D1795685

Topics in Applied Physics

Volume 121

Series Editors

Young Pak Lee, Physics, Hanyang University, Seoul, Korea (Republic of)
Paolo M. Ossi, NEMAS - WIBIDI Lab, Politecnico di Milano, Milano, Italy

Topics in Applied Physics is a well-established series of review books, each of which presents a comprehensive survey of a selected topic within the broad area of applied physics. Edited and written by leading research scientists in the field concerned, each volume contains review contributions covering the various aspects of the topic. Together these provide an overview of the state of the art in the respective field, extending from an introduction to the subject right up to the frontiers of contemporary research.

Topics in Applied Physics is addressed to all scientists at universities and in industry who wish to obtain an overview and to keep abreast of advances in applied physics. The series also provides easy but comprehensive access to the fields for newcomers starting research.

Contributions are specially commissioned. The Managing Editors are open to any suggestions for topics coming from the community of applied physicists no matter what the field and encourage prospective editors to approach them with ideas.

More information about this series at http://www.springer.com/series/560

Nianjun Yang

Editor

Novel Aspects of Diamond

From Growth to Applications

Second Edition

 Springer

Editor
Nianjun Yang
Institute of Materials Engineering
University of Siegen
Siegen, Germany

ISSN 0303-4216   ISSN 1437-0859   (electronic)
Topics in Applied Physics
ISBN 978-3-030-12468-7   ISBN 978-3-030-12469-4   (eBook)
https://doi.org/10.1007/978-3-030-12469-4

Library of Congress Control Number: 2018968532

1st edition: © Springer International Publishing Switzerland 2015
2nd edition: © Springer Nature Switzerland AG 2019
This work is subject to copyright. All rights are reserved by the Publisher, whether the whole or part of the material is concerned, specifically the rights of translation, reprinting, reuse of illustrations, recitation, broadcasting, reproduction on microfilms or in any other physical way, and transmission or information storage and retrieval, electronic adaptation, computer software, or by similar or dissimilar methodology now known or hereafter developed.
The use of general descriptive names, registered names, trademarks, service marks, etc. in this publication does not imply, even in the absence of a specific statement, that such names are exempt from the relevant protective laws and regulations and therefore free for general use.
The publisher, the authors and the editors are safe to assume that the advice and information in this book are believed to be true and accurate at the date of publication. Neither the publisher nor the authors or the editors give a warranty, expressed or implied, with respect to the material contained herein or for any errors or omissions that may have been made. The publisher remains neutral with regard to jurisdictional claims in published maps and institutional affiliations.

This Springer imprint is published by the registered company Springer Nature Switzerland AG
The registered company address is: Gewerbestrasse 11, 6330 Cham, Switzerland

In honor of Prof. Xin Jiang's contribution to the field of diamond research

Dedication to Xin Jiang

Dedication to Prof. Xin Jiang

A few months ago, I was asked by the editor of this book "Novel Aspects of Diamond: From Growth to Applications (2nd version)", Dr. Nianjun Yang (Siegen, Germany), to write a dedication note about Prof. Dr. Xin Jiang (University of Siegen, Siegen, Germany), who is one of my friends since we met the first time in 1994. At this time, we both got involved into diamond research and development, financed by the German Ministry of Education and Research (BMBF) with the title "Diamond as Electronic Material". At that time, Xin was employed at Fraunhofer Institute for Surface Engineering and Thin Films (IST), Braunschweig, and recognized as outstanding scientist on diamond nucleation and growth. Our first joint paper was published in the Journal of Diamond and Related Materials (1998, 7, 879–883) about the synthesis and electrical properties of CVD diamond films.

His vita is exciting. Prof. Jiang was born in 1960 in Shenyang, the largest City in Northern China. He was a student at Jilin University and got his Bachelor of Science in 1983. After that he matured at the Institute for Metal Research (CAS) to get his Master of Science on Titanium-Nitride layers in 1986. In 1987, he changed to Germany becoming a Ph.D. student at the Institute for Solid Sate Research (now: Peter Grünberg Institut (PGI)), Forschungszentrum Jülich (FZJ). His topic was Brillouin scattering and ultralow load indentation of amorphous carbon and silicon thin films, supervised by Prof. Peter Grünberg and Prof. Klaus Reichelt. For the first time, he combined the Brillouin scattering with nano-indentation methods to study the elastic mechanical properties of amorphous thin film materials systematically (Journal of Applied Physics, 1989, 66, 4729–4735). After obtaining his Ph.D. (Dr. rer. nat.) in experimental physics in 1990 from the Technical University Aachen (RWTH-Aachen, Germany), he was promoted to become a project leader and senior scientist at the Fraunhofer Institute IST, Braunschweig, Germany. In January 1998, he habilitated in the field of "Materials Science" at the Technical University of Braunschweig (Braunschweig, Germany). Finally, he was appointed to become a Full Professor at the University of Siegen (Siegen, Germany) in March 2003, chairing the Institute of Surface and Material Technology.

With about thirty years of experience in material science, he is one of the leading experts in CVD diamond nucleation and growth. Meanwhile, he is also covering other topics like deposition of super-hard materials, formation of new wide bandgap thin films, surface and interface properties, and nanomaterial science.

During his period at Fraunhofer IST, Prof. Jiang was in charge for five BMBF and two European Union (EU) projects in the field of diamond and ceramic materials. In this time, he collaborated with several world leading companies such as Daimler-Benz, Siemens, and Philips. Here the focus of research was optimization of CVD diamond growth where he also published several basic papers on the heteroepitaxial growth of diamond on silicon (Applied Physics Letters 1993, 62, 3438–3440; Diamond and Related Materials 1993, 2, 1112–1113). Up to now, these papers are recognized as milestones in the field of diamond CVD growth. This work has been cited several hundred times. The reported process has been widely applied for the preparation of highly oriented CVD diamond films. His electron microscopy investigations on the diamond/Si interface revealed that a direct epitaxial growth of diamond on Si is possible, in spite of extremely large differences of lattice parameter (Applied Physics Letters 1995, 67, 1197–1199). He invented the ion-assisted growth of CVD diamond films using a combination of CVD and ion bombardment to prepare new materials such as nanodiamond and (001)-textured diamond films. Extending these findings, he developed a new method for the formation of diamond nanocone tip arrays for sensor and field emission applications (Applied Physics Letters 1996, 68, 1927–1929). He invented novel selective deposition processes for diamond and silicon carbide, triggering the synthesis of diamond/ß-SiC composite films (Applied Physics Letters 1992, 61, 1629–1631). In collaboration with Phillips, he employed newly developed transparent conductive oxide films for display technologies, replacing indium tin oxide by aluminum-doped zinc oxide for applications in organic light-emitting devices (Applied Physics Letters 2003, 83, 1875–1877).

Since March 2003, Prof. Jiang is Full Professor and Chair of the Institute for Surface Engineering and Materials Technology (LOT), University Siegen, Germany. He acquired a variety of projects from DFG (German Research Foundation), BMBF (Bundesministerium für Bildung und Forschung, Germany), and EU. In the course of these activities, he improved the understanding as well as applications of new materials in different fields. He investigated adhesion problems and developed graded diamond/β-SiC nanocomposites (Applied Physics Letters, 2006, 88, 073109) as well as the growth of monochiral tubular graphite cones (Science 2003, 300, 472; Journal of American Chemical Society, 2007, 129, 8907). He also addressed chemical/biochemical sensors from diamond and related composite films and carbon nanomaterials to extend his activities. Diamond surfaces were functionalized by photochemical approaches (Applied Physics Letters 3009, 95, 143703). More recently, he focused on the growth of nanomaterials and multi-component composites (e.g., diamond, SiC, graphene). He reported for the first time about the preparation of porous diamond networks with enlarged surfaces (ACS Applied Materials and Interfaces 2015, 7, 5384–5390) and applications of architectural films for catalytic reduction of carbon dioxide (Nature communications 2017, 8, 1828), as well as their applications in high-performance supercapacitors (Advanced Energy Materials 2018, 8, 1702947).

Professor Dr. Jiang has invented 32 patents and more than 345 publications, predominantly in highly impacted journals such as Science, Nature Communications, Physical Review Letters, Nano Letters, and Advanced Energy Materials. These papers have been cited more than 8000 times. He is frequently invited to give plenary and keynote presentations at leading international conferences. He served as editorial board member and guest editor for a variety of international journals as well as reviewer of distinguished academic journals and science foundations such as Alexander von Humboldt (AvH), DFG, and Chinese Academy of Sciences (CAS). Additionally, he is an active member of several scientific organizations and well experienced in scientific administration. The recognition of his work led to numerous international collaborations. He is the awardee of two prestige scholarships: the "Changjiang Chair professor" (Ministry of Education, China) and the "1000 Talented Overseas Researcher" (Organization Department of the Central Committee of China, China). He was in charge of Joint International Laboratories at the Dalian University of Technology (Dalian, China) and the Institute of Metal Research, Chinese Academy of Sciences (Shenyang, China), both giving rise to an active scientific exchange between Germany and China.

Under the supervision of Prof. Dr. Jiang, 21 international students have received their Ph.D. degree since 2003. Currently, 16 Ph.D. students are members of his team. His graduated Ph.D. students achieved leading scientific positions either as professors or as senior scientists. He hosted two research awardees of the Humboldt Foundation (Prof. Wenjun Zhang and Prof. Ruiqing Zhang, both with City University of Hong Kong, China) and several DAAD scholars thereby building up an international network for young researchers.

Professor Dr. Jiang enjoys science and shares his knowledge with numerous colleagues and students. In his private time, he takes care of his family and spent his free time to read Chinese history books. I wish Prof. Dr. Xin Jiang all the best for the future, to stay in good health and to enjoy sciences also in the upcoming years.

Freiburg, Germany   Christoph E. Nebel
                                                                              Fraunhofer Institute for Applied
                                                                                        Solid State Physics

Preface

Diamond, as a wide bandgap semiconducting material, has been extensively studied for years. It is known more than 25 years as a perfect material for mechanical, optical, thermal, and electronic applications because diamond offers excellent physical and chemical properties. High-temperature electronic devices, radiation detectors, high-voltage switches, X-ray windows, audio speaker diaphragms, and protective coatings are examples of diamond-based devices. In the initial stage of diamond research, natural diamond crystals and single-crystalline diamond synthesized by the high temperature/high pressure (HPHT) were mostly utilized. About 40 years ago, the realization of the low-pressure chemical vapor deposition (CVD) of diamond triggered novel aspects of diamond research in laboratories and in industries all over the world. CVD-grown diamond films offer advantages for electronic applications with respect to crystal purification and doping for p-type or n-type conductivity. For example, the diamond films become electrically conductive when they are heavily doped with boron. Such boron-doped diamond films possess wide potential window, low background currents, and long stability. They are therefore recognized as the perfect electrodes in the fields of electroanalysis, pollution degradation, electrosynthesis, electrochemical biosensing, and so on.

The book "Novel Aspects of Diamond: From Growth to Applications (2nd version)" is in honor of Prof. Xin Jiang (Institute of Materials Engineering, University of Siegen, Germany). The objective of this book is to familiarize the reader with the scientific and engineering aspects of CVD diamond films and to provide experienced researchers, scientists, and engineers in academia and industry with the latest developments and achievements in this rapidly growing field. This second edition provides an updated, systematic review of diamond research, ranging from its growth and properties up to applications. The growth of single-crystalline and doped diamond films is included. The physical, chemical, and engineering properties of these films are discussed from theoretical and experimental aspects. The applications of various diamond films in the fields of chemistry, biology, medicine, physics, and engineering are presented. The book contains *seven updated and seven* new chapters dealing with new topics of diamond research.

The topics and the chapters are organized in the following way. The first part consists of ten chapters (from Chaps. 1 to 8) and is essentially devoted to growth, properties, and applications of diamond films. The first three chapters are about the growth, surface, and mechanical properties of diamond films. From Chaps. 4 to 8, the applications of various diamond films for micro- and nanomechanical resonators, field electron emission, organic electrosynthesis, electrochemical energy storage and conversion, and electrochemical devices are summarized. The second part of this book is about preparation, surface properties and chemistry, and applications of diamond micro- and nanostructures and particles. Chapters 9 and 10 focus on nanostructures generated on diamond films and from diamond films, respectively. Spectroscopy of nanodiamond surface, surface modification of diamond nanoparticles, and electrochemical applications of diamond particles are included in Chaps. 11, 12, 13, and 14, respectively.

From novel aspects and recent achievement of diamond researches presented in these chapters, it is quite clearly that some diamond activities are well established, while other new and promising opportunities for diamond science and industries are rapidly growing, together with upcoming and existing challenges for the whole diamond scientific community. It is hoped that this book will simulate more researchers, especially a new generation of scientists to contribute and promote diamond-related researches in different fields.

I certainly want to express my sincere thanks to all colleagues who delivered great chapters to this book. The efficient help and valuable suggestions from Claus Ascheron, Viradasarani Natarajan, Adelheid Duhm, and Elke Sauer in Springer Publisher Office, Germany, are highly appreciated. Last but not least, I want to thank my family, especially my wife, Dr. Xiaoxia Wang, and my children Zimo and Chuqian Luisa for their always and strong support to let me finalize such a book.

Siegen, Germany   Nianjun Yang

About This Book

This book is in honor of the contribution of Prof. Xin Jiang (Institute of Materials Engineering, University of Siegen, Germany) to the field of diamond research. The objective of this book is to familiarize readers with the scientific and engineering aspects of CVD diamond films and to provide experienced researchers, scientists, and engineers in academia and industry with the latest developments and achievements in this rapidly growing field. This second edition consists of 14 chapters, providing an updated, systematic review of diamond research, ranging from its growth and properties up to applications. The growth of single-crystalline and doped diamond films is included. The physical, chemical, and engineering properties of these films and diamond nanoparticles are discussed from theoretical and experimental aspects. The applications of various diamond films and nanoparticles in the fields of chemistry, biology, medicine, physics, and engineering are presented.

Contents

Contributors

J. C. Arnault CEA, LIST, Diamond Sensors Laboratory, Gif sur Yvette, France

Johnathan Ash School of Physics and Astronomy, Cardiff University, Cardiff, UK;
Department of Physics, Aberystwyth University, Aberystwyth, UK

Ashek-I-Ahmed Department of Physics, National Dong Hwa University, Hualien, Taiwan

Emmanuel B. Brousseau School of Engineering, Cardiff University, Cardiff, UK

Chia-Liang Cheng Department of Physics, National Dong Hwa University, Hualien, Taiwan

Patricia Rachel Fernandes da Costa Institute of Chemistry, Federal University of Rio Grande do Norte, Lagoa Nova, Natal, RN, Brazil

Djalma Ribeiro da Silva Institute of Chemistry, Federal University of Rio Grande do Norte, Lagoa Nova, Natal, RN, Brazil

Elisama Vieira dos Santos Institute of Chemistry, Federal University of Rio Grande do Norte, Lagoa Nova, Natal, RN, Brazil

John S. Foord Department of Chemistry, Oxford University, Oxford, UK

Atsuhiro Fujimori Graduate School of Science and Engineering, Saitama University, Sakura-ku, Saitama, Japan

Soliu Oladejo Ganiyu Institute of Chemistry, Federal University of Rio Grande do Norte, Lagoa Nova, Natal, RN, Brazil

Ken Haenen IMOMEC, IMEC vzw, Diepenbeek, Belgium

Artashes Karmenyan Department of Physics, National Dong Hwa University, Hualien, Taiwan

Georgina M. Klemencic School of Physics and Astronomy, Cardiff University, Cardiff, UK

Yasuo Koide Research Center for Functional Materials, National Institute for Materials Science (NIMS), Tsukuba, Ibaraki, Japan

Takeshi Kondo Department of Pure and Applied Chemistry, Faculty of Science and Technology, Tokyo University of Science, Noda, Chiba, Japan

Karin Larsson Department of Chemistry-Ångström Laboratory, Uppsala University, Uppsala, Sweden

Meiyong Liao Research Center for Functional Materials, National Institute for Materials Science (NIMS), Tsukuba, Ibaraki, Japan

Mailis M. Lounasvuori Department of Chemistry, Oxford University, Oxford, UK

Soumen Mandal School of Physics and Astronomy, Cardiff University, Cardiff, UK

Carlos A. Martínez-Huitle Institute of Chemistry, Federal University of Rio Grande do Norte, Lagoa Nova, Natal, RN, Brazil;
National Institute for Alternative Technologies of Detection, Toxicological Evaluation and Removal of Micropollutants and Radioactives (INCT-DATREM), Institute of Chemistry, Unesp, Araraquara, SP, Brazil;
Institut für Organische Chemie, Johannes Gutenberg-Universität Mainz, Mainz, Germany

Geoffrey W. Nelson Department of Chemistry, Oxford University, Oxford, UK

Elena V. Perevedentseva Department of Physics, National Dong Hwa University, Hualien, Taiwan;
P.N. Lebedev Physics Institute of Russian Academy of Science, Moscow, Russia

Liwen Sang Research Center for Functional Materials, National Institute for Materials Science (NIMS), Tsukuba, Ibaraki, Japan

Kamatchi Jothiramalingam Sankaran Institute for Materials Research (IMO), Hasselt University, Diepenbeek, Belgium

Evan L. H. Thomas School of Physics and Astronomy, Cardiff University, Cardiff, UK

Norio Tokuda Nanomaterials Research Institute, Kanazawa University, Kanazawa, Japan

Siegfried R. Waldvogel Institut für Organische Chemie, Johannes Gutenberg-Universität Mainz, Mainz, Germany

Jessica M. Werrell School of Physics and Astronomy, Cardiff University, Cardiff, UK

Oliver A. Williams School of Physics and Astronomy, Cardiff University, Cardiff, UK

Nianjun Yang Institute of Materials Engineering, University of Siegen, Siegen, Germany

Yuan Yu Institute of Atomic and Molecular Science, Shaanxi University of Science & Technology, Xi An, China;
Key Laboratory of Photochemical Conversion and Optoelectronic Materials, Technical Institute of Physics and Chemistry, Chinese Academy of Sciences, Beijing, People's Republic of China

Takeru Yunoki Graduate School of Science and Engineering, Saitama University, Sakura-ku, Saitama, Japan

Jinfang Zhi Key Laboratory of Photochemical Conversion and Optoelectronic Materials, Technical Institute of Physics and Chemistry, Chinese Academy of Sciences, Beijing, People's Republic of China

Chapter 1
Homoepitaxial Diamond Growth by Plasma-Enhanced Chemical Vapor Deposition

Norio Tokuda

Abstract Both carbon and silicon are group IV members, but carbon has the smaller atomic number. Diamond, with the same crystalline structure as that of silicon, is expected to act as the basic material for the next generation of high-power electronic, optoelectronic, bio/chemical electronic, quantum computing devices, etc. This is because diamond exhibits electrical properties similar to those of silicon, while having superior physical properties. In this chapter, the author reviewed and discussed the homoepitaxial growth of high-quality single-crystal diamond films with atomically flat surfaces, by using plasma-enhanced chemical vapor deposition (PECVD).

1.1 Introduction

Growth of diamond films by chemical vapor deposition (CVD), which has been studied since the 1950s, must be conducted under nonequilibrium conditions. This is because under normal conditions graphite is a more stable phase of carbon than diamond. Furthermore, during the CVD process, hydrogen radicals (atomic hydrogen) must be present to remove nondiamond carbon, including graphite which is formed on the diamond surface. The hydrogen radicals are generated either by thermal dissociation on a hot filament of W or Ta, or in plasma by electron impact, collisional energy transfer, etc. In plasma, the external energy input couples directly to free electrons, producing hydrogen radicals via

$$H_2 + e^- \rightarrow 2H + e^-. \tag{1.1}$$

N. Tokuda (✉)
Nanomaterials Research Institute, Kanazawa University, Kanazawa 920-1192, Japan
e-mail: tokuda@ec.t.kanazawa-u.ac.jp

© Springer Nature Switzerland AG 2019
N. Yang (ed.), *Novel Aspects of Diamond*, Topics in Applied Physics 121,
https://doi.org/10.1007/978-3-030-12469-4_1

Methane is commonly used as the carbon source for CVD diamond growth. Activated CH_x ($x = 0, 1, 2, 3$) species are formed by hydrogen abstraction reactions; for example, hydrogen radicals may produce methyl radicals from methane:

$$H + CH_4 \rightarrow CH_3 + H_2. \tag{1.2}$$

Then, recombination of the methyl radicals induces to form activated C_2H_y ($y = 0, 1, 2, 3, 4, 5, 6$) species:

$$CH_3 + CH_3 + M \rightarrow C_2H_6 + M \tag{1.3}$$

$$C_2H_y + H + M \leftrightarrow C_2H_{y-1} + H_2 + M, \tag{1.4}$$

where M is a third body. The CH_x and C_2H_y radicals are regarded as precursors for diamond growth during the CVD process, as shown in Fig. 1.1 [1–8]. Thus, radicals play an important role in CVD diamond growth; this differs from the other semiconductor films' growth conducted by nonplasma processes such as thermal CVD and molecular beam epitaxy.

Hot-filament CVD has been applied to large-scale industrial processes because of its simple system configuration and ability to coat large areas and complex

Fig. 1.1 Schematic of CVD diamond processes

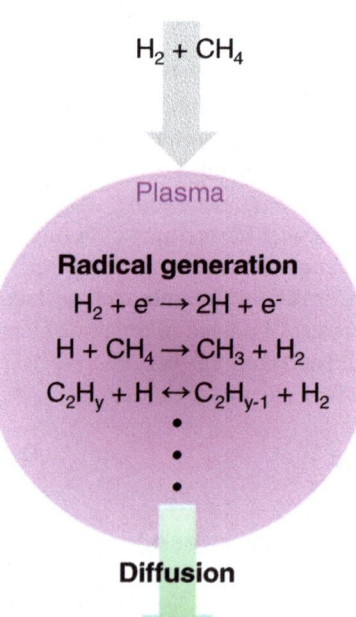

shapes. However, hot-filament CVD of diamond films must be carried out at lower gas temperatures and pressures than those of plasma CVD because of the upper temperature limit of the filament materials and the low production rate of hydrogen radicals. This leads to relatively low growth rates of diamond films compared to diamond growth by plasma-enhanced CVD (PECVD). Recently, homoepitaxial diamond growth rates of >100 μm/h by PECVD have been reported [9–15]. Additionally, both p- and n-type diamond films have been reproducibly grown by PECVD [16–42]. Therefore, the diamond films used in diamond electronic devices were grown by PECVD [43–77].

In this chapter, recent studies on homoepitaxial diamond growth by PECVD are reviewed. Additionally, impurity doping into diamond and the growth of atomically flat diamond surfaces are described.

1.2 Growth Mechanism

More than a decade has passed since PECVD of diamonds was established. Since then, many experimental and theoretical studies have been reported. Diamond growth by PECVD is not driven by thermodynamics but by the chemistry and kinetics of vapor phase and surface reactions. To elucidate the diamond growth mechanism during PECVD, both vapor-phase and surface reactions need to be understood. Evaluations of the vapor phase have been based on optical emission spectroscopy (OES) and mass spectrometry (MS) [78–85]. Here, as described in Sect. 1.1, the production and diffusion of hydrogen, CH_x radicals, and C_2H_y radicals are key processes, as shown in Fig. 1.1. Recently, the distributions of the radical, gas, and electron temperatures in plasmas have been simulated [86–96]. The simulation results provide some information on vapor-phase reactions in the CVD diamond process, but microscopic experimental results are still needed. Those radicals that arrive at diamond surfaces migrate and react with hydrogen, terminating the surface and/or carbon. It is extremely challenging to identify the involved processes because of the difficulty of conducting in situ characterizations in plasma environments. In this section, the author reviewed only those aspects of diamond surface chemistry that pertain to chemical reactions of hydrogen radicals and diamond precursors (CH_x and C_2H_y radicals).

1.2.1 Hydrogen

During diamond growth by PECVD, the diamond surfaces are continuously bombarded with hydrogen radicals. While under typical growth conditions the hydrogen concentration is 95% or higher (the hydrogen concentration is defined as the ratio of hydrogen flow rate to total gas flow rate). Consequently, most diamond surfaces are terminated by hydrogen and cannot react with diamond precursors.

However, hydrogen radicals abstract the hydrogen from terminated diamond surfaces, C_d–H, to form an active site, $C_d{}^\bullet$:

$$\text{(hydrogen abstraction)} \quad C_d{-}H + H \rightarrow C_d{}^\bullet + H_2. \qquad (1.5)$$

Then, the active site reacts with a hydrogen radical:

$$\text{(hydrogen adsorption)} \quad C_d{}^\bullet + H \rightarrow C_d{-}H. \qquad (1.6)$$

During diamond growth by PECVD, the fraction of active sites is determined by the dynamic equilibrium between chemical reactions (1.5) and (1.6). The diamond surfaces after the hydrogen plasma treatment and diamond growth by PECVD are terminated by hydrogen, as shown in Fig. 1.2.

Hydrogen radicals also play a role in the growth of high-quality diamond films by removing nondiamond carbon. Diamond is etched by reactions with hydrogen radicals, although the etching rate is lower than that of nondiamond carbon. The diamond etching rates by hydrogen radicals depend on the structures on the diamond surface: monohydride (CH), dihydride (CH$_2$), and trihydride (CH$_3$). Chen et al. proposed that the diamond etching rates, R, by hydrogen radicals are $R_{monohydride} < R_{dihydride} < R_{trihydride}$ [97]. They also reported that {111}-oriented facets form on both single-crystal diamond {110} and {100} surfaces by anisotropic etching. The anisotropic diamond etching by the hydrogen plasma treatment can selectively form an atomically flat diamond {111} surface on the trench bottom [98]. Thus, diamond growth by PECVD is accompanied by the reactions of hydrogen abstraction (1.5) and adsorption (1.6) and by anisotropic etching on diamond surfaces, which limits chemisorption of diamond precursors and diamond nucleation.

1.2.2 Carbon

As described in Sect. 1.2.1, diamond surfaces are nearly fully terminated by hydrogen during diamond growth by PECVD. Chemisorption by diamond precursors occurs not at hydrogen-terminated sites but at active sites, which are hydrogen-abstracted sites:

$$\text{(CH}_3 \text{ radical chemisorption)} \quad C_d{}^\bullet + CH_3 \rightarrow C_d{-}CH_3. \qquad (1.7)$$

The chemisorbed structure is a trihydride, which is readily etched by hydrogen radicals. Structures composed of monohydrides and/or dihydrides may need to nucleate on diamond surfaces during PECVD. Observation of the growth surface is crucial for elucidation of the growth mechanism because the growth process influences the structure of the growth surface. Scanning probe microscopy (SPM), low-energy electron diffraction (LEED), Fourier transform-infrared spectroscopy

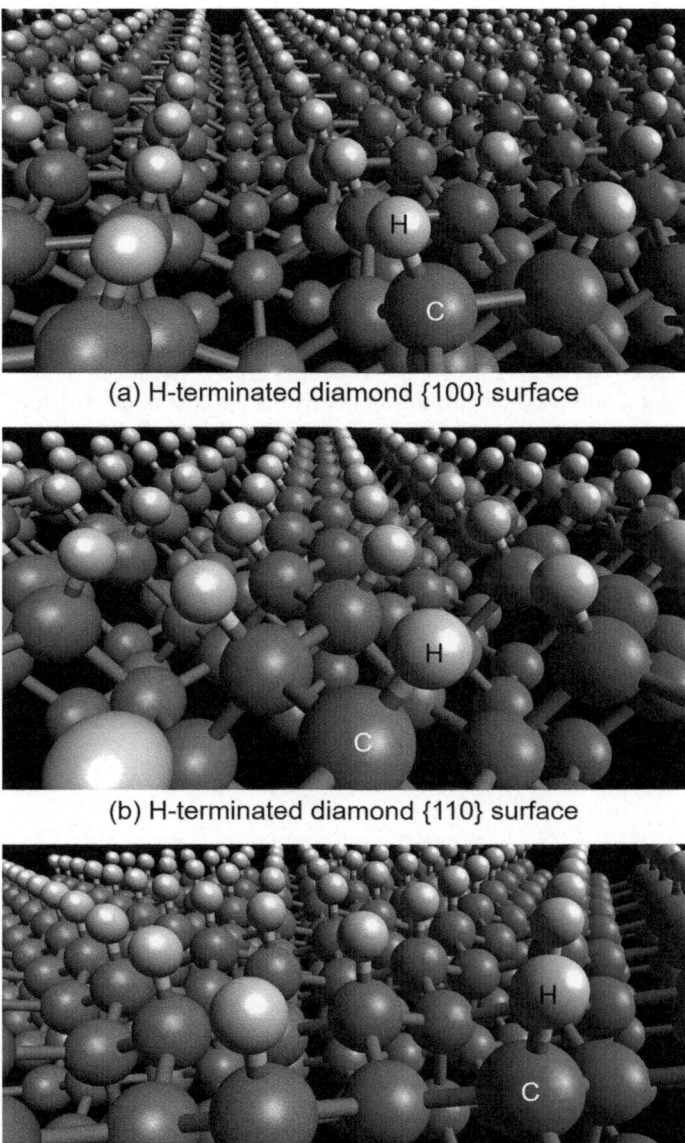

(a) H-terminated diamond {100} surface

(b) H-terminated diamond {110} surface

(c) H-terminated diamond {111} surface

Fig. 1.2 Hydrogen-terminated diamond {100}, {110}, and {111} surfaces

(FT-IR), and electron energy loss spectroscopy (EELS) provide physical and chemical information on surfaces at the atomic level, and are powerful tools for the study of diamond CVD growth. Results from such techniques reveal that as-grown diamond {100} and {111} surfaces have 2×1:H reconstructed structures with carbon dimer rows and 1×1:H structures, respectively [99–111]. Nevertheless, at present, the mechanism of diamond growth by PECVD is still not well-understood because of the difficulty of in situ observations.

1.3   Growth Modes

To realize diamond-based electronics, a growth technique is needed for producing device-grade diamond films. As described in Sect. 1.2, the growth mechanism of diamond films by PECVD remains poorly understood because of the difficulty of in situ characterization in plasma environments [112]. Additionally, the control of dynamic characterizations on well-defined surfaces, such as a scanning electron or probe microscope-molecular beam epitaxy system used for Si [113], GaAs [114], and GaN [115], is needed to elucidate the growth mode of PECVD diamond films.

Figure 1.3 illustrates a simplified model for the determination of growth modes by an alternative to in situ characterizations [116, 117]. Figure 1.3 illustrates the surface steps present on an ideal surface of an as-received diamond substrate after the formation of a mesa structure due to a misoriented angle between the basal plane and polished surface. Then, homoepitaxial growth is carried out under a

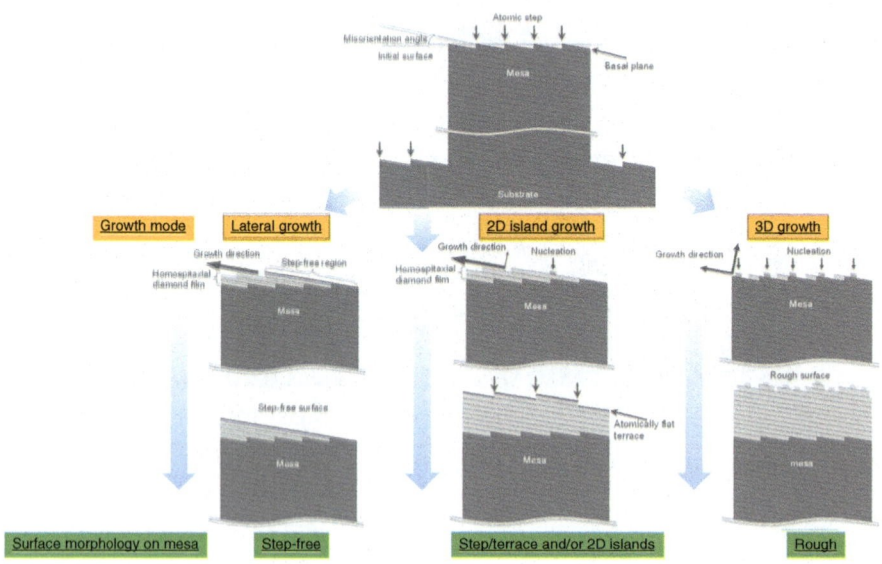

Fig. 1.3 Simplified models for the determination of growth modes by using mesa structures

lateral growth mode without 2D nucleation on terraces, as described below. Each atomic step on the mesa surface grows laterally. Under ideal conditions in which 2D terrace nucleation is completely suppressed, there would be no further growth perpendicular to the basal plane. Finally, a step-free surface is formed over the mesa surface, leaving the basal plane surface. In 2D island growth, new steps are formed by nucleation on the terraces during lateral growth. Finally, a surface with single atomic steps and atomically flat terraces is formed on the mesa surface. The interval between the formed islands is wider than the terrace width estimated from the misoriented angle of the substrate. In 3D growth, the interval between the formed islands is narrower than the terrace width estimated from the misoriented angle. As a result, the surface is very rough. In addition, this mesa structure eliminates the influence of abnormal growth, such as spiral growth induced by screw dislocations from trench bottoms. Thus, diamond growth modes can be determined from ex situ surface observations of diamond films grown on mesa surfaces.

Figure 1.4 shows AFM images of diamond {111} mesa surfaces before and after homoepitaxial growth by PECVD. For diamond growth at low methane concentrations (0.005–0.025% $CH_4/(H_2 + CH_4)$ ratio), a step-free surface, that is, a perfectly flat surface without any atomic steps, was formed on the mesa. This result shows that the growth mode of the homoepitaxial diamond {111} films was an ideal lateral growth without 2D terrace nucleation. For diamond growth at middle methane concentrations (0.05–0.25% $CH_4/(H_2 + CH_4)$ ratio), equilateral-triangular islands and/or single bi-atomic layer step/terrace structures on atomically flat surfaces were formed on the mesa. This shows that the growth mode of the

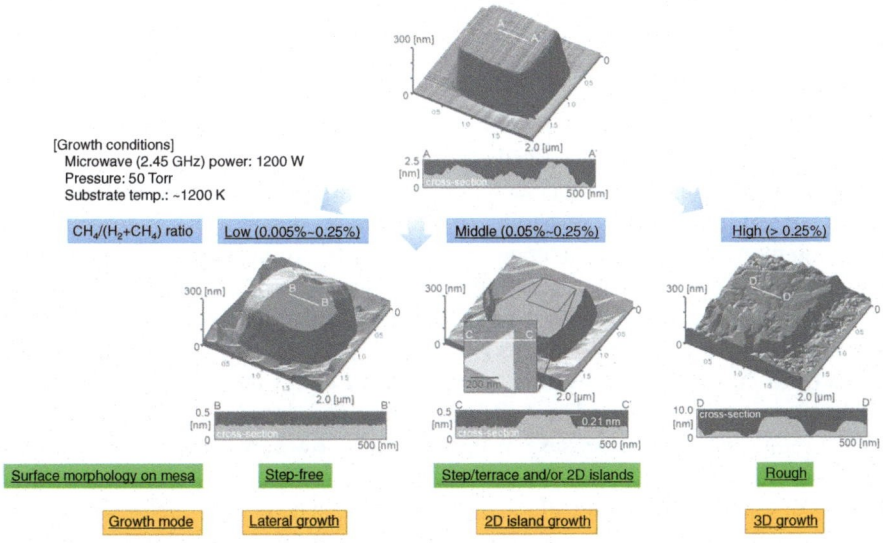

Fig. 1.4 AFM images of diamond {111} mesa surfaces before and after homoepitaxial growth by PECVD

homoepitaxial diamond {111} films is 2D island growth with 2D terrace nucleation and lateral growth. Additionally, the formation of equilateral-triangular islands shows that the diamond growth had extremely high selectivity. For diamond growth at high methane concentrations (>0.25% $CH_4/(H_2 + CH_4)$ ratio), the film surface, whose RMS value was 0.84 nm, is much rougher than the initial surface before growth (RMS = 0.44 nm). This shows that the growth mode for the homoepitaxial diamond {111} films is 3D growth.

Variations in methane concentrations give rise to different fluxes of hydrocarbon precursors arriving at the surface. This is because growth rates increase with methane concentration. Since the substrates used for growth have the same misoriented angle, the concentrations of adatoms on terraces increase with higher methane concentrations. When the flux is low, adatoms on a terrace remain below the critical size for 2D terrace nucleation. Adatoms, which are adsorbed precursors on the diamond surface, arriving at steps crystallize, resulting in step-edge growth (no 2D terrace nucleation).

In the middle methane concentration region, the adatom concentration increases and passes the critical value needed for 2D nucleation. Adatoms can cluster more easily on terraces because of their higher population. This causes 2D nucleation on terraces. As the methane concentration is increased further (into the high methane concentration region), adatoms form clusters as soon as they land on surfaces from the gas phase. This causes surface roughening because of 3D growth. This growth mechanism is a common process in thermal CVD and MBE. Despite the extremely short migration length, the common mechanism may also apply to PECVD diamond {111} growth because of the formation of atomically flat surfaces of diamond films.

1.4 Doping

For the realization of diamond-based electronic devices, doping acceptor and donor impurities into diamond is necessary to control the carrier type and concentration and to control the electrical resistivity of diamond semiconductors. Nitrogen is the most common impurity in diamond. Nitrogen is likely to form several types of complexes with vacancies. Recently, the nitrogen-vacancy (N-V) center in diamond has attracted much attention as a promising solid-state spin system for quantum information and sensing applications [71, 118–126]. However, as a donor, nitrogen in diamond has a high activation energy of 1.7 eV, which is higher than the bandgap of silicon (1.1 eV). The resistivity of nitrogen-doped diamond is extremely high at room temperature because of the extremely low concentration of thermally activated electrons. Generally, boron and phosphorus are used as p- and n-type dopants of diamond semiconductors, respectively. The activation energies of boron- and phosphorus-doped diamond are 0.37 eV and 0.57 eV, respectively, which are lower than that of nitrogen-doped diamond. However, compared with boron- and phosphorus-doped silicon, the resistivity of doped diamond is still too high,

Fig. 1.5 Room-temperature resistivities of Si and diamond as functions of impurity concentration. The resistivities of *p*- and *n*-type diamond were calculated when the compensation ratio was zero and mobility was constant

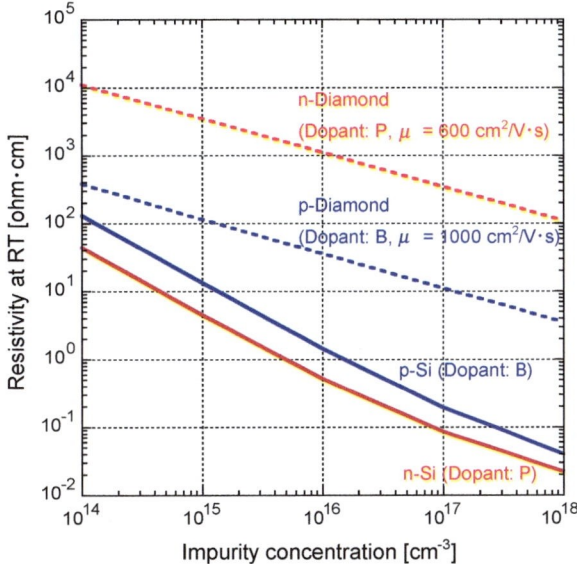

as shown in Fig. 1.5. Although other dopants with lower activation energies have been investigated, reproducibility has not yet been obtained.

Doping into diamond is carried out by HPHT, PECVD, and ion implantation [127–132]. Recently, Bormashov et al. reported that boron-doped {100} diamond without any extended defect. It was synthesized by HPHT and showed the high Hall hole mobility of 2200 cm^2/V s at 300 K and 7200 cm^2/V s at 180 K [133]. For device fabrication, doped diamond films are mostly grown on HPHT or CVD diamond substrates by PECVD because doping by PECVD provides both *p*- and *n*-type diamond with controlling concentrations of impurities. During diamond homoepitaxial growth by PEDVD boron and phosphorus doping is carried out by introducing diborane (or trimethylboron) and phosphine (or tertiarybutylphosphine) gases, respectively. The highest Hall hole and electron mobility of PECVD diamond films are 1860 cm^2/V s at 290 K [134] and 660 cm^2/V s at 300 K [34], respectively. Carrier mobility decreases with increasing boron or phosphorus concentrations in diamond films, reducing the resistivity of diamond. For [B] < 10^{19} cm^{-3}, conduction is dominated by free holes in the valence band. At higher doping concentrations, variable-range hopping conduction appears, and then the metal-insulator transition and superconductivity arise around 3×10^{20} B atoms/cm^3 [135–141]. The resistivity of heavily boron-doped diamond {100} film with 3×10^{20} B atoms/cm^3 is 10 mΩ cm or less at room temperature [135, 142–144]. In contrast, the resistivity of heavily phosphorus-doped diamond {111} film with 10^{20} P atoms/cm^3 is around 70 Ω cm at room temperature [145]. These can lead to the fabrication of chemical/bio and electronic devices, such as Schottky barrier diodes, *pn*-junction diodes, Schottky *pn* diodes, JFETs, and bipolar transistors.

1.5 Growth of Atomically Flat Diamond

To realize electronic devices with proper performance characteristics, one of the most important issues is the roughness of surfaces and interfaces in semiconductor materials. Device applications of diamond semiconductors require sharp interfaces at diamond homo- and hetero-junctions. Generally, growth of p- and n-type diamond semiconductors for Schottky contacts and pn junctions, etc., is carried out by boron and phosphorus doping, respectively, during homoepitaxial diamond growth. However, growth hillocks, which are macroscopic defects, are often observed on as-grown diamond surfaces after homoepitaxial diamond growth by PECVD, as shown in Fig. 1.6. In most cases, even macroscopic flat surfaces excluding hillocks are not atomically flat, as shown in Fig. 1.7. Therefore, surface flattening of homoepitaxial diamond films is extremely important.

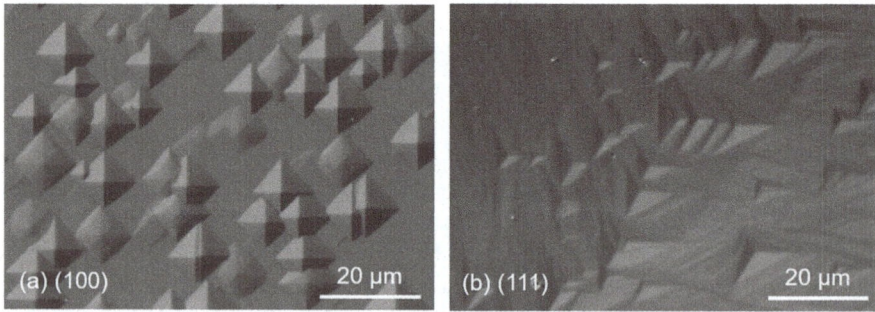

Fig. 1.6 OM images of hillocks formed on diamond surfaces after homoepitaxial growth by PECVD. **a** Quadrangular hillocks were observed on diamond {100} surfaces and **b** triangular hillocks on diamond {111} surfaces

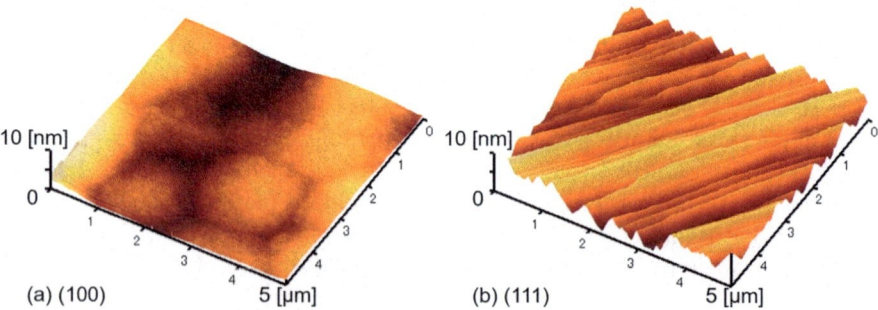

Fig. 1.7 3D AFM images of diamond surfaces, excluding hillocks, after homoepitaxial growth by PECVD. **a** The surface roughness (RMS) of the homoepitaxial diamond {100} film was 0.54 nm. **b** The surface roughness (RMS) of the homoepitaxial diamond {111} film was 1.68 nm

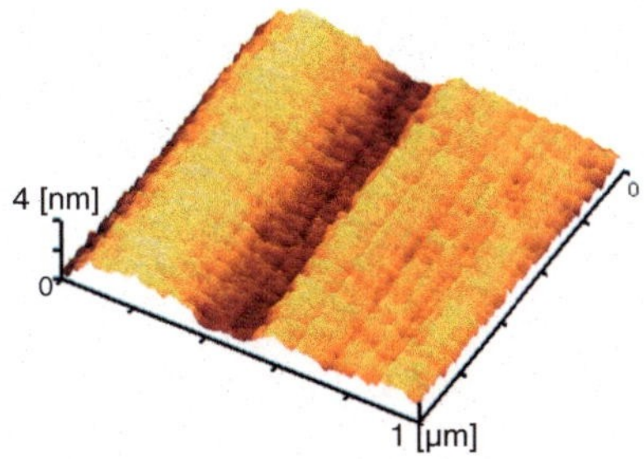

Fig. 1.8 3D AFM image of as-polished single-crystal diamond surface after acid treatment. The surface roughness (RMS) was 0.44 nm

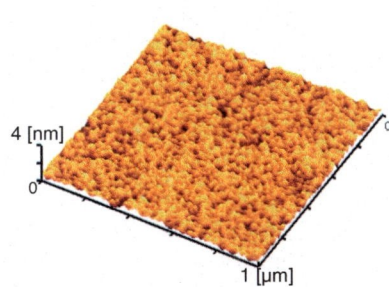

Fig. 1.9 3D AFM image of as-polished single-crystal Si surface after modified RCA cleaning. The surface roughness (RMS) was 0.16 nm

Surface flattening can be carried out by a polishing process. For example, defect-free Si wafers with surface flatness at the atomic level can be obtained by chemical–mechanical polishing. However, as-received diamond substrate surfaces are not atomically flat; instead, they have a roughness of several nanometers or more after mechanical polishing, as shown in Fig. 1.8. Compared with as-received Si wafer surfaces after chemical–mechanical polishing, as shown in Fig. 1.9, diamond surfaces are much rougher. Recently, a new technique for diamond surface polishing has been reported [146–148]. It is expected that surface roughness of diamond substrates will be reduced to that of Si wafers via some breakthrough diamond polishing technique. In this section, the author describes the growth of hillock-free, atomically step/terrace, and step-free diamond films.

1.5.1 Hillock-Free Surfaces

Growth hillocks are often observed on homoepitaxial diamond {100} and {111} surfaces, as shown in Fig. 1.6. A growth hillock is formed by single and double spiral growth centered on one and two screw dislocation cores, respectively, as shown in Fig. 1.10 [117, 149, 150], and the diamond surface is increasingly roughened by the growth of hillocks. The spiral growth rate at a screw dislocation is higher than the normal growth rate on the surface, excluding such crystal defects. Generally, growth hillocks make device fabrication difficult; hillocks are related to dielectric breakdown and current leakage in electronic devices such as *pn* junctions,

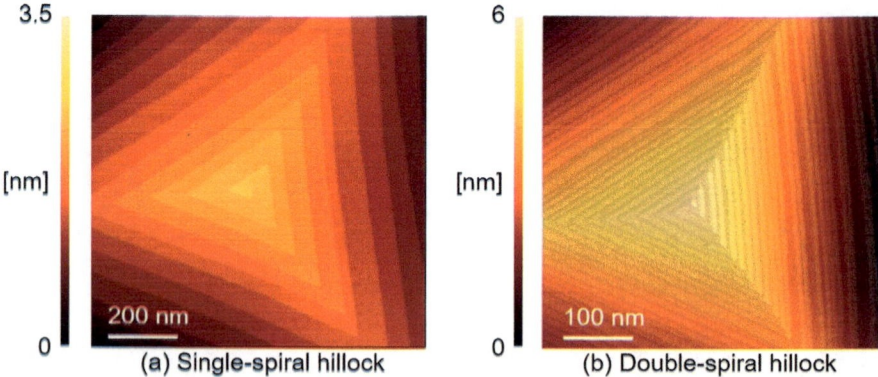

(a) Single-spiral hillock   (b) Double-spiral hillock

Fig. 1.10 AFM image of **a** single and **b** double spirally grown diamond {111} film due to screw dislocations. Each step height was approximately 0.21 nm, which is consistent with the single BL step height of {111} diamond

Schottky contacts, and MIS structures. Therefore, it is extremely important to eliminate hillocks from epitaxial diamond films. The most effective method for achieving hillock-free diamond is to completely eliminate dislocations from single-crystal diamond. However, dislocation-free diamond substrates are very expensive compared to common single-crystal diamond substrates with dislocation densities of 10^4–10^5 cm^{-2}. Alternatively, it is possible to eliminate hillocks by homoepitaxial lateral growth on highly misoriented diamond substrates. Figure 1.11 shows a simple model for suppressing the growth of hillocks [144]. The mechanism of hillock formation during diamond growth is considered to be as follows. The origins of hillocks have a growth rate higher than those of other areas.

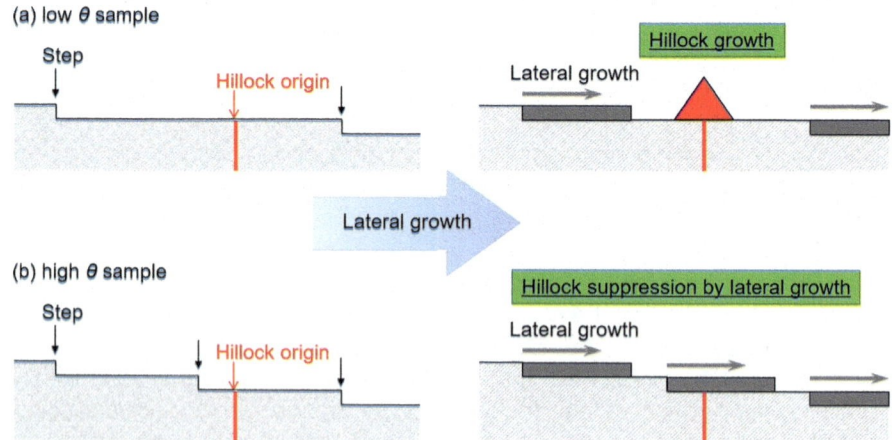

Fig. 1.11 Simplified models of hillock growth and suppression on samples having **a** low and **b** high misorientation angles θ by lateral growth

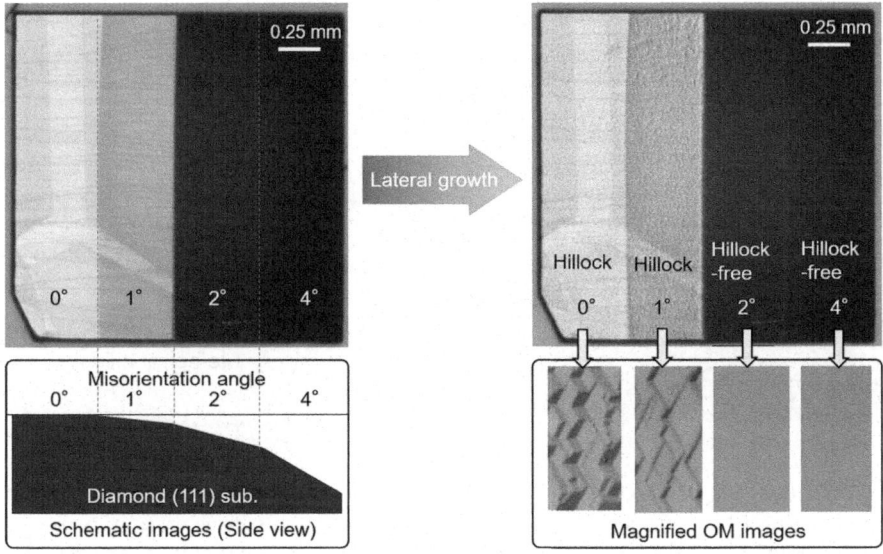

Fig. 1.12 OM and schematic cross-sectional images of diamond (111) surfaces with misorientation angles of 0°, 1°, 2°, and 4° before and after homoepitaxial lateral growth [150]

A local increase in growth rate due to the defects observed by transmission electron microscopy, which are dislocations [151] and twinning structures [152], has been reported. Hence, growth at the origins of hillocks could be suppressed when the growth rate of a normal epitaxial area exceeds that of hillocks, indicating an enhancement of lateral growth, as shown in Fig. 1.12 [150].

Figure 1.13 shows the OM images of homoepitaxial diamond films grown on diamond {001} substrates with (a) $\theta = 0.2°$ and (b) $\theta = 2.2°$. Hillocks were observed to spread over the entire surface of the low-θ sample, as shown in Fig. 1.13a. The density and size of the hillocks were 2×10^3 cm^{-2} and below 200 μm, respectively. The Hillock size increases with film thickness, and thus, the film surface becomes rough macroscopically. This roughening is a fatal issue for the additional growth of homoepitaxial diamond films intended for device fabrication. However, suppression of hillock growth can be achieved by increasing θ to above 2° [144]. The OM image of the high-θ sample is shown in Fig. 1.13b. Judging from this image, a hillock-free diamond film with a macroscopically flat surface was obtained over the entire surface by homoepitaxial lateral growth on highly misoriented substrates.

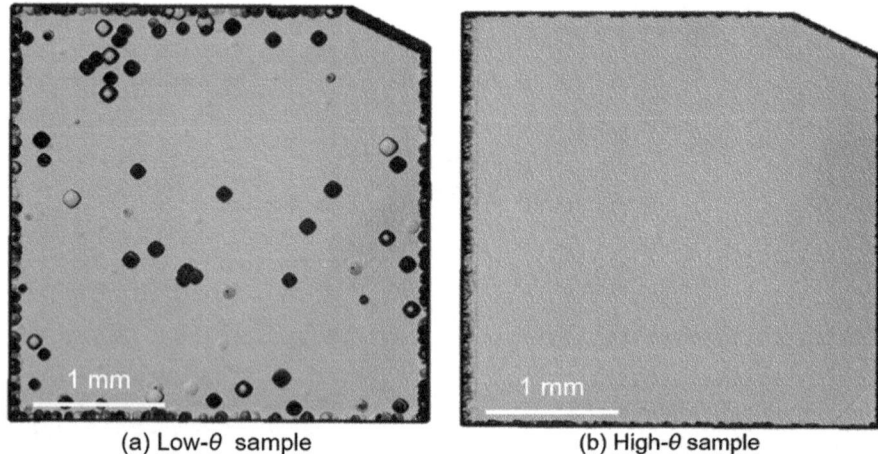

(a) Low-θ sample   (b) High-θ sample

Fig. 1.13 OM images of homoepitaxial diamond film surfaces on single-crystal diamond {100} substrates with **a** low and **b** high misorientation angles, θ. The low and high values of θ were 0.2 and 2.2, respectively. The homoepitaxial diamond films were grown under the same conditions (1200 W, 50 Torr, 0.6% $CH_4/(H_2 + CH_4)$ ratio, 1.6% B/C ratio, 70 h) [144]

1.5.2 Step/Terrace Structures

As described in Sect. 1.5.1, it is possible to obtain a hillock-free diamond film with a macroscopically flat surface by homoepitaxial lateral growth on a highly misoriented substrate. In most cases, the homoepitaxial lateral growth of diamond films accompanies 2D terrace nucleation. Therefore, macroscopically flat diamond surfaces, excluding hillocks, after homoepitaxial growth are not atomically flat but roughened because of 2D terrace nucleation, as shown in Fig. 1.7. Therefore, it is necessary to suppress 2D terrace nucleation during homoepitaxial growth by PECVD.

Watanabe et al. successfully formed an atomically flat diamond {100} film over the entire substrate by homoepitaxial lateral growth at extremely low CH_4/H_2 ratios [153]. Figure 1.14 shows an AFM image of an atomically flat diamond {100} surface by such a growth. The step heights were approximately 0.1 nm, which is consistent with the single atomic step of {100} diamond (0.089 nm), or 0.2 nm, which is consistent with the bi-atomic step of {100} diamond (2 × 0.089 nm). The average interval between steps (66 nm) is consistent with the estimated terrace width from the misorientation angle of the diamond {100} substrate (65 nm).

Atomically flat diamond {111} surfaces were also formed by homoepitaxial lateral growth, as shown in Fig. 1.15. The step height was approximately 0.2 nm, which is consistent with the single atomic step of {111} diamond (0.206 nm). The atomically flat diamond {111} surfaces with step/terrace structures were selectively formed by lateral growth on a diamond {111} substrate with mesa structures [154, 155].

Fig. 1.14 AFM image of atomically flat diamond {100} surfaces after homoepitaxial lateral growth at a low methane concentration (0.05%)

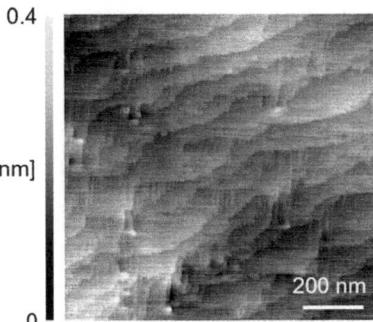

Fig. 1.15 AFM image of atomically flat diamond {111} surfaces after homoepitaxial lateral growth at a low methane concentration (0.05%)

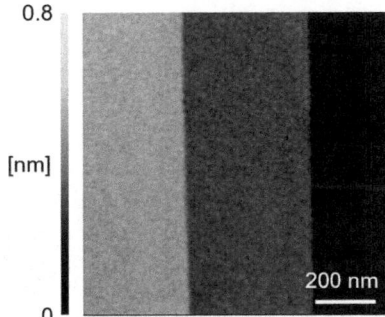

1.5.3 Atomically Step-Free Surfaces

Much effort has been expended on flattening the surfaces of Si, GaAs, SiC, and GaN at the atomic level; those efforts have led to the achievement of perfectly flat surfaces without any atomic steps (atomically *step-free* surfaces) through step-flow growth without 2D nucleation on terraces [156–160]. The aluminum nitride (AlN)/ diamond heterostructure is expected to combine the features of both wide-bandgap materials, thereby providing a new scheme for both nitride and diamond devices because they have opposite tendencies in doping characteristics [161–165]. Hirama et al. reported that single-crystal AlN{0001} growth on a diamond {111} surface was achieved [164], but the AlN layer on a diamond {100} surface had a multidomain structure consisting of tilted and rotated domains [166]. To realize useful devices, high-quality AlN films on {111} diamond and sharp AlN/diamond {111} interfaces are essential. For GaN/SiC heterostructures, previous studies have revealed that surface steps promote extended crystal defects in heteroepitaxial films grown on SiC [167–169]. Bassim et al. reported that very low dislocation densities were achieved in GaN films on step-free SiC mesa surfaces [170], and ultraviolet luminescence of GaN *pn*-junction diodes fabricated on the step-free SiC{0001} surfaces was improved relative to that on atomically flat SiC surfaces with atomic

Fig. 1.16 Simplified models
of step-free surface growth
using **a** mask and **b** maskless
processes. The step-free
surfaces of GaAs and GaN
were grown using the mask
process, while Si, SiC, and
diamond were grown using
the maskless process

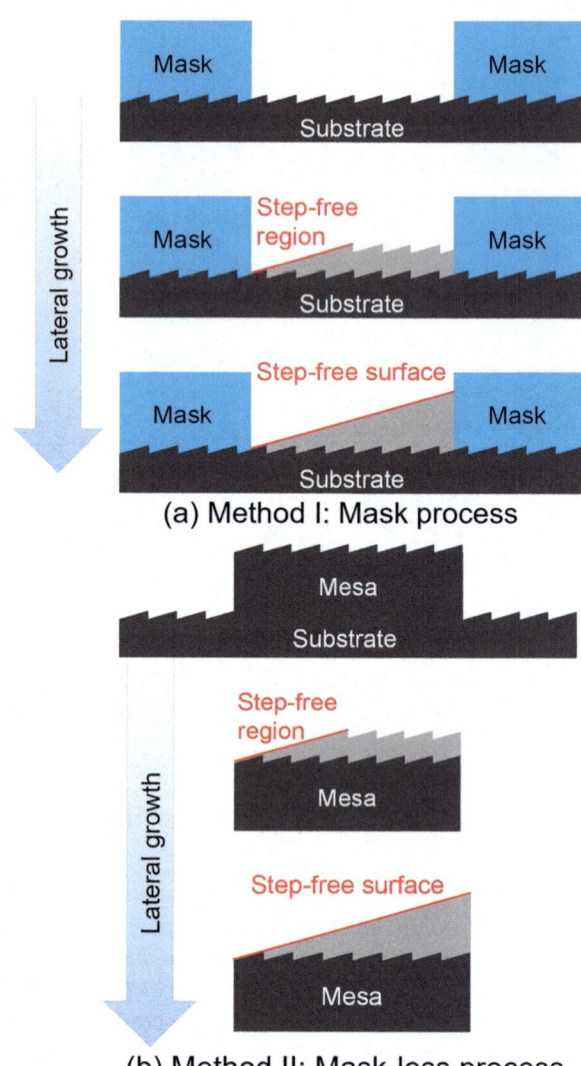

(a) Method I: Mask process

(b) Method II: Mask-less process

steps [171]. Thus, the formation of step-free diamond surfaces is a promising technique for improving the performance of devices that use diamond heterostructures.

Two methods for the formation of *step-free* surfaces have been proposed, as shown in Fig. 1.16. Both methods utilize an ideal lateral growth mode without 2D terrace nucleation. In method I, a mask, which is not etched and on which no nucleation occurs in the growth environment, is used for selective growth. Method II, in which a mesa structure is used, is a maskless process. Method II should be utilized for diamond growth by PECVD because, in plasma

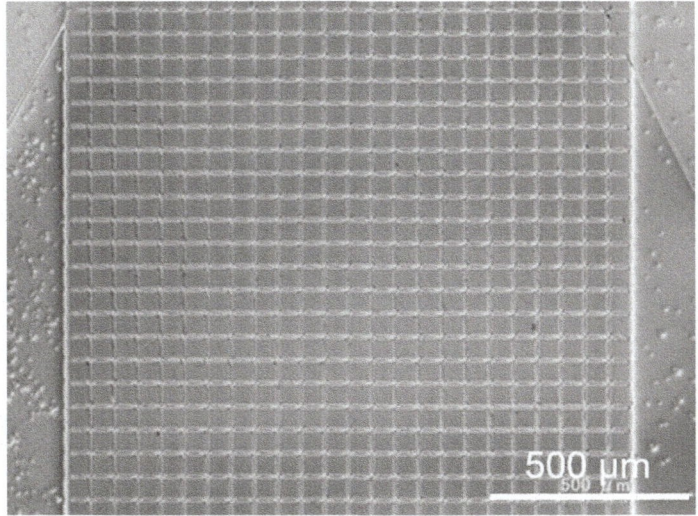

Fig. 1.17 OM image of single-crystal diamond substrates with $50 \times 50 \ \mu m^2$ mesas

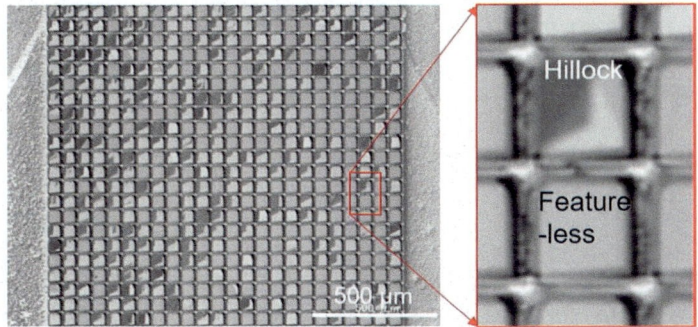

Fig. 1.18 OM image of diamond {111} films on single-crystal diamond substrates with $50 \times 50 \ \mu m^2$ mesas after homoepitaxial lateral growth

environments, most materials do not meet the conditions necessary for the mask. Figure 1.17 shows an OM image of a {111} diamond substrate surface after mesa fabrication. The array formation of $50 \times 50 \ \mu m^2$ mesas on single-crystal diamond substrates was carried out through a conventional lithographic technique. Figure 1.18 shows an OM image of a diamond {111} surface after homoepitaxial lateral growth by MPCVD. The mesas exhibit one of the following two characteristics in the OM observations shown in Fig. 1.18: (1) the surface was featureless; or (2) the surface contained at least one hillock, which grew spirally on the substrate, as shown in Fig. 1.10. The hillock was induced by a screw dislocation, as shown in Fig. 1.19 [117, 149]. The featureless mesa surface is an atomically step-free surface, as shown in Fig. 1.20a. Additionally, the diamond {111} films

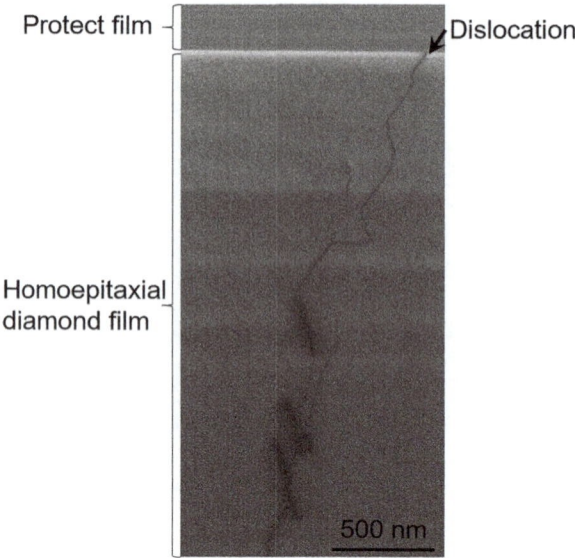

Fig. 1.19 Cross-sectional transmission electron microscope (XTEM) image of homoepitaxial diamond {111} film obtained from the mesa with hillock

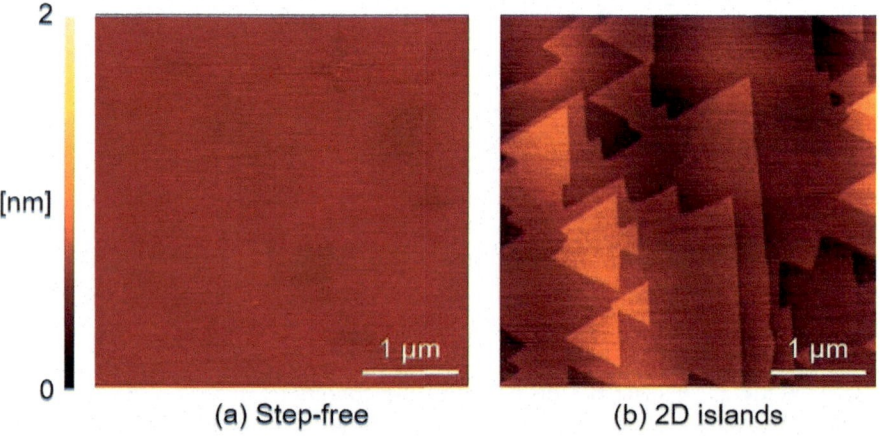

Fig. 1.20 AFM images of **a** atomically step-free diamond {111} surface and **b** atomically flat diamond {111} surface with 2D islands. The 2D islands were composed of atomically flat surfaces and a single BL step

with step-free surfaces contained no dislocations, as shown in Fig. 1.21. Recently, the growth of a $100 \times 100 \ \mu m^2$ step-free diamond {111} surface has been reported [116, 117]. This indicates the possibility of forming a step-free diamond surface over a substrate if the substrate contains no dislocation.

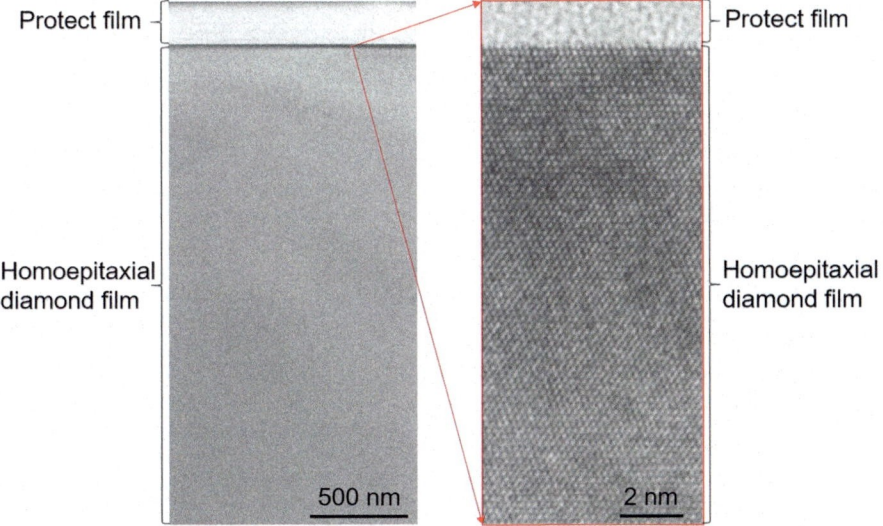

Fig. 1.21 XTEM images of homoepitaxial diamond {111} film obtained from featureless mesas

1.6 Conclusions

In this paper, the mechanism and control of homoepitaxial diamond growth by PECVD was reviewed. Recently, homoepitaxial diamond films have been used to fabricate many diamond devices. Judging from the reported performance of such devices, the technology of homoepitaxial diamond growth seems to have reached a certain level. However, because of the difficulty of in situ characterization in plasma environments, the growth mechanism remains unclear at present, especially regarding formation of precursors and their surface reactions. In the future, further developments of epitaxial diamond growth technology, including impurity doping and defect control, will be necessary to realize diamond-based electronics.

Acknowledgements The author sincerely thanks Dr. Satoshi Yamasaki, Dr. Hideyo Okushi, Dr. Daisuke Takeuchi, Dr. Masahiko Ogura, Dr. Toshiharu Makino, Dr. Hiromitsu Kato, Dr. Hitoshi Umezawa, Dr. Takehide Miyazaki of the National Institute of Advanced Industrial Science and Technology; Dr. Sung-Gi Ri of the National Institute for Materials Science; and Professor Takao Inokuma, Assistant Professor Tsubasa Matsumoto of Kanazawa University for fruitful discussions. This study was partly supported by JSPS KAKENHI Grant Numbers JP24686074, JP 17H02786, JP 17K18980, 18H03870 and the Adaptable and Seamless Technology Transfer Program through target-driven R&D, JST, and Kanazawa University SAKIGAKE Project 2018.

References

1. B.V. Spitsyn, L.L. Bouilov, B.V. Derjaguin, Vapor growth of diamond on diamond and other surfaces. J. Cryst. Growth **52**, 219–226 (1981). https://doi.org/10.1016/0022-0248(81)90197-4
2. S. Matsumoto, Y. Sato, M. Kamo, N. Setaka, Vapor deposition of diamond particles from methane. Jpn. J. Appl. Phys. **21**, L183–L185 (1982). https://doi.org/10.1143/JJAP.21.L183
3. M. Kamo, Y. Sato, S. Matsumoto, N. Setaka, Diamond synthesis from gas phase in microwave plasma. J. Cryst. Growth **62**, 642–644 (1983). https://doi.org/10.1016/0022-0248(83)90411-6
4. M. Kamo, H. Yurimoto, Epitaxial growth of diamond on diamond substrate by plasma assisted CVD. Appl. Surf. Sci. **33**(34), 553–560 (1988). https://doi.org/10.1016/0169-4332(88)90352-2
5. D.G. Goodwin, J. E. Butler, in *Handbook of Industrial Diamond and Diamond Films*, ed. by M.A. Prelas, G. Popovici, L.K. Biglow (Marcel Dekker, Inc., NY, 1997), p. 527
6. T. Teraji, in *Physics and Applications of CVD Diamond*, ed. by S. Koizumi, C.E. Nebel, M. Nesladek (Wiley-VCH Verlag GmbH & Co. KGaA, Weinheim, 2008), p. 29
7. J.E. Butler, A. Cheesman, M.N.R. Ashfold, in *CVD Diamond for Electronic Devices and Sensors*, ed. by R.S. Sussmann (Wiley, UK, 2009), p. 103
8. J.E. Butler, Y.A. Mankelevich, A. Cheesman, J. Ma, M.N.R. Ashfold, Understanding the chemical vapor deposition of diamond: recent progress. J. Phys. Cond. Mat. **21**, 364201 (2009). https://doi.org/10.1002/pssa.200777501
9. O.A. Williams, R.B. Jackman, High growth rate MWPECVD of single crystal diamond. Diam. Relat. Mater. **13**, 557–560 (2004). https://doi.org/10.1016/j.diamond.2004.01.023
10. J. Achard, F. Silva, O. Brinza, A. Tallaire, A. Gicquel, Coupled effect of nitrogen addition and surface temperature on the morphology and the kinetics of thick CVD diamond single crystals. Diam. Relat. Mater. **16**, 685–689 (2007). https://doi.org/10.1016/j.diamond.2006.09.012
11. H. Yamada, A. Chayahara, Y. Mokuno, S. Shikata, Numerical and experimental studies of high growth-rate over area with 1-inch in diameter under moderate input-power by using MWPCVD. Diam. Relat. Mater. **17**, 1062–1066 (2008). https://doi.org/10.1016/j.diamond.2008.01.045
12. Q. Liang, C.Y. Chin, J. Lai, C. Yan, Y. Meng, H. Mao, R.J. Hemley, Enhanced growth of high quality single crystal diamond by microwave plasma assisted chemical vapor deposition at high gas pressures. Appl. Phys. Lett. **94**, 024103 (2009). https://doi.org/10.1063/1.3072352
13. Y. Gu, J. Lu, T. Grotjohn, T. Schuelke, J. Asmussen, Microwave plasma reactor design for high pressure and high power density diamond synthesis. Diam. Relat. Mater. **24**, 210–214 (2012). https://doi.org/10.1016/j.diamond.2012.01.026
14. Y. Su, H.D. Li, S.H. Cheng, Q. Zhang, Q.L. Wang, X.Y. Lv, G.T. Zou, X.Q. Pei, J.G. Xie, Effect of N_2O on high-rate homoepitaxial growth of CVD single crystal diamonds. J. Cryst. Growth **351**, 51–55 (2012). https://doi.org/10.1016/j.jcrysgro.2012.03.041
15. J. Lu, Y. Gu, T.A. Grotjohn, T. Schuelke, J. Asmussen, Experimentally defining the safe and efficient, high pressure microwave plasma assisted CVD operating regime for single crystal diamond synthesis. Diam. Relat. Mater. **37**, 17–28 (2013). https://doi.org/10.1016/j.diamond.2013.04.007
16. N. Fujimori, T. Imai, A. Doi, Characterization of conducting diamond films. Vacuum **36**, 99–102 (1986). https://doi.org/10.1016/0042-207X(86)90279-4
17. N. Fujimori, H. Nakahata, T. Imai, Properties of boron-doped epitaxial diamond films. Jpn. J. Appl. Phys. **29**, 824–827 (1990). https://doi.org/10.1143/JJAP.29.824
18. S. Yamanaka, D. Takeuchi, H. Watanabe, H. Okushi, K. Kajimura, Low-compensated boron-doped homoepitaxial diamond films using trimethylboron. Phys. Stat. Sol. (a) **174**, 59–64 (1999). https://doi.org/10.1002/(SICI)1521-396X(199907)

19. T. Tsubota, T. Fukui, M. Kameta, T. Saito, K. Kusakabe, S. Morooka, H. Maeda, Effect of total reaction pressure on electrical properties of boron doped homoepitaxial (100) diamond films formed by microwave plasma-assisted chemical vapor deposition using trimethylboron. Diam. Relat. Mater. **8**, 1079–1082 (1999). https://doi.org/10.1016/S0925-9635(99) 00096-5

20. S. Ri, H. Kato, M. Ogura, H. Watanabe, T. Makino, S. Yamasaki, H. Okushi, Electrical and optical characterization of boron-doped (111) homoepitaxial diamond films. Diam. Relat. Mater. **14**, 1964–1968 (2005). https://doi.org/10.1016/j.diamond.2005.06.032

21. C. Baron, M. Wade, A. Deneuville, F. Jomard, J. Chevallier, Cathodoluminescence of highly and heavily boron doped (100) homoepitaxial diamond films. Diam. Relat. Mater. **15**, 597–601 (2006). https://doi.org/10.1016/j.diamond.2006.01.015

22. T. Teraji, H. Wada, M. Yamamoto, K. Arima, T. Ito, Highly efficient doping of boron into high-quality homoepitaxial diamond films. Diam. Relat. Mater. **15**, 602–606 (2006). https://doi.org/10.1016/j.diamond.2006.01.011

23. T. Teraji, Chemical vapor deposition of homoepitaxial diamond films. Phys. Stat. Sol. (a) **203**, 3324–3357 (2006). https://doi.org/10.1002/pssa.200671408

24. V. Mortet, M. Daenen, T. Teraji, A. Lazea, V. Vorlicek, J. D'Haen, K. Haenen, M. D'Olieslaeger, Characterization of boron doped diamond epilayers grown in a NIRIM type reactor. Diam. Relat. Mater. **17**, 1330–1334 (2008). https://doi.org/10.1016/j.diamond.2008. 01.087

25. J. Barjon, N. Habka, C. Mer, F. Jormard, J. Chevallier, P. Bergonzo, Resistivity of boron doped diamond. Phys. Stat. Sol. RRL **3**, 202–204 (2009). https://doi.org/10.1002/pssr. 200903097

26. J. Pernot, P.N. Volpe, F. Omnès, P. Muret, Hall hole mobility in boron-doped homoepitaxial diamond. Phys. Rev. B **81**, 205203 (2010). https://doi.org/10.1103/PhysRevB.81.205203

27. F. Omnès, P. Muret, P.N. Volpe, M. Wade, J. Pernot, F. Jomard, Study of boron doping in MPCVD grown homoepitaxial diamond layers based on cathodoluminescence spectroscopy, secondary ion mass spectroscopy and capacitance–voltage measurements. Diam. Relat. Mater. **20**, 912–916 (2011). https://doi.org/10.1016/j.diamond.2011.05.010

28. M. Ogura, H. Kato, T. Makino, H. Okushi, S. Yamasaki, J. Misorientation-angle dependence of boron incorporation into (0 0 1)-oriented chemical-vapor-deposited (CVD) diamond. J. Cryst. Growth **317**, 60–63 (2011). https://doi.org/10.1016/j.jcrysgro.2011.01.010

29. M.E. Belousov, Y.A. Mankelevich, P.V. Minakov, A.T. Rakhimov, N.V. Suetin, R.A. Khmelnitskiy, A.A. Tal, A.V. Khomich, Boron-doped homoepitaxial diamond CVD from microwave plasma-activated ethanol/trimethyl borate/hydrogen mixtures. Chem. Vap. Depos. **18**, 302–308 (2012). https://doi.org/10.1002/cvde.201206993

30. J. Achard, R. Issaoui, A. Tallaire, F. Silva, J. Barjon, F. Jomard, A. Gicquel, Freestanding CVD boron doped diamond single crystals: a substrate for vertical power electronic devices? Phys. Stat. Sol. (a) **209**, 1651–1658 (2012). https://doi.org/10.1002/pssa. 201200045

31. A. Lazea, Y. Garino, T. Teraji, S. Koizumi, High quality p-type chemical vapor deposited {111}-oriented diamonds: growth and fabrication of related electrical devices. Phys. Stat. Sol. (a) **209**, 1978–1981 (2012). https://doi.org/10.1002/pssa.201228162

32. S. Koizumi, M. Kamo, Y. Sato, H. Ozaki, T. Inuzuka, Growth and characterization of phosphorous doped 111 homoepitaxial diamond thin films. Appl. Phys. Lett. **71**, 1065–1067 (1997). https://doi.org/10.1063/1.119729

33. S. Koizumi, T. Teraji, H. Kanda, Phosphorus-doped chemical vapor deposition of diamond. Diam. Relat. Mater. **9**, 935–940 (2000). https://doi.org/10.1016/S0925-9635(00)00217-X

34. M. Katagiri, J. Isoya, S. Koizumi, H. Kanda, Lightly phosphorus-doped homoepitaxial diamond films grown by chemical vapor deposition. Appl. Phys. Lett. **85**, 6365–6367 (2004). https://doi.org/10.1063/1.1840119

35. M. Suzuki, H. Yoshida, N. Sakuma, T. Ono, T. Sakai, S. Koizumi, Electrical characterization of phosphorus-doped n-type homoepitaxial diamond layers by Schottky barrier diodes. Appl. Phys. Lett. **84**, 2349–2351 (2004). https://doi.org/10.1063/1.1695206

36. M. Suzuki, S. Koizumi, M. Katagiri, H. Yoshida, N. Sakuma, T. Ono, T. Sakai, Electrical characterization of phosphorus-doped n-type homoepitaxial diamond layers. Diam. Relat. Mater. **13**, 2037–2040 (2004). https://doi.org/10.1016/j.diamond.2004.06.022

37. H. Kato, S. Yamasaki, H. Okushi, n-type doping of (001)-oriented single-crystalline diamond by phosphorus. Appl. Phys. Lett. **86**, 222111 (2005). https://doi.org/10.1063/1. 1944228

38. S. Koizumi, M. Suzuki, n-Type doping of diamond. Phys. Stat. Sol. (a) **203**, 3358–3366 (2006). https://doi.org/10.1002/pssa.200671407

39. H. Kato, T. Makino, S. Yamasaki, H. Okushi, n-type diamond growth by phosphorus doping on (0 0 1)-oriented surface. J. Phys. D Appl. Phys. **40**, 6189–6200 (2007). https://doi.org/10. 1088/0022-3727/40/20/s05

40. J. Perot, S. Koizumi, Electron mobility in phosphorous doped {111} homoepitaxial diamond. Appl. Phys. Lett. **93**, 052105 (2008). https://doi.org/10.1063/1.2969066

41. H. Kato, D. Takeuchi, N. Tokuda, H. Umezawa, S. Yamasaki, H. Okushi, Electrical activity of doped phosphorus atoms in (001) n-type diamond. Phys. Stat. Sol. (a) **205**, 2195–2199 (2008). https://doi.org/10.1002/pssa.200879722

42. M.-A. Pinault-Thaury, B. Berini, I. Sternger, E. Chikoidze, A. Lusson, F. Jomard, J. Chevallier, J. Barjon, High fraction of substitutional phosphorus in a (100) diamond epilayer with low surface roughness. Appl. Phys. Lett. **100**, 192109 (2012). https://doi.org/ 10.1063/1.4712617

43. S. Koizumi, K. Watanabe, M. Hasegawa, H. Kanda, Ultraviolet emission from a diamond pn junction. Science **292**, 1899–1901 (2001). https://doi.org/10.1126/science.1060258

44. H. Okushi, High quality homoepitaxial CVD diamond for electronic devices. Diam. Relat. Mater. **10**, 281–288 (2001). https://doi.org/10.1016/S0925-9635(00)00399-X

45. T. Makino, N. Tokuda, H. Kato, M. Ogura, H. Watanabe, S. Ri, S. Yamasaki, H. Okushi, High-efficiency excitonic emission with deep-ultraviolet light from (001)-oriented diamond p-i-n junction. Jpn. J. Appl. Phys. **45**, L1042–L1044 (2006). https://doi.org/10.1143/jjap.45. l1042

46. D. Shin, N. Tokuda, B. Rezek, C.E. Nebel, Periodically arranged benzene-linker molecules on boron-doped single-crystalline diamond films for DNA sensing. Electrochem. Commun. **8**, 844–850 (2006). https://doi.org/10.1016/j.elecom.2006.03.014

47. D. Shin, B. Rezek, N. Tokuda, D. Takeuchi, H. Watanabe, T. Nakamura, T. Yamamoto, C. E. Nebel, Photo- and electrochemical bonding of DNA to single crystalline CVD diamond. Phys. Stat. Sol. (a) **203**, 3245–3272 (2006). https://doi.org/10.1002/pssa.200671402

48. H. Umezawa, N. Tokuda, M. Ogura, S. Ri, S. Shikata, Characterization of leakage current on diamond Schottky barrier diodes using thermionic-field emission modeling. Diam. Relat. Mater. **15**, 1949–1953 (2006). https://doi.org/10.1016/j.diamond.2006.08.030

49. K.-S. Song, T. Hiraki, H. Umezawa, H. Kawarada, Miniaturized diamond field-effect transistors for application in biosensors in electrolyte solution. Appl. Phys. Lett. **90**, 063901 (2007). https://doi.org/10.1063/1.2454390

50. E. Kohn, A. Denisenko, Concepts for diamond electronics. Thin Solid Films **515**, 4333–4339 (2007). https://doi.org/10.1016/j.tsf.2006.07.179

51. M. Liao, Y. Koide, J. Alvarez, Single Schottky-barrier photodiode with interdigitated-finger geometry: application to diamond. Appl. Phys. Lett. **90**, 123507 (2007). https://doi.org/10. 1063/1.2715440

52. T. Makino, N. Tokuda, H. Kato, M. Ogura, H. Watanabe, S. Ri, S. Yamasaki, H. Okushi, Electrical and light-emitting properties of (001)-oriented homoepitaxial diamond p–i–n junction. Diam. Relat. Mater. **16**, 1025–1028 (2007). https://doi.org/10.1016/j.diamond. 2007.01.024

53. C.E. Nebel, D. Shin, B. Rezek, N. Tokuda, H. Uetsuka, H. Watanabe, Diamond and biology. J. R. Soc. Interface **4**, 439–461 (2007). https://doi.org/10.1098/rsif.2006.0196

54. H. Umezawa, T. Saito, N. Tokuda, M. Ogura, S. Ri, H. Yoshikawa, S. Shikata, Leakage current analysis of diamond Schottky barrier diode. Appl. Phys. Lett. **90**, 073506 (2007). https://doi.org/10.1063/1.2643374

55. T. Makino, N. Tokuda, H. Kato, S. Kanno, S. Yamasaki, H. Okushi, Electrical and light-emitting properties of homoepitaxial diamond p-i-n junction. Phys. Stat. Sol. (a) **205**, 2200–2206 (2008). https://doi.org/10.1002/pssa.200879717
56. T. Makino, S. Tanimoto, Y. Hayashi, H. Kato, N. Tokuda, M. Ogura, D. Takeuchi, K. Oyama, H. Ohashi, H. Okushi, S. Yamasaki, Diamond Schottky-pn diode with high forward current density and fast switching operation. Appl. Phys. Lett. **94**, 262101 (2009). https://doi.org/10.1063/1.3159837
57. T. Makino, S. Ri, N. Tokuda, H. Kato, S. Yamasaki, H. Okushi, Electrical and light-emitting properties from (111)-oriented homoepitaxial diamond p–i–n junctions. Diam. Relat. Mater. **18**, 764–767 (2009). https://doi.org/10.1016/j.diamond.2009.01.016
58. K. Oyama, S. Ri, H. Kato, M. Ogura, T. Makino, D. Takeuchi, N. Tokuda, H. Okushi, S. Yamasaki, High performance of diamond p[sup +]-i-n[sup +] junction diode fabricated using heavily doped p^+ and n^+ layers. Appl. Phys. Lett. **94**, 152109 (2009). https://doi.org/10.1063/1.3120560
59. P.-N. Volpe, P. Muret, J. Pernot, F. Omnès, T. Teraji, Y. Koide, F. Jomard, D. Planson, P. Brosselard, N. Dheilly, B. Vergne, S. Scharnholz, Extreme dielectric strength in boron doped homoepitaxial diamond. Appl. Phys. Lett. **97**, 223501 (2010). https://doi.org/10.1063/1.3520140
60. R. Hoffmann, A. Kriele, H. Obloh, N. Tokuda, W. Smirnov, N. Yang, C.E. Nebel, The creation of a biomimetic interface between boron-doped diamond and immobilized proteins. Biomaterials **32**, 7325–7332 (2011). https://doi.org/10.1016/j.biomaterials.2011.06.052
61. T. Kawae, Y. Hori, T. Nakajima, H. Kawasaki, N. Tokuda, S. Okamura, Y. Takano, A. Morimoto, Structure and electrical properties of (Pr, Mn)-codoped $BiFeO_3$/B-doped diamond layered structure. Electrochem. Solid-State Lett. **15**, G31–G34 (2011). https://doi.org/10.1149/1.3568838
62. H. Kawarada, A.R. Ruslinda, Diamond electrolyte solution gate FETs for DNA and protein sensors using DNA/RNA aptamers. Phys. Stat. Sol. (a) **208**, 2005–2016 (2011). https://doi.org/10.1002/pssa.201100503
63. R. Hoffmann, H. Obloh, N. Tokuda, N. Yang, C.E. Nebel, Fractional surface termination of diamond by electrochemical oxidation. Langmuir **28**, 47–50 (2012). https://doi.org/10.1021/la2039366
64. T. Iwasaki, Y. Hoshino, K. Tsuzuki, H. Kato, T. Makino, M. Ogura, D. Takeuchi, T. Matsumoto, H. Okushi, S. Yamasaki, M. Hatano, Diamond junction field-effect transistors with selectively grown n^+-side gates. Appl. Phys. Express **5**, 091301 (2012). https://doi.org/10.1143/apex.5.091301
65. H. Kato, K. Oyama, T. Makino, M. Ogura, D. Takeuchi, S. Yamasaki, Diamond bipolar junction transistor device with phosphorus-doped diamond base layer. Diam. Relat. Mater. **27–28**, 19–22 (2012). https://doi.org/10.1016/j.diamond.2012.05.004
66. T. Kawae, H. Kawasaki, T. Nakajima, N. Tokuda, S. Okamura, A. Morimoto, Y. Takano, Fabrication of (Bi,Pr)(Fe,Mn)O_3 thin films on polycrystalline diamond substrates by chemical solution deposition and their properties. Jpn. J. Appl. Phys. **51**, 09LA08 (2012). https://doi.org/10.1143/jjap.51.09la08
67. R. Edgington, A.R. Ruslinda, S. Sato, Y. Ishiyama, K. Tsuge, T. Ono, H. Kawarada, R.B. Jackman, Boron delta-doped (111) diamond solution gate field effect transistors. Biosens. Bioelectron. **33**, 152–157 (2012). https://doi.org/10.1016/j.bios.2011.12.044
68. H. Kawarada, High-current metal oxide semiconductor field-effect transistors on H-terminated diamond surfaces and their high-frequency operation. Jpn. J. Appl. Phys. **51**, 090111 (2012). https://doi.org/10.1143/jjap.51.090111
69. T. Makino, H. Kato, D. Takeuchi, M. Ogura, H. Okushi, S. Yamasaki, Device design of diamond Schottky-pn diode for low-loss power electronics. Jpn. J. Appl. Phys. **51**, 090116 (2012). https://doi.org/10.1143/jjap.51.090116
70. S. Cheng, L. Sang, M. Liao, J. Liu, M. Imura, H. Li, Y. Koide, Integration of high-dielectric constant Ta_2O_5 oxides on diamond for power devices. Appl. Phys. Lett. **101**, 232907 (2012). https://doi.org/10.1063/1.4770059

71. N. Mizuochi, T. Makino, H. Kato, D. Takeuchi, M. Ogura, H. Okushi, M. Nothaft, P. Neumann, A. Gali, F. Jelezko, J. Wrachtrup, S. Yamasaki, Electrically driven single-photon source at room temperature in diamond. Nat. Photon. **6**, 299–303 (2012). https://doi.org/10.1038/nphoton.2012.75

72. M. Liao, L. Sang, T. Teraji, M. Imura, J. Alvarez, Y. Koide, Comprehensive investigation of single crystal diamond deep-ultraviolet detectors. Jpn. J. Appl. Phys. **51**, 090115 (2012). https://doi.org/10.1143/jjap.51.090115

73. D. Takeuchi, T. Makino, H. Kato, M. Ogura, H. Okushi, H. Ohashi, S. Yamasaki, High-voltage vacuum switch with a diamond p–i–n diode using negative electron affinity. Jpn. J. Appl. Phys. **51**, 090113 (2012). https://doi.org/10.1143/jjap.51.090113

74. H. Umezawa, M. Nagase, Y. Kato, S. Shikata, High temperature application of diamond power device. Diam. Relat. Mater. **24**, 201–205 (2012). https://doi.org/10.1016/j.diamond.2012.01.011

75. G. Chicot, A. Maréchal, R. Motte, P. Muret, E. Gheeraert, J. Pernot, Metal oxide semiconductor structure using oxygen-terminated diamond. Appl. Phys. Lett. **102**, 242108 (2013). https://doi.org/10.1063/1.4811668

76. T. Matsumoto, H. Kato, K. Oyama, T. Makino, M. Ogura, D. Takeuchi, T. Inokuma, N. Tokuda, S. Yamasaki, Inversion channel diamond metal-oxide-semiconductor field-effect-transistor with normally off characteristics. Sci. Rep. **6**, 31585 (2016). https://doi.org/10.1038/srep31585

77. T. Matsumoto, T. Mukose, T. Makino, D. Takeuchi, S. Yamasaki, T. Inokuma, N. Tokuda, Diamond Schottky-pn diode using lightly nitrogen-doped layer. Diam. Relat. Mater. **75**, 152–154 (2017). https://doi.org/10.1016/j.diamond2017.03.018

78. A. Gicquel, K. Hassouni, S. Farhat, Y. Breton, C.D. Scott, M. Lefebvre, M. Pealat, Spectroscopic analysis and chemical kinetics modeling of a diamond deposition plasma reactor. Diam. Relat. Mater. **3**, 581–586 (1994). https://doi.org/10.1016/0925-9635(94)90229-1

79. C. Benndorf, P. Joeris, R. Kröger, Mass and optical emission spectroscopy of plasmas for diamond synthesis. Pure Appl. Chem. **66**, 1195–1205 (1994). https://doi.org/10.1351/pac199466061195

80. T. Fujii, M. Kareev, Mass spectrometric studies of a CH_4/H_2 microwave plasma under diamond deposition conditions. J. Appl. Phys. **89**, 2543–2546 (2001). https://doi.org/10.1063/1.1346655

81. P. Deák, A. Kováts, P. Csíkváry, I. Maros, G. Hárs, Ethynyl (C_2H): a major player in the chemical vapor deposition of diamond. Appl. Phys. Lett. **90**, 051503 (2007). https://doi.org/10.1063/1.2437718

82. H. Zhou, J. Watanabe, M. Miyake, A. Ogino, M. Nagatsu, R. Zhan, Optical and mass spectroscopy measurements of $Ar/CH_4/H_2$ microwave plasma for nano-crystalline diamond film deposition. Diam. Relat. Mater. **16**, 675–678 (2007). https://doi.org/10.1016/j.diamond.2006.11.074

83. J. Ma, M.N.R. Ashfold, Y.A. Mankelevich, Validating optical emission spectroscopy as a diagnostic of microwave activated $CH_4/Ar/H_2$ plasmas used for diamond chemical vapor deposition. J. Appl. Phys. **105**, 043302 (2009). https://doi.org/10.1063/1.3078032

84. A. Gicquel, N. Derkaoui, C. Rond, F. Benedic, G. Cicala, D. Moneger, K. Hassouni, Quantitative analysis of diamond deposition reactor efficiency. Chem. Phys. **398**, 239–247 (2012). https://doi.org/10.1016/j.chemphys.2011.08.022

85. J.C. Richley, M.W. Kelly, M.N.R. Ashfold, Y.A. Mankelevich, Optical emission from microwave activated C/H/O gas mixtures for diamond chemical vapor deposition. J. Phys. Chem. A **116**, 9447–9458 (2012). https://doi.org/10.1021/jp306191y

86. P. Bou, J.C. Boettner, L. Vandenbulcke, Kinetic calculations in plasmas used for diamond deposition. Jpn. J. Appl. Phys. **31**, 1505–1513 (1992). https://doi.org/10.1143/JJAP.31.1505

87. M.C. McMaster, W.L. Hsu, M.E. Coltrin, D.S. Dandy, C. Fox, Dependence of the gas composition in a microwave plasma-assisted diamond chemical vapor deposition reactor on the inlet carbon source: CH_4 versus C_2H_2. Diam. Relat. Mater. **4**, 1000–1008 (1995). https://doi.org/10.1016/0925-9635(95)00270-7

88. J.M. Larson, M.T. Swihart, S.L. Girshick, Characterization of the near-surface gas-phase chemical environment in atmospheric-pressure plasma chemical vapor deposition of diamond. Diam. Relat. Mater. **8**, 1863–1874 (1999). https://doi.org/10.1016/S0925-9635(99)00143-0

89. O. Aubry, J.-L. Delfau, C. Met, L. Vandenbulcke, C. Vovelle, Precursors of diamond films analysed by molecular beam mass spectrometry of microwave plasmas. Diam. Relat. Mater. **13**, 116–124 (2004). https://doi.org/10.1016/j.diamond.2003.09.009

90. J. Achard, F. Silva, A. Tallaire, X. Bonnin, G. Lomvardi, K. Hassouni, A. Gicquel, High quality MPACVD diamond single crystal growth: high microwave power density regime. J. Phys. D **40**, 6175–6188 (2007). https://doi.org/10.1088/0022-3727/40/20/S04

91. H. Yamada, A. Chayahara, Y. Mokuno, Simplified description of microwave plasma discharge for chemical vapor deposition of diamond. J. Appl. Phys. **101**, 063302 (2007). https://doi.org/10.1063/1.2711811

92. J. Ma, J.C. Richley, M.N.R. Ashfold, Y.A. Mankelevich, Probing the plasma chemistry in a microwave reactor used for diamond chemical vapor deposition by cavity ring down spectroscopy. J. Appl. Phys. **104**, 103305 (2008). https://doi.org/10.1063/1.3021095

93. F. Silva, J. Achard, O. Brinza, X. Bonnin, K. Hassouni, A. Anthonis, K.D. Corte, J. Barjon, High quality, large surface area, homoepitaxial MPACVD diamond growth. Diam. Relat. Mater. **18**, 683–697 (2009). https://doi.org/10.1016/j.diamond.2009.01.038

94. K. Hassouni, F. Silva, A. Gicquel, Modelling of diamond deposition microwave cavity generated plasmas. J. Phys. D **43**, 153001 (2010). https://doi.org/10.1088/0022-3727/43/15/153001

95. H. Yamada, A. Chayahara, Y. Mokuno, S. Shikata, Model of reactive microwave plasma discharge for growth of single-crystal diamond. Jpn. J. Appl. Phys. **50**, 01AB02 (2011). https://doi.org/10.1143/jjap.50.01ab02

96. H. Yamada, Numerical simulations to study growth of single-crystal diamond by using microwave plasma chemical vapor deposition with reactive (H, C, N) species. Jpn. J. Appl. Phys. **51**, 090105 (2012). https://doi.org/10.1143/jjap.51.090105

97. C.-L. Cheng, H.-C. Chang, J.-C. Lin, K.-J. Song, J.-K. Wang, Direct observation of hydrogen etching anisotropy on diamond single crystal surfaces. Phys. Rev. Lett. **78**, 3713–3716 (1997). https://doi.org/10.1103/PhysRevLett.78.3713

98. H. Kuroshima, T. Makino, S. Yamasaki, T. Matsumoto, T. Inokuma, N. Tokuda, Mechanism of anisotropic etching on diamond (111) surfaces by a hydrogen plasma treatment. Appl. Surf. Sci. **422**, 452–455 (2017). https://doi.org/10.1016/j.apsusc.2017.06.005

99. T. Tsuno, T. Imai, Y. Nishibayashi, K. Hamada, N. Fujimori, Epitaxially grown diamond (001) $2 \times 1/1 \times 2$ surface investigated by scanning tunneling microscopy in air. Jpn. J. Appl. Phys. **30**, 1063–1066 (1991). https://doi.org/10.1143/JJAP.30.1063

100. H. Sasaki, H. Kawarada, Structure of chemical vapor deposited diamond (111) surfaces by scanning tunneling microscopy. Jpn. J. Appl. Phys. **32**, L1771–L1774 (1993). https://doi.org/10.1143/JJAP.32.L1771

101. L.F. Sutcu, C.J. Chu, M.S. Thompson, R.H. Hauge, J.L. Margrave, M.P. D'Evelyn, Atomic force microscopy of (100), (110), and (111) homoepitaxial diamond films. J. Appl. Phys. **71**, 5930–5940 (1992). https://doi.org/10.1063/1.350443

102. T. Tsuno, T. Tomikawa, S. Shikata, T. Imai, N. Fujirmori, Diamond(001) single-domain 2×1 surface grown by chemical vapor deposition. Appl. Phys. Lett. **64**, 572–574 (1994). https://doi.org/10.1063/1.111107

103. T. Tsuno, T. Tomikawa, S. Shikata, N. Fujimori, Diamond homoepitaxial growth on (111) substrate investigated by scanning tunneling microscope. J. Appl. Phys. **75**, 1526–1529 (1994). https://doi.org/10.1063/1.356389

104. M. McGonigal, J.N. Russell Jr., P.E. Pehrsson, H.G. Maguire, J.E. Butler, Multiple internal reflection infrared spectroscopy of hydrogen adsorbed on diamond(110). J. Appl. Phys. **77**, 4049–4053 (1995). https://doi.org/10.1063/1.359487

105. H. Kawarada, H. Ssaki, A. Sato, Scanning-tunneling-microscope observation of the homoepitaxial diamond (001) 2×1 reconstruction observed under atmospheric pressure. Phys. Rev. B **52**, 11351–11358 (1995). https://doi.org/10.1103/PhysRevB.52.11351

106. Y. Kuang, Y. Wang, N. Lee, A. Badzian, T. Badzian, T.T. Tsong, Surface structure of homoepitaxial diamond (001) films, a scanning tunneling microscopy study. Appl. Phys. Lett. **67**, 3721–3723 (1995). https://doi.org/10.1063/1.115361

107. C.-L. Cheng, J.-C. Lin, H.-C. Chang, J.-K. Wang, Characterization of CH stretches on diamond C(111) single- and nanocrystal surfaces by infrared absorption spectroscopy. J. Chem. Phys. **105**, 8977–8978 (1996). https://doi.org/10.1063/1.472938

108. T. Takami, K. Suzuki, I. Kusunoki, I. Sakaguchi, M. Nishitani-Gamo, T. Ando, RHEED and AFM studies of homoepitaxial diamond thin film on C(001) substrate produced by microwave plasma CVD. Diam. Relat. Mater. **8**, 701–704 (1999). https://doi.org/10.1016/S0925-9635(98)00391-4

109. T. Takami, I. Kusunoki, M. Nishitani-Gamo, T. Ando, Homoepitaxial diamond (001) thin film studied by reflection high-energy electron diffraction, contact atomic force microscopy, and scanning tunneling microscopy. J. Vac. Sci. Technol. B **18**, 1198–1202 (2000). https://doi.org/10.1116/1.591360

110. A. Heerwagen, M. Strobel, M. Himmelhaus, M. Buck, Chemical vapor deposition of diamond: an in situ study by vibrational spectroscopy. J. Am. Chem. Soc. **123**, 6732–6733 (2001). https://doi.org/10.1021/ja016056q

111. L.K. Bigelow, M.P. D'Evelyn, Role of surface and interface science in chemical vapor deposition diamond technology. Surf. Sci. **500**, 986–1004 (2002). https://doi.org/10.1016/S0039-6028(01)01545-X

112. L. Ackermann, W. Kulisch, Investigation of diamond etching and growth by in situ scanning tunneling microscopy. Diam. Relat. Mater. **8**, 1256–1260 (1999). https://doi.org/10.1016/S0925-9635(99)00119-3

113. B. Voigtländer, M. Kästner, P. Šmilauer, Magic islands in Si/Si(111) homoepitaxy. Phys. Rev. Lett. **81**, 858–861 (1998). https://doi.org/10.1103/PhysRevLett.81.858

114. H. Yamaguchi, Y. Homma, Imaging of layer by layer growth processes during molecular beam epitaxy of GaAs on (111)A substrates by scanning electron microscopy. Appl. Phys. Lett. **73**, 3079–3081 (1998). https://doi.org/10.1063/1.122678

115. M. H. Xie, S.M. Seutter, W.K. Zhu, L.X. Zheng, H. Wu, S.Y. Tong, Anisotropic step-flow growth and island growth of GaN(0001) by molecular beam epitaxy. Phys. Rev. Lett. **82**, 2749–2752 (1999). https://doi.org/10.1103/physrevlett.82.2749

116. N. Tokuda, T. Makino, T. Inokuma, S. Yamasaki, Formation of step-free surfaces on diamond (111) mesas by homoepitaxial lateral growth. Jpn. J. Appl. Phys. **51**, 090107 (2012). https://doi.org/10.1143/JJAP.51.090107

117. N. Tokuda, T. Makino, T. Inokuma, S. Yamasaki, Formation of step-free diamond (111)surfaces by plasma-enhanced CVD. J. Jpn. Assoc. Cryst. Growth **39**, 185–189 (2012)

118. F. Jelezko, T. Gaebel, I. Popa, A. Gruber, J. Wrachtrup, Observation of coherent oscillations in a single electron spin. Phys. Rev. Lett. **92**, 076401 (2004). https://doi.org/10.1103/PhysRevLett.92.076401

119. L. Childress, M.V. Gurudev Dutt, J.M. Taylor, A.S. Zibrov, F. Jelezko, J. Wrachtrup, P.R. Hemmer, M.D. Lukin, Coherent dynamics of coupled electron and nuclear spin qubits in diamond. Science **314**, 281–285 (2006). https://doi.org/10.1126/science.1131871

120. M.V. Gurudev Dutt, L. Childress, L. Jiang, E. Togan, J. Maze, F. Jelezko, A.S. Zibrov, P.R. Hemmer, M.D. Lukin, Quantum register based on individual electronic and nuclear spin qubits in diamond. Science **316**, 1312–1316 (2007). https://doi.org/10.1126/science.1139831

121. J.R. Maze, J.M. Taylor, M.D. Lukin, Electron spin decoherence of single nitrogen-vacancy defects in diamond. Phys. Rev. B **78**, 094303 (2008). https://doi.org/10.1103/PhysRevB.78.094303

122. P. Neumann, N. Mizuochi, F. Rempp, P. Hemmer, H. Watanabe, S. Yamasaki, V. Jacques, T. Gaebel, F. Jelezko, J. Wrachtrup, Multipartite entanglement among single spins in diamond. Scinece **320**, 1326–1329 (2008). https://doi.org/10.1126/science.1157233

123. G. Balasubramanian, P. Neumann, D. Twitchen, M. Markham, R. Kolesov, N. Mizuochi, J. Isoya, J. Achard, J. Beck, J. Tissler, V. Jacques, P.R. Hemmer, F. Jelezko, J. Wrachtrup, Ultralong spin coherence time in isotopically engineered diamond. Nat. Mater. **8**, 383–387 (2009). https://doi.org/10.1038/nmat2420

124. B.B. Buckley, G.D. Fuchs, L.C. Bassett, D.D. Awschalom, Spin-light coherence for single-spin measurement and control in diamond. Science **330**, 1212–1215 (2010). https://doi.org/10.1126/science.1196436

125. X. Zhu, S. Saito, A. Kemp, K. Kakuyanagi, S. Karimoto, H. Nakano, W.J. Munro, Y. Tokura, M.S. Everitt, K. Nemoto, M. Kasu, N. Mizuochi, K. Semba, Coherent coupling of a superconducting flux qubit to an electron spin ensemble in diamond. Nature **478**, 221–224 (2011). https://doi.org/10.1038/nature10462

126. K.C. Lee, M.R. Sprague, B.J. Sussman, J. Nunn, N.K. Langford, X.-M. Jin, T. Champion, P. Michelberger, K.F. Reim, D. England, D. Jaksch, I.A. Walmsley, Entangling macroscopic diamonds at room temperature. Science **334**, 1253–1256 (2011). https://doi.org/10.1126/science.1211914

127. J.F. Prings, Activation of boron-dopant atoms in ion-implanted diamonds. Phys. Rev. B **38**, 5576–5584 (1988). https://doi.org/10.1103/PhysRevB.38.5576

128. C. Uzan-Saguy, R. Kalish, R. Walker, D.N. Jamieson, S. Prawer, Formation of delta-doped, buried conducting layers in diamond, by high-energy, B-ion implantation. Diam. Relat. Mater. **7**, 1429–1432 (1998). https://doi.org/10.1016/S0925-9635(98)00231-3

129. K. Ueda, M. Kasu, T. Makimoto, High-pressure and high-temperature annealing as an activation method for ion-implanted dopants in diamond. Appl. Phys. Lett. **90**, 122102 (2007). https://doi.org/10.1063/1.2715034

130. N. Tsubouchi, M. Ogura, Enhancement of dopant activation in B-implanted diamond by high-temperature annealing. Jpn. J. Appl. Phys. **47**, 7047–7051 (2008). https://doi.org/10.1143/JJAP.47.7047

131. N. Tsubouchi, M. Ogura, N. Mizuochi, H. Watanabe, Electrical properties of a B doped layer in diamond formed by hot B implantation and high-temperature annealing. Diam. Relat. Mater. **18**, 128–131 (2009). https://doi.org/10.1016/j.diamond.2008.09.013

132. A.K. Ratnikova, M.P. Dukhnovsky, Y.Y. Fedorov, V.E. Zemlyakov, A.B. Muchnikov, A. L. Vikharev, A.M. Gorbachev, D.B. Radishev, A.A. Altukhov, A.V. Mitenkin, Homoepitaxial single crystal diamond grown on natural diamond seeds (type IIa) with boron-implanted layer demonstrating the highest mobility of 1150 cm^2/V s at 300 K for ion-implanted diamond. Diam. Relat. Mater. **20**, 12343–1245 (2011). https://doi.org/10.1016/j.diamond.2011.07.007

133. V.S. Bormashov, S.A. Tarelkin, S.G. Buga, M.S. Kuznetsov, S.A. Terentiev, A.N. Semenov, V.D. Blank, Electrical properties of the high quality boron-doped synthetic single-crystal diamonds grown by the temperature gradient method. Diam. Relat. Mater. **35**, 19–23 (2013). https://doi.org/10.1016/j.diamond.2013.02.011

134. S. Yamanaka, H. Watanabe, S. Masai, D. Takeuchi, H. Okushi, K. Kajimura, High-quality B-doped homoepitaxial diamond films using trimethylboron. Jpn. J. Appl. Phys. **37**, L1129–L1131 (1998). https://doi.org/10.1143/JJAP.37.L1129

135. J.-P. Lagrange, A. Deneuville, E. Gheeraert, Activation energy in low compensated homoepitaxial boron-doped diamond films. Diam. Relat. Mater. **7**, 1390–1393 (1998). https://doi.org/10.1016/S0925-9635(98)00225-8

136. E.A. Ekimov, V.A. Sidrov, E.D. Bauer, N.N. Mel'nki, N.J. Curro, J.D. Thompson, S.M. Stishov, Superconductivity in diamond. Nature **428**, 542–545 (2004). https://doi.org/10.1038/nature02449

137. Y. Takano, M. Nagao, I. Sakaguchi, M. Tachiki, T. Hatano, K. Kobayashi, H. Umezawa, H. Kawarada, Superconductivity in diamond thin films well above liquid helium temperature. Appl. Phys. Lett. **85**, 2851–2853 (2004). https://doi.org/10.1063/1.1802389

138. T. Yokoya, T. Nakamura, T. Matsushita, T. Muro, Y. Takano, M. Nagao, T. Takenouchi, H. Kawarada, T. Oguchi, Origin of the metallic properties of heavily boron-doped superconducting diamond. Nature **438**, 647–650 (2005). https://doi.org/10.1038/nature04278

139. E. Bustarret, Superconducting diamond: an introduction. Phys. Stat. Sol. (a) **205**, 997–1008 (2008). https://doi.org/10.1002/pssa.200777501

140. T. Klein, P. Achatz, J. Kacmarcik, C. Marcenat, F. Gustafsson, J. Marcus, E. Bustarret, J. Pernot, F. Omnes, B.E. Sernelius, C. Persson, A. Silva, C. Cytermann, Metal-insulator transition and superconductivity in boron-doped diamond. Phys. Rev. B **75**, 165313 (2007). https://doi.org/10.1103/PhysRevB.75.165313

141. A. Kawano, H. Ishiwata, S. Iriyama, R. Okada, T. Yamaguchi, Y. Takano, H. Kawarada, Superconductor-to-insulator transition in boron-doped diamond films grown using chemical vapor deposition. Phys. Rev. B **82**, 085318 (2010). https://doi.org/10.1103/PhysRevB.82.085318

142. N. Tokuda, T. Saito, H. Umezawa, H. Okushi, S. Yamasaki, The role of boron atoms in heavily boron-doped semiconducting homoepitaxial diamond growth—study of surface morphology. Diam. Relat. Mater. **16**, 409–411 (2007). https://doi.org/10.1016/j.diamond.2006.08.013

143. N. Tokuda, H. Umezawa, T. Saito, K. Yamabe, H. Okushi, S. Yamasaki, Surface roughening of diamond (001) films during homoepitaxial growth in heavy boron doping. Diam. Relat. Mater. **16**, 767–770 (2007). https://doi.org/10.1016/j.diamond.2006.12.024

144. N. Tokuda, H. Umezawa, K. Yamabe, H. Okushi, S. Yamasaki, Hillock-free heavily boron-doped homoepitaxial diamond films on misoriented (001) substrates. Jpn. J. Appl. Phys. **46**, 1469–1470 (2007). https://doi.org/10.1143/JJAP.46.1469

145. H. Kato, D. Takeuchi, N. Tokuda, H. Umezawa, H. Okushi, S. Yamasaki, Characterization of specific contact resistance on heavily phosphorus-doped diamond films. Diam. Relat. Mater. **18**, 782–785 (2009). https://doi.org/10.1016/j.diamond.2009.01.033

146. T. Yatsui, W. Nomura, M. Naruse, M. Ohtsu, Realization of an atomically flat surface of diamond using dressed photon-phonon etching. J. Phys. D **45**, 475302 (2012). https://doi.org/10.1088/0022-3727/45/47/475302

147. A. Kubota, S. Fukuyama, Y. Ichimori, M. Touge, Surface smoothing of single-crystal diamond (100) substrate by polishing technique. Diam. Relat. Mater. **24**, 59–62 (2012). https://doi.org/10.1016/j.diamond.2011.10.022

148. Y. Kato, H. Umezawa, S. Shikata, M. Touge, Effect of an ultraflat substrate on the epitaxial growth of chemical-vapor-deposited diamond. Appl. Phys. Express **6**, 025506 (2013). https://doi.org/10.7567/APEX.6.025506

149. N. Tokuda, H. Umezawa, K. Yamabe, H. Okushi, S. Yamasaki, Growth of atomically step-free surface on diamond {111} mesas. Diam. Relat. Mater. **19**, 288–290 (2010). https://doi.org/10.1016/j.diamond.2009.11.015

150. N. Tokuda, M. Ogura, T. Matsumoto, S. Yamasaki, T. Inokuma, Influence of substrate misorientation on the surface morphology of homoepitaxial diamond (111) films. Phys. Status Solidi A **213**, 2051–2055 (2016). https://doi.org/10.1002/pssa.201600082

151. H. Sawada, H. Ichinose, H. Watanabe, D. Takeuchi, H. Okushi, Cross-sectional TEM study of unepitaxial crystallites in a homoepitaxial diamond film. Diam. Relat. Mater. **10**, 2030–2034 (2001). https://doi.org/10.1016/S0925-9635(01)00477-0

152. T. Tsuno, T. Imai, N. Fujimori, Twinning structure and growth hillock on diamond (001) epitaxial film. Jpn. J. Appl. Phys. **33**, 4039–4043 (1994). https://doi.org/10.1143/JJAP.33.4039

153. H. Watanabe, D. Takeuchi, S. Yamanaka, H. Okushi, K. Kajimura, T. Sekiguchi, Homoepitaxial diamond film with an atomically flat surface over a large area. Diam. Relat. Mater. **8**, 1272–1276 (1999). https://doi.org/10.1016/S0925-9635(99)00126-0

154. N. Tokuda, H. Umezawa, S. Ri, M. Ogura, K. Yamabe, H. Okushi, S. Yamasaki, Atomically flat diamond (111) surface formation by homoepitaxial lateral growth. Diam. Relat. Mater. **17**, 1051–1054 (2008). https://doi.org/10.1016/j.diamond.2008.01.089

155. N. Tokuda, H. Umezawa, H. Kato, M. Ogura, S. Gonda, K. Yamabe, H. Okushi, S. Yamasaki, Nanometer scale height standard using atomically controlled diamond surface. Appl. Phys. Express **2**, 055001 (2009). https://doi.org/10.1143/APEX.2.055001

156. D. Lee, J.M. Blakely, T.W. Schroeder, J.R. Engstrom, A growth method for creating arrays of atomically flat mesas on silicon. Appl. Phys. Lett. **78**, 1349–1351 (2001). https://doi.org/10.1063/1.1352656

157. T. Nishida, N. Kobayashi, Step-free surface grown on GaAs (111)B substrate by selective area metalorganic vapor phase epitaxy. Appl. Phys. Lett. **69**, 2549–2550 (1996). https://doi.org/10.1063/1.117735

158. T. Nishida, N. Kobayashi, Formation of a 100-μm-wide stepfree GaAs (111)B surface obtained by finite area metalorganic vapor phase epitaxy. Jpn. J. Appl. Phys. **37**, L13–L14 (1997). https://doi.org/10.1143/JJAP.37.L13

159. J.A. Powell, P.G. Neudeck, A.J. Trunek, G.M. Beheim, L.G. Matus, R.W. Hoffman Jr., L.J. Keys, Growth of step-free surfaces on device-size (0001)SiC mesas. Appl. Phys. Lett. **77**, 1449–1451 (2000). https://doi.org/10.1063/1.1290717

160. T. Akasaka, Y. Kobayashi, M. Kasu, Step-Free GaN hexagons grown by selective-area metalorganic vapor phase epitaxy. Appl. Phys. Express **2**, 091002 (2009). https://doi.org/10.1143/APEX.2.091002

161. C.E. Nebel, C.R. Miskys, J.A. Garrido, M. Hermann, O. Ambacher, M. Eickoff, M. Stutzmann, AlN/diamond np-junctions. Diam. Relat. Mater. **12**, 1873–1876 (2003). https://doi.org/10.1016/S0925-9635(03)00313-3

162. C.R. Miskys, J.A. Garrido, C.E. Nebel, M. Hermann, O. Ambacher, M. Eickhoff, M. Stutzmann, AlN/diamond heterojunction diodes. Appl. Phys. Lett. **82**, 290–292 (2003). https://doi.org/10.1063/1.1532545

163. Y. Taniyasu, M. Kasu, MOVPE growth of single-crystal hexagonal AlN on cubic diamond. J. Cryst. Growth **311**, 2828–2830 (2009). https://doi.org/10.1016/j.jcrysgro.2009.01.021

164. K. Hirama, Y. Taniyasu, M. Kasu, Heterostructure growth of a single-crystal hexagonal AlN (0001) layer on cubic diamond (111) surface. J. Appl. Phys. **108**, 013528 (2010). https://doi.org/10.1063/1.3452362

165. M. Imura, K. Nakajima, M. Liao, Y. Koide, Growth mechanism of c-axis-oriented AlN on (1 1 1) diamond substrates by metal-organic vapor phase epitaxy. J. Cryst. Growth **312**, 1325–1328 (2010). https://doi.org/10.1016/j.jcrysgro.2009.09.020

166. K. Hirama, Y. Taniyasu, M. Kasu, Hexagonal AlN(0001) heteroepitaxial growth on cubic diamond (001). Jpn. J. Appl. Phys. **49**, 04DH01 (2010). https://doi.org/10.1143/jjap.49.04dh01

167. S. Tanaka, R.S. Kern, R.F. Davis, Initial stage of aluminum nitride film growth on 6H-silicon carbide by plasma-assisted, gas-source molecular beam epitaxy. Appl. Phys. Lett. **66**, 37 (1995). https://doi.org/10.1063/1.114173

168. J.A. Powell, J.B. Petit, J.H. Edgar, I.G. Jenkins, L.G. Matus, J.W. Yang, P. Pirouz, W.J. Choyke, L. Cleman, M. Yoganathan, Controlled growth of 3C-SiC and 6H-SiC films on low-tilt-angle vicinal (0001) 6H-SiC wafers. Appl. Phys. Lett. **59**, 333–335 (1991). https://doi.org/10.1063/1.105587

169. T. Ouisse, Electron transport at the SiC/SiO$_2$ interface. Phys. Status Solidi A **162**, 339–368 (1997). https://doi.org/10.1002/1521-396X(199707)162:1%3c339:AID-PSSA339%3e3.0.CO;2-G

170. N.D. Bassim, M.E. Twigg, C.R. Eddy Jr., J.C. Culbertson, M.A. Mastro, R.L. Henry, R.T. Holm, P.G. Neudeck, A.J. Trunek, J.A. Powell, Lowered dislocation densities in uniform GaN layers grown on step-free (0001) 4H-SiC mesa surfaces. Appl. Phys. Lett. **86**, 021902 (2005). https://doi.org/10.1063/1.1849834

171. J.D. Caldwell, M.A. Mastro, K.D. Hobart, O.J. Glembocki, C.R. Eddy Jr., N.D. Bassim, R.T. Holm, R.L. Henry, M.E. Twigg, F. Kub, P.G. Neudeck, A.J. Trunek, J.A. Powell, Improved ultraviolet emission from reduced defect gallium nitride homojunctions grown on step-free 4H-SiC mesas. Appl. Phys. Lett. **88**, 263509 (2006). https://doi.org/10.1063/1.2218045

Chapter 2
The Effect of Dopants on Diamond Surface Properties and Growth

Karin Larsson

Abstract The purpose with the present studies has been to support and explain the experimental observations made regarding the effect by N-, P-, S- and B-doping on the diamond (111), (100)−2 × 1 and (110) growth rate, respectively. All surfaces were assumed to be H-terminated. Density functional theory calculations were used, based on a plane wave approach under periodic boundary conditions. It was shown that the surface H abstraction reaction is most probably the rate-limiting step during diamond growth. Moreover, the results showed that it is N, substitutionally positioned within the upper diamond surface, that will cause the growth rate improvement, and not nitrogen chemisorbed onto the growing surface in the form of either NH (or NH$_2$). These results coupled very strongly to experimental counterparts. For the situation with P doping, there were no visible energy barrier obtained for the approaching H radical to any of the diamond surface planes. Hence, the growth rate must be appreciably increased as a function of doping with P. It was furthermore observed that S and B doping will lead to anomalous changes in the diamond growth rate (i.e., either increase or decrease), depending on the position of these two dopants in the lattice. These phenomena are also strongly supported by experimental observations where there are both increasing and decreasing effects of the diamond growth rate by S and B doping.

2.1 Introduction

The diamond material possesses very attractive properties, such as high transparency, high thermal conductivity at room temperature, radiation hardness, as well as an extreme mechanical hardness. In addition, diamond also exhibits superior electronic properties (including high carrier mobility), large electrochemical potential window, low dielectric constant, controllable surface termination, and a high breakdown voltage. Furthermore, when considering the well-known combination of chemical

K. Larsson (✉)
Department of Chemistry-Ångström Laboratory, Uppsala University, Uppsala, Sweden
e-mail: karin.larsson@kemi.uu.se

© Springer Nature Switzerland AG 2019
N. Yang (ed.), *Novel Aspects of Diamond*, Topics in Applied Physics 121,
https://doi.org/10.1007/978-3-030-12469-4_2

inertness and high degree of biocompatibility [1], diamond became recently a promising candidate for applications like artificial photosynthetic water-splitting, where interfaces between a photo-electrode surface and a redox protein is of utterly importance [2]. Boron-doped diamond surfaces, with attached Pt nanoparticles as the catalytic surface, are nowadays working as a new class of electrode materials. The boron-doped diamond electrode is a semiconducting material with very promising properties like (i) a wider potential window in aqueous solution (approximately −1.35 to +2.3 V versus the normal hydrogen electrode), (ii) low background current, and (iii) corrosion stability in aggressive environments [3].

More generally, the intrinsic reactivity of a surface atom (for a solid material) depends of various factors, such as (i) number of binding atoms, (ii) surface reconstructing, (iii) type and degree of chemisorption to the surface, and (iv) elemental doping within the upper surface region. The number of bonding atoms to a surface atom (factor i) is always smaller than the binding situation for a bulk atom. Moreover, a surface atom has the driving force to become more bulk-like (i.e., the more preferred surrounding electron density is the one for a bulk atom). This is one explanation to the larger surface reactivity relative the bulk. This is also the explanation to the specific surface electronic properties, which differ from the corresponding bulk scenario. Factors (ii)–(iv) do all represent changes in the surrounding electron density for the surface atoms, of which all are different from the situations for the bulk atoms.

Theoretical modelling based predominantly on Density Functional Theory (DFT) has during the last decades proven to become highly valuable in the explanation and prediction of experimental results. The simulation and theoretical analysis of surface reactivity has been shown to aid important information about thin film growth mechanisms, as well as about surface reconstruction, modification and functionalization.

2.2 Methods and Methodologies

Theoretical modelling and simulations are tools that are often necessary in (i) the interpretation of experimental results, (ii) within precursor design of interface materials (i.e. the prediction of experimental parameters), and (iii) for theoretical "experiments" where it is not possible to perform practical experiments. First-principle DFT calculations have become a useful tool for many inorganic materials of practical importance. It is well known that electronic structures and, especially, band gaps are sensitive to choice of theoretical method.

Moreover, chemical reactions where bonds are formed and/or broken need methods that consider the electrons in the system. Hence, the quantum mechanical DFT method is very useful for studies of e.g. surface stability, termination, functionalization and other types of surface (or interfacial) chemical processes.

The DFT calculations done for solid surfaces are most often performed under periodic boundary conditions. More specifically, ultrasoft pseudopotential [4] plane-wave approaches are most often used, based on the Perdew-Wang(PW91) generalized gradient approximation (GGA) [5] for the exchange-correlation functional. The GGA method usually gives a better overall description of the electronic subsystem, compared to the more simple LDA (local density approximation) corrections. The reason is that LDA, which is based on the known exchange-correlation energy of a uniform electron gas, is inclined to overbind atoms and to overestimate the cohesive energy in the system under study. On the contrary, GGA takes into account the gradient of the electron density, which gives a much better energy evaluation [6]. The total energy obtained from a DFT-calculation, and for a specific model, can further on be used in e.g. the calculations of reaction energies.

The atomic charges and bond electron populations (i.e. electron densities within the bond), might be estimated by using the methods of Mulliken analysis, which can be performed by using a projection of the plane wave states onto the localized basis by a technique described by Sanchez-Portal et al. [7] The calculated atomic charges and information about bond electron populations are important tools in the interpretation of the underlying causes to the observed doping-effects on diamond growth and surface properties.

2.3   General

2.3.1   Diamond Doping Using Nitrogen, Phosphorous, Sulphur or Boron

2.3.1.1   N-type Doping Using Nitrogen

Although diamond has been proven to be a genuine multifunctional material, thereby useful for quite many applications, it is still quite expensive to grow it using vapour phase methods like chemical vapour deposition (CVD). However, it has been experimentally proven that e.g. N doping will enhance the diamond growth rate, and thereby reduce the production costs. The influence of nitrogen on the deposition rate has already been reported for both poly- and mono-crystalline diamond [8–11]. The effect by N on the CVD diamond growth rate enhancement has been studied using laser reflection interferometry [12]. In addition, gaseous N concentrations as low as a few ppm were found to strongly boost up the crystal growth rate in the (100) orientation. As an explanation to this observation, Bar-Yam and Moustakas [13] proposed a defect-induced stabilization of diamond. However, this model does not explain the crystal orientation dependence of the increase in growth rates. Frauenheim et al. [14] suggested an alternative model, in which donor electrons originating from sub-surface N will lead to a lengthening of the

(100) surface reconstruction bond. Nitrogen (with its extra electron compared to C) is a well-known n-type dopant, with a very deep donor level in diamond (1.7 eV below the conduction band) [15].

2.3.1.2   N-type Doping Using Phosphorous or Sulphur

The effects of substitutional doping by sulfur, or phosphorous, have experimentally been studied thoroughly [16, 17]. It is not only the changes in the electronic properties that have been observed, but it is clear that these dopant elements can have a great impact on the diamond surface morphology and growth rate when using chemical vapor deposition methods (CVD) [16–18].

Regarding the influence on the diamond growth rate, various results have though been obtained when using an addition of hydrogen sulfide (H_2S) during the CVD process. It has been observed that the doping reagent could cause both a raise in growth rate, as well as a decrease, depending on the substrate temperature used [16]. While it was concluded that hydrogen sulfide will reduce both diamond nucleation and growth of individual crystals onto silicon substrates, the addition of this dopant resulted in an increase of the diamond layer growth rate [18]. Similarly, during the addition of phosphine (H_3P) to the CVD reaction chamber, the diamond growth rate has been shown to be heavily dependent on the substrate temperature. Opposite to the situation with non-doped diamond, a significant increase in diamond growth rate was obtained by using phosphine [18].

The improvement of the diamond growth rate could be one step in the development of a less expensive synthesis method for diamond thin film growth of different dimensions and thicknesses. For example, there is a strong need for a cost effective method whereby it would be possible to grow large single-crystalline diamond of high quality. This method is to be compared with high-pressure high-temperature (HPHT) synthesize techniques, which will restrict both the size and the quality of the diamond material [19]. A well-controlled growth process using dopants is also of utmost importance for the growth of specific crystalline planes. It is well known that the doping effect is different for the different low-index surface planes; (111), (100), and (110) [20–25].

The very complex and dynamic nature of the basic mechanism for diamond growth, has earlier been highlighted in several theoretical investigations [26]. However, only a small number of theoretical studies have been focused on the CVD growth of sulfur- or phosphorus-doped diamond [27, 28]. One report stated that phosphorus would decrease the growth rate due to the high energy associated with the strains caused by phosphorus atoms in the diamond lattice [29]. Another report concluded that sulfur atoms are spontaneously incorporated in a hydrogenated (100) diamond surface, and that they will also enhance the hydrogen desorption from the surface [30].

In choosing a suitable donor (or acceptor) for diamond, one has, however, not only to consider its donor (or acceptor) level but also its solubility and mode of incorporation (e.g. incorporation during growth by in-diffusion or by ion implantation). Impurities have been shown to become introduced into diamond during CVD or HPHT growth [31]. A review article has, in addition, been reporting doping by implantation [32]. A substitutional n- (or p-) type doping of diamond during growth has been of a special interest to study in the present report. The resulting substitutional doping will, hence, depend not only on the solubility of the impurity, but will also depend on the kinetics of the growth. Kinetic trapping may then be possible although the final n- (or p-) type doped product is thermodynamically somewhat unfavourable.

2.3.1.3 P-Type Doping Using Boron

The growth process of boron-doped (B-doped) diamond material, with its extraordinary electrochemical properties, has recently been well developed. It has been found that the B dopant will increase the electronic conductivity, as compared with the intrinsic non-doped diamond [33]. For different doping levels, the diamond material will show a metallic-type conductivity [34, 35], p-type semi-conductivity [36, 37], and superconductivity [38]. Especially the superconductivity property makes diamond very attractive for electronic applications. In order to achieve a good quality of B-doped diamond films and to better control the effects of density of boron, the chemical vapour deposition (CVD) technique has been shown to be a very useful technique. Thus, substitutional doping with boron (using CVD) will not only change the electronic properties, but it will also have a great impact on the diamond surface morphology and growth rate [38–41].

It has been shown, that when using a boron precursor such as trimethylborate, boron trioxide, diborane or trimethylboron during the CVD process, the growth rate will decrease compared to non-doped diamond. On the other hand, it is also observed that the doping reagent might cause a slight increase in growth rate, which depend on the substrate temperature [42] or pressure [43]. In the study by Hartmann et al. [44] and Miyayta [45], it was found that at very low boron concentration will increase the diamond growth rate. It is though experimentally very difficult to explain how B will affect the growth processes.

2.3.2 Growth Mechanism

The chemical vapour deposition (CVD) of diamond is a very complex and dynamic process [45]. The growing surface has to be terminated with e.g. H atoms, since it will otherwise collapse to the graphitic phase. However, these terminating species must be abstracted away from the surface to leave room for a growth species (e.g. CH_3). Moreover, in the initial part of the diamond growth process, the abstraction of two surface H species (terminating the surface and one of the H ligands on CH_3) has to take place before any surface migrating can take place, ending with a final incorporation at a step edge. Hence, the process of H abstraction is a very important reaction step during the overall diamond growth mechanism [46]. Since it is an endothermic reaction, whilst the adsorption of e.g. CH_3 highly exothermic (+17 vs. −348 kJ/mol), it is here assumed that the H abstraction is one of the rate-limiting steps for the diamond growth process [47].

2.3.3   H Abstraction Rates for Growth of Non-doped Diamond

The process of H abstraction was modelled by initially positioning an H radical 4 Å above one of the surface H adsorbates. This H radical was thereafter approached towards the surface in smaller steps (from 4.0 to 1.0 Å, in steps of 0.2 Å), thereby mimicking the initial growth step of surface H abstraction by gaseous H. This mechanism is demonstrated in Fig. 2.1 for a doped surface. For each step, the introduced H radical was fixed during the geometrical optimization of the whole system (in addition to the lowest C layer with its terminated H). When the H radical approached the surface-binding H (at an approximate distance of 1.0 Å), a reaction between these two H atoms started to take place. The processes of desorption of this hydrogen molecule (from the surface) was thereafter studied by monitoring a step-wise removal of the H–H molecule from the surface.

As was the situation with the approaching H, one of the H atoms (in H_2) was kept fixed during the geometry optimization processes. By using this procedure, the energy barriers for both the approaching H atom (to the most reactive surface H), and the outgoing H_2 molecule, could be calculated.

(a)   **(b)**

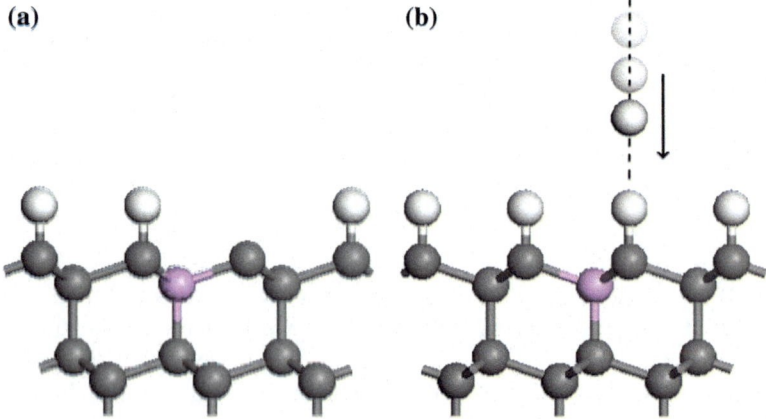

Fig. 2.1 a Representation of a radical surface C as a result of an H-abstraction. A dopant element is positioned in the second C layer. **b** Illustration of the simulated abstraction process

The energy barriers for the incoming H in the H abstraction process (i.e. the barrier for the approaching H atom to the most radical surface-binding H) has in [48] been calculated for three different H-terminated diamond planes; (100)−2 × 1, (110) and (111). The second step in this overall process (i.e. the energy barriers for H–H molecule desorption from the diamond surface) has also been calculated. Figure 2.2 shows the energy evolution for an approaching radical to an adsorbed H species on the surface for the three different surface planes of diamond; (110), (111) and (100)−2 × 1. The corresponding desorption of H_2 from the surfaces is shown in Fig. 2.2b. The respective barrier energies, as well as the distances of the transitions states (i.e. energy maxima) from the H adsorbate at the surface, are especially shown in the Figures. As can be seen in Fig. 2.2a, at a distance of about 3.0 Å from the H adsorbate, an energy barrier is starting to appear and the total energy of the systems starts to increase to a maximum.

In order to ensure the accuracy in the position of these maxima, shorter steps (0.05–0.10 Å instead of 0.2 Å) were chosen in the regions surrounding these

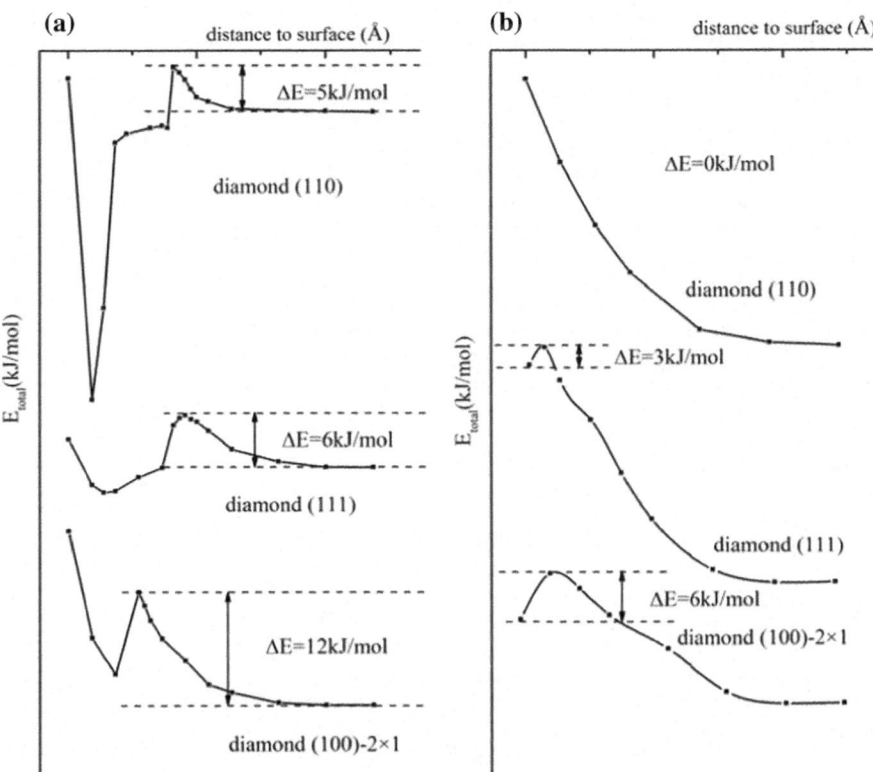

Fig. 2.2 The energy evolution curve for **a** an approaching H to a surface terminated H species, and for **b** the corresponding desorption of H_2 molecule from the surfaces. This is demonstrated for different diamond surface planes; (110), (111) and (100)−2 × 1

stationary points. The position of the transition states with respect to the H adsorbates was found to differ appreciably; 1.6, 2.0 and 1.9 for Å for diamond (100) -2×1, (111) and (110), respectively. And this observation is not due to the fact that the H adsorbates have undergone relaxation to various extents. In addition, the numerical values of the respective energy barriers were calculated using $\Delta E = E_{maximum} - E_{furthest}$ away, which also gave quite different values for the various surface planes (12, 6 vs. 5 kJ/mol, respectively). When continuing the calculation after the energy barriers, and hence with H approaching even closer to the H adsorbate at the diamond surface, the results also looks different depending on the surface plane.

As can be seen in Fig. 2.2a, for the (100)-2×1 plane, the energy value directly drops to energy minima. For the situation with the diamond (111) plane, the energy value reduces gradually. This behaviour is even more pronounced for the (110) plane, where the total energy value will first decrease rapidly, followed by a slow decrease that will end by a more dramatic decline to the lowest energy minima. These variations in energy evolution are strongly coupled to structural changes at the surface as a result of the approaching H species. Despite the variations in energy evolution for the incoming H radical to the surfaces, all of these three planes showed the lowest total energy minimum for a distance of 1.2–1.4 Å from the surface-binding H species. At this distance, the H radical is interacting with this H adsorbate, but to various extents depending on surface plane. When the approaching H comes close enough to the adsorbed H (for all three diamond planes), an electron population (as the result of a bond population analysis) was observed between the approaching H and the terminating one, being a strong indication of an H–H interaction in the form of bond covalence. For the (110) surface, this bond population will continue to increase as the H–H distance becomes smaller (and a simultaneous decrease in bond population for the C–H bond), with the final formation of gaseous H_2 [48]. The resulting value of bond population for H–H in H_2 is much larger than the electron bond population for the final C–H entity (0.73 vs. -0.04), indicating the C–H bond has been broken completely.

2.4 Nitrogen-Induced Effect on Diamond Growth

2.4.1 Introduction

It has been experimentally proven that e.g. N doping will enhance the diamond growth rate and thereby reduce the production costs [49]. Dunst et al. [5] were studying the effect of diamond growth by introducing nitrogen into a microwave plasma-activated CVD reactor in depositing single crystalline diamond (100) at 800 C. The results of that study state that the introduction of N_2 will improve the diamond growth rate with a factor of approximately 2.5 at a CH_4 concentration of

2%. However, it is experimentally very difficult to determine in which form N will be efficient in this process. To be more specific, it is experimentally not possible to discriminate between the effect of nitrogen in the form of an N-containing species adsorbed on the surface, or as N substitutionally positioned within the upper part of the diamond surface. An issue that has to be highlighted is the catalytic effect by the N dopant. The problem is that a very low concentration of N dopants shall be able to give a more pronounced effect on the overall diamond growth rate. However, it has earlier been shown that the surface hydrogens are very mobile on the diamond surface [48]. Hence, the desorption of an H species nearby the N dopant will immediately be replaced by a neighbouring H species on the surface, causing radical surface C sites at longer distances from the N dopant. In short, the substitutional N dopant will improve both the H abstraction rate and concentration of radical surface C sites, the latter being distributed evenly over the surface.

Three different super cells, modelling the diamond surface planes $(100)-2 \times 1$, (110) and (111), respectively, have been constructed. The corresponding models are shown in Fig. 2.3, together with the different ways, N is incorporated in the surface region. All of these surfaces are terminated by H atoms to simulate an as-grown CVD diamond surface. And for modelling the (100) surface, the 2x1 reconstructed mono-hydride surface, $(100)-2 \times 1$:H, was used.

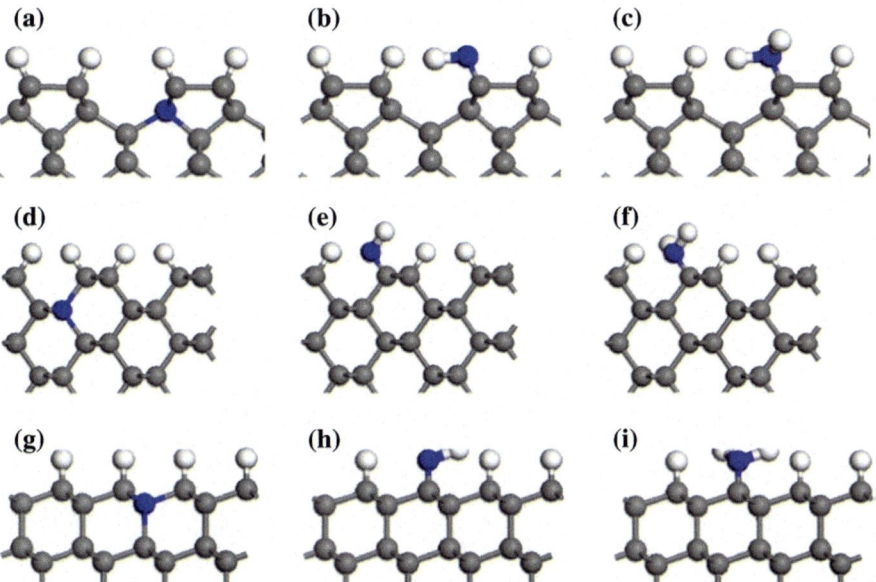

Fig. 2.3 Representations of nitrogen doping in different diamond surface structures; **a–c** (100) -2×1, **d–f** (110), and **g–i** (111). Figure **a, d** and **g** demonstrate substitutional doping into the 2nd C atomic layer. Furthermore, **b, e** and **h** demonstrate an adsorbed NH radical. Similarly, **c, f**, and **i** demonstrate adsorbed NH_2 species

2.4.2 N Substitutionally Positioned into Various C Atomic Layers

At first, the effect by substitutionally positioning N within different C atomic layers close to the surface had to be evaluated. For this purpose, the surface H abstraction energy was chosen to be calculated for the $(100)-2 \times 1$, (110) and (111) surface planes. As can be seen in Fig. 2.4, for these three surfaces there is a decrease in the abstraction curves for the position of N within C layers 2 and 3 in the upper diamond surface. Hence, the H abstraction becomes much more exothermic. Based on all of these calculations, it has been proven reasonable to continue the investigation in the present study by positioning the N dopant within the 2nd atomic surface layers.

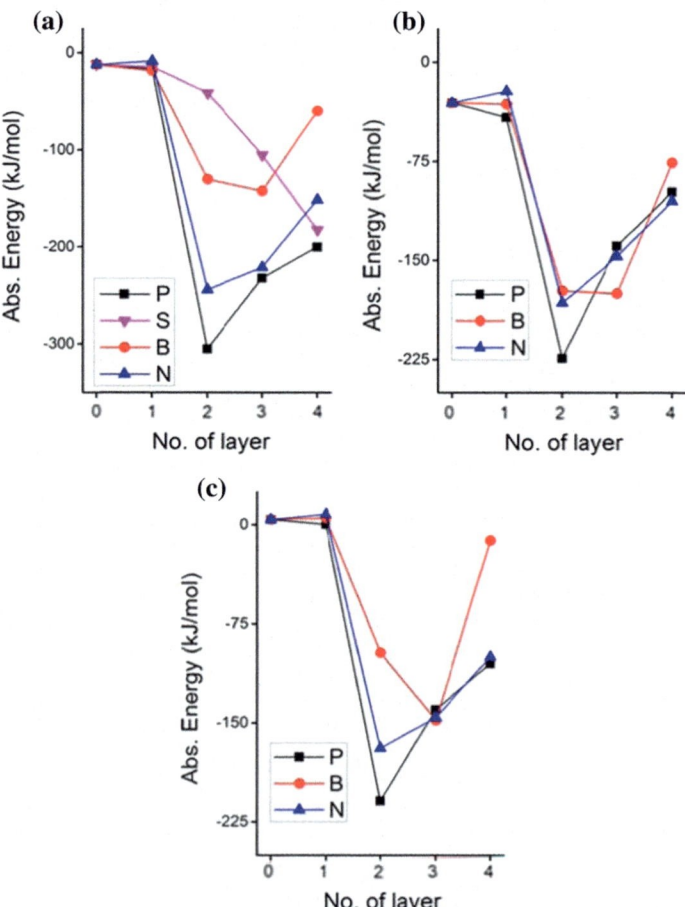

Fig. 2.4 Surface H abstraction energy for different dopants positions within different atomic C layers for diamond **a** (111), **b** (110) and **c** $(100)-2 \times 1$ surface planes

2.4.3   N Substitutionally Positioned Within the C Atomic Layer 2

Nitrogen, substitutionally positioned within the second C layer of various diamond surface planes, was here chosen since it was found that this is a position that has a large effect on the H abstraction reaction.

An identical methodology, as for the non-doped diamond surfaces, was followed in trying to estimate the energy barriers for the H abstraction processes for the doped situations. The values of H abstraction energy barrier for the different diamond surface planes, and for different kinds of nitrogen doping, are shown in Table 2.1. The most interesting result is obtained for substitutionally nitrogen doping in the second C atomic layer, for which there is no energy barrier (for the approaching H radical) for any of the surface planes investigated.

2.4.4   N Chemisorbed onto the Surface in the Form NH or NH$_2$

Since it was not experimentally clear in what form nitrogen will aid for the improved diamond growth rate, nitrogen was attached to the surface in the form of a radical NH group (i.e. chemisorbed onto the diamond surface). As can be seen in Fig. 2.5, the non-paired electron will stay locally at the adsorbate NH place. This local position of the extra electron is further supported by the deformation density results in Fig. 2.6. The extra electron in NH does not show any tendency to interact

	Energy barrier (eV)	Relative growth rate
(110)		
Non-doped	4.7	1.0
N in 2nd layer	0	1.7
NH$_2$ in adsorbate	5.7	0.9
(111)		
Non-doped	6.1	1.0
N in 2nd layer	0	1.9
NH$_2$ in adsorbate	6.0	1.1
(100)		
Non-doped	11.8	1.0
N in 2nd layer	0	3.7
NH$_2$ in adsorbate	8.2	1.5

Table 2.1 Effect of N on the surface H abstraction energy and relative growth rate (in relation to the respective non-doped surface planes: (110), (111) versus (100) -2×1

(a)   **(b)**

N(layer 2)   NH(rad)

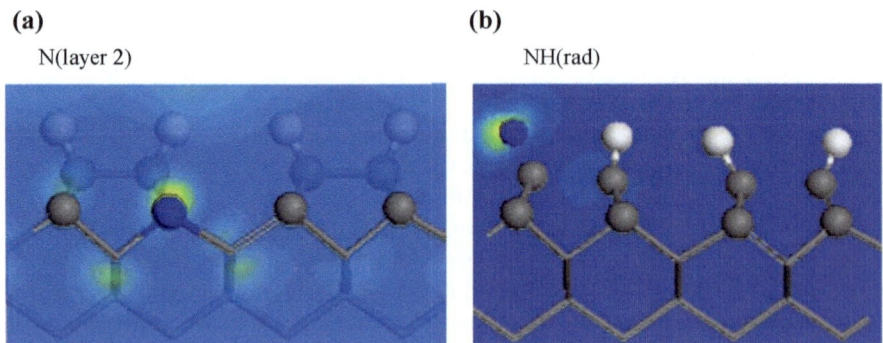

Fig. 2.5 Demonstration of the electron spin density for diamond (100)−2 × 1, with N substitutional positioned within the 2nd C atomic layer (**a**), or chemisorbed in the form of a radical NH species (**b**)

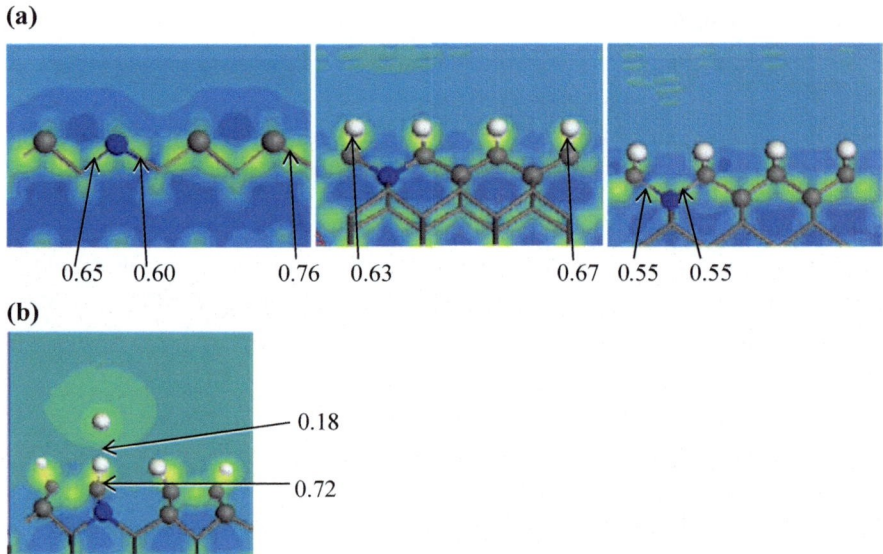

Fig. 2.6 Electron density difference maps showing the dopant-induced lattice weakening (**a**) and the interaction between the approaching H and the surface H atom (**b**)

with the surface in weakening bonds by filling anti-bonding orbitals. Due to this reason, the barrier energy for the H abstraction is expected to be very similar as for the non-doped diamond surfaces. As can be seen in [50], when nitrogen is chemisorbed in the form of NH_2 onto the diamond surface, the situation becomes more complex (even though NH_2 is not a radical species when adsorbed to the diamond surface).

Similar to the situation with NH adsorbates, the nitrogen doping in the form of NH_2 adsorbed do not show any larger effect as compared to the non-doped scenario. To be more specific, the H abstraction energy barriers for the diamond $(100)-2 \times 1$ and (111) planes were observed to decrease, whilst the barrier for the (110) plane increased; NH_2 doping (8, 6, 6) versus non-doping (12, 6.5) (see Table 2.1). However, it must be stressed that these difference are marginal for diamond (111) and (110).

2.5   Phosphorous- or Sulfur-Induced Effect on Diamond Growth

2.5.1   Introduction

Doping with either sulphur, or phosphorous, has experimentally been shown to have an effect on the diamond growth rate [23, 25]. However, the underlying reason to the fact that phosphorus doping will cause an increased growth rate, while sulphur has been observed to cause either an increased, or decreased, growth rate, has not yet been clarified. One of the questions is how P (or S) will affect the diamond structure when incorporated into the lattice. Another question is how the H abstraction energy and barrier will be influenced by this specific doping. To be able to outline these various influences by the P (or S) dopants, theoretical DFT calculations have become very useful. Within the present report, the dopants (S or P) have been substitutionally positioned into the upper diamond lattice. More precisely, the dopants were positioned within C layer 1, 2, 3 and 4, respectively.

In a thermodynamic study, the influence by the dopant P (or S) on the driving force for the H abstraction reaction, has been especially highlighted. In a kinetic study, the influence by the dopant P (or S) on the H abstraction rate, has been more thoroughly investigated by performing transition state calculations. With the assumption that the H abstraction is a rate limiting step, the rate of H abstraction will be more or less identical to the diamond growth rate.

2.5.2   Thermodynamics—H Abstraction Energies

When it comes to the H abstraction energy, the effect by P (or S), positioned within different C atomic layers close to the surface, had to be evaluated. As can be seen in Fig. 2.4, these two dopants show completely different behaviours.

For the P dopant, there is a strong effect when positioning P within the carbon layers 2–4. The corresponding reaction energy values for the surface H abstraction processes are −12 kJ/mol (non-doped), and −16, −305, −232, and −200 kJ/mol for P in the 1st, 2nd, 3rd and 4th diamond surface C layer, respectively. It is clear that the most exothermic abstraction energy is obtained when positioning P in the atomic layer two, whilst the position of P in layer one will not lead to any significant difference compared to the non-doped scenario.

Although the H abstraction energy curve is continuously decreasing (towards more exothermic values), the situation for S-doping is more complex than it looks. The corresponding reaction energy values are −12 kJ/mol (non-doped), and −14, −41, −105, −182 kJ/mol for S in the 1st, 2nd, 3rd and 4th diamond surface C layer, respectively. (These values are the obtained ones for the abstraction of H next-nearest to the S dopant).

2.5.3   Kinetics—H Abstraction Barriers

In order to investigate the kinetics of the H abstraction process (and thereby the diamond growth rate), the energy barriers of this process was calculated. For the different dopants (P and S) and their positions in the different atomic C layers, large variations in energy barriers of were observed. These calculations can be seen in Table 2.2. For example, when P is substitutionally positioned within the 1st atomic C layer, the corresponding energy barrier has increased almost twice larger than non-doped diamond which is found to be 6 kJ/mol (See Sect. 2.3.3.). However the bond population of C (surface) – H bond (0.90) is somewhat increased just a little when compared to non-doped diamond (0.89). Thus, more detailed calculations have been done in order to analyse the energy of the transition state for the approaching H radical to the surface H atom. In the transition state, the bond population of C (surface) – H increases a lot to 0.95 (from 0.90) which just increases a little to 0.90 (from 0.89) for non-doped diamond. It was thereby found that when P is substitutionally positioned within the 1st atomic C layer, it have influenced on the incoming H radical and hence enforce the energy barrier. More attractive result comes though from a deeper positioned doping with P, as there is no energy barrier detected for the approaching H radical to the surface. When analyzing the transition state for P positioned within the layers 2–4, the bond population of the C (surface) – H without any increasing but decreases approximate 0.1 directly.

Table 2.2 Effect of P and S on the H abstraction energy for the surface planes: (110), (111) versus (100)−2 × 1, respectively

No. of layer	P-doped energy barrier, kJ/mol	S-doped energy barrier, kJ/mol
1	11.1	6.6
2	0	13.4
3	0	11.7
4	0	0

The result of S doping is also here a bit more complicated. The first layer substitution makes almost no change of energy barrier, while substitution within 2nd and 3rd layer increases a lot compared to non-doped diamond (6 kJ/mol) (see Table 2.2). In addition, there is no energy barrier for the hydrogen radical approaching the surface when sulphur is substitutional positioned into the lattice within the fourth carbon atomic layer.

The experimental effect of doping by phosphorus, or sulphur, on the diamond growth rate, can be estimated on the basis of this kinetic study. Since there is always no energy barrier present for the approaching H radical when phosphorus is present in the lattice, the growth rate must be appreciably increased as a function of doping. In contrast, sulphur doping leads to anomalous changes in the diamond (111) growth rate. It can both increase, or decrease, depending on the position of S in the lattice. As compared to the non-doped situation for which the energy barrier is 6 kJ/mol, it can be concluded that when sulphur is present within the 4th C layer the growth rate will be increased, and when S is positioned within the upper layer it will lead to a decrease in growth rate.

2.6   Boron-Induced Effect on Diamond Growth

2.6.1   Introduction

As presented in Sect. 2.3.2, doping with boron has experimentally been shown to have an effect on the diamond growth rate. As was the situation with N-, P- and S-doping, it is experimentally very difficult to find the underlying the reason to the fact that boron doping will either cause an increase, or decrease, of the diamond growth rate. To be more specific, it is experimentally very difficult to understand how B will affect the diamond structure when incorporated into the lattice, and how the H abstraction energy and barrier will be influenced by this specific doping. For this purpose, theoretical DFT calculations will become very useful. As was the situation with N-, P- and S-doping, the B dopants has here been substitutionally positioned into the upper diamond lattice. More precisely, the B-dopant was positioned within C layer 1, 2, 3 or 4.

The H abstraction reaction for B-doped diamond surfaces has been focused in the present study, from both a thermodynamic and a kinetic point of view. In comparison with experimental results, an identity in trends, in addition to a large similarity in numerical values, for the diamond growth rates, were thereby observed. It has experimentally been proven that nitrogen doping will enhance the single crystalline diamond (100) CVD growth rate by a factor of 3.5 where the theoretical result was shown to be 3.7.

2.6.2 Thermodynamics—H Abstraction Energies

The effect by boron positioned within different C atomic layers close to the surface (i.e., 1st, 2nd, 3rd, 4th, or 5th C layer), on the H abstraction energy, has here been evaluated. As can be seen in Fig. 2.4, the three different diamond planes [i.e., (111), (100)−2 × 1 and (110)] show almost the same behaviour. Although the numerical values of the H abstraction energy are different, the trends for the different diamond planes are very similar. There is a particularly strong effect when positioning B within the 2nd, or 3rd, carbon layers. For deeper positioned B, such as within the 4th or 5th atomic C layer, this effect on the H abstraction energy becomes minor (especially for the (100)−2 × 1 surface). Moreover, it will not lead to any significant difference, as compared to the non-doped scenario, when boron is substitutionally positioned within the 1st C layer.

It is big interest to also investigate the effect by doping on the lateral distance to the dopant. The results of the H abstraction energy for B positioned within the 2nd layer in different diamond surfaces show almost the same behaviours. It is found that the abstraction energy will increase (i.e., the reaction process will be less exothermic) when the distance between the dopant and the H abstraction site becomes longer, as can be seen in Fig. 2.7. This result supports a local effect by B doping since the influence by B decreases for longer distances. Thus, the atomic charges of the H atoms which are further away from the dopant show the same value as for the non-doped scenario. Similar investigations have also been carried out for B positioned within the 3rd C layer. The H abstraction energy thereby increase from −142 to −35 kJ/mol (levelling out at −21 kJ/mol) for diamond (111), from −176 to −94 kJ/mol (levelling out at 75 kJ/mol) for diamond (110), and from −148 to −6 kJ/mol (levelling out at 4 kJ/mol) for diamond (100)−2 × 1, when laterally moving the H reaction sites further away from the B dopant. Thus, the B doping shows also here a very local effect, as was the situation with B doping within the 2nd C layer. Moreover, when B is positioned within the 3rd or 4th C layer in diamond (100)−2 × 1 another doping position needs to be considered.

2.6.3 Kinetics—H Abstraction Barriers

Another purpose with this study was to investigate the kinetics of the H abstraction process, and thereby to compare the CVD growth rate of the non-doped and boron-doped diamond. The most interesting results were obtained for B doping with B positioned within the 2nd C layer in diamond (see Table 2.3). For this situation, no energy barrier was observed for the approaching H radical to the surface. For other doping positions in the diamond lattice, the energy barriers are somehow increased as compared to the non-doped scenario. The corresponding energy barriers for non-doped diamond surfaces are 6.1, 4.7 and 11.8 kJ/mol for diamond (111), (110) and (100)−2 × 1 surface, respectively.

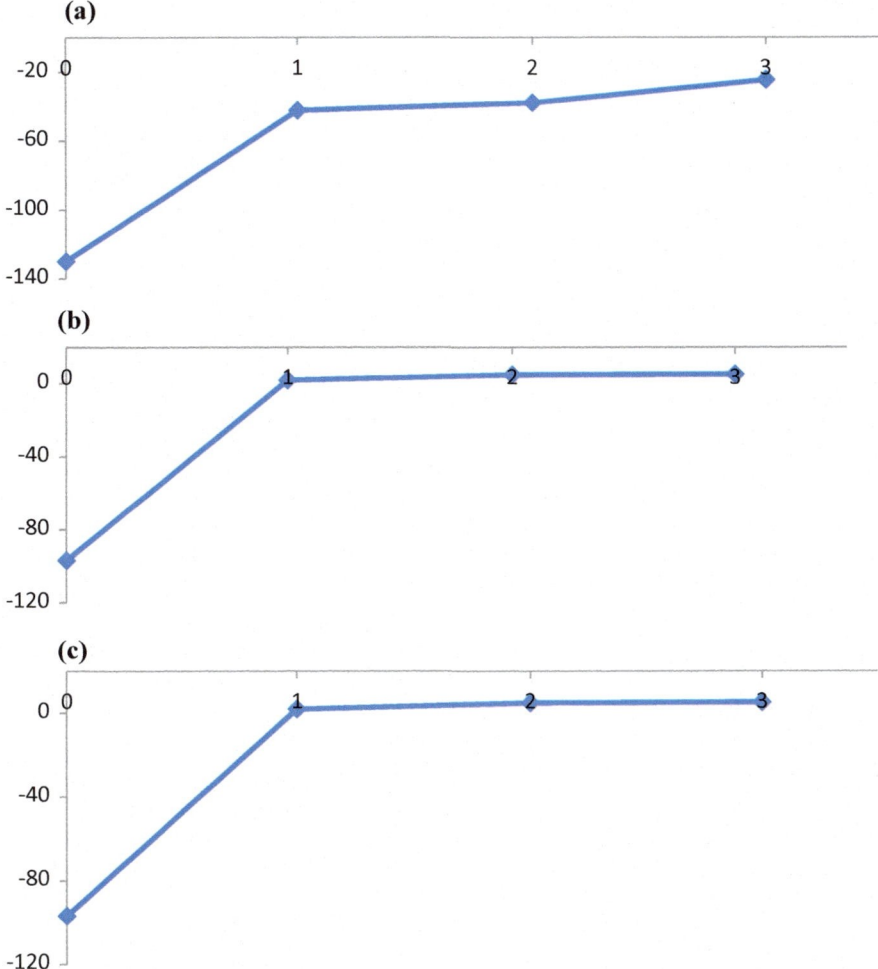

Fig. 2.7 Surface H abstraction energies for B positioned within different lateral positions relative the H abstraction site in diamond (111) (**a**), (110) (**b**) and (100)−2 × 1 (**c**)

The experimental effect by doping with boron, on the diamond growth rate, can be estimated on the basis of the present kinetic study. The Arrhenius equation $(v = v_0 \cdot e^{-\Delta E/RT})$ can be used to find the relation between the energy barrier and the diamond growth rate, with the assumption that the H abstraction process is a rate-limiting growth step. If the difference in energy barrier between non-doped and boron-doped diamond is positive (i.e., if the value of the energy barrier for the non-doped scenario is larger than for the doped situation), the growth rate will increase as an effect by B doping. It was here found that the growth rate of diamond will only increase when boron is positioned within the 2nd C layer for diamond

Table 2.3 Effect of B (substitutionally positioned within different atomic C layers) on the energy barriers for H abstraction processes for various diamond surface planes

No. of layer	Diamond (111)	Diamond (110)	Diamond (100)−2 × 1
1	7.7	6.1	14.7
2	0	0	0
3	6.8	5.9	12.6
4	9.1	8.5	19.5

(111), (110) and (100)−2 × 1 surface (since there is no energy barrier observed). Moreover, when boron is positioned within the other C layers (i.e., 3rd and 4th) it is found that the extent of the increased energy barrier for diamond (111) surface is the smallest compared to the (110) and (100)−2 × 1 surface (i.e., 0.5 vs. 1.2 vs. 0.8, and 3.0 vs. 3.8 vs. 7.7 for the energy barrier differences between B positioned within 3rd or 4th and non-doped situation, for (111), (110) and (100)−2 × 1 respectively). Thus, the growth rate of the diamond (111) surface will decrease to a smaller extent as compared to the diamond (110) and (100)−2 × 1 surface, respectively. That means that the effect of B doping on the slowing the abstraction process is less for diamond (111) plane than others.

As mentioned above, there is a lateral effect by the boron doping on the abstraction energy. Other attack sites for the incoming H radical have thereby been chosen in order to also study the lateral effect on the energy barrier. The energy barrier did in fact increase when laterally moving the attack site further away from the B dopant. As an example, the energy barrier was observed to increase to 8.6 kJ/mol, and furthermore to 11.9 kJ/mol, when lateral moving the attack site away from the dopant, which was positioned within the 3rd C layer in diamond (110). Even for the situation with B positioned within the 2nd C layer in diamond (111), when the attack site is far away from the dopant there is an energy barrier of 3.8 kJ/mol that levels out to 6.6 kJ/mol. This result is in accordance with the observations of abstraction energy which will decrease (i.e., less exothermic) when laterally moving the abstraction site further away from the B dopant. These results do strongly indicate that the boron doping has a very local effect on the H abstraction process.

2.7 Summary

The presented studies have been mainly focused onto the effect by the various dopants on the growth process of diamond (111), (100)−2 × 1 and (110) surfaces. Gas phase abstraction of terminating H species was then assumed to be one of the dominating rate-determining reactions.

At first, the effect by different dopants, positioned within different C atomic layers close to the surface, on the thermodynamics (i.e., H abstraction energy) had to be evaluated to investigate the driving force for the H abstraction reaction. It was shown that various dopants show different behaviours for each diamond plane. Moreover, all of them showed a more exothermic abstraction energy, which means to be more preferable for the H abstraction reaction. Hence, is thereby possible that the growth rate of diamond may increase due to these dopants incorporation within the diamond lattice.

In order to investigate the kinetics of the H abstraction process and thereafter to compare the growth rate between the non-doped and doped situation, in addition to the experimental observations during the CVD process, the energy barriers for a free H radical attacking the surface had to be evaluated. As a result, there is a remarkable increase in growth rate when introducing N within the 2nd C layer, which has been calculated with a factor of 3.7 (for diamond (100)−2 × 1 plane), compared to the non-doped scenario. Almost the same ratio enhancement of growth rate (i.e., with a maximum value of 3.5) was also found in an experiment within a MWPA-CVD reaction chamber at 1073 K surface temperature between with N_2 doping and without.

For the situation with P doping, there were no visible energy barrier obtained for the approaching H radical to any of the diamond surface planes (111), (100)−2 × 1 and (110), respectively. Hence, the growth rate must be appreciably increased as a function of doping with P.

It was furthermore observed that S and B doping will lead to anomalous changes in the diamond growth rate (i.e., either increase or decrease), depending on the position of these two dopants in the lattice. More specifically, the growth rate will increase when S is present in the 4th C layer or B in the 2nd C layer, since there is no energy barrier obtained for these situations. For other doping situations, there are much larger energy barriers as compared to the non-doped scenario. Thus, there will be a decrease in diamond growth rate for these situations. These phenomena are also strongly supported by experimental observations where there are both increasing and decreasing effects of the diamond growth rate by S and B doping.

Acknowledgements This work was supported by the Faculty of Uppsala University, and the Swedish Research Council (VR). The computational results were obtained using CASTEP from BIOVIA.

References

1. M. Amaral, A.G. Dias, P.S. Gomes, M.A. Lopes, R.F. Silva, J.D. Santos, M.H. Fernando, Nanocrystalline diamond *in vitro* biocompatibility assessment by MG63 and human bone marrow cell cultures. J. Biomed. Mater. Res. A **87**(1), 91–99 (2008). https://doi.org/10.1002/jbm.a.31742
2. J.P. McEvoy, G.W. Brudvig, Water splitting of photosystem II. Chem. Rev. **106**(11), 4455–4483 (2006). https://doi.org/10.1021/cr0204294
3. M. Panizza, G. Cerisola, Application of diamond electrodes to electrochemical processes. Electrochim. Acta **51**(2), 191–199 (2005). https://doi.org/10.1016/j.electacta.2005.04.023
4. D. Vanderbilt, Soft self-consistent pseudopotentials in a generalized eigenvalue formalism. Phys. Rev. B **41**(11), 7892–7895 (1990). https://doi.org/10.1103/PhysRevB.41.7892

5. J.P. Perdew, K. Burke, M. Ernzerhof, Generalized gradient approximation made simple. Phys. Rev. Lett. **77**(18), 3865–3868 (1996). https://doi.org/10.1103/PhysRevLett.77.3865

6. D. Petrini, K. Larsson, Theoretical study of the thermodynamic and kinetic aspects of terminated (111) diamond surfaces. J. Phys. Chem. C **112**(37), 14367–14376 (2008). https://doi.org/10.1021/jp709625a

7. H.J. Monkhorst, J.D. Pack, Special points for Brillouin-zone integrations. Phys. Rev. B **13** (12), 5188–5192 (1976). https://doi.org/10.1103/PhysRevB.13.5188

8. G.Z. Cao, J.J. Schermer, W.J.P. van Enckevort, W.A.L.M. Elst, L.J. Giling, Growth of 100 textured diamond films by the addition of nitrogen. J. Appl. Phys. **79**(3), 1357–1364 (1996). https://doi.org/10.1063/1.361033

9. W. Müller-Sebert, E. Wörner, F. Fuchs, C. Wild, P. Koidl, Nitrogen induced increase of growth rate in chemical vapor deposition of diamond. Appl. Phys. Lett. **68**, 759–760 (1996). https://doi.org/10.1063/1.116733

10. C.S. Yan, Y.K. Vohra, Multiple twinning and nitrogen defect center in chemical vapor deposited homoepitaxial diamond. Diam. Relat. Mater. **8**(2022), 2022–2031 (1999). https://doi.org/10.1016/S0925-9635(99)00148-X

11. T. Liu, D. Raabe, Influence of nitrogen doping on growth rate and texture evolution of chemical vapor deposition diamond films. Appl. Phys. Lett. **94**, 211191–211193 (2009). https://doi.org/10.1063/1.3072601

12. S. Dunst, H. Sternschulte, M. Schreck, Growth rate enhancement by nitrogen in diamond chemical vapor deposition—a catalytic effect. Appl. Phys. Lett. **94**, 224101–224103 (2009). https://doi.org/10.1063/1.3143631

13. Y. Bar-Yam, T.D. Moustakas, Defect-induced stabilization of diamond films. Nature **342**, 786 (1989). https://doi.org/10.1038/342786a0

14. T. Frauenheim, G. Jungnickel, P. Sitch, M. Kaukonen, F. Weich, J. Widany, D. Porezag, A molecular dynamics study of N-incorporation into carbon systems: doping, diamond growth and nitride formation. Diam. Relat. Mater. **7**, 348–355 (1998). https://doi.org/10.1016/S0925-9635(97)00186-6

15. G.B. Bachelet, D.R. Hamann, M. Scluter, Pseudopotentials that work: from H to Pu. Phys. Rev. B **26**, 4199–4219 (1982). https://doi.org/10.1103/PhysRevB.26.4199

16. H. Sternschulte, M. Schreck, B. Stritzker, A. Bergmaier, G. Dollinger, Growth and properties of CVD diamond films grown under H2S addition. Diam. Relat. Mater. **12**(3–7), 318–323 (2003). https://doi.org/10.1016/S0925-9635(02)00312-6

17. R. Haubner, D. Sommer, Hot-filament diamond deposition with sulfur addition. Diam. Rel. Mater. **12**(3), 298–305 (2003). https://doi.org/10.1016/S0925-9635(02)00342-4

18. S. Bohr, R. Haubner, B. Lux, Influence of phosphorus addition on diamond CVD. Diam. Rel. Mater. **4**(2), 133–144 (1995). https://doi.org/10.1016/0925-9635(94)00235-5

19. Q. Liang, C. Yan, Y. Meng, J. Lai, S. Krasnicki, H. Mao, R.J. Hemley, Recent advances in high-growth rate single-crystal CVD diamond. Diam. Rel. Mater. **18**(5), 698–703 (2009). https://doi.org/10.1016/j.diamond.2008.12.002

20. H. Kato, J. Barjon, N. Habka, T. Matsumoto, D. Takeuchi, H. Okushi, S. Yamasaki, Energy level of compensator states in (001) phosphorus-doped diamond. Diam. Rel. Mater. **20**(7), 1016–1019 (2011). https://doi.org/10.1016/j.diamond.2011.05.021

21. H. Kato, T. Makino, S. Yamasaki, H. Okushi, n-type diamond growth by phosphorus doping on (0 0 1)-oriented surface. J. Phys. D Appl. Phys. **40**(20), 6189–6200 (2007). https://doi.org/10.1088/0022-3727/40/20/S05

22. M. Nishitani-Gamo, C. Xiao, Y. Zhang, E. Yasu, Y. Kikuchi, I. Sakaguchi, T. Suzuki, Y. Sato, T. Ando, Homoepitaxial diamond growth with sulfur-doping by microwave plasma-assisted chemical vapor deposition. Thin Solid Films **382**(1), 113–123 (2001). https://doi.org/10.1016/S0040-6090(00)01770-3

23. S.C. Eaton, A.B. Anderson, J.C. Angus, Y.E. Evstefeeva, Y.V. Pleskov, Diamond growth in the presence of boron and sulfur. Diam. Rel. Mater. **12**(10), 1627–1632 (2003). https://doi.org/10.1016/S0925-9635(03)00202-4

24. S. Koizumi, M. Kamo, Y. Sato, H. Ozaki, T. Inuzuka, Growth and characterization of phosphorous doped {111} homoepitaxial diamond thin films. Appl. Phys. Lett. **71**(8), 1065–1067 (1997). https://doi.org/10.1063/1.119729

25. T. Miyazaki, H. Kato, H. Okushi, S. Yamasaki, Ab initio energetics of phosphorus impurity in subsurface regions of hydrogenated diamond surfaces. Surf. Sci. Nanotech. **4**, 124–128 (2006). https://doi.org/10.1380/ejssnt.2006.124

26. H. Wada, T. Teraji, T. Ito, Growth and characterization of P-doped CVD diamond (111) thin films homoepitaxially grown using trimethylphosphine. Appl. Surf. Sci. **244**(1), 305–309 (2005). https://doi.org/10.1016/j.apsusc.2004.10.137

27. A. Mainwood, Theoretical modelling of dopants in diamond. J. Mater. Sci. Mater. Electron. **17**(6), 453–458 (2006). https://doi.org/10.1007/s10854-006-8091-x

28. T. Miyazaki, H. Okushi, Theoretical modeling of sulfur–hydrogen complexes in diamond. Diam. Rel. Mater. **11**(3), 323–327 (2002). https://doi.org/10.1016/S0925-9635(01)00543-X

29. E. Gheeraert, N. Casanova, A. Tajani, A. Deneuville, E. Bustarret, J.A. Garrido, C.E. Nebel, M. Stutzmann, n-Type doping of diamond by sulfur and phosphorus. Diam. Rel. Mater. **11**(3), 289–295 (2002). https://doi.org/10.1016/S0925-9635(01)00683-5

30. Z.M. Shah, A. Mainwood, A theoretical study of the effect of nitrogen, boron and phosphorus impurities on the growth and morphology of diamond surfaces. Diam. Relat. Mater. **17**(7), 1307–1310 (2008). https://doi.org/10.1016/j.diamond.2008.03.028

31. R. Kalish, The search for donors in diamond. Diam. Relat. Mater. **10**, 1749–1755 (2001). https://doi.org/10.1016/S0925-9635(01)00426-5

32. S.A. Kajihara, A. Antonelli, J. Bernholc, R. Car, Nitrogen and potential *n*-type dopants in diamond. Phys. Rev. Lett. **66**, 21101 (1991). https://doi.org/10.1103/PhysRevLett.66.2010

33. R.L. McCreery, Advanced carbon electrode materials for molecular electrochemistry. Chem. Rev. **108**(7), 2646–2687 (2008). https://doi.org/10.1021/cr068076m

34. Y.V. Pleskov, A.Y. Sakharova, M.D. Krotova, L.L. Bouilov, B.V. Spitsyn, Photoelectrochemical properties of semiconductor diamond. J. Electroanalyt. Chem. **228**(1–2), 19–27 (1987). https://doi.org/10.1016/0022-0728(87)80093-1

35. G.M. Swain, R. Ramesham, The electrochemical activity of boron-doped polycrystalline diamond thin-film electrodes. Analyt. Chem. **65**(4), 345–351 (1993). https://doi.org/10.1021/ac00052a007

36. A.W. Williams, E. Lightowl, A.T. Collins, Impurity conduction in synthetic semiconducting diamond. J. Phys. Part C Solid State Phys. **3**(8), 1727 (1970). https://doi.org/10.1088/0022-3719/3/8/011

37. R.M. Chrenko, Boron, dominant acceptor in semiconducting diamond. Phys. Rev. B **7**(10), 4560–4567 (1973). https://doi.org/10.1103/PhysRevB.7.4560

38. H.D. Li, T. Zhang, L. Li, X. Lü, B. Li, Z. Jin, G. Zou, Investigation on crystalline structure, boron distribution, and residual stresses in freestanding boron-doped CVD diamond films. J. Cryst. Growth **312**(12–13), 1986–1991 (2010). https://doi.org/10.1016/j.jcrysgro.2010.03.020

39. N.G. Ferreira, E. Abramof, E.J. Corat, V.J. Trava-Airoldi, Residual stresses and crystalline quality of heavily boron-doped diamond films analysed by micro-Raman spectroscopy and X-ray diffraction. Carbon **41**(6), 1301–1308 (2003). https://doi.org/10.1016/s0008-6223(03)00071-x

40. K. Miyata, K. Kumagai, K. Nishimura, K. Kobashi, Morphology of heavily B-doped diamond films. J. Mater. Res. **8**(11), 2845–2857 (1993). https://doi.org/10.1557/jmr.1993.2845

41. J. Cifre, J. Puigdollers, M.C. Polo, J. Esteve, Trimethylboron doping of Cvd diamond thin-films. Diam. Rel. Mater. **3**(4–6), 628–631 (1994). https://doi.org/10.1016/0925-9635(94)90238-0

42. R. Haubner, S. Bohr, B. Lux, Comparison of P, N and B additions during CVD diamond deposition. Diam. Rel. Mater. **8**(2–5), 171–178 (1999). https://doi.org/10.1016/S0925-9635(98)00270-2

43. L. Wang, B. Shen, F. Sun, Z. Zhang, Effect of pressure on the growth of boron and nitrogen doped HFCVD diamond films on WC-Co substrate. Surf. Interface Anal. **47**(5), 572–586 (2015). https://doi.org/10.1002/sia.5748
44. P. Hartmann, S. Bohr, R. Haubner, B. Lux, P. Wurzinger, P. Wurzinger, M. Griesser, A. Bergmaier, G. Dolling, H. Sternschulte, R. Sauer, Diamond growth with boron addition. Inter. J. Refr. Metals Hard Mater. **16**(3), 223–232 (1998). https://doi.org/10.1016/S0263-4368(98)00022-5
45. H. Liu, D. Dandy, *Diamond Chemical Vapor Deposition* (Elsevier, Amsterdam, 1996)
46. K. Spear, J. Dismukes, *Synthetic Diamond: Emerging CVD Science and Technology* (Wiley, London, 1994)
47. T. Van Regemorter, K. Larsson, Effect of substitutional N on important chemical vapor deposition diamond growth steps. J. Phys. Chem. A **113**(13), 3274–3284 (2009). doi: CVDEFX
48. Z. Yiming, F. Larsson, K. Larsson, Effect of CVD diamond growth by doping with nitrogen. Theor. Chem. Acc. **133**(2), 1432. https://doi.org/10.1007/s00214-013-1432-y
49. C.J. Chu, R.H. Hauge, J.L. Margrave, M.P. D'Evelyn, Growth kinetics of (100), (110), and (111) homoepitaxial diamond films. Appl. Phys. Lett. **61**, 1393 (1992). https://doi.org/10.1063/1.107548
50. K. Larsson, J.-O. Carlsson, Surface migration during diamond growth studied by molecular orbital calculations. Phys. Rev. B **59**, 8315 (1999). https://doi.org/10.1103/PhysRevB.59.8315

Chapter 3
Chemical Mechanical Polishing
of Nanocrystalline Diamond

Soumen Mandal, Evan L. H. Thomas, Jessica M. Werrell, Georgina
M. Klemencic, Johnathan Ash, Emmanuel B. Brousseau and Oliver
A. Williams

Abstract In this chapter the chemical mechanical polishing of nanocrystalline diamond film is presented. It is shown that it is possible to polish a superhard material like nanocrystalline diamond with a much softer material like silica. It has also been demonstrated that this technique can be used for removing polishing marks on single crystal diamond. Experiments with other oxides like ceria and alumina showed polishing action on nanocrystalline diamond films. Surface roughness reduction rate was found to be inversely proportional to the size of abrasive material. Addition of redox agents to the polishing slurry accelerated the roughness reduction of nanocrystalline diamond films. Based on the experimental results and theoretical studies by other groups we have proposed a polishing mechanism for chemical mechanical polishing of diamond. Lastly, we have applied this technique to study its effect on superconducting diamond films. It was found that even after 14 h of polishing, superconductivity in diamond remained unchanged.

3.1 Introduction

Advances in growth technologies have made it possible to grow diamond thin films on large-area, non-diamond substrates [1]. Such large area diamond thin films have superlative properties rivalling smaller single crystals [2, 3], making them excellent candidates for a range of diverse applications such as micro-electro-mechanical

S. Mandal (✉) · E. L. H. Thomas · J. M. Werrell · G. M. Klemencic · J. Ash · O. A. Williams
School of Physics and Astronomy, Cardiff University, Cardiff, UK
e-mail: mandals2@cardiff.ac.uk

O. A. Williams
e-mail: williamso@cardiff.ac.uk

J. Ash
Department of Physics, Aberystwyth University, Aberystwyth, UK

E. B. Brousseau
School of Engineering, Cardiff University, Cardiff, UK

© Springer Nature Switzerland AG 2019
N. Yang (ed.), *Novel Aspects of Diamond*, Topics in Applied Physics 121,
https://doi.org/10.1007/978-3-030-12469-4_3

systems (MEMS), surface acoustic wave devices (SAW), and optical coatings to name a few [1, 4–6]. However, the growth of diamond on non-diamond substrates is not trivial. Due to the large differences in surface energies between diamond and the most commonly used substrates, and the low sticking coefficient of methyl growth precursors, it is difficult to grow diamond directly on non-diamond substrates. Hence, to grow coalesced, thin film diamond on non-diamond substrate a seeding step is required [7]. The seeded substrates are then exposed to chemical vapour deposition conditions where growth initially follows the Volmer-Weber growth model [8] until coalescence, and then transitions to a competitive columnar growth regime in accordance with the Van der Drift model [9]. Such competitive growth then results in a surface roughness that is proportional to film thickness, precluding/ hampering the use of thin film diamond in many of the applications stated above.

In the literature there are several methods to reduce roughness in nanodiamond films by altering the growth process. One of them is to limit the crystal size of the grains in the nanodiamond film. This leads to an increase in non-diamond carbon rich grain boundaries and a corresponding reduction in the Young's modulus [1], a key parameter for MEMS application. However, the roughness values achieved using this technique are still in excess of 5 nm root mean square (RMS), exceeding the acceptable roughness values for many applications [10, 11]. One can also use the smoother seed side of the wafer by etching the carrier substrates [6]. However, the quality of diamond on the seed side is poor as a result of the large surface/ volume ratio of the crystallites. Up to 1 μm of the film may then need to be lapped away from the underside before device fabrication [12, 13], ruling out the technique for thin diamond films.

Another approach to reducing the surface roughness of the films is through polishing of the rough as-grown surface. Historically, polishing has predominantly relied on pressing the samples against a fast rotating cast iron scaife charged with diamond grit at pressures of the order of a MPa [14]. Loose grit particles atop the scaife then degrade the surface through either fracture along the {111} plane or phase conversion to less dense forms of carbon [15]. Through the use of a technique reliant on fracture however, cracks initiated at the surface propagate microns deep into the samples being polished. The increase in dislocation defects brought about by these cracks then prevent use of the technique for samples to be used for device fabrication [15–18]. In addition, the growth of diamond on non-diamond substrates can introduce wafer bow in the thin films. This is shown in Fig. 3.1 as an exaggerated schematic. The wafer bow arises due to differences in thermal expansion coefficient between the substrate and diamond layer. The growth of diamond generally takes place at elevated temperatures. After the growth when the wafer is cooled a significant wafer bow will occur. In the cases where the wafer bow is larger than the film, pressing the film against the fast rotating hard scaife at high pressures will shatter the film or at best lead to uneven polishing [19].

Therefore, a gentler technique is required to produce smooth diamond samples without the inherent damage of mechanical polishing, while also being able to account for the wafer bow. This seems contrary to common logic, diamond being the hardest material known yet it can be polished using a gentle polishing

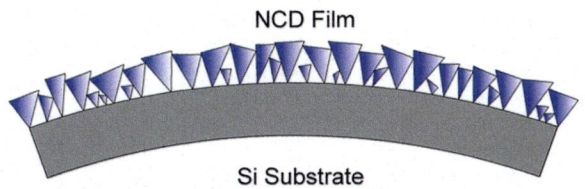

Fig. 3.1 Oversimplified schematic of wafer bow resulting from cooling of wafer after growth of diamond. The bow occurs due to difference in thermal expansion coefficient between diamond and silicon (Reprinted from Thomas et al. [20] under a Creative Commons License)

technique. Such a technique was developed by Thomas et al. [20]. The technique is an adaption of the chemical mechanical polishing used for polishing dielectric and metal interconnects in the IC fabrication industry [21, 22]. It is to be noted that the technique is able to polish a material as hard as diamond using a much softer silica powder and polyester pad.

In this chapter we will focus on the use of chemical mechanical polishing (CMP) for removing surface roughness from thin diamond films. First, the technique and its usage on both nanocrystalline (NCD) and single crystal diamond (SCD) are discussed. Subsequent studies into the effect of slurry composition as well as use of other oxide materials for polishing of diamond in an attempt to optimise the adaption are then introduced. Finally, the efficacy of the technique is shown through testing the superconducting properties of thin diamond films before and after being subjected to CMP.

3.2 Method for Chemical Mechanical Polishing of Diamond

For all our polishing experiments thin diamond films were grown on Silicon (100) p-type wafers, which on many occasions were buffered with 500 nm silicon dioxide layer [20, 23–25]. The wafers were cleaned with standard SC-1 cleaning process wherein 30% H_2O_2 and ammonia solution was mixed with de-ionised (DI) water in the ratio of 1:1:5 and the wafer was placed in the solution for 10 min at 75 °C. After that the wafers were rinsed with DI water in ultrasonic bath for 10 min and spun dry. The cleaned wafers were then seeded with a monodispersed nanodiamond seed solution resulting in nucleation densities in excess of 10^{11} cm^{-2} [7]. These seeded wafers were then exposed to diamond growth conditions and ~ 350 nm thick films were grown for polishing [20, 23–25].

The chemical mechanical polishing of diamond films were done using a Logitech Tribo polishing system. A SUBA-X polishing pad with Logitech supplied Syton SF1 alkaline slurry was used for polishing the wafers. The wafers were

Fig. 3.2 Schematic of the polishing setup used. The sample is held in the rotating carrier which swept across the polishing cloth. The polishing cloth is attached to a counter rotating plate (Reprinted from Thomas et al. [20] under a Creative Commons License)

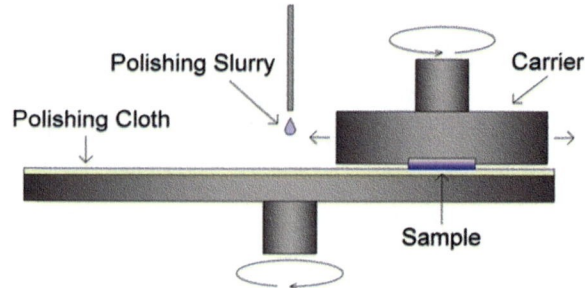

mounted on the polishing setup using a wafer carrier. The schematic of the polishing setup is shown in Fig. 3.2. The Syton SF1 slurry is made up of 15–50% silicon dioxide, 4–5% ethylene glycol mixed in water. The pH of the unaltered slurry is between 9.2 and 10.1. The pads were conditioned before use to increase the surface roughness of pads to maximize the polishing action and slurry distribution. The conditioning of the pads was done by a conditioning chuck consisting of a nickel plate embedded with diamond grit. The conditioning was done for 30 min using either the polishing slurry or DI water. The conditioner was rotated at 60 rpm with a down pressure of 2 psi on the counter rotating cloth at 60 rpm. The conditioner was also used in conjunction with the carrier during polishing to avoid clogging up the polishing pad.

The polishing was done with the carrier and pad rotating at 60 rpm in opposite direction to each other. The down pressure was varied between 2 and 4 psi and a backing pressure of 20 psi on the carrier chuck was used to roughly compensate the bow in the wafer. The polishing slurry was fed onto the pad at the rate of 40 ml/min as shown in the schematic. After, polishing the sample was thoroughly washed in water followed by a SC-1 clean. The polished surfaces were checked by scanning electron microscope and atomic force microscopy images were taken to estimate the roughness values of the polished wafers. X-ray photoelectron spectroscopy were also done on some of the wafers to explain the plausible mechanism behind this polishing technique.

3.3 Chemical Mechanical Polishing of Diamond

In this section we will discuss the results from our study on nanocrystalline and single crystal diamond. We will discuss the effect of size of particles on polishing rate and also the effect of using other oxides like ceria and alumina. The effect of pH and addition of redox agents to slurry will also be discussed. Finally, we will discuss the application of this technique to study the effect of polishing on superconducting diamond.

3.3.1 Polishing of Nanocrystalline Diamond Films

The polishing of diamond thin films was carried out on CVD grown nano diamond films. The films were polished for 4 h and the morphology of the films recorded after each hour of polishing. In Fig. 3.3 the scanning electron microscope (SEM) images of such a polished film is shown. The images were taken using a Raith e-line SEM operated at 10–20 kV, 10 mm working distance and using either the SE2 or in-Lens detector. Panel a of the figure shows the as-grown film before polishing. The image shows clear faceting on the surface with crystal sizes between 100 and 250 nm. As this film is polished the peaks of the crystals come in contact with the polishing pad and they are removed first. This is evident from Fig. 3.3b which shows the surface of the same sample after 1 h of polishing.

As the polishing progresses the crystal peaks and plateaus are reduced and the neighbouring crystals are next to each other without any voids. This is evident from panel c and d in Fig. 3.3 which shows the SEM image of the surface after 2 and 4 h of polishing respectively. Also to be noted, is the absence of any visible contamination from the polishing slurry. This shows that the SC-1 clean after polishing is enough to remove any contamination from the polishing process. Atomic force microscopy (AFM) images were taken on the unpolished and polished samples to estimate the surface roughness. The AFM images were taken using a TESPA tip

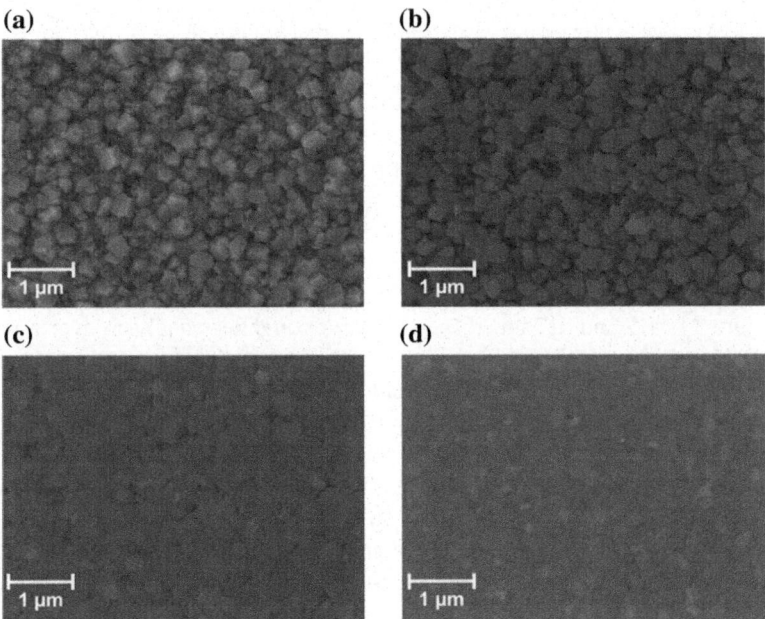

(a) **(b)** **(c)** **(d)**

Fig. 3.3 SEM images of unpolished and polished nanocrystalline diamond film. **a** As-grown **b–d** 1 h, 2 h and 4 h CMP respectively (Reprinted from Thomas et al. [20] under a Creative Commons License)

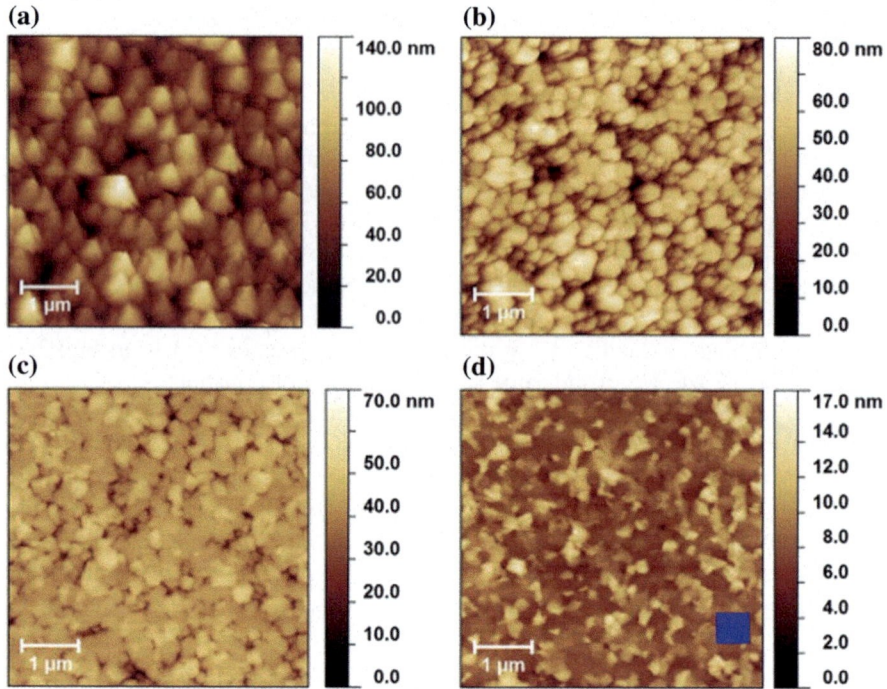

Fig. 3.4 AFM images of as-grown and polished nanocrystalline diamond film. **a** As-grown, **b–d** 1, 2 and 4 h CMP film respectively (Reprinted from Thomas et al. [20] under a Creative Commons License)

with 320 kHz resonant frequency, 8 nm tip radius and a spring constant of 42 N/m. In Fig. 3.4 we have shown the AFM images of an unpolished film (Panel a) and polished film (Panel b–d) after 1, 2 and 4 h polishing respectively.

The as-grown film in Fig. 3.4a has a root mean square (RMS) roughness of 18.3 nm over an area of 25 μm^2. Similarly, for panels b–d the RMS roughness over 25 μm^2 are 11, 4.5 and 1.7 nm respectively. A roughness analysis over a smaller area ~ 0.25 μm^2 (marked by blue square in Fig. 3.4d) yields a roughness value of 0.42 nm. Considering the individual grain size of the as-grown film ranges from 100 to 250 nm, an area represented by the blue square will have multiple grains, showing the effectiveness of the CMP process in reducing surface roughness of nanocrystalline diamond films. It is clear from the AFM data that a material like diamond with hardness of 90 GPa [26] can be polished with silica whose hardness value is 7.5 GPa [27] and in literature there are examples where harder materials have been polished by softer material [28–30]. X-ray photoelectron spectroscopy (XPS) was done on the polished and unpolished samples to identify the possible mechanism for polishing.

XPS studies on the thin films presented in this section was done by a VG ESCA Lab XPS spectrometer at 10^{-9} Torr with an Al Kα radiation source (1486.3 eV). The

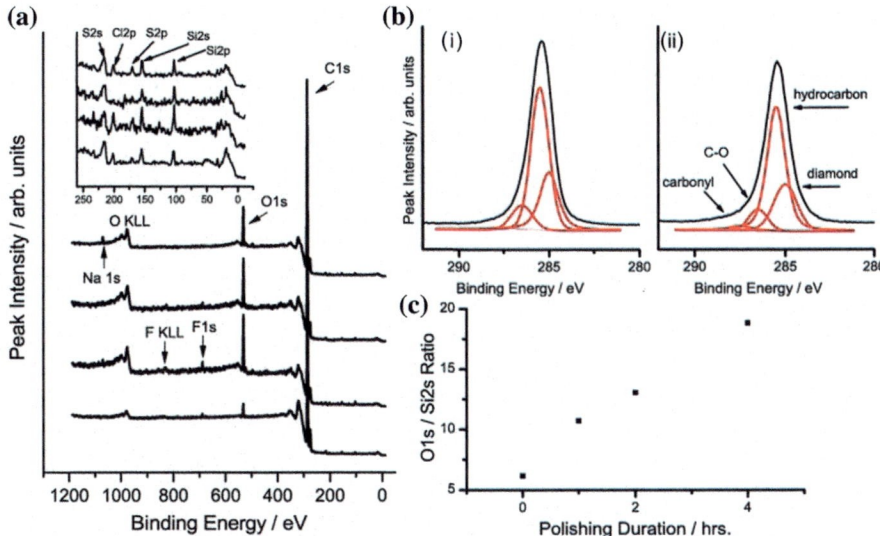

Fig. 3.5 **a** Survey XPS spectra of unpolished and polished diamond films are shown. The inset shows the spectrum at low binding energies. Spectra from as-grown, 1 h, 2 h and 4 h CMP films are shown from top to bottom respectively. **b** C1s spectra of the surface (i) before polishing and (ii) after 4 h CMP. The red curves are a fit to the data for various carbon containing species. **c** Ratio of O1s to Si2s peaks are plotted as function of polishing duration (Reprinted from Thomas et al. [20] under a Creative Commons License)

anode voltage was 10 kV with an emission current of 10 mA. The spectra were obtained using Fixed Analyser Transmission (FAT) mode with 50 eV or 25 eV pass energy for survey and narrow scans respectively. XPS Peak Fit was used for data analysis and normalization of peak areas were done with respect to XPS cross-section of the F1s photoelectron signal by using atomic sensitivity factors [31]. The survey and narrow XPS scans on the polished and unpolished nanocrystalline diamond are shown in Fig. 3.5.

In Fig. 3.5a the survey XPS spectra of diamond films are shown. The main photoelectron and Auger signals have been indicated on the same figure. The curves from bottom to top represent the as-grown, 1 h, 2 h and 4 h polished films respectively. The inset shows the spectra of the films at low binding energies. All the diamond films, polished and unpolished, show strong C1s ($\approx$285 eV) and O1s ($\approx$531 eV) peaks. There are also signals showing the presence of F, S, Cl and Si on the surface of the polished films, which is probably due to the polishing process. The contamination introduced by the process is not significant enough to effect any potential application of polished diamond thin films.

In Fig. 3.5b we have shown the C1s spectra of the diamond surface before and after 4 h of CMP (Fig. 3.5b (i) and (ii) respectively). The deconvolution of the C1s peaks into diamond (C–C, 285.0 eV), hydrocarbon (C–H, 285.5 eV), ether (C–O, 286.5 eV) and carbonyl (C=O, 287.5 eV) does not reveal any significant difference between the surfaces before and after polishing. Such deconvolution into different species have been used before to study CVD diamond surfaces [32, 33]. Another significant aspect of the chemical analysis is the absence of graphite or graphite related defects on the surface of polished diamond films. The other aspect of the data which is to be noted is the ratio of O1s to C1s peaks in the samples. On an as-grown film the ratio is 0.022 which increases significantly to 0.142 just after one hour of polishing which comes down to 0.1 after 4 h CMP. Assuming that the increase in oxygen content is mainly due to silica molecules getting attached to the diamond surface one would expect a constant O1s/Si2s ratio, which is not the case as is seen from Fig. 3.5c. This clearly indicates that the surface oxygen is not only due to silica but can be part of other elements detected on the surface. So in short, the polishing process introduces a range of strongly bonded chemical species on the diamond surface. To conclude, we have shown that thin diamond nanocrystalline films can be polished using silica colloidal slurry and the polishing slurry leads to an increase in oxygen containing species on the diamond surface indicating a chemical and mechanically assisted bonding and cleaving polishing mechanism.

3.3.2   Polishing of Single Crystal Diamond

In the previous section we have discussed the technique of chemical mechanical polishing of nanocrystalline diamond films. In this section we will discuss the use of this technique as a final polishing step for single crystal diamond. In the literature, many ways of polishing of single crystal diamond are detailed [16, 34–37], but traditionally mechanical polishing using fast rotating metal scaife and diamond grit has been preferred [38]. It is to be noted that polishing of diamond crystal is highly anisotropic and is dependent on the crystal face and direction of polishing [18, 39–43]. There are three principal crystal lattice plane groups in diamond. These are the cubic {100}, dodecahedral {110} and octahedral {111} plane groups. These facets or plane groups are also known as 4-point, 2-point and 3-point planes respectively. In Fig. 3.6a we have shown the preferred polishing direction on the various faces of a diamond crystal. The various letters c, d and o refer to cubic, dodecahedron and octahedron faces respectively. In Fig. 3.6b we have represented the individual faces and indicated the hard and soft polishing directions. It has been observed that the difference in polishing rates between hard and soft direction can be as high as 10:1 [44].

From Fig. 3.6b it is clear that while cubic and dodecahedral faces have at least one soft polishing direction, the octahedral face has only softer polishing direction making it the hardest faces to polish [45]. The wear mechanism along the hard and soft direction is significantly different leading to variation in wear rates [42, 46]. Along the soft direction polishing is mainly due to phase conversion to non-sp^3 material mediated shearing between the diamond charged scaife and the sample [15]. This results in formation of nanogrooves of varying lengths and depths depending on the grit size used [47]. Previous experiments have shown that the

(a)

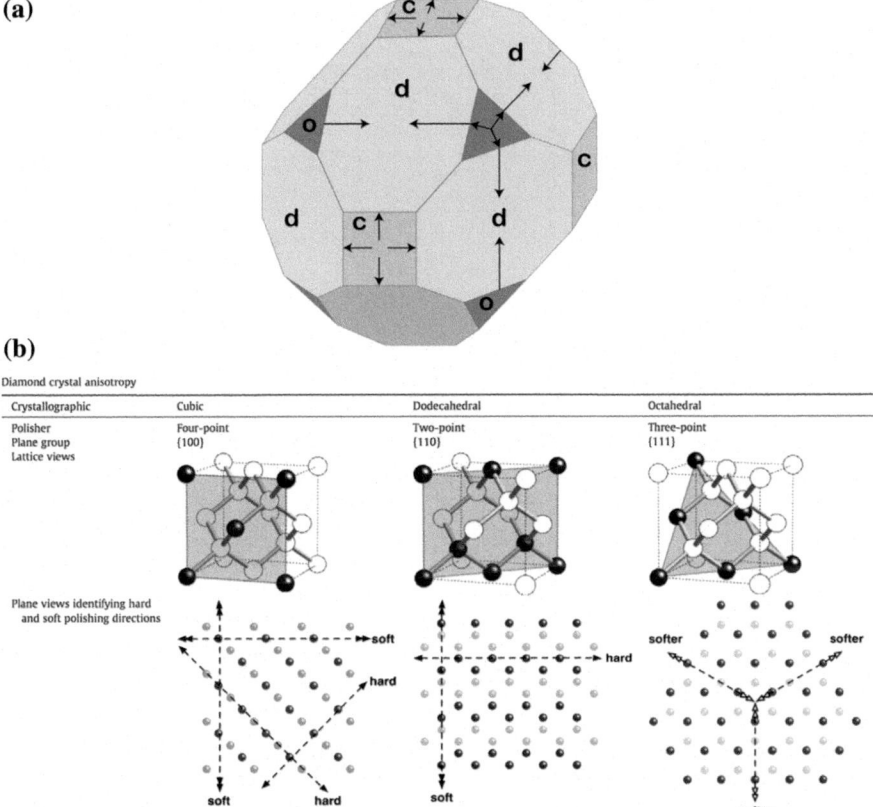

(b)

Diamond crystal anisotropy

Crystallographic	Cubic	Dodecahedral	Octahedral
Polisher	Four-point	Two-point	Three-point
Plane group	{100}	{110}	{111}
Lattice views			

Plane views identifying hard and soft polishing directions

Fig. 3.6 **a** In this figure we have shown the various faces of a diamond crystal with its preferred direction of polishing. The c, d and o represent cubic, dodecahedron and octahedron faces respectively. **b** Atomic arrangements of carbon atom on various faces of diamond crystal are shown along with hard and soft polishing directions (Reprinted from Schuelke and Grotjohn [18], with permission from Elsevier)

mechanical nature of the technique can lead to considerable subsurface damage propagating into the bulk of the material making it unsuitable for applications [48, 49]. So, it is essential that an alternative soft technique is implemented for polishing of single crystal diamond. In this regard chemical mechanical polishing is an excellent method but the polishing rates achieved on nanocrystalline diamond is extremely slow to remove subsurface damage in single crystal diamond. Hence, CMP should be used as a finishing technique after removal of subsurface damage by alternative methods. Here we will present our data on polishing on {100} and {111} single crystal diamond and show that we are able to remove polishing grooves from the diamond surface. The diamonds used for the study were high pressure high temperature single crystals obtained from Element Six.

Fig. 3.7 Plastic holder for
mounting small single crystal
diamond samples. The sample
was fixed to the holder within
the recess using
cyanoacrylate. Crystalbond
509 was used to fill the rest of
the recess as well as cap the
sample to stop it from moving
during the polishing process

The polishing of the small single crystal diamonds was done using the same
setup as the nanocrystalline diamond. However, the sample mount had to be
modified to hold such small samples on a machine designed for 2 in. wafers. In
Fig. 3.7 we have shown the image of our sample holder which is a polymer plastic
with a square recess in the middle. The sample was attached to the holder by
cyanoacrylate. Crystalbond 509 was then used to fill the recess and cap the sample.
When the capped sample comes in contact with the polishing pad the cap is easily
removed leaving only the surface of the diamond protruding from the sample
holder.

To check the effectiveness of the polishing process the sample surfaces were
checked before and after polishing using a Park Systems XE-100 AFM with
NT-MDT NSG30 tips operating in non-contact mode. The results from the AFM
measurements on the {100} diamond crystal are shown in Fig. 3.8. In Fig. 3.8a the
AFM image of the surface is shown before CMP polishing. The figure clearly
shows the presence of nanogrooves on the diamond surface. We have taken line
traces along and perpendicular to the nanogrooves and the traces are shown in
Fig. 3.8c. Plot 1 in the figure shows a line trace perpendicular to the grooves. The
widths of the grooves vary from 100 to 500 nm and has average depth of 3 nm. Plot
2 is the trace along the grooves showing far less variation in depth. The RMS values
of roughness along and perpendicular to the nanogrooves are 0.34 and 0.92 nm
respectively. Figure 3.8b shows the AFM image of the same diamond crystal after
3 h of CMP. The nanogrooves seen in the original sample has been reduced to the
point where there is little difference in RMS roughness along and perpendicular to
the groove. The RMS roughness values along and perpendicular to nanogrooves
after polishing are 0.19 and 0.23 nm respectively. Line traces along and

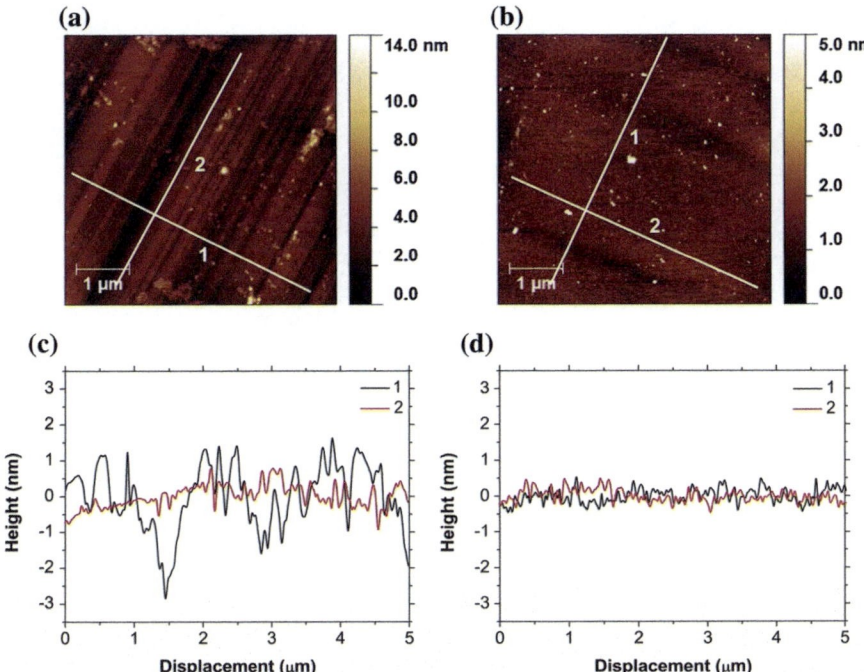

Fig. 3.8 The AFM images of {100} single crystal diamond is shown in panels **a** and **b**. Panel **a** is the surface before polishing and Panel **b** is the image of the surface after 3 h CMP. Panels **c** and **d** are the line traces along and perpendicular to nanogrooves created by the mechanical polishing of the commercial diamonds. Trace 1 is perpendicular to the grooves and trace 2 is parallel to the grooves (Reprinted from Thomas et al. Silica based polishing of {100} and {111} single crystal diamond, *Sci. Tech. Adv. Mater.* 2014, **15**, 035013 under a Creative Commons License)

perpendicular to the nanogrooves after polishing are plotted in Fig. 3.8d showing no clear distinction between the two directions. We did similar polishing process with a {111} oriented single crystal diamond and to check the polishing process we studied the surface of the crystal before and after polishing by AFM. The results of the AFM experiments are shown in Fig. 3.9.

From Fig. 3.9a, b it is clear that CMP is able to remove the nanogrooves created by mechanical polishing process. When we take the line profile perpendicular to the nanogrooves (Trace 1 in Fig. 3.9c), the depth of the grooves is much smaller than what is seen for {100} surface. The roughness values calculated from the line profiles parallel and perpendicular to the grooves are 0.23 and 0.31 nm RMS respectively. After polishing for 7 h it becomes difficult to differentiate between the two directions and the RMS roughness value for the line profiles were 0.09 nm. The line profiles after polishing are shown in Fig. 3.9d. When we compare polishing times for {100} (3 h) and {111} (7 h) diamond crystals it is evident that {111} surfaces are much harder to polish as is expected from the crystal profiles seen in Fig. 3.6. In conclusion, CMP is a good process to remove polishing grooves from

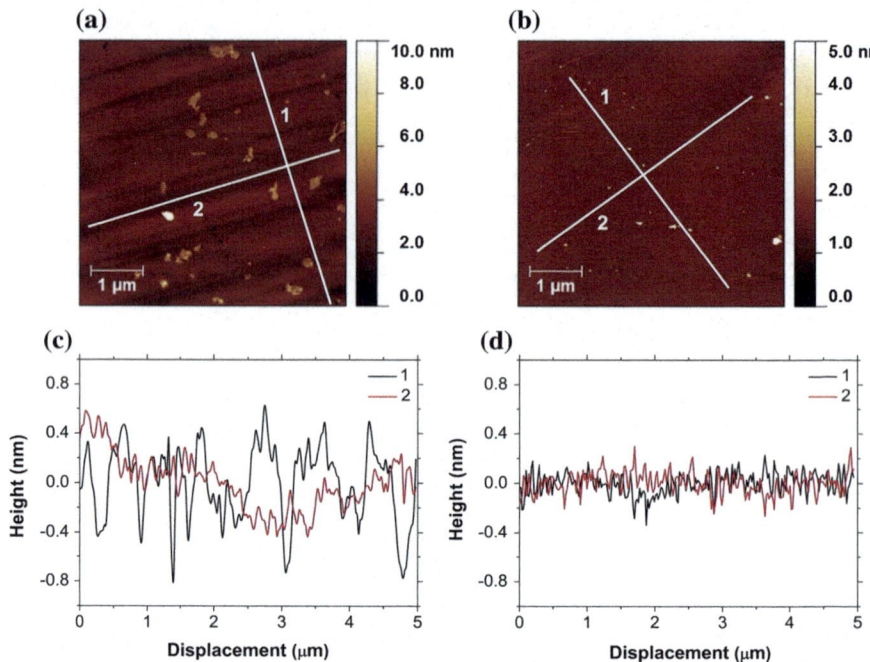

Fig. 3.9 The AFM images of {111} single crystal diamond is shown in panels **a** and **b**. Panel **a** is the surface before polishing and Panel **b** is the image of the surface after 7 h CMP. Panels **c** and **d** are the line traces along and perpendicular to nanogrooves created by the mechanical polishing of the commercial diamonds. Trace 1 is perpendicular to the grooves and trace 2 is parallel to the grooves (Reprinted from Thomas et al. Silica based polishing of {100} and {111} single crystal diamond, *Sci. Tech. Adv. Mater.* 2014, **15**, 035013 under a Creative Commons License)

the crystal surface. But, the polishing rates seen during our experiments are too slow to remove subsurface damage which can be microns deep in the crystal. Hence, CMP can be used as a finishing process for creating damage free diamond surfaces.

3.3.3 Polishing with Ceria and Alumina

In the previous sections we have discussed the CMP of nanocrystalline and single crystal diamond. So far, we have only used a silica based slurry for polishing the diamonds. It is clear that the CMP of diamond is an extremely slow process. In this section, we will try to investigate the process with other commonly used oxides in the polishing industry like ceria and alumina. Finally, we will compare the results with silica polishing. We have taken two different slurries for each particle, one acidic (pH $\sim$ 6) and one basic (pH $\sim$ 9), to check if the change in pH introduces any change in the polishing process.

As before, for this study we have grown a series of ~360 nm nanocrystalline diamond thin films on 500 nm silicon dioxide buffered 500 µm thick (100) silicon wafer. The as-grown films were imaged using a Park system Park XE-100 and their RMS roughness were measured over an area of 25 µm². The average roughness of the films was 25.1 nm. For polishing, four commercially available slurries were taken. While the basic and acidic alumina slurries were commercially available, silica and ceria slurries available commercially were basic in nature and phosphoric acid was added to the slurries to make them acidic. The silica slurry was SF1 polishing slurry from Logitech Ltd. The original slurry is basic in nature and it was made acidic in-house. The ceria slurry was Ultra-Sol® Optiq from Eminess. Like SF1 this slurry is also basic and had to be made acidic for our studies. The alumina slurries were Polycrystalline Alumina Slurries 9240 and 9245 from Saint Gobain Ltd. While 9240 is alkaline in nature, 9245 is acidic. The primary properties like, particle size, pH were measured in the lab, other properties like particle content, hardness of particle or density were taken from manufacturers documents or literature. The summary of all these properties for the slurries are listed in Table 3.1. The particle size was measured using dynamic light scattering (DLS) and compared with manufacturer provided values. Polishing was done with each slurry for a maximum of 3 h and the roughness was recorded after each hour of polishing. In Fig. 3.10 we have shown the AFM images of as-grown films along with the 3 h polished films for various slurries. The average RMS values of roughness were calculated by averaging roughness values over three 25 µm² areas on each film.

In Fig. 3.10a the AFM image of as-grown film is shown. The average RMS roughness of the film is ~25 nm over 25 µm². The value is slightly higher when compared to the as-grown value in Fig. 3.4a. This can be due to the extra silicon dioxide buffer layer present in sample shown in Fig. 3.10a. Panels b–g in Fig. 3.10 show the films polished with basic alumina, acidic alumina, basic silica, acidic silica, basic ceria and acidic ceria, respectively, for three hours. In Fig. 3.10d some

Table 3.1 Primary properties of slurry used for studying the use of other oxides. The hardness and density values taken from various sources [27, 50]. Other properties were measured in house and taken from Werrell et al. [24] (Reprinted from Werrell et al. [24] under a Creative Commons License)

Property	Silica	Alumina	Ceria
Density (g/cm³)	1.95	3.96	7.1
Hardness (GPa)	7.5	20	7.5
Particle size from DLS (µm)	0.1 (basic)	1.0 (basic)	0.5 (basic)
	0.1 (acidic)	0.2 (acidic)	0.5 (acidic)
Particle size from manufacturer (µm)	Unavailable	0.4 (basic)	0.4 (basic)
		0.3 (acidic)	
Particle content (%)	15–50	20 (basic)	20
		20 (acidic)	
pH	9.6	9.1	8.9
	5.6	5.8	5.8

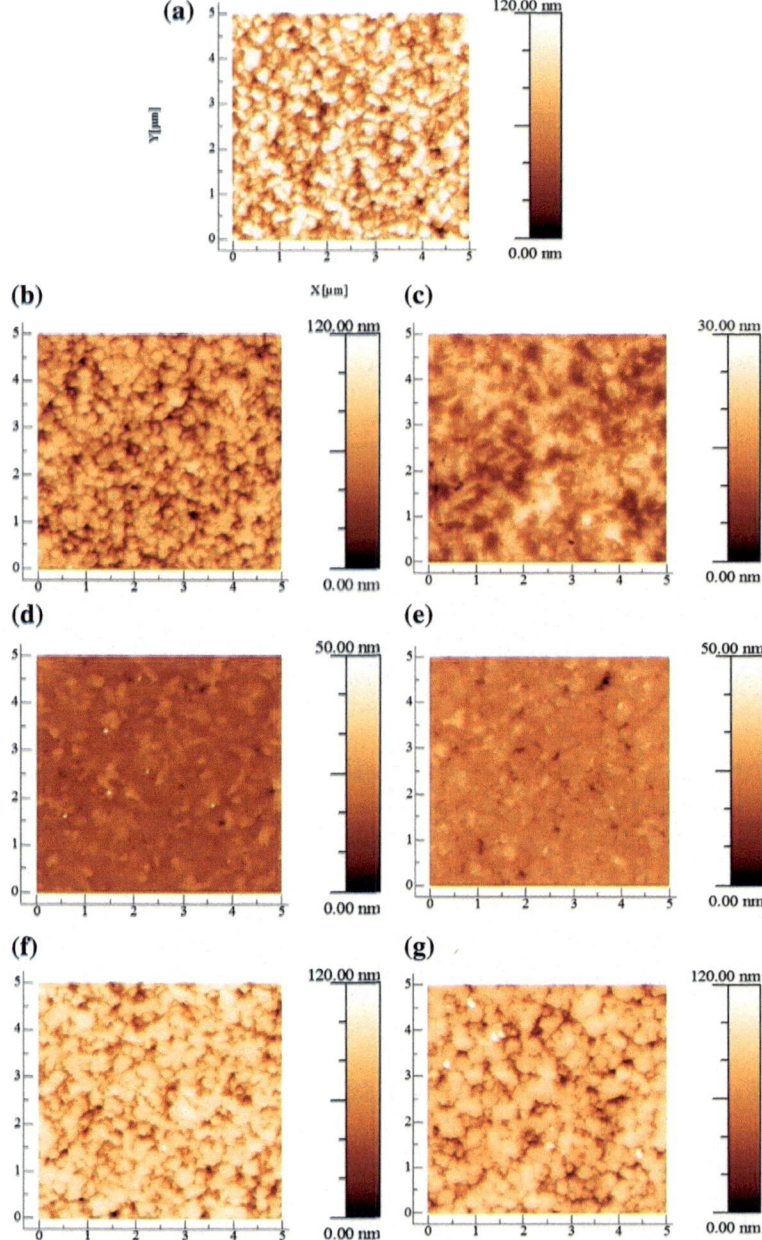

◄**Fig. 3.10** AFM images of as-grown and polished nanocrystalline diamond films. **a** Image of as-grown film. **b, c** Images of NCD films polished by basic and acidic alumina slurry respectively. The acidic slurry shows some polishing effect but the basic slurry shows very little polishing on the films. **d, e** Films polished with basic and acid silica slurry are shown here respectively. As expected polishing action is seen in these films. **f, g** Images of films polished with basic and acidic ceria slurry respectively. In case of ceria slurries no polishing action was observed (Reprinted from Werrell et al. [24] under a Creative Commons License)

minor white dots are seen on the film surface. This is mainly due to small dust particles on the surface. If we compare the six polished films with the as-grown film, it is quite evident that the films polished by silica slurries (Panel d (basic) and e (acidic)) are significantly smoother than the as-grown film. This result is expected as has been seen in previous sections. If we look at panels f and g which shows the films polished by basic and acidic ceria respectively we do not see much polishing on these films. They look similar to the as-grown films. Finally, if we look at films polished by basic alumina (panel b) and acidic alumina (panel c) there is considerable difference between the two slurries. While the image for basic slurry looks similar to that of as-grown film, considerable polishing can be seen for acidic alumina slurry.

We have also taken SEM images of the as-grown and polished films and the images are shown in Fig. 3.11. Figure 3.11a shows the image of as-grown film with peaks and valleys on the surface clearly visible. In Panels b–g images of the polished films are shown. In all the images some polishing action is seen as the peaks seen in panel a have disappeared. Only for films polished with basic silica slurry (Fig. 3.11d) we see that the peaks have completely disappeared and they have merged with their neighbouring crystals forming a smooth surface. For the rest though, the crystal peaks have formed into plateaus but the complete merger with the neighbouring crystals have not happened. To analyse further we have plotted the change in RMS roughness with polishing time in Fig. 3.12.

From Fig. 3.12 we can see that the slurries polish by varying amounts. The acidic silica slurry reduced surface roughness by 23.7 ± 0.4 nm while the basic slurry reduced it by 20.8 ± 0.6 nm after three hours of polishing. If these values are compared with alumina the numbers are 19.9 ± 0.4 nm and 9.5 ± 0.6 nm for acidic and basic slurries respectively. The difference in value is 10.4 nm compared to 2.9 nm for silica. Similarly, for ceria the numbers are 15.1 ± 0.6 nm and 9.3 ± 0.9 nm for acidic and basic slurries respectively giving a difference of 5.8 nm. The numbers clearly show that acidic slurries polish with greater rate then basic slurries but the variations in differences mean that apart from pH and particle type some other factors are also responsible for polishing. XPS measurements were done on the polished and unpolished films to study the surface chemistry of the films before and after polishing. A Thermo Scientific™ K-Alpha + spectrometer operating a monochromatic Al source at 72 W was used to record the data. Pass energies of 40 and 150 eV were used for high resolution and survey scans respectively over an area of 400 μm^2. The results of the XPS measurements are shown in Fig. 3.13.

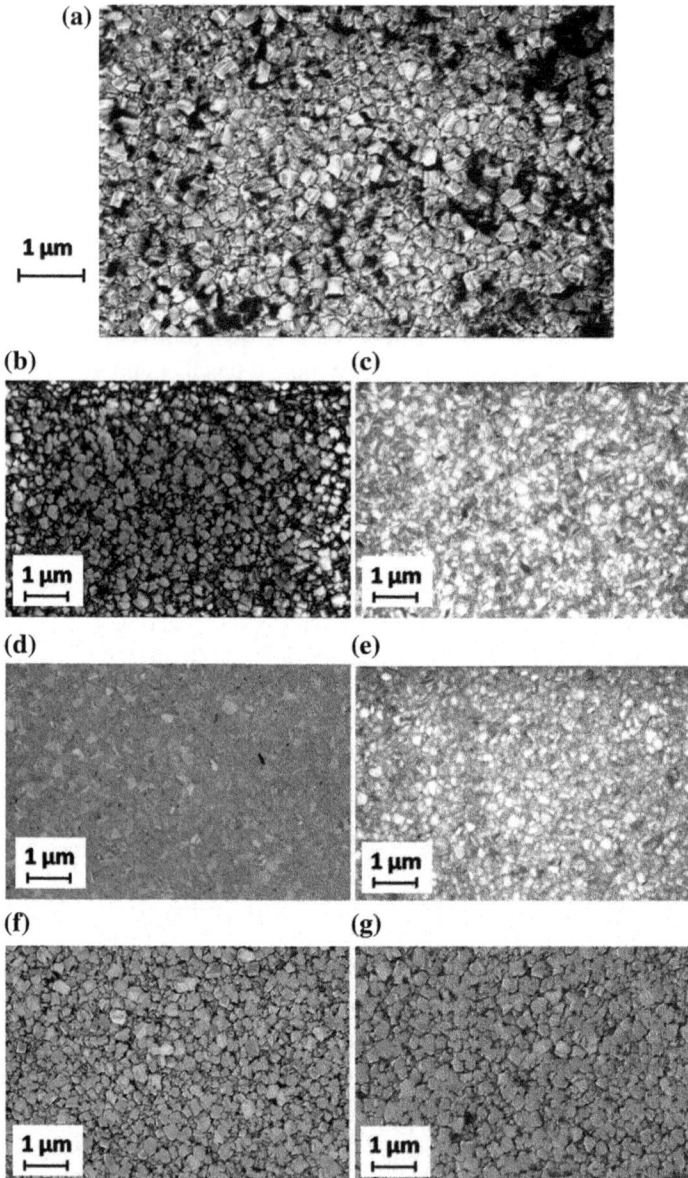

Fig. 3.11 SEM images of the polished films. **a** Unpolished as-grown film **b, c** NCD films polished by basic and acidic alumina slurry respectively. **d, e** Films polished with basic and acid silica slurry respectively. **f, g** Images of films polished with basic and acidic ceria slurry respectively (Reprinted from Werrell et al. [24] under a Creative Commons License)

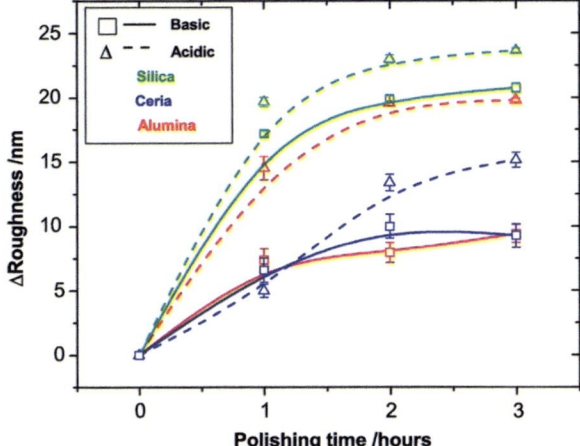

Fig. 3.12 The change in RMS roughness for the films after each hour of polishing. The RMS roughness is averaged over three 25 μm^2 area on each film. The standard deviations of the roughness values are also shown in the figure. The zero mark in the figure is the as-grown film (Reprinted from Werrell et al. [24] under a Creative Commons License)

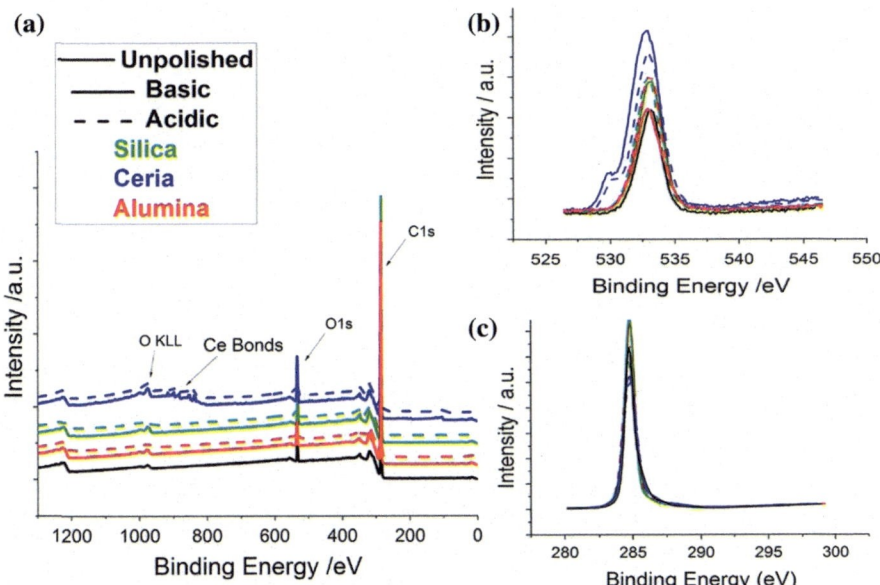

Fig. 3.13 Data from the XPS measurements are shown. **a** Shows the spectrum in the energy range of 0–1200 eV. The C1s, O1s and O KLL peaks are marked along with Ce peaks. **b** The O1s peaks from all the samples are presented in this panel. **c** The C1s peaks are represented in this panel (Reprinted from Werrell et al. [24] under a Creative Commons License)

Fig. 3.14 O1s/C1s ratio
calculated from the XPS data
from the polished and
unpolished films are shown
(Adapted from Werrell et al.
[24] under a Creative
Commons License)

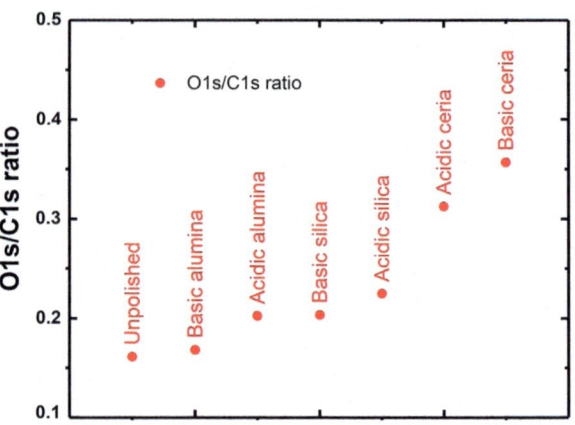

Figure 3.13a shows the broad spectrum of the sample surfaces between 0 and 1200 eV binding energy. Only the films polished with ceria show a significant change after polishing when compared with unpolished film surface. A clear signature of Ce on the sample surface around 900 eV is present. Small signatures of Al are seen for films polished with basic alumina but they are completely absent for acidic alumina polished films. Silica slurry polished films show much smaller Si signal when compared with Ce. In Fig. 3.13b we have expanded on the O1s peak in the range of 520–550 eV. The C–O peak at 532 eV is clearly visible. For the ceria polished samples a second peak at 530 eV belonging to Ce–O is also present. Finally, in Fig. 3.13c we have plotted the spectrum data around the C1s peak between 280 and 300 eV. No change has been observed between the polished and unpolished films in the C1s peaks. We have calculated the ratio of O1s to C1s for each films to see the change in oxygen content before and after polishing on the sample surface. The ratios have been plotted in Fig. 3.14.

The ratios in Fig. 3.14 show some interesting facts. We have already seen before that irrespective of slurry type acidic slurries performed slightly better than basic slurries. For alumina and silica slurries we see that O1s/C1s ratio is slightly higher for acidic slurry than basic slurry but for ceria the ratio is other way around, even though acidic slurry shows better polishing rate. This again confirms the fact that apart from pH and type of polishing particle some other parameters are also responsible for polishing. So, if we go back to Table 3.1 and compare the basic properties of each slurry an interesting relation between polishing rate and size of particle emerges. We will discuss this variation of rate with abrasive particle size in the next section. To conclude this section, we have seen that polishing of diamond can be done by other oxides like alumina and ceria. Furthermore, not only particle composition or pH but other factors can also lead to change in polishing rates.

3.3.4   Abrasive Particle Size Dependence of Polishing

From Figs. 3.12 and 3.14 it is clear that apart from pH and particle type some other parameters are also important for polishing of diamond. If we compare the primary properties of all the particles used in the previous section as listed in Table 3.1, a dependence of polishing rate on particle size emerges. This variation in polishing rate with particle size is plotted in Fig. 3.15. We have plotted the reduction in roughness after 2 h of polishing versus the particle size. The time 2 h was chosen because looking at Fig. 3.12 it is clear that after 2 h the rate of change is extremely slow and it saturates after that time.

From Fig. 3.15 it is clear that, larger the particle size of the abrasive material, lower is the polishing rate. The higher rate for smaller particle may be mainly due to increased surface area due to small particle size. Since the particle content of all the slurries are close to each other, smaller particles mean more available surface for polishing action. Apart from the above discussed results on particle size we have also done some experiments taking only silica slurries with various particle sizes. We took three slurries from Levasil CS45-38P, SP1039 and SP1049. The particle sizes and contents are listed in Table 3.2.

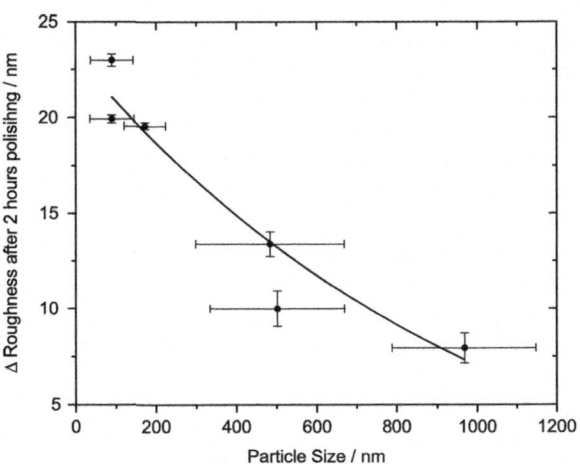

Fig. 3.15 Plot for change in surface roughness after two hours of polishing versus particle size of abrasive particle (Reprinted from Werrell et al. [24] under a Creative Commons License)

Table 3.2 Basic properties of slurries from Levasil. While the first two slurries have particle sizes close to each other CS45 particle size is smaller than the other two

Slurry	Silica content (%)	Particle size (nm)	pH
SP1039	30–60	97.0	9.23
SP1049	30–60	90.1	9.29
CS45-38P	30–60	58.4	9.93

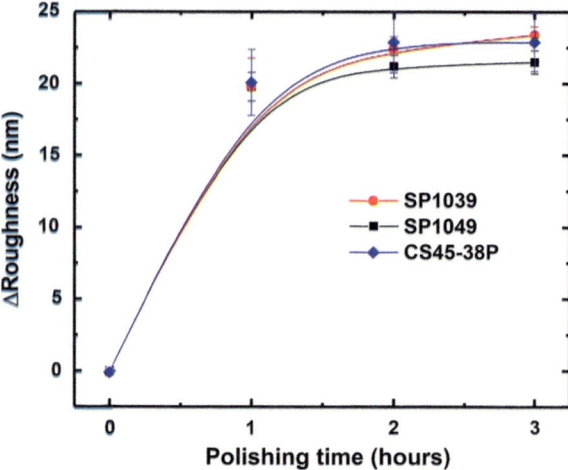

Fig. 3.16 Change in roughness for the films polished with three different silica slurry. As with SF1 slurry these slurries are able to polish the films equally efficiently

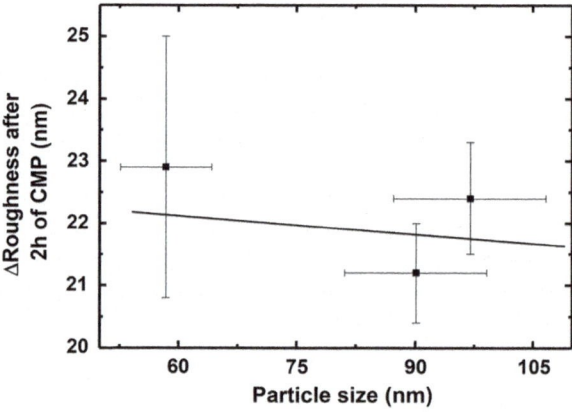

Fig. 3.17 Change in roughness of NCD films after two hours of polishing is plotted against particle size. The particle sizes were measured using dynamic light scattering

As before, two NCD films were polished with each slurry and the film was monitored after each hour of polish by AFM. RMS roughness were calculated over six areas of 25 μm^2 on each sample and the average change in roughness was calculated. The result of the polishing for the three slurries are plotted in Fig. 3.16. It is clear from the figure that these slurries were able to polish the NCD films as efficiently as SF1. To investigate whether the particle size has any effect on polishing we will plot the change in roughness of the films after two hours of polishing. The result of the comparison is plotted in Fig. 3.17. The line in the plot is a straight line fit to the data. It is clear that within the range of particles chosen there is very little change in polishing rates. But looking at the straight line fit we can see a negative slope indicating that higher the particle size lower is the polishing rate. If we put these data points in Fig. 3.15 the points fall within the trend seen in the figure.

In this case we have not seen large changes in polishing rates and this may be due to very small variation in particle size being selected for the experiments. This is mainly due to the availability of slurries commercially. Further experiments can be done by making slurries from silica of different sizes with size variation over a larger range. Here we would like to mention that this is not the first time that variation of polishing rate due to particle size variation is seen. For example, Bielman [51] saw the reduction in polishing rate of tungsten with increase in alumina particle size. On the contrary Tamboli et al. [52] saw higher polishing rates of tantalum as the particle size was increased. But then higher removal does not mean a smoother surface, because, in the same work we see that larger particles lead to rougher surface finish. Apart from that studies have been conducted on effects of surfactant on CMP [53]. Many groups have also tried to model these effects theoretically to study the CMP process [54–57]. To conclude this section, we have studied the effects of particle size on polishing and found that the smaller the particle size, the smoother the finish after polishing. In the next section we will detail the effects of redox agents on the polishing of NCD films.

3.3.5 Effect of Redox Agents on the CMP of NCD Films

So far we have studied the mechanical nature of CMP. We have also seen the effects of different abrasive particles on the process. In this section we would present our studies on changing the chemical nature of slurry by adding oxidising and reducing agents to the silica slurry SF1. Hocheng et al. [58] studied the mechanical polishing of diamond by adding strong oxidising agent to the slurry. In our study we have used three different oxidisers and two reducing agents. The oxidising agents were hydrogen peroxide (H_2O_2), ferric nitrate ($Fe(NO_3)_3$) and potassium permanganate ($KMnO_4$). The reducing agents were oxalic acid ($C_2H_2O_4$) and sodium thiosulfate ($Na_2S_2O_3$). Out of these hydrogen peroxide is liquid and the rest are solids. The concentration of hydrogen peroxide in SF1 slurry was 20% v/v. The solid additives had a concentration of 2.5 gm/l. The addition of oxalic acid made the resulting slurry mixture highly acidic making it unusable with the Suba-X pad used for polishing. Hence, sodium hydroxide was added to the slurry to bring the pH value within the specified range of Suba-X pad. For the rest of the additives no pH modification was necessary.

We have polished two wafers with each slurry mixture. The roughness of the polished wafers was monitored after each hour of polish using AFM and the roughness is the average RMS value over three 25 μm^2 area. All the results on roughness are over an area of 25 μm^2. The result from the AFM study is presented in Fig. 3.18. The polishing was done for maximum of 4 h or till the point where the RMS roughness dropped below 2 nm. Figure 3.18a is the AFM image of an as-grown film. The average RMS roughness of all the as-grown films taken together for this experiments was 24.5 $\pm$ 1.5 nm. Figure 3.18b presents the film polished with SF1 for 4 h. The average roughness after 4 h polishing was 2.7 $\pm$ 0.4 nm

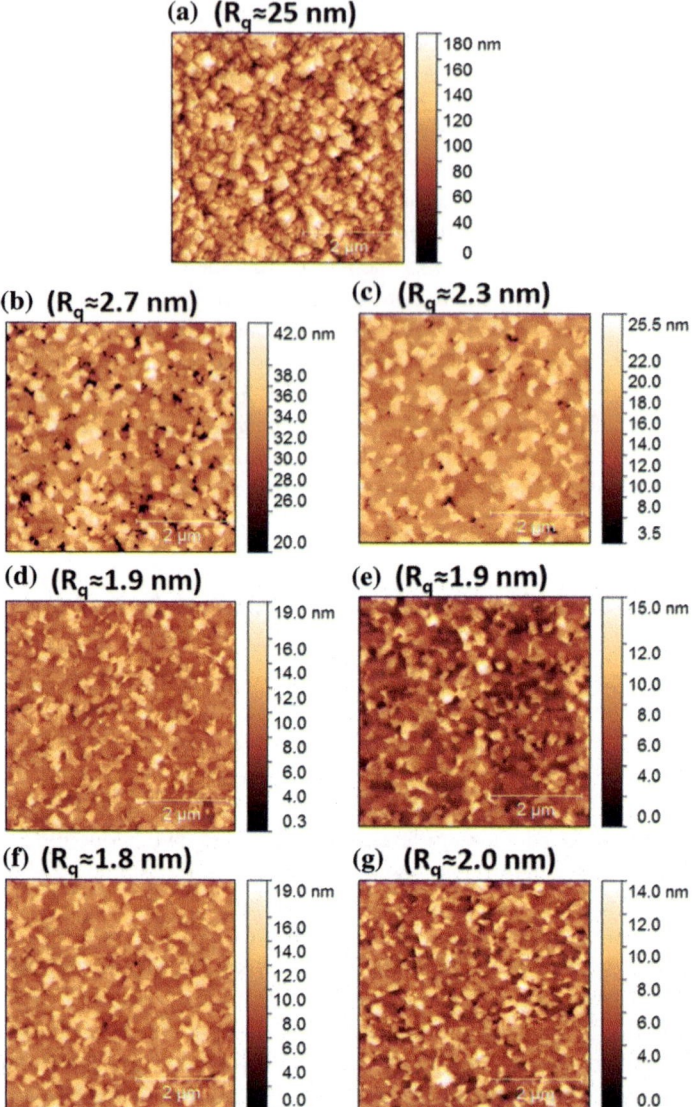

Fig. 3.18 AFM images of **a** as-grown film, **b** film polished with SF1 and films polished with SF1 mixed with **c** hydrogen peroxide, **d** ferric nitrate, **e** potassium permanganate, **f** oxalic acid and **g** sodium thiosulfate. The final roughness values are indicated within brackets (Reprinted from Mandal et al. [25], with permission from Elsevier)

RMS. In panels c–g we have presented the AFM images of films polished with SF1 mixed with hydrogen peroxide, ferric nitrate, potassium permanganate, oxalic acid and sodium thiosulfate respectively. The film polished with hydrogen peroxide and SF1 mixture was polished for 4 h and the rest were polished for only 3 h. The final

Table 3.3 Final RMS roughness values of the polished films and the total duration of polishing for each slurry (Reprinted from Mandal et al. [25], with permission from Elsevier)

Slurry	Polishing duration (h)	Final roughness (RMS nm)
SF1	4	2.7 ± 0.4
SF1 + H_2O_2 (20% v/v)	4	2.3 ± 0.3
SF1 + $Fe(NO_3)_3$ (2.5 gm/l)	3	1.9 ± 0.1
SF1 + $KMnO_4$ (2.5 gm/l)	3	1.9 ± 0.1
SF1 + $C_2H_2O_4$ (2.5 gm/l)	3	1.8 ± 0.1
SF1 + $Na_2S_2O_3$ (2.5 gm/l)	3	2.0 ± 0.1

roughness values and the total polishing duration for each slurry has been summarized in Table 3.3.

From Fig. 3.18 and Table 3.3 it is clear that the solid additives have accelerated the polishing process. Hydrogen peroxide addition results in slightly smoother films compared to unaltered slurry but its performance is not as good as the solid additives. This can be due to volatile nature of hydrogen peroxide. For the rest, oxalic acid produces marginally better finish then the other slurries. Here it is to be noted that the slurry with oxalic acid addition is the most acidic and we have seen in Sect. 3.3 that acidic slurry can produce marginally better performance. We have plotted the change in roughness after each hour of polish for all the slurries in Fig. 3.19. It is quite clear that slurries with solid additives reach their limiting value of roughness within 2 h of polishing. In contrast SF1 and SF1 with hydrogen peroxide do not show any plateauing until 4 h of polishing. As we have seen from Fig. 3.19 that most of the slurries reach their limiting value within the first two hours we have tried to compare the reduction in roughness an hour before the plateau sets in. Figure 3.20 shows the reduction in roughness after first hour of polishing for all the slurries. It seems that the slurry with hydrogen peroxide performs the worst after the first hour even though the final polish is better than pure

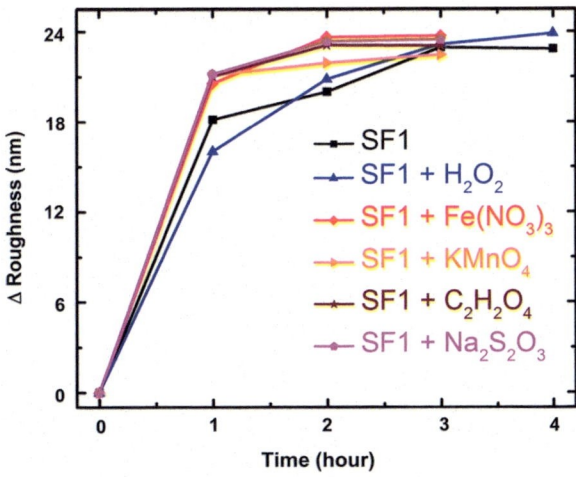

Fig. 3.19 Change in roughness after each hour of polishing for all the slurries. While the slurries with solid additives reach their limiting roughness values with 2 h of polishing, unaltered and hydrogen peroxide slurry do not show any saturation till 4 h of polishing (Reprinted from Mandal et al. [25], with permission from Elsevier)

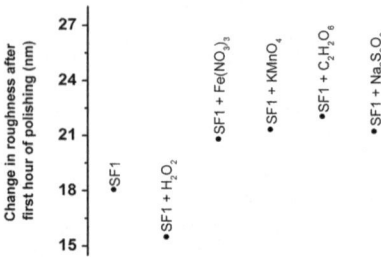

Fig. 3.20 Roughness reduction for various slurries after the first hour of polishing. The solid additives show ∼17% increased roughness reduction when compared with SF1 (Reprinted from Mandal et al. [25], with permission from Elsevier)

SF1. The solid additives though show a ∼17% increase in roughness reduction when compared with SF1. And, if we just compare the solid additives very little distinction can be made between the slurries with oxalic acid showing marginally better roughness reduction.

We have measured the particle size of silica in each slurry to confirm if the increased roughness reduction is due to change in particle size, as we have seen in Sect. 3.4 that smaller particles can lead to increased roughness reduction rate. Dynamic light scattering measurements of all the slurries are presented in Fig. 3.21. We can see that for hydrogen peroxide, oxalic acid and sodium thiosulfate mixed slurries there is no change in particle size. The particle size change seen for ferric nitrate and potassium permanganate is mostly due to agglomeration on addition of redox agents and as such the agglomerates will break apart upon mechanical action during polishing. Also, the roughness reduction rates do not correlate with particle size clearly proving that the increased rate is purely due to chemical nature of the slurry.

From the above data it is clear that the enhancement in polishing rate, on addition of redox agent, is more to do with the chemical nature of the process rather than the mechanical nature. We have done XPS measurements on the polished samples to see whether the surface of the samples polished with unaltered and

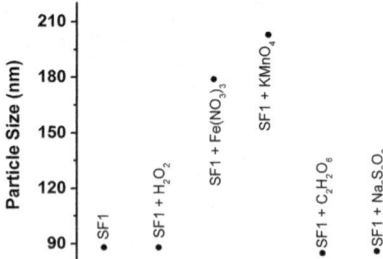

Fig. 3.21 Particle sizes of various slurries as measured by dynamic light scattering. For ferric nitrate and potassium permanganate increase in particle size is seen due to agglomeration. For the rest of the additives no change in particle size is observed (Reprinted from Mandal et al. [25], with permission from Elsevier)

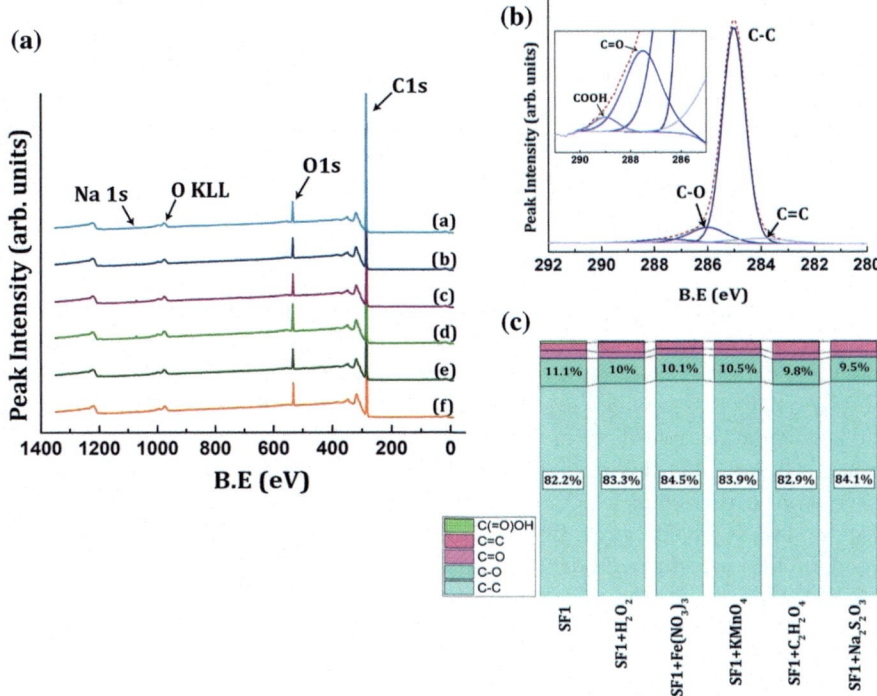

Fig. 3.22 a XPS data for the polished films. **b** deconvoluted C1s peak from the NCD film polished with oxalic acid added SF1. **c** fractions of different species of the C1s peak from the polished films (Reprinted from Mandal et al. [25], with permission from Elsevier)

altered slurries show any considerable difference. In Fig. 3.22a we have shown the survey XPS spectra of samples polished with (a) SF1, SF1 with addition of (b) hydrogen peroxide, (c) ferric nitrate, (d) potassium permanganate, (e) oxalic acid and (f) sodium thiosulfate. As has been seen in previous sections the polished films have significant amount of C1s (285 eV) and O1s (533 eV) character. Signatures of Si2s (103 eV) and Si2p (153 eV) are also visible signifying the presence of bonded silica from the slurry, resistant to post CMP cleaning using SC-1 process. Apart from that, sodium and calcium contaminants are also visible, probably coming from the parent slurry. For samples polished with slurries containing ferric nitrate and potassium permanganate, additional iron and manganese peaks are visible. This suggests that the additives have some chemical interaction with the NCD surface. Such contamination can be removed by using various techniques detailed in the literature [22, 59].

We have further deconvoluted the C1s peaks into its various components. It may be noted here that due to very small shifts in binding energy between various components it is difficult to precisely determine the components on the surface [33]. After deconvolution the main component of the peak at 285 eV was attributed to

Fig. 3.23 The O1s/C1s for
polished samples are shown
here. The slurry mixture is
indicated next to the data
point (Adapted from Mandal
et al. [25] with permission
from Elsevier)

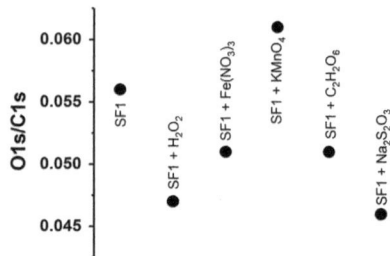

bulk and surface sp^3 C–C bond with mono-hydrate surface dimers having some
minor contribution [60]. Other species like graphitic carbon, ether/hydroxyl, carbonyl and carboxyl groups with relative binding energy shifts of −1, +1, 2.5 and
4 eV, respectively, provided a good fit to the XPS data [61]. We also tried taking
into consideration contributions from CH$_x$ (x = 2, 3) species with binding energy
shift of +0.5 eV, but this resulted in bad fit to the data. Such behaviour has been
observed in earlier studies [33, 61] and hence these components were not considered for the fit in this case. In Fig. 3.22b we have shown an example of peak
deconvolution used for XPS data analysis. The data in the figure is from the surface
polished with oxalic acid added SF1 slurry. Figure 3.22c shows the fractions of
each components in the C1s peak from each of the polished sample.

From Fig. 3.22c it is clear that there is very little difference between fractions of
various species across all samples. We also took the ratios of O1s peak area to C1s
peak area. As seen before [20, 24], polished films showed larger ratios indicating an
increase in oxygen content on the surface. We have plotted the O1s/C1s ratios in
Fig. 3.23 for various slurries. There is no correlation between the ratios and rates
seen in Fig. 3.20 and between unaltered and altered slurries there is little variation
in the oxygen content. This is an indication that addition of redox agents to SF1 do
not increase the oxygen content on the surface of NCD. Hence, higher rate in
roughness reduction cannot be attributed to higher density of oxygen containing
species on the sample surface. Also there is no signature of change in graphitic
carbon peak between the samples indicating the absence of phase conversion to less
dense form of carbon being responsible for the rise in roughness removal rate [56].
To conclude this section, we have seen an increase in roughness reduction rate with
addition of redox agents but that cannot be attributed to any increase in surface
species like oxygen containing species or conversion to less dense forms of carbon.

3.3.6 Possible Mechanism for Polishing

From the previous sections it is clear that the polishing of NCD starts with the
removal of peaks and formation of plateaus till all the plateaus connect with each
other. This clearly shows that polishing occurs only at places where the sample is in
contact with the pad. This is a clear indication of the mechanical nature of the

process. Further, the crack free finish produced by this process points to the chemical nature of CMP. Mechanical polishing relies on fracturing of diamond peaks using either diamond particles or particles of materials with hardness values within an order of magnitude, for example, alumina plates with hardness of ~ 20 GPa is routinely used in polishing of diamond with hardness of ~ 90 GPa [19]. In comparison, the pad used in our CMP process has a hardness value of 30–100 MPa and the silica particles have hardness value of ~ 7.5 GPa. Clearly, the pad will be unable to generate any substantial frictional force to cause any polishing action on NCD. It should also be noted here that the conditioner used in all our polishing experiments to roughen the pad is a diamond embedded nickel plate and it is common to use diamond embedded metal or ceramic plates for conditioning the polishing pads [62, 63].

From all our XPS data it is also clear that in all the processes the conversion of sp^3 bonded carbon to less dense forms are also unlikely. Further, solely chemical based etching of diamond [64–66] as seen with hot KOH or KNO$_3$ (600–1100 °C) is also not possible since the operating temperatures in our case are much below the boiling point of water. Hence, it is clear that traditional models used to describe polishing of diamond will not be able to explain the polishing seen with CMP. But, the XPS data on various polished films do throw some light onto the possible causes of material removal. In all the polishing experiments a marked increase in oxygen containing species on the surface of polished diamond is seen. Also, the presence of silicon signature in the XPS data points to chemically bonded silicon atoms on the surface.

So, to explain the polishing process we try to draw parallels between CMP of diamond and CMP of silicon dioxide using silica particles. In the oxide polishing process, silanol (Si(OH)$_4$) groups are formed on the surface through hydroxide ion based conversion of surface siloxane bonds (Si–O–Si). The silanol groups in turn facilitate the bonding of silica particles to the wafers surface [67, 68]. This bonded silica particle along with the silanol group is then sheared off by the shearing force from the pad. In the case of diamond polishing a similar mechanism has been proposed. The as-grown NCD film is largely hydrogen terminated. The alkaline slurry converts this hydrogen terminated surface partly to carbonyl and hydroxyl rich surface. On the converted surface silica from the slurries are able to attach themselves creating a Si–O–C bridge. A sufficient shearing force from the pad will be able to remove the whole bridge thus removing the surface carbon atom. The polishing process has been schematically shown in Fig. 3.24.

More recently Peguiron et al. [56] performed density functional theory calculations to study the bond breakage and role of silica in polishing of diamonds. They studied the polishing of diamond surface in contact with both silicon and silica. The configuration of their simulation is shown in Fig. 3.25a, b. On the timescale of simulation, it was found that for silicon/diamond interface a plastic deformation of silicon was observed while the (110) diamond surface remained intact. On the other hand, C-C bond breakage was seen occasionally for silica/diamond interface. It was also observed that the C–C bond breakage was seen at the bonds (marked by red in Fig. 3.25c) that connected the surface aromatic carbon chains (marked by black in Fig. 3.25c) to the underlying bulk diamond. Pastewka et al. [15] had seen that these

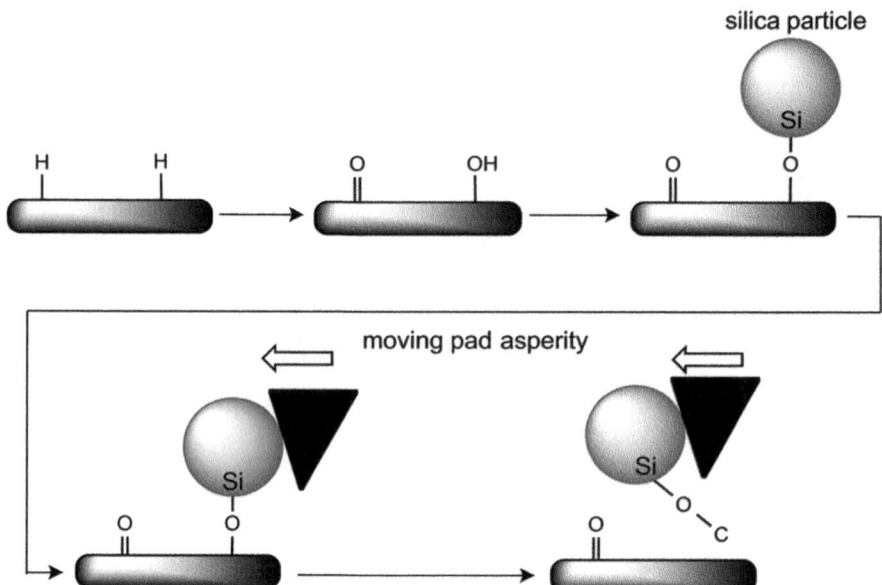

Fig. 3.24 Schematic of the polishing process. The alkaline slurry converts the hydrogen terminated surface to hydroxyl and carbonyl rich surface facilitating the attachment of silica particles to the surface. Shearing force from the pad on the silica particle then removes carbon atom providing polishing (Reprinted from Thomas et al. [20] under a Creative Commons License)

bonds are more likely to fail when compared to other C–C bonds. This C–C bond breakage results in the reconstruction of the C–C–C carbon chains (Fig. 3.25d, e) into C–O–C ether groups or C–C bonds between an uncoordinated C atom in neighbouring aromatic chain and top atom of C–C–C unit. Such a reconstruction leads to local destruction of the surface structure which can lead to both amorphization [15] or oxidative wear [69, 70] of diamond surface. If we consider the data from XPS measurements amorphization seems most unlikely and the polishing seems to progress through the oxidative wear mechanism brought about by the breakage of C–C bonds (red line in Fig. 3.25c) and subsequent reconstruction.

It is clear from the simulations and the results from our experiments that larger the diamond/silica interface, higher is the probability of wear from the diamond surface. This may be one of the reasons why smaller particles show higher roughness removal rates as they have larger surface to volume ratios when compared with bigger particles. Secondly, the acceleration seen due to addition of redox agents can be attributed to the acceleration of the surface reconstruction which leads to the oxidative wear process. Once the top carbon atoms are removed the underlying carbon atoms go through surface deconstruction to form another layer of aromatic chain and the process repeats itself. To conclude this section, we have given a tentative mechanism, based on our experiments and simulation from other groups, for polishing of superhard diamond surface by using a much softer material like silica.

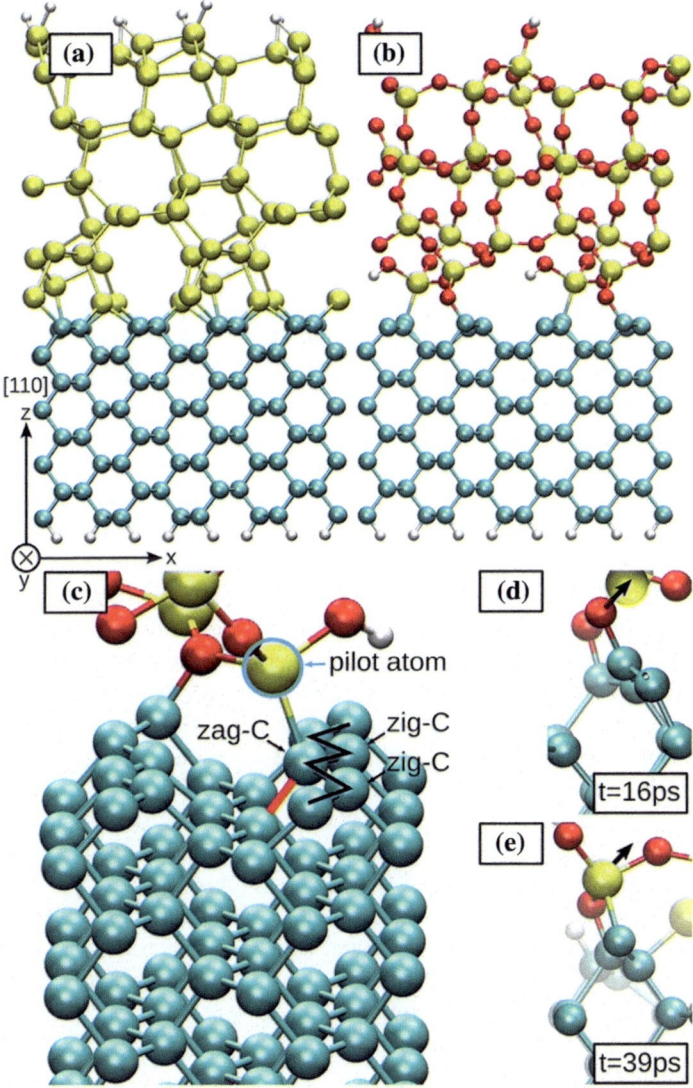

Fig. 3.25 **a** and **b** The configuration of the initial sliding surfaces are shown in these panels. Panel **a** shows the silicon/diamond interface and panel **b** shows the silica/diamond interface. In panel **c** the aromatic carbon chain is represented by the black lines and the red line represents the underlying bond that connects the aromatic chain to the bulk. The pilot atom that tries to break this bond is marked with blue circle. Both O and Si atoms can form the pilot atom and during sliding action the pilot atoms move away from the diamond/silica interface (shown by black arrows in panels **d** and **e**). Panels **d** and **e** show that both C–O and C–Si bonds respectively can lead to bond breakage shown by red line in Panel **c** (Reprinted from Peguiron et al. [56], with permission from Elsevier)

3.4 CMP of Superconducting NCD Thin Films

We have detailed the CMP of diamond in the previous sections. We have also given a plausible explanation to the process. But, the process is only effective if it is able to retain the superlative properties of diamond after polishing. There have been reports of loss of superconductivity in diamond after mechanical polishing [71]. Pure diamond is an insulator but when doped with boron it can behave like a semiconductor [72, 73], metal [74] or even superconductor [75]. Since the discovery of superconductivity in boron doped diamond (BDD) numerous researchers have studied its superconducting properties [76–83]. Attempts have also been made to make quantum devices [84–86] and MEMS [5, 87] out of BDD. All the devices in literature are either made from single crystal diamond samples or unpolished boron doped NCD films. The presence of surface roughness can lead to inferior device performances [88]. In this section we will discuss the superconductivity in planarized boron doped diamond NCD films.

Boron doped diamond films were grown on 500 nm silicon dioxide buffered silicon wafers. The B/C ratio in the gas mixture during growth was kept at 12800 ppm. The film thickness after 150 min growth was found to be 520 nm. The films were polished down to RMS roughness of 1.5 nm over 100 μm^2 from 44 nm RMS for as-grown film in 14 h. The transition temperature of the wafer was measured after each intermittent polishing step. The surface morphology was monitored using AFM. The transition temperature was measured by clamping the whole 2″ wafer to a variable temperature stage on the cold plate of a ^{4}He cryostat. Temperature was measured by a diode thermometer glued to the underside of the wafer. A surface mount heater was also fixed to the underside for varying the temperature of the wafer. Four terminal measurement was done using a centrally positioned spring loaded push pin setup directly in contact with the sample. Each pin in the setup was at the corner of a 1 cm^2 square. An AVS-47B AC resistance bridge was used to monitor the resistance of the sample as the temperature was varied by using the surface mounted heater. After each measurement the sample was removed from the cryostat, the thermometer and heater detached, and the wafer was polished till the limiting roughness was reached.

Figure 3.26 shows the AFM images of the as-grown BDD (Panel a) and 14 h polished film (Panel b). The as-grown film had a RMS roughness of 44 nm over an area of 100 μm^2. That is considerably rougher than the films we have seen in previous sections. This is due to the fact that the film used for this experiment is much thicker (520 nm) then the ones used before ($\sim$350 nm). The increased roughness means that longer polishing duration is needed to achieve the limiting surface roughness. In Fig. 3.26b we have shown the AFM image of the wafer polished for 14 h. Finally, in Fig. 3.26c we have shown the line profiles of AFM images at various intermediate polishing times. The evolution of the surface roughness is more clearly presented in the inset of Fig. 3.26c where we can see the surface roughness going down after each polishing step.

Fig. 3.26 AFM images showing the surface morphology of **a** as-grown film, **b** 14 h polished film. Panel **c** shows the line trace across AFM images at various polishing times. The inset shows the evolution of surface roughness with polishing time (Reprinted from Klemencic et al. [23] under a Creative Commons License)

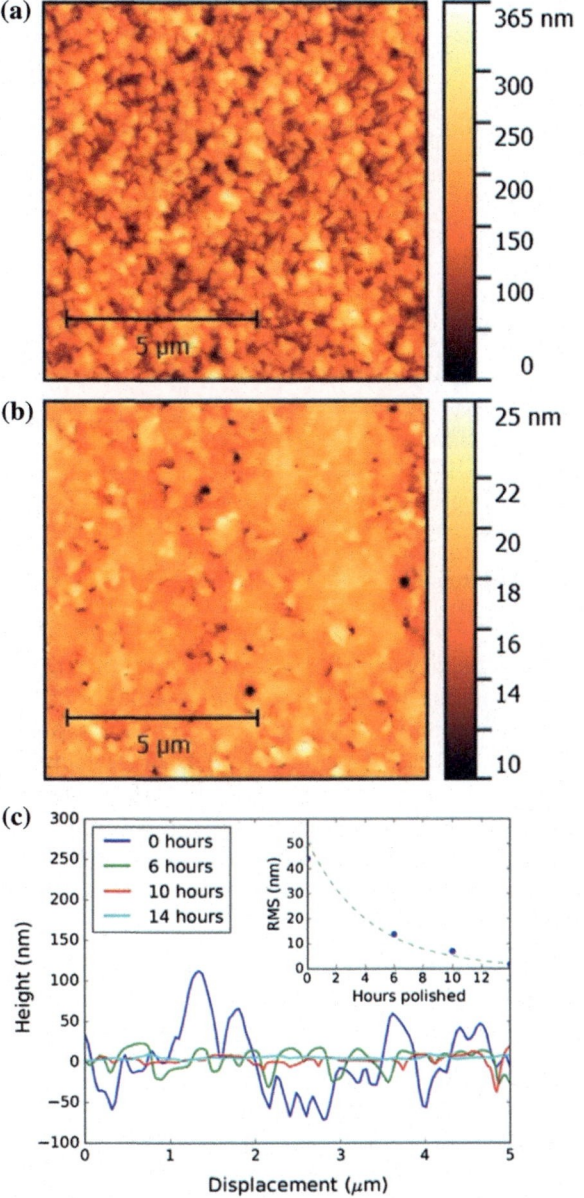

In Fig. 3.27 we have presented the normalized resistance (R/R_{10K}) for the polished and unpolished BDD wafer. The inset is the normalized resistance (R/R_{300K}) as a function of temperature between 1.6 and 300 K. The sample resistance increases as the temperature is decreased till the BDD reaches the superconducting transition. Earlier studies on free standing BDD films have seen similar behaviour

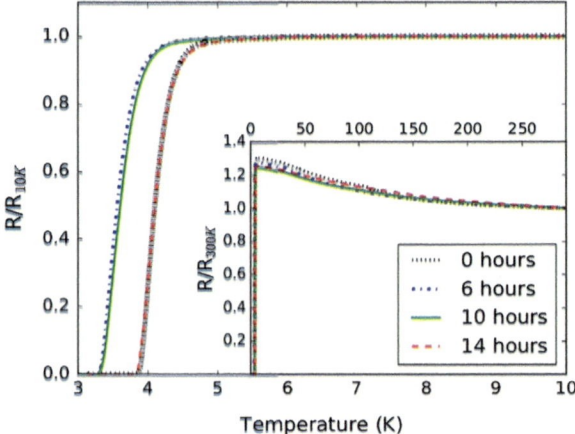

Fig. 3.27 Normalized resistance (R/R$_{10K}$) for BDD films after various intervals of CMP. The inset represents the variation of normalized resistance (R/R$_{300K}$) as a function of temperature between 1.6 and 300 K. The dashed red line showing the superconducting transition for 14 h polished film lies exactly on top of the curve for as-grown film shown with black dotted line (Reprinted from Klemencic et al. [23] under a Creative Commons License)

[75, 78] and also our film was grown on conducting silicon which makes it difficult to separate the wafer resistivity from BDD resistivity. The superconducting transition temperature of the film was $T_c = 4.2$ K with a transition width of 0.5 K. The behaviour of the 14 h polished BDD film is similar to the unpolished wafer with transition temperature and width remaining unchanged. The slight variation seen in transition temperature after 6 and 10 h polishing is down to the imprecision of the pressed contacts, as it is impossible to press the contacts at the same points after each polishing cycle. In conclusion, we have clearly shown that superconducting properties of BDD films remain unchanged even after 14 h of CMP unlike mechanical polishing [71].

3.5 Conclusion

In conclusion, we have presented our work on CMP of nanocrystalline diamond in this chapter. We have shown that a mixture of mechanical and chemical process is able to polish superhard diamond thin films (hardness $\sim$90 GPa) with much softer silica (hardness $\sim$7.5 GPa) particles. The technique is also applicable for single crystal diamond where polishing tracks from mechanical polishing were completely removed by CMP. Not only silica but it is also possible to CMP diamond with other oxides like alumina and ceria. It was found that roughness reduction rate was inversely proportional to the particle size. Addition of redox agents to the silica slurry accelerated the roughness reduction rate by almost 17% but still the CMP process is extremely slow to be used for meaningful removal of diamond. This process should be used for generating defect free diamond surfaces as a final polishing step or for polishing thin diamond films which will not survive the extreme conditions of mechanical polishing. Based on our experiments and theoretical studies by other groups we have proposed a mechanism where the oxidative wear of diamond surface takes place due to bond breakage between surface aromatic chains and the bulk diamond. Finally, we have applied our polishing technique to boron doped diamond where we studied its effect on superconductivity. We found that even after extensive polishing BDD films were able to retain their superconducting properties.

Acknowledgements The results presented in this chapter was supported by EPSRC under the grant 'Nanocrystalline diamond for Micro-Electro-Mechanical-Systems' reference number EP/J009814/1 and European Research Council (ERC) Consolidator Grant for the development of 'Superconducting Diamond Quantum Nano-Electro-Mechanical Systems', Project ID: 647471.

References

1. O.A. Williams, Nanocrystalline diamond. Diam. Relat. Mater. **20**, 621–640 (2011). https://doi.org/10.1016/j.diamond.2011.02.015
2. O.A. Williams, A. Kriele, J. Hees, et al., High young's modulus in ultra thin nanocrystalline diamond. Chem. Phys. Lett. **495**, 84–89 (2010). https://doi.org/10.1016/j.cplett.2010.06.054
3. M.A. Angadi, T. Watanabe, A. Bodapati et al., Thermal transport and grain boundary conductance in ultrananocrystalline diamond thin films. J. Appl. Phys. **99**, 114301 (2006). https://doi.org/10.1063/1.2199974
4. M. Imboden, O. Williams, P. Mohanty, Nonlinear dissipation in diamond nanoelectromechanical resonators. Appl. Phys. Lett. **102** (2013). https://doi.org/10.1063/1.4794907
5. T. Bautze, S. Mandal, O.A. Williams et al., Superconducting nano-mechanical diamond resonators. Carbon **72**, 100–105 (2014). https://doi.org/10.1016/j.carbon.2014.01.060
6. J.G. Rodriguez-Madrid, G.F. Iriarte, J. Pedros et al., Super-high-frequency SAW resonators on AlN/diamond. IEEE Electron Device Lett. **33**, 495–497 (2012). https://doi.org/10.1109/LED.2012.2183851
7. O.A. Williams, O. Douhéret, M. Daenen, et al., Enhanced diamond nucleation on monodispersed nanocrystalline diamond. Chem. Phys. Lett. **445**, 255–258 (2007). https://doi.org/10.1016/j.cplett.2007.07.091
8. X. Jiang, K. Schiffmann, C.-P.P. Klages, Nucleation and initial growth phase of diamond thin films on (100) silicon. Phys. Rev. B **50**, 8402–8410 (1994). https://doi.org/10.1103/PhysRevB.50.8402
9. P. Smereka, X. Li, G. Russo, D.J. Srolovitz, Simulation of faceted film growth in three dimensions: microstructure, morphology and texture. Acta Mater. **53**, 1191–1204 (2005). https://doi.org/10.1016/j.actamat.2004.11.013
10. O. Ergincan, G. Palasantzas, B.J. Kooi, Influence of surface modification on the quality factor of microresonators. Phys. Rev. B **85**, 205420 (2012). https://doi.org/10.1103/PhysRevB.85.205420
11. G.F. Iriarte, J.G. Rodríguez, F. Calle, Synthesis of c-axis oriented AlN thin films on different substrates: a review. Mater. Res. Bull. **45**, 1039–1045 (2010). https://doi.org/10.1016/j.materresbull.2010.05.035

12. R.B. Simon, J. Anaya, F. Faili, et al., Effect of grain size of polycrystalline diamond on its heat spreading properties. Appl. Phys. Express **9** (2016). https://doi.org/10.7567/apex.9. 061302

13. J. Anaya, T. Bai, Y. Wang et al., Simultaneous determination of the lattice thermal conductivity and grain/grain thermal resistance in polycrystalline diamond. Acta Mater. **139**, 215–225 (2017). https://doi.org/10.1016/j.actamat.2017.08.007

14. Y. Chen, L. Zhang, *Polishing of Diamond Materials* (Springer, London, 2013)

15. L. Pastewka, S. Moser, P. Gumbsch, M. Moseler, Anisotropic mechanical amorphization drives wear in diamond. Nat. Mater. **10**, 34–38 (2011). https://doi.org/10.1038/nmat2902

16. A. Malshe, B. Park, W. Brown, H. Naseem, A review of techniques for polishing and planarizing chemically vapor-deposited (CVD) diamond films and substrates. Diam. Relat. Mater. **8**, 1198–1213 (1999). https://doi.org/10.1016/S0925-9635(99)00088-6

17. I. Friel, S.L. Clewes, H.K. Dhillon et al., Control of surface and bulk crystalline quality in single crystal diamond grown by chemical vapour deposition. Diam. Relat. Mater. **18**, 808–815 (2009). https://doi.org/10.1016/j.diamond.2009.01.013

18. T. Schuelke, T.A. Grotjohn, Diamond polishing. Diam. Relat. Mater. **32**, 17–26 (2013). https://doi.org/10.1016/j.diamond.2012.11.007

19. C.D. Ollison, W.D. Brown, A.P. Malshe, et al., A comparison of mechanical lapping versus chemical-assisted mechanical polishing and planarization of chemical vapor deposited (CVD) diamond. Diam. Relat. Mater. **8**, 1083–1090 (1999). https://doi.org/10.1016/s0925-9635(99)00091-6

20. E.L.H. Thomas, G.W. Nelson, S. Mandal et al., Chemical mechanical polishing of thin film diamond. Carbon **68**, 473–479 (2014). https://doi.org/10.1016/j.carbon.2013.11.023

21. J. Luo, D.A. Dornfeld, Material removal regions in chemical mechanical planarization for submicron integrated circuit fabrication: coupling effects of slurry chemicals, abrasive size distribution, and wafer-pad contact area. IEEE Trans. Semicond. Manuf. **16**, 45–56 (2003). https://doi.org/10.1109/TSM.2002.807739

22. P.B. Zantye, A. Kumar, A.K. Sikder, Chemical mechanical planarization for microelectronics applications. Mater. Sci. Eng. R Rep. **45**, 89–220 (2004). https://doi.org/10.1016/j.mser.2004. 06.002

23. G.M. Klemencic, S. Mandal, J.M. Werrell et al., Superconductivity in planarised nanocrystalline diamond films. Sci. Technol. Adv. Mater. **18**, 239–244 (2017). https://doi.org/10.1080/ 14686996.2017.1286223

24. J.M. Werrell, S. Mandal, E.L.H. Thomas et al., Effect of slurry composition on the chemical mechanical polishing of thin diamond films. Sci. Technol. Adv. Mater. **18**, 654–663 (2017). https://doi.org/10.1080/14686996.2017.1366815

25. S. Mandal, E.L.H. Thomas, L. Gines et al., Redox agent enhanced chemical mechanical polishing of thin film diamond. Carbon **130**, 25–30 (2018). https://doi.org/10.1016/j.carbon. 2017.12.077

26. G.M. Pharr, D.L. Callahan, S.D. McAdams et al., Hardness, elastic modulus, and structure of very hard carbon films produced by cathodic-arc deposition with substrate pulse biasing. Appl. Phys. Lett. **68**, 779–781 (1996). https://doi.org/10.1063/1.116530

27. B. Hussey, J. Wilson, *Advanced Technical Ceramics Directory and Databook* (Springer, US, Boston, MA, 1998)

28. N. Elbel, Tungsten Chemical Mechanical Polishing. J. Electrochem. Soc. **145**, 1659 (1998). https://doi.org/10.1149/1.1838533

29. C. Li, I.B. Bhat, R. Wang, J. Seiler, Electro-chemical mechanical polishing of silicon carbide. J. Electron. Mater. **33**, 481–486 (2004). https://doi.org/10.1007/s11664-004-0207-6

30. C. Spiro, G. Steuer, F.B. Kaufman, Method of polishing a tungsten carbide, US8162723B2 (2006)

31. C.D. Wagner, L.E. Davis, M.V. Zeller et al., Empirical atomic sensitivity factors for quantitative analysis by electron spectroscopy for chemical analysis. Surf. Interface Anal. **3**, 211–225 (1981). https://doi.org/10.1002/sia.740030506

32. J.I. Wilson, J. Walton, G. Beamson, Analysis of chemical vapour deposited diamond films by X-ray photoelectron spectroscopy. J. Electron Spectrosc. Relat. Phenom. **121**, 183–201 (2001). https://doi.org/10.1016/S0368-2048(01)00334-6

33. S. Ferro, M. Dal Colle, A. De Battisti, Chemical surface characterization of electrochemically and thermally oxidized boron-doped diamond film electrodes. Carbon **43**, 1191–1203 (2005). https://doi.org/10.1016/j.carbon.2004.12.012

34. S. Jin, J.E. Graebner, G.W. Kammlott et al., Massive thinning of diamond films by a diffusion process. Appl. Phys. Lett. **60**, 1948–1950 (1992). https://doi.org/10.1063/1.107133

35. A. Hirata, H. Tokura, M. Yoshikawa, Smoothing of chemically vapour deposited diamond films by ion beam irradiation. Thin Solid Films **212**, 43–48 (1992). https://doi.org/10.1016/0040-6090(92)90498-Z

36. A.M. Ozkan, A.P. Malshe, W.D. Brown, Sequential multiple-laser-assisted polishing of free-standing CVD diamond substrates. Diam. Relat. Mater. **6**, 1789–1798 (1997). https://doi.org/10.1016/S0925-9635(97)00141-6

37. R.H. Olsen, D.K. Aspinwall, R.C. Dewes, Electrical discharge machining of conductive CVD diamond tool blanks. J. Mater. Process. Technol. **155–156**, 1227–1234 (2004). https://doi.org/10.1016/j.jmatprotec.2004.04.355

38. Y. Chen, L. Zhang, Mechanical Polishing, *Polishing of Diamond Materials: Mechanisms, Modeling and Implementation* (Springer, London, 2013), pp. 25–44

39. R.M. Denning, Directional grinding hardness in diamond. Am. Min. **38**, 108–117 (1953)

40. H. Whittaker, C.B. Slawson, Vector hardness in diamond tools. Am. Min. **31**, 143–149 (1946)

41. E.H. Kraus, C.B. Slawson, Variation of hardness in the diamond. Am. Min. **24**, 661–676 (1939)

42. S.E. Grillo, J.E. Field, The polishing of diamond. J. Phys. D Appl. Phys. **30**, 202–209 (1997). https://doi.org/10.1088/0022-3727/30/2/007

43. S.E. Grillo, J.E. Field, F.M. Van Bouwelen, Diamond polishing: the dependency of friction and wear on load and crystal orientation. J. Phys. D Appl. Phys. **33**, 985–990 (2000). https://doi.org/10.1088/0022-3727/33/8/315

44. M.P. Hitchiner, E.M. Wilks, J. Wilks, The polishing of diamond and diamond composite materials. Wear **94**, 103–120 (1984). https://doi.org/10.1016/0043-1648(84)90169-8

45. W.J. Huisman, J.F. Peters, S.A. de Vries et al., Structure and morphology of the as-polished diamond(111) −1 × 1 surface. Surf. Sci. **387**, 342–353 (1997). https://doi.org/10.1016/S0039-6028(97)00369-5

46. M.R. Jarvis, R. Perez, F.M. van Bouwelen, M.C. Payne, Microscopic mechanism for mechanical polishing of diamond (110) surfaces. Phys. Rev. Lett. **80**, 3428–3431 (1998). https://doi.org/10.1103/PhysRevLett.80.3428

47. J.R. Hird, J.E. Field, Diamond polishing. Proc. R Soc. Math. Phys. Eng. Sci. **460**, 3547–3568 (2004). https://doi.org/10.1098/rspa.2004.1339

48. P.N. Volpe, P. Muret, F. Omnes et al., Defect analysis and excitons diffusion in undoped homoepitaxial diamond films after polishing and oxygen plasma etching. Diam. Relat. Mater. **18**, 1205–1210 (2009). https://doi.org/10.1016/j.diamond.2009.04.008

49. A. Gaisinskaya, R. Edrei, A. Hoffman, Y. Feldheim, Morphological evolution of polished single crystal (100) diamond surface exposed to microwave hydrogen plasma. Diam. Relat. Mater. **18**, 1466–1473 (2009). https://doi.org/10.1016/j.diamond.2009.09.014

50. A.B. Shorey, K.M. Kwong, K.M. Johnson, S.D. Jacobs, Nanoindentation hardness of particles used in magnetorheological finishing (MRF). Appl. Opt. **39**, 5194 (2000). https://doi.org/10.1364/AO.39.005194

51. M. Bielmann, Effect of particle size during tungsten chemical mechanical polishing. Electrochem. Solid-State Lett. **2**, 401 (1999). https://doi.org/10.1149/1.1390851

52. D. Tamboli, G. Banerjee, M. Waddell, Novel interpretations of CMP removal rate dependencies on slurry particle size and concentration. Electrochem. Solid-State Lett. **7**, F62 (2004). https://doi.org/10.1149/1.1795033

53. Z. Zhang, W. Liu, Z. Song, Particle size and surfactant effects on chemical mechanical polishing of glass using silica-based slurry. Appl. Opt. **49**, 5480 (2010). https://doi.org/10.1364/AO.49.005480

54. C. Liu, B. Dai, W. Tseng, C. Yeh, Modeling of the wear mechanism during chemical-mechanical polishing. J. Electrochem. Soc. **143**, 716 (1996). https://doi.org/10.1149/1.1836507

55. Y. Wang, Y. Zhao, W. An et al., Modeling effects of abrasive particle size and concentration on material removal at molecular scale in chemical mechanical polishing. Appl. Surf. Sci. **257**, 249–253 (2010). https://doi.org/10.1016/j.apsusc.2010.06.077

56. A. Peguiron, G. Moras, M. Walter et al., Activation and mechanochemical breaking of C–C bonds initiate wear of diamond (110) surfaces in contact with silica. Carbon **98**, 474–483 (2016). https://doi.org/10.1016/j.carbon.2015.10.098

57. T. Kuwahara, G. Moras, M. Moseler, Friction regimes of water-lubricated diamond (111): role of interfacial ether groups and tribo-induced aromatic surface reconstructions. Phys. Rev. Lett. **119**, 96101 (2017). https://doi.org/10.1103/PhysRevLett.119.096101

58. H. Hocheng, C.C.C. Chen, H.H. Cheng, C.C.C. Chen, Chemical-assisted mechanical polishing of diamond film on wafer. Mater. Sci. Forum **505–507**, 1225–1230 (2006). https://doi.org/10.4028/www.scientific.net/MSF.505-507.1225

59. Y. Liu, K. Zhang, F. Wang, Y. Han, Study on the cleaning of silicon after CMP in ULSI. Microelectron. Eng. **66**, 433–437 (2003). https://doi.org/10.1016/S0167-9317(02)00906-1

60. S. Ghodbane, D. Ballutaud, F. Omnès, C. Agnès, Comparison of the XPS spectra from homoepitaxial 111}, {100 and polycrystalline boron-doped diamond films. Diam. Relat. Mater. **19**, 630–636 (2010). https://doi.org/10.1016/j.diamond.2010.01.014

61. F. Klauser, S. Ghodbane, R. Boukherroub et al., Comparison of different oxidation techniques on single-crystal and nanocrystalline diamond surfaces. Diam. Relat. Mater. **19**, 474–478 (2010). https://doi.org/10.1016/j.diamond.2009.11.013

62. M.Y. Tsai, S.T. Chen, Y.S. Liao, J. Sung, Novel diamond conditioner dressing characteristics of CMP polishing pad. Int. J. Mach. Tools Manuf **49**, 722–729 (2009). https://doi.org/10.1016/j.ijmachtools.2009.03.001

63. Y.C. Kim, S.J.L. Kang, Novel CVD diamond-coated conditioner for improved performance in CMP processes. Int. J. Mach. Tools Manuf **51**, 565–568 (2011). https://doi.org/10.1016/j.ijmachtools.2011.02.008

64. F.K. De Theije, O. Roy, N.J. Van Der Laag, W.J.P. Van Enckevort, Oxidative etching of diamond. Diam. Relat. Mater. **9**, 929–934 (2000). https://doi.org/10.1016/S0925-9635(99)00239-3

65. F.K. De-Theije, E. van-Veenendaal, W.J.P. van-Enckevort, E. Vlieg, Oxidative etching of cleaved synthetic diamond {111} surfaces. Surf. Sci. **492**, 91–105 (2001). https://doi.org/10.1016/s0039-6028(01)01398-x

66. Y. Yao, Y. Ishikawa, Y. Sugawara et al., Fast removal of surface damage layer from single crystal diamond by using chemical etching in molten KCl + KOH solution. Diam. Relat. Mater. **63**, 86–90 (2016). https://doi.org/10.1016/j.diamond.2015.10.003

67. M. Krishnan, J.W. Nalaskowski, L.M. Cook, Chemical mechanical planarization: slurry chemistry, materials, and mechanisms. Chem. Rev. **110**, 178–204 (2010). https://doi.org/10.1021/cr900170z

68. H. Hocheng, H.Y. Tsai, Y.T. Su, Modeling and experimental analysis of the material removal rate in the chemical mechanical planarization of dielectric films and bare silicon wafers. J. Electrochem. Soc. **148**, G581 (2001). https://doi.org/10.1149/1.1401087

69. G. Moras, L. Pastewka, M. Walter et al., Progressive shortening of sp-hybridized carbon chains through oxygen-induced cleavage. J. Phys. Chem. C **115**, 24653–24661 (2011). https://doi.org/10.1021/jp209198g

70. G. Moras, L. Pastewka, P. Gumbsch, M. Moseler, Formation and oxidation of linear carbon chains and their role in the wear of carbon materials. Tribol. Lett. **44**, 355–365 (2011). https://doi.org/10.1007/s11249-011-9864-9

71. D. Wu, Y.C. Ma, Z.L. Wang et al., Optical properties of boron-doped diamond. Phys. Rev. B **73**, 12501 (2006). https://doi.org/10.1103/PhysRevB.73.012501
72. V.S. Vavilov, Semiconducting diamond. Phys. Status Solidi **31**, 11–26 (1975). https://doi.org/10.1002/pssa.2210310102
73. M.I. Eremets, Semiconducting diamond. Semicond. Sci. Technol. **6**, 439–444 (1991). https://doi.org/10.1088/0268-1242/6/6/004
74. T. Tshepe, J. Prins, M. Hoch, Metal–insulator transition in boron-ion implanted type IIa diamond. Diam. Relat. Mater. **8**, 1508–1510 (1999). https://doi.org/10.1016/S0925-9635(99)00066-7
75. E.A. Ekimov, V.A. Sidorov, E.D. Bauer et al., Superconductivity in diamond. Nature **428**, 542–545 (2004). https://doi.org/10.1038/nature02449
76. E. Bustarret, J. Kačmarčik, C. Marcenat et al., Dependence of the superconducting transition temperature on the doping level in single-crystalline diamond films. Phys. Rev. Lett. **93**, 237005 (2004). https://doi.org/10.1103/PhysRevLett.93.237005
77. F. Dahlem, P. Achatz, O.A. Williams, et al., Nanocrystalline boron-doped diamond films, a mixture of BCS-like and non-BCS-like superconducting grains. Phys. Status Solidi. **207**, 2064–2068 (2010). https://doi.org/10.1002/pssa.201000013
78. Z.L. Wang, Q. Luo, L.W. Liu et al., The superconductivity in boron-doped polycrystalline diamond thick films. Diam. Relat. Mater. **15**, 659–663 (2006). https://doi.org/10.1016/j.diamond.2005.12.035
79. Y. Takano, M. Nagao, T. Takenouchi et al., Superconductivity in polycrystalline diamond thin films. Diam. Relat. Mater. **14**, 1936–1938 (2005). https://doi.org/10.1016/j.diamond.2005.08.014
80. S. Mandal, C. Naud, O.A. Williams et al., Nanostructures made from superconducting boron-doped diamond. Nanotechnology **21**, 195303 (2010). https://doi.org/10.1088/0957-4484/21/19/195303
81. S. Mandal, C. Naud, O.A. Williams, et al., Detailed study of superconductivity in nanostructured nanocrystalline boron doped diamond thin films. Phys. Status Solidi. **207**, 2017–2022 (2010). https://doi.org/10.1002/pssa.201000008
82. G.M. Klemencic, J.M. Fellows, J.M. Werrell et al., Fluctuation spectroscopy as a probe of granular superconducting diamond films. Phys. Rev. Mater. **1**, 44801 (2017). https://doi.org/10.1103/PhysRevMaterials.1.044801
83. N. Titova, A.I. Kardakova, N. Tovpeko et al., Slow electron-phonon cooling in superconducting diamond films. IEEE Trans. Appl. Supercond. **27**, 1–4 (2017). https://doi.org/10.1109/TASC.2016.2638199
84. M. Watanabe, A. Kawano, S. Kitagoh et al., Stacked SNS Josephson junction of all boron doped diamond. Phys. C Supercond. Appl. **470**, S613–S615 (2010). https://doi.org/10.1016/j.physc.2009.11.061
85. S. Mandal, T. Bautze, O.A. Williams et al., The diamond superconducting quantum interference device. ACS Nano **5**, 7144–7148 (2011). https://doi.org/10.1021/nn2018396
86. M. Watanabe, R. Kanomata, S. Kurihara et al., Vertical SNS weak-link Josephson junction fabricated from only boron-doped diamond. Phys. Rev. B Condens. Matter. Mater. Phys. **85**, 3–4 (2012). https://doi.org/10.1103/PhysRevB.85.184516
87. J. Zhang, J.W. Zimmer, R.T. Howe, R. Maboudian, Characterization of boron-doped micro- and nanocrystalline diamond films deposited by wafer-scale hot filament chemical vapor deposition for MEMS applications. Diam. Relat. Mater. **17**, 23–28 (2008). https://doi.org/10.1016/j.diamond.2007.09.010
88. M. Imboden, P. Mohanty, Dissipation in nanoelectromechanical systems. Phys. Rep. **534**, 89–146 (2014). https://doi.org/10.1016/j.physrep.2013.09.003

Chapter 4
Single Crystal Diamond Micromechanical and Nanomechanical Resonators

Meiyong Liao, Yasuo Koide and Liwen Sang

Abstract Micro- or nanoelectromechanical system (MEMS/NEMS) has witnessed explosive growth with applications spanning from automotive, consumer, industry, military, to biotechnology in the past decades. Presently, MEMS are dominated by Si material due to the mature CMOS technology. However, the intrinsic weakness of Si such as poor mechanical or tribological properties and poor thermal stability limit the device performance and hinder the applications of Si MEMS in harsh environments. Diamond is an outstanding material for MEMS/NEMS under harsh environments with diverse and much better performance than Si in terms of the excellent properties such as high Young's modulus, high thermal conductivity, hydrophobic surface, and tailorable electronic configuration of diamond. In this chapter, we review our recent progress in the batch fabrication of single crystal diamond (SCD) mechanical resonators on SCD. The energy loss mechanism in the SCD mechanical resonators were discussed and the strategies to improve quality factors of the SCD resonators were described.

4.1 Introduction

4.1.1 Background of MEMS

Microelectromechanical system (MEMS), coined by Prof. R. Howe around 1989, is a highly interdisciplinary science and technology including mechanics, electronics, physics, chemistry, and biology with nanotechnology involved [1]. A key feature of MEMS is that it integrates mechanical moving parts with semiconductor integrated circuits, which creates new functionalities and applications with miniaturized devices. Therefore, MEMS is recognized as an attractive strategy for "More-than-Moor" [2]. The feature size of MEMS ranges from 1 to 100 μm. The

M. Liao (✉) · Y. Koide · L. Sang
Research Center for Functional Materials, National Institute for Materials Science (NIMS),
1-1 Namiki, Tsukuba, Ibaraki 305-0044, Japan
e-mail: meiyong.liao@nims.go.jp

© Springer Nature Switzerland AG 2019
N. Yang (ed.), *Novel Aspects of Diamond*, Topics in Applied Physics 121,
https://doi.org/10.1007/978-3-030-12469-4_4

history of MEMS can be traced back in 1958, when the first gauge sensor based on Si piezoresistivity effect [3] was demonstrated. In 1968, Harvey Nathanson from Westinghouse invented the first microelectromechanical system (MEMS) device, which combined a mechanical resonator component and an electronic transistor by using surface micromachining batch fabrication technology for the first time [4]. Explosive growth in MEMS applications was experienced from 1980s to 2000s with progress in the combination of semiconductor integrated circuits with resonant microsystems. The MEMS devices vary from mechanical sensors and actuators, to microanalysis and chemical sensors, to microoptical systems and bioMEMS for microscopic surgery. Many MEMS products such as pressure, chemical, biological, and inertial sensors, high-frequency filters, and atomic force probes, can be found in automotive industry, communications, defense systems, security, health care, information technology, avionics, instruments, and environmental monitoring. Here are the milestones of MEMS.

Milestones of MEMS

1954 Piezoresistive effect in Germanium and Silicon (C.S. Smith)

1958 First silicon pressure sensor demonstrated (Kulite)

1967 Anisotropic deep silicon etching (H.A. Waggener et al.)

1968 Resonant Gate Transistor Patented (Surface Micromachining Process) (H. Nathanson, et al.)

1970s Bulk etched silicon wafers used as pressure sensors (Bulk Micromaching Process)

1970 First silicon accelerometer demonstrated

1979 Micromachined ink-jet nozzle (HP)

1982 "Silicon as a Mechanical Material" (K. Petersen)

1982 LIGA process (the German acronym for "Lithography Galvanik Abformung" = Lithography electroplating and molding, KfK, Germany, W. Ehrfeld et al.)

1982 Disposable blood pressure transducer (Honeywell)

1983 Integrated pressure sensor (Honeywell)

1985 Sensonor Crash sensor (Airbag)

1986 Silicon wafer bonding (M. Shimbo)

1988 Batch fabricated pressure sensors via wafer bonding (Nova Sensor)

1992 Bulk micromachining (SCREAM process, Cornell)

1993 Digital mirror display (Texas Instruments)

1994 Commercial surface micromachined accelerometer (Analog Devices)

1999 Optical network switch (Lucent)

2000 VLSI NEMS (IBM)

The development of MEMS strongly relies on the materials selection, process, device concept, fabrication, and characterization. The potential of MEMS is fulfilled with integration of sensors and actuators with semiconductor integrated circuits. The dominated material for MEMS is Si due to the mature complementary metal-oxide-semiconductor (CMOS) technology. However, Si has intrinsic weakness of poor mechanical properties, such as a low Young's modulus, a weak mechanical strength and low wear resistance. In addition, Si also has poor chemical inertness such as the formation of oxide on the surface and poor resistance to acid and erosion chemicals. The low thermal conductivity of Si limits the ultimate device performance and the application of Si-MEMS for high power handling. Furthermore, Si has a bandgap as narrow as 1.1 eV, which suffers from electrical leakage problem at elevated temperatures. In many cases, such as automobile, aerospace, nuclear plant, and chemical industries, MEMS devices able to operate in harsh environments are necessary. Due to the intrinsic material limitations mentioned above, Si cannot work well as semiconductor electronic devices above 250 ° C and the mechanical yield strength of Si also losses at 500 °C. The seeking for semiconductor materials as an alternative to Si for MEMS able to operate in harsh environments is in demand.

When the feature size of the MEMS devices is reduced to the scale of nanometer, the terminology nanoelectromechanical system (NEMS) is usually utilized [5, 6]. The physical properties for NEMS is dominated by the surface effects due to the large surface-to-volume ratio. The advantage of NEMS is the extremely high sensitivity when it is used as sensors due to the light mass of the resonator (i.e. 10^{-18} g) [7].

4.1.2 Diamond for MEMS/NEMS

Wide bandgap (WBG) semiconductors such as diamond, silicon carbide (SiC), and group-III nitrides attractive candidates for MEMS that can work under harsh environments because of their superior electrical properties, mechanical strength, and chemical inertness in harsh environments. Table 4.1 lists the physical, electronic and mechanical properties of diamond semiconductors compared with Si and other WBG semiconductors. As can be seen, diamond has the highest Young's modulus as high as 1200 GPa among the bulk materials among these WBG semiconductors, providing a high acoustic velocity as high as 18,000 m/s. The tensile strength of diamond was observed to be as high as 60 GPa and was predicted to be more than 90 GPa [8, 9]. Diamond has the highest mechanical hardness of more than 100 GPa and a low friction coefficient (<0.05). Therefore, it exhibits excellent tribological properties with an ideal wear life 10,000 times longer than Si, providing the foundation of high reliability for MEMS devices. Diamond is also an excellent thermal conductor because of the strong covalent bonding and low phonon scattering. The thermal conductivity of diamond is as high as 2200 W/(m·K) [10], five times higher than copper. Isotopically ^{12}C enriched (99.9%) synthetic

Table 4.1 Properties of diamond compared with Si and other WBG semiconductors with respect to MEMS applications

Properties	Diamond	Si	SiC	GaN	AlN
Density (g cm^{-3})	3.52	2.33	3.21	6.10	3.3
Young's modulus (GPa)	1200	130	450	210	100–400
Hardness (GPa)	100	10	33	12	11.8
Coefficient of frication	<0.05	0.4–0.6	0.2–0.5	0.2	0.4
Fracture strength (GPa)	5.3	1.0	5.2	1.73	0.29
Thermal conductivity (Wcm^{-1}K^{-1})	24	1.5	5	1.95	1.75
Bangap (eV)	5.5	1.1	3.3	3.4	6.2
Electron mobility (cm V^{-1} S^{-1})	4500	1450	900	1000	426
Hole mobility (cm V^{-1} S^{-1})	3800	480	120	200	14
Breakdown field (MV cm^{-1})	10	0.3	3.5	3.3	1.5
Dielectric constant	5.5	11.8	9.7	9.5	8.9

single crystal diamond has the highest thermal conductivity of 3320 W/(m·K) at room temperature, the highest value among the known solids [11]. Due to the high thermal conductivity, the intrinsic loss such as the thermoelastic damping (TED) limited by material properties can be reduced for MEMS/NEMS applications [12, 13]. The chemical inertness of diamond offers it the attractive candidate for NEMS, in which the native oxides do not exist on diamond surface. In addition, diamond has a wide-bandgap of 5.5 eV and a high radiation hardness, ensuring the applications under harsh environments. Therefore, diamond is an ideal material for MEMS/NEMS [14, 15].

The achievement of diamond MEMS/MEMS devices mostly relies on the fabrication of mechanical resonators such as cantilevers, and bridges or membranes, etc. Great progress on MEMS/NEMS devices based on polycrystalline or nanocrystalline diamonds have been achieved [16–21]. In those cases, high-performance MEMS/NEMS resonators were demonstrated by utilizing diamond grown on foreign substrates having a sacrificial layer. The quality factor as high as 365 000 was obtained [21]. However, polycrystalline or nanocrystalline diamond usually has the disadvantages of (i) a high volume fraction of grain boundaries and impurities and (ii) difficulty in semiconductor properties control. These drawbacks bring forward the fatal problem of reliability. Therefore, it is desirable to develop MEMS/NEMS devices made from single crystal diamond (SCD), so that the extreme properties of diamond for MEMS/NEMS can be exploited.

In this chapter, we will review our recent results on SCD MEMS/NEMS. These contents include the fabrication of SCD mechanical resonators, the energy dissipation mechanism analysis of the mechanical resonators, and the improvement of the quality factors of the SCD resonators.

4.2 Fabrication of Single Crystal Diamond Mechanical Resonators

Due to the chemical inertness and extremely high mechanical strength, the conventional micromachining process for Si and other WBG semiconductors cannot be simply transferred to SCD. Unique processing strategies should be developed for SCD mechanical resonators since the heteroepitaxial growth of SCD wafers on foreign substrates has not been established yet. Up to now, there are three approaches for the fabrication of SCD mechanical resonators: ion-implantation assisted lift-off (IAL) technique [22], diamond-on-insulator (DOI) method [23, 24], and angled etching [25].

Ovartchaiyapong et al. obtained high quality-factor SCD resonator over 330 000 with the DOI method by using an oxide bonding technique to form a diamond/SiO_2/silicon multilayer structure [23]. The DOI method was started from a commercial SCD plate with a thickness of 20 µm and a smooth surface (RMS < 0.5 nm) on both side of the SCD plate. In the fabrication, a 30 nm-thick SiO_2 layer was deposited on one side of the diamond plate by plasma-enhanced chemical vapor deposition technique. The SCD plate with oxide layer was mounted on a Si substrate and then bonded to an oxidized Si chip using a low-temperature oxide bonding technique. The diamond layer was thinned to be 1–2 µm by using an $ArCl_2$ reactive ion etch/inductively coupled plasma (RIE/ICP) etch with an etch rate of 3 µm/h. Subsequently, The DOI sample was patterned to form the cantilever patterns by photolithography. A 300 nm-thick gold was deposited on the patterned DOI substrate as a mask for O_2 plasma etching. Finally, SCD cantilevers with various dimensions were released by dipping DOI substrate in a buffered hydrofluoric acid to selectively remove the 1 µm-thick SiO_2 layer underneath the cantilevers. Figure 4.1 shows the cross section of the multilayer DOI structure on Si and the obtained SCD cantilevers. One advantage of the DOI method is the ability to maintain the high crystal quality of the initial diamond plate.

A simpler fabrication method for freestanding SCD nanoscale mechanical resonators was proposed and demonstrated by Burek et al. by using angled etching technique in a plasma [26]. In this approach, they firstly (i) deposited an etch mask on the SCD plate on the pattern diamond plate by standard electron beam lithography, then (ii) the SCD plate patterned with the etch mask was etched by conventional top down plasma etching, and (iii) finally, a second anisotropic etch step was performed at an oblique angle to the substrate surface to release the nanostructures. Such a simple method could yield the suspended nanobeams, as shown in Fig. 4.2. The authors employed a Faraday cage to perform the second angled etching in order to shield the electromagnetic fields. For example, triangular prism Faraday cages with θ ∼ 45° were used to fabricate suspended nanobeam mechanical structure. Other freestanding structures such as nanocavities could also be fabricated by properly designing the etching mask and the geometries of the Faraday cages.

Fig. 4.1 SCD cantilever fabricated by DOI method. **a** Scanning electron microscopy (SEM) image of DOI cross section. **b** SEM of patterned and etched SCD cantilevers on SiO$_2$. **c** Released SCD over Si. **d** Optical image of suspended SCD cantilevers [23]. Copyright © 2012, American Institute of Physics

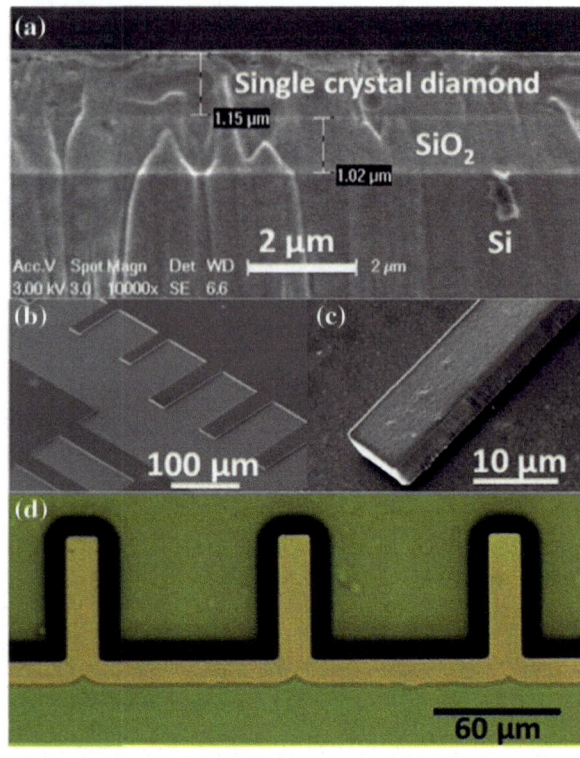

We developed a batch fabrication strategy for SCD mechanical resonators by using ion implantation assisted lift-off (IAL) technique (smart-cut) by borrowing the idea for the fabrication of SCD plates [27, 28]. Our method provides all SCD MEMS/NEMS with controlled dimensions from nanoscale to microscale and is highly reproducible. The fabrication was begun with a high-pressure high-temperature (HPHT) type-Ib SCD substrate. The HPHT SCD substrate was selectively implanted by carbon ions at an energy of 180 keV, corresponding to a penetration depth of around 217 nm evaluated by SRIM2003. The ion beam current density was 180 nA/cm^2 and the dose was 10^{16} cm^{-2}. To reduce the channeling effect, the implantation was conducted at an incident beam angle of 7° deviated from the normal direction of the diamond substrate. The implanted HPHT type-Ib SCD substrate was processed in a boiling acid solution before the homoepitaxial diamond growth. The homoepitaxial growth of diamond was conducted by a microwave plasma chemical vapor deposition (MPCVD) apparatus. During the MPCVD growth, the CH$_4$/H$_2$ flow ratio was 0.08–0.1% with a hydrogen flow rate of 500 sccm. For growing p-type diamond, hydrogen diluted B(CH$_3$)$_3$ (TMB) was used as the boron source and the flow ratio of TMB/CH$_4$ was from 5–200 ppm. The growth temperature was from 850 to 950 °C. After growth, the diamond epilayer was annealed at 900–1100 °C for 3 h in a UHV chamber (base pressure 10^{-7} Pa) to

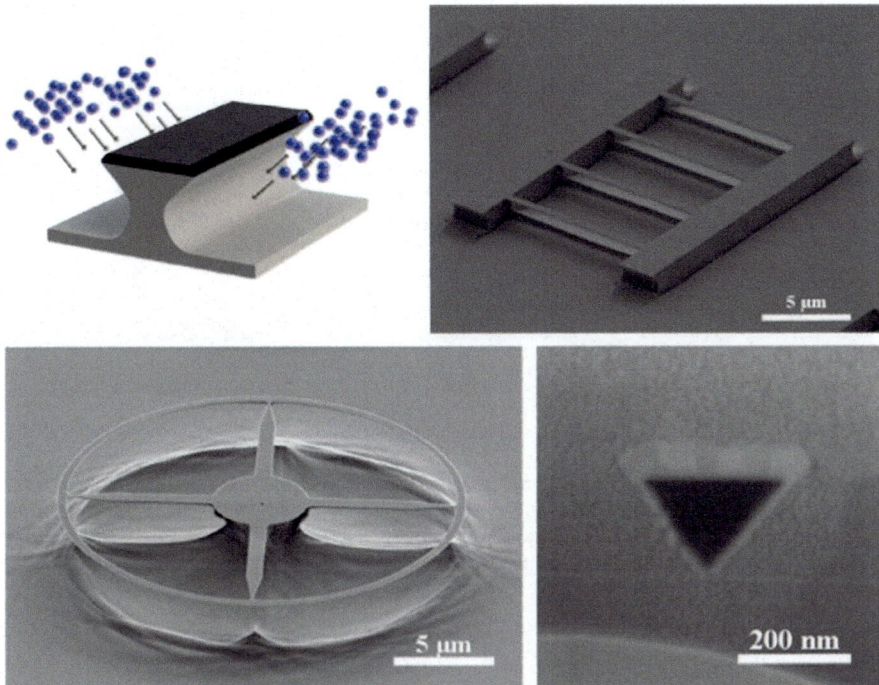

Fig. 4.2 Schematic of angled etching technique for SCD MEMS and the resulted SCD suspended structures [26]. Copyright © 2012, American Chemical Society

reduce the ion impinged defects within the diamond. During the growth and annealing process, the damaged layer induced by the ion implantation was transferred into graphite. The buried graphite layer acted as the sacrificial layer for the formation of free-standing SCD mechanical structures. After boiling the diamond in a HNO_3/H_2SO_4 acid solution, a photolithographic process was performed to pattern the homoepitaxial layer with cantilever or bridge shapes. A metallic layer such as aluminum or tungsten carbide or gold with a thickness of 100–200 nm was deposited on the patterned diamond epilayer as a mask for dry etching. The ICP-RIE etching was conducted at 800 W with oxygen gas at a pressure of 0.5 Pa and a flow rate of 90 sccm. The etching rate was around 60 nm/min. After etching, the metallization was removed, subsequently followed by a non-contact acid or electrochemical (EC) etching in water [29]. This process led to the formation of free-standing SCD mechanical resonators on SCD substrate. By defining the feature size or geometries of the patterns, mechanical resonators from microscale to nanoscale were achieved in a reproducible behavior. The thickness of the SCD resonators was controlled by the thickness of the SCD homoepitaxial layer or by tuning the ion energy in the ion implantation process. Figure 4.3 illustrates the batch fabrication process for the SCD mechanical resonators.

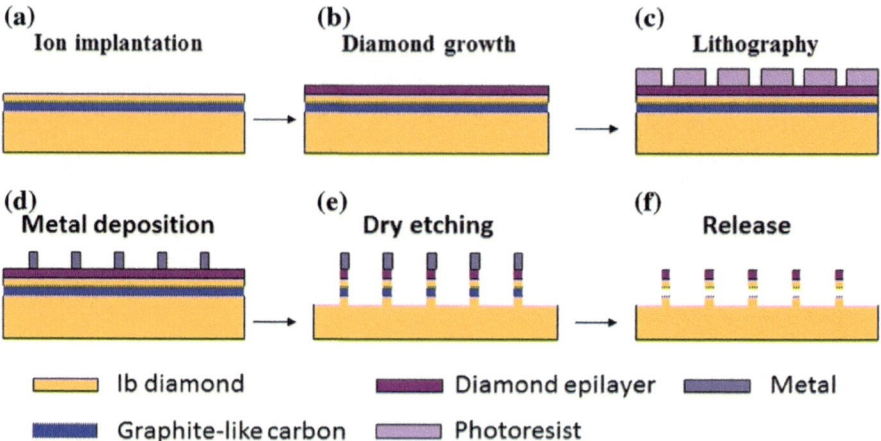

Fig. 4.3 Batch fabrication process of SCD mechanical resonators. **a** Ion implantation into the type-Ib diamond substrate. **b** MPCVD growth of SCD homoepitaxial layer. **c** Pattern formation by photo or e-beam lithography. **d** Formation of metal etch mask. **e** Mesa formation by reactive ion etching. **f** Finally, removal of the metal mask and graphite in an acid solution (H_2SO_4 + HNO_3) or electrochemical etching in water

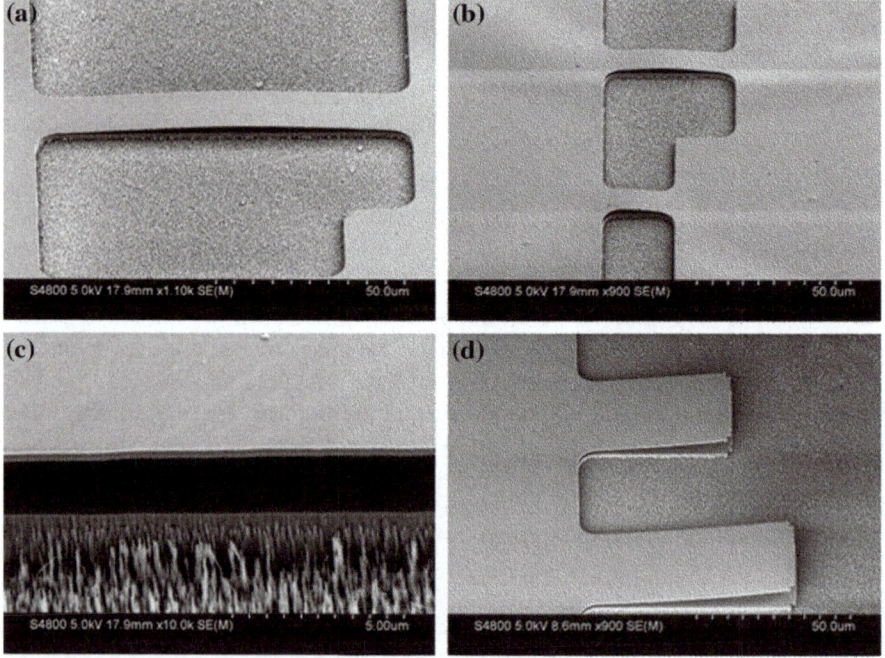

Fig. 4.4 SEM images of various SCD bridges and cantilevers [22] **a** a bridge with a length of 80 μm and a width of 15 μm, **b** bridges with lengths of 20 and 40 μm and a width of 15 μm, **c** enlarged image of the center part of the bridge, and **d** cantilevers with lengths of 50 and 70 μm and a width of 40 μm. © 2010 IOP Publishing Ltd

Figure 4.4 illustrates the scanning electron microscopy (SEM) images of the SCD bridges and cantilevers with microscale lengths and widths on the Ib-type diamond substrate. The diamond bridges and cantilevers are constituted of two layers: the homoepitaxial diamond layer and the type-Ib diamond substrate layer separated from the original substrate. The gap between the diamond bridge/cantilever and the substrate, which was formed by removing the graphite induced by ion implantation and annealing, can be clearly identified. In some cases, the air gap was larger than the thickness of the buried graphite layer (~ 200 nm) due to strain in the resonators. The gap at the center of a bridge with a length of 80 µm exceeds 1 µm, as disclosed from the enlarged SEM image (Fig. 4.4c). A bending was also observed for the cantilevers (Fig. 4.4d). The large gap of the bridges/cantilevers provides the motional operation as MEMS devices.

Nanoscale mechanical resonators were fabricated by using e-beam lithography. Figure 4.5 shows SCD cantilevers/bridges with widths from 100 to 400 nm and

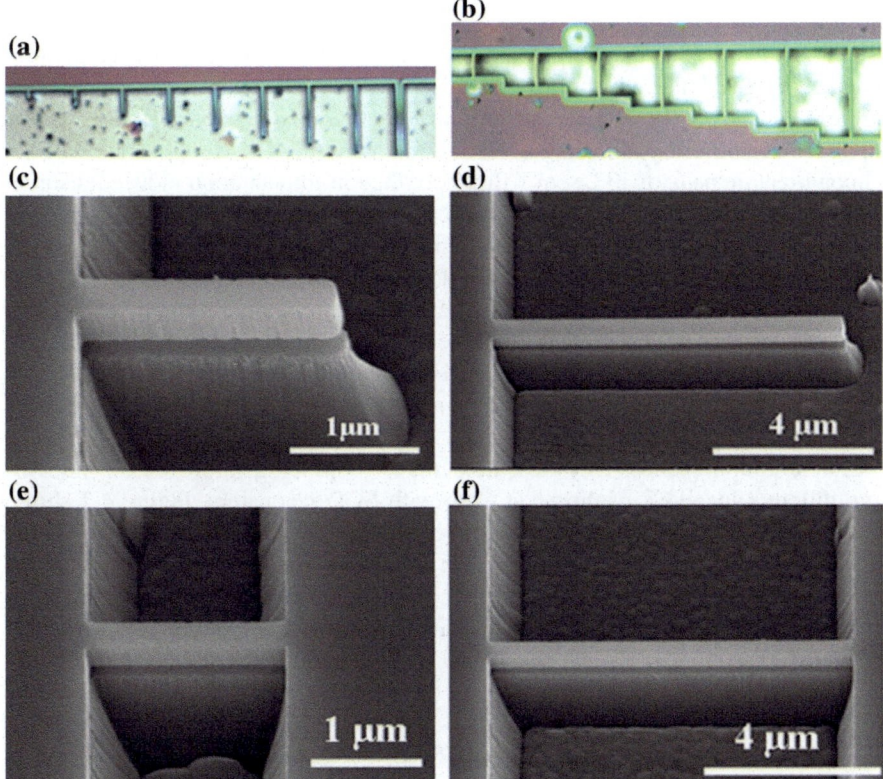

Fig. 4.5 **a** and **b** Optical images of the nanoscale cantilevers and bridges. SEM images of **c** a cantilever with a length of 2 µm and a width of 100 nm, **d** a cantilever with a length of 8 µm and a width of 200 nm, **e** a bridge with a length of 2 µm and a width of 200 nm, and **f** a bridge with a length of 6 µm and a width of 200 nm

lengths from 2 to 16 μm. Figure 4.5a, b are optical images of SCD cantilevers and bridges, respectively. Figure 4.5c–f are SEM images of these nano-beams. The SEM images reveal the SCD-on-SCD structure of the nanoscale mechanical resonators. The edge near the fixed end of the nanobeam was only slightly removed (~ 1 μm) during etching and the etching process automatically stopped. No clear bending was observed for these nanoscale resonators. These features ensured the control of the resonators dimensions and successful applications in NEMS switch [30]. In principle, suspended nanoscale SCD resonators in any shape can be produced using the IAL technique.

4.3 Structure of SCD Mechanical Resonators Fabricated by IAL Method

Since the method of IAL includes the ion impingement on the diamond crystal lattice, it is necessary to characterize the crystal quality of the SCD mechanical resonators. The resonators were characterized by Raman spectroscopy, cathodoluminescence, and high-resolution transmission electron microscopy (HRTEM). Figure 4.6a shows the Raman spectra of the diamond epilayer grown on the implanted region after annealing at 900 °C in vacuum. The disappearance of the diamond feature peak of 1332 cm^{-1} discloses the ion-implantation induced disorder and graphitization. After releasing the cantilever, the diamond 1332 cm^{-1} LO phonon peak with a full-width at half maximum (FWHM) of 3.6 cm^{-1} can be observed, as shown in Fig. 4.6b. Fortunately, during the releasing of the SCD resonators, one cantilever was broken and unintentionally bridged across two neighbor cantilevers. This broken cantilever was used for Raman characterization of the diamond cantilever alone. The FWHM is almost the same as the Raman spectrum from the diamond epilayer grown on the un-implanted region, suggesting the good quality of the diamond beam.

Cathodoluminescence (CL) measurements at room temperature were conducted from different locations of the SCD plate with SCD resonators. Figure 4.7 shows the CL spectra measured from (a) the Ib-type diamond substrate, (b) homoepitaxial diamond layer on the ion-implanted region with the buried graphite layer remained, and (c) a SCD diamond cantilever after removing the graphite layer. The free exciton peak at 235 nm was clearly observed at room temperature from the SCD cantilever, revealing the high crystal quality. The surface morphology of the epilayer above the implanted region was also the same as that of the epilayer grown on the non-implanted region.

We employed selective area electron (SAE) diffraction and HRTEM further to characterize the structure quality of the diamond resonators fabricated by the IAL method. The diamond specimen was prepared by focused-ion-beam

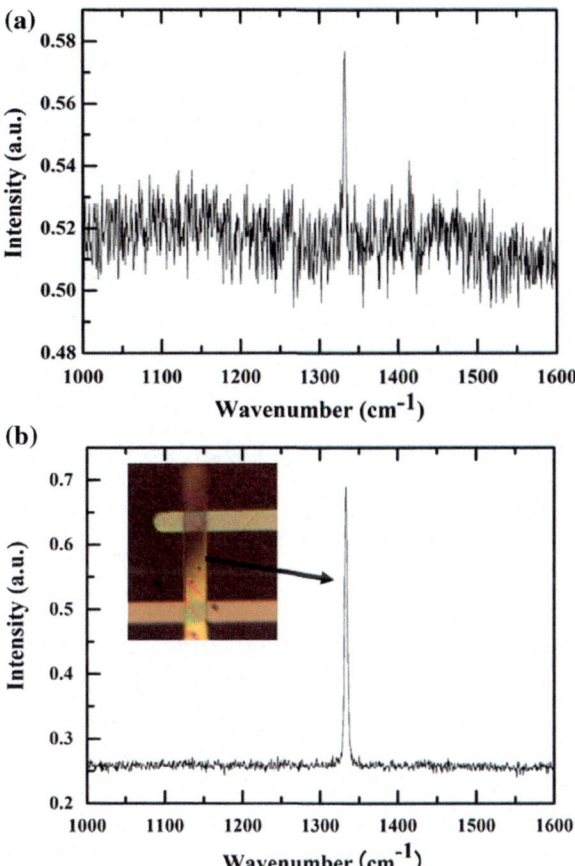

Fig. 4.6 Raman spectra from **a** the SCD epilayer (boron-doped) grown on the ion-implanted region and **b** a SCD beam relaxed from the diamond substrate. Adjusted from [22]. © 2010 IOP Publishing Ltd

(FIB) technique from the ion-implanted region on the diamond substrate. The SAE diffraction and the TEM images were shown in Fig. 4.8. The graphite-like carbon layer with a thickness around 200 nm was clearly observed. The SAE diffraction revealed the single crystal nature of the diamond epilayer grown on the substrate and the graphite-like structure within diamond due to ion irradiation. The single crystal nature of the diamond epilayer was also observed when boron ($>10^{20}cm^3$) was induced inside diamond during growth. These results suggest that diamond mechanical resonators fabricated by the IAL method remain the single crystal nature. However, it is difficult to rule out the point defects within the type-Ib SCD substrate originated from ion irradiation.

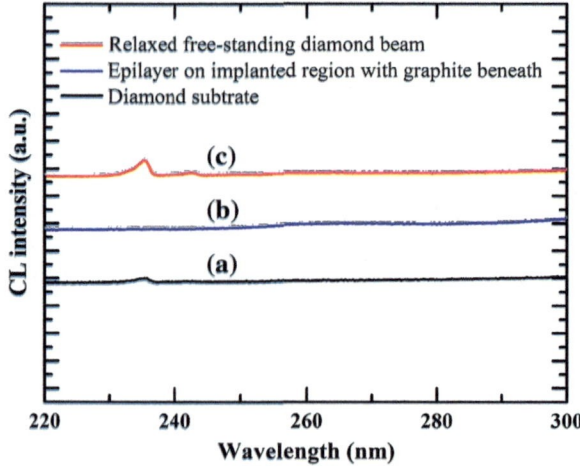

Fig. 4.7 Cathodoluminescence spectra taken from **a** the diamond substrate, **b** diamond epilayer grown on the ion-implanted region with graphite remained, and **c** a freestanding SCD beam [22]. © 2010 IOP Publishing Ltd

4.4 Nanoindentation of SCD Resonators

The spring constant and the Young's modulus of the SCD resonator were measured by Nanoindentation. To obtain large-size SCD resonators, we released the resonators by using EC method. This method led to a serious overhang effect. We measured force-displacement (F-d) of the SCD cantilevers/bridges by using atomic force microscopy (AFM) (JEOL JSPM-5200 scanning probe) in air [31]. Silicon cantilevers (Micromash Inc, tip radius <10 nm) with nominally known spring constants k_{Si} from 0.6 to 14 N m^{-1} were used. The spring constants were calibrated by using Sader method by obtaining the resonant frequency, quality factor, and geometric dimensions of the silicon cantilevers [32]. Such a way gave an error within 15% when the overhang space of the target cantilever was not considered [33].

Figure 4.9 shows the AFM images of the SCD bridge with a length of 40 μm and a width of 15 μm and the corresponding force-displacement (F-d) curve in comparison with that of the diamond base, where the measurement was conducted at the bridge center. The F-d curves of a SCD cantilever with a length of 50 μm were also measured at certain positions from the base. The F-d curve from the diamond substrate was taken as a reference, as shown in Fig. 4.10. The slope of the F-d curve of the cantilever decreased as the distance (defined as x) from the base increased. Slight non-linearity and hysteresis were observed.

The load-unload hysteresis was attributed to the friction effect during the nanoindentation. This hysteresis was inevitable in our measurement even though the minimum loading speed was used. The slopes of the load and unload curves

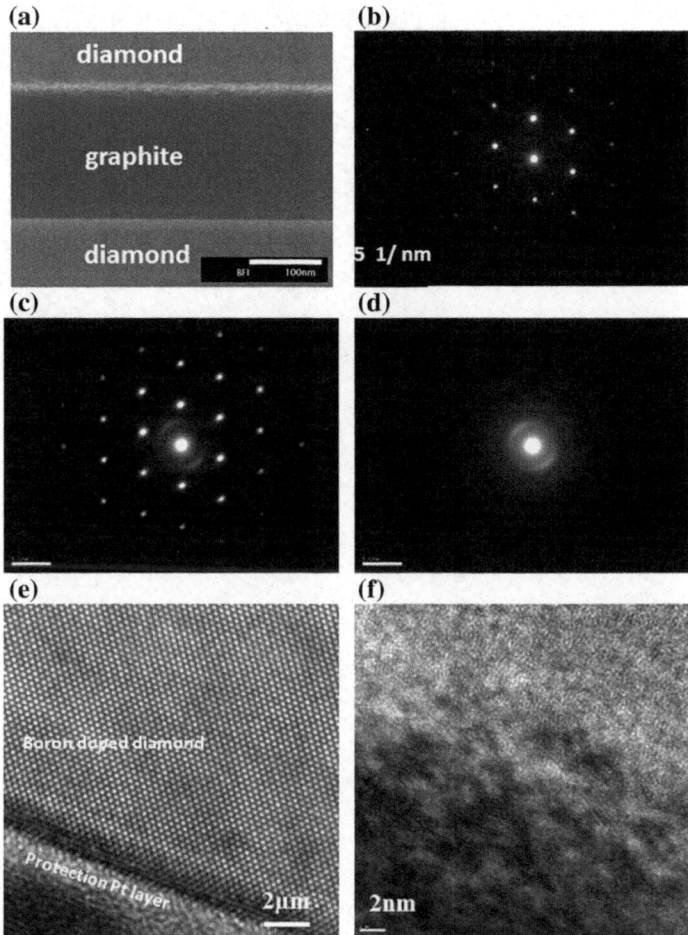

Fig. 4.8 **a** TEM image showing the sandwich structure comprised of diamond-graphite-diamond. **b**, **c** and **d** SAE diffractions from the diamond epilayer, the interface between diamond/graphite, and graphite, respectively. **e** and **f** High-resolution TEM images confirming the single crystal nature of the epilayer and the graphite nature of the implanted region

were similar, which may not affect so much about the data interpretation [34]. The F-d curves were recorded more than 5 times at each fixed sample position and most of the F-d curves were reversible. Therefore, the deformation of the diamond bridge/cantilevers was nearly elastic. Since the F-d curves were acquired in air, there was adhesion force exerted by thin layers of water vapor in addition to the van der Waals interaction between the Si cantilever and the diamond sample. The adhesion effect may be responsible for the nonoverlap behavior and the slight nonlinearity of the F-d curves [35, 36].

Fig. 4.9 a The AFM image
of a SCD bridge and b the
corresponding
force-displacement curve of
the bridge at the center
refereed to that of the
substrate [22]. © 2010 IOP
Publishing Ltd

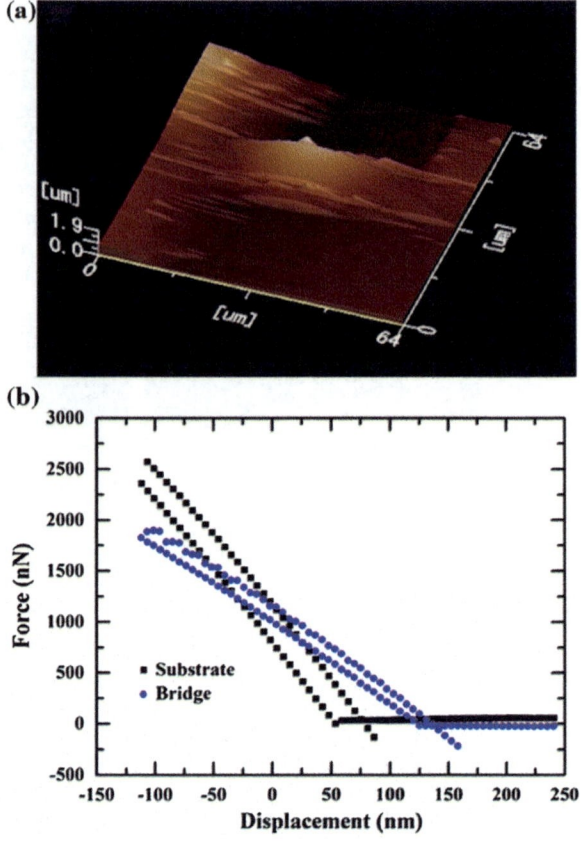

The stiffness of the diamond structure can be derived by the stiffness of the silicon cantilever and the slope k_m of the measured F-d curve, which is

$$k_{dia} = \frac{k_m k_{Si}}{k_{Si} - k_m}. \tag{4.1}$$

To avoid the slight nonlinearity and hysteresis of the F-d curves, the fitting of the slope was done at the high load part of the load curve in order to exclude the influence of surface energies according to Hertz-Sneddon limit [34]. The measured spring constant k_{dia} for the diamond bridge was 18.1 N/m. The spring constants for the diamond cantilever at different locations were different, which were 2.9, 1.0, and 0.35 N/m, respectively, for $x = 21$, 30, and 38 μm from (4.1). The Young's modus E of the diamond bridge was calculated from $E = k_{dia}L^3/192I$ for a center position, while for that of the cantilever $E = k_{dia}x^3/3I$. Here, I is the moment of the inertia and L is the length of the bridge. The resulting Young's modulus was calculated to be 800 ± 200 GPa. The measured values are lower than that of the bulk diamond (1143 GPa) [37]. The Young's modulus of these diamond

Fig. 4.10 **a** The AFM image
of a single crystal diamond
cantilever and **b** the
force-displacement curve of a
single crystal diamond
cantilever at a certain position
refereed to that of the
substrate [22]. © 2010 IOP
Publishing Ltd

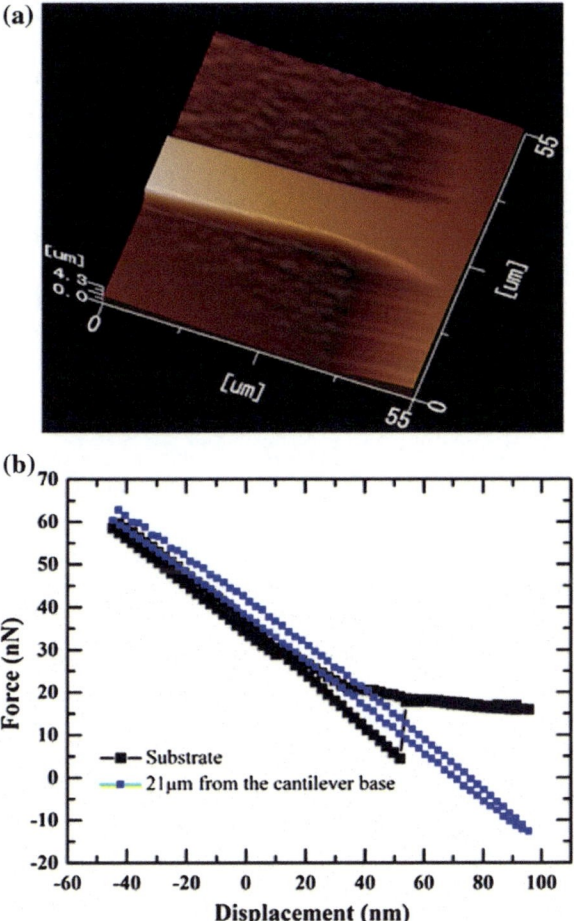

freestanding structures was underestimated. A main reason is the serious overhang effect of the SCD cantilevers fabricated by EC release. More accurate estimation of the SCD Young's modulus was described next by the resonance frequency measurement.

4.5   Mechanical Resonance Properties of SCD Mechanical Resonators

The mechanical resonance provides not only the mechanical Young's modulus of the resonators but also the energy dissipation in the resonator that fundamentally affects the device performance. Either optical or electrical method can be adopted to measure the resonance properties of the SCD mechanical resonators. The Laser Doppler Velocimetery (LDV) is the mostly employed optical technique to measure the mechanical resonator frequency, which is based on the measurement of the instantaneous velocity the target object. The LDV is non-instructive and can measure all the three velocity components. In this technique, the velocimeter sends a monochromatic laser signal toward the target and records the reflected optical signal. According to the Doppler effect, the change in wavelength of the reflected light is related to the targeted object's relative velocity. Therefore, the velocity of the moving object, which can be transferred into frequency, is obtained by measuring the change in the wavelength of the reflected laser light.

We measured the out-of-plane resonance frequencies of the SCD cantilevers by an optical interferometric velocity and displacement technique (MSA-500, Polytec. Ltd or Onosoki LV-1710) by utilizing the Doppler effect of a focused laser (He–Ne laser, 633 nm, <1 mW) at a vertical incidence to the substrate. The measurement was performed in a vacuum chamber to circumvent the air damping loss. The actuation of the resonators was performed by using a radio-frequency signal to drive a piezoceramic element placed at the bottom of the diamond samples. The laser light reflected from the cantilever interfered with the incident laser light, which modulated the detected light intensity. The laser spot was controlled by a computer to identify the target cantilevers precisely in the monitor. The frequency spectra were generated by the FFT procedure (Fast Fourier Transformation) of the velocity signals.

4.5.1 Resonance Frequency of the SCD Mechanical Resonators

We focused on the resonance frequency of the SCD cantilevers fabricated by the IAL technique here [38]. The SCD cantilevers were released by acid etching, so that the overhang could nearly be neglected. The high-pressure high-temperature (HPHT) type-Ib SCD (100) substrates were implanted by ions with high energies. Two kinds of ion implantation experiments were performed: carbon ions at an energy of 180 keV with a dose of 10^{16} cm^{-2} and helium ions at an energy of 1 meV with a dose of 5×10^{16} cm^{-2}. The SCD cantilevers fabricated by carbon ion implantation were denoted as C-SCD cantilevers and those fabricated by He ions implantation as He-SCD cantilevers. The cantilever width was varied from around 3.1–10 μm and the length was from around 30–206 μm. The thickness of the SCD cantilevers was varied from 0.68 to 1.7 μm, controlled by the ion implantation conditions or the epilayer thickness.

The resonance frequency of the SCD cantilevers was measured in the linear regime, namely, the driving voltage of the piezoceramics was carefully controlled not to affect the energy dissipation in the cantilevers. The overhang space at the base was around 1–3 μm, much smaller than the length (>30 μm) of the cantilevers when the cantilevers were released by acids. The typical fundamental resonance frequency spectra of the SCD cantilever were shown in Fig. 4.11. The resonance frequency strongly depended on the cantilever length for the SCD cantilever with the same thickness. The SCD cantilever was a typical harmonic oscillator, the spectrum of which followed a square-root of Lorentzian profile with a centred resonant frequency.

Fig. 4.11 Resonance frequency of SCD cantilevers. **a** optical image of the SCD cantilevers. **b** Resonance frequency spectra of the SCD cantilevers with different lengths. **c** A typical resonance frequency spectrum

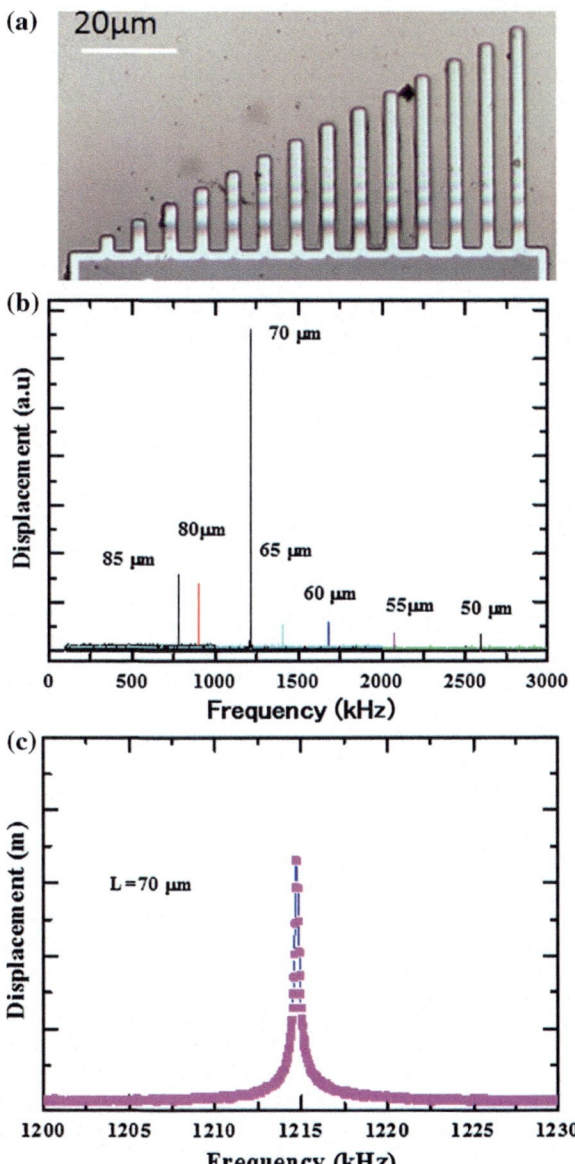

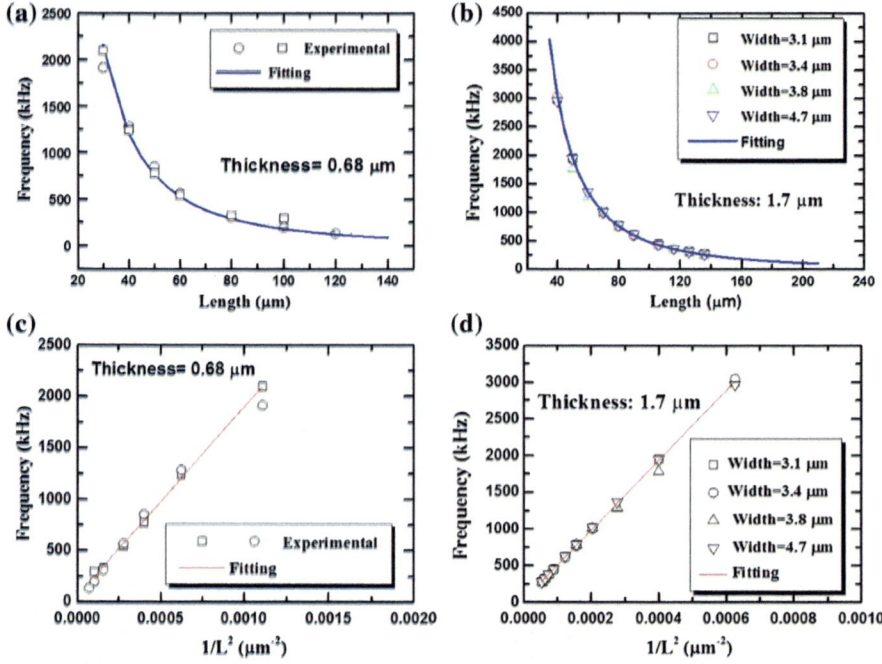

Fig. 4.12 The resonance frequency dependence of the cantilevers length: **a** the *C*-SCD cantilevers, **b** the *He*-SCD cantilevers. **c** The variation of the resonance frequency with $1/L^2$, calculated from (**a**). (**d**) The variation of the resonance frequency with $1/L^2$, calculated from **b** [38]. © 2014, American Institute of Physics

The detailed dependence of the fundamental resonance frequencies on the cantilever length is displayed in Fig. 4.12a for the *C*-SCD cantilevers. Figure 4.12b illustrates the characteristics of the resonance frequencies versus length for the *He*-SCD cantilevers.

We analyzed the resonance frequency of the SCD cantilevers by the Euler-Bernoulli theory as [39].

$$f = k \frac{t}{L^2} \sqrt{\frac{E}{\rho}} \tag{4.2}$$

where κ is 0.162 for a cantilever in the first vibration mode, E *is* the Young's modulus, ρ *is* the mass density, t *is* the thickness and L *is* the length of the cantilever. As revealed in Fig. 4.12c, d, the resonance frequency scales well as the law of $1/L^2$, regardless of the cantilever width and thickness. Possibly due to the large stress in the *C*-SCD cantilevers, a slight deviation between the experimental and theoretical frequency was observed, as shown in Fig. 4.12c. While for the *He*-SCD cantilevers, the resonance frequency obeyed the scaling law of $1/L^2$ in an almost perfect manner. The consistence of the experiments with the theoretical fitting confirmed the validity of the thin-beam approximation and also suggested the high reproducibility of the IAL method for the fabrication of the SCD mechanical resonators. The Young's modulus of the SCD-cantilevers was extracted by fitting the curve of the resonance frequency *versus* cantilever length, which was around 1100 GPa. This value is more precise and higher than that of obtained by nanoindentation. The evaluated Young's modulus is similar to that calculated from the diamond NEMS switch [40] and was also very close to that of bulk SCD (1143 GPa). When the overhang effect was considered, the Young's modulus increased to be around 1160 GPa [41]. Note that the strain in the cantilevers was not taken into account for the fitting.

4.5.2 Energy Dissipation in SCD Cantilevers and Quality Factor

A key figure-of-merit of a mechanical resonator is the quality factor (Q-factor), which ultimately determines the device response sensitivity (e.g. force, mass, displacement etc.). For a normal rectangular cantilever, the minimum detectable force can be expressed as [42].

$$F_{\min} = (\frac{wt^2}{LQ})^{1/2}(k_B TB)^{1/2}(E\rho)^{1/4} \qquad (4.3)$$

where E is the Young's modulus, ρ is the mass density of the resonator, B is the bandwidth and w, L, t stand for the width, length, and thickness of the cantilever, respectively. Therefore, a high Q-factor resonator is generally pursued in MEMS/NEMS.

The Q-factor was technically calculated from the full-width at half maximum (FWHM) of the resonance spectrum from Lorentzians fit, which is defined by

$$Q = \frac{f_r}{\Delta f} \qquad (4.4)$$

where f_r is the resonance frequency and Δf is the FWHM of the resonance spectrum. The Q-factor is determined by the main energy dissipation. Generally, the overall Q-factor can be expressed as

$$\frac{1}{Q} = \frac{1}{Q_{air}} + \frac{1}{Q_{clamp}} + \frac{1}{Q_{TED}} + \frac{1}{Q_{bulk}} + \frac{1}{Q_{surf}} \qquad (4.5)$$

where Q_{air}, Q_{clamp}, Q_{TED}, Q_{bulk}, and Q_{surf} denote the quality factors related to the loss mechanisms of air damping, support loss, thermoelastic damping (TED), bulk, and surface effect, respectively. The dominated energy dissipations determining the ultimate Q-factor was deduced by excluding other mechanisms.

For the C-SCD cantilevers with a thickness of 680 nm fabricated by carbon ion implantation at 180 keV, the Q factors were in the range of 3 000–7 000, as shown in Fig. 4.13a. The Q factor decreased with the L decreasing from 80 to 30 µm. When L was longer than 80 µm, the Q factors decreased. For the He-SCD cantilevers fabricated by He ion implantation at 1 meV, the Q factors were slightly lower than those of C-SCD cantilevers, as presented in Fig. 4.13b, c. Similar to the C-SCD, the Q factors of the He-SCD cantilevers decreased with L shortened from 176 to 35 µm. However, when L was longer than 176 µm, the Q-factors tended to be saturated. We also observed that the Q-factors slightly decreased for the He-SCD cantilevers with a width of 4.7 µm.

We explored the energy dissipation by excluding each mechanism one by one. In vacuum, the air damping was neglected. TED is often regarded as one of the most important energy dissipation mechanism at room temperature [43]. The TED dissipation was concluded not to govern the energy dissipation due to the following facts. As shown in Fig. 4.14, the measured Q factors at elevated temperatures did not show a reduction as increasing the temperature from RT to 373 K, which contradicted to the TED model.

The dominated energy dissipation was considered as a result of the clamping and the bulk or surface losses for the SCD cantilevers. The bulk or surface loss mechanism, which does not depend on the dimensions of the cantilevers, was not the main factor for the short cantilevers here. At shorter length, clamping loss governed the energy loss. We considered two models to describe the clamping loss of out-of-plane flexural cantilevers. For a cantilever of infinite width attached to a semi-infinite base, the Q factor due to clamping loss can be express as [44].

$$Q_{\text{clamp}} = \alpha \left(\frac{L}{t} \right)^3 \tag{4.6}$$

where α is around 2.17 when the cantilever and the base are fabricated using the same material. Clamping loss based on this model predicts a Q factor in the order of magnitude of above 10^6 for the current experiments. This model was believed to be effective for the NCD double-clamped beams [45]. In our case, the variation of the Q factor with L deviated from the L^3 law. Actually, the predicted Q factors are much larger than those of the SCD cantilevers in the present study. Another model for flexural cantilever with a finite base was proposed by Photiadis and Judge as [46]

$$Q_{\text{clamp}} = \beta \frac{L}{w} \left(\frac{t_b}{t} \right)^2 \tag{4.7}$$

where β is a constant, w is the cantilever width, and t_b is the basement thickness. As shown in Fig. 4.14, the Q factors show a nearly linear dependence on the cantilever length, which can be well fitted by (4.7) for each set of the SCD-cantilevers.

Fig. 4.13 Dependence of Q-factor on the cantilevers length: **a** the C-SCD cantilevers. **b**, **c** the He-SCD cantilevers with thicknesses of 1.4 and 1.7 μm, respectively [38]. © 2014, American Institute of Physics

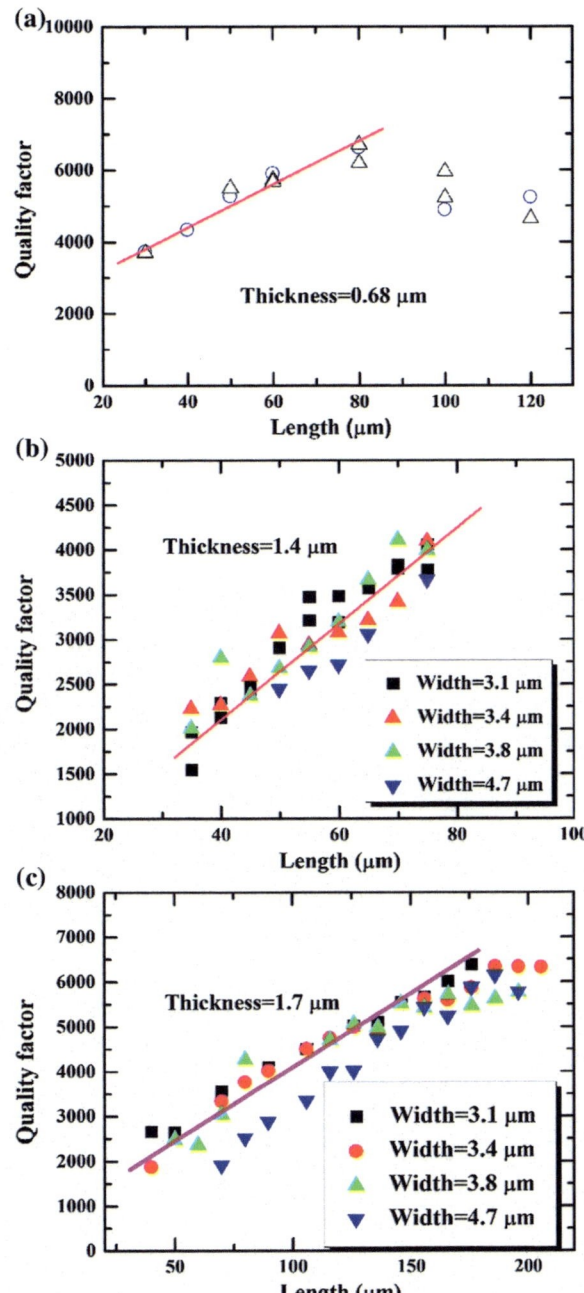

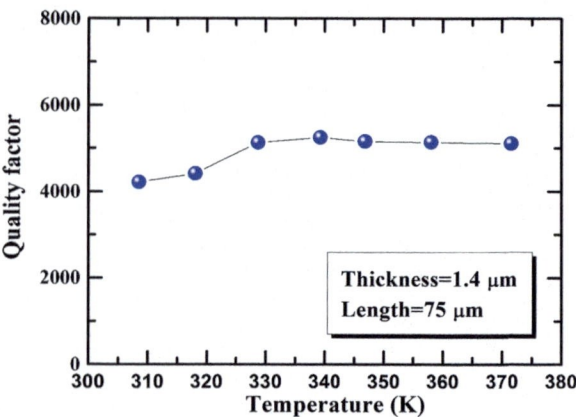

Fig. 4.14 Quality factor dependence on the measurement temperature [38]. © 2014, American Institute of Physics

A decrease of the Q factor was also observed for the cantilevers with a width of 4.7 μm, consistent with (4.7). The clamping loss mechanism also indicated that the thinner cantilevers exhibited higher Q factors. We plotted the variation of Q factors vs $L/(wt^2)$ in Fig. 4.15, which confirmed the clamping loss for the short cantilevers. The linear dependence of the Q factor on the cantilever length implied that the base for the SCD cantilevers behaved like a membrane. The deviation of Q factors from (4.7) for long cantilevers was due to surface loss.

The Q factor tended to be saturated for a certain value of length. Bulk defects and surface effect dissipation [47, 48], which are independent on the length, were considered as the dominant mechanism for the longer cantilevers. Note that the reported Q-factors of the SCD resonators fabricated by wafer bonding or direct dry etching are much larger than those of the as-fabricated SCD cantilevers by IAL method. Therefore, the ion implantation induced bulk defects limits the ultimate Q-factors of the IAL-SCD cantilevers. The energy loss induced by ion-irradiation

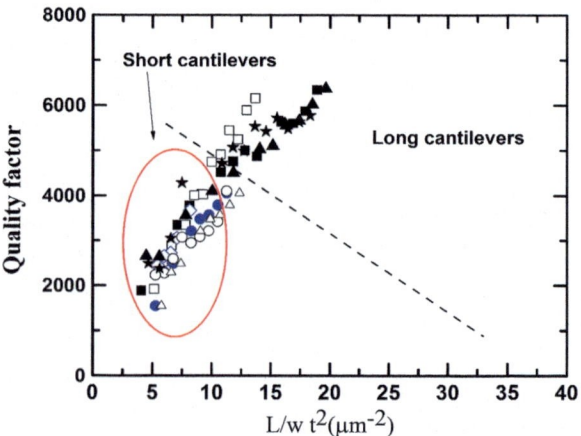

Fig. 4.15 Quality factor as a function of $L/(wt^2)$. The triangles, squares, circles, and stars indicate different cantilevers [38]. © 2014, American Institute of Physics

Fig. 4.16 Resonance frequency spectra from the type-Ib diamond substrate alone implanted by carbon ion. The Q-factor is around 200. The Vpp is the actuation voltage applied to the piezoceramics below the SCD plate [49]. Copyright (2017) The Japan society of Applied Physics

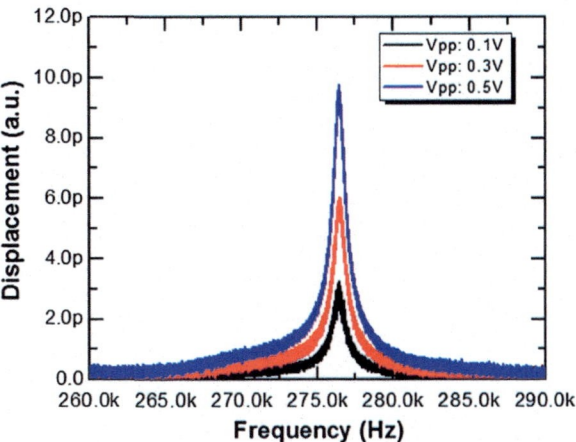

induced defects was realized by measuring the Q-factor of the ion implanted type-Ib substrate alone. The Q-factor from such diamond cantilevers had a Q-factor at the level of 1000 or less, as shown in Fig. 4.16. On the other hand, the surface effects, such as ion implantation induced damage at the bottom of the cantilever, RIE induced defects at the edge of the cantilever and surface stress also generate energy dissipation. The reduction of the Q factors for the C-SCD cantilevers with L longer than 80 μm may be explained by surface stress effect. In contrast, the surface stress was not as significant for the thicker He-SCD cantilevers as for the C-SCD cantilevers.

4.5.3 Strategies Toward High Quality-Factors

We note again that the SCD mechanical resonators fabricated by the IAL method contains two layers: the homoepitaxial layer and the ion implanted substrate diamond layer. Based on the above analysis, we proposed two strategies to improve the Q-factor of the IAL fabricated SCD mechanical resonators. One is the deposition of thick diamond epilayer on the ion implanted diamond substrate [49]. In Fig. 4.17, one can see that as the thickness ratio of the epilayer to the substrate increases, the Q-factor also increases. We interpret the dependence of the Q-factor on the epilayer thickness by a bilayer model as

$$\frac{1}{Q} = \frac{1}{1+\beta}\left(\frac{1}{Q_{\text{epi}}} + \beta\frac{1}{Q_{\text{ion}}}\right) \tag{4.8}$$

where Q is the quality factor of the overall cantilever, Q_{epi} represents the quality factor of the epilayer alone and Q_{ion} denotes the effective quality factor of the ion-implantation damaged layer. β is an dimensionless factor, described as

Fig. 4.17 Quality factor dependence on the thickness ratio of the ion damaged layer to the epilayer [49]. Copyright (2017) The Japan Society of Applied Physics

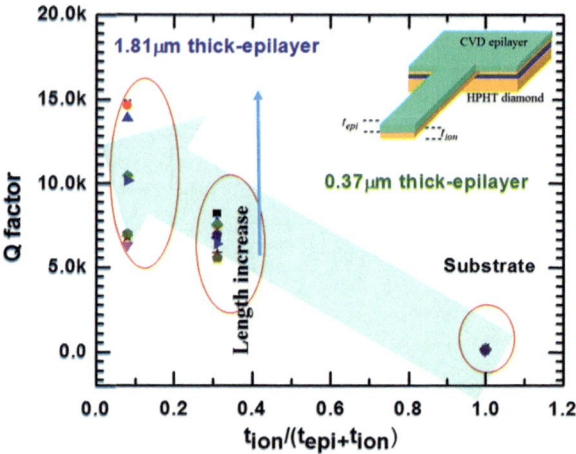

$\beta = E_{ion}t_{ion}/E_{epi}t_{epi}$. E_{ion} and E_{epi} are the Young's modulus of the ion damaged layer and epilayer, respectively. t_{ion} and t_{epi} are the thickness of the ion damaged layer and epilayer, respectively. It was assumed that $E_{ion} = E_{epi}$, then, β was simplified to be t_{ion}/t_{epi}. The behavior in Fig. 4.17 was found to be consistent with the bilayer model. The Q-factor was improved up to 39 417 by simply increasing the epilayer thickness to 1.81 μm (Fig. 4.18).

We employed the second strategy to improve the Q-factor by reducing the defects at the bottom of the SCD cantilever through a surface etching effect. We did the annealing experiments in oxygen ambient at 430 and 500 °C for different time durations [50]. A thin SCD cantilever with a thickness of 340 nm was investigated since such a thin cantilever was sensitive to the surface effect. Shown in Fig. 4.19 is the Q-factor evolution as the annealing temperature and time durations. It was observed that the annealing temperature at 430 °C did not affect the Q-factor so much. A significant improvement of the Q-factors from 3500 to 7000 in average occurred after annealing at 500 °C for 7 h. We note that the frequency-length dependence followed well the Euler-Bernoulli theory after annealing.

In order to clarify the mechanism of the Q-factor enhancement, the variation of the resonance frequency as the annealing temperature and time duration was examined, as illustrated in Fig. 4.20. Although little shift in the resonance frequency was observed for the annealing at 430 °C, a marked shift was observed after annealing at 500 °C.

We claimed from another viewpoint that the SCD cantilever were thermally stable at 430 °C even under oxygen ambient. The frequency reduced almost linearly with the annealing time with a slope of around 100 Hz/h. The reduction in the resonance frequency was attributed to the uniform etching of the diamond. From the resonance shift, an etching rate of about 0.4–0.5 nm/h was estimated at 500 °C in oxygen ambient. This etching rate corresponded to a carbon mass loss of 10^{-12} g/h, and a mass sensitivity of around 3×10^{-14} g/Hz was deduced. Therefore, the improvement of the Q-factor by annealing was concluded to the

Fig. 4.18 Resonance frequency spectra of the cantilevers with a thickness of 1.98 μm with a 1.81 μm-thick SCD epilayer. **a** At different piezo driving voltage V_p and **b** at zero driving voltage [49]. The quality factor was calculated to be around 30 417. Copyright (2017) The Japan Society of Applied Physics

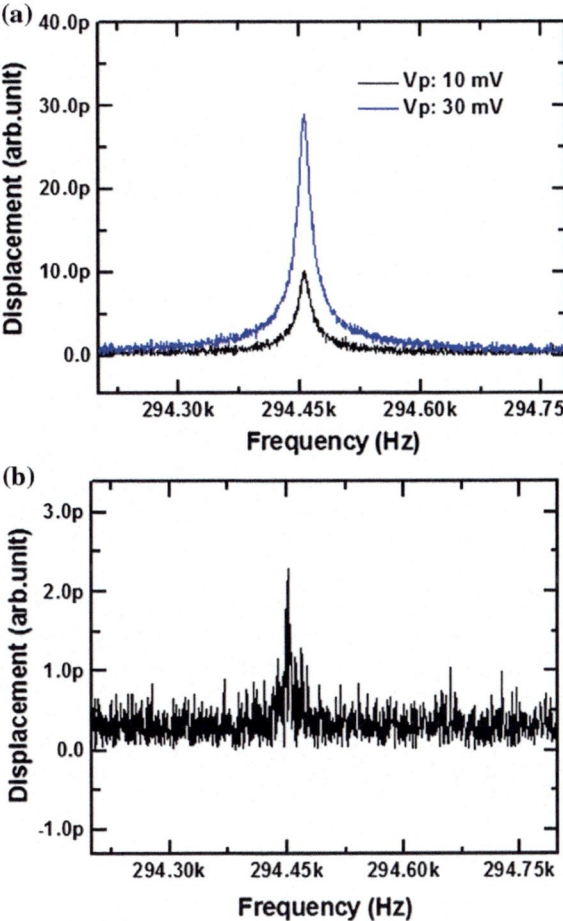

Fig. 4.19 Dependence of Q-factor on the annealing temperature and time durations. © 2017 Elsevier Ltd

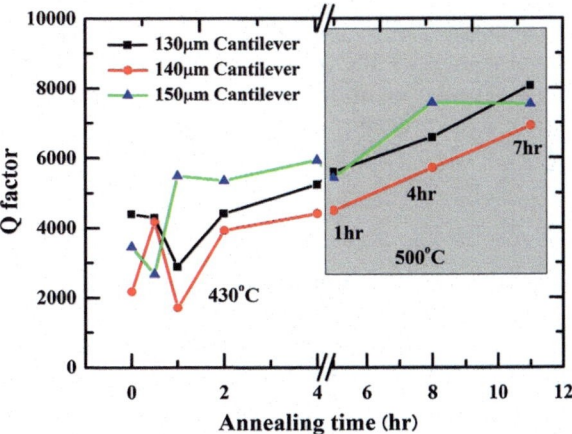

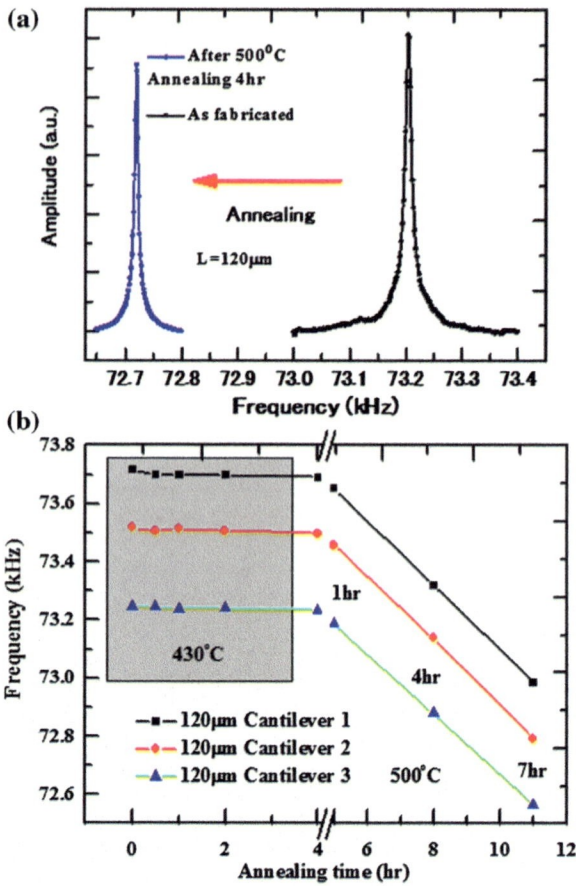

Fig. 4.20 **a** Resonance frequency shift after 500 °C annealing and **b** frequency shift as annealing temperature and time duration [50]. © 2017 Elsevier Ltd

removal or reduction of the surface defects. In order to support this viewpoint, we investigated the Q-factor dependence on the SCD cantilever length before and after annealing at 500 °C for 7 h. The results are shown in Fig. 4.21. It was disclosed that in both cases, the Q-factor exhibited little dependence on the cantilever length. The independence of the Q-factor on the cantilever length could be explained by a surface loss mechanism as written [51].

$$Q_{\text{surf}} = \frac{wt}{2\delta(3w+t)} \frac{E_1}{E_1^S} Q_S \tag{4.9}$$

where $Q_S = E_1^S/E_2^S$. The surface defects were mostly from the bottom of the SCD cantilever induced by ion implantation. The annealing in oxygen ambient reduced the bottom defects through a surface effect. The submicron scale feature of the SCD cantilevers provides the variation of the Q-factor sensitive to the surface treatment. The force sensitivity of the submicron SCD cantilever was evaluated to be as high

Fig. 4.21 Dependence of Q-factors on the C-SCD cantilever length **a** before and **b** after annealing at 500 °C for 7 h for different cantilever sets [50]. © 2017 Elsevier Ltd

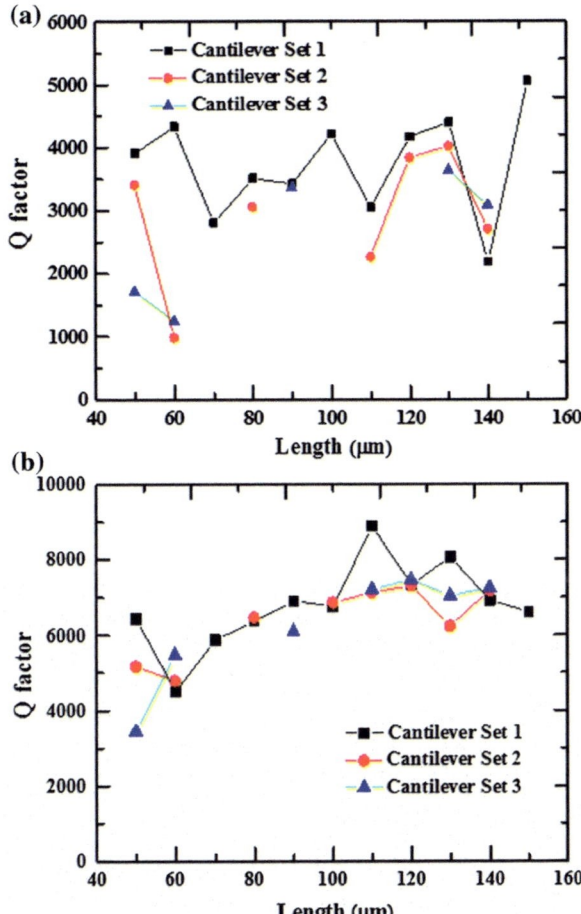

as 10^{-16}–10^{-15} N/Hz$^{1/2}$, comparable with those NEMS cantilevers with ultra-high sensitivity [52].

Another annealing experiment was conducted on a SCD MEMS cantilever with an original thickness of 1.6 µm fabricated by the IAL method. After annealing for 380 h in oxygen ambient at 500 °C, the thickness of the SCD MEMS cantilever was reduced to be around 1.44 µm. The long time annealing in oxygen ambient markedly improved the Q-factor up to over 410 000 measured by a ring-down method [53], as illustrated in Fig. 4.22. The Q-factor is higher than those values reported ever from any polycrystalline or nanocrystalline diamonds. Therefore, in the case of SCD NEMS, the ion-implantation induced defects at the bottom surface of the cantilever degrade the Q-factor through a surface effect.

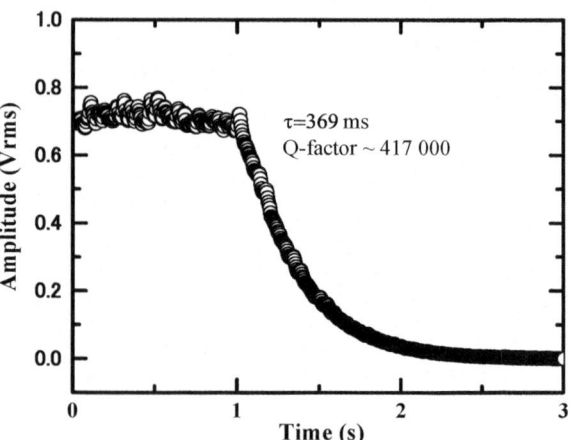

Fig. 4.22 Ring-down curve of the SCD cantilever after 380 h annealing in oxygen ambient [50]. © 2017 Elsevier Ltd

4.6 Summary and Outlook

In summary, we demonstrated the batch fabrication of SCD mechanical resonators by using an IAL method with well controlled manner in device dimensions. The IAL method provides the diamond-on-diamond device concept for MEMS/NEMS able to operate under harsh environments. The structural quality, mechanical bending, resonance properties, and energy dissipation in the SCD mechanical resonators were investigated. The freestanding diamond structures exhibited good single crystal quality. The maximum Young's modulus of the diamond bridges/cantilevers was as large as 1100 GPa. In order to examine the energy dissipation mechanism, SCD cantilevers with different dimensions of length ranging from 30 to 206 µm, width from 3.1 to 4.7 µm, and thickness from 0.68 to 1.7 µm were examined by different ion implantation conditions. The mechanical resonance frequencies of these cantilevers well followed the law of $1/L^2$ regardless of the width and thickness, revealing the high reproducibility of the present fabrication method. For the shorter cantilevers, clamping loss governed the energy dissipation. As the cantilever length increased to a certain value, the bulk defects and surface effect became dominant. In order to reduce the energy loss due to surface/bulk defects in the SCD mechanical resonators induced by ion implantation, the effect of annealing in oxygen ambient at 430 and 500 °C on the resonance performance of the SCD NEMS cantilevers was investigated. It was observed that both the resonance frequency and Q-factor changed little at 430 °C for different time durations. In contrast, at 500 °C, there were marked linear red-shift in the resonance frequency and obvious improvement in the Q-factor. The frequency shift and Q-factor improvement were resulted from the etching of the bottom surface defective layer of the SCD cantilever. A high Q-factor more than 400 000 was obtained, which is promising for MEMS/NEMS sensors, scanning microscopy probes, and quantum information coupled with spin centers in MEMS/NEMS resonators under harsh environments. In the future, on-chip characterization of SCD MEMS/NEMS should be developed and the applications under harsh environments should be explored.

Acknowledgements This work was partially supported by JSPS KAKENHI (Grant Number 15H03999, 15H03980, 26220903) and Nanotechnology Platform projects sponsored by the Ministry of Education, Culture, Sports, and Technology (MEXT) in Japan.

References

1. F. Chollet, H. Liu, A (not so) short introduction to MEMS (2008). http://memscyclopedia.org/introMEMS.html
2. C. Fischer, F. Forsberg, M. Lapisa, S.J. Bleiker, G. Stemme, N. Roxhed, F. Niklaus, Integrating MEMS and ICs. Microsyst. Nanoeng. **1**, 15005 (2015). https://doi.org/10.1038/micronano.2015.5
3. C.S. Smith, Piezoresistance effect in germanium and silicon. Phys. Rev. **94**(1), 42–49 (1954). https://doi.org/10.1103/PhysRev.94.42
4. N. Harvey, D. John, K. Terence, Resonant gate transistor with fixed position electrically floating gate electrode in addition to resonant member. U.S. 3,590,43 (A) (1971)
5. M.L. Roukes, Nanoelectromechanical systems face the future. Phys. World **14**, 25–31 (2001). https://doi.org/10.1088/2058-7058/14/2/29
6. K.L. Ekinci, M.L. Roukes, Nanoelectromechanical systems. Rev. Sci. Instr. **76**, 061101 (2005). https://doi.org/10.1063/1.1927327
7. K.L. Ekinci, X.M.H. Huang, M.L. Roukes, Ultrasensitive nanoelectromechanical mass detection. Appl. Phys. Lett. **84**(22), 4469–4472 (2004). https://doi.org/10.1063/1.1755417
8. A.L. Ruoff, J. Wanagel, High pressures on small areas. Science **198**(4321), 1037–1038 (1977). https://doi.org/10.1126/science.198.4321.1037
9. R.H. Telling, C.J. Pickard, M.C. Payne, J.E. Field, Theoretical strength and cleavage of diamond. Phys. Rev. Lett. **84**(22), 5160–5163 (2000). https://doi.org/10.1103/PhysRevLett.84.5160
10. J.E. Graebner, S. Jin, G.W. Kammlott, J.A. Herb, C.F. Gardiner, Large anisotropic thermal conductivity in synthetic diamond. Nature **359**, 401–403 (1992). https://doi.org/10.1038/359401a0
11. T.R. Anthony, W.F. Banholzer, J.F. Fleischer, L.H. Wei, P.K. Kuo, R.L. Thomas, R.W. Pryor, Thermal conductivity of isotopically enriched ^{12}C diamond. Phys. Rev. B **42**(2), 1104–1111 (1990). https://doi.org/10.1103/PhysRevB.42.1104
12. R. Lifshitz, M. Roukes, Thermoelastic damping in micro- and nanomechanical systems. Phys. Rev. B **61**(8), 5600–5609 (2000). https://doi.org/10.1103/PhysRevB.61.5600
13. A. Duwel, R.N. Candler, T.W. Kenny, M. Varghese, Engineering MEMS resonators with low thermoelastic damping. J. Microelectromech. Syst. **15**(6), 1437–1445 (2006). https://doi.org/10.1109/JMEMS.2006.883573
14. A.V. Sumant, O. Auciello, M. Liao, O.A. Williams, MEMS/NEMS based on mono-, nano-, and ultrananocrystalline diamond films. MRS Bull. **39**(6), 511–516 (2014). https://doi.org/10.1557/mrs.2014.98
15. M. Liao, Y. Koide, Carbon-based materials: growth, properties, MEMS/NEMS technologies, and MEM/NEM switches. Crit. Rev. Solid State Mater. Sci. **36**(2), 66–101 (2011). https://doi.org/10.1080/10408436.2011.572748
16. J.K. Luo, Y.Q. Fu, H.R. Le, J.A. Williams, S.M. Spearing, W.I. Milne, Diamond and diamond-like carbon MEMS. J. Micomech. Microeng. **17**, S147 (2007). https://doi.org/10.1088/0960-1317/17/7/S04

17. V.P. Adiga, A.V. Sumant, S. Sumant, S. Suresh, C. Gudeman, O. Auciello, J.A. Carlisle, R. W. Carpick, Mechanical stiffness and dissipation in ultrananocrystalline diamond microresonators. Phys. Rev. B **79**, 245403 (2009). https://doi.org/10.1103/PhysRevB.79.245403

18. A. Orlando, P. Sergio, S. Anirudha, G. Chris, S. Suresh, D. Arindom, C. Robert, A. Vivekananda, P. Zurcher, Z. Ma, H.C. Yuan, J.A. Carlisle, B. Kabius, J. Hiller, S. Srinivasan, Are diamonds a MEMS' best friend? IEEE Microw. Mag. **8**(6), 61–75 (2007). https://doi.org/10.1109/MMM.2007.907816

19. S. Srinivasan, J. Hiller, B. Kabius, O. Auciello, Piezoelectric/ultrananocrystalline diamond heterostructures for high-performance multifunctional micro/nanoelectromechanical systems. Appl. Phys. Lett. **90**, 134101 (2007). https://doi.org/10.1063/1.2679209

20. A. Gaidarzhy, M. Imboden, P. Mohanty, J. Rankin, B.W. Sheldon, Piezoelectric/ ultrananocrystalline diamond heterostructures for high-performance multifunctional micro/ nanoelectromechanical systems. Appl. Phys. Lett. **91**, 203503 (2007). https://doi.org/10.1063/ 1.2679209

21. H. Najar, C. Yang, A. Heidari, L.W. Lin, D. Horsley, Quality factor in polycrystalline diamond micromechanical flexural resonators. J. Microelectromech. Syst. **24**(6), 2152–2160 (2015). https://doi.org/10.1109/JMEMS.2015.2478802

22. M.Y. Liao, C. Li, S. Hishita, Y. Koide, Batch production of single-crystal diamond bridges and cantilevers for microelectromechanical systems. J. Micromech. Microeng. **20**, 085002 (2010). https://doi.org/10.1088/0960-1317/20/8/085002

23. P. Ovartchaiyapong, L.M.A. Pascal, B.A. Myers, P. Lauria, A.C. Bleszynski Jayich, High quality factor single-crystal diamond mechanical resonators. Appl. Phys. Lett. **101**, 163505 (2012). https://doi.org/10.1063/1.4760274

24. Y. Tao, J.M. Boss, B.A. Moores, C.L. Degen, Single-crystal diamond nanomechanical resonators with quality factors exceeding one million. Nat. Commun. **5**, 3638 (2014). https:// doi.org/10.1038/ncomms4638

25. M.J. Burek, D. Ramos, P. Patel, I.W. Frank, M. Lončar, Nanomechanical resonant structures in single-crystal diamond. Appl. Phys. Lett. **103**(13), 131904 (2013). https://doi.org/10.1063/ 1.4821917

26. M.J. Burek, N.P. de Leon, B.J. Shields, B.J.M. Hausmann, Y. Chu, Q. Quan, A.S. Zibrov, H. Park, M.D. Lukin, M. Lončar, Free-standing mechanical and photonic nanostructures in single-crystal diamond. Nano Lett. **12**, 6084–6089 (2012). https://doi.org/10.1021/nl302541e

27. N.R. Parikh, J.D. Hunn, E. McGucken, M.L. Swanson, C.W. White, R.A. Rudder, D. P. Malta, J.B. Posthill, R.J. Markunas, Single-crystal diamond plate liftoff achieved by ion implantation and subsequent annealing. Appl. Phys. Lett. **61**, 3124–3126 (1992). https://doi. org/10.1063/1.107981

28. M. Marchywka, P.E. Pehrsson, D.J. Vestyck, Daniel Moses, Low energy ion implantation and electrochemical separation of diamond films. Appl. Phys. Lett. **63**, 3521 (1993). https://doi. org/10.1063/1.110089

29. C.F. Wang, E.L. Hu, J. Yang, J.E. Bulter, Fabrication of suspended single crystal diamond devices by electrochemical etch. J. Vac. Technol. B **25**, 730 (2007). https://doi.org/10.1116/1. 2731327

30. M.Y. Liao, S. Hishita, E. Watanabe, S. Koizumi, Y. Koide, Suspended single-crystal diamond nanowires for high-performance nanoelectromechanical switches. Adv. Mater. **22**(47), 5393 (2010). https://doi.org/10.1002/adma.201003074

31. C. Li, Y. Bando, C. Zhi, Y. Huang, D. Golberg. Thickness-dependent bending modulus of hexagonal boron nitride nanosheets. Nanotechnology **20**(38), 385707 (2009). https://doi.org/ 10.1088/0957-4484/20/38/385707

32. J.E. Sader, J.W. Chon, P. Mulvaney, Calibration of rectangular atomic force microscope cantilevers. Rev. Sci. Instr. **70**, 3967–3979 (1999). https://doi.org/10.1063/1.1150021

33. G. Moeller. AFM Nanoindentation of viscoelastic materials with large end-radius probes. J. Poly. Sci. Part B Poly. Phys. **47**(16), 1573–1587 (2009). https://doi.org/10.1002/polb.21758

34. J.H. Hoh, A. Engel, Friction effects on force measurements with an atomic force microscope. Langmuir **9**(11), 3310–3312 (1993). https://doi.org/10.1021/la00035a089
35. B. Cappella, G. Dietler, Force-distance curves by atomic force microscopy. Sur. Sci Rep. **34** (1–3), 1–104 (1999). https://doi.org/10.1016/S0167-5729(99)00003-5
36. Q. Xiong, Force-deflection sepctroscopy: a new method to determine the Young's modulus of nanofilaments. Nano Lett. **6**(9), 1904–1909 (2006). https://doi.org/10.1021/nl060978f
37. S.F. Wang, Y.F. Hsu, J.C. Pu, J.C. Sung, L.G. Hwa, Determination of acoustic wave velocities and elastic properties for diamond and other hard materials. Mater. Chem. Phys. **85** (2–3), 432(2004). https://doi.org/10.1016/j.matchemphys.2004.02.003
38. M.Y. Liao, M. Toda, L.W. Sang, S. Hishita, S. Tanaka, Y. Koide, Energy dissipation in micron- and submicron-thick single crystal diamond mechanical resonators. Appl. Phys. Lett. **105**(25), 251904 (2014). https://doi.org/10.1063/1.4904990
39. K.B. Gavan, H.J.R. Westra, E.W.J.M. van der Drift, W.J. Venstra, H.S.J. van der Zant, Size-dependent effective young's modulus of silicon nitride cantilevers. Appl. Phys. Lett. **94** (23), 233108 (2009). https://doi.org/10.1063/1.3152772
40. M. Liao, Z. Rong, S. Hishita, M. Imura, S. Koizumi, Y. Koide, Nanoelectromechanical switch fabricated from single crystal diamond: experiments and modeling. Diam. Relat. Mater. **24**, 69–73 (2012). https://doi.org/10.1016/j.diamond.2011.10.026
41. J. Mencik, E. Quandt, Determination of elastic modulus of thin films and small specimens using beam bending methods. J. Mater. Res. **14**(5), 2152 (1999). https://doi.org/10.1557/JMR.1999.0291
42. K. Yasumura, T. Stowe, E. Chow, T. Pfafman, T. Kenny, B. Stipe, D. Rugar, Quality factors in micron- and submicron-thick cantilevers. J. Microelectromech. Syst. **9**(1), 117–125 (2000). https://doi.org/10.1109/84.825786
43. R. Lifshitz, M.L. Roukes, Thermoelastic damping in micro- and nanomechanical systems. Phys. Rev. B **61**(8), 5600 (2000). https://doi.org/10.1103/PhysRevB.61.5600
44. Y. Jimbo, K. Itao, Energy loss of a cantilever vibrator. J. Horol. Inst. Jpn. **47**, 1–15 (1968)
45. M. Imboden, P. Mohanty, A. Gaidarzhy, J. Rankin, B.W. Sheldon, Scaling of dissipation in megahertz-range micromechanical diamond oscillators. Appl. Phys. Lett. **90**, 173502 (2007). https://doi.org/10.1063/1.2732163
46. D.M. Photiadis, J.A. Judge, Attachment losses of high oscillators. Appl. Phys. Lett. **85**(3), 482–485 (2004). https://doi.org/10.1063/1.1773928
47. J. Yang, T. Ono, M. Esashi, Energy dissipation in submicrometer thick single-crystal silicon cantilevers. J. Microelectromech. Syst. **11**(6), 775–783 (2002). https://doi.org/10.1109/JMEMS.2002.805208
48. V.P. Adiga, Mechanical stiffness and diffipation in ultrananocrystalline diamond films. Publicly Accessible Penn Dissertations. p. 413 (2010)
49. M. Liao, M. Toda, L.W. Sang, T. Teraji, M. Imura, Y. Koide, Improvement of the quality factor of single crystal diamond mechanical resonators. Jpn. J. Appl. Phys. **56**(2), 024101 (2017). https://doi.org/10.7567/JJAP.56.024101
50. H. Wu, L.W. Sang, T. Teraji, T. Li, K. Wu, M. Imura, J. You, Y. Koide, M. Liao, Reducing energy dissipation and surface effect of diamond nanoelectromechanical resonators by annealing in oxygen ambient. Carbon **124**, 181–187 (2017). https://doi.org/10.1016/j.carbon.2017.08.069
51. J. Yang, T. Ono, M. Esashi, Investigating surface stress: Surface loss in ultrathin single-crystal silicon cantilevers. J. Vac. Sci. Technol. B **19**(2), 551–556 (2001). https://doi.org/10.1116/1.1347040
52. M. Li, H.X. Tang, M.L. Roukes, Ultra-sensitive NEMS-based cantilevers for sensing, scanned probe and very high-frequency applications. Nat. Nanotech. **2**(2), 114–120 (2007). https://doi.org/10.1038/nnano.2006.208
53. S. Ozdemir, S. Akhtar, O.E. Gunal, M.E. Khater, R. Saritas, E.M. Abdel-Rahman. M. Yavuz, Measuring the quality factor in MEMS devices. Micromachines **6**(12), 1935–1945 (2015). https://doi.org/10.3390/mi6121466

Chapter 5
Nitrogen Incorporated (Ultra) Nanocrystalline Diamond Films for Field Electron Emission Applications

Kamatchi Jothiramalingam Sankaran and Ken Haenen

Abstract Diamond is eminent to own a series of outstanding physical and chemical properties, thus rendering it to be a strong cold cathode material for field electron emission (FEE) applications. FEE from diamond materials comprises the supply of electrons to the conduction band of the materials, transport through bulk, and lastly emission at the surface. In this chapter, the enhancement in the FEE characteristics of nitrogen incorporated (ultra)nanocrystalline diamond ((U)NCD) films and their related nanostructures is discussed. Ion implantation, in situ impurity doping in plasma, and post plasma treatment processes have been employed to incorporate nitrogen in (U)NCD. The possible mechanism through which the change in microstructure enhances the FEE characteristics of nitrogen incorporated (U)NCD films/nanostructures is discussed.

5.1 Introduction

Field emission electron sources are vital building blocks in an array of appliances, including electron beam induced light sources, flat panel displays, microwave amplifiers and travelling wave tubes [1–5]. Field electron emission (FEE) is a unique quantum-mechanical effect of electrons tunneling from a solid surface into vacuum. This phenomenon is characterized by getting electrons to overcome the natural energy barrier existing at the surface and emit out. Fowler-Nordheim (F-N) theory is used to explain the FEE behavior of materials [6]. According to F-N theory,

$$J = AE^2 \exp[(-B\varphi^{3/2})/E] \tag{5.1}$$

K. J. Sankaran
Institute for Materials Research (IMO), Hasselt University, 3590 Diepenbeek, Belgium
e-mail: sankaran.kamatchi@uhasselt.be

K. Haenen (✉)
IMOMEC, IMEC vzw, 3590 Diepenbeek, Belgium
e-mail: ken.haenen@uhasselt.be

© Springer Nature Switzerland AG 2019
N. Yang (ed.), *Novel Aspects of Diamond*, Topics in Applied Physics 121,
https://doi.org/10.1007/978-3-030-12469-4_5

where J is current density, E is applied field, A and B are constants, and φ is the work function of the emitting material. Field emission based cold cathode devices are characterized by their instantaneous switching on properties, capability to operate at high frequency/high current densities, reduced weight and operation without any outside heating element for the emission process different to their thermionic based electron sources. Nanostructured materials are further used to enrich the performance of such FEE based cold cathode devices. Over the last few years, the design, realization, and applications of such nanostructured material based cold cathode devices have been the object of remarkable attention by the research communities.

Among the most systematically examined nanostructured materials for cold cathode applications, carbon nanotubes (CNTs) [7–9], graphene [10] and graphdiyne [13, 14] have a prominent place, owing to their superior FEE properties. But these nanocarbon materials face the challenge of insufficient lifetime stability during field emission [7–14]. Contrary to sp^2-bonded carbon based FEE materials, diamond with a strong covalently sp^3-bonded crystal structure is a potential candidate as a field electron source with high lifetime and reliability. Nevertheless because of the wide bandgap nature of diamond, the electrical conductivity in bulk diamond is low, which decreases its electron emission. Consequently, the high potential of diamond as a cold cathode field emitting material necessitates the material to be conductive.

Ultrananocrystalline diamond (UNCD) is a special form of diamond scientifically examined for its application as field electron emitters owing to its negative electron affinity (NEA) and low effective work function [15]. The term "UNCD" has been adopted [16] to differentiate from what is commonly termed nanocrystalline diamond (NCD) in the literature with 50–100 nm grain size [17]. In a microwave plasma enhanced chemical vapor deposition (MPECVD) process, when H is reinstated with Ar, which results in hydrogen deficient (CH$_4$/Ar) plasmas, leads to the growth of UNCD films [18]. The grains of UNCD films have an sp^3 character on a 2–5 nm scale and a large proportion of grain boundaries with a mixture of sp^2, hydrocarbon and amorphous carbon (a-C) [17]. FEE properties of UNCD films were measured with an electrometer using a parallel plate cathode–anode setup [18]. The turn-on field (E_0) is designated as the interception of the straight lines extrapolated from the low field and high field segments of the F-N plots. Because of the low E_0 and large J values from UNCD, as compared to other forms of diamond (microcrystalline diamond (MCD) or NCD), this material shows tremendous potential for applications such as cold cathode emitters and other vacuum microelectronic devices. The ultra-nano grain size and the sp^2 rich grain boundaries are believed to be responsible for the superior FEE properties of UNCD films.

Many groups proposed the possibilities of conduction channel and/or grain-boundary conduction mechanism for the enhanced conductivity/FEE properties for UNCD films [19–21]. The conductivity/FEE properties are believed to correlate with the number density of grain boundaries comprising non-diamond sp^2 phases [22]. To elucidate this, Cui et al. [23] described a mechanism that the threshold for FEE is lowered due to a local reduction of the electron affinity at the nanostructured diamond surface surrounded by the sp^2-bonded carbon grain boundaries. They conveyed that electron emission takes place from the grain boundaries, but the emission barrier is controlled by the surrounding diamond surface. Ilie et al. [24] and Robertson [25] also proposed a similar model for the improved FEE from other forms of carbon, containing flat, smooth diamond-like carbon. Hence, the non-diamond carbon phases such as a-C and $trans$-poly-acetylene (t-PA) that occur in the grain boundaries of UNCD films act as conducting channels, assisting easy tunneling of electrons and hence enhance the FEE characteristics of UNCD films through a 'grain boundary conduction emission' mechanism [22].

Whereas the physical properties of UNCD films depend on the intrinsic structure of the materials, their electrical and optical properties are more closely related to the microstructure of the films [26–28]. So, the ability to control the microstructure and surface morphology of UNCD films could tailor this material for field electron emitters. A decrease in diamond grain size increases the proportion of grain boundaries, but the low conductive capacity of the a-C and t-PA phases at the grain boundaries limits the FEE properties attainable for UNCD films [18, 29]. Therefore, for the development of UNCD based field emitters there is a necessity to control the microstructure. Particularly the modification of non-diamond carbon phases at the grain boundaries is necessary to enhance the electrical conductivity of the materials. Previous studies disclosed that doping by foreign dopants, variation of chemical bonding structure, interlayer modifications, surface hydrogenation, metallic film coating/annealing, and synthesizing hybrid-nanostructured diamond play an important role in improving the conductivity/FEE properties for diamond and related materials [30–40]. Sulphur-doped nanocrystalline diamond (NCD) films were synthesized by Baranauskas et al. and achieved a lowest threshold value for FEE of 13.20 V/μm [30]. Yamada et al. reported a low threshold field of 16.0 V/μm from reconstructed (annealing diamond at 950 °C) phosphorus-doped diamond surfaces [31]. Koinkar et al. [32] reported the synthesis of boron doped NCD films and observed that the field emission properties of NCD were improved upon increasing boron concentration. Metallic species such as Li, Cs, Ca, Sr, Ba, Cu, Ag, Pt and Au have also been incorporated into the diamond matrix by ion implantation to lower the E_0 value for electron emission [33–37]. Sankaran et al. reported high conductivity of 185 (S/cm) and superior FEE properties, viz. low E_0 of 4.88 V/μm with high J of 6.52 mA/cm^2 in UNCD films due to gold ion implantation [36]. Enhanced FEE characteristics of low E_0 of 3.2 V/μm and J of 2.086 mA/cm^2 are observed in NCD films upon irradiation with 100 meV Ag^{9+} ions [37]. The formation of interconnected sp^2 clusters in the NCD films assists in the enhancement of FEE properties of Ag ion irradiated NCD films. Moreover, Sankaran et al. observed enhanced FEE properties viz. a low E_0 of 5.3 V/μm and J of 3.6 mA/cm^2 at an applied field of 11.7 V/μm, of UNCD films due to hydrogen gas-treatment at 600 °C [38]. H_2 gas treatment induces graphitic phases along the grain boundaries throughout the thickness of the UNCD films, resulting in creation of conduction channels for the electrons to transport from the bottom of the films to the top and hence the superior FEE properties. Utilizing a Au interlayer between UNCD films and Si, the FEE properties of UNCD films are improved by attaining a E_0 of 8.2 V/μm and J of 4.5 mA/cm^2, which are superior to those of UNCD films grown directly on Si [39, 40]. The diffusion of Au in the interface layer results in the introduction of nanographitic phases in the interface, lowering the resistivity of the interfacial layer. The electrons can therefore be transferred effortlessly from Si substrates across the interfacial layer to the diamond and can consequently be field emitted. From all these observations, the sp^2 bonded nanographitic phase at the grain boundaries is the key factor on improving the FEE characteristics of diamond films. This chapter investigates several methods of nitrogen incorporation in NCD/UNCD films for the induction of nanographitic phases at the grain boundaries for improving the FEE characteristics of NCD/UNCD films.

5.2 Nitrogen in Diamond

Nitrogen is one of the most common impurities encountered both in natural and synthetic diamond and has a deep influence on diamond's physical/chemical properties [41–44]. It matches the crystal lattice of diamond well because of its similar atomic radius. It is present in primarily single-substitutional form with a relatively deep donor level located 1.7 eV below the conduction band minimum [44–46]. Distortion by bigger dopants is usually a problem with diamond due to its tight lattice and it can lead to graphitization. Another important advantage of nitrogen is its easy handling during the doping and deposition processes. As it is gaseous at room temperature and then also at typical fabrication temperatures, it can merely be added to the gas phase. A small amount of nitrogen prefers a (100) growth texture [47, 48], which has higher wear resistance, lower roughness, and higher heat conductivity [49–51] than other crystallographic directions. Also, a limited incorporation of nitrogen appears to encourage the deposition rate [52, 53]. Pertaining the disadvantages, nitrogen decreases the thermal conductivity and the optical transparency [54, 55]. In addition, nitrogen has a high solubility in diamond and is found to be either concentrated in small precipitates (type Ia diamond) or dispersed on substitutional atomic sites (type Ib diamond) [56]. Addition of nitrogen gives diamond a yellow tarnish and increases the defect content, ensuing in a negative influence on its physical properties [57].

Several groups have attempted to incorporate nitrogen into diamond and studied the FEE properties [58–60]. Nitrogen incorporated diamond films are obtained mainly by two approaches. The first approach is N ion implantation after deposition of diamond films, which is commonly used to modify the film surface, like inducing defects or a-C in the films [61]. The second one is to introduce nitrogen in the form of N_2, urea, NH_3, N_2O, $N(CH_3)_3$, $C_3H_6N_6$ into the chamber during the deposition process, which can achieve nitrogen atom doping in diamond films [62–65]. Okano et al. used urea as a dopant and achieved low-threshold emission from heavily nitrogen doped polycrystalline diamond films [64]. Triethylamine dissolved in the methanol was used as a liquid nitrogen source to synthesize nitrogen-doped UNCD films at 760 °C and achieved superior FEE properties of low E_0 of 3.4 V/μm and large J of 8.0 mA/cm^2 [65]. Recently, Okumura et al. also fabricated n-type semiconducting diamond in an acetlyene flame with addition of nitrogen [66].

Conversely, the mechanism based on the incorporation of nitrogen in diamond is still an open question. How nitrogen responds at the diamond surface and why the presence of nitrogen in the gas phase increases the growth rate and modifies the surface morphology are still not completely understood. This is mainly due to the complexity of the N-incorporation processes involved, depending not only on experimental parameters (e.g. type and concentration of the N-containing gas, substrate temperature etc.) but also on the diamond crystal orientation and its surface termination. Several nitrogen-containing species have been offered as contributors in gas–surface reactions and CN radicals are the major gas species contributing to diamond growth [67–69]. The CN adsorption on a diamond {111} surface has been proposed as a way to nucleating new layer growth [70]. Larsson and co-workers [71, 72] theoretically explored how the pre-adsorbed NH_x species affect gas-surface reactions and the ways in which previously incorporated substitutional N atoms can affect the energetics, and thus the rates, of the elementary reactions involved in CH_x incorporation [73, 74]. Cao et al. [75] also described the possible contributions from a range of gas-phase NH_x and CNH_x species.

Tight binding molecular dynamics simulations showed that incorporation of nitrogen into the diamond grain boundary is favored by 3–5 eV relative to incorporation into the diamond grains [76]. Moreover, nitrogen increases the amount of threefold-coordinated carbon atoms in the grain boundary because the nitrogen substitution energy for the grain boundary is 2.6–5.6 eV lower than that for the bulk diamond. Nitrogen induces percolation paths in the grain boundary regions and there is an increase in the density of states at the Fermi level [76]. Hence, grain boundary conduction involving carbon π states in the grain boundary is responsible for the high electrical conductivities via possible hopping or thermally activated conduction mechanisms in the diamond films [77].

This chapter is aimed at looking into the effect of nitrogen content in the UNCD film on its FEE properties. Different nitrogen incorporation methods like ion implantation, in situ doping in the plasma, and post plasma treatments have been employed to alter the microstructure of the UNCD films to achieve high conductivity and enhanced FEE properties of UNCD films. Diamond nanostructures fabricated from nitrogen doped UNCD films to further improve the FEE properties are also discussed. In addition, formation of heterostructures by a combination of nitrogen doped NCD/UNCD with other field emitters such as CNTs, graphene, ZnO and hBN, in which nitrogen doped NCD/UNCD films help to improve the FEE properties of other field emitters, is discussed. To shed light on the influence of superior FEE characteristics of nitrogen doped NCD/UNCD films/nanostructures for practical device applications, the plasma illumination (PI) performance of microplasma devices, which employed nitrogen doped diamond materials as cathode is conversed.

5.3 Nitrogen Ion Implanted UNCD Films

Ion implantation has been a common technique to engineer the structure and to incorporate dopant species at a specific depth, thereby improving the diverse properties of different materials. It can be used to tailor the sp^3/sp^2 ratio for diamond or related carbon materials by properly selecting the dose and energy of implantations [78, 79]. For the purpose of incorporating foreign ions into the diamond matrix by replacing the carbon, the energy of incident ions must be asserted at a few keV level. Furthermore, a thermal post-implantation annealing process is expected to eliminate or reduce defects and sometimes activate the dopants to have useful effects. The occurrence of severe damage during ion implantation in diamond materials can be minimized by performing the ion implantation at or above 300 °C, which can effectually anneal out irradiation-induced damages through dangling bond saturation. Such a method has been developed by Prins [80], and improved by Kalish et al. [81], to optimize the electrical properties of diamond by boron ion implantation. Much work has been carried out for microcrystalline diamond (MCD) and NCD films to modify their properties [79], which include the efforts to improve the FEE characteristics of diamond by ion implantation process [82]. The enhancement in FEE properties of diamond is due to ion implantation introduced defects and the removal of defects as well as enhanced emission properties with the post implantation annealing process [83, 84].

Diamond films were implanted by many ions for improving their FEE behavior but nitrogen ion implantation showed a significant modification in the microstructure that explains the enhancement in the FEE properties of diamond films. Li et al. [85] investigated the FEE properties of diamond films by implanting 10 keV nitrogen ions in the dose range from 1×10^{16} to 5×10^{17} ions/cm^2 at room temperature. Increasing the implantation dose could lower the threshold field of the emission of the diamond film from 18 to 4 V/µm and the effective work function was estimated to be in the range of 0.01–0.1 eV. The enhancement of FEE for nitrogen ion implanted diamond films was attributed to the increase of the sp^2 bonded carbon fraction and the formation of defect bands within the bulk diamond bandgap induced by nitrogen ion implantation, which could alter the work function of the diamond films.

Nitrogen ion implantation in UNCD can lead to desorption of hydrogen from the hydrocarbons at the grain boundaries. This results in depassivation of dangling bonds and a rise in the number of gap states, which ensues in increased conductivity and a reduction in bandgap [78, 86], all above factors lending to the enhanced field emission behavior. Joseph et al. made a systematical investigation on the enhancement of FEE properties of UNCD films by high dose N-ion implantation and annealing processes [87]. These studies showed that with increase of N ion implantation dose in the UNCD films, the implantation first expelled H$^-$, induced the formation of disordered carbon and a defect complex, and then brought about a transformation into a-C. Post implantation annealing healed the atomic defects, converting the disordered carbon to a stable defect complex and a-C into a stable graphitic phase. For films implanted at doses $<10^{14}$ ions/cm^2, the FEE properties reverted to the original values after the annealing process. A critical implantation dose of $>10^{15}$ ions/cm^2 followed by annealing is required for huge improvement of the FEE properties. High dose N-ion implantation ($>10^{15}$ ions/cm^2) followed by annealing at 600 °C improve the FEE properties with E_0 and J values of 7.0 V/μm and 5.5 mA/cm^2 (at an applied field of 20 V/μm), respectively. Such an improvement of FEE is attributed to the formation of nanographitic phases, which mediates the FEE process. At higher dosages N gets implanted into the grain boundaries replacing the C species, injecting an electron into the UNCD lattice, which facilitates the enhancement of FEE.

To locally locate the field emission sites from nitrogen ion implanted UNCD films, Panda et al. [88] used scanning tunneling spectroscopic (STS) measurements and showed substantial difference in current–voltage curves at the grain and grain boundary. For non-ion implanted UNCD, the electronic band structure estimation discloses that the bandgap is ~ 4.8 eV at the grain and ~ 3.8 eV at the grain boundary. The N-ion implantation decreases the bandgap for the grain to ~ 3.8 eV, while annealing further decreases the bandgap to ~ 3.2 eV. A density of states is introduced in the bandgap and the grain boundaries display metallic behavior with higher conductivity than the grains, then acting as the eminent electron emitters. The decrease in bandgap and the increase in density of states in the bandgap due to the sp^2 content and new bonds at the diamond grains, as well as an increase in conductivity at the thicker grain boundaries, lead to the enhanced FEE properties in the nitrogen ion implanted UNCD films.

Additionally, Joseph et al. [89, 90] compared the structural and FEE properties of N ion implantation with B- and C-ion implantations in UNCD films. The fluence of the ions to implant N-, B- and C-ions (10^{12} or 10^{15} ions cm^2) was chosen either lower or higher than a critical dose to create reversible or irreversible structural change. This was again only due to the indication that ion implantation over critical fluence is necessary for inducing enhanced FEE properties. The nature of the implant species seemed unimportant. Low-dose (10^{12} ions/cm^2) B ion implantation and annealing processes insignificantly changed the FEE properties, but high-dose (10^{15} ions/cm^2) ion implantation and subsequent annealing resulted in surface graphitization for UNCD films ensued in inferior FEE properties. The B ions react with carbon forming covalent bonds and are not transferring the charges to the UNCD grains. The high dose C-ion implantation degraded the FEE properties for the films as like B-ions did. While high dose N-ion implantation induced the amorphous phase, which converted to the graphitic phase after annealing, and thereby improving the FEE properties for the UNCD films. They proposed the possible mechanism that N-ions reside in the grain boundaries (i.e., inducing defects) and render UNCD grains semiconducting by a charge-transfer process. The B and C-ions do not have a similar effect as N-ions did. In addition, Sankaran et al. [91] studied the effect of high dose (10^{16} ions/cm^2) N-ion implantation with energies of 100 and 200 keV on improving the FEE properties of surface treated UNCD films and found that N-ion implantation on UNCD films showed a dominating role on enhancing the FEE properties in par with the surface treatment.

Fig. 5.1 Field emission data of UNCD films before and after the SENII and MENII processes carried out at room temperature [92]

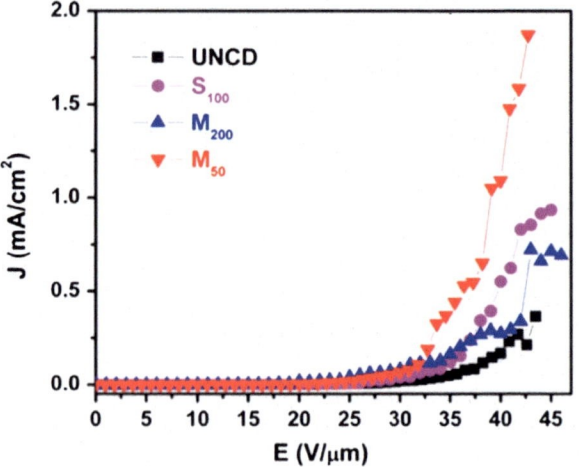

Following the previous work, Joseph et al. [92] displayed improvement of FEE of UNCD films owing to structural transformations after single- or multi-energy nitrogen ion implantation (S/MENII) processes (Fig. 5.1). Initially a dosage of 4×10^{14} ions/cm^2, just below the critical dose of 1×10^{15} ions/cm^2, is used, wherein the structure remains intact. The FEE properties are better for films undergoing 100 keV N-ions SENII process at 300 °C ($E_0 = 7.1$ V/μm) than the pristine UNCD film ($E_0 = 13.9$ V/μm) (Fig. 5.2). On the other hand, the MENII processes at 300 °C using the 50–100–150–200 keV N-ions enhance the FEE significantly with $E_0 = 4.5$ V/μm and $J = 2.0$ mA/cm^2 at 24.5 V/μm. Notably, in 50–100–150–200 keV MENII process the films were first ion implanted with 50 keV N-ions, followed by 100, 150, and 200 keV N-ion implantation consecutively.

Fig. 5.2 Field emission data of UNCD films before and after the SENII and MENII processes carried out at 300 °C [92]

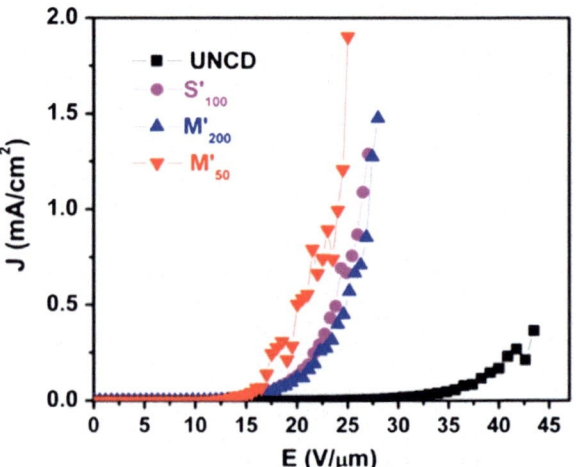

Taking one step further, Joseph et al. [93] fabricated UNCD pyramidal tips followed by 100 keV N ion (10^{15} ions/cm^2) implantation on the tips and by using transmission electron microscopy (TEM) revealed the evidence of complex nanostructural changes due to ion implantation, such as, graphitic or amorphous type transition, volume expansion and nanostair type surface formation on the pyramidal tips. The nanostructural modifications conceivably enhanced the grain boundary incorporation of N and the electron transport through a transfer doping mechanism, which in turn improved the FEE properties of the ion implanted pyramid tips, viz. the E_0 value is reduced from 3.6 V/μm (for unimplanted tips) to 2.8 V/μm and the J value is increased from 0.1 (unimplanted tips) to 5.4 mA/cm^2 at an applied field of 12.5 V/μm. The leading factors for bettering the FEE properties of N-ion implanted UNCD films are supposed to be the grain boundary incorporation of N, activation of the implanted N by the annealing effect and the healing of induced defects, which are clarified based on a surface charge transfer doping mechanism [92].

Electron energy loss spectroscopy along with TEM studies are employed by Joji et al. [94] to comprehend the phase transformation induced by N-ion implantation (1×10^{16} ions/cm^2) followed by an annealing process. They suggested a model for structural modifications in the N-ion implanted and annealed films. Pristine UNCD films contain chiefly the diamond phase with a small proportion of a-C or graphitic phases. In the N-ion implanted UNCD films, the layer where the implantation occurred saw a conversion into an amorphous like structure while where the N-ions do not reach a minor increase in the a-C content was perceived, possibly owing to coalescence of nano-sized diamond grains. The annealing after implantation healed defects in the diamond lattice, transforming the a-C phase into a nanographite phase. The existence of such a nanographite phase improved the conductivity and FEE properties of the UNCD films.

5.4 In situ Doping Using N₂ in a CH₄/H₂ Plasma

While the N-ion implantation/annealing process was revealed to be able to effectively enhance the FEE properties of diamond films, the elucidation about the mechanism is not unambiguous, as N-ions possibly replace C atoms, ensuing in a donor-doping effect. Moreover, the ion implantation process in diamond includes many steps and is also time consuming. Hence, addition of N_2 in the diamond growth plasma would be a better solution for modifying the microstructure of diamond films.

As effects of N_2 on the film properties are so large in comparison to its fraction in the feed gas, it is indispensable to have a perfect control over the process parameters, e.g. the vacuum leakage of the deposition system, as nitrogen is the major constituent of air. The presence of nitrogen in the reactant gases can seriously alter the morphology of deposited diamond films which restricts their usefulness for some applications, especially those that require high-quality electronic-grade diamond [95]. On the other hand, nitrogen-doped diamond films have very useful semiconducting properties, which are particularly important for microelectronics and field emission display applications [96–98]. These discoveries have attracted widespread research interest, and prompted many researchers to investigate the effects of nitrogen on the growth of diamond films using CH_4/H_2 and CH_4/Ar gas mixtures [99–104].

Small amounts of nitrogen in the process gas have also been found to have a tremendous effect on the nucleation density, the average diamond grain size and the growth rate in diamond films [105] and induces drastic changes in the optical, thermal, electrical and morphological properties of the diamond films [42, 48, 52, 57, 75, 100, 104, 106, 107]. Locher et al. [48] have studied the effect of very low concentrations of nitrogen in a CH_4/H_2 plasma. For a low methane concentration of 0.5% they do not state changes in morphology and texture of diamond films. For higher methane concentrations (1–2%), films prepared with 10–100 ppm of N_2 exhibit a (100) texture and a high diamond (sp^3) phase purity. Vandevelde et al. studied the influence of minute nitrogen addition in a CH_4/H_2 plasma on the morphology of the diamond film and on the optical emission intensities of the various emitting species in the bulk plasma [108]. They also displayed a link between the intensity of the emitting species, their concentration in the plasma, and the texture and preferred orientation of the synthesized diamond films [109]. Jin and Moustakas [42] encountered that the growth habit and deposition rate using $CH_4/H_2/N_2$ mixtures depended strongly on the ratio of carbon-to-nitrogen in the feed gas. They noticed that the overall growth rate was little affected by low N_2 addition (up to 1%). But at high N_2 concentrations an improved film quality, larger growth rates and a change in the surface morphology from (111) to predominantly (100) facets were observed. Recently, Krecmarová et al. [110] reported the variation of crystalline structure and morphology of diamond layers in H_2 rich plasmas with a variable addition of N_2 in the gas phase. The addition of nitrogen, first, increases the deposition rate to a maximum at $N_2 = 2.6\%$ and then decreases it, due to enhanced etching. A tiny addition of N_2 dramatically changes the surface morphology from well facetted diamond crystals to UNCD-like morphology. Ma et al. [111] synthesized NCD films in a gas mixture of $CH_4/H_2/N_2$ with the nitrogen concentration varying in a wide range from 0 to 75%. The introduction of nitrogen caused a significant change on the morphology of NCD films. It was found that the sp^2/sp^3 ratio of carbon bonds increased, and the grain size decreased with the addition of nitrogen in the plasma. The nitrogen incorporation reduces dramatically the resistivity by six orders of magnitude from 10^{11} to 10^5 Ω-cm, which is associated with the increased sp^2/sp^3 ratio and graphitic phase in the NCD films.

The addition of N_2 in the CH_4/H_2 plasma modifies the microstructure of the diamond films markedly, which, in turn, alters the electrical and FEE properties of the films. Wu et al. [112] observed an enhanced FEE behavior of $E_0 = 1$ V/μm and J of 10 mA/cm^2 for the NCD films synthesized using $CH_4/H_2/N_2$ gas mixtures and reported that a conducting–tunneling mechanism occurs, in which graphite phases play an important role in the field emitting process. Enhanced FEE properties of a low threshold field of 1.6 V/μm and a high emission current of 19 mA at 6 V/μm

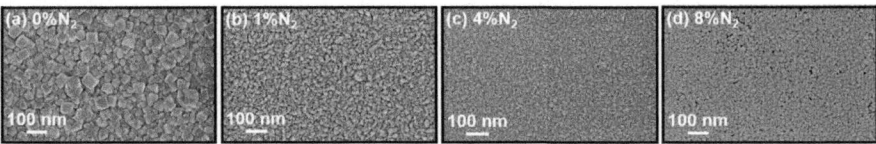

Fig. 5.3 SEM micrographs of diamond films grown using different N_2 concentrations in CH_4/H_2 plasmas **a** 0% N_2, **b** 1% N_2, **c** 4% N_2, and **d** 8% N_2 [114]

are observed from micropatterned pyramidal diamond tips fabricated with $CH_4/H_2/N_2$-deposited NCD by Subramanian et al. [113]. The improved FEE behavior from the NCD tip array is attributed to its better geometrical enhancement factor ($\sim 32,000$), increased sp^2-carbon content, and higher electrical conductivity by incorporation of the nitrogen impurity in the diamond.

Recently, Sankaran et al. [114] investigated the influence of N_2 concentration in $CH_4/H_2/N_2$ plasmas on microstructural evolution and FEE properties of diamond films and found that a systematic change in the ratio of the reactant gases is the way to control surface morphology, microstructure, and bonding structure of diamond films. The optical emission spectroscopy (OES) studies support the idea that CN species are the main criterion to judge the modifications of microstructure with the N_2 concentration. While the diamond films grown in CH_4/H_2 plasmas comprise large diamond grains (Fig. 5.3a), for the diamond films grown using $CH_4/H_2/(4\%)$ N_2 plasmas, the microstructure significantly changed due to the presence of a large proportion of CN (and C_2) species in the plasma, ensuing in ultra-nanosized diamond grains (Fig. 5.3b), along with the formation of so-called *n*-diamond and *i*-carbon clusters [114]. In addition, these films contain wide grain boundaries enclosing *a*-C phases. Further raising the N_2 concentration from 6 to 8 and 10%, the size of the diamond grains increased, accompanied by the presence of porosity in the films (Fig. 5.3c, d). The grains are shown to agglomerate, forming large elongated aggregates of a few hundred nanometers in size for diamond films grown using a $CH_4/H_2/(8\%)N_2$ plasma. Moreover, the films grown using a $CH_4/H_2/(4\%)$ N_2 plasma show a low resistivity of 74 Ω-cm with a low E_0 value of 14.3 V/μm and a high J value of 0.37 mA/cm^2 at an applied field of 25.4 V/μm [114]. It seems that the grain boundaries can provide electron conduction networks to transport expeditiously the electrons to emission sites for field emission, as long as they have adequate thickness.

5.5 Doping Using N_2 in a CH_4/Ar Plasma

Incorporating N_2 into conventional CH_4/H_2 plasmas was not efficacious since N forms a deep donor level [115] and does not enhance the electrical conductivity and FEE of diamond films [114]. In contrast, the conductivity of UNCD films grown using CH_4/Ar plasmas has been improved by incorporating N into grain boundaries of diamond films, so as to increase the electrical conductivity that enhances the FEE properties of the films [115, 116]. Preferred changes in the morphology of the films and, in turn, other properties of the films can be accomplished by changing the composition or changing the film deposition parameters [77, 116–119]. A few earlier reports mention the effect of substrate temperature on the properties of N_2 doped diamond films [57, 120–124]. Bhattacharyya et al. [125] mentioned that the electrical conductivity of the nitrogen-doped UNCD (N-UNCD) films increased by five orders of magnitude up to 143 S/cm with increasing the nitrogen content to about 20% in a $CH_4/Ar/N_2$ plasma using a high substrate temperature of 800 °C. Hall effect measurements were carried out by Williams et al. [126] and n-type conductivity with mobility values of 1.5 cm^2 V^{-1} s^{-1} and a sheet carrier concentration of 2×10^{17} cm^{-2} were observed for the UNCD films grown with 10% nitrogen added into a $CH_4/Ar/N_2$ plasma. By low-temperature conductivity measurements, the mechanism of high conductivity and the change of the electronic structure by nitrogen incorporation is investigated by Bhattacharyya [77]. The 0.5% N_2 (low N concentration) in a $CH_4/Ar/N_2$ plasma grown UNCD shows predominantly Arrhenius-type behavior with a different change in activation energy at different temperatures. A Coulomb gap was ascertained for the 5% N_2 grown UNCD sample. Moreover, the sample obeys Pollok's model in the low temperature range (85–15 K) and Mott's model in the range 300–160 K. For higher N_2 % grown UNCD, conductivity can be conveyed as a combination of delocalized state conduction and hopping conduction where a shift of Fermi level towards the conduction band has been observed. A model has also been discussed by Corrigan et al. [15] and Birrell et al. [117] that the nitrogen favorably enters the grain boundaries and encourages sp^2 bonding in the neighboring carbon atoms, therefore increases the electrical conductivity.

For the purpose of improving the FEE properties of N-UNCD films, Chen et al. made efforts to grow films at high substrate temperature using 10% N_2 in Ar/CH$_4$ plasmas [118]. The FEE properties of the films were also noticeably improved, that is, the E_0 value decreased from 20 to 10 V/μm and the J value increased from less than 0.05 to 15 mA/cm^2 when the substrate temperature increased from 600 to 830 °C. Furthermore, Shalini et al. [127] made a systematic investigation on the microstructural evolution of N-UNCD films due to N_2 incorporation in Ar/CH$_4$ plasma grown at constant substrate temperature of 800 °C. The N-UNCD films grown in Ar/CH$_4$ plasmas without N_2 show spherical grains. In contrast, the films grown in $CH_4/Ar/N_2$ plasmas contain wire-like grains. The size of the wire-like grains is small when grown in 3.4% N_2. The grain size increased monotonously with the content of N_2 in the plasma, reaching a size as large as 20 nm $\times$ 400 nm for those synthesized in a 10% N_2 plasma. The morphology of N-UNCD grains changed from a wire-like to plate-like shape when the N_2 content in the plasma further increased to 25%. The electrical conductivity and the fraction of sp^2 bonding in the grain boundaries increased proportionally with the amount of nitrogen incorporation. The N-UNCD films grown using 10% N_2 in CH$_4$/Ar plasmas were conductive and exhibited a markedly lower E_0 value of 13.0 V/μm with larger J value of 0.28 mA/cm^2 [128]. TEM investigations revealed that the main factor resulting in high conductivity and better FEE properties for these samples was the formation of a nanographitic phase that encased the wire-like diamond grains. The formation of wire-like diamond grains was seemingly due to the existence of CN species that adhered to the subsisting nanodiamond clusters, which inhibited radial growth of the nanodiamond crystals, encouraging anisotropic growth and the formation of wire-like nanodiamond. Arenal et al. also showed the transformation from randomly oriented 3–5 nm diamond crystallites to diamond nanowires surrounded by a largely sp^2-bonded carbon sheath when N-UNCD films were grown using 10% N_2 in a CH_4/Ar plasma at 800 °C [129]. The beneficial effect of the incorporated nitrogen on the FEE characteristics of the N-UNCD films was credited to the formation of additional electron gap states in the grain boundaries of the diamond [116] or formation of a diamond/graphite mixed phase [112]. Frolov et al. [130] also described more detailed measurements using charge-based deep level transient spectroscopy and complex scanning probe microscopy to validate that the shallow donor center is chiefly introduced by nitrogen incorporation into grain boundaries, which is then proposed as the potential mechanism of low-field emission from N-UNCD films. Ikeda and Teii [131] re-examined the FEE behavior of N-UNCD films, using FEE and ultraviolet photoelectron spectroscopy, demonstrating the delocalization of carriers, linking this phenomenon to a higher probability of electron injection into upper defect levels during the transport process. Consequently, the possible mechanism is supposed to be that the nitrogen incorporated in UNCD films are residing at grain boundary regions, converting sp^3-bonded carbons into sp^2-bonded ones. The nitrogen ions inject electrons into the grain boundary carbons, thereby increasing the electrical conductivity of the grain boundary regions, which in its turn improves the efficiency for electron transport from the substrate to the emission sites, i.e. the diamond grains. It is also to be noted here that, without heating the substrate Liu et al. observed poor FEE properties for the films grown using 40% N_2 in CH_4/Ar plasmas [132]. Hence substrate temperature is the major key factor for the modification of microstructure that influences the electrical conductivity and the FEE properties of UNCD films.

5.6 Diamond Nanowire Films from a CH_4/N_2 Plasma

5.6.1 Substrate Temperature Effect

To further improve the electrical conductivity and the FEE characteristics, efforts were also made to grow highly conducting N-UNCD films using CH_4/N_2 plasmas without any Ar or H_2 addition. Sankaran et al. systematically investigated the effect of substrate temperature on the microstructural evolution of N-UNCD films deposited in such CH_4/N_2 plasmas and explained the grain growth mechanism of nanowire morphology of diamond grains [133]. The morphology of the films grown at 550 °C shows clustered grains with random/spherical like shapes. With an

Fig. 5.4 SEM micrograph of
the N-UNCD films, which are
grown in a CH$_4$/N$_2$ plasma by
a MW PE CVD process at
700 °C [133]

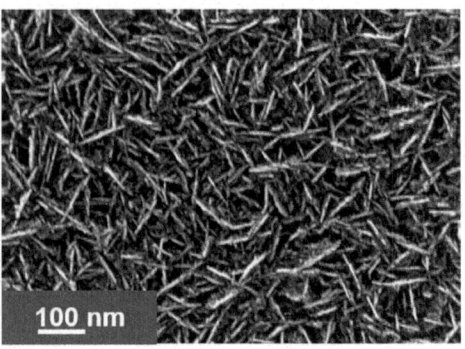

increase of substrate temperature to 600 °C an evident change in the grain structure
from random/spherical shape to acicular shaped grains is observed. On further
increase of substrate temperature to 700 °C, the grains evidently gain a discrete
wire-like structure (Fig. 5.4). For substrate temperatures ≥750 °C though, the
acicular grains grow larger in lateral dimension, while the boundaries describing
each grain get fainter. The substrate temperature dependent morphological changes
of the films reflect in the FEE and conductivity behavior of the films. It is noticed
that the N-UNCD films grown at 700 °C display the best FEE behavior i.e., lowest
E_0 of 6.13 V/μm, the highest J of 3.36 mA/cm^2 at an applied field of 8.8 V/μm and
the lowest effective work function value of 0.0124. The electrical conductivity of
N-UNCD films assessed by van der Pauw configuration first increases monotonously with increasing substrate temperature, from 1.2 S/cm for N-UNCD films
grown at 550 °C to about 186 S/cm for N-UNCD films grown at 700 °C. The
conductivity returns to lower values for films grown at substrate temperature
>700 °C, demonstrating a similar behavior to that of the FEE results.

To understand the mechanism that alters the electrical properties of the films
deposited at different substrate temperatures, the microstructural evolution of the
films is studied using TEM. Figure 5.5 depicts a TEM micrograph of a N-UNCD

Fig. 5.5 TEM micrograph of
a typical N-UNCD film
grown in a CH$_4$/N$_2$ plasma by
a MW PE CVD process at
700 °C [133]

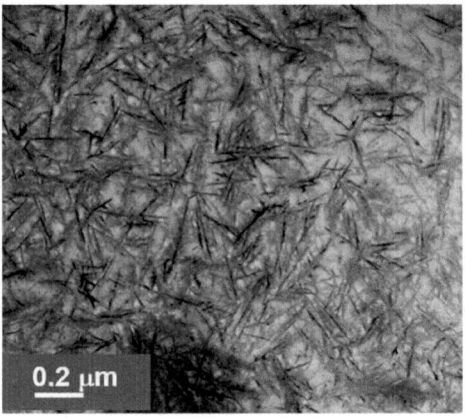

Fig. 5.6 High magnification
TEM inverse Fourier
transform image of a typical
N-UNCD film grown in a
CH_4/N_2 plasma by a MW PE
CVD process at 700 °C [133]

film grown at 700 °C with clear confirmation of wire-like acicular structure formation. The branching features reveal the anisotropic grain growth forming the
unique granular structure of N-UNCD films growth at 700 °C. Figure 5.6 exhibits
the merged high resolution dark field TEM image of the films displaying evidently
the growth of diamond into wire-like structures (yellow feature) with graphite
domains surrounding it (green feature). On the other hand, the acicular grains of the
N-UNCD films grown at 800 °C are much shorter in length and wider in diameter,
ensuing in much smaller aspect ratio, as associated with those in the films grown at
700 °C [133]. Furthermore, the growth of acicular grains along the longitudinal
direction (parallel to the growth axis) is somehow blocked and the growth in
circumferential direction (perpendicular to the growth axis) of the grains is started at
a substrate temperature of 800 °C. In case of N-UNCD films grown at substrate
temperature of 550 °C, the acicular granular structure for the diamond grains cannot
be detected leading to the formation of clusters of different domains prevailed by
diamond and amorphous phases and/or graphitic phases which have just started
forming, respectively.

The prominent, characteristic temperature depicts growth of diamond nanowires
surrounded by graphite domains, probably yielding the films more conducting
[134]. The OES of the CH_4/N_2 plasmas were recorded (Fig. 5.7) to gain information of the constituents of the plasma bombarding the substrate and the leading
factor on the formation of acicular structures and the graphitic phase lending to the
conductivity. This possibly will also help in understanding the grain-growth evolution. The OES spectra for the N-UNCD films grown at different substrate temperatures remain the same. Apart from the peak by N_2 in the plasma at ~ 358 nm,
the CN violet system (~ 386, ~ 418 nm) and the C_2 Swan system
(~ 468, ~ 516 nm), are also noticed in the spectra showing that N_2, CN and C_2 are
the key components in the plasma, which could possibly be the factors associated to
the microstructural evolution of the diamond and graphitic phases [134].

Fig. 5.7 Optical emission spectra of CH$_4$/N$_2$ plasmas used to grow UNCD films by a MW PE CVD process at substrate temperatures 550–800 °C [133]

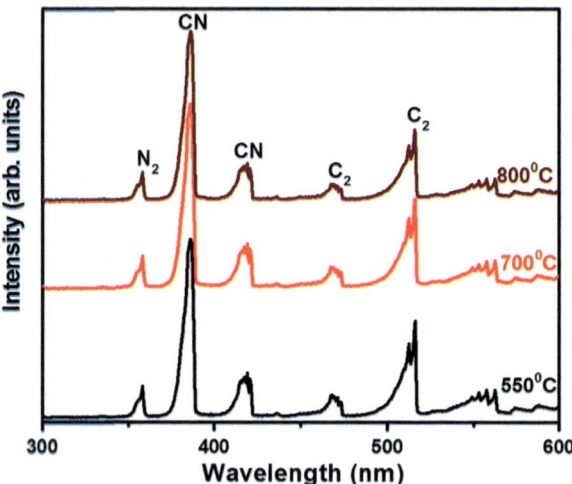

Models based on the growth mechanism of nano-carbon materials emphasize the role of catalytic nanoparticles, nucleation centers, substrate type, etc. [135–138]. Shang et al. elucidated that ultrathin diamond nanorods are grown by using catalysts of diamond nanoclusters that nucleate on graphene caps or nano onions and on the surfaces of which carbon radicals adsorb [135]. Sun et al. modeled the growth of diamond nanorods using carbon nanotubes: initially diamond crystallizes in the core of amorphous carbon clusters complied by faceting of the diamond nanocrystallites, after which diamond nanorods begin to grow at the tip of the facetted face [136]. Singh used the Ostwald ripening model to explain the growth of diamond grains [138]: carbon clusters, which are converted to the sp^3 diamond phase through an etching and recrystallization process, act as the nucleus for diamond growth. The deposited carbon diffuses inwards till a critical size is reached, beyond which secondary growth occurs. Nevertheless, none of these models obviously account for the growth of acicular diamond grains due to the increase in substrate temperature for the films grown in the presence of CN species in the plasma.

The variations in morphology of nitrogen doped diamond films has been conveyed earlier signifying that it is not the nitrogen incorporation in the diamond lattice but nitrogen related surface processes by the modifications in gas phase chemistry and surface kinetics which causes such changes [75, 129, 135, 139]. The N-UNCD films's morphology grown at different substrate temperatures infers that the nano-diamond grows anisotropically in the CH$_4$/N$_2$ plasma. The growth rates along the longitudinal and circumferential directions differ significantly for films grown with substrate temperature of 700 °C, leading to acicular grains,. Moreover, it has been mentioned that when growing N-UNCD films in CH$_4$/Ar plasmas, H and C$_2$ species prevail [134]. Earlier studies [140] perceive that for adequately high substrate temperature (700 °C), the granular structure changes from equiaxed for plasmas comprising a small proportion of N$_2$ (<5%, with small CN peaks in OES) to acicular one for plasmas containing a large proportion of N_2 (>10%, with large CN peaks in OES). These explanations, in combination with the above-described results, suggest strongly that the presence of CN species in the plasma is of critical prominence, besides a suitable substrate temperature, to form the acicular granular structure for the N-UNCD grains. From these observations, a model is proposed for the formation of the acicular structure of the grains. The role played by the substrate temperature and the constituents of the plasma, especially the CN species, in defining the grain morphology is emphasized. The grain growth process can be picturized as such: after the nucleation of nano-clusters (~ 5 nm in size), the initial growth of grains at 550 °C is by the Ostwald ripening process, which is a thermodynamically driven process of combining dissolutioned smaller crystals, to form bigger grains. For a substrate temperature >600 °C, particularly for 700 °C, the high in-plane temperature gives rise to directionally preferential growth of grains. This kind of growth may be similar to that explained by the van der Drift model [141], which highlights textured surface development due to directionality of the growth of grains along any particular direction, i.e., the anisotropic growth of grains.

Figure 5.8 shows a schematic diagram of the possible grain growth mechanism. Ions from the continuous plasma get deposited on the substrate. By the Ostwald ripening process, clusters of nano-diamond (~ 5 nm in size) are primarily formed on the substrate, with the C_2 dimers endorsing growth of diamond [142]. For a substrate temperature >600 °C the bigger grains begin to coalesce along any

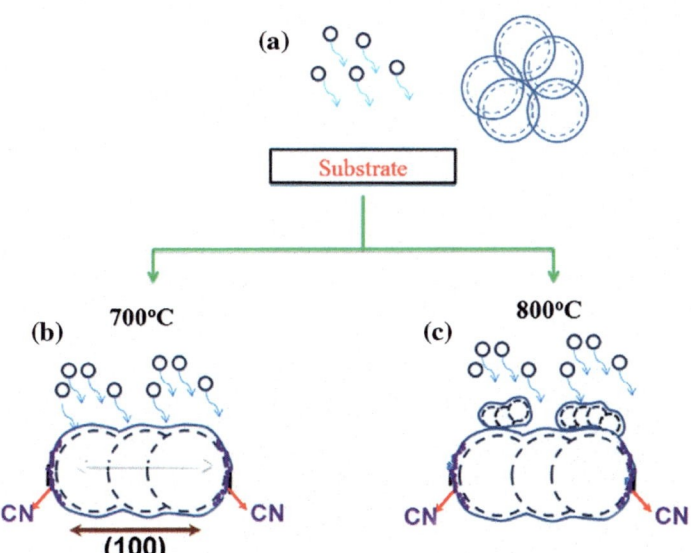

Fig. 5.8 Schematic diagram of grain-growth evolution: **a** formation of equiaxed grains, **b** formation of acicular grains at 700 °C and **c** the growth of nano-diamond clusters on the acicular grains at 800 °C [133]

preferred direction, ensuing in marginally elongated grains. The aspect ratio of the acicular grains increases with substrate temperature, attaining the largest value at 700 °C. Theoretical studies have shown that growth of elongated grains is persuaded by the preferential attachment of CN species on certain faces [e.g. (100)] of the nano-diamond clusters [143]. The adsorption of C_2 dimers is enhanced at the surface, which contains the adhered CN species, resulting in an anisotropic growth of the grains in that direction, e.g. [100], with the CN molecule staying on top of the growing surface, i.e. the (100) face, as this is energetically favorable [76, 143, 144]. For a substrate temperature of 800 °C, during the formation of the diamond nanowire, the surplus energy supplied by the substrate temperature leads to secondary nucleation in the circumferential direction of the nanowires. It is evidently demonstrated that the substrate temperature can play a significant role in the formation of the granular morphology. Moreover, it is observed that acicular grain growth has ceased in the films deposited at 800 °C, recognized by the "capping" feature at the tip of the nanowires. However for the films grown at 700 °C the grain growth ceased only due to termination of the deposition process. This indicates that upon continuation of film deposition the nanowires will elongate further to contribute even longer structures along the [100] direction [143]. A substrate temperature of 700 °C is thus the most promising condition for diamond nanowire growth in CH_4/N_2 plasmas, yielding high electrical conductivity and superior FEE characteristics of N-UNCD films.

5.6.2 Localized Electron Emission

Current imaging tunneling spectroscopy and STS measurements were executed on N-UNCD films in order to examine the main conducting paths for electrons from a microscopic viewpoint [145]. To scrutinize the microstructure on the surface of the diamond nanowire, the STM image in a highly contrast mode is presented in Fig. 5.9a for an easy recognition of diamond grains and graphitic grain boundaries. The N-UNCD film is made of smaller diamond clusters of equiaxed geometry. The cluster size changes between 2.0 and 5.0 nm which is reliable with that of the diamond grains, incorporated in the conventional UNCD films grown by CH_4/Ar

Fig. 5.9 a High contrast STM image to discern diamond grains (Grains) and grain boundaries (GBs) on N-UNCD films. **b** STS spectra of tunneling I–V curves at the marked position in (a) [145]

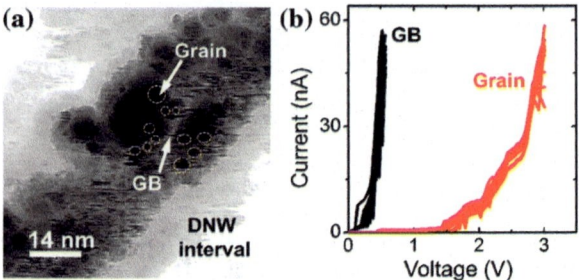

plasma [16]. Furthermore, Fig. 5.9b shows the STS spectra of tunneling I–V curves at the diamond grain (Grain) and graphitic grain boundary (GB) which are marked on Fig. 5.9a. Interestingly, the tunneling current at the graphitic GB position is much larger than that at the diamond Grain position. From such a microscopic inspection, this result proves again that the electron conducting path is along the graphitic GB rather than the diamond grain on the surface of N-UNCD films. Moreover, the GB regimes surrounding the isolated grains help to create conducting channels covering the whole surface of N-UNCD films. The size of GBs is roughly equal to the grain size of ∼3 nm and they are randomly dispersed and form three-dimensional conducting channels; similarly the tunneling I–V curve at GB establishes a very low threshold voltage lower than 0.5 V and a tunneling current higher than 50 nA. Such a special property of GB on the surface of N-UNCD films may provide enhanced FEE properties. May and co-workers [146] also showed grain boundary electron emission by using intentional high input bias to burn out the emission sites in diamond films. Ring-like craters detected in the damaged emission sites surrounding the diamond grains, displayed that the grain boundaries are the electron emission sites, damaged due to the passage of excess current.

5.6.3   Hydrogen Treatment Effect

To further increase the FEE properties of N-UNCD films, the N-UNCD films were treated with hydrogen plasma for different durations [147]. Due to a H_2 plasma treatment, the E_0 value decreases constantly and the best FEE properties are observed for N-UNCD films treated with 10 min of H_2 plasma with a lower E_0 value of 4.2 V/μm and a higher J value of 5.1 mA/cm^2 at an electric field of 8.5 V/μm. However, increasing the hydrogen treatment time further to 15 min surprisingly reduces the FEE properties, i.e., E_0 value increased to 10.1 V/μm with a sudden decrease in J value to 0.01 mA/cm^2 at an applied field of 8.5 V/μm. It is noticed that, the wire-like granular structure in N-UNCD films was transformed to smaller diamond nanowires with rod-like geometry, i.e., shorter in length and wider in diameter after hydrogen plasma treatment for 5 and 10 min (Fig. 5.10a, b). In

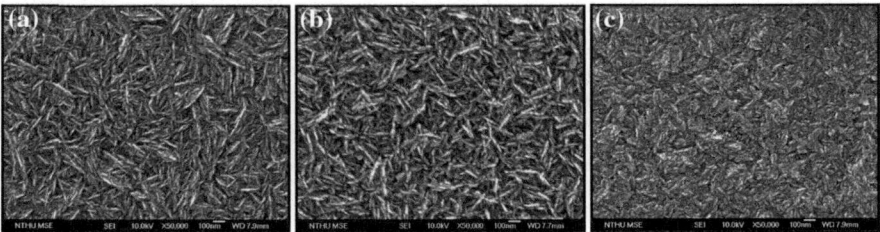

Fig. 5.10 SEM micrographs of hydrogen plasma-treated N-UNCD films for **a** 5, **b** 10, and **c** 15 min [147]

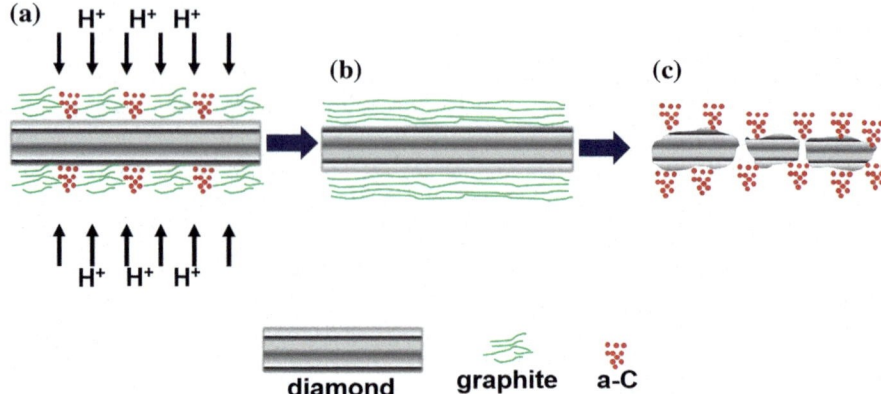

Fig. 5.11 Schematic diagrams showing the effect of hydrogen plasma treatment on modifying the microstructure of N-UNCD films **a** N-UNCD, **b** 10 min hydrogen plasma treated, and **c** the extended (>15 min) hydrogen plasma-treated N-UNCD films [147]

contrast, the wire-like granular structure for the diamond grains cannot be observed when the films were subjected to a 15 min hydrogen plasma treatment (Fig. 5.10c). The grains of N-UNCD films become random/spherical shaped due to the latter plasma treatment. Consequently, it can be decided that the graphitic phase around the N-UNCD films has been evidently destroyed after a 15 min hydrogen plasma treatment. These observations disclose that short time ($\sim$10 min) hydrogen plasma treatments increased significantly the graphitic content of the N-UNCD films, advancing the FEE properties, but extended hydrogen plasma treatment destroyed the graphitic content of these films, thus leading to a poor FEE behavior.

To comprehend the phenomenon why hydrogen plasma changes the characteristics of the graphitic sheath of the N-UNCD films, altering their FEE properties, a model is proposed (Fig. 5.11). In the sp^2-bonded phase encasing the wire-like diamond grains for the N-UNCD films, there are plentiful a-C phases, co-existing with the graphitic layers. When these diamond nanowires were exposed to the hydrogen plasma, the hydrogen terminating the a-C phases were abstracted by the atomic hydrogen in the plasma, leaving behind a reactive dangling bond that then immediately forms double bonds with other dangling bonds, ensuing in the formation of a graphitic phase that enriched significantly the electrical conductivity and FEE behavior of treated N-UNCD films. The increase in graphitic phase in diamond films after hydrogen beam irradiation is also observed by Kushita et al. [148]. Atomic hydrogen in a plasma is a very aggressive species capable of etching both the sp^2- and sp^3-bonded carbon. Hence, when the diamond nanowires were kept in a hydrogen plasma for a longer time, not only did the graphitic phases disappear but also the wire-like diamond started to break down into smaller pieces, explaining the degradation of the FEE properties for 15 min H$_2$ plasma treated N-UNCD films.

5.7 Bias-Enhanced Grown N-UNCD Films

As described in Sect. 5.6, N-UNCD films provide lower turn-on voltages, low resistivity, high mobility, high electrical conductivity and stable electron emission behavior compared to other conventional diamond films [145]. The N-UNCD films offer unique properties due to the nanocomposite microstructure of sp^2–sp^3 graphite carbon and diamond materials [133]. However, the growth process entails a relatively high substrate temperature (700 °C) to acquire such wire-like diamond grains, which is not very well compativle with Si-based device technology. Additionally, the conventional MW PE CVD process for growing diamond films on Si-substrates can result in non-optimal interfacial layers prior to the formation of diamond films, e.g. SiC. Such interfacial layer can be rather resistive, obstructing the transport of electrons from Si to the diamond film, thus restraining the FEE properties achievable [39]. The nucleation of diamond grains is successfully accomplished with the following methods: mechanical scratching of the substrate surface using diamond powder, ultrasonic agitation of the substrate surface in a solution containing nanodiamond powder, deposition of a carbide interlayer for assisting diamond nucleation, and bias enhanced nucleation and growth (BEN-BEG) process. Among the effective diamond seeding and nucleation techniques, the BEN-BEG method is more proficient in reducing the size of the diamond grains [149, 150]. A few earlier reports [150–154] observed that the application of bias voltage in the CH_4/H_2 plasma-based MW PE CVD process not only promoted the growth of diamond, but also proficiently reduced the size of the grains, resulting in diamond films with nanosized granular structure. Teng et al. [155] reported the enhanced FEE behavior of BEG NCD films using CH_4/H_2 plasma and attributed this phenomenon to the formation of nanographite phases. Moreover, the UNCD films were grown in CH_4/Ar plasmas by BEG process using different bias voltages. While both the positive and the negative biases enhanced growth processes encouraged high growth rate and transformed the grains from nano-size (100 nm) into ultranano size (~ 5–10 nm), only the application of large negative bias voltages can lead to the formation of nanographitic filaments, thereby resulting in the improvement of the FEE properties. The films grown with negative bias have a higher growth rate and a larger nucleation density than those of films grown with positive bias or without bias [156]. Hence, the application of negative bias voltage encourages the conversion of a-C phases into nanographitic filaments in the grain boundaries, besides the enhancement on the re-nucleation process on forming a nano-grain granular structure. The electrons can be transported easily across along the nanographitic filaments from the bottom side to the top side, and are then emitted to vacuum without any difficulty. Henceforth, the BEN-BEG processes have pronounced prospective for prompting the formation of nano-graphitic phase in diamond films. This is also a promising method for preparing highly conducting N-UNCD films without the need of a relatively high growth temperature (700 °C).

Saravanan et al. reported the growth of N-UNCD films on silicon substrates by the BEG process with the application of a negative biased voltage of -250 V using a CH_4/N_2 gas mixture at a low substrate temperature of 450 °C [157]. It is observed that by controlling the growth time, the microstructure of the N-UNCD films and, thus, the electrical conductivity and the FEE properties of the films can be manipulated. In 10 min growth, the films retain rod-shaped diamond grains, whereas in the films grown for 30 min, wire-like diamond grains are formed, which comprises a diamond core encased in a sheath of sp^2-bonded graphite phase. These films achieved high conductivity of 900 S/cm and superior FEE properties, namely, low E_0 of 2.9 V/μm and high J of 3.8 mA/cm^2 at an applied field of 6.0 V/μm. On increasing the growth time to 60 min, the growth of acicular grains ceases and a large proportion of graphite clusters or defective diamond clusters (n-diamond) is formed. Even though films grown at 60 min possess higher electrical conductivity of 1549 S/cm than the films grown for 30 min, the FEE properties degraded. The consequence of this result is that higher conductivity by itself does not promise better FEE properties. This is seemingly due to the fact that for the films growth at 60 min, the emission sites are mostly nanographite clusters or defective diamond cluster (n-diamond clusters), which are not suitable emitting sites.

A model is proposed by Sankaran et al. based on the evolution of microstructure of the N-UNCD films while the films grow thicker [158]. The growth of diamond films starts with the formation of a-C on a Si surface prior to the presence of diamond nuclei [159]. Abundant spherical diamond clusters were nucleated because of the abundant C_2-species in the CH_4/N_2 plasma [133]. The application of -250 V bias voltage increased prominently the kinetic energy of C_2^+ and CN^+, further improving the anisotropic growth for diamond grains [157]. As the nanosized diamond rods grow longer, they ensnare one another, resulting in a dendrite-like granular structure, schematically depicted in Fig. 5.12b. Particularly, in this region, every wire-like diamond grain is capsulized with a graphene-like phase of a few layers thick. The presence of the graphene-like layer frustrates the formation of diamond bonds, resulting in the formation of partly crystallized diamond, such as n-diamond with an equiaxed granular structure (Fig. 5.12c). The wire-like diamond grains are no more observable. Consequently, for a thick film, such as 60 min grown N-UNCD films, there occur three layers in which each layer has a different microstructure. As schematically shown in (Fig. 5.12d), after the formation of a thin a-C, the first layer of rod-like diamond was formed (region I), followed by the sequential transformation of granular structure to wire-like diamond (region II) and to equiaxed granular structured material, which mainly consist of n-diamond and graphite phases with diamond as a minor phase (region III). Consequently, the most favorable geometry for the emission sites are nanosized diamond grains with wire-like geometry, especially when they are encased with graphene-like layers. These encapsulating layers can transport electrons efficiently to the diamond for field emitting sites.

Fig. 5.12 Schematic diagrams showing the evolution of microstructure as the diamond films grew thicker with time in a CH$_4$/N$_2$ plasma under an applied voltage of −250 V: the a-C layer was formed first, followed by the formation of diamond grains with (I) short-rod geometry and (II) needle-like geometry sequentially, and finally (III) a layer dominated by the n-diamond clusters [158]

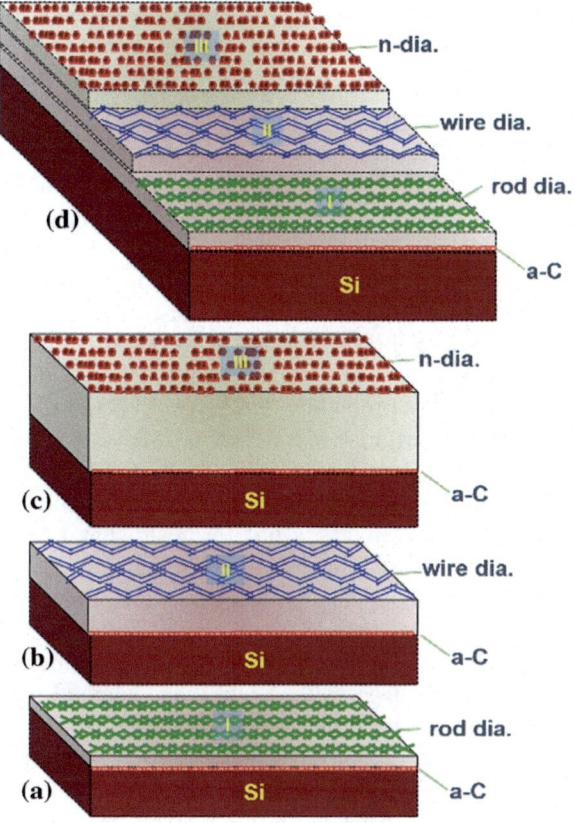

In addition to this work, Yeh et al. [160] reported a plasma post-treatment process in CH$_4$/N$_2$ plasma with −200 V applied bias to efficiently convert the microstructure of CH$_4$/Ar plasma grown UNCD films into wire-like granular structure, resulting in N-UNCD films with high conductivity and excellent FEE properties. The plasma post-treatment process is a diffusional process. It begins from the surface of CH$_4$/Ar plasma grown UNCD films and gradually continues towards the depth of the film. Long enough plasma post-treatment process is necessary (>60 min) to completely convert the granular structure of UNCD films into a wire-like diamond grain one over the complete film thickness. To conclude this section, the application of a negative bias voltage efficiently forms N-UNCD films at a low substrate temperature of only 450 °C. Such a low temperature growth process for synthesizing highly conducting diamond films is compatible with Si-based device technology and is exclusively beneficial for the exploitation of multifunctional field emission devices [161].

5.8 Nanostructured Nitrogen Doped Diamond

5.8.1 Vertically Aligned Diamond Nanorods

In previous Sects. 5.6 and 5.7, the synthesis, growth mechanism and efficient FEE behavior of highly conducting N-UNCD films was thoroughly discussed. Diamond nanostructuring is an efficient method for enhancing these peculiar properties even more. Generally, one dimensional (1D) nanostructures show high surface-to-volume ratio and tunable electron transport properties owing to quantum confinement effects. In two dimensional thin films, accumulation of charge carriers befalls only on the surface, while, the charge accumulation or depletion in the 1D nanostructure takes place in the "bulk" structure, thus giving rise to great changes in their electrical properties [162, 163]. Nanostructures of extremely hard and chemically inert materials such as diamond have been acquired by top-down methods, for example reactive ion etching (RIE), and also by bottom-up approaches [164, 165]. Shiomi et al. described the fabrication of porous diamond nanofilms by RIE using oxygen gas [166]. Nanostructured honeycomb diamond films were synthesized by etching diamond through a porous anodic alumina mask [167]. Moreover, diamond nanopillar arrays were produced using self-aligned Au particles as the etching mask in a biased assisted RIE with H_2/Ar plasma [168]. Sankaran et al. [169] fabricated vertically aligned diamond nanorods (DNRs) from N-UNCD films using the RIE method with nanodiamond particles as mask. The N-UNCD films were immersed in a stable suspension of nanodiamond particles of 8–10 nm in diameter in deionized water and sonicated for 10 min to seed nanodiamond particles on the N-UNCD film surface. The ND particle layer on the N-UNCD films is dense, which depends on the suspension quality and time of sonication. After masking, the N-UNCD films were then etched using the RIE process in an O_2/CF_4 gas mixture. In the process, nanodiamond particles acted as etching mask for fabricating vertically aligned DNRs. The DNRs are of diameters about 15–20 nm and lengths of about 460 nm (Fig. 5.13).

The proposed formation mechanism of DNRs is as follows: generally, the N-UNCD film contains nanodiamond grains separated by the sp^2-bonded grain

Fig. 5.13 SEM morphology of vertically aligned diamond nanorods (DNRs), which are fabricated from N-UNCD films grown using a CH_4/N_2 plasma at 700 °C, then using a RIE etching process with nanodiamond particles as masking materials [169]

200 nm

boundaries [145]. The grain boundaries contain a large proportion of sp^2-graphitic phases, which are more susceptible to O_2 plasma etching than the diamond material. At the initial stage of the etching process, nanodiamond particles sitting on top of the N-UNCD films mask the N-UNCD films from the ion bombardment. The chemical etching of oxygen ions starts from the grain boundaries of non-masked regions of the N-UNCD films. Because of the ease of etching on the sp^2-bonded carbon from the boundaries of N-UNCD films, an etching path for shaping the nanorods is created, resulting in vertically aligned nanorods [170, 171]. The etching process was stopped once the masking nanodiamond particles were also etched away by the O_2 plasma so as to prevent the plasma damage on the formed nanorods. Such a technique is similar to Yang's process [172], which utilized nanodiamond particles as a mask in a RIE process for fabricating the diamond nanorods from UNCD films.

These DNRs require only $E_0 = 2.04$ V/μm to turn on the FEE process and reach a J value of 4.84 mA/cm^2 at an applied field of 3.2 V/μm. Such FEE properties are markedly superior to those of the N-UNCD films [169]. Moreover, the field enhancement factor (β) is calculated by assuming the work function of diamond as 5.0 eV [173] and the value of DNRs is 1945, which is larger than that of the N-UNCD films (624), the enhanced value being due to the electrical field at the nanorod tips. The excellent FEE performance of the DNRs is mainly ascribed to the unique granular structure of nanorods and a high proportion of graphitic phase surrounding each nanorod. By using self-assembled Au nanodots as a mask on the N-UNCD films, Srinvasu et al. also fabricated grass-like nanostructures from N-UNCD films in the RIE process and achieved better FEE properties, viz, an E_0 value of 2.6 V/μm and a J value of 2.38 mA/cm^2 at an applied voltage of 5.8 V/μm. Recently, a "patterned-seeding technique" for selective area growth of nitrogen doped NCD (N-NCD) films (synthesized using $CH_4/H_2/5\%N_2$ plasma mixtures at 540 °C) in combination with a "nanodiamond masked RIE process" is demonstrated for fabricating patterned DNR arrays (Fig. 5.14) [174]. The patterned DNR arrays show a E_0 value of 5.26 V/μm and need an applied field of 10.55 V/μm to reach a J value of 1.0 mA/cm^2 with a high β value of 3270.

Fig. 5.14 SEM image of patterned DNR arrays, which are fabricated using a combination of "patterned-seeding technique" for selective area growth of nitrogen doped NCD films and a "nanodiamond masked RIE process" [174]

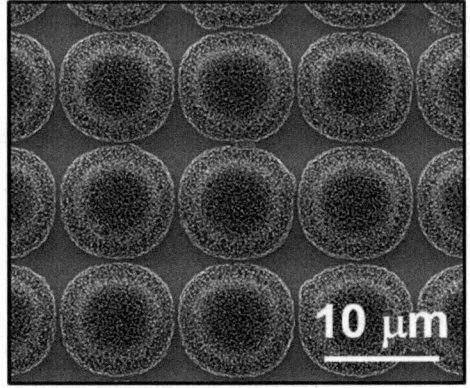

5.8.2 *Flexible N-UNCD Pyramidal Microtips*

Obviously, the DNRs fabricated using RIE method enhance the FEE properties to a certain extent but there are a few difficulties faced during the RIE process in controlling the etching rate of N-UNCD film and the induction of sp^2 bonded carbon in the nanorods [169]. The induction of more sp^2 bonded carbon in the nanorods leads to a lowerering of the lifetime of these DNRs. To overcome these issues, an efficient method of fabricating a flexible field emission device by incorporating arrays of pyramidal shaped N-UNCD microtips onto a plastic substrate is demonstrated in Fig. 5.15 [175]. First, N-UNCD films were synthesized on inverted pyramidal microcavity Si arrays. Then a thick polynorbornene (PNB)-based polymer layer was spin coated onto the N-UNCD films. After the spin coating process, the PNB coated N-UNCD substrates were baked at 100 °C, followed by annealing at 300 °C. The backside of the substrates was etched by acid to remove the Si completely by which the pyramidal shaped DNW microtips were transferred to the PNB substrate. Figure 5.16 show SEM micrographs of the flexible N-UNCD pyramidal microtips with sizes of 2, 4, 6 and 8 μm, respectively. The 2 μm sized array of flexible N-UNCD pyramidal microtips in bent configuration reveal a very low E_0 value of 1.80 V/μm with high J value of 5.8 mA/cm^2 at an applied field of 4.20 V/μm. Moreover, this flexible field emitting device exhibits a large β value of 4580 and good emission current stability of 210 min. Such flexible N-UNCD pyramidal microtips can be prospective candidates for a proficient field emitter material for flat panel displays and high brightness electron sources.

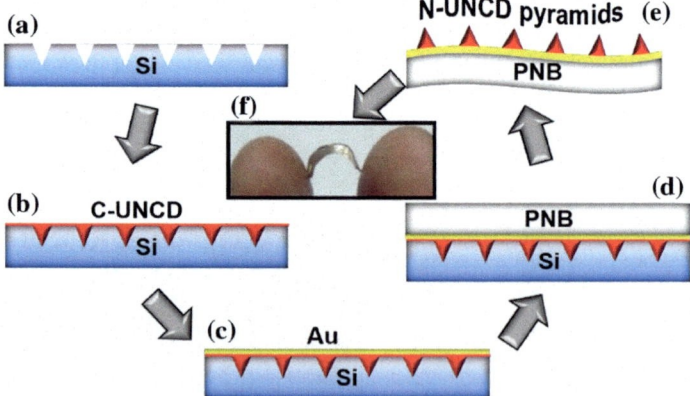

Fig. 5.15 Schematics of the process for fabricating flexible N-UNCD pyramidal microtips: **a** anisotropic etching of the Si substrates using a potassium hydroxide:normal propanol:deionized water solution to form inverted pyramidal microcavities; **b** deposition of N-UNCD films on inverted pyramidal microcavities using MW PE CVD; **c** sputtered deposition of Cr (5 nm) and Au (100 nm) on N-UNCD films; **d** spin coating of PNB; **e** arrays of flexible N-UNCD pyramidal microtips after etching of Si; **f** the digital photograph of typical hand bent, flexible N-UNCD pyramidal microtips [175]

Fig. 5.16 SEM micrographs of fabricated flexible N-UNCD pyramidal microtips after the back side etching of silicon, with tip sizes of **a** 2 μm, **b** 4 μm, **c** 6 μm, and **d** 8 μm [175]

5.9 Nitrogen Doped Diamond-Based Heterostructures

Combination of two different nanostructured materials in a heterostructure is a versatile building method for modern nanodevices. Compared to single component structures, heterostructures habitually reveal enhanced characteristics, such as emission efficiency and high electron mobility [176, 177], which are significant factors for many device performances [178–180]. A myriad of heterostructured materials have been established [181–188]. Fabrication of such heterostructures is not only critical for fundamental studies but also for assorted advanced functional devices, for instance, interconnects and emitters, etc. In this section, the enhanced FEE properties of heterostructures formed by the combination of nitrogen doped NCD/UNCD films with other field emitting materials like ZnO, CNTs, graphene, and hBN are discussed. It is to be noted that as single materials ZnO, CNTs, graphene, and hBN are not stable for long time field emitting operation. Interestingly, nitrogen doped NCD/UNCD in combination with these materials as a heterostructure assists in improving the FEE lifetime stability.

5.9.1 N-UNCD Coated ZnO Core–Shell Heterostructured Nanorods

A facile and reproducible way of fabricating zinc oxide nanorod (ZNR)-N-UNCD core–shell heterostructures on Si substrates (Fig. 5.17a) is demonstrated with excellent FEE performances [189]. Overcoming the problems of ZNRs field emitters, which generally have high E_0 and low J, the ZNR-N-UNCD core–shell heterostructures display outstanding FEE performances with low E_0 of 2.08 V/µm and high J of 5.5 mA/cm^2 at an applied field of 4.25 V/µm. Such an enhancement in the field emission originates from the unique materials combination, ensuing in good electron transport from ZNRs to N-UNCD nanoneedles and efficient field emission of electrons from the N-UNCD nanoneedles. The enhanced FEE behavior of the core–shell heterostructures can be clarified as follows: initially, the good electron transport properties of ZnO can advance the FEE by providing adequate electrons to the N-UNCD nanoneedles. The presence of graphitic phases in the interface region of ZNR-N-UNCD core–shell heterostructure drops the resistivity of the interfacial layer (Fig. 5.17b). The electrons can then be transferred enthusiastically from ZNRs across the interfacial layer to the N-UNCD nanoneedles. Under the action of an electric field, the N-UNCD nanoneedles act as radial electrodes and the electrons emitted from the ZNR core are collected radially along the diameter and transferred to the N-UNCD shell; the N-UNCD entrenched in graphitic sheaths present proficient electron transport paths for the emitted electrons to reach the tip of the nanoneedles from which they escape into vacuum. In short, the thin sheath of graphitic phase surrounding each N-UNCD nanoneedle collects the electrons from all over the surface of the nanoneedles and reduces the E_0 by lowering the barrier

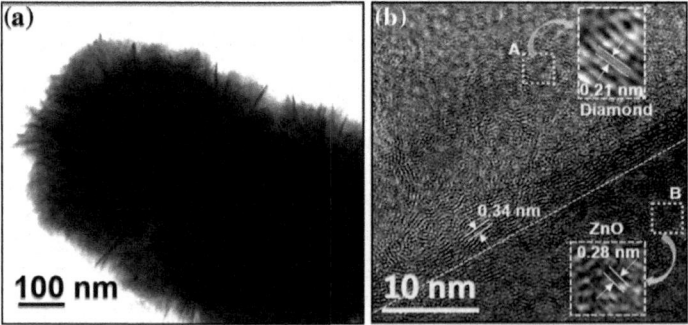

Fig. 5.17 a A low-magnification TEM image of the ZNR-N-UNCD core–shell heterostructures; **b** typical HRTEM image of a ZNR-N-UNCD interface. The graphitic phases are present in the interface with a lattice spacing of 0.34 nm; high magnification HRTEM image of the region A shows the diamond structure with a lattice spacing of 0.21 nm and region B shows the ZnO structure with a lattice spacing of 0.28 nm [189]

for the emitting electrons that enhances the J value. The vertically aligned ZNR-N-UNCD structure pointing the anode can be conceived as a supplementary reason for enhancing the FEE properties of these core–shell heterostructures.

5.9.2   Combination of N-UNCD Films and CNTs

Coating N-UNCD films on top of CNTs can enhance noticeably the robustness of CNTs. Chang et al. [190] demonstrated a possible way to fabricate an electron field emitter with superior FEE properties and improved lifetime stability via the combination of CNTs and the N-UNCD films. The resistance of the CNTs to CH_4/N_2 plasma ion bombardment during the growth of N-UNCD films is improved by the formation of carbon nanocones on the side walls of the CNTs, thus forming strengthened CNTs. The N-UNCD films can thus be grown on strengthened CNTs, forming N-UNCD-CNTs nanocomposite materials. This nanocomposite material possesses marvelous FEE properties, such as low E_0 of 3.58 V/μm with large J of 1.86 mA/cm^2 at an applied field of 6.0 V/μm. Moreover, the FEE emitters can be operated under 0.19 mA/cm^2 for more than 350 min without showing any sign of degradation. TEM investigations (Fig. 5.18) indicated that the N-UNCD films contain wire-like diamond grains encased in a few layers of nanographitic phase, which enhanced markedly the transport of electrons in the N-UNCD films. Moreover, the wire-like diamond grains were nucleated from the strengthened CNTs without the necessity of forming an interlayer, facilitating the transport of electrons crossing the diamond-to-Si interface (Fig. 5.18b, c). Both these factors contributed to the improved FEE behavior of the N-UNCD-CNT nanocomposites.

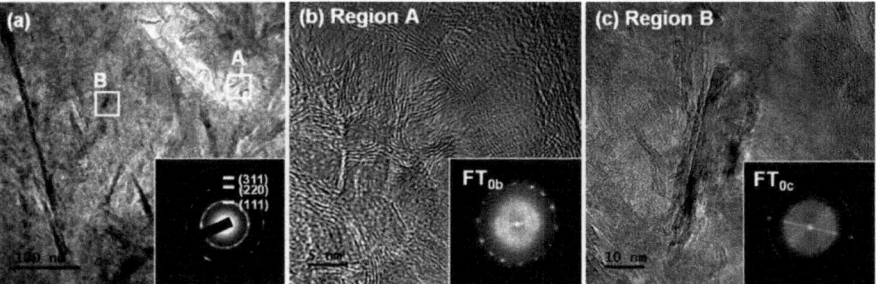

Fig. 5.18 a TEM bright field, **b** TEM structure images corresponding to region A designated in "a" and **c** TEM structure image corresponding to region B designated in "a" for the interface region of N-UNCD-CNTs films. The inset in "a" is the corresponding selected area electron diffraction pattern. The FT$_{0b}$ and FT$_{0c}$ are the Fourier-transformed diffractogram corresponding to the entire structure image in "b" and "c", respectively [190]

5.9.3 N-NCD-Graphene Hybrids

Graphene is a two-dimensional honeycomb lattice entailing of hexagonally arrayed sp^2-bonded carbon atoms [191]. The open surface and sharp edge of graphene presents a large aspect ratio therefore creating a fascinating candidate for FEE applications [192, 193]. Recent reports designate the E_0 value for flat graphene sheets to be high [10]. An alternative configuration is the use of graphene nano-flakes (GNFs) owing to a plentiful presence of sharp edge planes in GNFs, which can be a high density source of individual field emission sites. The existence of emissive sites in GNFs is predominantly beneficial and applicable for high effi-ciency graphene-based field electron emitters [194]. GNFs are comprised of ver-tically stacked graphene sheets which are unlike carbon nanowalls that are associated with highly defective nanostructured graphite [195]. Nevertheless, the short lifetime and the poor stability of the graphene emitters in a plasma environ-ment are foremost obstacles precluding their advantageous integration. Decorating GNFs with N-NCD results in superior functioning and enhanced lifetime of FEE devices associated to bare GNFs as cathodes [196]. The N-NCD grains were grown using $CH_4/H_2/N_2$ plasma mixtures at 540 °C. The N-NCD-GNFs possess better FEE characteristics with a low E_0 of 9.36 V/μm to induce the field emission, a high J of 2.57 mA/cm^2 and a large β value of 2380. The superior FEE of the N-NCD-GNFs are ascribed to the unique combination of N-NCD and graphene. Detailed structural characterizations through TEM reveal that the GNFs are homogeneously covered with N-NCD grains (Fig. 5.19). The nanographitic phases in the grain boundaries of the N-NCD grains form electron transport networks that lead to improvement in the FEE characteristics of the N-NCD-GNFs.

On the other hand, few-layer graphene (FLG) was catalytically formed on ver-tically aligned diamond nanorods (DNRs) by a high temperature annealing process [197]. The DNRs were fabricated from N-NCD films by O_2-plasma based RIE

Fig. 5.19 TEM micrograph of the N-NCD-GNFs [196]

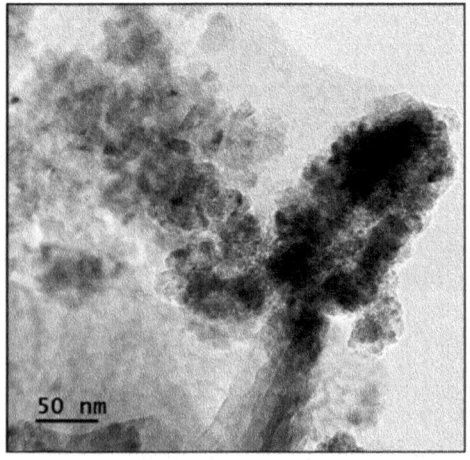

50 nm

Fig. 5.20 High resolution TEM micrograph of FLG-DNRs [197]

process [169]. A copper layer of about 150 nm was coated on the DNRs and thermal annealed at 800 °C for 1 h, followed by cooling down to room temperature. The copper coating was then etched away using nitric acid, resulting in FLG-DNRs hybrids. The FLG conformally covers the DNRs with 4–5 layer sharp edges (Fig. 5.20). The FLG-DNRs hybrid displays excellent FEE characteristics with a low E_0 of 4.21 V/µm, $J = 2.24$ mA/cm^2 at an applied field of 10 V/µm and a large β value of 3480, respectively [197].

5.9.4 Hexagonal Boron Nitride Nanowalls-N-NCD Heterostructures

Principally hexagonal boron nitride (hBN) is a structure analogous to graphite but reveals a large bandgap of 5.95 eV [198], viz. hBN is an insulating layered material. To make a precise application of hBN in nanoelectronics, it is imperative to change the electronic properties of hBN by doping it with appropriate dopant elements. Predominantly, when doped with carbon, hBN showed semiconducting properties due to the appearance of dopant or defect related states in the bandgap. Additionally, for a high-quality electron field emitter, a suitable electron supply from the substrate, e.g. Si, to the emitting sites, i.e. hBN, is critical, in addition to the need for a low work function for the emitting surface.

FEE properties of vertically aligned hexagonal boron nitride nanowalls (hBNNWs) grown on Si have been significantly enhanced through the use of N-NCD films as an interlayer. The FEE properties of hBNNWs-N-NCD heterostructures show a low E_0 of 15.2 V/µm, a high J of 1.48 mA/cm^2 and lifetime up to a period of 248 min [199]. These values are far superior to those for hBNNWs grown directly on Si substrates without the N-NCD interlayer, which have a E_0 of 46.6 V/μm with 0.21 mA/cm^2 current density and lifetime of 27 min. Cross-sectional TEM investigation exposes that the application of the N-NCD interlayer avoided the formation of amorphous boron nitride preceding the growth of hBN, henceforth hBN nucleated and grew directly on the N-NCD surfaces. The addition of N_2 in the N-NCD growth plasma encourages sp^2-bonded graphite phases in the grain boundaries of the N-NCD films that advance the electron transport from diamond to the hBNNWs. Then, the direct growth of hBN on the diamond surface lowers the resistivity of the interfacial layer and consequently the electrons can be transferred readily from N-NCD films across the interfacial layer to the hBNNWs. Lastly, the incorporation of C in the hBNNWs affords effective electron transport paths for the emitted electrons to reach the tip of the nanowalls from which they escape into vacuum without any difficulty as the hBN surfaces are NEA in nature [200, 201] ensuing in the superior FEE properties of hBNNWs-N-NCD heterostructures.

Further improving the FEE characteristics of hBN, hierarchical hBN-DNR were fabricated via a combination process of chemical vapor deposition synthesis for N-NCD film, RIE process for fabricating DNRs and the RF sputtering synthesis for hBN nanowalls (Fig. 5.21). Covering 1D DNRs with two dimensional hBNNWs is an effective approach to utilize the advantages of both nanostructured materials in field emission device applications. FEE measurements of the material show a low E_0 value of 6.0 V/μm, a high β value of 5870, and high lifetime stability of 435 min (under $J = 1.56$ mA/cm^2) [202]. Such an enhancement in the FEE properties of hBN-DNR derives from the distinctive material combination. The relatively large aspect ratio of the DNR and the sharp layer edges in the hBNNWs jointly contribute to the field enhancement. A large number of layer edges in the hBNNWs function as the emission sites.

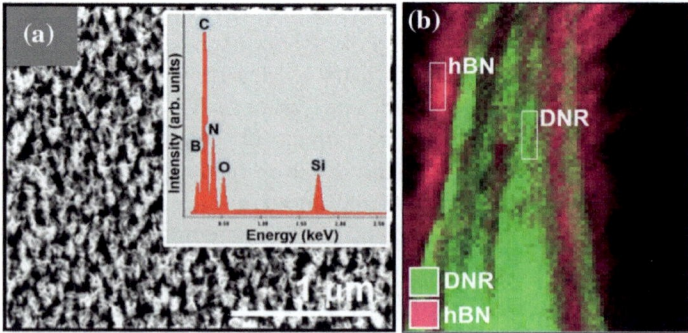

Fig. 5.21 **a** Tilted view SEM micrograph for hBN-DNR with an EDX spectrum shown as inset. **b** EELS elemental mapping for hBN (pink) and DNR (green) [202]

5.10 Plasma Illumination Cathodic Device

Microplasma science and technology, a new photonics technology, is a juncture of plasma science, optoelectronics and materials science that offers not only a hierarchy of plasma phenomenology but also device functionality [203–206]. Microplasma sources are attaining huge interest owing to their effectiveness in terms of portability, cost, and robustness in applications [207–211] in various fields such as surface treatment of materials for enhancing the wetting and adhesion properties of polymers, extremely proficient excimer ultra-violet (UV) light sources, ozone plasma display panels, production for air cleaning, and CO_2 lasers [212–218]. Microplasmas have been engaged in the field of biotechnology for biological decontamination, sterilization and diffuse discharges. Microplasma devices were efficaciously exploited in the fabrication of nanomaterials. Development of atmospheric pressure microplasmas has excessive prospective for material synthesis, processing and material modifications for bio-related applications. Due to these possible applications, a new area of research explores different candidate cathode materials that can increase the plasma characteristics as well as enhance the long-term sustainability of micro-discharges. High efficiency in making secondary electrons via plasma ion bombardment, i.e. having a large secondary electron emission coefficient (γ-coefficient), which can enhance the sustainability of micro-discharges, is of major importance in the choice of cathode materials for microplasma devices.

Many different types of materials have been successfully applied as electrodes in microplasmas, ranging from refractive metals to semiconductors [219–223]. Owing to ease of microfabrication of Si, it was the most promising material as electrode in microplasma applications [203–205, 210]. In case of microplasma applications at high pressures, electrode materials with high melting point are required, such as molybdenum, alumina, and boron nitride for operating at high temperature [224, 225]. Nevertheless, to withstand the plasma discharges, cathode materials with large γ-coefficient are required. MgO is typically used as a coating in plasma cells as it retains such a large γ-coefficient [226]. Diamond generally displays a large γ-coefficient owing to diamond's wide band gap (5.5 eV), negative electron affinity (NEA), low sputtering yield, and insensitivity to various processing conditions [227]. The Paschen-curve based γ-measurement [228] designates that diamond films synthesized via MW PE CVD process have a higher γ-coefficient than that of single crystal MgO. Additionally, the observations by Chakrabarti et al. [229] suggest that the high γ-coefficient is associated to the enhanced FEE properties of the material. As a consequence, high FEE diamond material is being explored for the development of a high γ-coefficient diamond material with enhanced plasma illumination (PI) characteristics.

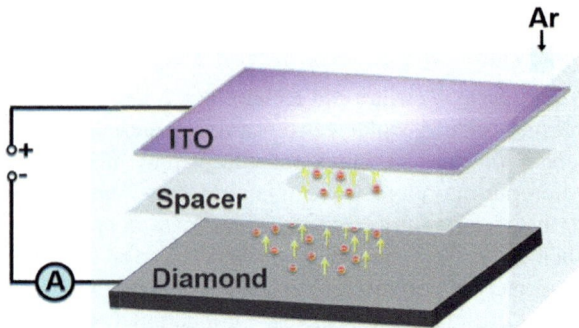

Fig. 5.22 Schematic of the microplasma device measurement

This section briefly describes the utilization of the high FEE nitrogen doped UNCD/NCD films and their related nanostructures as cathodes in a microplasma device and their improved PI properties. To investigate the PI behavior of these nitrogen doped UNCD/NCD films, a cylindrical type microplasma device was made by using the nitrogen doped UNCD/ NCD film as cathode and indium tin oxide coated glass as anode, which were separated by a Teflon spacer ($\sim$1 mm in thickness). Figure 5.22 shows a schematic of the plasma illumination measurement set-up. A circular hole about 8.0 mm in diameter was cut out from the spacer to form a microcavity. The plasma was triggered using a pulsed direct current voltage in a bipolar pulse mode (20 ms square pulse, 6 kHz repetition rate). The chamber was evacuated to reach a base pressure of 0.1 mTorr (13 MPa) and then purged with Ar for 10 min. The Ar gas was channeled into the chamber at a flow rate of 10 sccm throughout the measurements. The plasma current versus applied voltage was evaluated using an electrometer.

Srinivas et al. demonstrated that the N-UNCD based microplasma device discloses better PI characteristics with a low threshold field of 0.39 and high current density of 3.98 mA/cm^2 at an applied field of 0.58 V/μm. Moreover, the lifetime measurements in Ar plasma for N-UNCD based devices reveals that the current density profile started to decay in 2.02 days showing thus a high robustness of the used N-UNCD films [230]. In case of vertically aligned DNRs fabricated from N-UNCD films using RIE [169], the microplasma cavities using DNRs as cathode show PI characteristics with a low threshold field of 0.21 V/μm and a plasma current density of 7.06 mA/cm^2 at an applied field of 0.35 V/μm. The intensity of the plasma increases monotonously with the applied voltage (Fig. 5.23). On the other hand, the microplasma devices using N-UNCD or N-NCD in combination with ZNRs [189], CNTs [190], graphene [196, 197], and hBN [199, 202] as heterostructures (as described in Sects. 5.8.1, 5.9.1–5.9.4) show enhanced PI characteristics and especially high robustness as compared with bare CNTs, ZNR, hBN, and graphene utilized alone as cathode in a microplasma device. Such superior PI properties of nitrogen doped diamond films and their nanostructures make a significant impact on the diamond-based microplasma display technology.

Fig. 5.23 Photographs of plasma illumination characteristics of a microplasma cavity utilizing DNRs as cathode and ITO glass as anode [169]

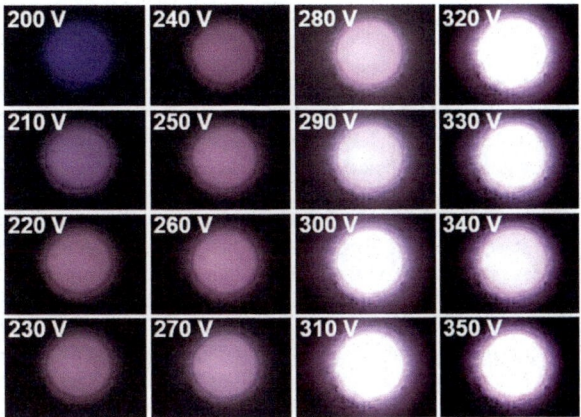

5.11 Conclusions

In this chapter, field electron emission and plasma illumination properties of nitrogen incorporated NCD/UNCD films were reviewed. The nitrogen incorporation was achieved by means of ion implantation and in situ doping processes, and recent achievements dealing with bias enhanced grown nitrogen doped UNCD films were presented. The deposited materials possess a high electrical conductivity and superior field electron emission properties, comparable with those of existing diamond-based field materials. A detailed look demonstrates that, whatever the approach employed to incorporate nitrogen in diamond, be it ion implantation, in situ doping in the plasma, or formation of nanostructures/heterostructures, the proven improvement of the electrical conductivity and field electron emission properties is principally connected with microstructural modification. Through a nitrogen-induced effect that increases the density of grain boundaries, sp^2 graphitic phases start to reside in said grain boundaries. These sp^2 phases, when finally interconnected, form excellent paths for the electrons to move easily through the material, thereby enhancing the conductivity as well as the field electron emission characteristics. As a consequence, the observed superior field electron emission and plasma illumination properties of nitrogen doped NCD/UNCD films and their nanostructures provide great prospective for the fabrication of high brightness display and multifunctional microplasma devices with long device lifetimes.

Acknowledgements The authors like to thank the financial support of the Research Foundation Flanders (FWO) via Research Grant 12I8416N and Research Project 1519817N, and the Methusalem "NANO" network. K. J. Sankaran is a Postdoctoral Fellow of FWO.

References

1. D.M. Trucchi, N.A. Melosh, Electron-emission materials: advances, applications, and models. MRS Bull. **42**, 488–492 (2017). https://doi.org/10.1557/mrs.2017.142
2. Y. Li, Y. Sun, J.T.W. Yeow, Nanotube field electron emission: principles, development, and applications. Nanotechnology **26**, 242001 (2015). https://doi.org/10.1088/0957-4484/26/24/242001
3. M.L. Terranova, S. Orlanducci, M. Rossi, E. Tamburri, Nanodiamonds for field emission: state of the art. Nanoscale **7**, 5094–5114 (2015). https://doi.org/10.1039/c4nr07171a
4. A.A. Talin, K.A. Dean, J.E. Jaskie, Field emission displays: a critical review. Solid State Electron. **45**, 963–976 (2001). https://doi.org/10.1016/S0038-1101(00)00279-3
5. K. Subramanian, W.P. Kang, J.L. Davidson, N. Ghosh, K.F. Galloway, A review of recent results on diamond vacuum lateral field emission device operation in radiation environments. Microelectron. Engg. **88**, 2924–2929 (2011). https://doi.org/10.1016/j.mee.2011.03.161
6. R.H. Fowler, L. Nordheim, Electron emission in intense electric fields. Proc. R. Soc. Lond. Ser. A 1928, **119**, 173–181. https://doi.org/10.1098/rspa.1928.0091
7. S. Sridhar, L. Ge, C.S. Tiwary, A.C. Hart, S. Ozden, K. Kalaga, S. Lei, S.V. Sridhar, R.K. Sinha, H. Harsh, K. Kordas, P.M. Ajayan, R. Vajtai, Enhanced field emission properties from CNT arrays synthesized on inconel superalloy. ACS Appl. Mater. Interfaces **6**, 1986–1991 (2014). https://doi.org/10.1021/am405026y
8. K.J. Sankaran, K. Srinivasu, K.C. Leou, N.H. Tai, I.N. Lin, High stability electron field emitters made of nanocrystalline diamond coated carbon nanotubes. Appl. Phys. Lett. **103**, 251601 (2013). https://doi.org/10.1063/1.4850525
9. K.S. Jung, S.H. Park, Enhanced field emission properties from carbon nanotube emitters on the nanopatterned substrate. J. Vac. Sci. Technol. B **35**, 011802–011808 (2017). https://doi.org/10.1116/1.4972119
10. S. Santandrea, F. Giubileo, V. Grossi, S. Santucci, M. Passacantando, T. Schroeder, G. Lupina, A.D. Bartolomeo, Field emission from single and few-layer graphene flakes. Appl. Phys. Lett. **98**, 163109–163112 (2011). https://doi.org/10.1063/1.3579533
11. C.X. Zhao, Y. Zhang, S.Z. Deng, N.S. Xu, J. Chen, Surface nitrogen functionality for the enhanced field emission of free-standing few-layer graphene nanowalls. J. Alloys Compd. **672**, 433–439 (2016). https://doi.org/10.1016/j.jallcom.2016.01.265
12. S. Pandey, S. Patole, F. Gunes, G.D. Kwon, J.B. Yoo, P. Nikolaev, S. Arepalli, Improved electron field emission from morphologically disordered monolayer graphene. Appl. Phys. Lett. **100**, 043104–042108 (2012). https://doi.org/10.1063/1.3679135
13. G. Li, Y. Li, X. Qian, H. Liu, H. Lin, N. Chen, Y. Li, Construction of tubular molecule aggregations of graphdiyne for highly efficient field emission. J. Phys. Chem. C **115**, 2611–2615 (2011). https://doi.org/10.1021/jp107996f
14. J. Zhou, X. Gao, R. Liu, Z. Xie, J. Yang, S. Zhang, G. Zhang, H. Liu, Y. Li, J. Zhang, Z. Liu, Synthesis of graphdiyne nanowalls using acetylenic coupling reaction. J. Am. Chem. Soc. **137**, 7596–7599 (2015). https://doi.org/10.1021/jacs.5b04057
15. T.D. Corrigan, D.M. Gruen, A.R. Krauss, P. Zapol, R.P.H. Chang, The effect of nitrogen addition to Ar/CH$_4$ plasmas on the growth, morphology and field emission of ultra-nanocrystalline diamond. Diam. Relat. Mater. **11**, 43–48 (2002). https://doi.org/10.1016/S0925-9635(01)00517-9
16. O. Auciello, A.V. Sumant, Status review of the science and technology of ultrananocrystalline diamond (UNCD™) films and application to multifunctional devices. Diam. Relat. Mater. **19**, 699–718 (2010). https://doi.org/10.1016/j.diamond.2010.03.015
17. O.A. Williams. M. Daenen, J. D'Haen, K. Haenen, J. Maes, V.V. Moschalkov, M. Nesladek, D.M. Gruen, Comparison of the growth and properties of ultrananocrystalline diamond and nanocrystalline diamond. Diam. Relat. Mater. **15**, 654–658 (2006). https://doi.org/10.1016/j.diamond.2005.12.009

18. K.J. Sankaran, P.T. Joseph, H.C. Chen, N.H. Tai, I.N. Lin, Investigation in the role of hydrogen on the properties of diamond films grown using Ar/H$_2$/CH$_4$ microwave plasma. Diam. Relat. Mater. **20**, 232–237 (2011). https://doi.org/10.1016/j.diamond.2010.12.018

19. W. Zhu, C.A. Randall, A.R. Badzian, R. Messier, Graphite formation in diamond film deposition. J. Vac. Sci. Technol. A **7**, 2315 (1989). https://doi.org/10.1116/1.575935

20. Y. Muto, T. Sugino, J. Shirafuji, Electrical conduction in undoped diamond films prepared by chemical vapor deposition. Appl. Phys. Lett. **59**, 843–845 (1991). https://doi.org/10.1063/1.105254

21. D. Hong, M. Aslam, Field emission from p-type polycrystalline diamond films. J. Vac. Sci. Technol. B **13**, 427–430 (1995). https://doi.org/10.1116/1.587962

22. D. Pradhan, I.N. Lin, Grain-size-dependent diamond-nondiamond composite films: characterization and field-emission properties. ACS Appl. Mater. Interfaces **1**, 1444–1450 (2009). https://doi.org/10.1021/am9001327

23. J.B. Cui, J. Ristein, L. Ley, Low-threshold electron emission from diamond. Phys. Rev. B **60**, 16135–16142 (1999). https://doi.org/10.1103/PhysRevB.60.16135

24. A.C. Ilie, A.C. Ferrari, T. Yagi, S.E. Rodil, J. Robertson, E. Barborini, P. Milani, Role of sp2/sp3 phase in field emission from nanostructured carbons. J. Appl. Phys. **90**, 2024–2032 (2001). https://doi.org/10.1063/1.1381001

25. J. Robertson, Mechanisms of electron field emission from diamond, diamond-like carbon and nanostructured carbon. J. Vac. Sci. Technol. B **17**, 659–665 (1999). https://doi.org/10.1116/1.590613

26. C. Popov, A. Gushterov, L. Lingys, C. Sippel, J.P. Reithmaier, Structural and optical properties of ultrananocrystalline diamond/InGaAs/GaAs quantum dot structures. Thin Solid Films **518**, 1489–1492 (2009). https://doi.org/10.1016/j.tsf.2009.09.097

27. V. Mortet, O. Elmazria, M. Nesladek, M.B. Assouar, G. Vanhoyland, J. D'Haen, M.D. Olieslaeger, P. Alnot, Surface acoustic wave propagation in aluminum nitride-unpolished freestanding diamond structures. Appl. Phys. Lett. **81**, 1720–1722 (2002). https://doi.org/10.1063/1.1503875

28. W. Zhu, G.P. Kochanski, S. Jin, Low-field electron emission from undoped nanostructured diamond. Science **282**, 1471–1473 (1998). https://doi.org/10.1126/science.282.5393.1471

29. J.P. Thomas, H.C. Chen, N.H. Tai, I.N. Lin, Freestanding ultrananocrystalline diamond films with homojunction insulating layer on conducting layer and their high electron field emission properties. ACS Appl. Mater. Interfaces **3**, 4007–4013 (2011). https://doi.org/10.1021/am200867c

30. V. Baranauskas, M.A. Sampaio, A.C. Peterlevitz, H.J. Ceragioli, J.C.R. Quispe, Field-emission properties of sulphur doped nanocrystalline diamonds. J. Phys. Conf. Series 61, 66–70 (2007). https://doi.org/10.1088/1742-6596/61/1/014

31. T. Yamada, H. Okano, H. Yamaguchi, H. Kato, S. Shikata, C.E. Nebel, Field emission from reconstructed heavily phosphorus-doped homoepitaxial diamond (111). Appl. Phys. Lett. **88**, 212114 (2006). https://doi.org/10.1063/1.2206552

32. P.M. Koinkar, S.S. Patil, T.G. Kim, D. Yonekura, M.A. More, D.S. Joag, R. Murakami, Enhanced field emission characteristics of boron doped diamond films grown by microwave plasma assisted chemical vapor deposition. Appl. Surf. Sci. **257**, 1854–1858 (2011). https://doi.org/10.1016/j.apsusc.2010.08.121

33. K.J. Sankaran, K. Srinivasu, C.J. Yeh, J.P. Thomas, S. Drijkoningen, P. Pobedinskas, B. Sundaravel, K.C. Leou, K.T. Leung, M.K. Van Bael, M. Schreck, I.N. Lin, K. Haenen, Field electron emission enhancement in lithium implanted and annealed nitrogen incorporated nanocrystalline diamond films. Appl. Phys. Lett. **110**, 261602 (2017). https://doi.org/10.1063/1.4990393

34. K. Panda, J.J. Hyeok, J.Y. Park, K.J. Sankaran, S. Balakrishnan, I.N. Lin, Nanoscale investigation of enhanced electron field emission for silver ion implanted/postannealed ultrananocrystalline diamond films. Sci. Rep. **7**, 16325 (2017). https://doi.org/10.1038/s41598-017-16395-1

35. J. Kurian, K.J. Sankaran, I.N. Lin, On the role of graphite in ultrananocrystalline diamond films used for electron field emitter applications. Phys. Status Solidi A **211**, 2223–2237 (2014). https://doi.org/10.1002/pssa.201431231

36. K.J. Sankaran, H.C. Chen, B. Sundaravel, C.Y. Lee, N.H. Tai, I.N. Lin, Gold ion implantation induced high conductivity and enhanced electron field emission properties in ultrananocrystalline diamond films. Appl. Phys. Lett. **102**, 061604 (2013). https://doi.org/10.1063/1.4792744

37. H.C. Chen, U. Palnitkar, B. Sundaravel, I.N. Lin, A.P. Singh, R.K. Rana, Enhancement of field emission properties in nanocrystalline diamond films upon 100 MeV silver ion irradiation. Diam. Relat. Mater. **18**, 164–168 (2009). https://doi.org/10.1016/j.diamond.2008.11.008

38. K.J. Sankaran, S. Kunuku, K.C. Leou, N.H. Tai, I.N. Lin, Enhancement of the electron Field emission properties of ultrananocrystalline diamond films via hydrogen post-treatment. ACS Appl. Mater. Interfaces **6**, 14543–14551 (2014). https://doi.org/10.1021/am503823n

39. K.J. Sankaran, K. Panda, B. Sundaravel, H.C. Chen, I.N. Lin, C.Y. Lee, N.H. Tai, Engineering the interface characteristics of ultrananocrystalline diamond films grown on Au-coated Si substrates. ACS Appl. Mater. Interfaces **4**, 4169–4176 (2012). https://doi.org/10.1021/am300894u

40. H.C. Chen, H.C. Chen, S.C. Lo, L.J. Lin, N.H. Tai, C.Y. Lee, I.N. Lin, Using an Au interlayer to enhance electron field emission properties of ultrananocrystalline diamond films. J. Appl. Phys. **112**, 103711 (2012). https://doi.org/10.1063/1.4766414

41. G. Sreenivas, S.S. Ang, W.D. Brown, Effects of nitrogen doping on the growth and properties of plasma-enhanced chemical-vapor-deposited diamond-like-carbon films. J. Electron. Mater. **23**, 569–575 (1994). https://doi.org/10.1007/BF02670661

42. S. Jin, T.D. Moustakas, Effect of nitrogen on the growth of diamond films. Appl. Phys. Lett. **65**, 403–405 (1994). https://doi.org/10.1063/1.112315

43. S.J. Breuer, P.R. Briddon, Energy barrier to reorientation of the substitutional nitrogen in diamond. Phys. Rev. B **53**, 7819–7822 (1996). https://doi.org/10.1103/PhysRevB.53.7819

44. J. Walker, Optical absorption and luminescence in diamond. Rep. Prog. Phys. **42**, 1605–1659 (1979). https://doi.org/10.1088/0034-4885/42/10/001

45. C.A. Ammerlaan, E.A. Burgemeister, Reorientation of nitrogen in type-Ib diamond by thermal excitation and tunneling. Phys. Rev. Lett. **47**, 954–957 (1981). https://doi.org/10.1103/PhysRevLett.47.954

46. P.K. Sitch, G. Jungnickel, M. Kaukonen, D. Porezag, Th. Frauenheim, M.R. Pederson, K.A. Jackson, A study of substitutional nitrogen impurities in chemical vapor deposition. J. Appl. Phys. **83**, 4642–4646 (1998). https://doi.org/10.1063/1.367249

47. Z. Yu, U. Karlsson, A. Flodström, Influence of oxygen and nitrogen on the growth of hot-filament chemical vapor deposited diamond films. Thin Solid Films **342**, 74–82 (1999). https://doi.org/10.1016/S0040-6090(98)01352-2

48. R. Locher, C. Wild, N. Herres, D. Behr, P. Koidl, Nitrogen stabilized <100> texture in chemical vapor deposited diamond films. Appl. Phys. Lett. **65**, 34–36 (1994). https://doi.org/10.1063/1.113064

49. Y. Avigal, O. Glozman, I. Etsion, G. Halperin, A. Hoffman, [100]-textured diamond films for tribological applications. Diam. Relat. Mater. **6**, 381–385 (1997). https://doi.org/10.1016/S0925-9635(96)00625-5

50. A.K. McCurdy, Phonon conduction in elastically anisotropic cubic crystals. Phys. Rev. B **26**, 6971–6986 (1982). https://doi.org/10.1103/PhysRevB.26.6971

51. A.K. McCurdy, H.J. Maris, C. Elbaum, Anistropic heat conduction in cubic crystals in the boundary scattering regime. Phys. Rev. B **2**, 4077–4083 (1970). https://doi.org/10.1103/PhysRevB.2.4077

52. W. Müller-Sebert, E. Wörner, F. Fuchs, C. Wild, P. Koidl, Nitrogen induced increase of growth rate in chemical vapor deposition of diamond. Appl. Phys. Lett. **68**, 759–760 (1996). https://doi.org/10.1063/1.116733

53. M. Guerino, M. Massi, H.S. Maciel, C. Otani, R.D. Mansano, P. Verdonck, J. Libardi, The influence of nitrogen on the dielectric constant and surface hardness in diamond-like carbon (DLC) films. Diam. Relat. Mater. **13**, 316–319 (2004). https://doi.org/10.1016/j.diamond. 2003.10.016

54. J. Mort, M.A. Machonkin, K. Okumura, Compenstaion effects in nitrogen doped diamond thin films. Appl. Phys. Lett. **59**, 3148–3150 (1991). https://doi.org/10.1063/1.105767

55. E. Boettger, A. Bluhm, X. Jiang, L. Schäfer, C.P. Klages, Investigation of the high-field conductivity and dielectric strength of nitrogen containing polycrystalline diamond films. J. Appl. Phys. **77**, 6332–6337 (1995). https://doi.org/10.1063/1.359103

56. G. Davies, The A nitrogen aggregate in diamond-its symmetry and possible structure. J. Phys. C Solid State Phys. **9**, L537–L542 (1976). https://doi.org/10.1088/0022-3719/9/19/005

57. S. Bohr, R. Haubner, B. Lux, Influence of nitrogen additions on hot-filament chemical vapor deposition of diamond. Appl. Phys. Lett. **68**, 1075–1077 (1996). https://doi.org/10.1063/1.115717

58. M.W. Geis, J.C. Twichell, N.N. Efremow, K. Krohn, T.M. Lyszczarz, Comparison of electric field emission from nitrogen-doped, type Ib diamond, and boron doped diamond. Appl. Phys. Lett. **68**, 2294–2296 (1996). https://doi.org/10.1063/1.116168

59. A.T. Sowers, B.L. Ward, S.L. English, R.J. Nemanich, Field emission properties of nitrogen-doped diamond films. J. Appl. Phys. **86**, 3973–3982 (1999). https://doi.org/10.1063/1.371316

60. K. Okano, T. Mine, I. Saito, H. Yamaguchi, T. Yamada, A. Sawabe, Electron emission from heavily nitrogen-doped heteroepitaxial chemical vapor deposition diamond. J. Vac. Sci. Tech. B **22**, 1327–1330 (2004). https://doi.org/10.1116/1.1756585

61. R. Kalish, C. Uzan-Saguy, B. Phiiosoph, V. Richter, J.P. Lagrange, E. Gheeraert, A. Deneuville, A.T. Collins, Nitrogen doping of diamond by ion implantation. Diam. Relat. Mater. **6**, 516–520 (1997). https://doi.org/10.1016/S0925-9635(96)00657-7

62. Sh. Cherf, M. Chandran, Sh. Michaelson, S. Elfimchev, R. Akhvlediani, A. Hoffman, Nitrogen and hydrogen content, morphology and phase composition of hot filament chemical vapor deposited diamond films from $NH_3/CH_4/H_2$ gas mixtures. Thin Solid Films **638**, 264–268 (2017). https://doi.org/10.1016/j.tsf.2017.07.060

63. B. Atakan, M. Beuger, K.K. Höinghaus, Nitrogen compounds and their influence on diamond deposition in flames. Phys. Chem. Chem. Phys. **1**, 705–708 (1999). https://doi.org/10.1039/A808850K

64. Y. Kudo, Y. Sato, T. Masuzawa, T. Yamada, I. Saito, T. Yoshino, W.J. Chun, S. Yamasaki, K. Okano, Field emission from N-doped diamond doped with dimethylureaa. J. Vac. Sci. Tech. B **28**, 506–510 (2010). https://doi.org/10.1116/1.3385784

65. W. Yuan, L. Fang, Z. Feng, Z. Chen, J. Wen, Y. Xiong, B. Wang, Highly conductive nitrogen-doped ultrananocrystalline diamond films with enhanced field emission properties: triethylamine as a new nitrogen source. J. Mater. Chem. C **4**, 4778–4785 (2016). https://doi.org/10.1039/c6tc00087h

66. Y. Okumura, K. Kanayama, H. Nishiguchi, Synthesis of n-type semiconducting diamond films in acetylene flame with nitrogen doping. Proc. Combust. Inst. **000**, 1–9 (2016). https://doi.org/10.1016/j.proci.2016.06.158

67. H. Yamada, A. Chayahara, Y. Mokuno, Effects of intentionally introduced nitrogen and substrate temperature on growth of diamond bulk single crystals. Jpn. J. Appl. Phys. **55**, 01AC07 (2016). iopscience.iop.org/1347-4065/55/1S/01AC07

68. C.S. Yan, Y.K. Vohra, Multiple twinning and nitrogen defect center in chemical vapour deposited homoepitaxial diamond. Diam. Relat. Mater. **8**, 2022–2031 (1999). https://doi.org/10.1016/S0925-9635(99)00148-X

69. D.S. Dandy, Influence of the gas phase on doping in diamond chemical vapor deposition. Thin Solid Films **381**, 1–5 (2001). https://doi.org/10.1016/S0040-6090(00)01355-9

70. J.E. Butler, I.A. Oleynik, Mechanism for crystal twinning in the growth of diamond by chemical vapour deposition. Philos. Trans. R. Soc. A **366**, 295–311 (2008). https://doi.org/10.1098/rsta.2007.2152

71. T.V. Regemorter, K. Larsson, Effect of coadsorbed dopants on diamond initial growth processes: CH_3 adsorption. J. Phys. Chem. A **112**, 5429–5435 (2008). https://doi.org/10.1021/jp711402e

72. T.V. Regemorter, K. Larsson, Effect of a NH coadsorbate on the CH_3 (or CH_2) adsorption to a surface step on diamond (100). J. Phys. Chem. C **113**, 19891–19896 (2009). https://doi.org/10.1021/jp900853a

73. T. Van Regemorter, K. Larsson, Effect of substitutional N on important chemical vapor deposition growth steps. J. Phys. Chem. A **113**, 3274–3284 (2009). https://doi.org/10.1021/jp811505w

74. Z. Yiming, F. Larsson, K. Larsson, Effect of CVD diamond growth by doping with nitrogen. Theor. Chem. Acc. **133**, 1432 (2014). https://doi.org/10.1007/s00214-013-1432-y

75. G.Z. Cao, J.J. Schermer, W.J.P. van Enckevort, W.A.L.M. Elst, L.J. Giling, Growth of {100} textured diamond films by the addition of nitrogen. J. Appl. Phys. **79**, 1357–1364 (1996). https://doi.org/10.1063/1.361033

76. P. Zapol, M. Sternberg, L.A. Curtiss, T. Frauenheim, D.M. Gruen, Tight-binding molecular-dynamics simulation of impurities in ultrananocrystalline diamond grain boundaries. Phys. Rev. B **65**, 045403 (2001). https://doi.org/10.1103/PhysRevB.65.045403

77. S. Bhattacharyya, Mechanism of high n-type conduction in nitrogen-doped nanocrystalline diamond. Phys. Rev. B **70**, 125412 (2004). https://doi.org/10.1103/PhysRevB.70.125412

78. S. Talapatra, J.Y. Cheng, N. Chakrapani, S. Trasobares, A. Cao, R. Vajtai, M.B. Huang, P.M. Ajayan, Ion irradiation induced structural modifications in diamond nanoparticles. Nanotechnology **17**, 305–309 (2006). https://iopscience.iop.org/article/10.1088/0957-4484/17/1/052

79. S. Prawer, R. Kalish, Ion-beam-induced transformation of diamond. Phys. Rev. B **51**, 15711–15722 (1995). https://doi.org/10.1103/PhysRevB.51.15711

80. J.F. Prins, Ion-implanted structures and doped layers in diamond. Mater. Sci. Rep. **7**, 275–364 (1992). https://doi.org/10.1016/0920-2307(92)90001-H

81. F. Fontaine, C. Uzan-Saguy, B. Philosoph, R. Kalish, Boron implantation/in situ annealing procedure for optimal p-type properties of diamond. Appl. Phys. Lett. **68**, 2264–2266 (1996). https://doi.org/10.1063/1.115879

82. E.J. Correa, Y. Wu, J.G. Wen, R. Chandrasekharan, M.A. Shannon, Electrical conduction in undoped ultrananocrystalline diamond thin films and its dependence on chemical composition and crystalline structure. J. Appl. Phys. **102**, 113706 (2007). https://doi.org/10.1063/1.2816214

83. W. Zhu, G.P. Kochanski, S. Jin, L. Seibles, D.C. Jacobson, M. McCormack, A.E. White, Electron field emission from ion-implanted diamond. Appl. Phys. Lett. **67**, 1157–1159 (1995). https://doi.org/10.1063/1.114993

84. P.M. Koinkar, R.S. Khairnar, S.A. Khan, R.P. Gupta, D.K. Avasthi, M.A. More, Influence of high energy ion irradiation on the field emission characteristics of CVD diamond films. Nucl. Instrum. Methods B **244**, 217–220 (2006). https://doi.org/10.1016/j.nimb.2005.11.020

85. J.J. Li, W.T. Zheng, C.Z. Gu, Z.S. Jin, G.R. Gu, X.X. Mei, Z.X. Mu, C. Dong, Field emission from nitrogen-implanted CVD diamond film grown on silicon wafer. Appl. Phys. A **81**, 357–361 (2005). https://doi.org/10.1007/s00339-004-2974-6

86. Y.C. Chen, X.Y. Zhong, B. Kabius, J.M. Hiller, N.H. Tai, I.N. Lin, Improvement of field emission performance on nitrogen ion implanted ultrananocrystalline diamond films through visualization of structure modifications. Diam. Relat. Mater. **20**, 238–241 (2011). https://doi.org/10.1016/j.diamond.2010.12.017

87. P.T. Joseph, N.H. Tai, C.Y. Lee, H. Niu, W.F. Pong, I.N. Lin, Field emission enhancement in nitrogen-ion-implanted ultrananocrystalline diamond films. J. Appl. Phys. **103**(4), 043720 (2008). https://doi.org/10.1063/1.2885348

88. K. Panda, B. Sundaravel, B.K. Panigrahi, P. Magudapathy, D. Nandagopala Krishna, K.G.M. Nair, H.C. Chen, I.N. Lin, Structural and electronic properties of nitrogen ion implanted ultra nanocrystalline diamond surfaces. J. Appl. Phys. **110**, 044304 (2011). https://doi.org/10.1063/1.3622517

89. P.T. Joseph, N.H. Tai, C.H. Chen, H. Niu, H.F. Cheng, W.F. Pong, I.N. Lin, On the mechanism of enhancement on electron field emission properties for ultrananocrystalline diamond films due to ion implantation. J. Phys. D Appl. Phys. **42**(10), 105403 (2009). https://doi.org/10.1088/0022-3727/42/10/105403

90. T.J. Palathinkal, N.H. Tai, C.Y. Lee, H. Niu, H.F. Cheng, W.F. Pong, I.N. Lin, Field emission enhancement in ion implanted ultra-nanocrystalline diamond films. Plasma Process. Polym. **6**, S834–S839 (2009). https://doi.org/10.1002/ppap.200930107

91. K.J. Sankaran, P.T. Joseph, N.H. Tai, I.N. Lin, High dose N ion implantation effects on surface treated UNCD films. Diam. Relat. Mater. **19**, 927–931 (2010). https://doi.org/10.1016/j.diamond.2010.02.027

92. P.T. Joseph, N.H. Tai, C.H. Chen, H. Niu, H.F. Cheng, U.A. Palnitkar, I.N. Lin, Field emission enhancement in ultrananocrystalline diamond films by in situ heating during single or multienergy ion implantation processes. J. Appl. Phys. **105**(12), 123710 (2009). https://doi.org/10.1063/1.3152790

93. P.T. Joseph, H.C. Chen, C.H. Chen, H. Niu, N.H. Tai, I.N. Lin, Nanostair formation and Field emission enhancement on high dose N-ion-implanted ultrananocrystalline diamond pyramid tips. Mater. Express **1**, 68–73 (2011). https://doi.org/10.1166/mex.2011.1009

94. J. Kurian, K.J. Sankaran, J.P. Thomas, N.H. Tai, H.C. Chen, I.N. Lin, The role of nanographitic phase on enhancing the electron field emission properties of hybrid granular structured diamond films: the electron energy loss spectroscopic studies. J. Phys. D Appl. Phys. **47**, 415303 (2014). https://doi.org/10.1088/0022-3727/47/41/415303

95. C. Pan, C.J. Chu, J.L. Margrave, R.H. Hague, Chlorine-activated diamond chemical vapor deposition. J. Electrochem. Soc. **141**, 3246–3249 (1994). https://doi.org/10.1149/1.2059312

96. K. Okano, S. Koizumi, S.R.P. Silva, G.A.J. Amaratunga, Low-threshold cold cathodes made of nitrogen-doped chemical-vapour-deposited diamond. Nature **381**, 140–141 (1996). https://doi.org/10.1038/381140a0

97. T. Masuzawa, Y. Sato, Y. Kudo, I. Saito, T. Yamada, A.T.T. Koh, D.H.C. Chua, T. Yoshino, W.J. Chun, S. Yamasaki, K. Okano, Correlation between low threshold emission and C–N bond in nitrogen-doped diamond films. J. Vac. Sci. Tech. B **29**, 02B119 (2011). https://doi.org/10.1116/1.3569821

98. P. Ball, Diamond films put on display. Nature **381**, 116 (1996)

99. R. Samlenski, C. Haug, R. Brenn, C. Wild, R. Locher, P. Koidl, Incorporation of nitrogen in chemical vapor deposition diamond. Appl. Phys. Lett. **67**, 2798–2800 (1995). https://doi.org/10.1063/1.114788

100. M. Ficek, K.J. Sankaran, J. Ryl, R. Bogdanowicz, I.N. Lin, K. Haenen, K. Darowicki, Ellipsometric investigation of nitrogen doped diamond thin films grown in microwave CH_4/H_2/N_2 plasma enhanced chemical vapor deposition. Appl. Phys. Lett. **108**, 241906 (2016). https://doi.org/10.1063/1.4953779

101. M.A. Lobaev, A.M. Gorbachev, S.A. Bogdanov, A.L. Vikharev, D.B. Radishev, V.A. Isaev, V.V. Chernov, M.N. Drozdov, Influence of CVD diamond growth conditions on nitrogen incorporation. Diam. Relat. Mater. **72**, 1–6 (2017). https://doi.org/10.1016/j.diamond.2016.12.011

102. S. Jin, T.D. Moustakas, Effect of nitrogen on the growth of diamond films. Appl. Phys. Lett. **63**, 2354–2356 (1993). https://doi.org/10.1063/1.112315

103. Q. Sun, J. Wang, J. Weng, F. Liu, Surface structure and electric properties of nitrogen incorporated NCD films. Vacuum **137**, 155–162 (2017). https://doi.org/10.1016/j.vacuum.2016.12.040

104. A. Badzian, T. Badzian, S.T. Lee, Synthesis of diamond from methane and nitrogen mixture. Appl. Phys. Lett. **62**, 3432–3434 (1993). https://doi.org/10.1063/1.109039

105. H. Chatei, J. Bougdira, M. ReÂmy, P. Alnot, C. Bruch, J.K. Kruger, Effect of nitrogen concentration on plasma reactivity and diamond growth in a H_2-CH_4-N_2 microwave discharge. Diam. Relat. Mater. **6**, 107–119 (1997). https://doi.org/10.1016/S0925-9635(96)00588-2

106. F. Silva, A. Gicquel, A. Tardieu, P. Cledat, Th. Chauveau, Control of an MPACVD reactor for polycrystalline textured diamond films synthesis: role of microwave power density. Diam. Relat. Mater. **5**, 338–344 (1996). https://doi.org/10.1016/0925-9635(95)00428-9

107. P.W. May, P.R. Burridge, C.A. Rego, R.S. Tsang, M.N.R. Ashfold, K.N. Rosser, R.E. Tanner, D. Cherns, R. Vincent, Investigation of the addition of nitrogen-containing gases to a hot filament diamond chemical vapour deposition reactor. Diam. Relat. Mater. **5**, 354–358 (1996). https://doi.org/10.1016/0925-9635(95)00379-7

108. T. Vandevelde, M. Nesladek, C. Quaeyhaegens, L. Stals, Optical emission spectroscopy of the plasma during CVD diamond growth with nitrogen addition. Thin Solid Films **290**, 143–147 (1996). https://doi.org/10.1016/S0040-6090(96)09189-4

109. T. Vandevelde, M. Nesladek, C. Quaeyhaegens, L. Stals, Optical emission spectroscopy of the plasma during microwave CVD of diamond thin films with nitrogen addition and relation to the thin film morphology. Thin Solid Films **308**, 154–158 (1997). https://doi.org/10.1016/S0040-6090(97)00408-2

110. M.J. Krecmarová, V. Petrák, A. Taylor, K.J. Sankaran, I.N. Lin, A. Jäger, V. Gärtnerová, L. Fekete, J. Drahokoupil, F. Laufek, J. Vacík, P. Hubík, V. Mortet, M. Nesládek, Change of diamond film structure and morphology with N_2 addition in MW PECVD apparatus with linear antenna delivery system. Phys. Status Solidi A **211**, 2296–2301 (2014). https://doi.org/10.1002/pssa.201431255

111. K.L. Ma, W.J. Zhang, Y.S. Zou, Y.M. Chong, K.M. Leung, I. Bello, S.T. Lee, Electrical properties of nitrogen incorporated nanocrystalline diamond films. Diam. Relat. Mater. **15**, 626–630 (2006). https://doi.org/10.1016/j.diamond.2005.11.017

112. K. Wu, E.G. Wang, Z.X. Cao, Z.L. Wang, X. Jiang, Microstructure and its effect on field electron emission of grain-size-controlled nanocrystalline diamond films. J. Appl. Phys. **88**, 2967–2974 (2000). https://doi.org/10.1063/1.1287602

113. K. Subramanian, W.P. Kang, J.L. Davidson, R.S. Takalkar, B.K. Choi, M. Howell, D.V. Kerns, Enhanced electron field emission from micropatterned pyramidal diamond tips incorporation CH_4/H_2/N_2 plasma-deposited nanodiamond. Diam. Relat. Mater. **15**, 1126–1131 (2006). https://doi.org/10.1016/j.diamond.2005.12.047

114. K.J. Sankaran, N.H. Tai, I.N. Lin, Microstructural evolution of diamond films from CH_4/H_2/N_2 plasma and their enhanced electrical properties. J. Appl. Phys. **117**, 075303 (2015). https://doi.org/10.1063/1.4913258

115. E. Rohrer, C.F.O. Graeff, R. Janssen, C.E. Nebel, H. Guettler, R. Zachai, Nitrogen-related dopant and defect states in CVD diamond. Phys. Rev. B. **54**, 7874–7880 (1996). https://doi.org/10.1103/PhysRevB.54.7874

116. D. Zhou, A.R. Krauss, L.C. Qin, T.G. McCauley, D.M. Gruen, T.D. Corrigan, R.P.H. Chang, H. Gnaser, Synthesis and electron field emission of nanocrystalline diamond thin films grown from N_2/CH_4 microwave plasmas. J. Appl. Phys. **82**(9), 4546–4550 (1997). https://doi.org/10.1063/1.366190

117. J.E. Birrell, O. Auciello, J.M. Gibson, D.M. Gruen, J.A. Carlisle, Bonding structure in nitrogen doped ultrananocrystalline diamond. J. Appl. Phys. **93**, 5606–5612 (2003). https://doi.org/10.1063/1.1564880

118. Y.C. Chen, N.H. Tai, I.N. Lin, Substrate temperature effects on the electron field emission properties of nitrogen doped ultra-nanocrystalline diamond. Diam. Relat. Mater. **17**, 457–461 (2008). https://doi.org/10.1016/j.diamond.2007.10.020

119. C.R. Lin, W.H. Liao, D.H. Wei, J.S. Tsai, C.K. Chang, W.C. Fang, Formation of ultrananocrystalline diamond films with nitrogen addition. Diam. Relat. Mater. **20**, 380–384 (2011). https://doi.org/10.1016/j.diamond.2010.12.015

120. N.M. Santos, T.M. Arantes, N.G. Ferreira, M.R. Baldan, Characterization of nitrogen doped diamond electrodes produced by hot filament chemical vapor deposition. Mat. Sci. Forum **802**, 180–185 (2014). https://doi.org/10.4028/www.scientific.net/MSF.802.180

121. H. Yoneda, K. Ueda, Y. Aikawa, K. Baba, N. Shohata, The grain size dependence of the mobility and lifetime in chemical vapor deposited diamond photoconductive switches. J. Appl. Phys. **83**, 1730–1733 (1998). https://doi.org/10.1063/1.366891

122. O.A. Shenderova, D.W. Brenner, L.H. Yang, Atomistic simulations of structures and mechanical properties of polycrystalline diamond: symmetrical <001> tilt grain boundaries. Phys. Rev. B **60**, 7043–7052 (1999). https://doi.org/10.1103/PhysRevB.60.7043

123. R.S. Tsang, C.A. Rego, P.W. May, M.N.R. Ashfold, K.N. Rosser, Examination of the effects of nitrogen on the CVD diamond growth mechanism using in situ molecular beam mass spectrometry. Diam. Relat. Mater. **6**, 247–254 (1997). https://doi.org/10.1016/S0925-9635(96)00647-4

124. A. Afzal, C.A. Rego, W. Ahmed, R.I. Cherry, HFCVD diamond grown with added nitrogen: film characterization and gas-phase composition studies. Diam. Relat. Mater. **7**, 1033–1038 (1998). https://doi.org/10.1016/S0925-9635(98)00148-4

125. S. Bhattacharyya, O. Auciello, J. Birrell, J.A. Carlisle, L.A. Curtiss, A.N. Goyette, D.M. Gruen, A.R. Krauss, J. Schlueter, A. Sumant, P. Zapol, Synthesis and characterization of highly-conducting nitrogen-doped ultrananocrystalline diamond films. Appl. Phys. Lett. **79**, 1441–1443 (2001). https://doi.org/10.1063/1.1400761

126. O.A. Williams, S. Curat, J.E. Gerbi, D.M. Gruen, R.B. Jackman, n-type conductivity in ultrananocrystalline diamond films. Appl. Phys. Lett. **85**, 1680–1682 (2004). https://doi.org/10.1063/1.1785288

127. J. Shalini, Y.C. Lin, T.H. Chang, K.J. Sankaran, H.C. Chen, I.N. Lin, C.Y. Lee, N.H. Tai, Ultra-nanocrystalline diamond nanowires with enhanced electrochemical properties. Electrochim. Acta **92**, 9–19 (2013). https://doi.org/10.1016/j.electacta.2012.12.078

128. Y.C. Lin, K.J. Sankaran, Y.C. Chen, C.Y. Lee, H.C. Chen, I.N. Lin, N.H. Tai, Enhancing electron field emission properties of UNCD films through nitrogen incorporation at high substrate temperature. Diam. Relat. Mater. **20**, 191–195 (2011). https://doi.org/10.1016/j.diamond.2010.11.026

129. R. Arenal, P. Bruno, D.J. Miller, M. Bleuel, J. Lal, D.M. Gruen, Diamond nanowires and the insulator-metal transition in ultrananocrystalline diamond films. Phys. Rev. B. **75**, 195431 (2007). https://doi.org/10.1103/PhysRevB.75.195431

130. V.D. Frolov, S.M. Pimenov, V.I. Konov, V.I. Polyakov, A.I. Rukovishnikov, N.M. Rossukanyi, J.A. Carlisle, D.M. Gruen, Electronic properties of low-field-emitting ultrananocrystalline diamond films. Surf. Interface Anal. **36**, 449–454 (2004). https://doi.org/10.1002/sia.1709

131. T. Ikeda, K. Teii, Origin of low threshold field emission from nitrogen-incorporated nanocrystalline diamond films. Appl. Phys. Lett. **94**, 143102 (2009). https://doi.org/10.1063/1.3115767

132. Y.K. Liu, P.L. Tso, D. Pradhan, I.N. Lin, M. Clark, Y. Tzeng, Structural and electrical properties of nanocrystalline diamond (NCD) heavily doped by nitrogen. Diam. Relat. Mater. **14**, 2059–2063 (2005). https://doi.org/10.1016/j.diamond.2005.06.012

133. K.J. Sankaran, J. Kurian, H.C. Chen, C.L. Dong, C.Y. Lee, N.H. Tai, I.N. Lin, Origin of a needle-like granular structure for ultrananocrystalline diamond films grown in a N_2/CH_4 plasma. J. Phys. D Appl. Phys. **45**, 365303 (2012). https://doi.org/10.1088/0022-3727/45/36/365303

134. C.S. Wang, G.H. Tong, H.C. Chen, W.C. Shih, I.N. Lin, Effect of N_2 addition in Ar plasma on the development of microstructure of ultra-nanocrystalline diamond films. Effect of N2 addition in Ar plasma on the development of microstructure of ultra-nanocrystalline diamond films. Diam. Relat. Mater. **19**, 147–152 (2010). https://doi.org/10.1016/j.diamond.2009.09.009

135. N. Shang, P. Papakonstantinou, P. Wang, A. Zakharov, U. Palnitkar, I.N. Lin, M. Chu, A. Stamboulis, Self-assembled growth, microstructure, and field-emission high-performance of ultrathin diamond nanorods. ACS Nano **3**, 1032–1038 (2009). https://doi.org/10.1021/nn900167p

136. L. Sun, J. Gong, D. Zu, Z. Zhu, S. He, Diamond nanorods from carbon nanotubes. Adv. Mater. **16**, 1849–1853 (2004). https://doi.org/10.1002/adma.200400429

137. Y. Li, W. Kim, Y. Zhang, M. Rolandi, D. Wang, H. Dai, Growth of single-walled carbon nanotubes from discrete catalytic nanoparticles of various sizes. J. Phys. Chem. B **105**, 11424–11431 (2001). https://doi.org/10.1021/jp012085b

138. J. Singh, Nucleation and growth mechanism of diamond during hot-filament chemical vapour deposition. J. Mater. Sci. **29**, 2761–2766 (1994). https://doi.org/10.1007/BF00356830

139. S.A. Rakha, G. Yu, J. Cao, S. He, X. Zhou, Influence of CH_4 on the morphology of nanocrystalline diamond films deposited by Ar rich microwave plasma. J. Appl. Phys. **107**, 114324 (2010). https://doi.org/10.1063/1.3410804

140. I.I. Vlasov, O.I. Lebedev, V.G. Ralchenko, Hybrid diamond-graphite nanowires produced by microwave plasma chemical vapor deposition. Adv. Mater. **19**, 4058–4062 (2007). https://doi.org/10.1002/adma.200700442

141. A. van der Drift, Evolutionary selection, a principle governing growth orientation in vapour-deposited Layers. Philips Res. Rep. **22**, 267–268 (1967)

142. P.C. Redfern, D.A. Horner, L.A. Curtiss, D.M. Gruen, Theoretical studies of growth of diamond (110) from dicarbon. J. Phys. Chem. **100**, 11654–11663 (1996). https://doi.org/10.1021/jp953165g

143. M. Sternbergt, P. Zapol, T. Frauenheimt, J.A. Carlisle, D.M. Gruen, L.A. Curtiss, Density functional based tight binding study of C_2 and CN deposition on (100) diamond surface. Mat. Res. Soc. Symp. Proc. **675**, W12 (2001). https://doi.org/10.1557/PROC-675-W12.11.1

144. S.A. Kajihara, A. Antonelli, J. Bernhol, R. Car, Nitrogen and potential n-type dopants in diamond. Phys. Rev. Lett. **66**, 2010–2013 (1991). https://doi.org/10.1103/PhysRevLett.66.2010

145. K.J. Sankaran, Y.F. Lin, W.B. Jian, H.C. Chen, K. Panda, B. Sundaravel, C.L. Dong, N.H. Tai, I.N. Lin, Structural and electrical properties of conducting diamond nanowires. ACS Appl. Mater. Interfaces **5**, 1294–1301 (2013). https://doi.org/10.1021/am302430p

146. R.L. Harniman, O.J.L. Fox, W. Janssen, S. Drijkoningen, K. Haenen, P.W. May, Direct observation of electron emission from grain boundaries in CVD diamond by PeakForce-controlled tunnelling atomic force microscopy. Carbon **94**, 386–395 (2015). https://doi.org/10.1016/j.carbon.2015.06.082

147. K. Panda, K.J. Sankaran, B.K. Panigrahi, N.H. Tai, I.N. Lin, Direct observation and mechanism for enhanced electron emission in hydrogen plasma-treated diamond nanowire films. ACS Appl. Mater. Interfaces **6**, 8531–8541 (2014). https://doi.org/10.1021/am501398s

148. K.N. Kushita, K. Hojou, S. Furuno, H. Otsu, In situ EELS observation of diamond during hydrogen-ion bombardment. J. Nucl. Mater. **191–194**, 346–350 (1992). https://doi.org/10.1016/S0022-3115(09)80063-9

149. Y.C. Chu, C.H. Tu, C. Liu, Y. Tzeng, O. Auciello, Ultrananocrystalline diamond nano-pillars synthesized by microwave plasma bias-enhanced nucleation and bias-enhanced growth in hydrogen-diluted methane. J. Appl. Phys. **112**, 124307 (2012). https://doi.org/10.1063/1.4769861

150. X.Y. Zhong, Y.C. Chen, N.H. Tai, I.N. Lin, J.M. Hiller, O. Auciello, Effect of pretreatment bias on the nucleation and growth mechanisms of ultrananocrystalline diamond films via bias-enhanced nucleation and growth: an approach to interfacial chemistry analysis via chemical bonding mapping. J. Appl. Phys. **105**, 034311 (2009). https://doi.org/10.1063/1.3068366

151. C.Z. Gu, X. Jiang, Deposition and characterization of nanocrystalline diamond films prepared by ion bombardment-assisted method. J. Appl. Phys. **88**, 1788–1793 (2000). https://doi.org/10.1063/1.1305460
152. N. Jiang, K. Sugimoto, K. Eguchi, T. Inaoka, Y. Shintani, H. Makita, A. Hatta, A. Hiraki, Reducing the grain size for fabrication of nanocrystalline diamond films. J. Cryst. Growth **222**, 591–594 (2001). https://doi.org/10.1016/S0022-0248(00)00972-6
153. M. Schreck, T. Baur, R. Fehling, M. Mialler, B. Stritzker, A. Bergmaier, G. Dollinger, Modification of diamond film growth by a negative bias voltage in microwave plasma chemical vapor deposition. Diam. Relat. Mater. **7**, 293–298 (1998). https://doi.org/10.1016/S0925-9635(97)00260-4
154. V. Mortet, L. Zhang, M. Eckert, J. D'Haen, A. Soltani, M. Moreau, D. Troadec, E. Neyts, J.D. Jaeger, J. Verbeeck, A. Bogaerts, G.V. Tendeloo, K. Haenen, P. Wagner, Grain size tuning of nanocrystalline chemical vapor deposited diamond by continuous electrical bias growth: Experimental and theoretical study. Phys. Status Solidi A **209**, 1675–1682 (2012). https://doi.org/10.1002/pssa.201200581
155. K.Y. Teng, H.C. Chen, G.C. Tzeng, C.Y. Tang, H.F. Cheng, I.N. Lin, Bias-enhanced nucleation and growth processes for improving the electron field emission properties of diamond films. J. Appl. Phys. **111**, 053701 (2012). https://doi.org/10.1063/1.3687918
156. A. Saravanan, B.R. Huang, K.J. Sankaran, G. Keiser, J. Kurian, N.H. Tai, I.N. Lin, Structural modification of nanocrystalline diamond films via positive/negative bias enhanced nucleation and growth processes for improving their electron field emission properties. J. Appl. Phys. **117**, 215307 (2015). https://doi.org/10.1063/1.4921875
157. A. Saravanan, B.R. Huang, K.J. Sankaran, N.H. Tai, I.N. Lin, Highly conductive diamond—graphite nanohybrid films with enhanced electron field emission and microplasma illumination properties. ACS Appl. Mater. Interfaces **7**, 14035–14042 (2015). https://doi.org/10.1021/acsami.5b03166
158. K.J. Sankaran, B.R. Huang, A. Saravanan, D. Manoharan, N.H. Tai, I.N. Lin, Nitrogen incorporated ultrananocrystalline diamond microstructures from bias-enhanced microwave N_2/CH_4-plasma chemical vapor deposition. Plasma Process. Polym. **13**, 419–428 (2016). https://doi.org/10.1002/ppap.201500079
159. T.H. Chang, K. Panda, B.K. Panigrahi, S.C. Lou, C. Chen, H.C. Chen, I.N. Lin, N.H. Tai, Electrophoresis of nanodiamond on the growth of ultrananocrystalline diamond films on silicon nanowires and the enhancement of the electron field emission properties. J. Phys. Chem. C **116**, 19867–19876 (2012). https://doi.org/10.1021/jp306086b
160. C.J. Yeh, D. Manoharan, H.T. Chang, K.C. Leou, I.N. Lin, Synthesis of ultra-nano-carbon composite materials with extremely high conductivity by plasma post-treatment process of ultrananocrystalline diamond films. Appl. Phys. Lett. **107**, 083104 (2015). https://doi.org/10.1063/1.4929587
161. A. Saravanan, B.R. Huang, D. Manoharan, D. Kathiravan, I.N. Lin, Engineered design and fabrication of long lifetime multifunctional devices based on electrically conductive diamond ultrananowire multifinger integrated cathodes. J. Mater. Chem. C **4**, 9727–9737 (2016). https://doi.org/10.1039/C6TC03340G
162. U. Yogeswaran, S.M. Chen, A review on the electrochemical sensors and biosensors composed of nanowires as sensing material. Sensors **8**, 290–313 (2008). https://doi.org/10.3390/s8010290
163. M.M. Rahman, A. Ahammad, J.H. Jin, S.J. Ahn, J.J. Lee, A comprehensive review of glucose biosensors based on nanostructured metal-oxides. Sensors **10**, 4855–4886 (2010). https://doi.org/10.3390/s100504855
164. C.H. Hsu, H.C. Lo, C.F. Chen, C.T. Wu, J.S. Hwang, D. Das, J. Tsai, L.C. Chen, K.H. Chen, Generally applicable self-masked dry etching technique for nanotip array fabrication. Nano Lett. **4**, 471–475 (2004). https://doi.org/10.1021/nl049925t

165. Q. Wang, P. Subramanian, M. Li, W.S. Yeap, K. Haenen, Y. Coffinier, R. Boukherroub, S. Szunerits, Non-enzymatic glucose sensing on long and short diamond nanowire electrodes. Electrochem. Commun. **34**, 286–290 (2013). https://doi.org/10.1016/j.elecom. 2013.07.014

166. H. Shiomi, Reactive ion etching of diamond in O_2 and CF_4 plasma, and fabrication of porous diamond for field emitter cathodes. Jpn. J. Appl. Phys. **36**, 7745 (1997). https://doi.org/10. 1143/JJAP.36.7745

167. H. Masuda, M. Watanabe, K. Yasui, D. Tryk, T. Rao, A. Fujishima, Fabrication of a nanostructured diamond honeycomb film. Adv. Mater. **12**, 444–447 (2000). https://doi.org/ 10.1002/(SICI)1521-4095(200003)12:63.3.CO;2-B

168. Y. Zou, Y. Yang, W. Zhang, Y. Chong, B. He, I. Bello, S. Lee, Fabrication of diamond nanopillars and their arrays. Appl. Phys. Lett. **92**, 053105 (2008). https://doi.org/10.1063/1. 2841822

169. K.J. Sankaran, S. Kunuku, S.C. Lou, J. Kurian, H.C. Chen, C.Y. Lee, N.H. Tai, K.C. Leou, C. Chen, I.N. Lin, Microplasma illumination enhancement of vertically aligned conducting ultrananocrystalline diamond nanorods. Nanoscale Res. Lett. **7**, 522 (2012). https://doi.org/ 10.1186/1556-276X-7-522

170. W.J. Zhang, Y. Wu, C.Y. Cha, W.K. Wong, X.M. Meng, I. Bello, Y. Lifshitz, S.T. Lee, Structuring single- and nano-crystalline diamond cones. Diam. Relat. Mater. **13**, 1037–1043 (2004). https://doi.org/10.1016/j.diamond.2003.10.007

171. J. Birgit, M. Hausmann, M. Khan, Y. Zhang, T.M. Babinec, K. Martinick, M. McCutcheon, P.R. Hemmer, M. Loncar, Fabrication of diamond nanowires for quantum information processing applications. Diam. Relat. Mater. **19**, 621–629 (2010). https://doi.org/10.1016/j. diamond.2010.01.011

172. N. Yang, U. Hiroshi, O.A. Williams, E. Osawa, N. Tokuda, C.E. Nebel, Vertically aligned diamond nanowires: fabrication, characterization, and application for DNA sensing. Phys. Status Solidi A **206**, 2048–2056 (2009). https://doi.org/10.1016/j.diamond.2010.01.011

173. J. Liu, V.V. Zhirnov, A.F. Myers, G.J. Wojak, W.B. Choi, J.J. Hren, S.D. Wolter, M.T. McClure, B.R. Stoner, J.T. Glass, Field emission characteristics of diamond coated silicon field emitters. J. Vac. Sci. Technol. B **13**(2), 422–426 (1995). https://doi.org/10.1116/1. 587961

174. R. Ramaneti, K.J. Sankaran, S. Korneychuk, C.J. Yeh, G. Degutis, K.C. Leou, J. Verbeeck, M.K. Van Bael, I.N. Lin, K. Haenen, Vertically aligned diamond-graphite hybrid nanorod arrays with superior field electron emission properties. APL Mat. **5**, 066102 (2017). https:// doi.org/10.1063/1.4985107

175. K.J. Sankaran, N.H. Tai, I.N. Lin, Flexible electron field emitters fabricated using conducting ultrananocrystalline diamond pyramidal microtips on polynorbornene films. Appl. Phys. Lett. **104**, 031601 (2014). https://doi.org/10.1063/1.4862891

176. M. Nirmal, L. Brus, Luminescence photophysics in seminconductor nanocrystals. Acc. Chem. Res. **32**, 407–414 (1999). https://doi.org/10.1021/ar9700320

177. X. Xia, J. Tu, Y. Zhang, X. Wang, C. Gu, X. Zhao, H.J. Fan, High-quality metal oxide core/ shell nanowire arrays on conductive substrates for electrochemical energy storage. ACS Nano **6**, 5531–5538 (2012). https://doi.org/10.1021/nn301454q

178. K. Wang, J. Chen, W. Zhou, Y. Zhang, Y. Yan, J. Pern, A. Mascarenhas, Direct growth of highly mismatched type II ZnO/ZnSe core/shell nanowire arrays on transparent conducting oxide substrates for solar cell applications. Adv. Mater. **20**, 3248–3253 (2008). https://doi. org/10.1002/adma.200800145

179. Y. Tak, H. Kim, D. Lee, K. Yong, Type-II CdS nanopartice-ZnO nanowire heterostructure arrays fabricated by a solution process: enhanced photocatalytic activity. Chem. Commun., 4585–4587 (2008). https://doi.org/10.1039/b810388g

180. J. Schrier, D.O. Demchenko, L.W. Wang, Optical properties of ZnO/ZnS and ZnO/ZnTe heterostructures for photovoltaic applications. Nano Lett. **7**, 2377–2382 (2007). https://doi. org/10.1021/nl071027k

181. U.K. Gautam, X. Fang, Y. Bando, J. Zhan, D. Golberg, Synthesis, structure, and multiply enhanced field-emission properties of branched ZnS nanotube–in nanowire core–shell heterostructures. ACS Nano **2**, 1015–1021 (2008). https://doi.org/10.1021/nn800013b

182. J. Lin, Y. Huang, Y. Bando, C. Tang, C. Li, D. Golberg, Boron nitride nanotubes and nanosheets. ACS Nano **4**, 2452–2458 (2010). https://doi.org/10.1021/nn1006495

183. G. Li, Y. Jiang, Y. Zhang, X. Lan, T. Zhai, G.C. Yi, High-performance photodetectors and enhanced field-emission of CdS nanowire arrays on CdSe single-crystalline sheets. J. Mater. Chem. C **2**, 8252–8258 (2014). https://doi.org/10.1039/C4TC01503G

184. Q. Xie, C. Wang, X. Xu, J. Liu, J. Zhang, Formation of Ge nanosheets decorated hierarchical ZnSe/GeSe nanowire heterostructures. Jpn. J. Appl. Phys. **49**, 025001 (2010). https://doi.org/10.1143/JJAP.49.025001

185. X. Jiang, Q. Xiong, S. Nam, F. Qian, Y. Li, C.M. Lieber, InAs/InP radial nanowire heterostructures as high electron mobility devices. Nano Lett. **7**, 3215–3218 (2007). https://doi.org/10.1021/nl072024a

186. J. Xiong, C. Han, W. Li, Q. Sun, J. Chen, S. Chou, Z. Li, S. Dou, Ambient synthesis of a multifunctional 1D/2D hierarchical Ag–Ag$_2$S nanowire/nanosheet heterostructure with diverse applications. CrystEngComm **18**, 930–937 (2016). https://doi.org/10.1039/C5CE02134K

187. J. Yang, J. Liang, G. Zhang, J. Li, H. Liu, Z. Shen, Heterostructures of MoS2 nanofilms on TiO2 nanorods used as field emitters. Vacuum **123**, 17–32 (2016). https://doi.org/10.1016/j.vacuum.2015.10.004

188. X. Yang, Z. Li, F. He, M. Liu, B. Bai, W. Liu, Z. Qiu, H. Zhou, C. Li, Q. Dai, Coaxial carbon@boron nitride nanotube arrays with enhanced thermal stability and compressive mechanical properties. Small **11**, 3710–3716 (2015). https://doi.org/10.1039/C6NR01199C

189. K.J. Sankaran, M. Afsal, S.C. Lou, H.C. Chen, C. Chen, C.Y. Lee, L.J. Chen, N.H. Tai, Electron field emission enhancement of vertically aligned ultrananocrystalline diamond-coated ZnO core–shell heterostructured nanorods. Small **10**, 179–185 (2014). https://doi.org/10.1002/smll.201301293

190. T.H. Chang, P.Y. Hsieh, S. Kunuku, S.C. Lou, D. Manoharan, K.C. Leou, I.N. Lin, N.H. Tai, High stability electron field emitters synthesized via the combination of carbon nanotubes and N2-plasma grown ultrananocrystalline diamond films. ACS Appl. Mater. Interfaces **7**, 27526–27538 (2015). https://doi.org/10.1021/acsami.5b09778

191. A.K. Geim, K.S. Novoselov, The rise of graphene. Nat. Mater. **6**, 183–191 (2007). https://doi.org/10.1038/namt1849

192. Z.S. Wu, S. Pei, W. Ren, D. Tang, L. Gao, B. Liu, F. Li, C. Liu, H.M. Cheng, Field emission of single-layer graphene films prepared by electrophoretic deposition. Adv. Mater. **21**, 1756–1760 (2009). https://doi.org/10.1002/adma.200802560

193. U.A. Palnitkar, R.V. Kashid, M.A. More, D.S. Joag, L.S. Panchakarla, C.N.R. Rao, Remarkably low turn-on field emission in undoped, nitrogen-doped, and boron-doped graphene. Appl. Phys. Lett. **97**, 063102 (2010). https://doi.org/10.1063/1.3464168

194. N. Soin, S.S. Roy, S. Roy, K.S. Hazra, D.S. Misra, T.H. Lim, C.J. Hetherington, J.A. McLaughlin, Enhanced and stable field emission from in situ nitrogen-doped few-layered graphene nanoflakes. J. Phys. Chem. C **115**, 5366–5372 (2011). https://doi.org/10.1021/jp110476m

195. A.T.H. Chuang, J. Robertson, B.O. Boskovic, K.K.K. Koziol, Three-dimensional carbon nanowall structures. Appl. Phys. Lett. **90**, 123107 (2007). https://doi.org/10.1063/1.2715441

196. K.J. Sankaran, T.H. Chang, S.K. Bikkarolla, S.S. Roy, P. Papakonstantinou, S. Drijkoningen, P. Pobedinskas, M.K. Van Bael, N.H. Tai, I.N. Lin, K. Haenen, Growth, structural and plasma illumination properties of nanocrystalline diamond-decorated graphene nanoflakes. RSC Adv. **6**, 63178–63184 (2016). https://doi.org/10.1039/C6RA07116C

197. K.J. Sankaran, C.J. Yeh, S. Drijkoningen, P. Pobedinskas, M.K. Van Bael, K.C. Leou, I.N. Lin, K. Haenen, Enhancement of plasma illumination characteristics of few-layer graphene-diamond nanorods hybrid. Nanotechnology **28**, 065701 (2017). https://doi.org/10.1088/1361-6528/aa5378

198. Z. Remes. M. Nesladek, K. Haenen, K. Watanabe, T. Taniguchi, The optical absorption and photoconductivity spectra of hexagonal boron nitride single crystals. Phys. Status Solidi A **202**, 2229–2233 (2005). https://doi.org/10.1002/pssa.200561902

199. K.J. Sankaran, D.Q. Hoang, S. Kunuku, S. Korneychuk, S. Turner, P. Pobedinskas, S. Drijkoningen, M.K. Van Bael, J.D. Haen, J. Verbeeck, K.C. Leou, I.N. Lin, K. Haenen, Enhanced optoelectronic performances of vertically aligned hexagonal boron nitride nanowalls-nanocrystalline diamond heterostructures. Sci. Rep. **6**, 29444 (2016). https://doi.org/10.1038/srep29444

200. M.J. Powers. M.C. Benjamin, L.M. Porter, R.J. Nemanich, R.F. Davis, J.J. Cuomo, Observation of a negative electron affinity for boron nitride. Appl. Phys. Lett. **67**, 3912–3914 (1995). https://doi.org/10.1063/1.115315

201. K.P. Loh, I. Sakaguchi, M.N. Gamo, Surface conditioning of chemical vapor deposited hexagonal boron nitride film for negative electron affinity. Appl. Phys. Lett. **74**, 28–30 (1999). https://doi.org/10.1063/1.123122

202. K.J. Sankaran, D.Q. Hoang, S. Korneychuk, S. Kunuku, J.P. Thomas, P. Pobedinskas, S. Drijkoningen, M.K. Van Bael, J.D. Haen, J. Verbeeck, K.C. Leou, K.T. Leung, I.N. Lin, K. Haenen, Hierarchical hexagonal boron nitride nanowall–diamond nanorod heterostructures with enhanced optoelectronic performance. RSC Adv. **6**, 90338–90346 (2016). https://doi.org/10.1039/C6RA19596B

203. J.G. Eden, S.J. Park, N.P. Ostrom, S.T. McCain, C.J. Wagner, B.A. Vojak, J. Chen, C. Liu, P. von Allmen, F. Zenhausern, D.J. Sadler, C. Jensen, D.L. Wilcox, J.J. Ewing, Microplasma devices fabricated in silicon, ceramic, and metal/polymer structures: arrays, emitters and photodetectors. J. Phys. D Appl. Phys. **36**, 2869–2877 (2003). https://doi.org/10.1088/0022-3727/36/23/001

204. S.J. Park, J. Chen, C.J. Wagner, N.P. Ostrom, C. Liu, J.G. Eden, Microdischarge arrays: a new family of photonic devices. IEEE J. Sel. Top. Quantum Electron. **8**, 387–394 (2002). https://doi.org/10.1109/2944.991409

205. J.G. Eden, S.J. Park, N.P. Ostrom, K.F. Chen, Recent advances in microcavity plasma devices and arrays: a versatile photonic platform. J. Phys. D Appl. Phys. **38**, 1644–1648 (2005). https://doi.org/10.1088/0022-3727/38/11/002

206. K.H. Becker, K.H. Schoenbach, J.G. Eden, Microplasmas and applications. J. Phys. D Appl. Phys. **39**, R55–R70 (2006). https://doi.org/10.1088/0022-3727/39/3/R01

207. J.M. Thores, R.S. Dhariwal, Electric field breakdown at micrometer separations. Nanotechnology **10**, 102–107 (1999). https://doi.org/10.1088/0957-4484/10/1/020

208. R. Dorai, M.J. Kushner, A model for plasma modification of polypropylene using atmospheric pressure discharges. J. Phys. D Appl. Phys. **36**, 666–685 (2003). https://doi.org/10.1088/0022-3727/36/6/309

209. T. Callegari, R. Gante, J.P. Boeuf, Diagnostics and modeling of a macroscopic plasma display panel cell. J. Appl. Phys. **88**, 3905–3913 (2000). https://doi.org/10.1063/1.1308094

210. S.J. Park, K.F. Chen, N.P. Ostrom, J.G. Eden, 40000pixel40000pixel arrays of ac-excited silicon microcavity plasma devices Appl. Phys. Lett. **86**, 111501 (2005). https://doi.org/10.1063/1.1880441

211. D. Mariotti, K. Ostrikov, Tailoring microplasma nanofabrication: from nanostructures to nanoarchitectures. J. Phys. D Appl. Phys. **42**, 092002 (2009). https://doi.org/10.1088/0022-3727/42/9/092002

212. S.S. Yang, J.K. Lee, S.W. Ko, H.C. Kim, J.W. Shon, Two-dimensional kinetic and three-dimensional fluid-radiation transport simulations of plasma display panel. Contrib. Plasma Phys. **44**, 536–541 (2004). https://doi.org/10.1002/ctpp.200410076

213. H. Wang, G. Li, L. Jia, L. Li, G. Wang, High-temperature anisotropic silicon-etching steered synthesis of horizontally aligned silicon-based Zn_2SiO_4 nanowires. Chem. Commun. **7**, 3786–3788 (2009). https://doi.org/10.1039/B906787F

214. P. Kurunczi, J. Lopez, H. Shah, K. Becker, Excimer formation in high-pressure microhollow cathode discharge plasmas in helium initiated by low-energy electron collisions. Int. J. Mass Spectrum. **205**, 277–283 (2001). https://doi.org/10.1016/S1387-3806(00)00377-8

215. T. Svensson, M. Andersson, L. Rippe, J. Johansson, S. Folestad, S.A. Engels, High sensitivity gas spectroscopy of porous, highly scattering solids. Opt. Lett. **33**, 80–82 (2008). https://doi.org/10.1364/OL.33.000080

216. M. Saito, T. Hiraga, M. Hattori, S. Murakami, T. Nakai, An investigation of pipeline materials for continuous hyperpolarized 129Xe gas spectroscopy. Magn. Reson. Imaging **23**, 607–610 (2005). https://doi.org/10.1016/j.mri.2005.02.004

217. L. Que, C.G. Wilson, Y.B. Gianchandani, Microfluidic electrodischarge devices with integrated dispersion optics for spectral analysis of water impurities. J. Microelectromech. Syst. **14**, 185–191 (2005). https://doi.org/10.1109/JMEMS.2004.839337

218. A.K. Chakraborty, A.J. Golumbfskie, Polymer adsorption-driven self-assembly of nanostructures. Annu. Rev. Phys. Chem. **52**, 537–573 (2001). https://doi.org/10.1146/annurev.physchem.52.1.537

219. K.H. Schoenbach, M. Moselhy, W. Shi, R. Bentley, Microhollow cathode discharges J. Vac. Sci. Technol. A **21**, 1260–1265 (2003). https://doi.org/10.1116/1.1565154

220. P. Kurunczi, H. Shah, K. Becker, Hydrogen Lyman-α and Lyman-β emission from high-pressure microhollow cathode discharges in Ne-H2 mixtures. J. Phys. B At. Mol. Opt. Phys. **32**, L651–L658 (1999). https://doi.org/10.1088/0953-4075/32/22/103

221. P. von Allmen, D.J. Sadler, C. Jensen, N.P. Ostrom, S.T. McCain, B.A. Vojak, J.G. Eden, Linear, segmented microdischarge array with an active length of ∼1 cm: cw and pulsed operation in the rare gases and evidence of gain on the 460.30 nm transition of Xe+. Appl. Phys. Lett. **82**, 4447–4449 (2003). https://doi.org/10.1063/1.1585137

222. P. von Allmen, S.T. McCain, N.P. Ostrom, B.A. Vojak, J.G. Eden, F. Zenhausern, C. Jensen, M. Oliver, Parallel vacuum arc discharge with microhollow array dielectric and anode Appl. Phys. Lett. **82**, 2562–2564 (2003). https://doi.org/10.1063/1.4890124

223. J.W. Frame, D.J. Wheeler, T.A. DeTemple, J.G. Eden, Microdischarge devices fabricated in silicon. Appl. Phys. Lett. **71**, 1165–1167 (1997). https://doi.org/10.1063/1.119614

224. R. Block, O. Toedter, K.H. Schoenbach, *Proceedings of the 30th AIAA Plasma Dynamics and Lasers Conference*, Norfolk, VA (1999), p. 3434

225. R. Block, M. Laroussi, F. Leipold, K.H. Schoenbach, in *Proceedings of 14th International Symposium on Plasma Chemistry*, Prague, Czech Republic (1999), p. 945

226. G. Auday, P.H. Guillot, J. Galy, Secondary emission of dielectrics used in plasma display panels. J. Appl. Phys. **88**, 4871–4874 (2000). https://doi.org/10.1063/1.1290461

227. L.S. Pan, D.R. Kania, *Diamond: Electronic Properties and Applications* (Klumer Academic, Boston, 1995)

228. P.K. Bachmann, V. van Elsbergen, D.U. Wiechert, G. Zhong, J. Robertson, CVD diamond: a novel high γ-coating for plasma display panels? Diam. Relat. Mater. **10**, 809–817 (2001). https://doi.org/10.1016/S0925-9635(01)00377-6

229. K. Chakrabarti, R. Chakrabarrti, K.K. Chattopadhyay, S. Chaudhrui, A.K. Pal, Nano-diamond films produced from CVD of camphor. Diam. Relat. Mater. **7**, 845–852 (1998). https://doi.org/10.1016/S0925-9635(97)00312-9

230. S. Kunuku, K.J. Sankaran, C.L. Dong, N.H. Tai, K.C. Leou, I.N. Lin, Development of long lifetime cathode materials for microplasma application. RSC Adv. **4**, 47865–47875 (2014). https://doi.org/10.1039/C4RA08296F

Chapter 6
Trends of Organic Electrosynthesis by Using Boron-Doped Diamond Electrodes

Carlos A. Martínez-Huitle and Siegfried R. Waldvogel

Abstract The electro-organic synthesis is currently experiencing a renaissance due to the tremendous contributions of various electrocatalytic materials as well as the use of electric current as an inexpensive and suitable reagent to drive the electrosynthetic transformations, avoiding conventional chemical oxidizers or reducing agents. Consequently, electrosynthesis has a significant technical impact, because these processes can be easily scaled up, benefiting from advantages such as versatility, environmental compatibility (possibility of recovering and recycling non-converted substrates), automation (switching on or off electric current), inherent safety and potential cost effectiveness among others. Although many novel electrode materials have been developed and established in electro-organic synthesis, diamond films emerge as a novel and sustainable solution in selective electrochemical transformations for value-added organic products. This chapter aims to offer an overview on the recent synthetic developments which represent hot topics in BDD electro-organic synthesis.

6.1 Introduction

Novel electrosynthetic approaches are of particular interest for technical innovations and future industrial applications because they enable the direct use of electricity to generate valuable compounds. The combination with other approaches in organic chemistry or concepts of contemporary synthesis allows the establishment of powerful synthetic tools. Additionally, the substitution of chemical redox

C. A. Martínez-Huitle (✉)
Institute of Chemistry, Federal University of Rio Grande do Norte,
Campus Universitário, Lagoa Nova, Natal, RN 59072-970, Brazil
e-mail: carlosmh@quimica.ufrn.br

C. A. Martínez-Huitle · S. R. Waldvogel (✉)
Institut für Organische Chemie, Johannes Gutenberg-Universität Mainz,
Duesbergweg 10-14, 55128 Mainz, Germany
e-mail: waldvogel@uni-mainz.de
URL: http://www.chemie.uni-mainz.de/OC/AK-Waldvogel/

© Springer Nature Switzerland AG 2019
N. Yang (ed.), *Novel Aspects of Diamond*, Topics in Applied Physics 121,
https://doi.org/10.1007/978-3-030-12469-4_6

reagents by electricity can be considered a substantial step towards green chemical processes [1, 2].

Recently, synthetic organic approaches by using electricity is experiencing a renaissance due to the generation of a variety of valuable synthetic pathways, new electrochemical reactions, synthesis of complex organic molecules, use of novel electrode materials, coupling conventional chemical procedures with electrochemistry as well as new protocols for controlling the selectivity of electrochemical transformations. In this frame, novel and value-added organic products can be synthetized and innovative applications can be also scaled-up [1–3].

The existence of direct or mediated electrochemical reactions has allowed the proposal of three main approaches for the synthetic pathways by electrosynthesis [2]:

(i)   Direct electro-conversion (or direct electron transfer) occurs at the electrode surface (by using inert electrode), adjusting the appropriate electrode potential to promote specific selectivity.

(ii) Mediated electrochemical transformations occur by using active electrode with electrocatalytically active sites on the surface (it is a compact and electrically conductive coating that can be regenerated in-situ), which provide a unique reactivity with minor applied potential-dependence due to the active electrode layer.

(iii) Mediated electrochemical reactions by using electro-active species within the supporting electrolyte are also feasible in which organic compounds are selectively transformed into products. The mediators (redox-active species) are considered as electrochemical generated/regenerated reagents, with unique reactivity, avoiding over-potentials.

In general, there are two different operational modes to conduct electrolytic conversions: controlled potential or constant current conditions. Electrolysis at controlled potential leads to selective conversion as a result of the applied potential at the working electrode since it is chosen to match that of the substrate. Unfortunately, this requires a three-electrode-arrangement and therefore, a much more costly electronic periphery. In addition, such conversions can take prolonged electrolysis times. The galvanostatic mode of conducting the electrolysis provides a technically simple two-electrode setup. In this frame, undivided or divided electrolysis cells are usually used, taking into account the electrochemical and chemical reversibility of the conversion as well as the stability desired product (it can be oxidized or reduced at the working electrode), in such cases, flow cells might help [1, 2, 4]. Based on coupled nature of anodic and cathodic processes in electrochemical transformations, both reactions can be employed for preparative and synthetic purposes [5], following the approaches described above.

In general, the use of non-active electrodes, which do not provide any catalytic active site or their adsorption properties, is recommended in electrosynthesis [1–3]. These only act as an inert electrode and as a sink for the removal of electrons. In principle, only outer-sphere reactions are possible with this kind of anode, however, mediators can be electrochemically produced and anhydrous electrolytes are used, respectively [3]. For the later, the suppression of electrolysis of or the loss of the applied current in the production of molecular hydrogen and oxygen in cathodic and anodic processes, respectively, is the main target.

The above model for electro-organic conversions presupposes that an electrode material with a high offset potential, non-corrosion phenomenon or fouling deactivation to be the best choice [1, 2].

Boron-doped diamond (BDD) thin films still represent relatively new electrode materials that have received great attention, because they exhibit several technologically important characteristics including an inert surface with low adsorption properties, remarkable corrosion stability even in strongly acidic media and show an utmost high over-potential for O_2 evolution [1, 2, 6, 7]. Based on these superior and outstanding features, BDD is excellent material for different electrochemical applications [1, 2, 7, 8]. BDD electrodes have been defined as non-active materials, since it is expected that they do not provide any catalytically active site for the adsorption of reactants and/or products in aqueous or non-aqueous media [6] as well as the electrogeneration of active species on its surface is feasible [7, 8]. In the past years, BDD electrodes became available in a variety of sizes, different support materials and with an enhanced stability in organic media [1, 2, 6]. BDD films thus can be used for new oxidation/reduction reactions which otherwise are not possible in water [9]. The desired non-destructive pathway towards the product competes with the electrochemical incineration at high currents [9]; however, buffering electrical conditions can be beneficial for a synthetic and non-destructive transformation [1, 2].

6.2 BDD Features in Aqueous and Non-aqueous Media

In aqueous media, BDD characteristics such as their large potential window, low species adsorption, corrosion stability, high efficiency and low double-layer capacitance and background current, among many others [7–9], promoted in particular the use of BDD for destruction of organic pollutants [9] and waterborne pathogens [10]. These exceptional BDD properties are clearly evident when comparing its potentiodynamic profile with that exhibited by a typical platinum electrode, as illustrated in Fig. 6.1a. Lower background current (less than 1 μA) is observed in acidic and alkaline media. In the former, the oxygen evolution peak appears about +2.0 V versus SCE Fig. 6.1a; while, a broad oxidation peak at ca. +0.9 V versus SCE is exhibited, in the latter (Fig. 6.1b). This peak is due to water decomposition. Meanwhile, the hydrogen evolution reaction is shifted to negative potentials (−2.0 V and −1.5 V vs. SCE for acidic and alkaline media, respectively) [8]. Thus, a significant potential window is exhibited for BDD electrodes than those other electrocatalytic materials.

Recently, the Waldvogel lab [6] found that when fluorinated alcohols are used as electrolytes, instead of water, the potential window is tremendously opened up. Not

Fig. 6.1 Potentiodynamic profiles **a** for the BDD and Pt electrodes in acidic media (0.5 M H₂SO₄) and **b** for the BDD electrode in alkaline media (0.25 M NaOH + 0.5 M Na₂SO₄)

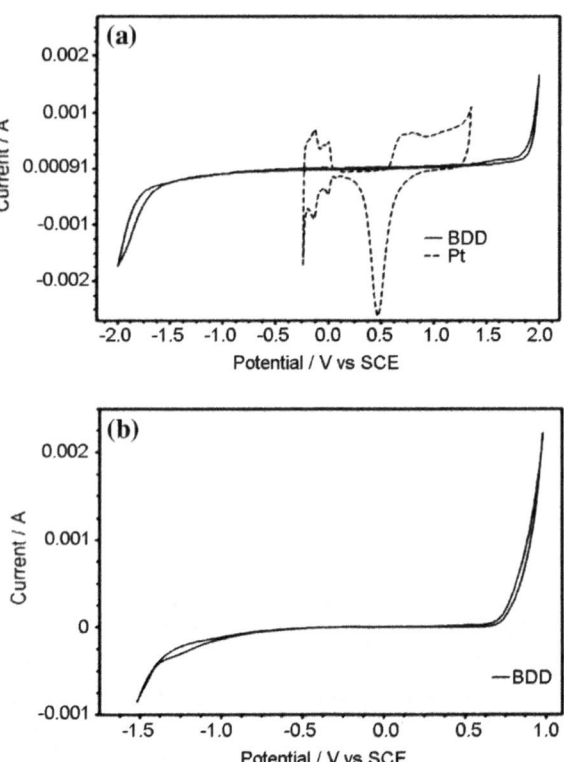

only on the anodic part, where the potential window is shifted to 3.2 V, but also in the reductive region, where almost 2 V were increased. Therefore, the use of 1,1,1,3,3,3-hexafluoroisopropanol (HFIP) as electrolyte represents the protic electrolyte with the largest electrochemical window of about 5 V when BDD is employed Fig. 6.2.

The supporting electrolyte used in the system is methyltriethylammonium methylsulfate [11, 12]. The limitation in the anodic part might be caused by the supporting electrolyte because slight conversions occur due to the impurities in the HFIP. In particular, BDD electrodes, for anodic electrochemical reactions, have very promising characteristics due to its unique reactivity which is based on the selective formation of oxyl radicals, highly reactive intermediates produced on its surface, no electrode fouling by coating of by-products or carbonization is anticipated. The resulting products consist of small molecules and will be easily removed from the electrolyte and the electrolysis cell. Since no deactivation of the anode surface is expected and a dramatically reduced effort for the maintenance of electrodes might be encountered the technical interest for these electrodes exists. Consequently, such self-cleaning electrodes based on BDD material may cause a small but important cost advantage for chemical processes [1, 2, 6, 11, 12].

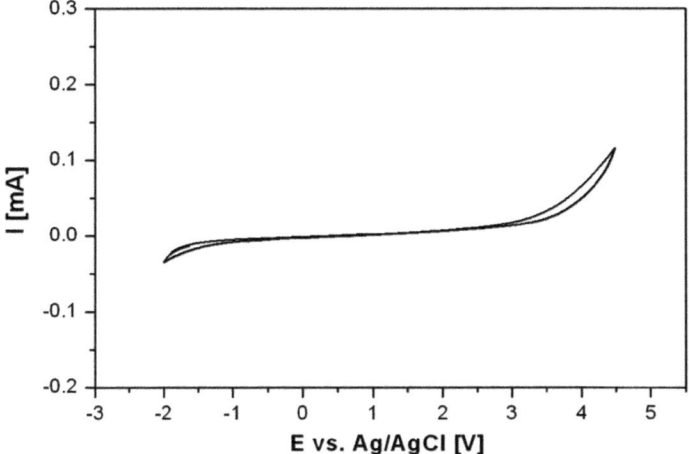

Fig. 6.2 Unusual broad electrochemical window for the system BDD/1,1,1,3,3,3-hexa-fluoroisopropanol (HFIP) with 0.1 M $Et_3NCH_3O_3SOCH_3$. Reproduced from [6]

6.3 Cathodic Synthesis on BDD Electrodes

The cathodic conversions are less elaborated than the anodic electro-synthetic approaches because the counter reactions are often not well controlled. However, the electric current is by far the less expensive reduction reagent which could be applied and few electrode materials with high offset potentials for the evolution of hydrogen can be used, as cathodes, in protic media [1, 2, 6]. Therefore, lead, mercury and cadmium show usually the best results. The high toxicity of those electrode materials and eventually formed organometallic species thereof lead to a reluctant application of such reductions [6]. Consequently, BDD can be an appropriate substitute for these heavy metal electrodes due to the protective cathodic polarization, no problems in stability as well as in the cathodic corrosion.

6.3.1 Electrochemical Approaches for Reducing CO_2

A chemical pathway for converting CO_2 and H_2 to methane was developed by Sabatier [13]. Since then, the utilization of CO_2 as the C1 feedstock for production of synthesis gas (e.g., CO/H_2) and valuable organic compounds (e.g., methanol, methane, ethylene, formic acid) have been promoted with various chemical, photochemical, biochemical, and electrochemical approaches [14]. The direct cathodic conversion and electrocatalytic reduction have been extensively investigated in aqueous and non-aqueous media during the past decades [14]. The most used cathodic electrodes for electrochemical approaches to convert CO_2 in valuable organic compounds have been mainly carbon materials [1, 14]. These cathodes exhibit a long lifetime during electrochemical experiments and a higher over-potential for hydrogen evolution reaction than most metallic cathodes, and consequently, suppressing the undesired reactions and achieving higher Faradaic efficiencies. In the case of *electrochemical reduction*, direct CO_2 conversion is attained at the electrode surface without any catalysts, and several carbon materials have been used, such as on graphite, glassy carbon, BDD, and CNTs [15–21]. Meanwhile, the *electrocatalytic reduction* of CO_2 employs homogeneous (in solution) or heterogeneous (at the electrode surface) catalysts to increase reduction efficiency, to improve the selectivity of reduction products, as well as to reduce the cost for CO_2 reduction. The employed catalysts involve various metals, metal alloys, metal oxides, metal complexes, organic and biomolecules.

6.3.1.1 Electrochemical Reduction of CO_2

Einaga and co-workers [16] investigated for the first time the electrochemical reduction of CO_2 at BDD cathode in a methanol solution containing tetrabutylammonium perchlorate by using voltammetry technique. A reduction peak of CO_2 was observed at a potential of -1.3 V (vs. Ag/Ag^+), achieving 97.5 mA cm^{-2} as a maximum current density. This result allows to perform bulk electrolysis at ambient conditions, obtaining as the main reaction products: HCHO, HCOOH and H_2. Higher Faradaic efficiency was obtained for HCHO (74%, at -1.7 (V vs. Ag/Ag^+)), while only 15% was achieved for HCOOH at -1.5 V. Regarding the H_2 formation, the Faradaic efficiency was less than 1.1% by applying potentials lower than -1.7 V. Another interesting feature is that, the electrochemical reduction of CO_2 was also performed by using different BDD electrodes with different amount of sp^2-impurities (graphite) in their surface, under the experimental conditions identified above [16]. According to the results, 15% of Faradaic efficiency was achieved for the HCHO generation with BDD cathodes with higher contents of sp^2-impurities. Conversely, high and stable Faradaic efficiencies were attributed to the use of BDD electrodes with lower amount of sp^2-impurities due to the sp^3-bonded carbon (high quality diamond grains). Additionally, the same group [16] demonstrated that the applicability of CH_3OH, NaCl aqueous solution and seawater as electrolytes for electrochemical CO_2 reduction on BDD, obtaining 36% of efficiency in seawater for the selective formation of formaldehyde.

Einaga and co-workers [22] investigated the electrochemical reduction of CO_2 in an aqueous ammonia solution with BDD electrodes, achieving a faradaic efficiency of about 24.3% for methanol production at -1.3 V (vs. Ag/AgCl), while 0.13% and 0.05% were also obtained for methane and CO, respectively as well as a 19.7% of hydrogen production. As the distribution of carbonaceous species in the aqueous solution depends on the pH conditions. This was studied in detail by the pH-dependence of electrochemical reduction of CO_2 by using BDD electrodes [22]. It is known that dissolved $CO_{2(aq)}$ is the dominant specie at pH < 5, while at pH levels from 7.5 to 9, HCO_3^- (bicarbonate ion) is dominant, and CO_3^{2-} (carbonate ion) is dominant above pH 12. Thus, in aqueous ammonium bicarbonate solution at pH 7.9 without CO_2 bubbling, methanol was produced, while no methanol production was observed at pH conditions over 10.6 and lower than 3.38. These results suggest that bicarbonate ions are reducible species and not CO_2 and CO_3^{2-} under these experimental conditions. Then, the proposed mechanism of methanol production in that system should be (1) [22]:

$$HCO_3^- + 5H_2O + 6e^- \rightarrow CH_3OH + 7OH^-$$ (1)

Based on the results reported in Jiwanti et al. [22], the selectivity for methanol production in aqueous ammonia solutions is due to the electrochemical reduction of bicarbonate ions which are formed by the reaction between ammonia and CO_2. In order to understand the effect of cations during the electrochemical reduction of CO_2 in aqueous solutions by using BDD electrodes, the same lab at Keio University [23] studied the production of formic acid when different cations, such as K^+, Na^+, Rb^+, and Cs^+, were present in alkaline solutions. By adjusting the pH to lower values, the presence of HCO_3^- becomes evident, which is deemed to be the species responsible for the formation of formic acid. At the same time, the concentration of the electrolyte influenced in the production of formic acid on BDD electrodes. A Faradaic efficiency of 71% was achieved in the case of a 0.075 M Rb^+ solution neutralized to pH 6.2 by the addition of HCl. In the case of a Cs^+ solution neutralized to pH 6.2, an efficient generation of formic acid was obtained with the more dilute concentration of 0.02 M (38%). Conversely, lower Faradaic efficiencies were achieved for K^+ (41%) and Na^+ (24%) solutions with solutions at around 0.5 M. These variations on the production of formic acid were attributed to the interaction between the alkali-metal cations and water molecules (solvation phenomena), which promotes a change on the structure of the electric double layer formed on the BDD electrode, and consequently, it favours the electrochemical conversions to formic acid [23].

The production of HCOOH by electrochemical reduction of CO_2 was investigated by using a circulation flow cell (Fig. 6.3) with BDD electrode, increasing the mass transport of CO_2 [24]. Then, a high faradaic efficiency for the production of HCOOH was achieved (94.7%) by applying 2 mA cm^{-2} at 200 mL min^{-1}. Moreover, CO and H_2 were also obtained in negligible amounts (0.6% and 4.1% for CO and H_2, respectively). This implies that the use of flow cell favours an increase on the mass-transport of the CO_2 towards BDD surface, and consequently, diamond surface seems to have a tendency for HCOOH production compared to the H_2 evolution reaction. On the other hand, the dependence of the faradaic efficiency and the production rate of HCOOH on current density were also investigated [24]. Electrochemical reduction experiments at BDD by applying 2, 5, 10, 15, and 20 mA cm^{-2} for 60 min with flow rates of 200 and 500 mL min^{-1} were found by the Einaga lab [24]. However, a decrease on the Faradaic efficiency was registered when an increase on the current density was achieved, decaying until 19% of HCOOH production at 20 mA cm^{-2}.

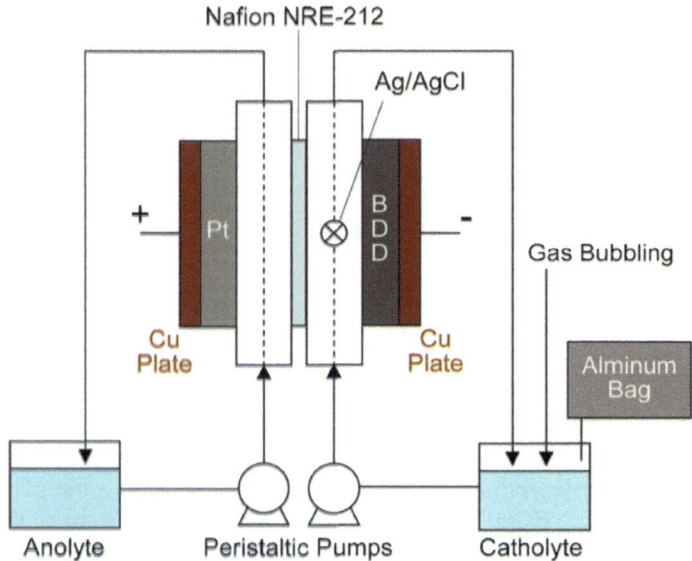

Fig. 6.3 Diagram of the two-compartment flow cell. Reproduced with permission from [24]. Copyright 2018, Wiley-VCH Verlag GmbH & Co

It is important to remark that, the selectivity for the production of HCOOH was more than 99%. The rate of the production was increased to 473 μmol m^{-2} s^{-1} at a current density of 15 mA cm^{-2} with a faradaic efficiency of 61%. The faradaic efficiency and the production rate are almost the same as or larger than those achieved using Sn and Pb electrodes [25]. Furthermore, the stability of the BDD electrodes was confirmed by a continuous 24 h operation.

6.3.1.2 Electrocatalytic Reduction of CO_2 on Metal Nanoparticles, Metal Oxides and Metal Complexes Using BBD Support Material

The electrocatalytic CO_2 reduction in different media (water, organic, and ionic liquid) was attained by Cu nanoparticles deposited on BDD surface Fig. 6.4. The variation of the sizes and the densities of copper nanoparticles did not alter the starting potential for CO_2 reduction (-0.1 V vs. NHE) if the surface coverage of copper nanoparticles is higher than 10%. Lower surface coverage of copper nanoparticles than 10% shifted the starting potential of CO_2 reduction in a negative direction, limiting the CO_2 reduction efficacy. However, the poor stability of Cu nanoparticles on its surface led to non-stable and repeatable CO_2 reduction current in these media [26].

The electrocatalytic reduction of CO_2 was also investigated by Panglipur et al. [27] and Jiwanti et al. [28]. In the first case [27], Cu-BDD electrodes were prepared

Fig. 6.4 SEM image of a planar diamond electrode decorated with copper nanoparticles. Reproduced with permission from [26]. Copyright 2013, American Chemical Society

by electrochemical reduction with various concentrations of $CuSO_4$ solutions (ranging from 2 to 10 mM). The electrocatalytic reduction of CO_2 was performed at various Cu-BDD electrodes using chronoamperometry technique by applying -1.2 V (vs. Ag/AgCl) during 60 min, detecting the formation of formaldehyde, formic acid, and acetic acid by HPLC. When BDD or Cu electrodes were used, the concentrations of by-products (formaldehyde and formic acid) were below 5 ppm. Similar results were obtained at all Cu-BDD electrodes, excepting at Cu-BDD electrode prepared by using 10 mM $CuSO_4$ solution where the concentration of formic acid was 8 ppm. Meanwhile, the amount of acetic acid increased with the increase of Cu amount deposited at BDD by using the $CuSO_4$ solutions ranging from 2 to 8 mM. Conversely, no acetic acid production was achieved by electrocatalytic reduction at BDD and Cu electrodes, as well as at Cu-BDD prepared by 10 mM $CuSO_4$. The large number of Cu deposited at BDD probably influences the Cu-BDD electrode to behave like Cu electrode. In the second case [28], modified BDD electrodes with Cu nanoparticles were prepared at different electrodeposition times (50, 100, 300 s) to produce electrocatalytically higher carbon-containing compounds from CO_2 reduction Fig. 6.5. The average sizes of the Cu particles on the Cu-BDD electrode surfaces were around 50–85 nm.

Higher faradaic efficiencies for ethanol (the main product), acetaldehyde and acetone (other C2 and C3 compounds) were obtained (42.4%, 13.7%, and 7%, respectively) by reduction at -1.0 V (vs. Ag/AgCl) for 2 h at Cu-BDD (100 s) electrode. Conversely, at Cu-BDD (50 s) electrode, the amount of Cu nanoparticles was not sufficient to react with CO_2 effectively. In fact, bare BDD electrode without Cu could not generate any products at -1.0 V (vs. Ag/AgCl), except H_2; meantime, for Cu-BDD (300 s), the efficiency dropped due to the instability of the Cu nanoparticles deposited on the BDD surface (see, Fig. 6.5) [28].

Meanwhile, when BDD was thermally coated with a RuO_2-deposit, the stability of the electrode material was preserved [29], obtaining as the main reduction products, HCOOH and CH_3OH, with Faradaic efficiencies of 40% and 7.7%, respectively, by applying -0.6 V (vs. SCE) in 0.4 M Britton−Robinson solutions (with adjusted pH values from 1.8 to 6.0) or in neutral media (0.3 M $NaClO_4$ in

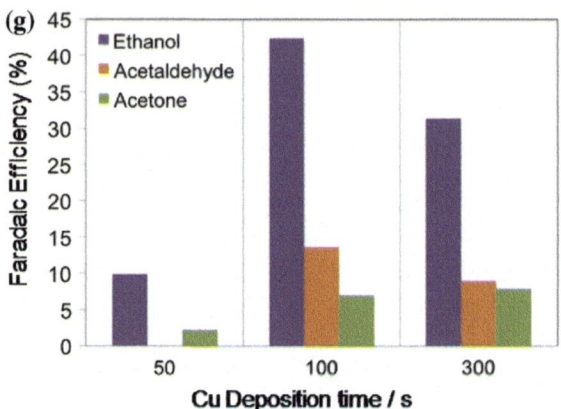

Fig. 6.5 **a**, **b**, **c** show SEM images of the Cu-BDD-50, Cu-BDD-100, Cu-BDD-300 electrodes before electrochemical reduction, and **d**, **e**, **f** show images of the same electrodes after electrochemical reduction of CO_2 at −1.0 V (vs. Ag/AgCl) for 2 h at room temperature and pressure. **g** Liquids produced at each electrode by electrochemical reduction for 2 h. Reproduced with permission from [28]. Copyright 2018, Elsevier

phosphate buffer). The production of H_2, CO and CH_4 was also achieved with 25.25%, 0.01% and 1.70%, respectively.

The use of BDD cathodes photochemically-functionalized with Co complex [30] has been also investigated for electrocatalytic CO_2 reduction to CO at a potential of −1.5 V (vs. Ag/AgCl 3 M), a close value to −1.89 V (vs. Fc/Fc$^+$) obtained on Co homogeneous catalysts in 0.1 M Bu_4NPF_6 in anhydrous acetonitrile. The catalytic current was consistent for more than 1000 cycles. The turnover frequency of this system was estimated to be 0.8 s^{-1} for a 16 h period under a constant potential of −1.8 V. This "smart" electrodes system thus exhibited good stability and electrocatalytic activity for electrocatalytic CO_2 reduction. Other report described the use of Ag nanoparticles deposited on oxygen-terminated BDD electrode to reduce selectively CO_2 by photoelectrochemical approach [31]. Excellent selectivity (estimated $CO:H_2$ mass ratio of 318:1) and recyclability (stable for five cycles of 3 h each) were obtained for the optimum Ag nanoparticle-modified BDD electrode at −1.1 V (vs. RHE) under 222 nm of irradiation. The high efficiency and stability of this catalyst were due to the in situ photoactivation of the BDD surface during the photoelectrochemical reaction, revealing that BDD can be used as a high-energy electron source with co-catalysts in photochemical conversion.

Then, as a promising approaches, physical trapping of metal catalysts, thermal formation of metal carbide as well as the covalent bonding of metal catalysts to carbon electrodes, can be considered as a solve, combined with other features of BDD, to the stability and electrocatalytic activity for CO_2 reduction.

6.3.2 Reduction of Oximes

The synthesis of amines starting from the corresponding oximes is a very common method since amines with the full carbon skeleton is provided. The reduction can be performed in a Bouveault-Blanc-type reaction by treatment with alkali metals in the presence of a proton source [32]. In order to avoid reagent waste and facing the safety issues of large quantity of alkali metals the conversion can be electrochemically conducted, wherein mainly reductions at a mercury pool are used [33]. The high over-potential for the H_2 evolution by electrolysis of protic electrolytes should be beneficial for cathodic transformations [34]. The electrochemical and chemoselective reduction of cyclopropyl phenylketone oxime (**1**) to *rac*-α-cyclopropyl benzylamine (**2**) was achieved with 96% yield on a BDD cathode with a niobium support (Scheme 6.1).

As electrolyte a MeOH/NaOMe mixture in a divided cell was applied. The results are similar to the conversion at the same conditions (1% $NaOCH_3$ in anhydrous MeOH, 40 °C, 3.4 A dm^{-2}) using lead cathodes [35]. The cathodic reduction is superior to a catalytic hydrogenation of oxime **1** to the desired compound **2**. The product of the conventional reduction by hydrogenation on noble

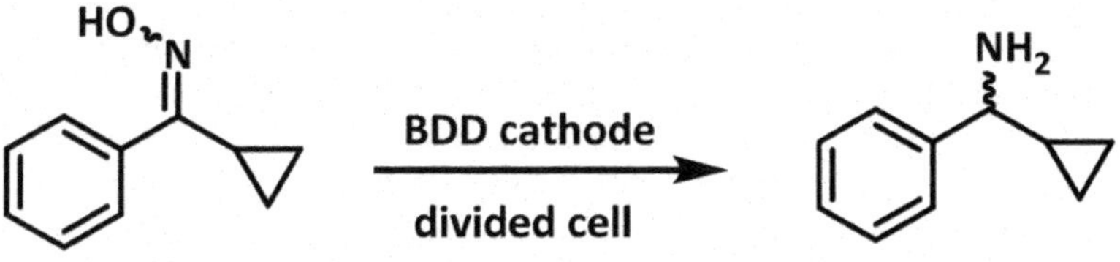

Scheme 6.1 Oxime reduction at BDD cathodes providing access to cyclopropyl amines

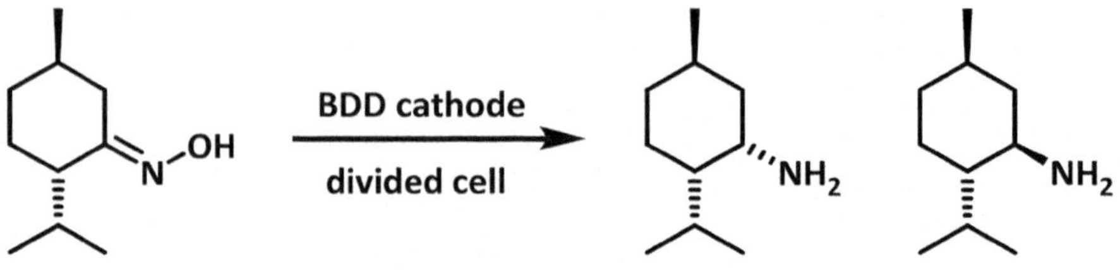

Scheme 6.2 Electro-organic synthesis of menthyl- and neomenthylamines

metal catalyst is concomitant with significant amounts of ring-opened compounds making purification by distillation very challenging.

Inspired by this work, the electroorganic reduction was tested on the sterically more demanding menthone oxime (**3**). The substrate **3** which is derived from optically pure L-menthone could provide upon reduction the both epimeric products, (+)-neomenthylamine (**4**) and (−)-menthylamine (**5**), respectively (Scheme 6.2).

For both diastereomers there exist unique applications [36]. The reduction was previously conducted by treatment with 30 equiv. of sodium metal leading safety concerns for a scale-up [37]. However, detailed studies with BDD cathodes revealed that **3** is not a useful substrate. Almost no conversion is observed and only traces of both epimeric amines could be detected. This indicates that the cathodic conversion occurs directly at the BDD surface and bulkiness of the substrate has a strong influence. In that conversion cathodic treatment using lead results in almost quantitative reduction and a splendid current efficiency (6 F mol^{-1}, 66%) [38].

6.3.3 Reduction of Nitro Arenes

Several synthetic approaches to nitrones have been established. However, these methods often require several synthetic steps and sacrificial metals, as well as a careful control of the reaction conditions to avoid mixture of products. In this context, the use of electroreduction approach with BDD electrodes for the synthesis of nitrones was proposed for first time by Waldvogel's group [39], avoiding the metals by using green solvent conditions with aldehydes and nitro compounds as reagents. It is well-known that the reduction of aldehydes under electrochemical conditions produces alcohols or pinacols, but the electroreduction of nitro derivatives can be attained in the presence of the aldehyde to yield the nitrone directly (Scheme 6.3), reducing the operational steps.

Moreover, since no metals are utilized, the subsequent work-up and purification are significantly facilitated. All the used enals afforded different nitrones in good yields (up to 72%) by using BDD cathode in 0.01 M NBu$_4$BF$_4$ as supporting

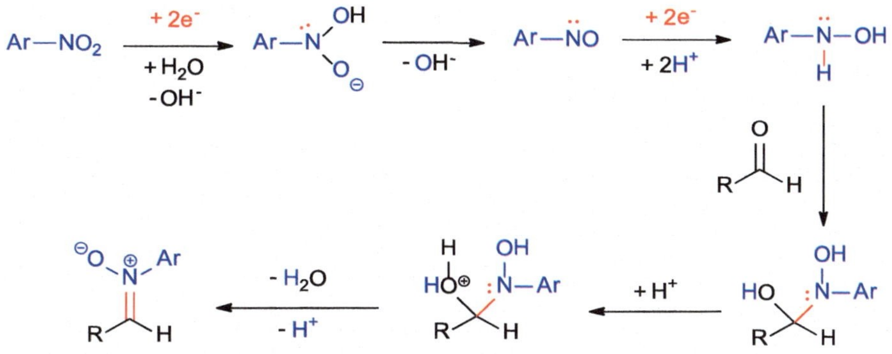

Scheme 6.3 Direct cathodic formation of nitrones starting from nitro arene

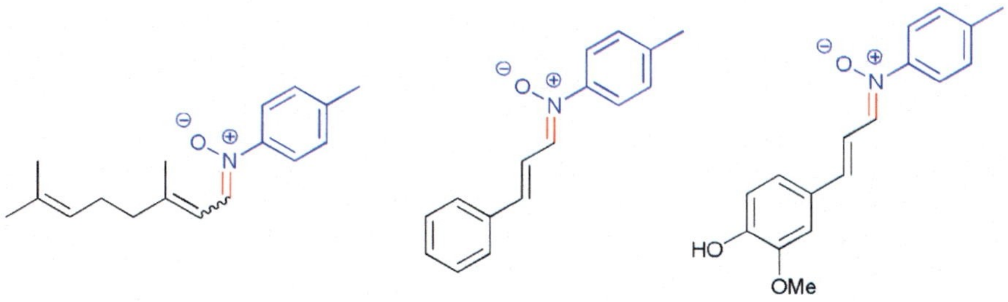

71%   72%   70%

Fig. 6.6 Selection of nitrones obtained by electrochemical cathodic conversions at BDD

electrolyte by applying 7.5 mA cm^{-2} (Fig. 6.6). This new pathway represents a very important feature for a technical scale-up and from a waste-avoiding point of view [39]. Interestingly, no pinacol formation of the aldehyde component was detected. It is noteworthy, that BDD provides not only the yield for nitrones but also cleanest and best conversion. In addition, this method proofed to be scalable as well.

6.3.4 Reductive Carboxylation

Carbon dioxide is good electron acceptor which can be employed for cathodic transformations [40]. For the carbon dioxide balance and due to ecological considerations the electrochemical fixation of this particular C1 building block into chemicals seems to be very attractive. 2-Hydroxy-4-methylsulfanylbutyric acid (**7**) often named as methionine hydroxy analogue (MHA) is an important technical product required animal feeding. **7** exhibits an improved bioavailability compared

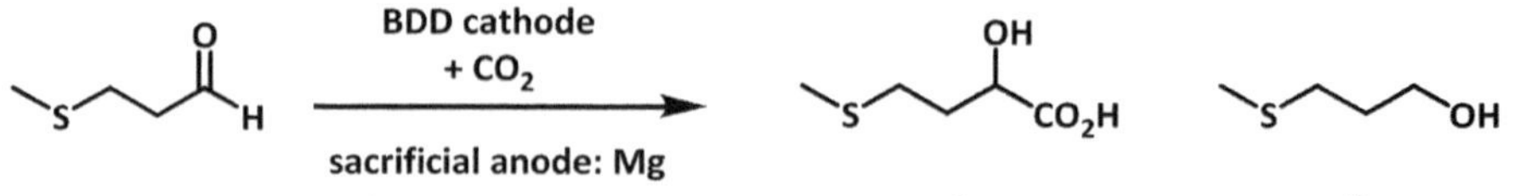

Scheme 6.4 Carboxylation of aldehydes yielding hydroxycarboxylates

to the essential amino acid methionine. MHA **7** is made on a several thousand ton scale by treatment of methylsulfanylpropionaldehyde (**6**) with cyanhydric acid to form the corresponding cyanohydrins. Subsequent nitrile hydrolysis provides **7**. An electrochemical approach to **7** exploits a reductive carboxylation of **6** with magnesium as sacrificial anode (Scheme 6.4).

The transformation requires a CO_2 atmosphere using for both electrodes magnesium. Electrolysis in a DMF/NBu$_4^+$BF$_4^-$ electrolyte resulted upon application of 5 F mol^{-1} a conversion of 90% with a selectivity of 75% for **7** [41]. The conversion can also be carried out on BDD electrodes. If a divided cell is used, less magnesium has to be applied. Unfortunately, the direct reduction of the aldehyde **6** providing alcohol **8** is a significant side reaction. The yield for **7** in this process seems to be significantly lower than on magnesium electrodes (conversion 66%, current efficiency 22%, divided flow cell, Nafion membrane, BDD on silicon support, 20–25 °C, 0.6 A dm^{-2}) [42]. However, this is an impressive example for the versatility of BDD electrodes in preparative electro-organic synthesis.

6.3.5 Reductive Dimerization

The use of the BDD electrode for the electrochemical reduction of methyl cinnamate was investigated by Nishiyama and co-workers [43] to obtain new neolignan-type bioactive substances. As shown in Fig. 6.7, the radical intermediate derived from cinnamate by a one-electron reduction is an interesting cathodic process. Therefore, the reductive dimerization of cinnamic acid derivatives could lead to unprecedented neolignan-type dimeric compounds.

Methyl cinnamate (**9**) was cathodically electrolyzed to the hydrodimer, dimethyl 3,4-diphenylhexanedioate (**10**), under constant current electrolysis conditions in a

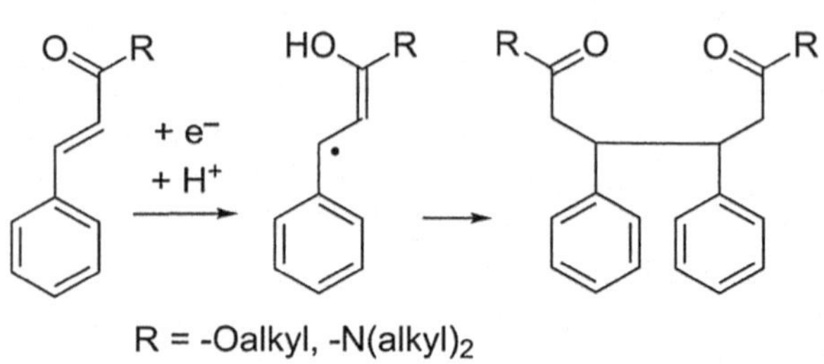

Fig. 6.7 Expected coupling products from one-electron reduction of C6–C3 substrates

R = -Oalkyl, -N(alkyl)$_2$

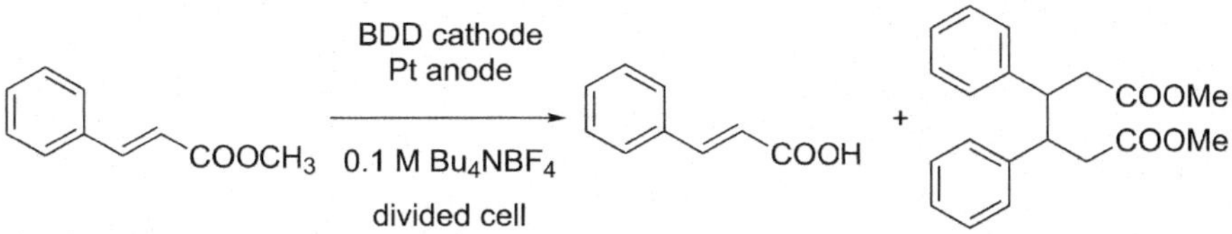

Scheme 6.5 Cathodic coupling of cinnamates

divided cell by applying 1.29 mA cm^{-2} in acetonitrile containing 0.33 M pH 7.0 phosphate buffer, obtaining 85% of yield (racemate/meso = 74:26, 2.5 F current). After separation of the diastereomeric mixture, ($\pm$)-dimethyl 3,4-diphenylhexanedioate (10) was subjected to the chemical conversion into the novel neolignan-type derivatives 4,5-diphenyloxepan-2-one and 4,5-diphenyloxepane (Scheme 6.5). BDD electrode gave predominantly the coupling product and behave superior to other electrode materials. These results underline the synthetic utility of the boron-doped diamond cathode.

6.4   Anodic Transformations on BDD Electrodes

The most important feature of BDD electrode is its widely electrochemical window in aqueous and protic media and the specific formation of oxyl radicals [7–10, 22], which can be created in higher concentration in aqueous media or alcoholic solutions [6, 22]. This opens unique reaction pathways because the generated hydroxyl or alkoxyl species may act either as radicals being implemented into the product or serve as strong oxidants, which transform the substrate. In the latter case, reaction sequence the oxyl spin centres play the role of a mediator. Therefore, anodic transformations are the most efficient at BDD anodes [6].

6.4.1   Electrochemical C–H Amination

The sustainable installation of electron-releasing groups such as hydroxyl or amino moieties is challenging and commonly considered to be the Holy Grail in C–H functionalization [44]. Consequently, tremendous efforts have been made to perform this transformation in a catalytic manner [45]. In modern versions, this type of reaction does not require leaving groups [46] or is viable with ammonia as the NH$_2$ source [47]. However, these approaches require significant loadings of expensive catalysts based on copper or noble metals. Recently, Yoshida and co-workers reported a powerful method for the electrochemical amination of electron-rich arenes (11) [48] via pyridinium derivatives (12) and Zincke salts (Scheme 6.6).

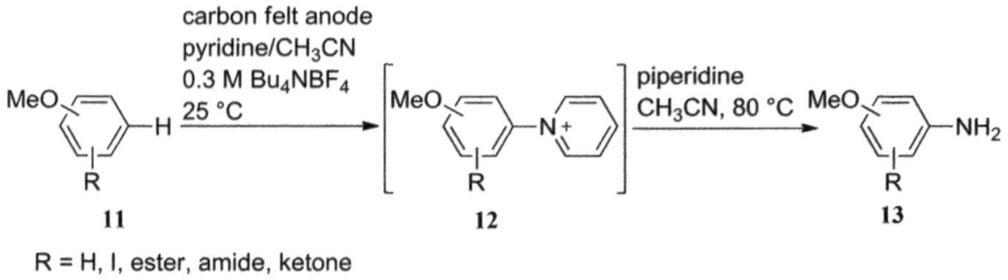
carbon felt anode
pyridine/CH3CN
0.3 M Bu4NBF4
25 °C
MeO
H
R
11
MeO
N+
R
12
piperidine
CH3CN, 80 °C
MeO
NH2
R
13
R = H, I, ester, amide, ketone

Scheme 6.6 Yoshida amination reaction by anodic conversion

A broad variety of arenes has been aminated in good to excellent yields of up to 99% by using carbon felt as the anode material. However, the approach is limited to activated, quite electron-rich substrates, e.g. methoxy-substituted arenes [44, 48].

Nevertheless, the employment of anodes with a higher performance at more-positive potentials, for example, BDD [49] might allow the electrolysis of less-activated substrates. Additionally, multiple conversions at large aromatic scaffolds might be viable. In fact, the method initially described by Yoshida et al. [48] for electron-rich arenes was expanded by the Waldvogel lab to such less-activated aromatic systems (simple alkylated benzene derivatives) [49] as well as for the twofold amination of naphthalene [50] by using BDD anodes. In the first case [49], BDD showed a significantly better performance for the electrochemical amination of m-xylene (**14**) than platinum (Scheme 6.7), glassy carbon, carbon felt, carbon fleece, or isostatic graphite, in all current densities applied (ranging from 2 to 12 mA cm^{-2}). Except BDD all other anodes suffer from carbonization of the substrate (anode fouling).

The highest isolated yield was 60% of 2, 4-dimethylaniline (**15**), at a current density of 8 mA cm^{-2}. The effect of supporting electrolyte concentration (tetrabutylammonium tetrafluoroborate at different concentrations ranging from 0.1 to 0.6 M) was also investigated, achieving 60% of 2, 4-dimethylaniline with 0.2 and 0.3 M electrolyte solutions. Consequently, a 0.2 M tetrabutylammonium tetrafluoroborate solution was used as the electrolyte for the anodic amination of alkylated arenes (see Table 6.1) using BDD at 22 °C by applying 4 mA cm^{-2} [49]. The 1,3-alkyl substitution pattern leads to the corresponding aniline derivatives, achieving 60% of conversion (Table 6.1, entry 1). Isopropyl substituents on the substrate **18** did not interfere with the amination process; obtaining 50% of the

Scheme 6.7 Electrochemical amination of less activated arenes

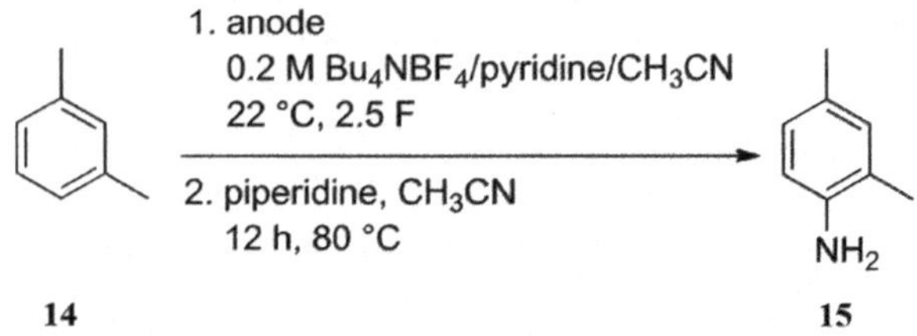
1. anode
0.2 M Bu4NBF4/pyridine/CH3CN
22 °C, 2.5 F
2. piperidine, CH3CN
12 h, 80 °C
NH2
14
15

Table 6.1 Scope of anodic amination at BBD electrodes of less activated substrates

Entry	Substrate	Product	Yield (%)[a]
1	**16**	**17**	60
2	**18**	**19**	50
3	**20**	**21**	61
4	**22**	**23**	50
5	**24**	**25**	32
		26	8
6	**27**	**28**	20
		29	35
7	**30**	**31**	30
8	**32**	**33**	24

corresponding aniline **19** (Table 6.1, entry 2). Substrates with non-branched alkyl substituents can also be successfully treated. Tetrahydronaphthalene **24** is preferentially aminated at the less-hindered position (Table 6.1, entry 5).

Interestingly, the functionalization *ortho* to a tert-butyl moiety is sterically hindered, leading to slightly lower yields (Table 6.1, entries 6 and 7). Most remarkably, the amino moiety is installed adjacent to a tert-butyl group in a significant amount forming **28** (Table 6.1, entry 6). Meanwhile, the steric demand from both adjacent positions are even more pronounced in the amination of mesitylene, in which the corresponding aniline **33** is obtained in only 24% (Table 6.1, entry 8) [49].

For the second case [50], the first direct and highly regioselective process for the twofold amination of naphthalene (**34**) to 1,5-diaminonaphthalene (**36**) by using BDD anodes at a current density of 10 mA cm^{-2} and 6 F was reported by Waldvogel group [50] with an initial yield of 10% (Scheme 6.8).

After that, screening of different supporting electrolytes (Bu$_4$NBF$_4$, Bu$_4$NPF$_6$, LiClO4, LiBF$_4$, LiB(C$_2$O$_4$)$_2$, Et$_3$NMeO$_3$SOMe, Bu$_4$NCF$_3$SO$_3$, Bu$_4$NC$_4$F$_9$SO$_3$) at different current densities (5–15 mA cm^{-2}, 6 F) was performed in a divided cell (separator: a porous glass frit or Thomapor) with BDD and platinum as anode and cathode, respectively, at 60 °C, achieving different yields ranging from traces to 18% of 1,5-diaminonaphthalene [50]. For tetrabutylammonium tetrafluoroborate and tetrabutylammonium hexafluorophosphate in acetonitrile comparable yields of 10% and 12%, respectively, were accomplished. Lithium perchlorate, lithium tetrafluoroborate, lithium bisoxalato borate (LiBOB) and triethylmethylammonium methylsulfate gave inferior results. Tetrabutylammonium triflate and tetrabutylammonium nonafluorobutane sulfonate provided the best yields with 18% and 15%, respectively. The study of the variation of applied charged was also performed by the authors, in the range of 7–11 F with BDD as anode by applying 10 mA cm^{-2} in 0.2 M Bu$_4$NBF$_4$/CH$_3$CN or 0.2 M Bu$_4$NBF$_4$/1,2-propylene carbonate solution, in order to increase the yield of 1,5-diaminonaphthalene as main product. However, only a 20% was produced as maximum yield in the propylene carbonate based electrolyte at 6 F. No significant increase in yield was achieved by using acetonitrile-based electrolyte. A detailed study of this reaction was carried out

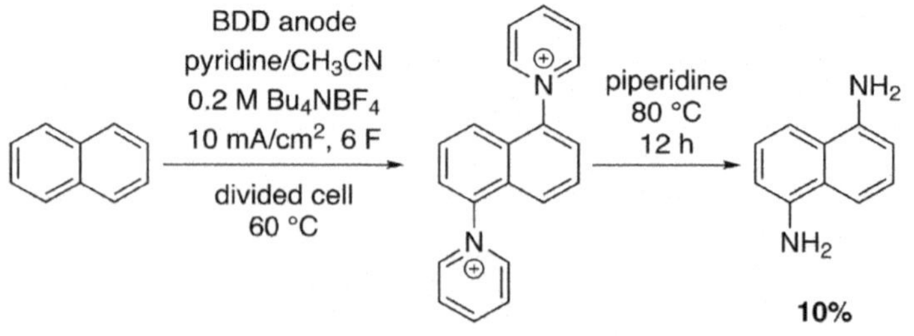

Scheme 6.8 Multiple electrochemical amination at napthtalene

[50] by varying other different electrolysis parameters, e.g. separators, undivided cell, electrosynthesis steps and different aromatic compounds. In the latter, diphenylmethane, triphenylmethane, biphenyl, and phenanthrene were subjected to this diamination method. In the case of diphenylmethane and triphenylmethane significant *ortho*-functionaliziation was observed due to π-π-interactions. As regioisomeric mixtures were obtained upon anodic amination of these substrates, a novel purification process was also developed. Finally, it is important to remark that, novel results obtained by Waldvogel group [50] demonstrated that the scope of the Yoshida amination reaction [48] could be expanded to diaminated products by using BDD anodes.

6.4.2 Clean Electrosynthesis

Many electro-organic protocols focus on the use of a three-electrode setup for constant potential conditions but these exhibit significant disadvantages (cell setup, depletion of the starting material, current decreases during conversion) over a galvanostatic two-electrode arrangement in an undivided cell. This configuration enables operation with a very simple DC power source and good electrochemical conversion in acceptable time scales. Based on the existing literature, the most electro-organic transformations are limited to a low current density of 0.5– 5 mA cm^{-2} or less [51]. Although few electrochemical transformations can be operated at high current densities of ≥ 20 mA cm^{-2}, the crucial requirement is the inherent electrochemical stability of products formed. In this frame, Waldvogel lab reported, for first time, a unique robust electrochemical synthesis of phenols and arenes, in high yields and selectivity, to non-symmetric biphenols, partially protected non-symmetric biphenols, and m-terphenyl-2,2″-diols by anodic C–H activating cross-coupling reaction by direct electrolysis of simple starting materials, and without the necessity of either a previous installation of leaving groups or the use of costly catalysts. The combination of electrochemically very stable HFIP and electrochemically robust BDD anodes were employed for electrochemical cross-coupling at 1, 5, 10, 20, and 25 mA cm^{-2} and even extended to 35, 50, 75, and 100 mA cm^{-2}. An example is displayed in Fig. 6.8.

In these clean electrochemical conversions, the selectivity and yield could be maintained in a range of current density of almost two orders of magnitude by using HFIP-based electrolytes. Meanwhile, the electrochemical reactions can be performed in markedly shorter times because the HFIP plays an important role in the reactions, by the stabilization of radical intermediates, by the prevention of the products from over-oxidation by solvation, and by the support of mass transport by the low viscosity and micro-heterogeneity of the HFIP/methanol mixture [52]. It is noteworthy, that other inert electrodes can be used as well but the yields are significantly lower in the electro-conversion [53]. The powerful combination of BDD anode and HFIP as stabilizing key electrolyte component is applicable to phenol-arene [12], phenol-phenol [54–56], phenol-protected phenol [57],

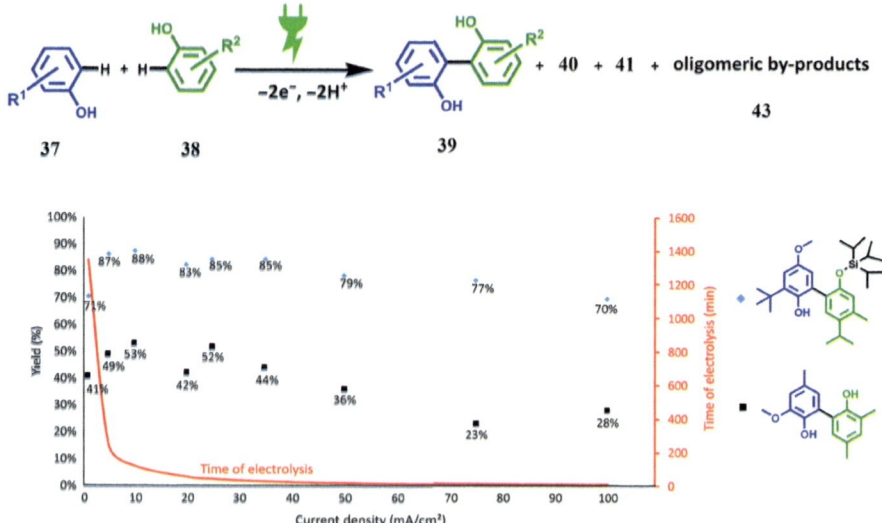

Fig. 6.8 Potential products for oxidative cross-coupling reaction of anodic phenol-phenol coupling and variation of current density for single anodic C–C cross-coupling reaction: synthesis of 2,2′-biphenols (black squares) and partially protected non-symmetric 2,2′-biphenols (blue diamonds)

phenol-thiophene [58], and phenol-benzofuran. The latter does not only show an arylation reaction but also a following up benzofuran metathesis [59]. Solvation by HFIP not only stabilized reactive intermediates, but also gave rise to a decoupling of nucleophilicity and the oxidation potential due to diverse solvation of the coupling partners, making cross-coupling possible. Yield and selectivity could be increased in many cases by additives like water or methanol manipulating the solvation.

6.5   Conclusions

The synthetic use of boron-doped diamond electrodes in preparative organic chemistry has just started. Consequently, only a small flavour of the synthetic possibilities was explored. However, in particular when applying very positive potentials this anode is the material of choice enabling novel electro-conversions or bringing electrosynthesis beyond current limits. Therefore, the unique reactivity of the diamond surface promises a fruitful field for future developments. Since BDD stands extremely strong oxidizing species the fluorination and even fluorine generation of BDD seems to be an interesting topic. A major issue for anodic transformations will be the reliable control of the reactive intermediates generated on BDD. Potential strategies in order to reduce mineralization involve ultra-stable electrolytes. Alternatively, the exchange between electrode surface and bulk should be enhanced for diluting the reactive intermediates. Such approaches can include electrolysis in micro reactors or exploiting ultrasonic methods.

Acknowledgements Carlos A. Martínez-Huitle acknowledges the funding provided by the Alexander von Humboldt Foundation (Germany) and Coordenação de Aperfeiçoamento de Pessoal de Nível Superior (Brazil) as a fellowship for experienced researcher. The authors highly appreciate the financial support by the Center for INnovative and Emerging MAterials (CINEMA). Support by the Advanced Lab of Electrochemistry and Electrosynthesis—ELYSION (Carl Zeiss Stiftung) is gratefully acknowledged.

References

1. S.R. Waldvogel, S. Möhle, M. Zirbes, E. Rodrigo, T. Gieshoff, A. Wiebe, Modern electrochemical aspects for the synthesis of value added organic products. Angew. Chem. Int. Ed. (2018) (in press). https://doi.org/10.1002/anie.201712732
2. S.R. Waldvogel, A. Wiebe, T. Gieshoff, S. Möhle, E. Rodrigo, M. Zirbes, Electrifying organic synthesis. Angew. Chem. Int. Ed. (2018) (in press). https://doi.org/10.1002/anie.201711060
3. M. Yan, Y. Kawamata, P.S. Baran, Synthetic organic electrochemical methods since 2000: on the verge of a renaissance. Chem. Rev. **117**(21), 13230–13319 (2017). https://doi.org/10.1021/acs.chemrev.7b00397
4. A.J. Fry, in *Synthetic Organic Electrochemistry*, 2nd edn. (Wiley, New York, 1989)
5. (a) H.-J. Schäfer, in *Encyclopedia of Electrochemistry*, ed. by A.J. Bard, M. Stratmann, H.-J. Schäfer (Wiley-VCH, Weinheim, 2004), pp. 125–170; (b) H.-J. Schäfer, in *Organic Electrochemistry*, ed. by H. Lund, O. Hammerich (Marcel Dekker, New York, 2001), pp. 207–222
6. S.R. Waldvogel, A. Kirste, S. Mentizi, in *Synthetic Diamond Films. Preparation, Electrochemistry, Characterization, and Applications*, ed. by C.A. Martínez-Huitle, E. Brillas (Wiley, Hoboken, N.J, 2011), pp. 483–510. https://doi.org/10.1002/9781118062364.ch19
7. S. Garcia-Segura, E. Vieira dos Santos, C.A. Martínez-Huitle, Role of sp^3/sp^2 ratio on the electrocatalytic properties of boron-doped diamond electrodes: a mini review. Electrochem. Commun. **59**, 52–55 (2015). https://doi.org/10.1016/j.elecom.2015.07.002
8. E. Brillas, C.A. Martínez-Huitle, *Synthetic Diamond Films: Preparation, Electrochemistry, Characterization, and Applications* (Wiley, Hoboken, N.J, 2011). https://doi.org/10.1002/9781118062364
9. C.A. Martínez-Huitle, M.A. Rodrigo, I. Sires, O. Scialdone, Single and coupled electrochemical processes and reactors for the abatement of organic water pollutants: a critical review. Chem. Rev. **115**(24), 13362–13407 (2015). https://doi.org/10.1021/acs.chemrev.5b00361
10. C.A. Martínez-Huitle, E. Brillas, Electrochemical alternatives for drinking water disinfection. Angew. Chem. Int. Ed. **47**(11), 1998–2005 (2008). https://doi.org/10.1002/anie.200703621
11. A. Kirste, M. Nieger, I.M. Malkowsky, F. Stecker, A. Fischer, S.R. Waldvogel, Ortho-selective phenol-coupling reaction by anodic treatment on boron-doped diamond electrode using fluorinated alcohols. Chem. Eur. J. **15**(10), 2273–2277 (2009). https://doi.org/10.1002/chem.200802556
12. A. Kirste, G. Schnakenburg, F. Stecker, A. Fischer, S.R. Waldvogel, Anodic phenol arene cross-coupling reaction on boron-doped diamond electrodes. Angew. Chem. Int. Ed. **49**(5), 971–975 (2010). https://doi.org/10.1002/anie.200904763

13. P. Sabatier, in *Nobel Lectures*, Chemistry (Elsevier Publishing, Amsterdam, 1966), pp. 1901 −1920
14. N. Yang, S.R. Waldvogel, X. Jiang, Electrochemistry of carbon dioxide on carbon electrodes. ACS Appl. Mater. Interfaces. **8**(42), 28357–28371 (2016). https://doi.org/10.1021/acsami. 5b09825
15. B.R. Eggins, E.M. Brown, E.A. O'Neill, J. Grimshaw, Carbon dioxide fixation by electrochemical reduction in water to oxaiate and glyoxylate. Tetrahedron Lett. **29**(8), 945–948 (1988). https://doi.org/10.1016/S0040-4039(00)82489-2
16. K. Nakata, T. Ozaki, C. Terashima, A. Fujishima, Y. Einaga, High-yield electrochemical production of formaldehyde from CO_2 and seawater. Angew. Chem., Int. Ed. **53**(3), 871–874 (2014). https://doi.org/10.1002/anie.201308657
17. Y. Liu, S. Chen, X. Quan, H. Yu, Efficient electrochemical reduction of carbon dioxide to acetate on nitrogen-doped nanodiamond. J. Am. Chem. Soc. **137**(36), 11631–11636 (2015). https://doi.org/10.1021/jacs.5b02975
18. S. Zhang, P. Kang, S. Ubnoske, M.K. Brennaman, N. Song, R.L. House, J.T. Glass, T.J. Meyer, Polyethylenimine-enhanced electrocatalytic reduction of CO_2 to formate at nitrogen-doped carbon nanomaterials. J. Am. Chem. Soc. **136**(22), 7845–7848 (2014). https://doi.org/10.1021/ja5031529
19. J. Wu, R.M. Yadav, M. Liu, P.P. Sharma, C.S. Tiwary, L. Ma, X. Zou, X.-D. Zhou, B.I. Yakobson, J. Lou, P.M. Ajayan, Achieving highly efficient, selective, and stable CO_2 reduction on nitrogen-doped carbon nanotubes. ACS Nano **9**(5), 5364–5371 (2015). https://doi.org/10.1021/acsnano.5b01079
20. P.A. Christensen, A. Hamnett, A.V.G. Muir, N.A. Freeman, CO_2 reduction at platinum, gold and glassy carbon electrodes in acetonitrile, an in-situ FTIR study. J. Electroanal. Chem. Interfacial Electrochem. **288**(1–2), 197–215 (1990). https://doi.org/10.1016/0022-0728(90) 80035-5
21. K. Hara, A. Kudo, T. Sakata, Electrochemical CO_2 reduction on a glassy carbon electrode under high pressure. J. Electroanal. Chem. **421**(1–2), 1–4 (1997). https://doi.org/10.1016/ S0022-0728(96)01028-5
22. P.K. Jiwanti, K. Natsui, K. Nakatab, Y. Einaga, Selective production of methanol by the electrochemical reduction of CO_2 on boron-doped diamond electrodes in aqueous ammonia solution. RSC Adv. **6**(104), 102214–102217 (2016). https://doi.org/10.1039/C6RA20466J
23. N. Ikemiya, K. Natsui, K. Nakata, Y. Einaga, Effect of alkali-metal cations on the electrochemical reduction of carbon dioxide to formic acid using boron-doped diamond electrodes. RSC Adv. **7**(36), 22510–22514 (2017). https://doi.org/10.1039/C7RA03370B
24. K. Natsui, H. Iwakawa, N. Ikemiya, K. Nakata, Y. Einaga, Stable and highly efficient electrochemical production of formic acid from carbon dioxide using diamond electrodes. Angew. Chem. Int. Ed. **57**(10), 2639–2643 (2018). https://doi.org/10.1002/anie.201712271
25. Y. Hori, H. Wakebe, T. Tsukamoto, O. Koga, Electrocatalytic process of CO selectivity in electrochemical reduction of CO_2 at metal electrodes in aqueous media. Electrochim. Acta **39** (11–12), 1833–1839 (1994). https://doi.org/10.1016/0013-4686(94)85172-7
26. N. Yang, F. Gao, C.E. Nebel, Diamond decorated with copper nanoparticles for electrochemical reduction of carbon dioxide. Anal. Chem. **85**(12), 5764–5769 (2013). https://doi.org/10.1021/ac400377y
27. H.S. Panglipur, T.A. Ivandini, R. Wibowo, Y. Einaga, Electroreduction of CO_2 using copper-deposited on boron-doped diamond (BDD). AIP Conf. Proc. **1729**, 020047 (2016). https://doi.org/10.1063/1.4946950
28. P.K. Jiwanti, K. Natsui, K. Nakata, Y. Einaga, The electrochemical production of C_2/C_3 species from carbon dioxide on copper-modified boron-doped diamond electrodes. Electrochim. Acta **266**, 414–419 (2018). https://doi.org/10.1016/j.electacta.2018.02.041
29. N. Spataru, K. Tokuhiro, C. Terashima, T.N. Rao, A. Fujishima, Electrochemical reduction of carbon dioxide at ruthenium dioxide deposited on boron-doped diamond. J. Appl. Electrochem. **33**(12), 1205–1210 (2003). https://doi.org/10.1023/B:JACH.0000003866. 85015.b6

30. S.A. Yao, R.E. Ruther, L.H. Zhang, R.A. Franking, R.J. Hamers, J.F. Berry, Covalent attachment of catalyst molecules to conductive diamond: CO_2 reduction using "smart" electrodes. J. Am. Chem. Soc. **134**(38), 15632–15635 (2012). https://doi.org/10.1021/ja304783j

31. N. Roy, Y. Hirano, H. Kuriyama, P. Sudhagar, N. Suzuki, K. Katsumata, K. Nakata, T. Kondo, M. Yuasa, I. Serizawa, T. Takayama, A. Kudo, A. Fujishima, C. Terashima, Boron-doped diamond semiconductor electrodes: efficient photoelectrochemical CO_2 reduction through surface modification. Sci. Rep. **6**(38010), 1–9 (2016). https://doi.org/10.1038/srep38010

32. (a) L. Bouveault, G. Blanc, Compt. Rend. **136**, 1676–1678 (1903); (b) B.I. Seo, L.K. Wall, H. Lee, J.W. Buttrum, D.E. Lewis, An improved practical synthesis of isomerically pure 3-endo-(p-Methoxybenzyl)isoborneol. Synth. Commun. **23**(1), 15–22 (2006). https://doi.org/10.1080/00397919308020396

33. J. Tafel, E. Pfeffermann, Elektrolytische reduktion von oximen und phenylhydrazonen in schwefelsaurer lösung. Ber. Dtsch. Chem. Ges. **35**, 1510–1518 (1902)

34. H.B. Martin, A. Argoita, U. Landau, A.B. Anderson, J.C. Angus, Hydrogen and oxygen evolution on boron-doped diamond electrodes. J. Electrochem. Soc. **143**(6), L133–L136 (1996). https://doi.org/10.1149/1.1836901

35. U. Griesbach, D. Zollinger, H. Pütter, C. Comninellis, Evaluation of boron doped diamond electrodes for organic electrosynthesis on a preparative scale. J. Appl. Electrochem. **35**(12), 1265–1270 (2005)

36. (a) M.C. Schopohl, C. Siering, O. Kataeva, S.R. Waldvogel, Reversible enantiofacial dicrimination of a single heterocyclic substrate by supramolecular receptors—a new concept for chiral induction. Angew. Chem. Int. Ed. **42**(23), 2620–2623 (2003). https://doi.org/10.1002/anie.200351102; (b) C. Siering, S. Grimme, S.R. Waldvogel, Direct assignment of enantiofacial discrimination on single heterocyclic substrates by self-induced CD (SICD). Chem. Eur. J. **11**(6), 1877–1888 (2005). https://doi.org/10.1002/chem.200401002; (c) M.C. Schopohl, A. Faust, D. Mirk, R. Fröhlich, O. Kataeva, S.R. Waldvogel, Synthesis of rigid receptors based on Triphenylen Ketals. Eur. J. Org. Chem. (14), 2987–2999 (2005). https://doi.org/10.1002/ejoc.200500108; (d) M. Bomkamp, C. Siering, K. Landrock, H. Stephan, R. Fröhlich, S.R. Waldvogel, Chem. Extraction experiments of radio-labelled xanthine derivatives by artificial receptors—deep insight into the association behaviour. Eur. J. **13**(13), 3724–3732 (2007). https://doi.org/10.1002/chem.200601231; (e) W. Schade, C. Bohling, K. Hohmann, C. Bauer, R. Orghici, S.R. Waldvogel, D. Scheel, Photonic sensors for security applications. Phot. Int. **1**, 32–34 (2007); (w) W. Schade, C. Bohling, K. Hohmann, C. Bauer, R. Orghici, S.R. Waldvogel, D. Scheel, Photonische Sensoren für die Sicherheitstechnik. Photonik **38**, 70–73 (2006); (f) R. Orghici, U. Willer, M. Gierszewska, S.R. Waldvogel, W. Schade, Fiber optic evanescent-field-sensor for detection of explosives and CO_2 dissolved in water. Appl. Phys. B **90**, 355 (2008). https://doi.org/10.1007/s00340-008-2932-7; (g) S. Börner, R. Orghici, S.R. Waldvogel, U. Willer, W. Schade, Evanescent field sensors and the implementation of waveguiding nanostructures. Appl. Opt. **48**(4), B183 (2009). https://doi.org/10.1364/ao.48.00b183; (h) U. Schwartz, R. Großer, K.-E. Piejko, B. Bömer, D. Arlt, Optisch aktive (Meth)-acrylamide, Polymere daraus, Verfahren zu ihrer Herstellung und ihre Verwendung zur Racematspaltung, DE3532356A1. Ger. Pat. Appl. (1987); (i) M. Grose-Bley, B. Bömer, R. Großer, D. Arlt, W. Lange, Optisch aktive schwefelhaltige aminosaeure-derivate, ihre herstellung, ihre polymerisation zu optisch aktiven polymeren und deren verwendung, DE4120695 Ger. Pat. Appl. (1992); (j) B. Bömer, R. Großer, W. Lange, U. Zweering, B. Köhler, W. Sirges, M. Grose-Bley, Chirale stationäre Phasen für die chromatographische Trennung von optischen Isomeren, DE19546136A1 Ger. Pat. Appl. (1997); (k) W. Lange, R. Grosser, B. Köhler, S. Michel, B. Bömer, U. Zweering, Chromatographic enantiomer of lactones, DE19714343A1 Ger. Pat. Appl. (1998); (l) J. Looft, T. Vössing, J. Ley, M. Backes, M. Blings, Substituted cyclopropane carbolic acid(3-methyl-cyclohexyl) amides as taste substances, EP1989944A1 Ger. Pat. Appl. (2008)

37. M.C. Schopohl, K. Bergander, O. Kataeva, R. Fröhlich, S.R. Waldvogel, Synthesis and characterization of enantiomerically pure menthylamines and their isocyanates. Synthesis **17**, 2689–2694 (2003). https://doi.org/10.1055/s-2003-42432

38. U. Griesbach, S.R. Waldvogel, J. Kulisch, I.M. Malkowsky, Process for the preparation of pantoprazole sodium. PCT Int. Appl. WO2008003620 A2 20080110 (2008)

39. E. Rodrigo, S.R. Waldvogel, Very simple one-pot electrosynthesis of nitrones starting from nitro and aldehyde components. Green Chem. (2018) (in press). https://doi.org/10.1039/c8gc00474a

40. J. Yoshida, K. Kataoka, R. Horcajada, A. Nagaki, Modern strategies in electroorganic synthesis. Chem. Rev. **108**(7), 2265–2299 (2008). https://doi.org/10.1021/cr0680843

41. (a) T. Lehmann, R. Schneider, C. Weckbecker, E. Dunach, S.Oliviero, Process for the production of 2-hydroxy-4-methylmercaptobutyric acid, WO 02/16671 (2002); (b) T. Lehmann, R. Schneider, C. Reufer, R. Sanzenbacher, in *Chemie und Biochemie*, ed. by J. Russow, H.J. Schäfer, GDCh-Monographie **23**, 251–258 (2001)

42. C. Reufer, M. Hateley, T. Lehmann, C. Weckbecker, R. Sanzenbacher, J. Bilz, EP 1 631702 (2006)

43. T. Kojima, R. Obata, T. Saito, Y. Einaga, S. Nishiyama, Cathodic reductive coupling of methyl cinnamate on boron-doped diamond electrodes and synthesis of new neolignan-type products. Beilstein J. Org. Chem. **11**, 200–203 (2015). https://doi.org/10.3762/bjoc.11.21

44. S.R. Waldvogel, S. Möhle, Versatile electrochemical C, H-amination via Zincke intermediates. Angew. Chem. Int. Ed. **54**(22), 6398–6399 (2015). https://doi.org/10.1002/anie.201502638

45. (a) J.F. Hartwig, Discovery and understanding of transition-metal-catalyzed aromatic substitution reactions. Synlett **9**, 1283–1294 (2006). https://doi.org/10.1055/s-2006-939728; (b) J.F. Hartwig, Carbon-heteroatom bond formation catalysed by organometallic complexes. Nature **455**(7211), 314–322 (2008). https://doi.org/10.1038/nature07369; (c) J.F. Hartwig, Evolution of a fourth generation catalyst for the amination and thioetherification of aryl halides. Acc. Chem. Res. **41**(11), 1534–1562 (2008). https://doi.org/10.1021/ar800098p; (d) D.S. Surry, S.L. Buchwald, Biaryl phosphine ligands in palladium-catalyzed amination. Angew. Chem. Int. Ed. **47**(34), 6338–6361 (2008). https://doi.org/10.1002/anie.200800497; D.S. Surry, S.L. Buchwald, Biarylphosphanliganden in der palladiumkatalysierten aminier- ung. Angew. Chem. **120**(34), 6438–6461 (2008). https://doi.org/10.1002/ange.200800497; (e) D.S. Surry, S.L. Buchwald, Diamine ligands in copper-catalyzed reactions. Chem. Sci. **1**, 13–31 (2010). https://doi.org/10.1039/c0sc00107d; (f) D.S. Surry, S.L. Buchwald, Dialkylbiaryl phosphines in Pd-catalyzed amination: a user's guide. Chem. Sci. **2**, 27–50 (2011). https://doi.org/10.1039/c0sc00331j

46. (a) T.W. Lyons, M.S. Sanford, Palladium-catalyzed ligand-directed C-H functionalization reactions. Chem. Rev. **110**(2), 1147–1169 (2010). https://doi.org/10.1021/cr900184e; (b) N. Kuhl, M.N. Hopkinson, J. Wencel-Delord, F. Glorius, Beyond directing groups: transition-metal-catalyzed C-H activation of simple arenes. Angew. Chem. Int. Ed. **51**(41), 10236–10254 (2012). https://doi.org/10.1002/anie.201203269; N. Kuhl, M.N. Hopkinson, J. Wencel-Delord, F. Glorius, Ohne dirigierende gruppen: übergangsmetallkatalysierte C-H- Aktivierung einfacher Arene. Angew. Chem. **124**(41), 10382–10401 (2012). https://doi.org/10.1002/ange.201203269

47. (a) Q. Shen, J.F. Hartwig, Palladium-Catalyzed coupling of ammonia and lithium amide with aryl halides. J. Am. Chem. Soc. **128**(31), 10028–10029 (2006). https://doi.org/10.1021/ja064005t; (b) G.D. Vo, J.F. Hartwig, Palladium-Catalyzed coupling of ammonia with aryl chlorides, bromides, iodides, and sulfonates: a general method for the preparation of primary arylamines. J. Am. Chem. Soc. **131**(31), 11049–11061 (2009). https://doi.org/10.1021/ja903049z

48. T. Morofuji, A. Shimizu, J.I. Yoshida, Electrochemical C-H amination: synthesis of aromatic primary amines via N-arylpyridinium ions. J. Am. Chem. Soc. **135**(13), 5000–5003 (2013). https://doi.org/10.1021/ja402083e

49. S. Herold, S. Möhle, M. Zirbes, F. Richter, H. Nefzger, S.R. Waldvogel, Electrochemical amination of less-activated alkylated arenes using boron-doped diamond anodes. Eur. J. Org. Chem. **2016**(7), 1274–1278 (2016). https://doi.org/10.1002/ejoc.201600048

50. S. Möhle, S. Herold, F. Richter, H. Nefzger, S.R. Waldvogel, Twofold electrochemical amination of naphthalene and related arenes. ChemElectroChem **4**(9), 2196–2210 (2017). https://doi.org/10.1002/celc.201700476

51. A. Wiebe, B. Riehl, S. Lips, R. Franke, S.R. Waldvogel, Unexpected high robustness of electrochemical cross-coupling for a broad range of current density. Sci. Adv. **3**(10), 1–7 (2017). https://doi.org/10.1126/sciadv.aao3920

52. O. Holloczki, A. Berkessel, J. Mars, M. Mezger, A. Wiebe, S.R. Waldvogel, B. Kirchner, The catalytic effect of fluoroalcohol mixtures depends on domain formation. ACS Catal. **7**(3), 1846–1852 (2017). https://doi.org/10.1021/acscatal.6b03090

53. A. Kirste, B. Elsler, G. Schnakenburg, S.R. Waldvogel, Efficient anodic and direct phenol-arene C, C cross-coupling—the benign role of water or methanol. J. Am. Chem. Soc. **134**(7), 3571–3576 (2012). https://doi.org/10.1021/ja211005g

54. B. Elsler, D. Schollmeyer, K.M. Dyballa, R. Franke, S.R. Waldvogel, Metal- and reagent-free highly selective anodic cross-coupling reaction of phenols. Angew. Chem. Int. Ed. **53**(20), 5210–5213 (2014). https://doi.org/10.1002/anie.201400627

55. B. Riehl, K.M. Dyballa, R. Franke, S.R. Waldvogel, Electro-organic synthesis as sustainable alternative for dehydrogenative cross-coupling of phenols and naphthols. Synthesis **49**(02), 252–259 (2017). https://doi.org/10.1055/s-0036-1588610

56. B. Elsler, A. Wiebe, D. Schollmeyer, K.M. Dyballa, R. Franke, S.R. Waldvogel, Source of selectivity in oxidative cross-coupling of aryls by solvent effect of 1,1,1,3,3,3-Hexafluoropropan-2-ol. Chem. Eur. J. **21**(35), 12321–12325 (2015). https://doi.org/10.1002/chem.201501604

57. A. Wiebe, D. Schollmeyer, K.M. Dyballa, R. Franke, S.R. Waldvogel, Selective synthesis of partially protected non-symmetric biphenols by reagent- and metal-free anodic cross-coupling reaction. Angew. Chem. Int. Ed. **55**(39), 11801–11805 (2016). https://doi.org/10.1002/anie.201604321

58. A. Wiebe, S. Lips, D. Schollmeyer, R. Franke, S.R. Waldvogel, Single and twofold metal- and reagent-free anodic c, c cross-coupling of phenols with thiophenes. Angew. Chem. Int. Ed. **56**(46), 14727–14731 (2017). https://doi.org/10.1002/anie.201708946

59. S. Lips, B.A. Frontana-Uribe, M. Dörr, D. Schollmeyer, R. Franke, S.R. Waldvogel, Metal- and reagent-free anodic C, C cross-coupling of phenols with benzofurans leading to a furan metathesis. Chem. Eur. J. **24**, 6057–6061 (2018). https://doi.org/10.1002/chem.201800919

Chapter 7
Diamond Films as Support for Electrochemical Systems for Energy Conversion and Storage

Patricia Rachel Fernandes da Costa, Elisama Vieira dos Santos, Djalma Ribeiro da Silva, Soliu Oladejo Ganiyu and Carlos A. Martínez-Huitle

Abstract Many efforts have been dedicated to develop and study different catalysts supported materials for energy storage and conversion. Polymer electrolyte membranes (PEM) and capacitors have been topics of special interest for the scientific community, then, the research to find excellent catalyst-supports has constantly increased. The use of conductive diamond films has been proposed due to their mechanical and chemical stability properties. In this context, the application of BDD-catalyst surfaces for PEM fuel cells as well as the production of electrochemical capacitors using BDD materials have been summarized and discussed in this chapter.

7.1 Introduction

The significant demand and use of petroleum in the near future have motivated scientists around the world to develop new clean and renewable energy sources. Several catalysts supported materials for energy storage and conversion have been investigated to comprehend the kinetic of the anodic and cathodic reactions in order to enhance the production or the accumulation of electrical energy [1]. In this frame,

P. R. F. da Costa · E. V. dos Santos · D. R. da Silva · S. O. Ganiyu ·
C. A. Martínez-Huitle (✉)
Institute of Chemistry, Federal University of Rio Grande do Norte, Lagoa Nova,
Natal, RN CEP 59078-970, Brazil
e-mail: carlosmh@quimica.ufrn.br

C. A. Martínez-Huitle
National Institute for Alternative Technologies of Detection, Toxicological Evaluation
and Removal of Micropollutants and Radioactives (INCT-DATREM),
Institute of Chemistry, Unesp, P.O. Box 355, Araraquara, SP 14800-900, Brazil

C. A. Martínez-Huitle
Institut für Organische Chemie, Johannes Gutenberg-Universität Mainz,
Duesbergweg 10-14, 55128 Mainz, Germany

© Springer Nature Switzerland AG 2019   199
N. Yang (ed.), *Novel Aspects of Diamond*, Topics in Applied Physics 121,
https://doi.org/10.1007/978-3-030-12469-4_7

the development of fuel technology has received great attention because this technology consists in directly converting chemical energy into electricity (e.g.: Direct Alcohol Fuel Cell (DAFC)) and after that, its accumulation is also other target.

The use of Pt–Ru catalyst is considered the most promising approach as anode material. However, the applicability of conductive diamond films as catalyst-support has been studied because of the chemical/electrochemical features of these materials, such as wide potential window, low background current, a high chemical and dimensional stability [2–11]. Several types of diamond electrodes can be prepared with different structural characteristics. Diamond films can be grown on different substrate materials (e.g.: Ti, Si, W, Mo, Nb and Pt) by chemical vapor deposition (CVD) technique. The conditions used to deposit diamond are specific to different reactor designs, influencing on many of their essential properties. Diamond films must be doped with boron or other dopants (e.g.: nitrogen, fluorine, phosphorous and sulfur) in order to have sufficient electrical conductivity for electrochemical measurements [1, 2]. Even when, different supports or dopants are used to synthetize diamond films; the most diamond electrode used is, boron-doped diamond (BDD). In this context, this chapter aims to give an overview about the techniques used to modify the BDD surfaces with catalysts for PEM fuel cells (methanol and ethanol oxidation) as well as for the fabrication of the electrochemical capacitors. Recent results about the use of the modified electrodes will be discussed.

7.2 Modification Procedures of BDD Surfaces

Diamond films have been used as supports to modify their surface with metal or metal oxide clusters on its surface, as nanoparticles [12]. Several methods have been used to prepare these catalysts; however, the synthesis technique influences strongly the shape and size distribution of particles. After that, the choice of the support material depends on its stability, chemical resistance, electrochemical properties as well as the interactions between the catalyst and its surface. Therefore, a significant effort, to improve particles adherence and dispersion, has been performed by investigating several deposition techniques. The most important procedures used to prepare nanoparticles and after that, deposit onto diamond surfaces are [13–46]: microemulsion synthesis, thermal deposition, electrodeposition and sol-gel modification: Each one of these method is summarized and described in Table 7.1, giving several examples of nanoparticles deposited as well as their application in fuel cells. Among these nanoparticles procedures, the most relevant results have been achieved by using microemulsion and sol-gel modification. In the last case, surface modifications of BDD electrodes have been performed by depositing different metal oxides and some mixed composites (see Table 7.1) [36–39]. The modification of the BDD electrode with other Pt–metal oxide catalysts or other oxide metal nanoparticles (IrO_2, PbO_2, SnO_2, Ta_2O_5, and some mixed

Table 7.1 Metal/metal-oxide nanoparticle procedures to synthetize and deposit onto BDD surfaces for application in fuel cells

Method	Definition	Metal/metal-oxide nanoparticles supported in diamond films	Application in Fuel cells
Microemulsion synthesis [13–15]	The synthesis of inorganic nanoparticles is usually carried out in water/oil microemulsions to produce nanoparticles with narrow size distribution. Water-in-oil microemulsion is the coexistence of an excess water phase and the surfactant molecules which produce reverse micelles. The water core of these aggregates is surrounded by surfactant molecules which have the non-polar part of their molecule towards the oil phase. In the water core of this aggregate, electrolytes may be solubilized for instance metal salts. These metals will be then transformed into inorganic precipitates by using an appropriate reducing or precipitating agent. There are two main ways of preparation in order to obtain nanoparticles from microemulsions: (i) by mixing two microemulsions, one is containing the precursor and the other the precipitating agent and (ii) by adding the precipitating agent directly into the microemulsion containing the metal precursor	Pt [16] (Nafion® films were used to mechanically stabilize the electrode) Pt–Ru [17] (nanoparticles of different compositions with a reducing agent were synthetized) Pt–Sn [18] Pt–Ru–Sn [19]	Pt, $Pt_{50}Ru_{50}$, and Ru nanoparticles deposited on BDD in pure supporting electrolyte (1 M $HClO_4$) achieved substantial decrements in the quality of the H adsorption–desorption characteristic and increase in the background current in the double-layer region when the Ru content in the particles was increased. The change in the electronic structure of the Pt component in the alloys (Pt/Sn and Pt/Ru) could modify the Pt work function and thus weakens bonding of adsorbed intermediates Pt–CO that could produce an enhancement in rates of methanol oxidation

(continued)

Table 7.1 (continued)

Method	Definition	Metal/metal-oxide nanoparticles supported in diamond films	Application in Fuel cells
Thermal deposition [20]	Appropriate precursors are dissolved in suitable solvents and spread on a metallic support before thermal decomposition. The particle size, non-stoichiometry, and morphology of the oxide layer depends on the nature of the precursor and the decomposition temperature, therefore, these must be controlled during the procedure	This method has been used to deposit IrO_2 nanoparticles (solutions of H_2IrCl_6 in 2-propanol was applied to the BDD surface (1 cm^2). After solvent evaporation at 80 °C, calcination was performed at 350 °C to oxidize the precursor acid to IrO_2) [17, 21], Au nanoparticles (Au-nanoparticles deposited on BDD surface by using a sputter deposition method with a heat treatment at 400–600 °C) [17, 22] and Pt nanoparticles (the application of 5 μL of 0.2–3 mM H_2PtCl_6 in 2-propanol on the BDD surface; evaporation of the solvent at 60 °C during 5 min, and finally, thermal decomposition of the precursor by treatment in an oven at 350 °C during 1 h) [23]	The agglomeration of platinum particles was observed due to the inhomogeneity of the interfacial surface tension with the BDD support. No significant efficiencies were attained on the methanol oxidation in acidic conditions at p-Si/BDD/Pt electrode
Electrodeposition [29]	The electrodeposition is one of the most widely used methods for the preparation and deposition of particles on BDD Advantages: process at room temperature, controlled growth of nanoparticles, and low cost methodology. The deposition and co-deposition of the different metals on have been demonstrated as feasible preparation procedure	The modification of a BDD electrode surface has been reported for a limited range of metal nanoparticles, including Ag, Au, Pt, Pd, Cu, Bi, Ni, Hg, Pb, Co, Ir, Ru, Te, Ti, and Fe [24–35]. For example: applying a potential step to a deaerated 2 mM H_2PtCl_6 solution in 1 M $HClO_4$ electrodeposition of Pt particles on a BDD electrode is generally performed. The potential is shifted from an equilibrium	The best electrocatalyst for methanol oxidation is Pt–Ru binary metallic, being oxidize according to a bifunctional mechanism [31]. Surface-sited Pt atoms oxidatively dehydrogenate the chemisorbed methyl moiety in consecutive steps to yield a residual Pt–CO fragment that cannot be oxidized to CO_2 at DMFC potentials, Pt adsorbed CO is removed via an oxygen-transfer step from

(continued)

Table 7.1 (continued)

Method	Definition	Metal/metal-oxide nanoparticles supported in diamond films	Application in Fuel cells
		potential (1 V, where no reduction of platinic ions takes place) to a potential at which the reduction of Pt^{4+} to metallic Pt occurs (0.02–0.15 V)	electrogenerated Ru–OH. Ru transfers oxygen more effectively than Pt due to its ability to oxidatively absorb water at less positive potentials [32]
Sol-gel modification [36–39]	The sol-gel method starts with a solution consisting of metal compounds, such as metal alkoxides, and acetylacetonates as source of oxides, water as hydrolysis agent, alcohol as solvent and acid or base catalyst. Metal compounds undergo hydrolysis and polycondensation at room temperature, giving rise to sol, in which polymers or fine particles are dispersed. Further reaction connects the particles, solidifying the sol into wet gel, which still contains water and solvents. Vaporization of water and solvents produces a dry gel (xerogel), an aerogel results from a supercritical drying process. Heating gels to several hundred degrees produces dense oxides as products. Coating films can be made by dip coating or spin coating of the sol. Unsupported films can be made by synthesizing the film at the interface between alkoxide solution and water. Membranes are prepared by pouring the sol on the porous oxide with coarse pores. Particles with sharp size distribution can be precipitated and grown in the sol	Pt–RuO$_2$ deposits [36] Platinum oxide particles PtO$_x$ and PtO$_{x-}$RuO$_2$, [41] Pt, Pt–RuO$_2$, and Pt–RuO$_2$–RhO$_2$ [42] IrO$_2$, PbO$_2$, SnO$_2$, Ta$_2$O$_5$, and some mixed composites (Pt–RuO$_2$, Pt–RuO$_x$, Pt–RuO$_2$, Pt–PbO$_x$, Pt–IrO$_2$ Pt–(RuO$_2$–IrO$_2$), Pt–(RuO$_2$–PbO$_x$), Pt–(IrO$_2$–PbO$_x$), Pt, Pt–SnO$_2$, Pt–Ta$_2$O$_5$) [36, 43–46]	

composites) prepared by sol-gel has been investigated with the aim to improve its electrocatalytic response towards methanol and/or ethanol oxidation reactions as a fuel cell anode [36, 43–46].

7.3 Modified BDD Films as Electrocatalytic Surfaces for Fuel Cells

Catalytic properties of Pt–RuO$_2$/C supported on BDD surfaces were evaluated by cyclic voltammetric technique [37] presented a higher current density than the composite supported on GC electrode. These studies indicated that the forward and backward lines of BDD substrate were almost coincident while a large difference was observed for GC electrode. However, other tests using the same Pt–RuO$_2$/C material showed that the differences are attributed to the participation of substrate, evidencing the capacitive effect of the GC. Conversely, BDD surfaces did not contribute in the catalyst response. In these studies, methanol and ethanol oxidation responses, for Pt–RuO$_2$/C catalyst on BDD surface, indicated that the oxidation of methanol started at 380 mV versus SHE, as already indicated by He et al. [32] where Pt–Ru nanoparticles were electrodeposited on carbon nanotubes. For the case of ethanol oxidation, the oxidation ethanol potential for Pt–RuO$_2$/C was much lower than for the Pt/C.

Salazar-Banda et al. [42] studied the oxidation of methanol and ethanol by modifying the BBD surfaces with Pt, alloys Pt–RuO$_2$, and Pt–RuO$_2$–RhO$_2$. Voltammetric assays showed that BDD modified surfaces improved the electroactive area in approximately five times with respect to original BDD surface (Fig. 7.1). BDD/Pt, BDD/Pt–RuO$_2$, BDD/Pt–RuO$_2$–RhO$_2$ electrodes also promoted a significant potential-shift in the potentials of oxygen reduction (OER) and hydrogen evolution (HER) reactions due to the catalytic effect of the deposited metals. The curve in Fig. 7.2 displays the typical electrochemical behavior of a polycrystalline Pt surface, such as hydrogen absorption/desorption and the oxide formation, evidencing that the Pt particles were successfully deposited onto the BDD surface with high purity and homogeneous distribution. Methanol and ethanol oxidation in acidic conditions, by cyclic voltammetric assays, revealed that the CO poisoning effect for both alcohols oxidation reactions was mainly inhibited at BDD/Pt–RuO$_2$–RhO$_2$ due to the Rh presence. Rh promotes a better catalytic effect for these reactions by either prompting the oxidation of the adsorbed intermediates to CO$_2$ or diminishing the absorption of CO as well as the others intermediates over Pt surface.

Another composite catalysts, such as Pt–PbO$_x$/C, Pt–IrO$_2$/C, Pt–(RuO$_2$–IrO$_2$)/C, Pt–(RuO$_2$–PbO$_x$)/C, and Pt–(IrO$_2$–PbO$_x$)/C, were also fixed on BDD substrate by sol-gel method. It favored the formation of the nanometric crystallite dimensions of the composites which can be responsible for the enhanced catalytic activity toward ethanol oxidation, in all cases [45]. González-González et al. [47] studied the

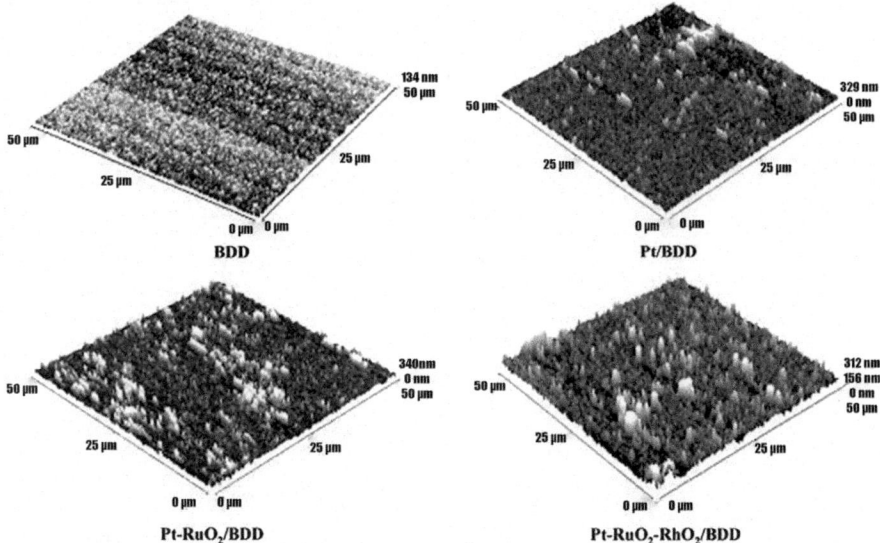

Fig. 7.1 AFM images for BDD, Pt/BDD, Pt–RuO$_2$/BDD, and Pt–RuO$_2$–RhO$_2$/BDD electrode surfaces. Reprinted with permission from [42]

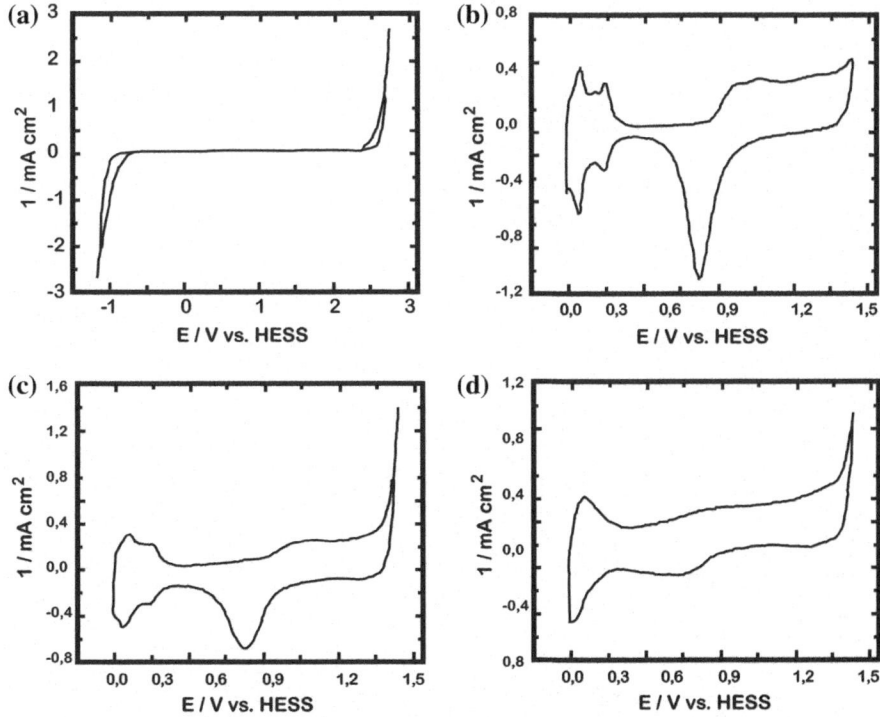

Fig. 7.2 Cyclic voltammetric studies in a 0.5 M of H$_2$SO$_4$ aqueous medium for **a** BDD, **b** Pt/BDD, **c** Pt–RuO$_2$/BDD, and **d** Pt–RuO$_2$–RhO$_2$/BDD. Reprinted with permission from [42]

Fig. 7.3 Cyclic
voltammograms recorded in
0.5 M H₂SO₄ for the BDD
(thin solid-line), Pt/BDD
(thick-solid line), Pt–SnO₂/
BDD (dashed line), and Pt–
Ta₂O₅/BDD (dotted line)
surfaces, v = 50 mV s⁻¹.
Reprinted with permission
from [46]

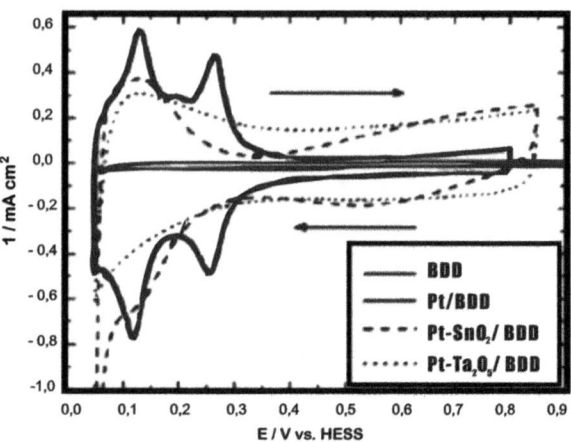

modified electrodes (Pt/BDD and Pt–Ru/BDD) for the oxidation of methanol in
acidic media by using cyclic voltammetry. Maximum currents densities for
methanol oxidation were obtained about 0.73 mA cm⁻² for Pt and 0.94 mA cm⁻²
for Pt–Ru deposited on BDD, respectively. However, as indicated by the authors,
the fact that Pt–Ru exhibited lower potentials than Pt may be expected in the basis
of previous studies [30] and hence, more investigation is necessary to completely
understand the composition and particle size effects.

Salazar-Banda et al. [43] also investigated the oxidation of methanol by using
Pt–RuO$_x$/BDD powder electrode, and its comparison with Pt–Ru/C; demonstrating
that the incorporation of ruthenium inhibited the hydrogen adsorption/desorption
processes. As a consequence, the BDD powder modification presented an important
enhancement of the catalytic activity to methanol oxidation respect to other
materials such as carbon modified composites. Similar modification was proposed
by Swope et al. [48] (conductive diamond powders electrochemically deposited),
achieving an increase on the surface area, approximately 100 m² g⁻¹, and good
corrosion resistance. Salazar-Banda et al. [46] also modified BDD film surfaces
with Pt, Pt–SnO₂, and Pt–Ta₂O₅ nanocrystallines to evaluate the methanol and
ethanol oxidation, but the presence of tantalum oxides produced an enhancement in
defective surface sites, which increases the interfacial capacitance and also raises
the ability of charge accumulation. Voltammograms carried out on BDD surfaces
showed that the oxidation of ethanol started at approximately 0.39, 0.35, and
0.61 V on the same materials (see Fig. 7.3), clearly indicating that, the enhancement of the catalytic activity of the Pt coatings with Sn or Ta oxide, was mainly due
to the presence of Sn. In addition, Pt–SnO₂/BDD electrode exhibited a faster kinetic
for ethanol oxidation avoiding the production of unwanted intermediates on the
surface.

The electrosynthesis of uniformly dispersed nickel hydroxide nanoparticles
(NPs) with narrow size distributions on pBDD electrodes was demonstrated for the
first time by Hutton and co-workers [49]. This was achieved by electrogenerating

OH$^-$ in the presence of Ni^{2+} to create highly supersaturated (S > 105) nickel hydroxide solutions close to the electrode for short time periods (approximately seconds). This resulted in the electrodeposition of nickel hydroxide via precipitation directly on the electrode surface was confirmed by XPS, FE-SEM, and AFM. The size of the NPs could be tuned by controlling the reaction conditions, particularly the concentration of OH$^-$ by the electrogeneration time. After 1 s, NPs with dimensions of 12 nm were produced with a surface coverage of 25 NPs μm^{-1}, increasing in size and surface coverage to ~ 39 nm and 87 NPs μm^{-2}, respectively, after 15 s. Longer times resulted in larger particles, which ultimately formed aggregates. After 100 s, the surface was dominated by such structures which were a few micrometers in height. The nickel hydroxide surface coverage was calculated by considering the charge passed during direct oxidation of nickel hydroxide. This value was in good agreement with that calculated on the basis of AFM images of the surface coverage, assuming spherical NPs. The close correlation of the two results suggests that during this solid-state electrooxidation process, the entire volume of nickel hydroxide was oxidized, not just the surface of the NPs. The electrocatalytic oxidation of this electrode toward methanol and ethanol was found to be very efficient, achieving very high density currents of ~ 1010 A g^{-1} for 0.5 Methanol and 990 A g^{-1} for 0.47 M methanol. Gonzalez-Gonzalez and co-workers also examined the possibility of the use of BDD as the support for electrocatalyst particles in a fuel cell. The electrochemical behavior of oxidized BDD films in 0.5 M H$_2$SO$_4$ using cyclic voltammetry showed a very wide electrochemical window. Platinum particles, which were deposited from different platinum concentrations, did not show a significant difference in the particle size; but particle distribution was considerably different. Pt and Pt–Ru particles deposited on BDD film substrate were characterized by SEM/EDS, XPS, and Auger, observing that the films were oxidized with C–O, C–OH, and C–O–C in their surfaces. Auger electron spectroscopy mapping showed that, no deposition of Ru is achieved at oxidized BDD surface. Meanwhile, particles with 5–10% of ruthenium with respect to platinum exhibited better performance for methanol oxidation in terms of oxidation peak current and current stability. But the insertion of ruthenium provokes a loss in the catalytic properties of Pt.

Steam activation of BDD electrodes was conducted by Ohashi et al. [50]. Triangular pits and islands, most likely reflecting the atomic arrangement of the {111} plane, were formed by steam-activation at 700 °C. Higher activation temperatures lead to rigorous corrosion of the BDD surface forming a highly porous structure with a columnar texture. The electrochemically active surface area of the steam-activated BDD was up to 20 times larger than the pristine BDD electrode owing to the porous texture. Additionally, a widening of the potential window was observed after steam activation, suggesting that the quality of BDD was enhanced due to oxidative removal of graphitic impurities during the activation process allowing to fabricate high quality porous BDD electrodes with high surface area and a wide potential window for a durable electrocatalyst support for energy conversion applications.

Pt and Pt–Ru nanoparticle sizes obtained ranged from 2 to 5 nm and deposited on BDD surfaces (BDDNPs) by La-Torre-Riveros and co-workers [51]. These results indicated that this fast reaction was convenient to obtain small nanoparticles that are desirable to increase the surface area of a catalyst system. The undoped DNPs and BDDNPs surfaces were successfully used as support systems for metallic catalyst particulates obtained by chemical reduction. The platinum and ruthenium compounds mainly deposit on the –OH sites were confirmed by XPS and FTIR analysis. The spectroscopic and surface characterization showed that a catalytic system based on undoped DNPs and BDDNPs supports was obtained, and platinum and ruthenium nanoparticles, which can be used in applications such as direct methanol fuel cells. Same research group [52] also investigated BDD in nanosize particles and they were used as a support for noble metal-based nanoparticles. Nanosize platinum oxide was deposited on diamond nanoparticles as well as on boron-doped diamond nanoparticles by the sol-gel method. The route used to prepare the suspensions (inks) of DNPs and BDDNPs for the electrochemical studies performed by La-Torre-Riveros and co-workers [53] demonstrated the excellent performances concerning the stability, electrical properties and distribution of nanoparticles in the diamond supported systems. The cyclic voltammetry results showed that there were significant differences between the undoped DNP and BDDNPs samples, with the latter exhibiting superior characteristics in terms of various electrochemical applications, including that of catalyst support. For the latter application, low capacitance and wide potential window were the most important consequence achieved. Chemical reduction of metal nanoparticles at the nanometer scale was successfully performed using DNPs and BDDNPs as support materials by using an excess of a mild reducing agent ($NaBH_4$) and a surfactant (SDBS). X-ray diffraction peaks for the metallic nanoparticles were clearly demonstrated this result. The XPS and DRIFTS results provided important information about the type of surface functional groups on the diamond involved in the metal deposition. These techniques indicated that platinum ions interact, become reduced, and are deposited as metal on sites containing mainly $-OH$ and CH_2 (or $-CH_3$) groups. TEM micrographs showed that the nanosize metals were crystals of less than 5 nm, which exhibited lattice fringes for the atomic planes of metal material, indicating good crystallinity. Anodic polarization results for methanol oxidation demonstrated that respectable current densities, in the range of mA cm^{-2}, could be obtained with both undoped DNPs and BDDNPs decorated with Pt and Pt–Ru catalysts. The current density (ca. 55 mW cm^{-2}) obtained with Pt–Ru/ BDDNP was comparable to that obtained with amorphous carbon-supported catalytic system, which is typically around 60 and 70 mW cm^{-2} at the same temperature. On the basis of the single fuel cell testing, it can be concluded that undoped DNPs and BDDNPs can be used as practical electrocatalyst supports for Pt and Pt–Ru in direct methanol fuel cells or hydrogen-fueled polymer electrolyte fuel cells.

The preparation of Pt–PrO_{2-x} electrocatalyst structures on diamond electrodes using electrodeposition techniques was explored by Chen et al. [54]. Pt was first electrodeposited on the diamond surface using a two-stage process involving PSD followed by a potentiostatic stage. Physically, the Pt deposited in this way consists of nanocrystals of the order of 5–30 nm diameters, depending on the Pt loading, aggregated into clusters of 200 nm of diameter. XPS and electrochemical methods were then employed to demonstrate that thin porous PrO_{2-x} preferentially deposited on these crystals, thus modifying their electrochemical properties, as assessed through a range of redox couples. Electrochemical measurements suggested that the main role of the PrO_{2-x} is to improve the tolerance to surface poisoning of the Pt structure.

The unsupported electrodeposition of Pt on carbon nano-onions (CNOs) produced from nanodiamond was investigated by Santiago et al. [55], using the rotating disk—slurry electrode (RoDSE) technique, obtaining mass production of platinum atoms and cluster-carbon-supported catalysts for fuel cell applications. CNOs were synthesized by annealing nanodiamond powder and subjected to acidic treatment to increase their surface oxygen functionalities in order to facilitate platinum electrodeposition. The electrodeposition process behaves similarly to that for Vulcan as a carbon support. The quantity of Pt electrodeposited (i.e., metal loading) on Pt/CNOs was similar to that for samples prepared using Vulcan (i.e., 11.5 wt%). Although both Pt/CNOs and Pt/VXC have the same metal loading, the CNO samples exhibited enhanced thermal stability. The electrochemical analysis showed active platinum electrodeposited on CNOs with defined hydrogen adsorption−desorption peaks, showing enhanced onset potentials for methanol electrooxidation. According to 3Pt/CNO geometric parameters obtained from the experimental study, Pt atoms are bonded to a hexagonal ring. These structural motifs explained the improvement on the thermal stability and lower methanol oxidation onset potentials.

Polytyramine (PTy) and cobalt oxide (Co_3O_4) were also used as substrates for platinum electrodeposition by Spătaru et al. [56] in order to obtain electrode systems with electrocatalytic properties. PTy and Co_3O_4 were previously electrochemically deposited on graphite and BDD supports, respectively. Anodic oxidation of methanol was used as a test-reaction for assessing possible functional effects of the substrate on the electrochemical behavior of the Pt particles. Then, when PTy or cobalt oxide was deposited, the electrocatalyst exhibited higher activity with lower fouling effect, via strong adsorption of reaction intermediates. X-ray photoelectron spectroscopy (XPS) measurements suggested that these features can be ascribed to the presence of −OH functional groups at the surface of the Pt particles. Electrodeposition of platinum on BDD resulted only in metallic Pt and $Pt(OH)_2$, whereas when graphite was used as substrate the presence of platinum oxides was also evident. During prolonged anodic oxidation of methanol, a higher stability of the −OH groups from the platinum surface was observed for graphite-supported Pt particles, compared to the case when bare BDD was used as support.

Although at first sight attractive, existing electro- and electroless routes for the deposition of Pt nanoparticles on diamond supports for fuel cell applications are problematic since they lead to very non-uniform deposits (electrodeposition) or require an activating coating (electroless deposition). Lyu et al. [57] demonstrated that a combined process, which involves simultaneous electroless and galvanic processes. The approach solves the uniformity problem associated with electrode-position on heterogeneous substrates like BDD, and eliminates the requirement of a preactivation process used for conventional electroless deposition. A range of differing reaction conditions were explored using differing reducing agents to drive the electroless reaction, reaction temperature, addition of a surfactant and the use of ultrasonic pre-treatment. Optimal deposition conditions involve reaction at 0 °C, and the use of ascorbic acid as the reducing agent along with an ultrasonic pre-treatment. After optimizing the deposition conditions, the Pt deposits were found to exhibit a uniform distribution of particles, and small particle size; they showed a high electrochemical activity in terms of a large electroactive area measured from hydrogen desorption, a low onset potential, and a high mass activity and surface activity for methanol oxidization. The method therefore represented a significant improvement in current electrochemical and electroless approaches for the deposition of Pt nanoparticles on diamond and other inert substrates for elec-trocatalysis applications.

7.4   BDD-Electrochemical Capacitors

Electrochemical capacitors (ECs) are an alternative as energy storage devices in many fields [58, 59]. ECs can charge and discharge much faster than a battery, and can have potentially much longer life than batteries. Typically they exhibit 20–200 times larger capacitance per unit volume or mass than conventional capacitors [60]. Therefore, a significant number of applications use ECs [61, 62].

New types of carbon materials, such as carbon nanotubes and nanofibers, have been studied as possible ECs because these materials are attractive due to their high surface area and good matrix conductivity. However, high capacitances as well as wide working potential range, in highly conductive aqueous electrolytes are the most important characteristics preferred in the materials for ECs. Therefore, BDD is a very attractive candidate.

Honda and coworkers demonstrated the suitability of BDD nanoporous hon-eycomb electrodes for the development of aqueous electrochemical capacitors [63]. Nanoporous honeycomb diamond films were fabricated from microwave plasma chemical vapor deposition by oxygen plasma etching through an alumina mask. These films exhibited a wide working potential range (ca. 2.5 V) in the aqueous electrolytes, just as in the case of unetched as-deposited diamond electrodes. The capacitance of the honeycomb diamond electrode was found to be 1.97×10^{-3} $F\,cm^{-2}$ (geometric area), which was 200-folds higher than the unetched counterpart (as-deposited surface). The capacitance values obtained from galvanostatic mea-surements were consistent with this value, thus indicating that significant electro-chemical features of honeycomb diamond electrodes as ECs. Meanwhile, the oxygen plasma-etched nanohoneycomb diamond electrodes were examined for electrochemical capacitor applications in non-aqueous electrolytes [64], achieving an increase on the potential window (up to 7.3 V) respect to the aqueous electrolytes. For pore type 400 nm × 1.8 μm in non-aqueous electrolyte, the power and energy densities could reach only similar values as those in aqueous electrolytes. However, the impedance behavior observed in non-aqueous electrolytes was significantly different from that in aqueous electrolyte and indicated that the AC signal cannot penetrate to the bottom of the honeycomb pores in the non-aqueous electrolytes due to their low conductivity, and that not all the surface may contribute to the double-layer capacitance. Therefore, the authors concluded that the combination of pore type 400 nm × 1.8 μm and aqueous electrolyte could be best for examined thus far.

Nanoporous BDD films, with various pore diameters (30–400 nm) and pore depths (50 nm to 3 μm), were fabricated by etching polished polycrystalline diamond films through porous alumina masks with oxygen plasma [65]. The capacitance values increased with increasing roughness factor, based on the pore dimensions. The honeycomb diamond electrode with pore dimensions 400 nm 3 μm exhibited a 400-fold increase in the capacitance (3.91×10^{-3} μF cm^{-2}, geometric area) in comparison to the as-deposited surface, and this value was 80 and 500-folds higher than that achieved at GC and HOPG, respectively. For the porous film with 30 nm diameter pores, there was only a very small effect of the pore structure on the capacitance due to the high pore impedance.

Almeida and co-workers [66] developed nanocrystalline diamond (NCD) grown on carbon fibers (CF) substrate to be used as electric double-layer capacitor. A high specific capacitance (2.6 mF cm^{-2}) and rectangular-shaped CV curves were obtained up to a high potential scan rate (100 mV s^{-1}) in 0.5 mol L^{-1} H$_2$SO$_4$ aqueous solution for NCD/CF-1300 (CF-1300 consists of felt disks with 0.15-cm thickness diameter). These results showed that the NCD/CF electrodes could be an excellent candidate for electrochemical double-layer capacitors by controlling deposition parameters and CF substrate microstructures. The morphologic and electrochemical characterization of carbon fibers (CF) and their hybrid material formed by BDD films grown on CF was also studied [67]. Figure 7.4 shows the composition of CF surfaces, treated at 1000 and 2000 °C, associated with their respective SEM micrographs. SEM images obtained for the BDD/CF-1000 (A, A1 and A2) and BDD/CF-2000 electrodes (B, B1 and B2), exhibited in Fig. 7.5, demonstrate that the CF substrates were completely covered by a polycrystalline diamond coating. These micrographs also showed that the grain size of BDD/CF-2000 electrode was larger than that for BDD/CF-1000 electrode, supporting the strong influence of CF structural parameters on diamond growth. Cyclic voltammetric curves demonstrated that BDD films grown on CF carbonized at 2000 °C (BDD/CF-2000), achieving higher capacitance values (1940 μF cm^{-2} (geometric area) which were significantly enhanced respect to the CF-2000 (Table 7.2). This capacitive behavior was attributed to the increase on its surface area as a result of the singular diamond film morphology formed on such carbon fiber.

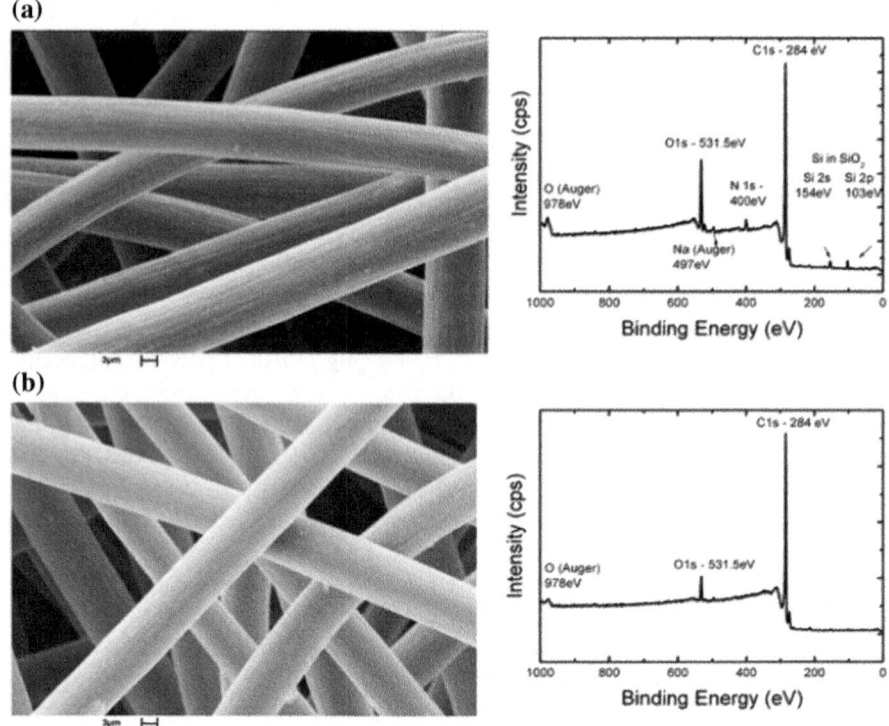

Fig. 7.4 Scanning electron microscopy images and X-ray photoelectron spectroscopy spectra of CF-1000 (**a**) and CF-2000 (**b**). Reprinted with permission from [67]

7.5　New Trends in Electrochemical Energy Conversion and Storage

7.5.1　Modification of BDD Surfaces for Fuel Cells

Hernández-Lebrón and Cabrera [68] used the square wave voltammetry (SWV) as a route for structure modification of platinum nanoparticles previously electrode-posited on BDD films. A solution of sulfuric acid was used with ascorbic acid and with acetone during the SWV treatment, generating structural and morphological changes in platinum surface and nanoparticle structure. Platinum shapes and some of the nanocubes were the specific alterations of platinum nanoparticles by using ascorbic acid or acetone during SWV produces. In some cases, the ammonia oxidation with Pt–BDD electrodes was achieved, and consequently, these platinum nanostructures could be used for the oxidation of other fuels. SWV can be also used for the synthesis and reconstruction of shaped controlled particles.

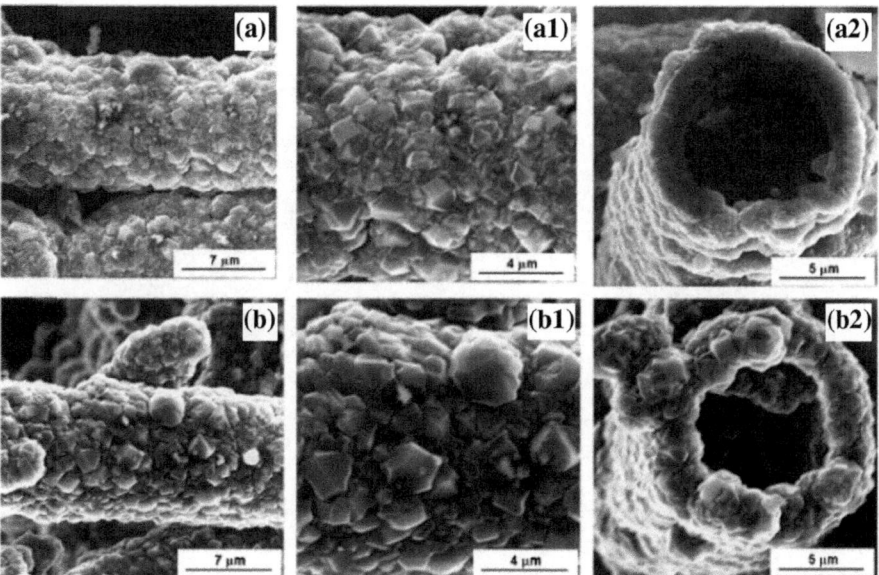

Fig. 7.5 Scanning electron microscopy images of the BDD films growth on carbon fibers: BDD/CF-1000 (**a, a1** and **a2**) and BDD/CF-2000 (**b, b1** and **b2**). **a, a1, b,** and **b1** images show the surface film morphology, while images **a2** and **b2** show the film thickness around each fiber. Reprinted with permission from [67]

Table 7.2 BET area, capacitance and parameters used for fitting the impedance results in both the Nyquist and Bode plots

Electrode	BET ($m^2 g^{-1}$)	C ($\mu F\ cm^{-2}$)	R_1 (Ω)	R_2 ($10^3\ \Omega$)	CPE_1 ($10^{-4}\ \Omega^{-1}\ s^n$)	n_1
CF-1000	1.51	266	142.6	0.474	0.06639	0.8427
CF-2000	0.33	245	16.26	1.338	0.1717	0.8493
BDD/CF-1000	4.05	459	305.4	4.41	0.1936	0.9004
BDD/CF-2000	14.00	1940	38.6	0.820	1.213	0.9086

In direct methanol fuel cells, conventional carbonaceous materials generally used as supports can undergo oxidation (particularly during fuel starvation and resulting polarity reversal) leading to the agglomeration of electrocatalyst particles and to a loss of electrical contact. Generally, platinum active sites are deactivated by CO poisoning during methanol anodic oxidation, which also considerably reduces the lifetime of the electrodes. For this reason, Spătaru et al. [69] have investigated the possibility of using BDD as a substrate for Pt–RuOx·nH$_2$O composites. BDD films were subjected to cathodic or anodic pre-treatments in order to achieve hydrogen-terminated or oxygen-terminated surfaces to understand the influence of the surface termination, prior to hydrous ruthenium oxide electrodeposition. SEM measurements performed after platinum deposition on the oxide-modified BDD-O electrodes demonstrated that with higher ruthenium oxide loadings (ca. 140 $\mu g\ cm^{-2}$),

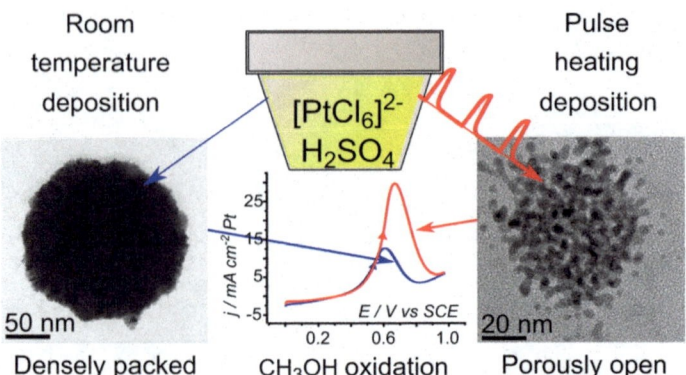

Room temperature deposition · $[PtCl_6]^{2-}$ H_2SO_4 · Pulse heating deposition

50 nm · 20 nm

Densely packed · CH_3OH oxidation · Porously open

$j / mA\ cm^{-2} Pt$ · $E / V\ vs\ SCE$ · 25 15 5 -5 · 0.2 0.6 1.0

Fig. 7.6 Scheme of nanoporous platinum nanoparticles synthesis. Reprinted with permission from [70]

P tparticles are more uniformly distributed and smaller in size, thus ensuring more efficient use of the noble metal. The activity for methanol oxidation of the Pt–$RuO_x\cdot nH_2O$/BDD-O electrodes was performed, and it was found that the overall process is strongly promoted by a thick hydrous ruthenium oxide layer. However, in terms of resistance to fouling, a lower amount of $RuO_x\cdot nH_2O$ appears to be more advantageous, probably due to the fact that it allows higher surface concentrations of oxidized species of both platinum and carbon.

A new approach to producing nanoporous Pt nanoparticles of high electrocatalytic activity supported on BDD films was proposed by Hussein [70] (Fig. 7.6). Injecting bursts of heat via IR laser modulation into the interface and then rapidly cooling acts to separate the nucleation and growth processes, leading to nanoparticles size homogeneity and the production of "porous", electrochemically accessible structures. Pt aggregates were obtained with 2–5 nm in size, containing high-index crystal planes. Investigation of nanoporous Pt nanoparticles activity toward MeOH oxidation revealed a significantly higher catalytic activity, however, these are susceptible to adsorbed CO poisoning.

Electrochemical deposition of Pd and bimetallic Pd-Ni nanoparticles on an oxygen-terminated BDD substrate were prepared and used in a direct ethanol fuel cell [71]. A potentiostatic two-step electrochemical method involving the electrodeposition of Ni nanoparticles on BDD followed by mono-dispersed Pd nanoparticles was used for the fabrication of a Pd-Ni/BDD electrode. The electrocatalytic activity of the bimetallic Pd-Ni nanoparticles was evaluated in an alkaline solution containing ethanol and compared to that of the Pd nanoparticles alone. The bimetallic Pd-Ni nanoparticles showed 2.4 times higher mass activity than similar systems in the literature as well as stability when operated in alkaline media via the synergistic effects of the electron interaction at the interface of the two metals. Loading of mono-dispersed Pd on a foreign Ni metal as nanoparticles was about 3.63×10^6 mA g^{-1} and specific (10.53 mA cm^{-2}) electrocatalytic activity of Pd towards ethanol electrooxidation in alkaline media.

7.5.2 BDD Electrodes as Supercapacitors

Gao et al. [72] fabricated Si nanowires covered by a 135 nm-thin nanocrystalline boron-doped diamond layer. The capacitive properties were tested in an ionic liquid (1-methyl-1-propylpyrrolidinium bis(trifluoromethylsulfonyl)imide, PMPyrrTFSI) in propylene carbonate (PC), which provided good conductivity and wettability. The electrochemical measurements allowed to determine the capacitance over a wide potential window of 4 V, reaching 105 mF cm^{-2} at 1 V s^{-1}, but dropped to 47 mF cm^{-2} upon increasing the scan rate to 100 V s^{-1}. The supercapacitor obtained has the capacity to retain 93.3% of its capacitance after 10,000 cycles at 5 V s^{-1}. Aradilla et al. [73] fabricated Si nanowires with lengths of 50 mm and diameters in the range 20–200 nm with a diamond coating of 100 nm thickness. The micro-supercapacitors achieved 1.9 mF cm^{-2} of capacitance at 2 mA cm^{-2} and subsequently dropped to 1.5 mF cm^{-2} for higher scan rates. This value is higher than those obtained with electric double-layer micro-supercapacitors based on carbon materials. Electrochemical impedance spectroscopy (EIS) measurements suggested a relaxation time of 165 ms, justifying the device's ability to deliver high power densities in a very short time. The micro-supercapacitors of BDD on the Si nanowires exhibited energy and power density values in the range of 11–15 mJ cm^{-2} and 3–25 mW cm^{-2}, respectively. Its capacitance was reduced (35%) after more than 1 million galvanostatic charge/discharge cycles. May and co-workers [74] reported a similar approach, Si needles were obtained with a length of almost 20 mm by inductively coupled plasma etching process. The diamond coating was reached, resulting in a heavily boron-doped micro- or nanocrystalline diamond films over the Si needles. The capacitance for partially overgrown BDD-coated long Si needles reached 638 mF cm^{-2} whereas the flat BDD exhibits only 2.9 mF cm^{-2}, and BDD-coated short and long needles with 7.6 and 279.4 mF cm^{-2}, respectively. The results discussed here indicate that the large surface-area-to-volume ratio (micro- or nanocrystalline diamond size) plays an important role in the double-layer charge, which is a crucial parameter for supercapacitor applications.

Zanin et al. [75, 76] investigated the electrochemical behavior of carbon/diamond electrodes in typical aqueous electrolytes. Tepee-BDD and honeycomb-BDD composite materials were fabricated by using vertically aligned carbon nanotubes with lengths of 5 and 40 mm, respectively, resulting in a significant difference in the surface. The electrochemical measurements showed that significant differences in the typical shapes for double-layer charging at both materials. Due to the higher real surface area, highest currents and hence higher capacitances were reached at honeycomb-BDD with carbon nanotubes material.

Highly corrugated diamond-like carbon films on the vertically aligned carbon nanotubes (40 mm long, 60 nm in diameter) were also studied, achieving capacitances of about 642 mF cm^{-2} for at 0.15 V, and this value is significantly higher than typical double-capacitance for a carbon electrode (5 mF cm^{-2}). The formation of oxygenated surface groups of the nanotubes improved the capacitive current [76].

Aradilla et al. [77] used silicon nanowires (50 mm long; 20–200 nm in diameter), and after that, these were covered by a diamond film, and finally a poly (3,4-ethylenedioxythiophene) (PEDOT) layer. Cyclic voltammetric profiles showed a pseudocapacitive behavior associated to the polymer coating, and the redox couple corresponded to the reduction and oxidation reactions of PEDOT. A decrease in the specific capacitance from 140 F g^{-1} at 1 mV s^{-1} down to 24 F g^{-1} at 200 mV s^{-1} was observed, which is caused by the slow diffusion of the electrolyte ions at high scan rates. The galvanostatic charge/discharge cycles carried out at different current densities exhibited an areal capacitance of 9.5 mF cm^{-2} at 0.1 mA cm^{-2}, which dropped to 8.5 mF cm^{-2} at 0.1 mA cm^{-2}. The energy and power densities were 29–25 mJ cm^{-2} and 0.1–3 mW cm^{-2}, respectively, at current densities varying in the range of 0.1–2 mA cm^{-2}. The device lifetime was over 15,000 charge/discharge cycles performed at 1 mA cm^{-2} within 2.5 V-wide potential windows, and the microcapacitors maintained 80% of its initial capacitance.

Gao et al. [78] fabricated diamond nanowires (15–20 nm in diameter) by a top-down etching approach. Anodization of the substrate resulted in the Ni/diamond electrode material. When the thickness of the deposited Ni increases, the nickel hydroxide layer becomes more corrugated.

Shi et al. [79] reported the BDD films grown on a Ta plate as a substrate for Ti layer (100 nm) deposition by magnetron sputtering, followed by anodization in fluoride-containing solution. The average crystal size of BDD was 30 mm, whereas the obtained TiO_2 increased the surface roughness and the number of electroactive sites. The electrochemical performance showed a significant increase on the pseudocapacitance (5.23 mF cm^{-2} at a scan rate of 5 mV s^{-1} and decreases when the scan rate increases), whereas nanostructured titania is responsible for the ion absorption enhancing the charge-transfer efficiency. After 250 and 500 cycles of galvanostatic charge/discharge, 96.4 and 89.3% of capacitance retention were observed.

Apart from titania, manganese oxide as a part of the diamond-based heterojunction was proposed by Yu et al. [80]. MnO_2 films were potentiostatic electrodeposition from $MnSO_4$ solution on the BDD. In this form, a porous MnO_2 layer was deposited (24 mg cm^{-2}) forming a coating on BDD surface without any cracks. Bare BDD exhibited a capacitance of 3.6 mF cm^{-2} at a scan rate of 10 mV s^{-1} and current density of 0.05 mA cm^{-2}, reducing 5% after 1000 charge/discharge cycles. Meanwhile, BDD/MnO_2 registered a high capacitance (7.9 mF cm^{-2}) when 24 mg cm^{-2} of MnO_2 was deposited, decreasing when the amount of oxide increases. After 1000 cycles, a loss of 34% of the initial capacitance was also attained, which is related to the weak binding between the MnO_2 film and BDD substrate, irreversible redox reaction during charging/discharging processes, fouling of the porous BDD/MnO_2 structure by impurities, and mechanical expansion of manganese oxide due to the intercalation/deintercalation of electrolyte cations lowering the structural stability. Conversely, highly ordered layer of titania was produced by anodization of a Ti metal plate serving as a substrate for BDD growth. This material exhibits different capacitances (7.46, 4.79, and 2.1 mFcm^{-2} for TiO_2

modified with BDD) depending on the B/C ratios (10,000, 5000, and 2000). Meanwhile, 0.11 mF cm^{-2} was registered for bare BDD. The capacitive behavior of titania-based diamond composites clearly showed that these materials could be considered as promising materials [81, 82].

7.6   Conclusions

BDD films can be modified with micro- and nanometric metallic and/or metallic oxide deposits by using different methods like electrodeposition, sol-gel, thermal deposition, microemulsion methods, among others. The depositions of metal or metal oxide clusters onto the BDD electrodes have been extensively investigated to understand the catalytic activity of nanoparticles. Different types of electrocatalysts have been employed to oxidize methanol or ethanol in order to produce electrical energy. Concerning the electrochemical capacitors using BDD as support, novel materials have been investigated, giving emphasis on the preparation of capacitive materials with high-area. As a consequence, the research about modified BDD materials as electrochemical capacitors systems has significantly increased in the last years [83]. Future developments will rely upon the close collaboration of analytical chemists, engineers and electrochemists to ensure effective application and exploitation of new catalysts to increase the efficiency of fuel cells using BDD anodes.

Acknowledgements   The authors acknowledge support from projects CNPq—465,571/2014-0; CNPq—446,846/2014-7 and CNPq—401,519/2014-7 and FAPESP—2014/50,945-4. Carlos A. Martínez-Huitle acknowledges the funding provided by the Alexander von Humboldt Foundation (Germany) and Coordenação de Aperfeiçoamento de Pessoal de Nível Superior (Brazil) as a fellowship for experienced researcher.

References

1. E. Antolini, Platinum-based ternary catalysts for low temperature fuel cells. Part II. Electrochemical properties. Appl. Cat. B Environ. **74**, 337–350 (2007). https://doi.org/10.1016/j.apcatb.2007.03.001
2. U. Griesbach, D. Zollinger, H. Pütter, Ch. Comninellis, Evaluation of boron doped diamond electrodes for organic electrosynthesis on a preparative scale. J. Appl. Electrochem. **35**, 1265–1270 (2005). https://doi.org/10.1007/s10800-005-9038-2
3. M. Panizza, E. Brillas, Ch. Comninellis, Application of boron-doped diamond electrodes For Wastewater Treatment. J. Environ. Eng. Manage. **18**, 139–153 (2008)
4. P. Canizares, A. Gadri, J. Lobato, B. Nasr, R. Paz, M.A. Rodrigo, C. Saez, Electrochemical oxidation of azoic dyes with conductive-diamond anodes. Ind. Eng. Chem. Res. **45**, 3468–3473 (2006). https://doi.org/10.1021/ie051427n
5. J. Iniesta, P.A. Michaud, M. Panizza, G. Cerisola, A. Aldaz, Ch. Comninellis, Electrochemical oxidation of phenol at boron-doped diamond electrode. Electrochim. Acta **46**, 3573–3578 (2001)

6. M.A. Rodrigo, P.A. Michaud, I. Duo, M. Panizza, G. Cerisola, Ch. Comninellis, Oxidation of 4-chlorophenol at boron-doped diamond electrode for wastewater treatment. J. Electrochem. Soc. **148**, D60–D64 (2001). https://doi.org/10.1149/1.1362545

7. B. Boye, P.A. Michaud, B. Marselli, M.M. Dieng, E. Brillas, Ch. Comninellis, Anodic oxidation of 4-chlorophenoxyacetic acid on synthetic boron-doped diamond electrodes. New Diam. Front. C. Technol. **12**, 63–72 (2002)

8. I. Sirés, N. Oturan, M.A. Oturan, R.M. Rodríguez, J.A. Garrido, E. Brillas, Electro-Fenton degradation of antimicrobials triclosan and triclocarban. Electrochim. Acta **52**, 5493–5503 (2007). https://doi.org/10.1016/j.electacta.2007.03.011

9. C. Flox, J.A. Garrido, R.M. Rodríguez, P.L. Cabot, F. Centellas, C. Arias, E. Brillas, Mineralization of herbicide mecoprop by photoelectro-Fenton with UVA and solar light. Catal. Today **129**, 29–36 (2007). https://doi.org/10.1016/j.cattod.2007.06.049

10. I. Sirés, F. Centellas, J.A. Garrido, R.M. Rodríguez, C. Arias, P.L. Cabot, E. Brillas, Mineralization of clofibric acid by electrochemical advanced oxidation processes using a boron-doped diamond anode and Fe2 + and UVA light as catalysts. Appl. Catal. B-Environ. **72**, 373–381 (2007). https://doi.org/10.1016/j.apcatb.2006.12.002

11. Y. Shao, J. Liu, Y. Wang, Y. Lin, Novel catalyst support materials for PEM fuel cells: current status and future prospects. J. Mater. Chem. **19**, 46–59 (2009). https://doi.org/10.1039/b808370c

12. A. Kraft, Doped diamond: A compact review on a new, versatile electrode material. Int. J. Electrochem. Sci. **2**, 355–385 (2007)

13. M. Boutonnet, S. Lögdberg, E.E. Svensson, Recent developments in the application of nanoparticles prepared from w/o microemulsions in heterogeneous catalysis. Curr. Opin. Colloid Interface Sci. **13**, 270–286 (2008). https://doi.org/10.1016/j.cocis.2007.10.001

14. S. Eriksson, U. Nylén, S. Rojas, M. Boutonnet, Preparation of catalysts from microemulsions and their applications in heterogeneous catalysis. Appl. Catal. A Gen. **265**, 207–219 (2004). https://doi.org/10.1016/j.apcata.2004.01.014

15. M. Boutonnet, J. Kizling, P. Stenius, G. Maire, The preparation of monodisperse colloidal metal particles from microemulsions. Colloids Surf. **5**, 209–225 (1982). https://doi.org/10.1016/0166-6622(82)80079-6

16. G. Siné, Ch. Comninellis, Nafion®-assisted deposition of microemulsion-synthesized platinum nanoparticles on BDD: Activation by electrogenerated OH radicals. Electrochim. Acta **50**, 2249–2254 (2005). https://doi.org/10.1016/j.electacta.2004.10.008

17. G. Siné, G. Fóti, Ch. Comninellis, Boron-doped diamond (BDD)-supported Pt/Sn nanoparticles synthesized in microemulsion systems as electrocatalysts of ethanol oxidation. J. Electroanal. Chem. **595**, 115–124 (2006). https://doi.org/10.1016/j.jelechem.2006.07.012

18. G. Siné, I. Duo, B. El Roustom, G. Fóti, Ch. Comninellis, Deposition of clusters and nanoparticles onto boron-doped diamond electrodes for electrocatalysis. J. Appl. Electrochem. **36**, 847–862 (2006). https://doi.org/10.1007/s10800-006-9159-2

19. G. Siné, D. Smida, M. Limat, G. Fóti, Ch. Comninellis, Microemulsion synthesized Pt/Ru/Sn nanoparticles on BDD for alcohol electro-oxidation. J. Electrochem. Soc. **154**, B170–B174 (2007). https://doi.org/10.1149/1.2400602

20. F. Cardarelli, P. Taxil, A. Savall, Ch. Comninellis, G. Manoli, O. Leclerc, Preparation of oxygen evolving electrodes with long service life under extreme conditions. J. Appl. Electrochem. **28**, 245–250 (1998). https://doi.org/10.1023/A:1003251329958

21. I. Duo, P.A. Michaud, W. Haenni, A. Perret, Ch. Comninellis, Activation of boron-doped diamond with IrO$_2$ clusters. Electrochem. Solid-State Lett. **3**, 325–334 (2000). https://doi.org/10.1149/1.1391137

22. B.E. Roustom, G. Fóti, Ch. Comninellis, Preparation of gold nanoparticles by heat treatment of sputter deposited gold on boron-doped diamond film electrod. Electrochem. Commun. **7**, 398–405 (2005). https://doi.org/10.1016/j.elecom.2005.02.014

23. F. Montilla, E. Morallón, I. Duo, Ch. Comninellis, J.L. Vázquez, Platinum particles deposited on synthetic boron-doped diamond surfaces. Application to methanol oxidation. Electrochim. Acta **48**, 3891–3897 (2003). https://doi.org/10.1016/s0013-4686(03)00526-7

24. A. De Battisti, S. Ferro, M. Dal Colle, Electrocatalysis at conductive diamond modified by noble-metal oxides. J. Phys. Chem. B **105**, 1679–1689 (2001)
25. J.S. Gao, T. Arunagiri, J.J. Chen, P. Goodwill, O. Chyan, J. Perez, D. Golden, Preparation and characterization of metal nanoparticles on a diamond surface. Chem. Mater. **12**, 3495–3504 (2000). https://doi.org/10.1021/cm000465o
26. M. Li, G. Zhao, R. Geng, H. Hu, Facile electrocatalytic redox of hemoglobin by flower-like gold nanoparticles on boron-doped diamond surface. Bioelectrochemistry **74**, 217–224 (2008). https://doi.org/10.1016/j.bioelechem.2008.08.004
27. V. Saez, J. Gonzalez-Garcia, M.A. Kulandainathan, F. Marken, Electro-deposition and stripping of catalytically active iron metal nanoparticles at boron-doped diamond electrodes. Electrochem. Commun. **9**, 1127–1133 (2007). https://doi.org/10.1016/j.elecom.2007.01.018
28. K.E. Toghill, G.G. Wildgoose, A. Moshar, C. Mulcahy, R.G. Compton, Fabrication and characterization of a bismuth nanoparticle modified boron doped diamond electrode and its application to the simultaneous determination of cadmium(II) and lead(II). Electroanalysis **20**, 1731–1739 (2008). https://doi.org/10.1002/elan.200804277
29. J. Wang, G.M. Swain, Fabrication and evaluation of platinum/diamond composite electrodes for electrocatalysis: Preliminary studies of the oxygen-reduction. J. Electrochem. Soc. **150**, E24–E32 (2003). https://doi.org/10.1149/1.1524612
30. O. Enea, B. Riedo, G. Dietler, AFM Study of Pt clusters electrochemically deposited onto boron-doped diamond films. Nano Lett. **2**, 241–249 (2002). https://doi.org/10.1021/nl0156661
31. C.A. Martínez-Huitle, M.A. Rodrigo, I. Sirés, O. Scialdone, Chem. Rev. **115**, 13362–13407 (2015)
32. Z. He, J. Chen, D. Lui, H. Zhou, Y. Kuang, Electrodeposition of Pt–Ru nanoparticles on carbon nanotubes and their electrocatalytic properties for methanol electrooxidation. Diam. Relat. Mater. **13**, 1764–1770 (2004). https://doi.org/10.1016/j.diamond.2004.03.004
33. I. Gonzalez-Gonzalez, D.A. Tryk, C.R. Cabrera, Polycrystalline boron-doped diamond films as supports for methanol oxidation electrocatalysts. Diam. Relat. Mater. **15**, 275–278 (2006). https://doi.org/10.1016/j.diamond.2005.08.037
34. H.A. Gasteiger, N. Markovic, P.N. Ross, E.J. Cairns, Methanol electrooxidation on well-characterized Pt–Ru alloys. J. Phys. Chem. **97**, 12020–12029 (1993)
35. X. Lu, J. Hu, J.S. Foord, Q. Wang, Electrochemical deposition of Pt–Ru on diamond electrodes for the electrooxidation of methanol. J. Electroanal. Chem. **654**, 38–43 (2011). https://doi.org/10.1016/j.jelechem.2011.01.034
36. H.B. Suffredini, G.R. Salazar-Banda, S.T. Tanimoto, M.L. Calegaro, S.A.S. Machado, L.A. Avaca, AFM studies and electrochemical characterization of boron-doped diamond surfaces modified with metal oxides by the sol-gel method. J. Braz. Chem. Soc. **17**, 257–264 (2006). https://doi.org/10.1590/S0103-50532006000200007
37. H.B. Suffredini, V. Tricoli, N. Vatistas, L.A. Avaca, Electro-oxidation of methanol and ethanol using a Pt–RuO2/C composite prepared by the sol-gel technique and supported on boron-doped diamond. J. Power Sources **158**, 124–128 (2006). https://doi.org/10.1016/j.jpowsour.2005.09.040
38. L.C. Klein, M. Aparicio, F. Damay, Sol-gel processing for battery and fuel cell applications, in *Applications of Sol-Gel Technology*, vol. III, ed. by S. Sakka (Kluwer Academic Publisher, 2004), p. 311
39. S. Sakka (ed.), *Applications of Sol-Gel Technology* (Kluwer Academic Publisher, 2004), p. 3
40. H.B. Suffredini, V. Tricoli, L.A. Avaca, N. Vatistas, Sol-gel method to prepare active Pt–RuO 2 coatings on carbon powder for methanol oxidation. Electrochem. Commun. **6**, 1025–1028 (2004). https://doi.org/10.1016/j.elecom.2004.08.008
41. G.R. Salazar-Banda, H.B. Suffredini, L.A. Avaca, Improved stability of PtOx Sol-gel-modified diamond electrodes covered with a Nafion® film. J. Braz. Chem. Soc. **16**, 903–906 (2005). https://doi.org/10.1590/S0103-50532005000600003

42. G.R. Salazar-Banda, H.B. Suffredini, M.L. Calegaro, S.T. Tanimoto, L.A. Avaca, Sol-gel-modified boron-doped diamond surfaces for methanol and ethanol electro-oxidation in acid medium. J. Power Sources 162, 9–20 (2006). https://doi.org/10.1016/j.jpowsour.2006.06.045

43. G.R. Salazar-Banda, K.I.B. Eguiluz, L.A. Avaca, Boron-doped diamond powder as catalyst support for fuel cell applications. Electrochem. Commun. 9, 59–64 (2007). https://doi.org/10.1016/j.elecom.2006.08.038

44. H.B. Suffredini, V. Tricoli, N. Vatistas, L.A. Avaca, J. Power Sources 158, 124–128 (2006)

45. H.B. Suffredini, G.R. Salazar-Banda, L.A. Avaca, Enhanced ethanol oxidation on PbOx-containing electrode materials for fuel cell applications. J. Power Sources 171, 355–362 (2007). https://doi.org/10.1016/j.jpowsour.2007.06.048

46. G.R. Salazar-Banda, H.B. Suffredini, L.A. Avaca, S.A.S. Machado, Mater. Chem. Phys. 117, 434–442 (2009)

47. I. González-González, D.A. Tryk, C.R. Cabrera, Diam. Relat. Mater. 15, 275–278 (2006)

48. V. M. Swope, I. Sasaki, A. Ay. G. M. Swain, Conductive diamond powder: A new catalyst support for the polymer electrolyte membrane fuel cell, ECS Trans. 3, 27–36 (2007). https://doi.org/10.1149/1.2753281

49. L.A. Hutton, M. Vidotti, A.N. Patel, M.E. Newton, P.R. Unwin, J.V. Macpherson, Electrodeposition of nickel hydroxide nanoparticles on boron-doped diamond electrodes for oxidative electrocatalysis. J. Phys. Chem. C 115, 1649–1658 (2011). https://doi.org/10.1021/jp109526b

50. I. Gonzalez-Gonzalez, C. Lorenzo-Medrano, C.R. Cabrera, Adv. Phys. Chem. 10, 679246 (2011), https://doi.org/10.1155/2011/679246

51. L. La-Torre-Riveros, E. Abel-Tatis, A.E. Mendez-Torres, D.A. Tryk, Mark Prelas, C.R. Cabrera, Synthesis of platinum and platinum-ruthenium-modified diamond nanoparticles, J. Nanopart. Res. 13, 2997–3009 (2011). https://doi.org/10.1007/s11051-010-0196-8

52. L. Cunci, C.R. Cabrera, Preparation and electrochemistry of boron-doped diamond nanoparticles on glassy carbon electrodes. Electrochem. Solid-State Lett. 14, K17–K19 (2011). https://doi.org/10.1149/1.3532943

53. L. La-Torre-Riveros, R. Guzman-Blas, A.E. Méndez-Torres, M. Prelas, D.A. Tryk, C.R. Cabrera, Diamond nanoparticles as a support for Pt and PtRu catalysts for direct methanol fuel cells. ACS Appl. Mater. Interfaces. 4, 1134–1147 (2012). https://doi.org/10.1021/am2018628

54. L. Chen, J. Hu, J.S. Foord, Electrodeposition of a Pt–PrO2-x electrocatalyst on diamond electrodes for the oxidation of methanol. Phys. Status Solidi A 209(9), 1792–1796 (2012). https://doi.org/10.1002/pssa.201200049

55. D. Santiago, G.G. Rodríguez-Calero, A. Palkar, D. Barraza-Jimenez, D.H. Galvan, G. Casillas, A. Mayoral, M. Jose-Yacamán, L. Echegoyen, C.R. Cabrera, Platinum electrode-position on unsupported carbon nano-onions. Langmuir 28, 17202–17210 (2012). https://doi.org/10.1021/la3031396

56. T. Spătaru, P. Osiceanu, M. Marcu, C. Lete, C. Munteanu, N. Spătaru, Jpn. J. Appl. Phys. 51, 090119 (2012)

57. X. Lyu, J. Hua, J.S. Foord, Q. Wang, A novel electroless method to prepare a platinum electrocatalyst on diamond for fuel cell applications. J. Power Sources 242, 631–637 (2013). https://doi.org/10.1016/j.jpowsour.2013.05.057

58. A. Lewandowski, M. Galinski, Practical and theoretical limits for electrochemical double-layer capacitors. J. Power Sources 173, 822–828 (2007). https://doi.org/10.1016/j.jpowsour.2007.05.062

59. H.D. Abruña, Y. Kiya, J.C. Henderson, Batteries and electrochemical capacitors. Phys. Today 61, 43–47 (2008). https://doi.org/10.1063/1.3047681

60. Y. Zhang, Y. Gui, X. Wu, H. Feng, A. Zhang, L. Wang, T.C. Xia, Preparation of nanostructures NiO and their electrochemical capacitive behaviors. Int. J. Hydrog Energy 34, 2467–2470 (2009). https://doi.org/10.1016/j.ijhydene.2008.12.078

61. J.R. Miller, Electrochemical capacitor thermal management issues at high-rate cycling. Electrochim. Acta 52, 1703–1708 (2006). https://doi.org/10.1016/j.electacta.2006.02.056

62. C. -Z. Yuan, B. Gao, X. -G. Zhang, Electrochemical capacitance of NiO/Ru0.35V0.65O2 asymmetric electrochemical capacitor. J. Power Sources 173, 606–612 (2007). https://doi.org/10.1016/j.jpowsour.2007.04.034

63. K. Honda, T.N. Rao, D.A. Tryk, A. Fujishima, M. Watanabe, K. Yasui, H. Masuda, Electrochemical characterization of the nanoporous honeycomb diamond electrode as an electrical double-layer capacitor. J. Electrochem. Soc. 147, 644–659 (2000). https://doi.org/10.1149/1.1393249

64. M. Yoshimura, K. Honda, R. Uchikado, T. Kondo, T.N. Rao, D.A. Tryk, A. Fujishima, Y. Sakamoto, K. Yasui, H. Masuda, Electrochemical characterization of nanoporous honeycomb diamond electrodes in non-aqueous electrolytes. Diam. Relat. Mater. 10, 620–626 (2001). https://doi.org/10.1016/S0925-9635(00)00381-2

65. K. Honda, T.N. Rao, D.A. Tryk, A. Fujishima, M. Watanabe, K. Yasui, H. Masuda, Impedance Characteristics of the Nanoporous Honeycomb Diamond Electrodes for Electrical Double-Layer Capacitor Applications. J. Electrochem. Soc. 148, A668–A679 (2001). https://doi.org/10.1149/1.1373450

66. E.C. Almeida, A.F. Azevedo, M.R. Baldan, N.A. Braga, J.M. Rosolen, N.G. Ferreira, Nanocrystalline diamond/carbon felt as a novel composite for electrochemical storage energy in capacitor. Chem. Phys. Lett. 438, 47–52 (2007). https://doi.org/10.1016/j.cplett.2007.02.040

67. E.C. Almeida, M.R. Baldan, J.M. Rosolen, N.G. Ferreira, Impedance characteristics of the diamond/carbon fiber electrodes for electrical double-layer capacitor. Diam. Relat. Mater. 17, 1529–1533 (2008). https://doi.org/10.1016/j.diamond.2008.03.006

68. Y. Hernández-Lebrón, C.R. Cabrera, Square wave voltammetry restructuring of platinum nanoparticle at boron doped diamond electrode for enhanced ammonia oxidation. J. Electroanal. Chem. 793, 174–183 (2017). https://doi.org/10.1016/j.jelechem.2016.12.036

69. T. Spătaru, L. Preda, P. Osiceanu, C. Munteanu, M. Marcu, C. Lete, N. Spătaru, A. Fujishima, Electrochemical deposition of Pt–RuOx·nH2O composites on conductive diamond and its application to methanol oxidation in acidic media. Electrocatalysis 7, 140–148 (2016). https://doi.org/10.1007/s12678-015-0292-8

70. H.E.M. Hussein, H. Amari, J.V. Macpherson, Electrochemical synthesis of nanoporous platinum nanoparticles using laser pulse heating: Application to methanol oxidation. ACS Catal. 7, 7388–7398 (2017). https://doi.org/10.1021/acscatal.7b02701

71. C.K. Mavrokefalos, M. Hasan, J.F. Rohan, J.S. Foord, Enhanced mass activity and stability of bimetallic Pd–Ni nanoparticles on boron-doped diamond for direct ethanol fuel cell applications. Chem. Electro. Chem. 5, 455–463 (2018). https://doi.org/10.1002/celc.201701105

72. F. Gao, G. Lewes-Malandrakis, M.T. Wolfer, W. Mgller-Sebert, P. Gentile, D. Aradilla, T. Schubert, C.E. Nebel, Diamond-coated silicon wires for supercapacitor applications in ionic liquids. Diam. Relat. Mater. 51, 1–6 (2015). https://doi.org/10.1016/j.diamond.2014.10.009

73. D. Aradilla, F. Gao, G. Lewes-Malandrakis, W. Mgller-Sebert, D. Gaboriau, P. Gentile, B. Iliev, T. Schubert, S. Sadki, G. Bidan, C.E. Nebel, A step forward into hierarchically nanostructured materials for high performance micro-supercapacitors: Diamond-coated SiNW electrodes in protic ionic liquid electrolyte. Electrochem. Commun. 63, 34–38 (2016). https://doi.org/10.1016/j.elecom.2015.12.008

74. P.W. May, M. Clegg, T.A. Silva, H. Zanin, O. Fatibello-Filho, V. Celorrio, D.J. Fermin, C.C. Welch, G. Hazell, L. Fisher, A. Nobbs, B. Su, Diamond-coated 'black silicon' as a promising material for high-surface-area electrochemical electrodes and antibacterial surfaces. J. Mater. Chem. B 4, 5737–5746 (2016). https://doi.org/10.1039/c6tb01774f

75. H. Zanin, P.W. May, D.J. Fermin, D. Plana, S.M.C. Vieira, W.I. Milne, E.J. Corat, Porous boron-doped diamond/carbon nanotube electrodes, ACS Appl. Mater. Interf. 6, 990–995 (2014), https://doi.org/10.1021/am4044344

76. H. Zanin, P.W. May, R.L. Harniman, T. Risbridger, E.J. Corat, D.J. Fermin, High surface area diamond-like carbon electrodes grown on vertically aligned carbon nanotubes. Carbon 82, 288–296 (2015). https://doi.org/10.1016/j.carbon.2014.10.073

77. D. Aradilla, F. Gao, G. Lewes-Malandrakis, W. Mgller-Sebert, P. Gentile, M. Boniface, D. Aldakov, B. Iliev, T.J.S. Schubert, C.E. Nebel, G. Bidan, Designing 3D Multihierarchical Heteronanostructures for High-Performance On-Chip Hybrid Supercapacitors: Poly(3,4-(ethylenedioxy)thiophene)-Coated Diamond/Silicon Nanowire Electrodes in an Aprotic Ionic Liquid. ACS Appl. Mater. Interf. **8**, 18069–18077 (2016). https://doi.org/10.1021/acsami. 6b04816

78. F. Gao, C.E. Nebel, Diamond nanowire forest decorated with nickel hydroxide as a pseudocapacitive material for fast charging–discharging. Phys. Status Solidi A **212**, 2533–2538 (2015)

79. C. Shi, H. Li, C. Li, M. Li, C. Qu, B. Yang, Preparation of TiO2/boron-doped diamond/Ta multilayer films and use as electrode materials for supercapacitors. Appl. Surf. Sci. **357**, 1380–1387 (2015). https://doi.org/10.1016/j.apsusc.2015.10.006

80. S. Yu, N. Yang, H. Zhuang, J. Meyer, S. Mandal, O.A. Williams, I. Lilge, H. Schönherr, X. Jiang, Electrochemical Supercapacitors from Diamond. J. Phys. Chem. C **119**, 18918–18926 (2015). https://doi.org/10.1021/acs.jpcc.5b04719

81. K. Siuzdak, R. Bogdanowicz, M. Sawczak, M. Sobaszek, Enhanced capacitance of composite TiO2 nanotube/boron-doped diamond electrodes studied by impedance spectroscopy. Nanoscale **7**, 551–558 (2015). https://doi.org/10.1039/c4nr04417g

82. M. Sobaszek, K. Siuzdak, M. Sawczak, J. Ryl, R. Bogdanowicz, Fabrication and characterization of composite TiO2 nanotubes/boron-doped diamond electrodes towards enhanced supercapacitors. Thin Solid Films **601**, 35–40 (2016). https://doi.org/10.1016/j.tsf. 2015.09.073

83. K. Siuzdak, R. Bogdanowicz, Nano-engineered diamond-based materials for supercapacitor electrodes: a review. Energy Technol. **6**, 223–237 (2018). https://doi.org/10.1002/ente. 201700345

Chapter 8
Diamond Electrochemical Devices

Nianjun Yang

Abstract Conductive boron-doped diamond is one of the best electrode materials and has been widely used for different electrochemical applications. Among them, the fabrication, properties, and applications of small-dimensional diamond electrochemical devices (e.g., diamond microelectrode, ultramicroelectrode, nanoelectrode and their arrays as well as scanning probe microscopy tips) have been paid much attention. In this chapter we summarize recent progress and achievements with respect to these small dimensional diamond electrochemical devices. The potential applications and future research directions of these devices are also discussed and outlined.

8.1 Introduction

Conductive boron-doped diamond (BDD) has been recognized as one of the best electrode materials after its introduction in 1983 by Iwaki et al. [1] and later in 1987 by Pleskov et al. [2]. Over passed decades, it has been extensively utilized in the various fields of electrochemistry for different applications [3–14]. This is because BDD films feature numerous unique physical and chemical properties over other solid electrode materials [7, 8, 14–18]. Being an electrode, the capacitive currents of BDD are low and stable in both aqueous and non-aqueous solutions. For example, the capacitive current of a hydrogen-terminated (HT) BDD electrode (boron concentration: 2×10^{20} cm^{-3}) in 0.1 M pH 7 phosphate buffer is 10 times lower than that of a gold electrode and about 100 times lower than that of a glassy carbon electrode. Moreover, BDD has wide electrochemical potential windows, depending on the used electrolyte (e.g., pH), its boron-concentration, its surface termination, and surface defects, etc. A heavily boron-doped single crystalline diamond exhibits a potential window of about 3.2 V in aqueous solutions, 4.6 V in organic solutions,

N. Yang (✉)
Institute of Materials Engineering, University of Siegen, 57076 Siegen, Germany
e-mail: nianjun.yang@uni-siegen.de

© Springer Nature Switzerland AG 2019
N. Yang (ed.), *Novel Aspects of Diamond*, Topics in Applied Physics 121,
https://doi.org/10.1007/978-3-030-12469-4_8

and 4.9 V at room temperature ionic liquid. These electrochemical potential windows are estimated when an absolution value of current density less than 1.0 mA cm^{-2} is defined. Furthermore, BDD is chemically inert, not swelling in electrolyte solutions, and does not show surface fouling, especially for HT BDD. BDD shows high chemical stability in harsh environments/media, at high current densities, and even under sun exposure. In additional, the BDD surface can be terminated with hydrogen, oxygen, or the mixtures of both. These terminations can be achieved by hydrogen or oxygen plasma treatment at high temperatures or by electrochemical reduction/oxidation at high current densities in acidic solutions. Altering the surface terminations of a BDD electrode can lead to the convenient optimization of its electronic properties of the solid/electrolyte interfaces. BDD is also biocompatible and can be bio-functionalized via rich carbon-carbon chemistry [12–14, 19, 20]. BDD is a hard material and thus its surface can be textured with dimensions of typically a few nanometers [21–23] to nanowires with lengths of a few micrometers [24].

Up to now, planar and macroscopic BDD electrodes have been frequently applied for fundamental electrochemistry, bio electrochemistry, sensor, and environmental related applications [3–14]. Under these situations, average electrochemical signals over the full electrode areas are detected. However, planar and macroscopic BDD electrodes suffer from non-uniform boron doping (especially for polycrystalline BDD films), boundary effects, and varied ratios of sp^2-graphite to sp^3-diamond for different samples.

To overcome these shortcomings and have more sensitive electrochemical signals for BDD electrodes, small dimensional BDD electrodes, especially those at the micro- and submicron- dimensions have been fabricated and widely investigated in past years [25–30]. Depending on their dimensions (e.g., diameters of electrodes), these small-dimensional BDD electrodes have been divided into BDD microelectrode (ME) when the diameters of BDD electrodes are in the range of 25–100 μm, BDD ultramicroelectrode (UME) when the diameters of BDD electrodes are in the range of 0.1–25 μm, and BDD nanoelectrode (NE) when the diameters of BDD electrodes are smaller than 100 nm [25, 26]. Different from planar and macroscopic BDD electrodes, these small-dimensional BDD electrodes have shown many advantages [25–30]. For example, they feature reduced Ohmic resistances, enhanced mass transport rates, decreased charging currents, decreased deleterious effects of solution resistance, and high possibility for fast voltammetric measurements [31–41].

It has to be pointed out that single ME/UME/NE generates only small currents. In some cases, these small currents are relatively difficult to be detected with conventional electrochemical setups. In this context, small-dimensional BDD electrodes are assembled and operated in parallel. When these electrodes are regularly arranged, one array is obtained, or else it is an ensemble. Provided that these arrays/ensembles are well designed (e.g., suitable diameter, optimized distance between electrodes, etc.), the signals of individual BDD ME/UME/NEs can be amplified. Meanwhile, the beneficial characteristics of individual small dimensional BDD electrodes will not lose [1–4]. Therefore, on these small-dimensional BDD electrode arrays, the performance with respect to the sensitivity, detection limit, lifetime, and reproducibility towards different targets, will be much improved. This is partially due to the employment of the most appropriate and optimized electrode material (namely BDD) for the fabrication of these small dimensional electrode arrays, and partially due to the features of small dimensional electrodes and their arrays.

Herein, we summarize in this chapter the fabrication, properties and applications of these small-dimensional BDD electrodes (including ME, UME, and NE) [31–41] as well as their ensembles/arrays [42–57]. As one kind of similar electrochemical device which can be used in air or vacuum or in solutions, scanning tunneling microscopy (STM) tips produced from conducting diamond are also included. Note that the progress and achievements on the fabrication, properties and applications of planar and macroscopic BDD electrodes do not belong to the content of this chapter and thus have been excluded. The potential applications and future research directions of these BDD ME/UME/NAs and their arrays/ensembles are also discussed and outlined.

8.2 Diamond Microelectrodes and Ultramicroelectrodes

8.2.1 Fabrication

Several approaches have been developed to fabricate diamond MEs and UMEs, including overgrowth of conformal diamond on small-diameter metal wires or direct coating of metal wires (e.g., produced by means of electrochemically etching processes) only exposed at the end of glass capillary columns, overgrowth of diamond layers on (lithographically) patterned substrates, and oxygen-based ion beam plasma etching of bulk diamond material. Among them, one frequently applied approach is to coat sharpened metal wires with a thin BDD film. The used metals are mainly tungsten [31–36] and platinum [37, 38]. Coating tungsten wires sealed in a quartz glass capillary is another approach [39, 40]. For example, in 1988 diamond ME was reported for the first time [31]. It was fabricated through overgrowth of an electrochemically etched tungsten wire with an electrically conducting microcrystalline BDD film using the microwave plasma enhanced chemical vapour deposition (MWCVD) technique. The overgrown tungsten wire was subsequently sealed in glass. Careful mechanical polishing or chemical etching of the glass with HF led to the exposure of electroactive diamond. In 2003, diamond MEs were prepared by overgrowing electrochemically sharpened platinum wires (76-, 25-, and 10-μm diameter) with polycrystalline diamond thin films. A MWCVD reactor was applied [37]. Focused ion beam (FIB) milling was further applied to improve the exposure areas of disc shaped MEs [36]. However, FIB exposure of BDD MEs unavoidably implants Ga^+ ions into the surface of BDD layers as well as the amorphization of BDD surface layers down to several tens of nanometers [58].

Although many approaches have been successfully developed to fabricate diamond MEs and UMEs with different geometries [31–40, 59, 60], there still exist several challenges [61]. For example, the reproducibility of highly conductive and conformal diamond films on fibrous substrates (e.g., metal wires, etc.) needs to be improved. Simple and reliable methods to insulate cylindrical portions of as-fabricated diamond MEs and UMEs are required. The precise control of exposure surface areas of diamond MEs and UMEs, their diameters, and shapes are still challenging.

8.2.2 Characterization

These MEs and UMEs have been characterized with various techniques such as scanning electron microscopy (SEM), optical microscopy, Raman spectroscopy, and electrochemical techniques (e.g., voltammetry, impedance, etc.) [31–40].

Figure 8.1 shows example SEM images of two electrochemically etched platinum wires with the diameter of 76 and 25 µm before and after BDD overgrowth

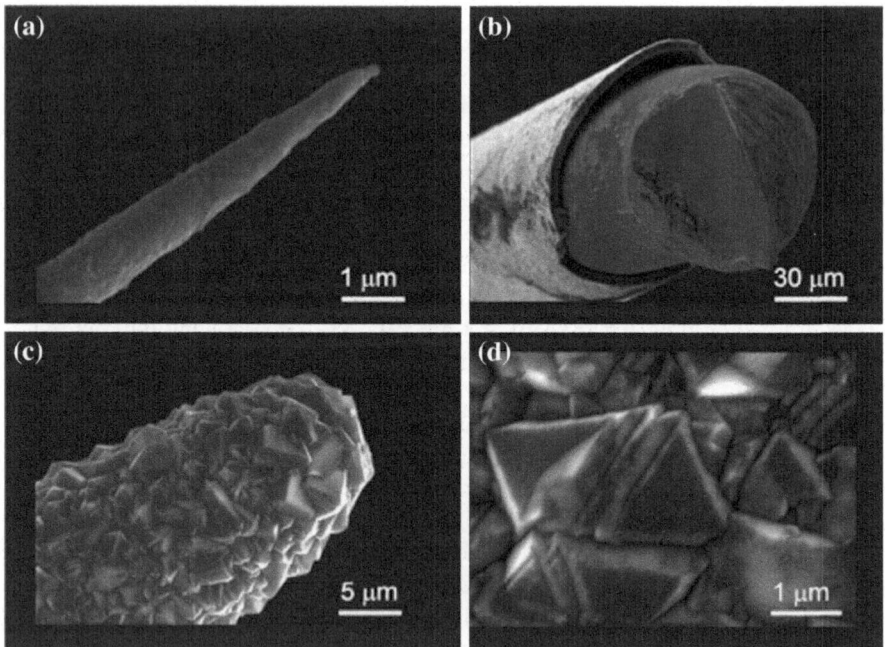

Fig. 8.1 SEM images of **a** the end of a 76-µm electrochemically etched platinum wire; **b** the cross section of a 76-µm wire covered with a BDD film, obtained by cutting the microelectrode with razor blade; **c** the tip of a 25-µm Pt wire covered with a BDD film; **d** the surface of the polycrystalline diamond film deposited on the platinum wire. Reprinted with permission from [37], Copyright ACS Publisher 2003

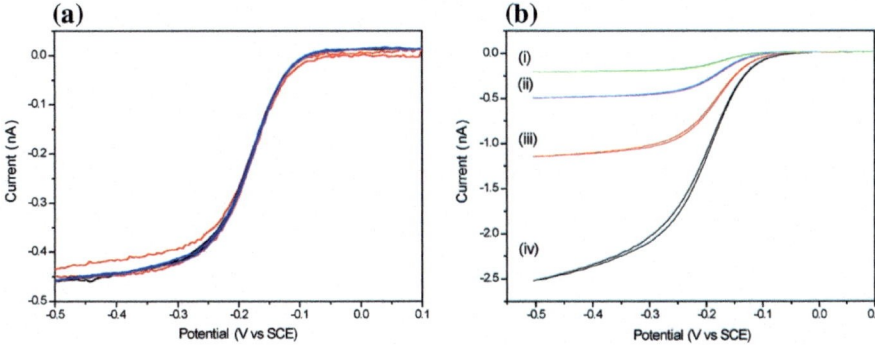

Fig. 8.2 CVs of the smallest diamond UME obtained from side-on FIB milling. **a** CVs at a scan rate of 25 (black), 50 (red), and 100 (blue) mV s^{-1}. **b** CVs in Ru(NH$_3$)$_6^{3+}$ with a concentration of 0.5 (i), 1 (ii), 2 (iii), and 4 (iv) mM. Reprinted with permission from [36]. Copyright ACS Publisher 2009

[37]. On diamond MEs, low and stable background currents were obtained [36, 37]. At low scan rates, the cyclic voltammograms (CVs) of redox probes (e.g., Ru(NH$_3$)$_6^{3+/2+}$ and Fe(CN)$_6^{3-/4-}$, etc.) are sigmoidally shaped. For example, on a diamond UME (radius < 1 μm) fabricated from side-on ion milling, its CVs in Ru(NH$_3$)$_6^{3+}$ solutions at the scan rates of 25, 50, and 100 mV s^{-1} are perfect sigmoidally shaped [36]. The limiting currents are independent of sweep rates up to values that are dependent on the sizes of diamond MEs (Fig. 8.2a), but linear with the concentration of Ru(NH$_3$)$_6^{3+}$ (Fig. 8.2b). The voltammetric behavior on diamond MEs is stable for several weeks [36].

Electrochemical characterization of diamond UMEs by means of cyclic voltammetry also revealed that the porosity of diamond films (more exactly diamond grain sizes) affects deleteriously the performance and lifetime of BDD UMEs [36]. For example, an increase of its limiting current from around 0.4 to 2.2 nA after repeated use reveals a possible leakage in the insulation and electrolyte ingress into the diamond layer. This speculation is confirmed by the drastic increase of the background signal of this diamond UME. In other words, the leakage current from underlying tungsten carbide is noticed. For diamond films with large grain sizes, the effect is much pronounced. In contrast, for those with small sizes it is neglectable [36].

Electrochemical impedance spectroscopy (EIS) was applied to investigate the size effect of diamond MEs on their electrochemical properties [62]. Figure 8.3 shows the related Nyquist plots recorded in 5 mM Fe(CN)$_6^{3-/4-}$ + 1 M KCl. The amplitude is 10 mV and the frequency range is from 0.1 to 100 kHz. By reducing the size (or diameter) of BDD ME/UME from 250 down to 10 μm, the shape of Nyquist plots changes from linear lines to two arcs [62]. The fitting of experimental data to electrochemical circuit model suggests that each arc likely corresponds to the grains of ultrananocrystalline diamond (UNCD) and grain boundary phases.

Fig. 8.3 Nyquist plot of diamond MEs with different sizes or diameters: **a** 250 (black dotted), 200 (green dotted), and 150 μm (brown dotted); **b** 100 (orange dotted) and 50 μm (red dotted); **c** 25 (gray dotted) and 10 μm (purple dotted). The inset in **c** shows the expanded view of a 25-μm plot. The solid curves are fitted results using electrical circuits inset in **b** and **c**. Reprinted with permission from [62]. Copyright Elsevier Publisher 2015

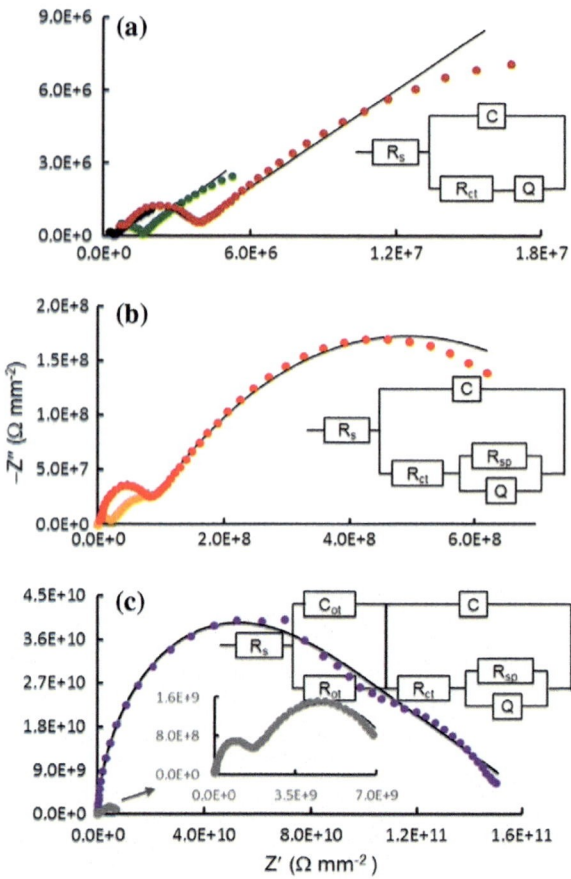

Two phases become separable as a result of the reduction of the sizes of BDD MU/UMEs. For MEs with their diameters less than 100 μm, microstructural and morphological defects/heterogeneities of grain boundaries and the presence of surface oxygen affect their EIS spectra. The size reduction of BDD MEs/UMEs specifically affects the impedance of the grain boundaries. For example, the grain-grain boundary properties are noticed more sensitively in their EIS spectra since the impedance of grain boundary is increased with a factor of 30 [62].

8.2.3   Applications

These diamond MEs and UMEs have been widely explored for electrochemical sensing applications in both non-aqueous [31] and aqueous solutions [32–40]. For example, they have been used in biological media to detect dopamine in mouse brain [33], to monitor norepinephrine release in mesenteric artery [38, 63], to investigate the role of adenosine in the modulation of breathing within animal tissue [39], to inspect serotonin as a neuromodulator [40, 64], to determine ozone [65, 66], and others [67–70]. In comparison to planar and macroscopic diamond electrodes as well as to carbon fibers and metal wires, these diamond MEs and UMEs have shown superior analytical figures of merit in terms of signal-to-noise ratios, linear dynamic ranges, limits of detection, and response precision towards different analytes in solutions [71–79].

Diamond MEs and UMEs have been integrated with other techniques such as capillary electrophoresis (CE) [37, 80] and flowing systems (e.g., microfluidic devices) [81, 82]. For example, using a 10 mM pH 6.0 phosphate buffer as the run buffer, a 30-cm-long fused-silica capillary (75-µm i.d.), and a diamond ME, reproducible separation (elution time) and detection (peak current or area) of dopamine, catechol, and ascorbic acid were achieved. The response precisions were less than 4.1%. The measurements were doable within a linear range over 4 orders of magnitude, up to a concentration between 0.1 and 1 mM. The detection limits for dopamine and catechol were as low as 1.7 and 2.6 fM, respectively (S/N = 3).

Recently, all-diamond MEs have been proposed for novel sensing applications [81, 83], such as being used as solid state probes for localized electrochemical sensing, more accurately to investigate localized corrosion [83]. It was conducted via amperometric measurement of oxygen distribution above an Al–Cu–CFRP (Carbon Fiber Reinforced Polymer) galvanic corrosion cell. Such kind of all-diamond microprobes was constructed with an inner BDD layer and an outer undoped diamond layer. Both layers were subsequently grown on a sharp tungsten tip by hot-filament CVD (HFCVD) in a stepwise manner within a single deposition run. FIB cutting of the apex of the ME was used to expose an electroactive BDD disk. Prior to localized sensing application, cyclic voltammetric behavior of ferrocenemethanol was tested on those MEs.

Diamond UMEs have been employed as the tips for scanning electrochemical microscopy (SECM). Spatially and temporally resolved electrochemical measurements have been successfully conducted [36, 41]. For example, on a diamond UME tip, fabricated through MWCVD overgrowth of an electrochemically etched tungsten wire with a BDD layer, various approach curves have been recorded towards conducting and insulating surfaces [34]. The fitting of these approach curves was conducted after taking two different electrode geometries of these BDD UMEs (e.g., disc-shape and hemi-sphere) into consideration during the course of mathematical fitting. Moreover, a disc-shaped diamond UME was employed to record SECM imaging of immobilized *E. coli*, namely to map the related respiratory activity immobilized onto a glass slide using poly-L-lysine. The cells were immobilized on the right-hand side of the slide as demarked by a white line in Fig. 8.4a. To record SECM images of these cells, ferricyanide was reduced by the respiratory chain enzymes to ferrocyanide, which was further detected at the diamond UME (radius ∼3 µm) biased at 0 V (vs. Ag/AgCl). The BDD UME tip was approximately 5 µm above the cells. The image was scanned in the $x–y$ plane at a

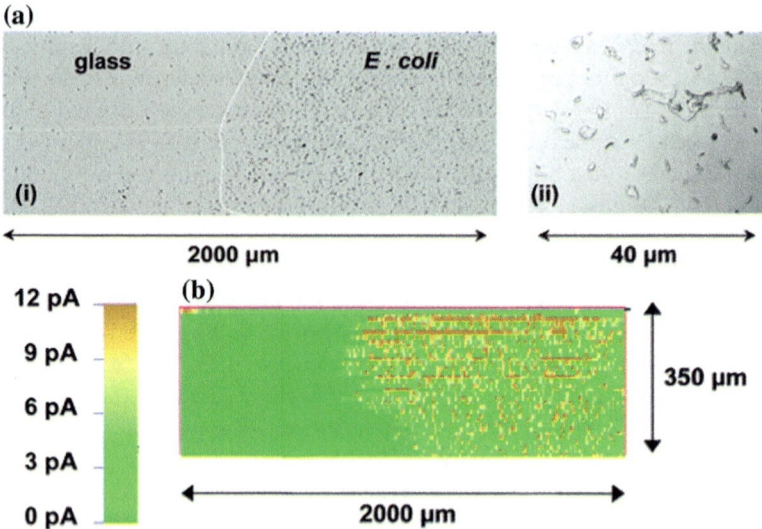

Fig. 8.4 **a** Optical microscope images using (i) × 5 objective, *E. coli* immobilized onto a glass slide using poly-L-lysine (cells are immobilized on the right-hand side of the slide as demarked by a white line); (ii) × 50 objective, immobilized *E. coli* showing isolated cells of about 2 μm in length as well as aggregated clumps of cells; **b** SECM image of immobilized *E. coli* cells. Reprinted with permission from [34]. Copyright ACS Publisher 2007

speed of 300 μm s^{-1}. The recorded SECM image (Fig. 8.4b) is comparable with the optical one (Fig. 8.4a).

Besides SECM tips, diamond MEs and UMEs have been integrated into AFM-SECM probes. For example, a conically shaped BDD-AFM-SECM probe has been fabricated from commercially available AFM probes. Reactive ion etching (RIE) and FIB processing were further applied [84]. Unfortunately, such a probe exhibited a relatively high capacitive current even at a scan rate of 10 mV s^{-1}. Later, BDD-AFM-SECM probes with recessed electrodes have been produced [85, 86]. These recessed AFM-SECM probes were fabricated according to the following steps: (i) overgrowth of commercially available Si AFM probes with a 500-nm thick nanocrystalline BDD layer by means of a MWCVD technique; (ii) subsequently coating with Parylene-C or an approx. 900-nm thick silicon nitride/silicon oxide layers; (iii) FIB milling; (iv) an electrochemical treatment of BDD-AFM-SECM probes in sulfuric acid to remove amorphous carbon from the FIB-milled surface as well as to achieve a HT BDD surface. On such a probe, sigmoidal-shaped CVs of redox probes (e.g., 10 mM K$_4$[Fe(CN)$_6$] in 0.1 M KCl) were obtained, as expected for a ME. SECM images have been conducted at model samples (e.g., the electroactive area of a Pt disc ME, at a structured Parylene C feature deposited at a conductive platinum surface). The images in AFM contact mode and SECM negative feedback mode were recorded. The visible feature of circular Parylene C in the AFM topography image corresponds well with the dark ring in SECM image [86].

8.3 Diamond Microelectrode Arrays and Ultramicroelectrode Arrays

8.3.1 Fabrication

In 2002, Fujishima and coworkers used structured silicon substrates to fabricate diamond microelectrode arrays (MEA) [42]. The array consists of 200 micro-disks with the diameters between 25 and 30 μm and with electrode spacing of 250 μm. Rychen and coworkers [43] produced diamond UME arrays (UMEAs) by depositing a BDD film with a thickness of 5 μm onto patterned silicon nitrite. The diameter of the UMEAs is 5 μm, the distance between UMEs is 150 μm and the number of UMEs is 106. Kang, Swain and coworkers [44–48] produced diamond UMEAs with different shapes, spacing, and number of UMEs. They utilized the "as-grown" diamond surface with randomly micro-structured topology as a planar diamond electrode. They also used a micro-patterning technique to produce a well-defined pyramidal tip-array with a controlled uniformity. In 2005, Compton and coworkers [49–51] realized for the first time all-diamond UMEAs. The diameters of diamond UMEs are between 10 and 25 μm. The separation of these UMEs is in the range of 100–250 μm. Bergonzo et al. [52]. and Carabelli et al. [53]. utilized nanocrystalline diamond (NCD) films to generate diamond MEAs. Interconnected and individually addressable diamond (U)MEAs have been also fabricated, including a 10-channel diamond array on polymer based MEA by Hess et al. [87], a 4-channel NCD MEA by Gao et al. [88], and a 8 × 8 multichannel addressable diamond MEA by Bergonzo et al. [89]. Recently, batch-production of integrated diamond UMEAs has been demonstrated by use of polycrystalline diamond (PCD) films [55, 56].

Taking the batch-production of integrated diamond UMEAs as an example, the involved fabrication steps are schematically shown in Fig. 8.5A [55, 56]. For these diamond UMEAs, insulting PCD (iD) films were grown on silicon (Si) wafers in an ellipsoidal shaped MWCVD system [56]. The temperature was in the range of 750–900 °C, the microwave power was 3 kW, the gas was 3% methane diluted in hydrogen, and the pressure was 60 mbar. Boron-doped PCD (B-PCD) films were grown by adding 7000 ppm trimethyl boron (TMB) to the gas phase. Prior to the fabrication of these UMEAs, as-grown diamond films were cleaned wet-chemically in the mixture of concentrated sulfuric acid (98%) and concentrated nitric acid (65%) (V:V = 3:1) at 200 °C for 1.5 h. The batch-production of diamond UMEAs includes three photolithography steps, two etching steps, one overgrowth step, one metal-deposition step, and one lift-off step. In the first step, a B-PCD film with a thickness 200–500 nm is deposited on a polished iD film with a thickness of 8–10 μm (Fig. 8.5A-a). Then, the first photolithography step is applied (Fig. 8.5A-b). After spin coating of the diamond wafer with the photoresist, a 350 nm SiO_2 layer is deposited. SiO_2 based patterns are generated by etching with SF_6 gas (Fig. 8.5A-c). To produce BDD based structures, RIE of the wafer (Fig. 8.5A-d) is conducted in a gas mixture of oxygen and hydrogen [58–90]. Please note that these

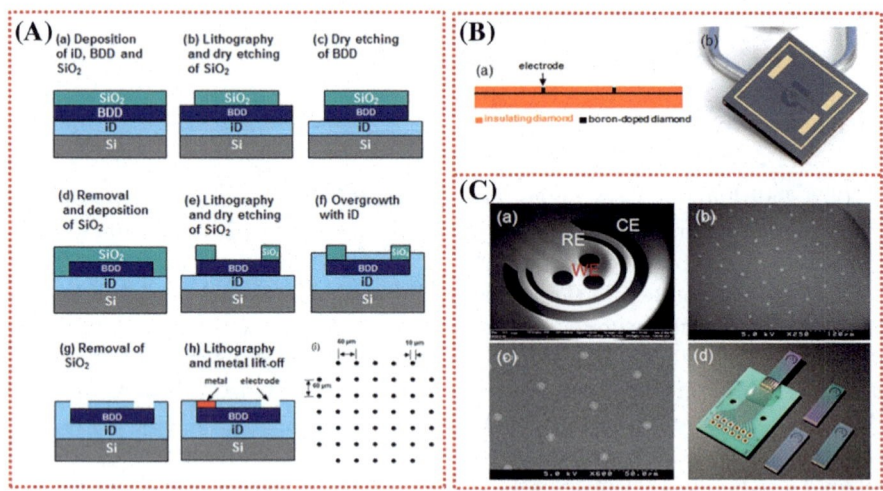

Fig. 8.5 **A** Schematic steps of the fabrication of integrated all-diamond UMEAs; **B** schematic plot of the structure of a diamond UMEA (**a**) and one photo of one integrated UMEA (**b**). Reprinted with permission from [55], Copyright ACS Publisher 2011; **C** SEM images of several addressable diamond UMEAs (**a, b, c**). WE, RE and CE are denoted for working electrode, reference electrode, and counter electrode, respectively; **d** the layout of an addressable diamond UMEAs after mounting the chips on a mother-board and wire-bonding

structures are protected. The conductivity of un-protected areas is checked frequently to make sure BDD is etched away. For the overgrowth of iD films, the second photolithography process is applied (Fig. 8.5A-e). The overgrowth (Fig. 8.5A-f) coats all area with iD except the parts (namely the counter, reference and UMEs) that are protected by SiO_2. The third photolithography step is finally applied (Fig. 8.5A-g) and the Ti/Pt/Au (20/60/200 nm) metal layers are deposited for electric connection. The last step is the application of the lift-off technique (Fig. 8.5A-h).

Depending on the design of the lithography masks, various diamond MEAs and UMEAs have been produced. The arrangement of one fabricated diamond UMEA is schematically shown in Fig. 8.5A-i. Figure 8.5B and C show one photo of as-fabricated diamond UMEA and their SEM images, respectively [56]. In Fig. 8.5B, the diamond MEA chip has a size of 5×5 mm^2. The yellow parts are metal contacts. The dark part with a semi-circle is the counter electrode, the dark rectangle is the reference electrode, and the center electrode is one diamond MEA. The counter and quasi-reference electrodes are oxidized BDD films. The diameter of UMEs is 10 μm and the total number of UMEs is 45. The vertical and horizontal center-to-center spacing in between UMEs is 60 μm. For one 2-inch wafer, more than 40 integrated chips were produced. Figure 8.5C-a shows SEM images of other diamond UMEAs where three working electrodes (WEs) are addressed, including one 600-μm macroscopic BDD electrode and two UMEAs. The counter electrode (CE) and reference electrode (RE) are made from BDD and integrated as well. The

UMEs in the arrays are arranged using a honeycomb structure. Their diameters are 20, 10, and 5 μm. Up to the distance of these UMEs, the number of these electrodes is varied from 37 to 85. For example, for a center-to-center distance of 60 μm between UMEs, the number of UMEs is 85; for a distance of 120 μm, the number of UMEs is 37. Figure 8.5C-b and c show the SEM images of the diamond UMEAs having 85 UMEs with the diameter of 20 and 10 μm, respectively. In both case, the center-to-center distance between two UMEs is 60 μm. These UMEAs were further mounted onto a motherboard through wire bonding. Such a demonstrator is presented in Fig. 8.5C-d.

8.3.2 Characterization

Various characterization techniques, including electron and optical microscopy, Raman spectroscopy, and electrochemical techniques have been applied to investigate the properties of these diamond MEAs/UMEAs. Some photos and SEM images of as-fabricated diamond MEAs and UMEAs are shown in Fig. 8.5B, C.

Voltammetry and impedance have been extensively applied to characterize diamond MEAs and UMEAs. For example, a 64-channel diamond MEA has been characterized using cyclic voltammetry and EIS [89], where a very fast electrode transfer rate up to 0.05 cm s^{-1} was achieved. Voltammetry has been also applied to investigate the effect of surface terminations of diamond electrodes, boron doping concentrations, used electrolytes, and scan rates on the Faradaic currents of redox couple of $Fe(CN)_6^{3-/4-}$ on diamond UMEAs [55, 56]. HT diamond UMEAs with boron-concentration of $4.2(\pm 2) \times 10^{20}$ cm^{-3} show the highest Faradaic current, indicating the fastest electron transfer process. The variation of supporting electrolyte does not change much capacitive currents of diamond UMEAs but alters dramatically obtained Faradaic currents [25–30]. Figure 8.6 shows the CVs of $Fe(CN)_6^{3-/4-}$ on diamond UMEAs (Fig. 8.5B) at different scan rates (solid lines). The magnitudes of the Faradaic currents increase and the shape of the CVs varies with an increase of the scan rates. At low scan rates (e.g., from 0.02 to 0.2 V s^{-1}) and at a fast scan rate of 20 V s^{-1}, peak-shaped CVs are obtained, indicating linear diffusion-limited transport processes of redox analytes. A sigmoidal-shaped CV is detected at a scan rate of 2 V s^{-1}, which is consistent with hemi-sphere diffusion to the UMEs on the array. The thickness of diffusion layer (δ) was calculated using the equation [25] of $\delta = (2D\Delta E/v)^{1/2}$, where v is the scan rate, D (= 7.6×10^{-6} cm s^{-1}) is the diffusion-coefficient of redox analytes, ΔE is the potential range over which electrolysis occurs. For example, to estimate the size of the diffusion layer thickness at $E = 0.4$ V, the value $\Delta E = 0.8$ V was used since significant electrolysis current started at $E = -0.4$ V. The center-to-center separation and the diameter of UMEs are known to be 60 μm and 10 μm, respectively. Namely, the separation between UMEs is 50 μm. At a scan rate of 0.02 V s^{-1}, $\delta = 250$ μm, which is much larger than the spacing between UMEs, indicating a complete overlap of redox

Fig. 8.6 Cyclic voltammograms of one diamond UMEA in 0.1 M KCl solution with (solid lines) and without (dashed lines) 1.0 mM $Fe(CN)_6^{3-/4-}$ at a scan rate of 0.02 (**a**), 0.2 (**b**), 2 (**c**), and 20 (**d**) V s^{-1}. Reprinted with permission from [55], Copyright ACS Publisher 2011

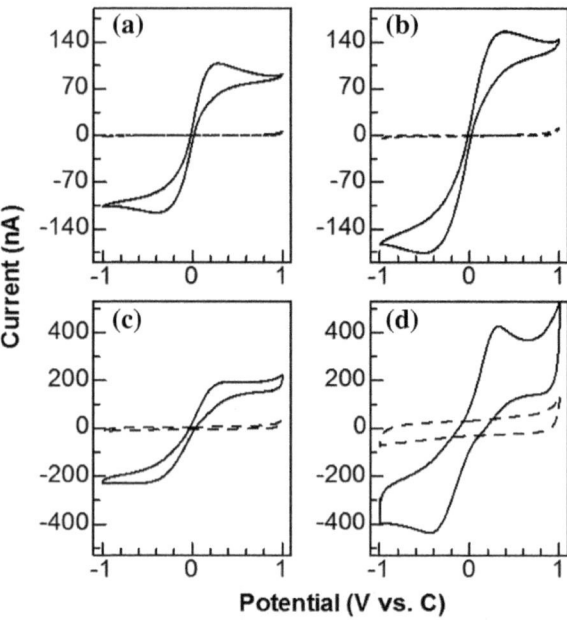

molecule diffusion on individual UMEs and subsequently a linear diffusion profile is seen. At a higher scan rate of 0.2 V s^{-1}, $\delta = 78$ μm, which is only slightly greater than the separation of UMEs, an overlap of adjacent diffusion profiles is still dominating. When δ becomes larger than the diameter of a UME but is still smaller than the separation between UMEs, the voltammetric response is the sum of an individual UME (sigmoidal curve) multiplied by the total number of UMEs in an array. This can be detected for example at a scan rate of 2 V s^{-1}. However, further increase of the scan rate (e.g., to 20 V s^{-1}) leads to even smaller values of δ (e.g., to 7.8 μm) than the size of UMEs (10 μm). In this case, the linear diffusion dominates the mass transport, again resulting in peak-shaped CVs.

Moreover, the capacitive currents of these diamond UMEAs (dashed lines in Fig. 8.6) were recorded in 0.1 M KCl. As expected, they increase linearly with the scan rates. It is known that to achieve sensitive detection of analytes in solutions with low detection limits and high sensitivity by use of MEAs/UMEAs, one has to optimize the ratio of Faradaic current (signal, S) to capacitive current (background, B). The estimated S/B ratios for these diamond UMEAs at the scan rate of 0.02, 0.2, 2 and 20 V s^{-1} are 1817 ± 40, 215 ± 18, 28 ± 6, and 10 ± 1, respectively. The highest ratio was achieved at the slowest scan rate. Namely, the largest S/B ratio can be obtained when maximum Faradaic current is achieved with minimum capacitive current. Note that both capacitive and Faradaic currents are known to be affected by the scan rates applied, the spacing between (U)MEs, and the concentration of supporting electrolyte. For diamond UMEAs, they are also affected by surface terminations and boron-doping concentrations. In this case the diffusion profiles at neighboring UMEAs are overlapped and thus peak-shaped CVs are detected. The magnitude of the Faradaic current is thus proportional to the geometric area, which is comprised of all UMEAs and the insulating parts. While the contribution of the capacitive current to the total current is small since the capacitive current is proportional to the scan rate and to the electrochemical active area, which is only the area for all UMEAs. This gives rise to the enhanced ratios of S/B, leading to increased sensitivity for analytes. Since the geometric area of an UMEA is always 50–1000 times larger than the electrochemical active area, a 50–1000 times better sensitivity is expected for such diamond UMEAs [55, 56].

Voltammetric response of addressable diamond UMEAs (Fig. 8.5C) was also tested. On a macroscopic diamond electrode (with a diameter of 600 μm) with an electrochemically HT surface, the peak difference of the anodic peak potential from the cathodic peak of surface sensitive redox couple $Fe(CN)_6^{3-/4-}$ is about 65 mV, measured at a scan rate of 100 mV s^{-1}. The CVs of $Fe(CN)_6^{3-/4-}$ were also recorded on several addressable diamond UMEAs and at different scan rates. One used UMEA has 37 UMEs and the center-to-center distance of the UMEs is 120 μm. Another one has 85 UMEs and the center-to-center distance of the UMEs is 60 μm. The diameters of these UMEs are 20, 10, and 5 μm. The shapes of recorded CVs are found to vary as a function of the diameters of UMEs and the applied scan rates. This is due to the changed thicknesses of diffusion layers at different scan rates as well as the distances in between the diamond UMEs. For example, peak-shaped CVs result from the overlapped diffusion domains of different UMEs; the sigmoidal curves arise from 3-dimensional diffusion properties toward the small dimensional electrodes. If the distance between UMEs is large, the diffusion properties of each UME do not affected by the neighboring electrodes. Please note that by optimizing scan rates and the center-to-center distance of UMEs, the Faraday current can be maximized and meanwhile the capacitive current is minimized.

SECM has been employed to characterize diamond UMEAs as well. As shown in Fig. 8.7a, a Pt UME with a diameter of 3 μm was used as the tip and one as-fabricated diamond UMEA (Fig. 8.5B) as the substrate. The solution was 5 mM ferrocene dissolved in 0.1 M TBAPF$_6$ + propylencarbonate. The distance between the tip and the substrate was about 1–2 μm. The tip-collection/substrate-generation mode was applied for SECM mapping. From the recorded SECM image (Fig. 8.7b), the local reactivity of these individual UMEs is seen. From the diameter of each dot, the electroactive diameter of these UMEs is confirmed to be 10 μm. From the distance of two dots, the distance of two UMEs is approved to be 60 μm. These values are well consistent with the design of these diamond UMEAs as shown in Fig. 8.5A.

Diamond MEAs with four different geometries (e.g., size and spacing between MEs) have been characterized using EIS. Together with the results from cyclic voltammetry, it is reported that the charge transfer resistance increases while the double layer capacitance decreases as the MEs are further spaced from one another. Diamond MEAs spaced further from one another gave better resolution from the background in fast scan cyclic voltammetric measurements of dopamine [91].

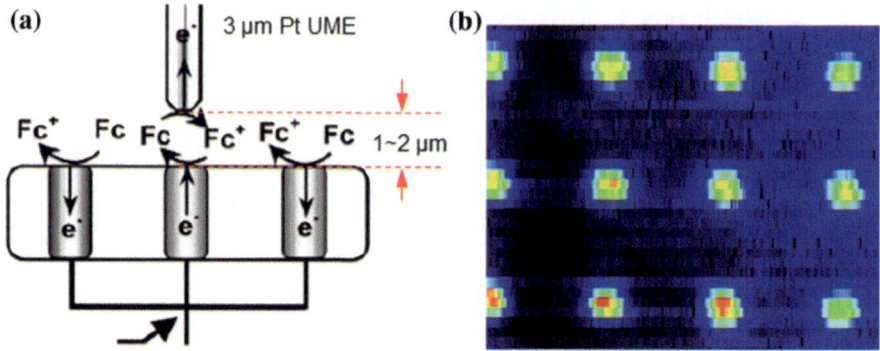

Fig. 8.7 **a** The setup for SECM mapping by use of a Pt UME at the tip and a diamond UMEA as the substrate; **b** one SECM image of a diamond UMEA using the tip-collection/substrate-generation mode

8.3.3 Applications

Diamond MEAs/UMEAs have been applied for many different sensing applications, e.g., for the detection of environmental analytes (e.g., nitrate, 4-nitrophenol [49–51], Cr(VI) ions, Ag(I) ions, sulphate, peroxodisulfate [43], hydrogen peroxide [55]), for bio-detections (e.g., detection of dopamine [44–48], neuronal activity measurements [52, 53], quantal catecholamine secretion from chromaffin cells [54]), and for SECM generation and detection of peroxidisulfate [71]. For example, dopamine has been sensitively and selectively detected on a diamond UMEA in the presence of ascorbic acid [55]. Compared with the results shown on other diamond electrode (including macroscopic diamond electrodes [72–77], MEA [78], UMEAs [47, 48], and diamond nanograss [79]), on diamond UMEAs the lowest detection limit (1.0 nM) was achieved for dopamine detection. This detection limit is 50–100 times lower than that reported [55]. Diamond UMEAs are thus promising for the detection of low concentrated dopamine (0.01–1 μM) in biological samples individually or in the presence of other similar compounds such as ascorbic acid. Moreover, the fabricated addressable diamond UMEAs are possible to be employed for simultaneous detection of at least three target compounds or the individual monitoring of different species.

It has to be highlighted that electrochemical and physical properties of various diamond MEAs and UMEAs as well as their applications for recording released neurotransmitter molecules and all-or-none action potentials from living cells have been recently reviewed [92]. Specifically, how high-density diamond MEAs/UMEAs resolve localized exocytotic events from subcellular compartments have been summarized, together with the application of low-density MEAs to monitor oxidizable neurotransmitter released from populations of cells in culture and tissue slices. It has been stated that interfacing diamond UMEs and MEAs with excitable cells is currently leading to the promising opportunity of recording electrical signals as well as creating neuronal interfaces through the same device. Recent newly available diamond MEAs and UMEAs featuring various geometries are expected to monitor electrical activity and neurotransmitter release in a variety of excitable and neuronal tissues [92].

Combination of EIS with multiplex diamond MEAs (3×3 format, 200 μm in diameter), a model Escherichia coli K12 bacterium was reliably detected together significantly improved signal reproducibility and sensitivity (e.g., by four orders of magnitude). A circuit model was constructed to explain impedance change upon bacteria binding. Moreover, a unique two-Q behavior was observed and explained by nano scale grains and grain boundaries of UNCD films [93].

8.4   Diamond Nanoelectrode Arrays

8.4.1   Fabrication

Several techniques (e.g., e-beam lithography [94], FIB milling [95, 96] nanoimprint [94], and nanosphere lithography [97]) have been applied for the fabrication of NE arrays (NEAs) and nanoelectrode ensembles (NEEs). The deposition of metals into pores of polycarbonate nanoporous membranes [98–100], nanosphere lithography [101], and block copolymer self-assembly [102–104] are alternative approaches. In addition, spatially separated carbon nanofibers and diamond nanograss as well as porous diamond film have been utilized as NEEs for electrochemical applications [105–114].

All-diamond NEAs and NEEs were fabricated for the first time in 2011 [57]. NCD films were used in that they have more features over PCD films. For example, their grain sizes vary only from a few tens of nanometers up to hundreds of nanometer. In other words, electrochemical activity obtained on NCD films are better and less affected than PCD films, of which grain sizes are in the range of micrometers [115, 116]. NCD films were grown on 3 inch Si substrates in an ellipsoid shaped MWCVD reactor [56]. Prior to NCD growth, diamond nanoparticle seeding was conducted by immersing Si wafers in nanodiamond suspension with average particle size of 5 nm [117]. The densities of seeds (or diamond nanoparticles) are more than 10^{11} cm^{-2}. The growth of insulating NCD (i-NCD) was performed using the H$_2$/CH$_4$ plasma with a methane admixture of 1 or 2%. Boron-doped NCD (B-NCD) films were grown when TMB was added into the gas phase with B/C ratios of 6000 ppm. The boron concentrations of B-NCD films measured by secondary ion mass spectroscopy were in the range of 1 to 4×10^{21} cm^{-3}.

Two approached have been developed to fabricate diamond NEAs and NEEs. As shown in the top of left column in Fig. 8.8, e-beam lithography was applied to fabricate NEAs [57]. In the first step, a 200 nm thick SiO$_2$ film is deposited on a 200 nm thick B-NCD film. It is then structured using e-beam lithography with

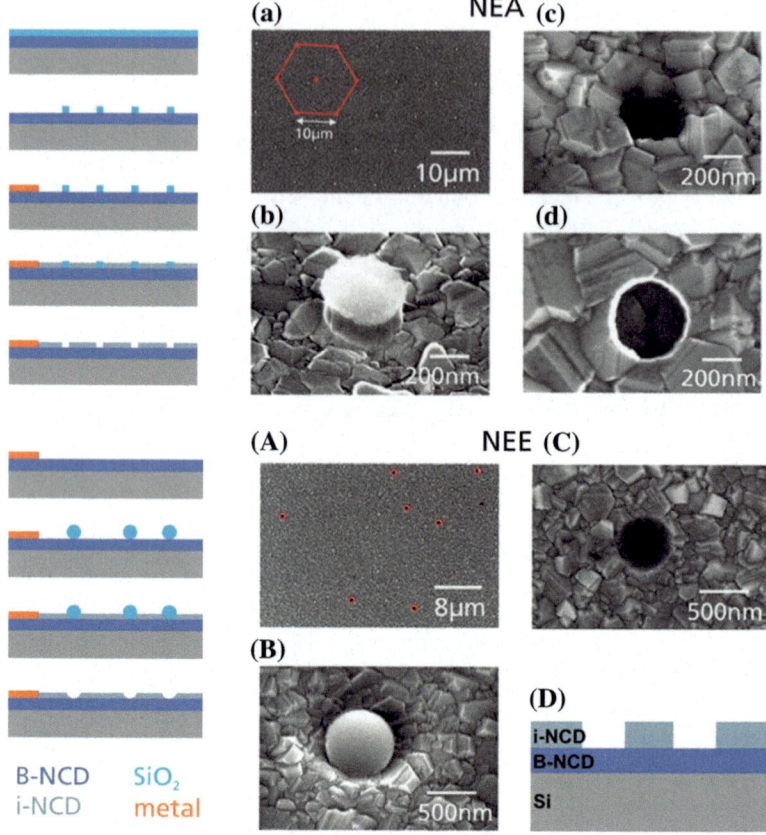

Fig. 8.8 Process steps for the fabrication of all-diamond NEAs (top in the left column) and NEEs (bottom in the left column); SEM images (right column) of a diamond NEA (**a–d**) and a diamond NEE (**A–D**) at different stages during the fabrication processes. Reproduced with permission from [57], Copyright ACS Publisher 2011

subsequent nickel deposition and SiO_2 etching by use of SF_6. In the next step, metal contacts are deposited using photolithography to allow electrical contacts for electrochemical characterization. In the crucial step, an i-NCD film with a thickness of 140 nm is grown on the part of the B-NCD layer that is exposed to the CVD plasma and not protected by SiO_2 islands. With the removal of SiO_2 with HF, recessed diamond NEAs surrounded by i-NCD are obtained. The SEM images at the top of the right column exhibit different stages during the course of the fabrication of such diamond NEAs. For example, SEM image in Fig. 8.8a is the design overview of a diamond NEA where the distance of neighboring NEs is 10 μm and the diamond NEs are arranged in a hexagonal order (indicated in red). The image in Fig. 8.8b is the structured SiO_2/Ni islands on a B-NCD layer; the one in Fig. 8.8c indicates the overgrowth of a i-NCD around SiO_2.

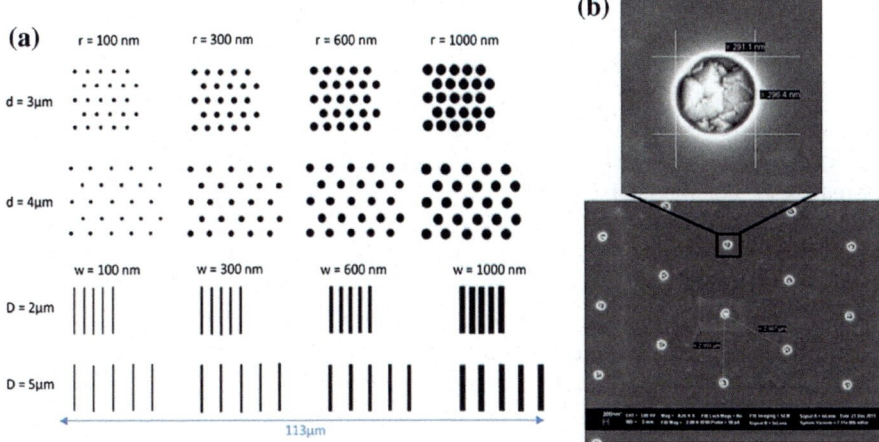

Fig. 8.9 a Schematic drawing of the fabricated diamond NEAs. The parameters *r*, *d*, *w*, and *D* correspond to the disc radius, the center-to-center distance between discs, the width of the bands and the pitch between the bands, respectively. **b** Two SEM images of a diamond NEA constituted by diamond nanodisc UMEs with *r* = 150 ± 10 nm and *d* = 3 μm at different magnifications. Reprinted with permission from [118], Copyright Springer Publisher, 2016

Nanosphere lithography was employed to fabricate all-diamond NEEs [57]. Initially, a photolithography step is used to deposit metal contacts on a B-NCD, which is then immersed in a solution of SiO_2 spheres (with a radius of 500 nm). To reach the equilibrium between adsorption and desorption of SiO_2 spheres on the B-NCD surface, ultra sonication treatment is applied. Obviously, the concentration of the SiO_2 sphere solution is directly correlated to the density of SiO_2 spheres on the B-NCD surface, namely the average distance of neighboring spheres. For example, a concentration of 9.55×10^8 cm^{-3} corresponds to a surface density of 9.7×10^5 cm^{-2} and an average distance of neighboring spheres of ∼ 10 μm. These values were derived from a large area (e.g., 50 × 50 $μm^2$). The next step involves the growth of i-NCD around SiO_2 spheres. In other words, selective growth of i-NCD layers is conducted. After wet-chemical removal of SiO_2 spheres with HF, concave-shaped diamond NEEs are obtained. The SEM images at the bottom of the right column exhibit different stages during the course of the fabrication of such diamond NEEs. For example, SEM image in Fig. 8.8A is overview of statistically distributed diamond NEs (indicated by red circles) on a B-NCD substrate. The image in Fig. 8.8B indicates one overgrown SiO_2 sphere with an i-NCD layer. The one in Fig. 8.8C is a final diamond NE after SiO_2 removal. The final electrode structure of these diamond NEAs and NEEs is schematically shown in Fig. 8.9D.

Recently, multiple diamond nanodisc and nanoband NEA platforms have been fabricated by e-beam lithography [118]. In the first step, markers were created on the BDD films (grown on Si substrates and with a thickness of 400 nm) to identify later the position of the arrays under a microscope. Prior to coating this BDD film with a polycarbonate (PC) PC film, the BDD film was treated with O_2 plasma for 15 s. Using 3% PC solutions in cyclopentanone, a spin coating speed of 2000 rpm, a bake at 180 °C for 30 min, a PC film with a thickness of 87 nm was deposited. The patterns of the UMEs were then exposed on the BDD film by a focused e-beam at 30 kV and developed for 1 min in 5 M NaOH at 35 °C. The platform structures were cleaned from residuals of resist by oxygen plasma in an inductively coupled plasma (ICP) reactor at a pressure of 4 mT. A bias of 30 V was also applied, generated by applying radio frequency powers of 200 and 10 W to the coil and to the platen, respectively. As shown in Fig. 8.9a, 16 kinds of as-fabricated NEA platforms have different geometries (diameters) and varied center-to-center distances in between diamond NEs. An example multiple platform is shown in Fig. 8.9b, where the NE has a diameter of 150 ± 10 nm and the distance in between NEs is 3 μm (first line and first column in Fig. 8.9a) [118].

8.4.2 Characterization

To investigate the properties of these diamond UMEAs, different characterization techniques including electron and optical microscopy, atomic force microscopy (AFM), Raman spectroscopy, and electrochemical methods have been applied. For example, some SEM images of fabricated diamond NEAs and NEEs are shown in Figs. 8.8 and 8.9. NEs in diamond NEAs are well arranged (e.g., in a hexagonal order) (Fig. 8.8a). In contrast, they are randomly distributed on a diamond NEEs (Fig. 8.8A). For diamond NEAs shown in Fig. 8.8a, it has a radius of 250 nm and a distance of 10 μm next to another NE. Its density on a NEA is 11×10^5 cm^{-2} and the number of NEs is 18 000. The diamond NEE shown in Fig. 8.8A has a radius of 175 nm and its density is 8.5×10^5 cm^{-2}.

Diamond NEEs were characterized further with conductive AFM (C-AFM) [57]. The density determined by C-AFM is 8.5×10^5 cm^{-2}, which is in good agreement with the expected value from SiO_2 sphere concentration in the solution. The size of individual NEs is about 175 nm in radius. The cross sections of the topography for two neighboring NEs clarify that the NE is surrounded by a 140-nm thick i-NCD layer.

These diamond NEAs and NEEs were characterized further with voltammetry [57]. The CVs were recorded in 1.0 mM $Fe(CN)_6^{3-/4-}$ dissolved in 0.1 M KCl at different scan rates (e.g., from few mV s^{-1} up to 10 V s^{-1}). At small scan rates (e.g., 20 mV s^{-1} for the NEA and 1 mV s^{-1} for the NEE), the CVs have mixed shapes, indicating partially overlapping diffusion hemispheres. Increasing the scan rate leads to typical steady-state sigmoidal CVs on both electrodes, indicating sphere-diffusion [119–123]. Moreover, the voltammetric response of $Ru(NH_3)_6^{2+/3+}$ and $IrCl_6^{2-/3-}$ was investigated on diamond NEAs and NEEs with different surface terminations. As shown in Fig. 8.10, the dependence of surface termination of diamond NEAs and NEEs on the charge transfer rates of redox electrolytes is revealed [57]. On a

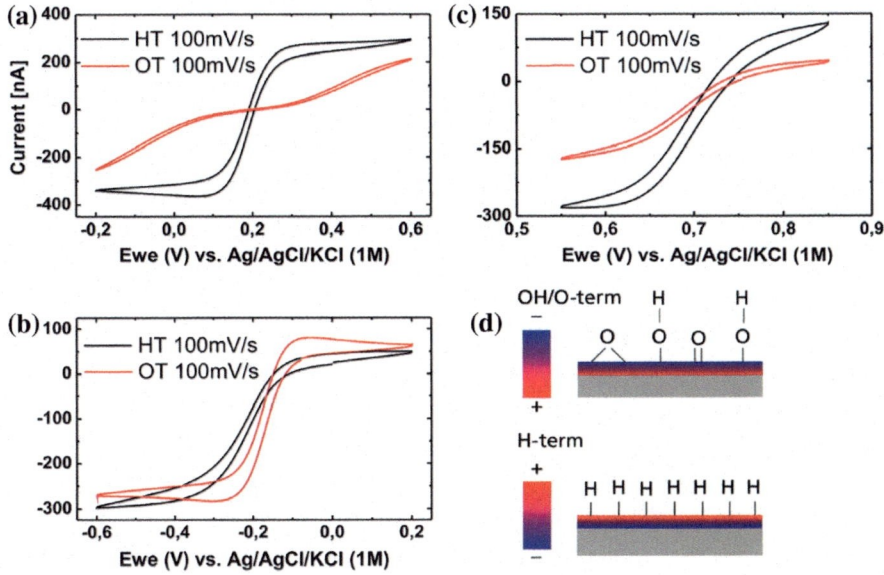

Fig. 8.10 Comparison of CVs of $Fe(CN)_6^{3-/4-}$ (**a**), $Ru(NH_3)_6^{2+/3+}$ (**b**), and $IrCl_6^{2-/3-}$ (**c**) on HT and OT diamond NEAs. A schematic of the charge distribution on a HT and OT diamond surface is given in (**d**). Reprinted with permission from [57], Copyright ACS Publisher 2011

diamond NEA, the CV of the anion $IrCl_6^{2-/3-}$ shows a fast electron transfer on the HT surface. While at the oxygen-terminated (OT) surface, the steady-state current as well as the slope of the transition from reduction to oxidation are reduced, indicative of a slower electron transfer. This tendency is similar as seen using another negatively charged redox couple of $Fe(CN)_6^{3-/4-}$. For the positively charged redox molecules $Ru(NH_3)_6^{2+/3+}$, the opposite effect is observed. That is to say, on an OT surface, the electron transfer rate for $Ru(NH_3)_6^{2+/3+}$ is faster than that on a HT surface. This effect for $IrCl_6^{2-/3-}$ and $Ru(NH_3)_6^{2+/3+}$ is smaller compared to that for $Fe(CN)_6^{3-/4-}$.

It is known that a HT surface has a positive surface dipole layer ("positive" refers to the interface of diamond to the liquid) and the OT surface results in a negative surface dipole layer (Fig. 8.10d). A macroscopic diamond electrode shows a higher degree of inhomogeneity with respect to boron-doping level and termination effects due to its macroscopic dimensions. Therefore, such a phenomena has not been shown on planar and macroscopic diamond electrodes for these redox probes [57, 124]. While for diamond NEAs, one would expect a homogenized behavior due to the small grains of the NCD films as well as a more effective termination of the small electrochemical active area. Therefore the possible effects responsible for the decrease of electron transfer rate constant on OT NEAs are either an electrostatic or a site blocking effect.

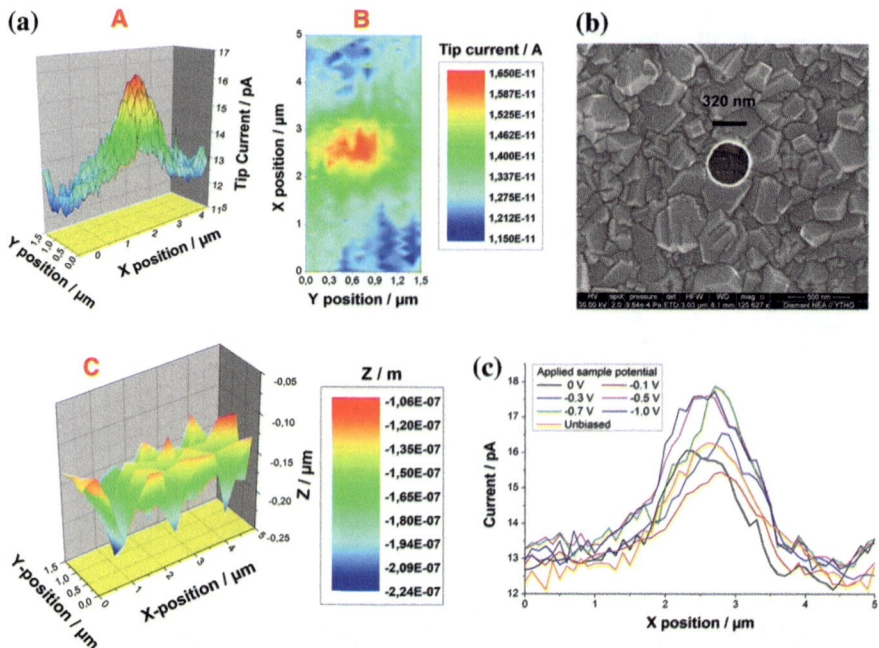

Fig. 8.11 **a** Amperometric (A: 3-D, B: top view) and topographic feedback (C: 3-D) images of a $5 \times 1.5 \ \mu m^2$ area of diamond NEAs. **b** SEM image of a diamond NE. **c** 2D-scan with SECM of a single diamond NE in feedback mode with different substrate potentials. Reprinted with permission from [125]. Copyright Elsevier Publisher, 2012

Besides voltammetry, EIS was performed to characterize diamond NEAs and NEEs [57]. Their impedance spectra show similar characteristic features. Both graphs display a large semicircle in the high-frequency regime. At low frequencies, a transition to linear diffusion with unity slope occurs, particularly observable for the NEA. A semicircle at high frequency regime is due to a three-dimensional hemispherical diffusion on a diamond NEA and NEE [95–97]. The transition at low frequencies represents the regime of overlapping diffusion hemispheres. The change from typical three-dimensional diffusion to overlapping diffusion hemispheres is very distinct for the NEA [57].

The phase-operated shear force technique and SECM feedback mode were also applied to characterize diamond NEAs. A Pt UME tip with an active radius of 167 nm was approached over a diamond NEA at a constant distance of 45 nm. The SECM measurements were accomplished with a constant tip potential of 700 mV (vs. Ag/AgCl) where diffusion limited oxidation of ferrocene occurs. The SECM images of a diamond NEA in an area of $5 \times 1.5 \ \mu m^2$, recorded in feedback mode are shown in Fig. 8.11a. A lateral resolution in the range of 100 nm is realized, leading to monitoring individual diamond NEs. The size of diamond NEs estimated from SECM images is consistent with that (320 nm in diameter)

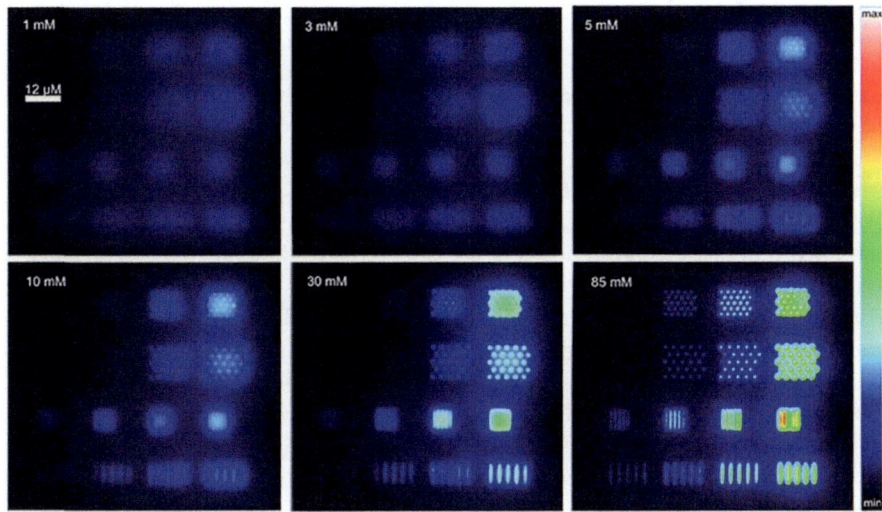

Fig. 8.12 ECL images of diamond NEAs obtained in a pH 7.4 phosphate buffer solution containing both 1 mM Ru(bpy)$_3^{2+}$ and increased concentrations of TPrA. All images were coded according to the same false color scale (right). Reprinted with permission from [118], Copyright Springer Publisher, 2016

estimated from their SEM image (Fig. 8.10b). The variation of the tip currents as a function of the distances during SECM mapping was also measured under different substrate potentials (namely the potentials applied on a diamond NEA). These 2D-SECM scans are shown in Fig. 8.10c [125].

Diamond NEAs were characterized by electrogenerated chemiluminescence (ECL) mapping simultaneously of different NEs on multiple platforms (Fig. 8.9). Ru(bpy)$_3^{2+}$ was used as the ECL luminophore and tri-n-propylamine (TPrA) as the co-reactant. Figure 8.12 displays the ECL images recorded on diamond NEAs when imposing a constant potential of 1.2 V (vs. Ag/AgCl/KCl) in a pH 7.4 phosphate buffer solution containing 1 mM Ru(bpy)$_3^{2+}$ and increased TPrA concentrations (ranging from 1 to 85 mM). The emitted ECL patterns are clearly detected but with different intensities when the TPrA concentrations are varied. At high concentrations of TPrA, the design (structure) of multiple NEA platforms is identified from well-separated individual ECL signals (spots or bands). These images also confirm that the thickness of the ECL-emitting zone at each NE scales inversely with the co-reactant concentration, while significantly stronger ECL signals are detected for NEAs. The roles of geometrical and mechanistic parameters of diamond NEAs on the ECL generation are thus clarified [118].

8.4.3 Applications

Diamond NEAs have been applied to investigate surface-sensitive adsorption phenomena [124]. The adsorption of neutral methyl viologen (MV^0) was used as a model system. The adsorption of MV^0 was examined with their different surface terminations, namely HT or OT. Diffusion-controlled processes manifest themselves as sigmoidal-shaped CVs on OT diamond NEAs, whereas adsorption-controlled processes result in peaks in the CVs for HT diamond NEAs. The change in the shapes of these CVs is due to the drastic changes that occur in the diffusion profiles of redox species, altering from hemispherical diffusion on an OT surface to thin layer electrochemistry upon adsorption on a HT surface. In this way the de-convolution of diffusion-controlled current from adsorption-controlled current was conducted. By further analysing anodic stripping process at high scan rates, the deposition of amorphous MV^0 was further approved on HT diamond NEAs. The types and the concentrations of the buffer solutions were then changed to alter the interaction of MV^0 with a HT diamond NE surface. The effect of guanidine, widely used for denaturation of proteins (which weakens hydrophobic interaction), on the stripping current of MV^0 was studied. The CVs were recorded in 1 mM MV^{2+} and 0.1 M KCl with increasing concentrations of guanidine. Initially adsorption occurs, whereas the stripping peak continuously vanishes with increasing concentrations of guanidine up to 1.5 M. Measurements with increased urea concentrations show the same impact on the adsorption of neutral MV as guanidine, which weakens hydrophobic interaction. Subsequently, the adsorption of MV^0 on a HT diamond NE is controlled by hydrophobic interaction. This effect of ions on the interaction of MV^0 and the hydrophobic diamond surface is correlated with the Hofmeister series. Therefore diamond NEA is ideal for the study of adsorption phenomena at the liquid-solid interface in voltammetry [124].

OT diamond NEAs were employed for sensitive and reproducible detection of dopamine in the presence of ascorbic acid by use of differential pulse voltammetry. The reported sensitivity for the detection of dopamine was 57.9 nA μM^{-1} cm^{-2}, the detection limit was less than 100 nM, and the linear range was up to a concentration of 20 μM. Namely, by use of diamond NEAs with the appropriate termination, a low-level detection of biogenic substances such as dopamine was achieved without the need for a selective membrane [126].

8.5 Scanning Tunneling Microscopy Tips

Besides commercially available diamond AFM cantilevers and newly developed diamond AFM-SECM tips (e.g., diamond UMEAs), conductive diamond has been utilized for the fabrication of scanning tunneling microscopy (STM) tips [127]. The reason of adopting diamond for the fabrication of various scanning probes mainly results from chemical stability of diamond, namely its robustness and chemical

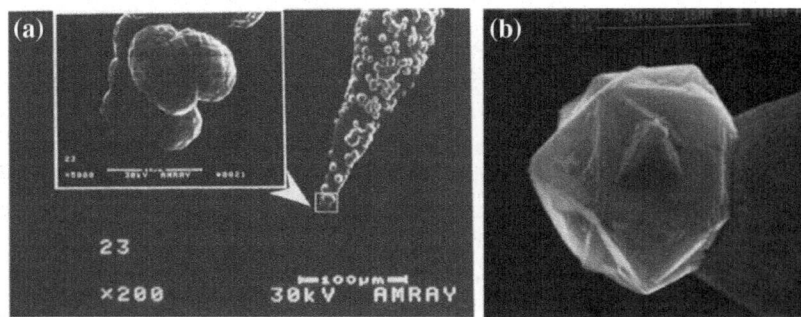

Fig. 8.13 **a** SEM image of a tungsten tip after growth of diamond grains; the insert is a magnified image of the top part (mark white rectangle). Reproduced with permission from [130], Copyright Elsevier Publisher, 1993; **b** SEM image of a diamond particle grown on a tungsten tip. Note that the apex of the diamond tip is along the axis of the tungsten probe. Reproduced with permission from [131]. Copyright AIP Publishing, 1997

inertness under different situations. Thanks to state-of-art nanotechnologies as well as super high hardness of diamond, diamond nanostructures with different sizes and radius are possible to be fabricated. This is another fact that accelerates the usage of diamond for scanning probe microscopy [127].

Figure 8.13 shows some example SEM images of as-fabricated STM tips [130, 131]. The first attempt to apply diamond STM tip was proposed in 1990 [128]. This diamond STM tip was fabricated through diamond polishing, and followed by boron ion implantation. An ion-implanted diamond tip was sharpened to a radius of about 100 nm and further approved to have sufficient conductivity for STM. This tip was used repeatedly even if it contacted with the sample surface. A diamond STM tip with a tip radius less than 12 nm was later fabricated [129]. Such a tip was obtained through conventional polishing of a heavily BDD layer epitaxially grown on an insulating (type IIa) natural diamond substrate using a HFCVD reactor. Atomic resolution on graphite surfaces was obtained under normal operating conditions for STM in air. The feasibility of using the diamond tip to create nanostructures on surfaces was also investigated. In 1993, diamond coated STM tips were shown (Fig. 8.13a) [130]. They were fabricated via MWCVD growth of diamond on an electrochemically etched tungsten probe, of which radius is less than 100 nm. Produced diamond STM probes showed sufficient conductivity, which was suggested to result from impurities, hydrogen surface termination, defects, or amorphous structures. With sharp apices, these diamond STM tips were used to image gold coated mica serving as a model sample. In 1997, diamond micro-particles were grown on etched tungsten wires using a MWCVD process (Fig. 8.13b) [131]. The apexes on cubo-octahedral particles bound by {100} and {111} facets were effectively used as STM tips. Atomically resolved surface images of highly oriented pyrolytic graphite were acquired. Tunneling characteristics revealed a higher electron emission from the diamond tips than that from the platinum-iridium tips. In 2007, STM was integrated with Berkovich diamond semiconductive tip [132]. Diamond was grown by the temperature gradient method at high pressure-high temperature (HPHT) conditions. Diamond tips were fabricated from these diamond crystals considering their zonal structure. The combined STM-nanoindentation instrument enabled submicron resolution indentation and in situ scanning of the gold film deposited on the silicon substrate, demonstrating the usability, acceptable resolution, and sensitivity of the fabricated diamond STM tips. Later, diamond single crystals grown by the temperature gradient method were used for STM [133, 134]. They decreased the probability of incidental tunneling channels with participation of the surface states caused by the presence of boron atoms in the diamond structure. Meanwhile, they increased the reliability of experimental data. Successful imaging of the atomic resolution of the (0001) graphite plane was thus conducted by use of these diamond STM tips. Very recently, single crystalline conductive BDD probes were proposed to record STM imaging with sub-Ångström lateral resolution [135]. A heavily boron-doped IIb diamond single crystal (with boron concentration from 5.0×10^{20} to 3.0×10^{21} cm^{-3}) synthesized by the temperature gradient method under HPHT conditions was used as a material for the fabrication of STM tips. By use of first-principles tunneling current calculations, the highest spatial resolution was achieved at tip-sample distances of 3–5 Å, if p-orbitals of the BDD tip maximally contribute to the tunneling current.

8.6   Summary and Outlook

Various small-dimensional diamond electrodes and electrode arrays have been produced, including microelectrodes, ultramicroelectrodes, nanoelectrodes, as well as their arrays. The technologies like photolithography, e-beam lithography, nanosphere lithography, and etching processes have been employed to fabricate these small dimensional diamond electrodes and electrode arrays. These small-dimensional diamond electrodes have shown advantages over planar and macroscopic diamond electrodes for sensor development, investigation of surface-sensitive reactions etc. For example, by use of diamond MEs and UMEs, various neurotransmitters have been in vivo and in vitro detected; Optimization of Faradaic and capacitive currents of diamond UMEAs led to 50–1000 times better sensitivity (over a planar and macroscopic diamond electrode) for voltammetric monitoring of target compounds in solutions; Diamond NEAs and NEEs with selected surface terminations provided new electrochemical interfaces to detect surface-sensitive adsorption at the nanometer scale.

With respect to scanning probe microscopy tips, diamond STM tips have been fabricated using different methods. Recording STM images at different resolutions have been realized. The utility of diamond STM tips is thus evident. In other words, diamond can be a promising candidate as the STM tip since these diamond STM tips features many advantages over conventional metallic STM probes.

It is quite reasonable to anticipate increased usage of these small dimensional diamond electrodes and electrode arrays as well as diamond STM tips in electrochemistry, corrosion science, materials research, and the life sciences in the near future. More and new research topics will be triggered. For example, future work can focus on in vivo applications of these arrays (e.g., fast and sensitive detection of biomolecules released from cells, in situ and on-line monitoring of cell/bacterial growth, etc.), surface treatment and properties at the nanoscale, and the formation of three-dimensional surface structures (e.g., nanotextures, nanowires, and porosity) on the arrays. The applications of these small dimensional electrodes, electrode arrays, STM tips in the fields of electrocatalysis, energy storage and conversion, electrosynthesis are much expected. For example, different kinds of electrocatalysts can be deposited on individual NEs or addressable NEAs to clarify their effect (e.g., composition, sizes, etc.) on the related electrocatalytic efficiencies (e.g., electrochemical carbon dioxide/nitrogen reduction, hydrogen evolutions, oxidation reduction, etc.). For energy storage and conversion (e.g., supercapacitors, etc.), diamond STM tips are possible to be employed to clarify the double layer structures of different electrode materials in various solutions.

However, the fabrication of these small-dimensional electrodes, electrode arrays, and STM tips are still difficult, expensive, complicated, and time consumable. For example, although several approaches have been demonstrated the possibility of generating suitable diamond STM tips and there are research groups routinely fabricating and using diamond STM tips, the appropriate methods to finally produce sharp and well-oriented STM tips at an industrial scale are actually still missing. Additionally, diamond exhibits varied electrochemical behavior, depending on dopant concentration, structural defects, non-diamond carbon impurities (mainly sp^2 hybridized carbon), crystallographic orientation, surface termination, and fraction of grain boundaries, etc. [136–138]. In other words, the appropriate facility for the growth of high quality diamond is still highly required. All these facts hinder seriously the fabrication, property investigation, and electrochemical applications of these diamond devices. Therefore, simplification of fabrication processes to produce price reduced and high quality small dimensional electrodes and electrode arrays will be the first key issue to be researched and solved in next years.

Acknowledgements The author thanks the financial support from German Research Foundation (DFG) under the project (grant no. YA344/1-1).

References

1. M. Iwaki, S. Sato, K. Takahashi, H. Sakairi, Electrical conductivity of nitrogen and argon implanted diamond. Nucl. Instrum. Methods Phys. Res. **209–210**(2), 1129–1133 (1983). https://doi.org/10.1016/0167-5087(83)90930-4
2. Y.V. Pleskov, A.Y. Sakharova, M.D. Krotova, L.L. Bouilov, B.V. Spitsyn, Photoelectrochemical properties of semiconductor diamond. J. Electroanal. Chem. **228**(1–2), 19–27 (1987). https://doi.org/10.1016/0022-0728(87)80093-1

3. R. Tenne, C. Levy-Clement, Diamond electrodes. Isr. J. Chem. **38**(1–2), 57–73 (1998). https://doi.org/10.1002/ijch.199800007

4. G.M. Swain, A.B. Andreson, J.C. Angus, Applications of diamond thin films in electrochemistry. MRS Bull. **23**(9), 56–60 (1998). https://doi.org/10.1557/S0883769400029389

5. Y.V. Pleskov, Synthetic diamond in electrochemistry. Russ. Chem. Rev. **68**(5), 381–392 (1999). https://doi.org/10.1070/RC1999v068n05ABEH000494

6. S.J. Cobb, Z.J. Ayres, J.V. Macpherson, Boron doped diamond: a designer electrode material for the twenty-first century. Annu. Rev. Anal. Chem. 11 (2018). https://doi.org/10.1146/annurev-anchem-061417-010107

7. T.A. Ivandini, Y. Einaga, Polycrystalline boron-doped diamond electrodes for electrocatalytic and electrosynthetic applications. Chem. Commun. **53**, 1338–1347 (2017). https://doi.org/10.1039/C6CC08681K

8. N. Yang, J.S. Foord, X. Jiang, Diamond electrochemistry at the nanoscale: a review. Carbon **99**, 90–110 (2016). https://doi.org/10.1016/j.carbon.2015.11.061

9. R.L. McCreery, Advanced carbon electrode materials for molecular electrochemistry. Chem. Rev. **108**(7), 2646–2687 (2008). https://doi.org/10.1021/cr068076m

10. O. Chailapakul, W. Siangproh, D.A. Tryk, Boron-doped diamond-based sensors: a review. Sens. Lett. **4**(2), 99–119 (2006). https://doi.org/10.1166/sl.2006.008

11. Y.L. Zhou, J.F. Zhi, The application of boron-doped diamond electrodes in amperometric biosensors. Talanta **79**(5), 1189–1196 (2009). https://doi.org/10.1016/j.talanta.2009.05.026

12. C.E. Nebel, B. Rezek, D. Shin, H. Uetsuka, N. Yang, Diamond for bio-sensor applications. J. Phys. D Appl. Phys. **40**(20), 6443–6466 (2007). https://doi.org/10.1088/0022-3727/40/20/S21

13. R. Linares, P. Doering, B. Linares, Diamond bio electronics. Stud. Health Technol. Inform. **149**, 284–296 (2009). https://doi.org/10.3233/978-1-60750-050-6-284

14. V. Vermeeren, S. Wenmackers, P. Wagner, L. Michiels, DNA sensors with diamond as a promising alternative transducer material. Sensor **9**(7), 5600–5636 (2009). https://doi.org/10.3390/s90705600

15. A. Argoitia, H.B. Martin, E.J. Rozak, U. Landau, J.C. Angus, Electrochemical studies of boron-doped diamond electrodes. MRS. Proc. **416**, 349 (1995). https://doi.org/10.1557/PROC-416-349

16. G.M. Swain, R. Ramesham, The electrochemical activity of boron-doped polycrystalline diamond thin film electrodes. Anal. Chem. **65**(4), 345–351 (1993). https://doi.org/10.1021/ac00052a007

17. G.M. Swain, The use of CVD diamond thin films in electrochemical systems. Adv. Mater. **6**(5), 388–392 (1994). https://doi.org/10.1002/adma.19940060511

18. R. Hoffmann, A. Kriele, H. Obloh, J. Hees, M. Wolfer, W. Smirnov, N. Yang, C.E. Nebel, Electrochemical hydrogen termination of boron-doped diamond. Appl. Phys. Lett. **97**(5), 052103 (2010). https://doi.org/10.1063/1.3476346

19. W. Yang, O. Auciello, J.E. Butler, W. Cai, J.A. Carlisle, J.E. Gerbi, D.M. Gruen, T. Knickerbocker, T.L. Lasseter, J.N. Russell Jr., J.M. Smith, R.J. Hamers, DNA-modified nanocrystalline diamond thin-films as stable, biologically active substrates. Nat. Mater. **1**(4), 253–257 (2002). https://doi.org/10.1038/nmat779

20. A. Hartl, E. Schmich, J.A. Garrido, J. Hernando, S.C.R. Catharino, S. Walter, P. Feulner, A. Kromka, D. Steinmuller, M. Stutzmann, Protein-modified nanocrystalline diamond thin films for biosensor applications. Nat. Mater. **3**(10), 736–742 (2004). https://doi.org/10.1038/nmat1204

21. Y.S. Zou, Y. Yang, W.J. Zhang, Y.M. Chong, B. He, I. Bello, S.T. Lee, Fabrication of diamond nanopillars and their arrays. Appl. Phys. Lett. **92**(5), 053105 (2008). https://doi.org/10.1063/1.2841822

22. N. Yang, H. Uetsuka, E. Osawa, C.E. Nebel, Vertically aligned nanowires from boron-doped diamond. Nano Lett. **8**(11), 3572–3576 (2008). https://doi.org/10.1021/nl801136h

23. C.E. Nebel, N. Yang, H. Uetsuka, E. Osawa, N. Tokuda, O. William, Diamond nano-wires, a new approach towards next generation electrochemical gene sensor platforms. Diam. Relat. Mater. **18**(5–8), 910–917 (2009). https://doi.org/10.1016/j.diamond.2008.11.024

24. W. Smirnov, A. Kriele, N. Yang, C.E. Nebel, Aligned diamond nano-wires: fabrication and characterisation for advanced applications in bio- and electrochemistry. Diam. Relat. Mater. **19**(2–3), 186–189 (2010). https://doi.org/10.1016/j.diamond.2009.09.001

25. A.J. Bard, L.R. Faulkner, *Electrochemical Methods, Fundamentals and Applications*, 2nd edn. (Wiley-VCH, New York, 2001)

26. J. Wang, *Analytical Electrochemistry*, 2nd edn. (Wiley-VCH, New York, 2000)

27. X.J. Huang, A.M. O'Mahony, R.G. Compton, Microelectrode arrays for electrochemistry: approaches to fabrication. Small **5**(7), 776–788 (2009). https://doi.org/10.1002/smll.200801593

28. D.W.M. Arrigan, Nanoelectrodes, nanoelectrode arrays and their applications. Analyst **129**(12), 1157–1165 (2004). https://doi.org/10.1039/B415395M

29. R.G. Compton, G.G. Wildgoose, N.V. Rees, I. Streeter, R. Baron, Design, fabrication, characterisation and application of nanoelectrode arrays. Chem. Phys. Lett. **459**(1–6), 1–17 (2008). https://doi.org/10.1016/j.cplett.2008.03.095

30. O. Ordeig, J. del Campo, F.X. Munoz, C.E. Banks, R.G. Compton, Electroanalysis utilizing amperometric microdisk electrode arrays. Electroanalysis **19**(19–20), 73–1986 (2007). https://doi.org/10.1002/elan.200703914

31. J.B. Cooper, S. Pang, S. Albin, J. Zheng, R.M. Johnson, Fabrication of boron-doped CVD diamond microelectrodes. Anal. Chem. **70**(3), 464–467 (1998). https://doi.org/10.1021/ac9707621

32. B.V. Sarada, T.N. Rao, D.A. Tryk, A. Fujishima, Electrochemical characterization of highly boron-doped diamond microelectrodes in aqueous electrolyte. J. Electrochem. Soc. **146**(4), 1469–1471 (1999). https://doi.org/10.1149/1.1391788

33. B. Duran, R.F. Brocenschi, M. France, J.J. Galligan, G.M. Swain, Electrochemical activation of diamond microelectrodes: implications for the in vitro measurement of serotonin in the bowel. Analyst **139**(12), 3160–3166 (2014). https://doi.org/10.1039/c4an00506f

34. K.B. Holt, J. Hu, J.S. Foord, Fabrication of boron-doped diamond ultramicroelectrodes for use in scanning electrochemical microscopy experiments. Anal. Chem. **79**(6), 2556–2561 (2007). https://doi.org/10.1021/ac061995s

35. J. Hu, J.S. Foord, K.B. Holt, Hot filament chemical vapour deposition of diamond ultramicroelectrodes. Phys. Chem. Chem. Phys. **9**(40), 5469–5475 (2007). https://doi.org/10.1039/B710241K

36. J. Hu, K.B. Holt, J.S. Foord, Focused ion beam fabrication of boron-doped diamond ultramicroelectrodes. Anal. Chem. **81**(14), 5663–5670 (2009). https://doi.org/10.1021/ac9003908

37. J. Cvacka, V. Quaiserova, J.W. Park, Y. Show, A. Muck, G.M. Swain, Boron-doped diamond microelectrodes for use in capillary electrophoresis with electrochemical detection. Anal. Chem. **75**(11), 2678–2687 (2003). https://doi.org/10.1021/ac030024z

38. J. Park, Y. Show, V. Quaiserova, J.J. Galligan, G.D. Fink, G.M. Swain, Diamond microelectrodes for use in biological environments. J. Electroanal. Chem. **583**(1), 56–68 (2005). https://doi.org/10.1016/j.jelechem.2005.04.032

39. J.M. Halpern, S. Xie, G.P. Sutton, B.T. Higashikubo, C.A. Chestek, H. Lu, H.J. Chiel, H.B. Martin, Diamond electrodes for neurodynamic studies in aplysia californica. Diam. Relat. Mater. **15**(2–3), 183–187 (2006). https://doi.org/10.1016/j.diamond.2005.06.039

40. S. Xie, G. Shafer, C.G. Wilson, H.B. Martin, In vitro adenosine detection with a diamond-based sensor. Diam. Relat. Mater. **15**(2–3), 225–228 (2006). https://doi.org/10.1016/j.diamond.2005.08.018

41. A.L. Colley, C.G. Williams, U. D'Haenens Johnsson, M.E. Newton, P.R. Uniwin, N.R. Wilson, J.V. Macpherson, Examination of the spatially heterogeneous electroactivity of boron-doped diamond microarray electrodes. Anal. Chem. **78**(8), 2539–2548 (2006). https://doi.org/10.1021/ac0520994

42. K. Tsunozaki, Y. Einaga, T.N. Rao, A. Fujishima, Fabrication and electrochemical characterization of boron-doped diamond microdisc array electrodes. Chem. Lett. **31**(5), 502–503 (2002). https://doi.org/10.1246/cl.2002.502

43. C. Provent, W. Haenni, E. Santoli, P. Rychen, Boron-doped diamond electrodes and microelectrode-arrays for the measurement of sulfate and peroxodisulfate. Electrochim. Acta **49**(22–23), 3737–3744 (2004). https://doi.org/10.1016/j.electacta.2004.02.047

44. K.L. Soh, W.P. Kang, J.L. Davidson, Y.M. Wong, A. Wisisoraat, G. Swain, D.E. Cliffel, CVD diamond anisotropic film as electrode for electrochemical sensing. Sens. Actuators B **91**(1–3), 39–45 (2003). https://doi.org/10.1016/S0925-4005(03)00064-9

45. K.L. Soh, W.P. Kang, J.L. Davidson, S. Basu, Y.M. Wong, D.E. Cliffel, A.B. Bonds, G.M. Swain, Diamond-derived microelectrodes array for electrochemical analysis. Diam. Relat. Mater. **13**(11–12), 2009–2015 (2004). https://doi.org/10.1016/j.diamond.2004.07.025

46. K.L. Soh, W.P. Kang, J.L. Davidson, Y.M. Wong, D.E. Cliffel, G. Swain, Ordered array of diamond ultramicroband electrodes for electrochemical analysis. Diam. Relat. Mater. **17**(3), 240–246 (2008). https://doi.org/10.1016/j.diamond.2007.12.023

47. K.L. Soh, W.P. Kang, J.L. Davidson, Y.M. Wong, D.E. Cliffel, G. Swain, Diamond-derived ultramicroelectrodes designed for electrochemical analysis and bioanalyte sensing. Diam. Relat. Mater. **17**(4–5), 900–905 (2008). https://doi.org/10.1016/j.diamond.2007.12.041

48. S. Raina, W.P. Kang, J.L. Davidson, Fabrication of nitrogen-incorporated nanodiamond ultra-microelectrode array for Dopamine detection. Diam. Relat. Mater. **19**(2–3), 256–259 (2010). https://doi.org/10.1016/j.diamond.2009.10.013

49. M. Pagels, C.E. Hall, N.S. Lawrence, A. Meredith, T.G.L. Jones, H.P. Godfried, C.S. J. Pickles, J. Wilman, C.E. Banks, R.G. Compton, L. Jiang, All-diamond microelectrode array device. Anal. Chem. **77**(11), 3705–3708 (2005). https://doi.org/10.1021/ac0502100

50. A.O. Simm, C.E. Banks, S. Ward-Jones, T.J. Davies, N.S. Lawrence, T.G.J. Jones, L. Jiang, R.G. Compton, Boron-doped diamond microdisc arrays: electrochemical characterisation and their use as a substrate for the production of microelectrode arrays of diverse metals (Ag, Au, Cu) via electrodeposition. Analyst **130**(9), 1303–1311 (2005). https://doi.org/10.1039/B506956D

51. N.S. Lawrence, M. Pagels, A. Meredith, T.G.J. Jones, C.E. Hall, C.S. Pickles, H. P. Godfried, C.E. Banks, R.G. Compton, L. Jiang, Electroanalytical applications of boron-doped diamond microelectrode arrays. Talanta **69**(4), 829–834 (2006). https://doi.org/10.1016/j.talanta.2005.11.020

52. M. Bonnauron, S. Saada, L. Rousseau, G. Lissorgues, C. Mer, P. Bergonzo, High aspect ratio diamond microelectrode array for neuronal activity measurements. Diam. Relat. Mater. **17**(7–10), 1399–1404 (2008). https://doi.org/10.1016/j.diamond.2007.12.065

53. M. Bonnauron, S. Saada, C. Mer, C. Gesset, O.A. Williams, L. Rousseau, E. Scorsone, P. Mailley, Transparent diamond-on-glass micro-electrode arrays for ex-vivo neuronal study. Phys. Status Solidi (a) **205**(9), 2126–2129 (2008). https://doi.org/10.1002/pssa.200879733

54. V. Carabelli, S. Gosso, A. Marcantoni, Y. Xu, E. Colombo, Z. Gao, E. Vittone, E. Kohn, A. Pasquarelli, E. Carbone, Nanocrystalline diamond microelectrode arrays fabricated on sapphire technology for high-time resolution of quantal catecholamine secretion from chromaffin cells. Biosens. Bioelectron. **26**(1), 92–98 (2010). https://doi.org/10.1016/j.bios.2010.05.017

55. W. Smirnov, N. Yang, R. Hoffmann, J. Hees, H. Obloh, W. Muller-Sebert, C.E. Nebel, Integrated all-diamond ultramicroelectrode arrays: optimization of Faradaic and capacitive currents. Anal. Chem. **83**(19), 7438–7443 (2011). https://doi.org/10.1021/ac201595k

56. N. Yang, W. Smirnov, J. Hees, R. Hoffmann, A. Kriele, H. Obloh, W. Müller-Sebert, C.E. Nebel, Diamond ultra-microelectrode arrays for achieving maximum Faradaic current with minimum capacitive charging. Phys. Status Solidi (a) **208**(9), 2087–2092 (2011). https://doi.org/10.1002/pssa.201100016

57. J. Hees, R. Hoffmann, A. Kriele, W. Smirnov, H. Obloh, K. Glorer, B. Raynor, R. Driad, N. Yang, O.A. Williams, C.E. Nebel, Nanocrystalline diamond nanoelectrode arrays and ensembles. ACS Nano **5**(4), 3339–3346 (2011). https://doi.org/10.1021/nn2005409

58. A. Eifert, P. Langenwalter, J. Higl, M. Lind|n, C. E. Nebel, B. Mizaikoff, C. Kranz, Focused ion beam (FIB)-induced changes in the electrochemical behavior of boron-doped diamond (BDD) electrodes. Electrochim. Acta **130**, 418–425 (2014). https://doi.org/10.1016/j. electacta.2014.03.029

59. E.L. Silva, M.A. Neto, A.J.S. Fernandes, A.C. Bastos, R.F. Silva, M.L. Zheludkevich, F. J. Oliveira, Fast coating of ultramicroelectrodes with boron-doped nanocrystalline diamond. Diam. Relat. Mater. **19**(10), 1330–1335 (2009). https://doi.org/10.1016/j.diamond.2010.06.023

60. M.B. Joseph, E. Bitziou, T.L. Read, L. Meng, N.L. Palmer, T.P. Mollart, M.E. Newton, J.V. Macpherson, Fabrication route for the production of coplanar, diamond insulated, boron doped diamond macro- and microelectrodes of any geometry. Anal. Chem. **86**(11), 5238–5244 (2014). https://doi.org/10.1021/ac501092y

61. A. Suzuki, T.A. Ivandini, K. Yoshimi, A. Fujishima, G. Oyama, T. Nakazato, N. Hattori, S. Kitazawa, Y. Einaga, Fabrication, characterization, and application of boron-doped diamond microelectrodes for in vivo dopamine detection. Anal. Chem. **79**(22), 8608–8615 (2007). https://doi.org/10.1021/ac071519h

62. G. Dutta, S. Siddqui, H. Zeng, J.A. Carlisle, P.U. Arumugam, The effect of electrode size and surface heterogeneity on electrochemical properties of ultrananocrystalline diamond microelectrode. J. Electroanal. Chem. **756**, 61–68 (2015). https://doi.org/10.1016/j.jelechem. 2015.08.016

63. J. Park, J.J. Galligan, G.D. Fink, G.M. Swain, In vitro continuous amperometry with a diamond microelectrode coupled with video microscopy for simultaneously monitoring endogenous norepinephrine and its effect on the contractile response of a rat mesenteric artery. Anal. Chem. **78**(19), 6756–6764 (2006). https://doi.org/10.1021/ac060440u

64. Y.S. Singh, L.E. Sawarynski, H.M. Michael, R.E. Ferrell, M.A. Murphey-Corb, G.M. Swain, B.A. Patel, A.M. Andrews, Boron-doped diamond microelectrodes reveal reduced serotonin uptake rates in lymphocytes from adult rhesus monkeys carrying the short allele of the 5-HTTLPR. ACS Chem. Neurosci. **1**(1), 49–64 (2010). https://doi.org/10.1021/ cn900012y

65. Y. Ishii, T.A. Ivandini, K. Murata, Y. Einaga, Development of electrolyte-free ozone sensors using boron-doped diamond electrodes. Anal. Chem. **85**(9), 4284–4288 (2013). https://doi. org/10.1021/ac400043b

66. T. Ochiai, Y. Ishii, S. Tago, M. Hara, T. Sato, K. Hirota, K. Nakata, T. Murakami, Y. Einaga, A. Fujishima, Application of boron-doped diamond microelectrodes for dental treatment with pinpoint ozone-water production. ChemPhysChem **14**(10), 2094–2096 (2013). https://doi.org/10.1002/cphc.201200845

67. K. Yoshimi, Y. Naya, N. Mitani, T. Kato, M. Inoue, S. Natori, T. Takahashi, A. Weitemier, N. Nishikawa, T. McHugh, Y. Einaga, S. Kitazawa, Phasic reward responses in the monkey striatum as detected by voltammetry with diamond microelectrodes. Neurosci. Res. **71**(1), 49–62 (2011). https://doi.org/10.1016/j.neures.2011.05.013

68. E.L. Silva, A.C. Bastos, M.A. Neto, R.F. Silva, M.L. Zheludkevich, M.G.S. Ferreira, F. J. Oliveira, Boron doped nanocrystalline diamond microelectrodes for the detection of Zn^{2+} and dissolved O_2. Electrochim. Acta **76**, 487–494 (2012). https://doi.org/10.1016/j.electacta. 2012.05.074

69. S.F. Peteu, B.W. Whitman, J.J. Galligan, G.M. Swain, Electrochemical detection of peroxynitrite using hemin–PEDOT functionalized boron-doped diamond microelectrode. Analyst **141**, 1796–1806 (2016). https://doi.org/10.1039/C5AN02587G

70. K. Asai, T.A. Ivandini, Y. Einaga, Continuous and selective measurement of oxytocin and vasopressin using boron-doped diamond electrodes. Sci. Rep. **6**, 32429 (2016). https://doi. org/10.1038/srep32429

71. D. Khamis, E. Mahe, F. Dardoize, D. Devilliers, Peroxodisulfate generation on boron-doped diamond microelectrodes array and detection by scanning electrochemical microscopy. J. Appl. Electrochem. **40**(10), 1829–1838 (2010). https://doi.org/10.1007/s10800-010-0114-x

72. E. Popa, H. Notsu, T. Miwa, D.A. Tryk, A. Fujishima, Selective electrochemical detection of dopamine in the presence of ascorbic acid at anodized diamond thin film electrodes. Electrochem. Solid-State Lett. **2**(1), 49–51 (1999). https://doi.org/10.1149/1.1390730

73. A. Fujishima, T.N. Rao, E. Popa, B.V. Sarada, I. Yagi, D.A. Tryk, Electroanalysis of dopamine and NADH at conductive diamond electrodes. J. Electroanal. Chem. **473**(1–2), 179–185 (1999). https://doi.org/10.1016/S0022-0728(99)00106-0

74. D. Sopchak, B. Miller, R. Kalish, Y. Avyigal, X. Shi, Dopamine and ascorbate analysis at hydrodynamic electrodes of boron doped diamond and nitrogen incorporated tetrahedral amorphous carbon. Electroanalysis **14**(7–8), 473–478 (2002). https://doi.org/10.1002/1521-4109(200204)

75. W.C. Poh, K.P. Loh, W.D. Zhang, S. Triparthy, J.-S. Ye, F.-S. Sheu, Biosensing properties of diamond and carbon nanotubes. Langmuir **20**(13), 5484–5492 (2004). https://doi.org/10.1021/la0490947

76. P.S. Siew, K.P. Loh, W.C. Poh, H. Zhang, Biosensing properties of nanocrystalline diamond film grown on polycrystalline diamond electrodes. Diam. Relat. Mater. **14**(3–7), 426–431 (2005). https://doi.org/10.1016/j.diamond.2004.11.016

77. G.-H. Zhao, M.-F. Li, M.-L. Li, Differential pulse voltammetric determination of dopamine with the coexistence of ascorbic acid on boron-doped diamond surface. Cent. Eur. J. Chem. **5**(4), 1114–1123 (2007). https://doi.org/10.2478/s11532-007-0049-1

78. A. Suzuki, T.A. Ivandini, K. Yoshimi, A. Fujishima, G. Oyama, T. Nakazato, N. attori, S. Kitazawa, Y. Einaga, Fabrication, characterization, and application of boron-doped diamond microelectrodes for in vivo dopamine detection. Anal. Chem. **79**(22), 8608–8615 (2007). https://doi.org/10.1021/ac071519h

79. M. Wei, G. Terashima, M. Lv, A. Fijishima, Z.-Z. Gu, Boron-doped diamond nanograss array for electrochemical sensors. Chem. Commun. **45**(24), 3624–3629 (2009). https://doi.org/10.1039/b903284c

80. G.W. Muna, V. Quaiserová-Mocko, G.M. Swain, Chlorinated phenol analysis using off-line solid-phase extraction and capillary electrophoresis coupled with amperometric detection and a boron-doped diamond microelectrode. Anal. Chem. **77**(20), 6542–6548 (2005). https://doi.org/10.1021/ac050473u

81. L.A. Hutton, M. Vidotti, J.G. Iacobini, C. Kelly, M.E. Newton, P.R. Unwin, J.V. Macpherson, Fabrication and characterization of an all-diamond tubular flow microelectrode for electroanalysis. Anal. Chem. **83**(14), 5804–5808 (2011). https://doi.org/10.1021/ac2010247

82. R. Oyobiki, T. Kato, M. Katayama, A. Sugitani, T. Watanabe, Y. Einaga, Y. Matsumoto, K. Horisawa, N. Doi, Toward high-throughput screening of NAD(P)-dependent oxidoreductases using boron-doped diamond microelectrodes and microfluidic devices. Anal. Chem. **86** (19), 9570–9575 (2014). https://doi.org/10.1021/ac501907x

83. E.L. Silva, C.P. Gouvêa, M.C. Quevedo, M.A. Neto, B.S. Archanjo, A.J.S. Fernandes, C.A. Achete, R.F. Silva, M.L. Zheludkevich, F.J. Oliveira, All-diamond microelectrodes as solid state probes for localized electrochemical sensing. Anal. Chem. **87**(13), 6487–6492 (2015). https://doi.org/10.1021/acs.analchem.5b00756

84. A. Avdic, A. Lugstein, M. Wu, B. Gollas, I. Pobelov, T. Wandlowski, K. Leonhardt, G. Denuault, E. Bertagnolli, Fabrication of cone-shaped boron doped diamond and gold nanoelectrodes for AFM–SECM. Nanotechnology **22**, 145306 (2011). https://doi.org/10.1088/0957-4484/22/14/145306

85. W. Smirnov, A. Kriele, R. Hoffmann, E. Sillero, J. Hees, O.A. Williams, N. Yang, C. Kranz, C.E. Nebel, Diamond-modified AFM probes: from diamond nanowires to atomic force microscopy-integrated boron-doped diamond electrodes. Anal. Chem. **83**(12), 4936–4941 (2011). https://doi.org/10.1021/ac200659e

86. A. Eifert, W. Smirnov, S. Frittmann, C. Nebel, B. Mizaikoff, C. Kranz, Atomic force microscopy probes with integrated boron doped diamond electrodes: fabrication and application. Electrochem. Commun. **25**, 30–34 (2012). https://doi.org/10.1016/j.elecom.2012.09.011

87. A.E. Hess, D.M. Sabens, H.B. Martin, C.A. Zorman, Diamond-on-polymer microelectrode arrays fabricated using a chemical release transfer process. J. Microelectromechanical Syst. **20**(4), 867–875 (2011). https://doi.org/10.1109/JMEMS.2011.2159099
88. Z. Gao, V. Carabelli, E. Carbone, E. Colombo, M. Dipalo, C. Manfredotti, A. Pasquarelli, A. Feneberg, K. Thonke, E. Vittone, Transparent microelectrode array in diamond technology. J. Micro-Nano Mechatronics **6**(1–2), 33–37 (2011). https://doi.org/10.1007/s12213-010-0032-3
89. R. Kiran, L. Rousseau, G. Lissorgues, E. Scorsone, A. Bongrain, B. Yvert, S. Picaud, P. Mailley, P. Bergonzo, Multichannel boron doped nanocrystalline diamond ultramicroelectrode arrays: design, fabrication and characterization. Sensors **12**(6), 7669–7681 (2012). https://doi.org/10.3390/s120607669
90. W. Smirnov, J.J. Hees, D. Brink, W. Muller-Sebert, A. Kriele, O.A. Williams, C.E. Nebel, Anisotropic etching of diamond by molten Ni particles. Appl. Phys. Lett. **97**(7), 073117 (2010). https://doi.org/10.1063/1.3480602
91. C.A. Rusinek, M.F. Becker, R. Rechenberg, T. Schuelke, Fabrication and characterization of boron doped diamond microelectrode arrays of varied geometry. Electrochem. Commun. **73**, 10–14 (2016). https://doi.org/10.1016/j.elecom.2016.10.006
92. V. Carabelli, A. Marcantoni, F. Picollo, A. Battiato, E. Bernardi, A. Pasquarelli, P. Olivero, E. Carbone, Planar diamond-based multiarrays to monitor neurotransmitter release and action potential firing: new perspectives in cellular neuroscience. ACS Chem. Neurosci. **8** (2), 252–264 (2017). https://doi.org/10.1021/acschemneuro.6b00328
93. S. Siddiqui, Z. Dai, C.J. Stavis, H. Zeng, N. Moldovan, R.J. Hamers, J.A. Carlisle, P.U. Arumugam, A quantitative study of detection mechanism of a label-free impedance biosensor using ultrananocrystalline diamond microelectrode array. Biosensor. Bioelectron. **15**, 284–290 (2012). https://doi.org/10.1016/j.bios.2012.03.001
94. M.E. Sandison, J.M. Cooper, Nanofabrication of electrode arrays by electron-beam and nanoimprint lithographies. Lab Chip **6**(8), 1020–1025 (2006). https://doi.org/10.1039/B516598A
95. Y.H. Lanyon, D.W.M. Arrigan, Recessed nanoband electrodes fabricated by focused ion beam milling. Sens. Actuators B **121**(1), 341–347 (2007). https://doi.org/10.1016/j.snb.2006.11.029
96. Y.H. Lanyon, G. De Marzi, Y.E. Watson, A.J. Quinn, J.P. Gleeson, G. Redmond, D.W.M. Arrigan, Fabrication of nanopore array electrodes by focused ion beam milling. Anal. Chem. **79**(8), 3048–3055 (2007). https://doi.org/10.1021/ac061878x
97. H. Li, N. Wu, A large-area nanoscale gold hemisphere pattern as a nanoelectrode array. Nanotechnology **19**(27), 275301 (2008). https://doi.org/10.1088/0957-4484/19/27/275301
98. R.M. Penner, C.R. Martin, Preparation and electrochemical characterization of ultramicroelectrodes ensembles. Anal. Chem. **59**(21), 2625–2630 (1987). https://doi.org/10.1021/ac00148a020
99. V.P. Menon, C.R. Martin, Fabrication and evaluation of nanoelectrode ensembles. Anal. Chem. **67**(13), 1920–1925 (1995). https://doi.org/10.1021/ac00109a003
100. M. Yang, F. Qu, Y. Lu, Y. He, G. Shen, R. Yu, Platinum nanowire nanoelectrode array for the fabrication of biosensors. Biomaterials **27**(35), 5944–5950 (2006). https://doi.org/10.1016/j.biomaterials.2006.08.014
101. T. Lohmuller, U. Muller, S. Breisch, W. Nisch, R. Rudorf, W. Schuhmann, S. Neugebauer, M. Kaczor, S. Linke, S. Lechner, J. Spatz, M. Stelzle, Nano-porous electrode systems by colloidal lithography for sensitive electrochemical detection: fabrication technology and properties. J. Micromechanics Microengineering **18**(11), 115011 (2008). https://doi.org/10.1088/0960-1317/18/11/115011
102. E. Jeoung, T.H. Galow, J. Schotter, M. Bal, A. Ursache, M.T. Tuominen, C.M. Stafford, T. P. Russell, V.M. Rotello, Fabrication and characterization of nanoelectrode arrays formed via block copolymer self-assembly. Langmuir **17**(21), 6396–6398 (2001). https://doi.org/10.1021/la010531g

103. C. Wang, X. Shao, Q. Liu, Y. Mao, G. Yang, H. Xue, X. Hu, One step fabrication and characterization of platinum nanopore electrode ensembles formed via amphiphilic block copolymer self-assembly. Electrochim. Acta **52**(2), 704–709 (2006). https://doi.org/10.1016/j.electacta.2006.06.003

104. C. Wang, Q. Liu, X. Shao, G. Yang, H. Xue, X. Hu, One step fabrication of nanoelectrodes ensembles formed via amphiphilic block copolymers self-assembly and selective voltammetric detection of uric acid in the presence of high ascorbic acid content. Talanta **71**(1), 178–185 (2007). https://doi.org/10.1016/j.talanta.2006.03.055

105. J. Li, J.E. Koehne, A.M. Cassell, H. Chen, H.T. Ng, Q. Ye, W. Fan, J. Han, M. Meyyappan, *Inlaid Multi-Walled Carbon Nanotube Nanoelectrode Arrays for Electroanalysis*, vol. 17 (Wiley-VCH, Weinheim, 2005), pp. 15–27. https://doi.org/10.1002/elan.200403114

106. J. Koehne, J. Li, A.M. Cassell, H. Chen, Q. Ye, H.T. Ng, J. Han, M. Meyyappan, The fabrication and electrochemical characterization of carbon nanotube nanoelectrode arrays. J. Mater. Chem. **14**(4), 676–684 (2004). https://doi.org/10.1039/B311728F

107. Y. Tu, Y. Lin, W. Yantasee, Z. Ren, *Carbon Nanotubes Based Nanoelectrode Arrays: Fabrication, Evaluation, and Application in Voltammetric Analysis, Electroanalysis*, vol. 17 (Wiley-VCH, Weinheim, 2005), pp. 79–84. https://doi.org/10.1002/elan.200403122

108. S. Siddiqui, P.U. Arumugam, H. Chen, J. Li, M. Meyyappan, Characterization of carbon nanofiber electrode arrays using electrochemical impedance spectroscopy: effect of scaling down electrode size. ACS Nano. **4**(2), 955–961 (2010). https://doi.org/10.1021/nn901583u

109. N. Yang, W. Waldemar, C.E. Nebel, Fabrication, properties and electrochemical applications of diamond nanostructures. MRS Proc. **1511**, mrsf12-1511-ee07-01 (2013). https://doi.org/10.1557/opl.2012.1661

110. D. Luo, L. Wu, J. Zhi, Fabrication of boron-doped diamond nanorod forest electrodes and their application in nonenzymatic amperometric glucose biosensing. ACS Nano. **3**(8), 2121–2128 (2009). https://doi.org/10.1021/nn9003154

111. M. Lv, M. wei, F. Rong, C. Terashima, A. Fujishima, Z.-Z. Gu, Electrochemical detection of catechol based on as-grown and nanograss array boron-doped diamond electrodes. Electroanalysis **22**(2), 199–203 (2010). https://doi.org/10.1002/elan.200900296

112. W. Wu, L. Bai, X. Lin, Z. Tang, Z. Gu, Nanograss array boron-doped diamond electrode for enhanced electron transfer from Shewanella loihica PV-4. Electrochem. Commun. **13**(8), 872–874 (2011). https://doi.org/10.1016/j.elecom.2011.05.025

113. D. Luo, J. Zhi, Fabrication and electrochemical behaviour of vertically aligned boron-doped diamond nanorod forest electrodes. Electrochem. Commun. **11**(6), 1093–1096 (2009). https://doi.org/10.1016/j.elecom.2009.03.011

114. Y. Yang, J.-W. Oh, Y.-R. Kim, C. Terashima, A. Fujishima, J.S. Kim, H. Kim, Enhanced electrogenerated chemiluminescence of a ruthenium tris(2,2′)bipyridyl/tripropylamine system on a boron-doped diamond nanograss array. Chem. Commun. **46**(31), 5793–5795 (2010). https://doi.org/10.1039/c0cc00773k

115. M.C. Granger, G.M. Swain, The influence of surface interactions on the reversibility of ferri/ferrocyanide at boron-doped diamond thin-film electrodes. J. Electrochem. Soc. **146**(12), 4551–4558 (1999). https://doi.org/10.1149/1.1392673

116. W. Gajewski, P. Achatz, O.A. Williams, K. Haenen, E. Bustarret, M. Stutzmann, J.A. Garrido, Electronic and optical properties of boron-doped nanocrystalline diamond films. Phys. Rev. B **79**(4), 045206 (2009). https://doi.org/10.1103/PhysRevB.79.045206

117. O.A. Williams, O. Douheret, M. Daenen, K. Haenen, E. Osawa, M. Takahashi, Enhanced diamond nucleation on monodispersed nanocrystalline diamond. Chem. Phys. Lett. **445**(4–6), 255–258 (2007). https://doi.org/10.1016/j.cplett.2007.07.091

118. M. Sentic, F. Virgilio, A. Zanut, D. Manojlovic, S. Arbault, M. Tormen, N. Sojic, P. Ugo, Microscopic imaging and tuning of electrogenerated chemiluminescence with boron-doped diamond nanoelectrode arrays. Anal. Bioanal. Chem. **408**(25), 7085–7094 (2016). https://doi.org/10.1007/s00216-016-9504-1

119. J. Guo, E. Lindner, Cyclic voltammograms at coplanar and shallow recessed microdisk electrode arrays: guidelines for design and experiment. Anal. Chem. **81**(1), 130–138 (2009). https://doi.org/10.1021/ac801592j

120. M. Fleischmann, S. Pons, J. Daschbach, The ac impedance of spherical, cylindrical, disk, and ring microelectrodes. J. Electroanal. Chem. **317**(1–2), 1–26 (1991). https://doi.org/10.1016/0022-0728(91)85001-6

121. M. Fleischmann, S. Pons, The behavior of microdisk and microring electrodes. Mass transport to the disk in the unsteady state: the ac response. J. Electroanal. Chem. **250**(2), 277–283 (1988). https://doi.org/10.1016/0022-0728(88)85169-6

122. L.M. Abrantes, M. Fleischmann, L.M. Peter, S. Pons, B.R. Scharifker, On the diffusional impedance of microdisc electrodes. J. Electroanal. Chem. **256**(1), 229–233 (1988). https://doi.org/10.1016/0022-0728(88)85023-X

123. O. Koster, W. Schuhmann, H. Vogt, W. Mokwa, Quality control of ultra-microelectrode arrays using cyclic voltammetry, electrochemical impedance spectroscopy and scanning electrochemical microscopy. Sens. Actuators B **76**(1–3), 573–581 (2001). https://doi.org/10.1016/S0925-4005(01)00637-2

124. J. Hees, R. Hoffmann, N. Yang, C.E. Nebel, Diamond nanoelectrode arrays for the detection of surface sensitive adsorption. Chem. Eur. J. **19**(34), 11287–11292 (2013). https://doi.org/10.1002/chem.201301763

125. C. Dincer, E. Laubender, J. Hees, C.E. Nebel, G. Urban, J. Heinze, SECM detection of single boron doped diamond nanodes and nanoelectrode arrays using phase-operated shear force technique. Electrochem. Commun. **24**, 123–127 (2012). https://doi.org/10.1016/j.elecom.2012.08.005

126. C. Dincer, R. Ktaich, E. Laubender, J.J. Hees, J. Kieninger, C.E. Nebel, J. Heinze, G. A. Urban, Nanocrystalline boron-doped diamond nanoelectrode arrays for ultrasensitive dopamine detection. Electrochim. Acta **185**, 101–106 (2015). doi:https://doi.org/10.1016/j.electacta.2015.10.113

127. C. Kranz, Diamond as advanced material for scanning probe microscopy tips. Electroanalysis **28**, 35–45 (2016). https://doi.org/10.1002/elan.201500630

128. R. Kaneko, S. Oguchi, Ion-implanted diamond tip for a scanning tunneling microscope. Jpn. J. Appl. Phys. **29**, 1854–1855 (1990)

129. E.P. Visser, J.W. Gerritsen, W.J.P. van Enckevort, H. van Kempen, Tip for scanning tunneling microscopy made of monocrystalline, semiconducting, chemical vapor deposited diamond. Appl. Phys. Lett. **60**, 3232–3234 (1992). https://doi.org/10.1063/1.106703

130. Z. Chang, Z. Ma, J. Shen, X. Chu, C. Zhu, J. Wang, S. Pang, Z. Xue, Diamond tips and nanometer-scale mechanical polishing. Appl. Surf. Sci. **70–71**, 407–412 (1993). https://doi.org/10.1016/0169-4332(93)90466-O

131. S. Albin, J. Zheng, J.B. Cooper, W. Fu, A.C. Lavarias, Microwave plasma chemical vapor deposited diamond tips for scanning tunneling microscopy. Appl. Phys. Lett. **71**, 2848–2850 (1997). https://doi.org/10.1063/1.120152

132. O. Lysenko, N. Novikov, A. Gontar, V. Grushko, A. Shcherbakov, Combined scanning nanoindentation and tunneling microscope technique by means of semiconductive diamond berkovich tip. J. Phys.: Conf. Ser. **61**, 740–744 (2007). https://doi.org/10.1088/1742-6596/61/1/148

133. O. Lysenko, N. Novikov, V. Grushko, A. Shcherbakov, A. Katrusha, S. Ivakhnenko, V. Tkach, A. Gontar, Fabrication and characterization of single crystal semiconductive diamond tip for combined scanning tunneling microscopy. Dia. Relat. Mater. **17**, 1316–1319 (2008). https://doi.org/10.1016/j.diamond.2008.02.013

134. A.P. Chepugov, A.N. Chaika, V.I. Grushko, E.I. Mitskevich, O.G. Lysenko, Boron-doped diamond single crystals for probes of the high-vacuum tunneling microscopy. J. Superhard Mater. **3**, 151–157 (2013). https://doi.org/10.3103/S1063457613030040

135. V. Grushko, O. Libben, A.N. Chaika, N. Novikov, E. Mitskevich, A. Chepugov, O. Lysenko, B.E. Murphy, S.A. Krasnikov, I.V. Shvets, Atomically resolved STM imaging with a diamond tip: simulation and experiment. Nanotechnology **25**, 025706 (2014). https://doi.org/10.1088/0957-4484/25/2/025706
136. A. Fujishima, Y. Einaga, T.N. Rao, D.A. Tryk, *Diamond Electrochemistry* (Elsevier Academic Press, Tokyo, 2005)
137. C.E. Nebel, J. Ristein, *Thin Film Diamond II: Semiconductors and Semimetals*, vol. 77 (Elsevier Academic Press, New York, 2004)
138. E. Brillas, C.A. Martinez-Huitle, *Synthetic Diamond Films: Preparation, Electrochemistry, Characterization, and Applications* (Wiley, New Jersey, 2011)

Chapter 9
Nanoparticle-Based Diamond Electrodes

Mailis M. Lounasvuori, Geoffrey W. Nelson and John S. Foord

Abstract This chapter reviews the construction, modification, and physical characteristics of two types of diamond electrodes: nanoparticle-modified diamond electrodes (NMDE) and detonation nanodiamond-based electrodes (DNDE). These particular types of diamond electrodes show great promise for improving the performance of diamond electrodes via the incorporation of nano-scale chemistry at their surfaces. The construction of both types of electrodes is reviewed, along with the resultant physical and electronic effects. The methods reviewed here are particularly applicable for electroanalytical and electrocatalytic applications of nanoparticle-based diamond electrodes. A brief review of progress on the interactions between metals and diamond at nanoparticle-based electrodes is also included. Finally, an outline of the present state-of-the art research in this field is presented.

9.1 Introduction

Carbon based materials have the potential to be an important class of materials for 21st century technologies. One particular carbon-based material has rather extreme properties—diamond. For several decades now, researchers have been excited by its transparency, extreme hardness, high thermal conductivity, bio-compatibility, and chemical resistance [1]. Examples of its utility abound. For instance, diamond is made conductive when doped, thus creating a useful electronic material. In this form, diamond can be used in biosensors [2–4], electroanalytical devices [5, 6], wastewater treatment [7, 8] and as a catalytic support in fuel cells [9]. At the cutting edge of diamond research are efforts to investigate how nitrogen vacancies in diamond could be crucial to a functioning quantum computing device [10–13]. The present interest in diamond remains high, as exemplified by the many reviews that

M. M. Lounasvuori · G. W. Nelson (✉) · J. S. Foord (✉)
Department of Chemistry, Oxford University, Oxford OX1 3TA, UK
e-mail: g.nelson@imperial.ac.uk

J. S. Foord
e-mail: john.foord@chem.ox.ac.uk

© Springer Nature Switzerland AG 2019
N. Yang (ed.), *Novel Aspects of Diamond*, Topics in Applied Physics 121,
https://doi.org/10.1007/978-3-030-12469-4_9

describe the historical development, properties, and synthesis of this unique functional material [1, 2, 14–18].

There exist many types of diamond, both natural and synthetic. However, it is the conductive forms of diamond which show the most promise as a functional material. Conductive diamond is counter-intuitive, as diamond is usually considered an insulator. However, when doped to a 1:1000 ratio of an intentionally added impurity (e.g. B or N), diamond can become a wide band-gap semiconductor [1]. In 1986, Fujimori et al. [19] were the first to develop a polycrystalline conductive diamond film using chemical vapour deposition (CVD) and boron-doping. Since then, boron-doped diamond (BDD) has become a commercially important material supplied by Element Six, U.K., among others, and is heavily used for electrochemical applications.

BDD, a p-type semiconductor, has desirable electrochemical properties, such as a low signal to background current ratio, wide potential window, robust nature, chemical resistance, and bio-compatibility [1]. BDD also does not suffer from extensive surface corrosion, oxide formation, or have unwanted electrical interactions with deposited catalyst, thus making it a useful substrate for fundamental chemical studies of electrocatalysis [15]. Its most unfavourable trait is its relative lack of electroactivity, compared to other carbon-based materials [20]. This is due to a reduced surface concentration of the following: sp^2 carbon, delocalised electrons, and oxygenated species. These characteristics reduce its electroactivity towards several chemical reactions of industrial and biological importance, such as the methanol oxidation reaction (MOR), oxygen reduction reaction (ORR), and neurotransmitter redox reactions. Thus, to make a practical diamond-based device, some modification of the diamond surface or structure is needed.

Many of the physical, wet chemical, and electrochemical routes leading to improved electrical properties are well-reviewed by others [18, 20–23]. Of the many possible modifications, the incorporation of nano-scale chemistry to the surfaces and structure of diamond is of tremendous recent interest. The two most interesting research pathways are the nanoparticle modification of diamond, and the use of unmodified or modified diamond nanoparticles. An example of each type of substrate and their modification is shown in Figs. 9.1 and 9.2.

Diamond could be modified by a film of metal or metal oxide, but this is an expensive process. It is also a process which does not benefit from the well-known catalytic properties of nano-particulate metals and metal oxides. For instance, bulk gold is relatively inert, but in nanoparticle form it is highly catalytic towards certain redox reactions, such as the ORR [28]. The properties of BDD make it an effective and stable support for these NPs. The modification of BDE with metal and metal oxide NPs has been recently reviewed by Toghill and Compton [20]. In their review, they detail how BDD electrochemistry is enhanced by the high surface area, improved reactivity, and zero-overlap of diffusion zones, thereby ensuring maximum mass transfer of analytes or molecules to the surface [20]. The adsorption of reactants is different at nano-sized particles, compared to their macro-sized counterparts [29]. Also, the exploitation of quantum effects to enhance electrochemical reactions is possible on nanoparticles-based electrodes [29]. Consensus suggests

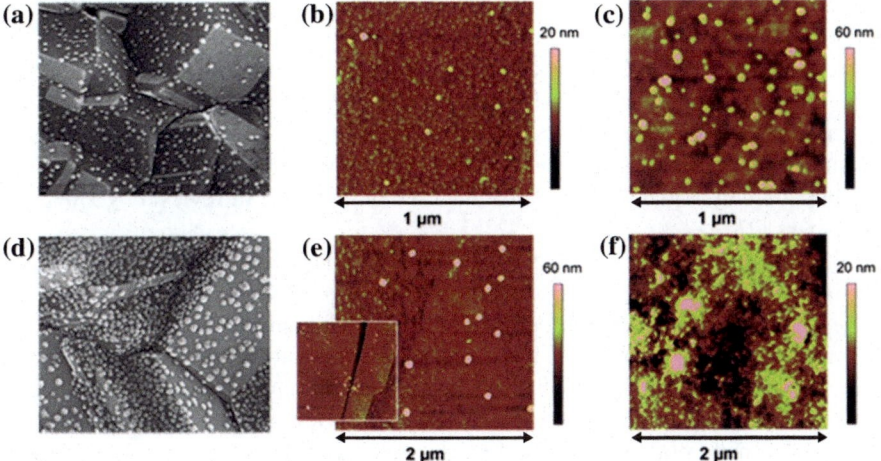

Fig. 9.1 SEM and AFM images of nanoparticles on boron-doped electrodes: SEM of BDD electrodes with Au nanoparticles after potentiostatic deposition at −0.4 V for **a** 10 s and **d** 60 s. AFM (tapping mode) of boron-doped diamond electrodes with Ni(OH)$_2$ after potentiostatic deposition in which Ni(OH)$_2$ was electro-precipitated onto polycrystalline boron-doped diamond electrode for **b** 1 s, **c** 15 s, **e** 30 s, and **f** 100 s. SEM images from Yamada et al. [24], with permission from Elsevier (Copyright © 2008); AFM images reprinted with permission from [25], Copyright (2011) American Chemical Society

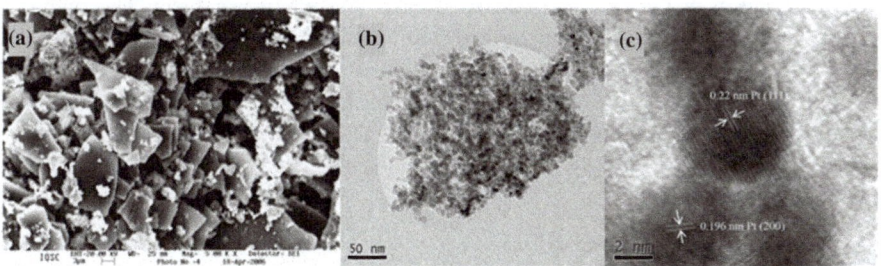

Fig. 9.2 Images of detonation nanodiamond particles decorated with nanoparticles: **a** SEM image of detonation nanodiamond powder surface decorated with Pt–RuO$_x$ deposited by the sol-gel method; TEM **b** and HRTEM **c** images of DND powder decorated with Pt via microwave-heating of ethylene glycol solutions containing H$_2$PtCl$_6$ and the detonation nanodiamond particles. Figure adapted from Salazar-Banda et al. [26], with permission from Elsevier (Copyright © 2006) and from Bian et al. [27], with permission from the International Journal of Electrochemical Science (Copyright © 2012)

that sub-monolayer coverage of nano-sized metals or metal oxides are effective catalysts for energy conversion and fuel cell applications.

Alternatively, diamond can be made nano-sized via detonation of explosives in oxygen-deficient environments. Unlike its bulk counterpart, detonation nanodiamond (DND) is electroactive in its undoped form. This remarkable property is due its "giant" surface area, coupled with a surface chemistry rich in reactive oxygen species, sp^2 carbon, unsaturated carbon bonds, and delocalised electrons [30]. The redox behaviour of ND has been found to be "molecule-like", with the reduction and oxidation of the surface occurring at discrete electrochemical potentials [30]. This relatively novel material is now used in advanced sensor and catalytical technologies. Further information can be found in reviews by Holt [10] and others [2].

While there exist reviews on the topics of nanoparticle-modified diamond electrodes (NMDE) [20] and DND-based electrodes (DNDE) [2, 10, 18] there are no reviews detailing the progress made for their construction. If one wishes to optimize the effects of nano-scale chemistry at diamond-based devices, it is clear from the literature that the conditions of construction have a tremendous effect on their physical and electrical properties. Thus, a review of the key considerations required to create practical diamond electrodes is warranted.

Discussion will revolve around the use of these electrodes for electroanalysis and electrocatalysis. These are the two most popular applications for boron-doped diamond electrodes (BDE). Non-electronic or electrochemical applications will not be discussed here; however, the methods discussed are applicable to the creation of nano-composite materials in which diamond enhances their physical properties (e.g. improved durability, coating technologies, etc.).

Of the many means to modify BDE, those concerning the construction of NMDE and DNDE will be discussed in depth. These two types of modification have found wide application in the fields of electroanalysis and electrocatalysis. The methods used to construct them have a variety of chemical and physical consequences, which need to be discussed in detail. Use of detonation nanodiamond particles as additives in electrolyte will be briefly mentioned. Other types of nanostructured diamond electrodes, such as nanostructured CVD diamond, will be touched upon. Finally, we will briefly outline the present state-of-the-art research concerning NMDE and DNDE.

9.2   Metal and Metal Oxide Nanoparticle Coated Diamond Electrodes

9.2.1   Choice of Material

Metals, metal alloys and metal oxides have all been used in electroanalytical and electrocatalytical applications [20], as summarized by Table 9.1 and seen in Fig. 9.1, in the case of $Ni(OH)_2$ and Pt deposition. Common reasons for choosing a particular material include: chemical aim, chemical environment, compatibility with deposition method, stability, and cost.

Table 9.1 Materials, methods, and applications of nanoparticles on BDE

Material	Method	References	Applications
Metals			
Pt	Spontaneous wet chemical	[31]	H_2O_2 evolution, alcohol oxidation, electroanalysis (arsenite, enzymes, environmental contaminants), ORR. fuel cell technologies
	Wet chemical assisted	[32]	
	Potential step	[33–35]	
	CV/Potential cycling	[36–41]	
	Multi-step electrodeposition	[38]	
	Potentiostatic	[33, 34, 36, 41–45]	
	Microemulsion	[46]	
	Sputtering/ion implantation	[47–51]	
	Sputtering	[52]	
	Thermal salt decomposition	[34]	
	Electroless deposition	[53]	
	Linear potential sweep	[37]	
	Galvanostatic	[54]	
	Pulsed galvanostatic	[55]	
	Sol-Gel	[56–58]	
	DC sputtering	[59]	
	Wet chemical	[60]	
Sb	Potentiostatic	[61]	Electroanalysis of heavy metals
	In situ linear sweep	[62]	
Sn	In situ linear sweep	[62]	Electroanalysis of heavy metals
Bi	Potentiostatic	[63]	Electroanalysis of heavy metals
	In situ linear sweep	[62]	
Co	Potentiostatic	[64]	H_2O oxidation catalyst
	Photoreduction	[65]	
Ni	Potentiostatic	[66]	Alcohol oxidation, biosensing
	Metal implantation	[5, 67]	
Fe	Potentiostatic	[68]	H_2O_2 detection, electroanalysis
Pb	Potentiostatic	[69]	Electroanalysis of heavy metals
Ir	Metal implantation	[48]	Arsenic detection, detection of pesticides
	Sputtering	[52]	
Ag	Spontaneous deposition	[31]	Perchlorate oxidation, biosensing; H_2O_2 detection, CO_2 reduction
	Wet chemical	[70]	
	Potentiostatic	[70–72]	
Cu	Spontaneous deposition	[31]	Nitrate detection, biosensing, organic catalysis, H_2O_2 detection, CO_2 reduction
	Potentiostatic	[73–77]	
	Wet-chemical		
	Potentiostatic coulometry	[78]	
	Metal implantation	[5]	

(continued)

Table 9.1 (continued)

Material	Method	References	Applications
Au	Spontaneous deposition	[31]	Biosensing (proteins, neurotransmitters), oxygen demand, ORR catalysis, arsenite detection, mercury detection
	Thermal salt decomposition	[35]	
	Sputtering	[79–81]	
	Potentiostatic	[82]	
	Self-assembly	[83–86]	
	Linear sweep voltammetry	[87]	
	Electro-aggregation	[88]	
	Dropcoating	[89]	
	Cyclic voltammetry	[89, 90]	
	Cyclic voltammetry with power ultrasound	[90]	
	Wet chemical	[91, 92]	
Pd	Spontaneous deposition	[31]	Proton detection, hydrazine detection, ORR catalysis, ethanol oxidation, neurotransmitter detection, C = C and C-C bond hydrogenation catalyst
	Potentiostatic	[93–95]	
Bimetallic			
Pt-Sn	Microemulsion	[96]	Alcohol oxidation, pollutant degradation
Pt-Ru	Microemulsion	[46, 97]	Alcohol oxidation for fuel cell technologies
	Potential cycling	[37]	
	Sequential and simultaneous electro-deposition	[98]	
Pt–Au	Sputtering of Au followed by potentiostatic electrodeposition of Pt	[99]	ORR, glucose detection
	Simultaneous potentiostatic electrodeposition	[100]	
Pt-Cu	Sequential potentiostatic electrodeposition	[74]	Methanol oxidation
Sn-Sb	In situ linear sweep	[62]	Electroanalysis of heavy metals
Bi-Sn	In situ linear sweep	[62]	Electroanalysis of heavy metals
Pd-Sn	Potentiostatic	[95]	Ethanol oxidation
Metal Oxides			
Pt-RuO$_x$	Sol-gel	[26, 101]	MOR catalysis, fuel cell technologies
Pt–RuO$_x$·nH$_2$O	Cyclic voltammetric deposition of RuO$_x$·nH$_2$O followed by pulsed potentiostatic electrodeposition of Pt	[102]	MOR catalysis

(continued)

Table 9.1 (continued)

Material	Method	References	Applications
Pt-PrO$_x$	Multi-step potentiostatic with electrogeneration of oxide	[103]	MOR catalysis, fuel cell technologies
Pt-SnO$_2$	Sol-gel	[58]	Methanol and ethanol oxidation
Pt-Ta$_2$O$_5$	Sol-gel	[58]	Methanol and ethanol oxidation
Fe$_3$O$_4$-Au	Dropcoating	[104]	Detection of biomarkers
IrO$_2$	Thermal salt decomposition	[35]	H$_2$O$_2$ detection, pH sensors
	Potential pulsing	[105]	
	Potential cycling	[105]	
	Galvanostatic	[105]	
TiO$_2$	Wet chemical adsorption	[106]	Solar cell technologies, H$_2$ evolution, detection of maleic acid, Ni^{2+}, cytochrome c, drinking water purification, wastewater treatment
	Potentiostatic	[107]	
	Electrophoretic	[108]	
RuO$_x$	Sol-gel/electrostatic	[109]	Cl evolution, H$_2$ evolution
FeO$_x$	Sol-gel	[109]	Oxidation of OH$^-$ to O$_2$
Ni(OH)$_2$	Electrogeneration of OH at Ni particles	[25, 110, 111]	Sensing (glucose, organic acids), MeOH and EtOH oxidation
PbO$_x$	Electro-deposition with power ultrasound	[112]	Ethylene glycol oxidation
Cu$_2$O	Amperometric electrodeposition	[113]	HER
NiO	Sol-gel	[113]	HER
Cu(II) phthalocyanine	Dropcoating	[114]	ORR
Composites			
Pt-Nafion	Microemulsion with Nafion	[97]	Biosensing
Pt-Dendrimer	Dendrimer encapsulation	[35]	Biosensing
Pt-electropolymers	Electrodeposition/ polymerization	[115–117]	Biosensing (proteins, neurotransmitters, H$_2$O$_2$)
Au-electropolymers	Electrodeposition/ polymerization	[118]	Biosensing
Ni-graphite	Thermal catalytic etching	[119]	Glucose sensing

The chemical aim is the most important parameter. Metals, metal oxides, and alloys/composites are useful in a variety of chemical systems, as listed in Table 9.1. Notably, Au and Pt are popular materials due to their versatility and catalytic properties. Metal oxides are useful in photocatalytic applications [113] and as alternatives to bimetallic alloys in direct methanol oxidation fuel cells [120]. Bimetallic alloys have advantages compared to their single metal counterparts, whereby metal-metal interactions (e.g. bifunctional mechanism, metal back-bonding, etc.) can increase reaction rates and decrease poisoning of reaction sites of a single metal catalyst (see Pt-and Pd-based alloys in Table 9.1) [74, 95, 99, 121, 122]. For example, Mavrokefalos et al. [95] compared the performance of Pd-modified BDD and bimetallic Pd-Sn-BDD in ethanol oxidation. The addition of Sn led to higher current density for the reaction as well as improved poisoning resistance [95]. The same group [74] examined the performance of Pt-modified BDD towards methanol oxidation and reported higher current densities and less fouling when Cu nanoparticles were present with Pt compared to Pt alone. XPS analysis showed a shift in the Pt4f peaks consistent with metallic bonding between Pt and Cu [74]. Codeposition of two metals is often employed for alcohol oxidation catalysis for fuel cell applications and it can be achieved by either simultaneous or sequential potentiostatic electrodeposition [74, 95, 98, 100], potential sweep techniques [37, 62], microemulsion methods [46, 96, 97] or sputtering followed by potentiostatic deposition [99].

Chemical aims may require other considerations. For instance, multi-step reactions may require more than one type of nanoparticle, or the presence of a specific facet, shape or set of interfaces to be present [123, 124]. The chemical effects or compatibility with stabilizers (e.g. Nafion), conductive, and bio-compatible coatings (e.g. SAMs on Au) also need to be considered [83, 116, 125, 126].

Some metals, such as gold and silver, can enhance Raman and infrared signals through surface plasmon resonance [127] and thus have been extensively deposited onto various substrates to achieve the enhancement effect in surface-enhanced infrared spectroscopy (SEIRAS). In recent years, boron-doped diamond has successfully been used as the working electrode in spectroelectrochemistry, where the boron-doped diamond film has been mounted in a transmission cell [128], pressed onto a ZnSe ATR crystal [129] or grown on top of a diamond ATR [130] in order to achieve a conductive, IR transparent electrode. Two different methods have been reported for depositing Au nanoparticles onto BDD electrodes for use in SEIRAS: wet chemical [91] and electrochemical [131]. The wet chemical route involved precipitation of Au NPs in solution and consequent gravity-assisted immobilisation onto amino-terminated BDD substrate and resulted in particle size of around 23 nm; however, this method did not lead to a reproducible enhancement effect [91]. The electrochemical route employed by Izquierdo et al. [131] involved pulsed nucleation-growth potential steps and gave a larger average particle size (ca. 80 nm), but no conclusions were made on the reproducibility of the enhancement effect.

The chemical environment to which the electrode is exposed will dictate the type of nanoparticles to be deposited. A crucial consideration is the use of toxic metals (e.g. Pb, Ag, etc.) on an electrode to be used in biological environments. It may be better to use a non-toxic metal, such as gold. The environment of fuel cells and batteries can be either acidic or alkaline. The dissolution of metal nanoparticles in acids makes their use in the former environment challenging. Metal oxides are an alternative material, due to their increased stability in those harsh environments. For biological applications, metal nanoparticles need to be modified with known organo-metallic chemistry to render it more bio-inert and/or stable at the surface (e.g. SAMs on Au, Ag, etc.).

The material must be compatible with the chosen deposition route. For instance, metal nanoparticles are easily deposited using various electrochemical methods. However, the deposition of metal oxides or alloys often requires more complex methods, such as multi-step electrochemical routes [102, 103] and sol-gel based methods [20, 26, 58, 101, 109, 113]. The deposition of metal oxides can occur by physical vapour deposition (PVD), with recent advances promising to enable the deposition of core-shell metal oxides onto surfaces (i.e. Mantis Deposition Ltd.). Some metal oxides can be deposited via a simple electrodeposition method provided that the metal cations are stabilised in the deposition solution during deposition. An example of this approach is the amperometric deposition of Cu_2O by using lactate ligands to stabilise the Cu^{2+} ions in solution that can then be reduced to form Cu_2O particles on BDE [113]. In the same work, NiO particles were subsequently deposited onto the Cu_2O particles via the sol-gel method from an alcoholic solution of nickel acetate followed by annealing in air [113].

Ideally, electrodes should be functional more than once. Therefore, the resultant NP-modification must be stable, both physically and chemically. Some nanoparticles, such as platinum, are known to have weak adhesion to diamond, thus motivating the search for their increased stability at BDE [33, 132]. Stability is of utmost importance in biological environments, where NPs are known to have inflammatory properties in solution. The instability of metals can be resolved by using alternative metal oxides, as exemplified by the replacement of Ti, Ta, Ti-Pd alloys on carbon-based electrodes by RuO_2 for electrocatalysis and chlorine evolution [101].

The cost of materials can be high, particularly those with the best catalytic versatility and performance (e.g. Pt, Au). Although these materials have been shown to be electroactive at extremely low loading rates (i.e. sub-monolayer), any further reduction in cost, without sacrificing performance is attractive [20]. Many strategies are used, including the use of cheaper materials (e.g. Ni, Pb, Sn, Fe, Cu, Co) and metal oxides [20], as outlined in Table 9.1. It is clear from this table that these alternatives can be used for the same reactions as those of Pt and Au. The use of an alternative material may offer superior performance, or reduce costs by increasing nanoparticle stability and longevity.

9.2.2 Methods of Deposition

Ideally, nanoparticles should be deposited in a chemically active form, having small size, uniform distribution, and low mass loading (<10% of surface coverage); these characteristics are associated with the best performing electrochemical and catalytic diamond-based electrodes [20, 98]. The deposition method chosen is paramount to achieving these results. The deposition methods in Table 9.1 can be generally categorized by whether they utilise electrochemistry or not. These two general approaches are discussed in this section.

9.2.2.1 Electrochemical Methods

Electrodeposition, in its various forms, is the most popular route towards metal nanoparticle deposition. These methods share a common feature, with metal cations from the appropriate salt (e.g. H_2PtCl_6) being reduced at the diamond electrode by some applied potential (e.g. Fig. 9.1b) [20]. The deposition is further controlled by the choice of electrochemical method, the deposition time, deposition potential, metal ion concentration, scan rate, and number of deposition phases/stages [20]. The advantages and disadvantages of popular electro-deposition methods are described here.

The potentiostatic method is the most popular technique. This is certainly clear from Table 9.1 and noted elsewhere [20]. Simply, it involves the exposure of the electrode to a metal salt solution, followed by deposition at a fixed potential [20]. The deposition conditions, such as applied potential, duration, and metal salt concentration are easily tuned. For instance, it is accepted that deposition time is proportional to mass loading for most nanoparticles [32, 36, 73, 78, 123]. An example of this phenomenon is demonstrated by comparing Fig. 9.1a, d, in the case of Pt deposition on BDE. The method has disadvantages. As noted in several reports, potentiostatic deposition leads to poor particle adhesion, large particle size (cf. Fig. 9.1a, d), low mass loading, and inhomogeneous distribution; all these characteristics could result in an impractical diamond electrode, despite decent electrocatalytic performance [33, 132–134].

Changing the electrochemical method is often a useful strategy. Potential step deposition methods, such as cyclic voltammetry (CV), linear sweep voltammetry (LSV), potentiostatic step deposition (PSD), chronoamperometry, and galvanostatic methods have all been used to control Pt deposition on diamond (see Table 9.1). The first three alternatives lead to more uniform size and distribution of Cu and Pt nanoparticles [32, 36, 73, 78, 123]. CV and LSV are slow techniques, but Welch et al. [78] have noted that slowly increasing or decreasing the potential during deposition offers better control over nanoparticle nucleation, growth, and mass loading. Chronoamperometry has the advantage of promoting high mass loading of Au particles via enhanced mass transfer effects initiated by its use [42, 134]. Galvanostatic and pulsed galvanostatic methods favour the growth of individual Pt particles, as these methods minimize the overlap of reactant depletion zones [55]; lower pulse times and higher applied currents lead to lower particle size and denser distribution [55, 132, 135]. Pulsed potentiometric method has been devised for Au nanoparticle growth, where short nucleation pulses at more negative potential are followed by longer growth pulses at less negative potential [131, 136]. The potentials are chosen such that no changes in the surface termination of the diamond can take place, hydrogen evolution is avoided, and small current is maintained in the growth phase [131, 136].

Multi-step and/or multi-technique deposition is a successful means to minimizing the disadvantages of any single method. The Foord group has shown that the combination of potentiostatic and potentiostatic step deposition enables particles to grow by both instantaneous and progressive nucleation stages [103]. Also, potentiostatic deposition followed by LSV increases and homogenizes particle size [73]. Combining electrodeposition with electropolymerization of a polymeric stabilizer (e.g. Nafion) or conductive polymer (e.g. polyaniline) is an effective route towards stabilizing nanoparticles on the diamond surface [137–139].

Deposition conditions

The conditions at which deposition occur are extremely important. The most important of which is that the potential applied to facilitate deposition must be sufficient to drive the reduction of metal cations at the electrode [20]. The most systematic studies have been conducted by the Compton group, notably their work on Cu [78, 140], Pt, Pd [93, 94], Ag [140], Au [140], Ir [141] and other metals, as reviewed here [20]. Vinokur et al. [142] demonstrated that the nucleation of Ag and Hg on BDE depends on overpotential. The appropriate applied potential(s) can provide a higher driving force for the deposition reaction, with the consequences of higher nanoparticle nucleation rates, higher nanoparticle densities, and smaller sizes [20, 33].

Caution and thought must be employed when choosing the applied potential(s). Depositing at potentials near that of H_2 or O_2 evolution is problematic, as gas bubbles can dislodge freshly deposited nanoparticles [33, 42]. Also, the potential can have desirable or undesirable *chemical* effects, such as H_2 cleansing of Pt nanoparticles [33], the co-deposition of bimetallic alloys [37], the pre-concentration of other ions at the interface (e.g. OH^-, H^+) [103], or the stripping of nanoparticles from the electrode [78]. In addition, the choice of potential has been found to affect the shape of Au [82], Cu [73], and Pt [98] nanoparticles. Examples of these effects can be seen by comparing Fig. 9.3a, c, as well as Fig. 9.3e, f, pairwise. In the first pair, it is clear that applied potential can affect shape, while in the second pair shows that applied potential can affect particle size. For the deposition of metal oxides via an electrochemical route, the choice of deposition potential is essential for the deposition of the nanoparticle *and* the electro-precipitation of the oxide, due to local pH changes at the nanoparticle surface during deposition [25, 103]. This can result in the deposition of nanoparticles with controllable size and morphology compared to sol-gel routes. An example of this electro-deposition is shown in Fig. 9.1a, for the case of $Ni(OH)_2$.

Other controllable conditions include the length of deposition, the scan rate and the nature of the deposition solution. The length of deposition is generally proportional to mass loading [103, 132] and is known to influence the morphology of deposition, with longer times exploited to create larger particle sizes [132]. In the case of CV, multiple scans can homogenize the size of nanoparticles, but not the particle distribution [78]. Adjustment of the scan rate has been shown to promote facet selective deposition of Pt on polycrystalline diamond [123]. At higher sweep rates, particles preferred to deposit on the (111) facet, as dramatically shown in Fig. 9.4. The sweep rate also affected shape and distribution, with the (111) facet having more numerous and smaller particles than the (110) facet. Oxygenated surface moieties are thought to concentrate on (111) facets. As mentioned in other sections, the presence of oxygen functionality increases nanoparticle stability, nucleation, and diamond is more electroactive in such regions [123].

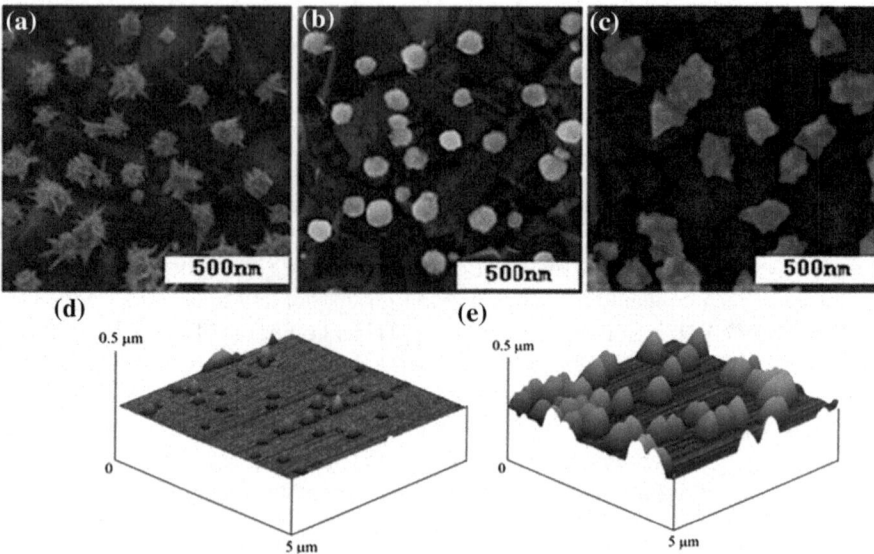

Fig. 9.3 Images of nanoparticles on diamond with electrodeposition influenced by applied potential or concentration of metal salt in solution. (Top row) SEM images of gold nanoparticles electrodeposited on boron-doped diamond surfaces: **a** Flower-like Au nanoparticles deposited at +0.5 V versus Ag/AgCl in 0.2 M H$_2$SO$_4$ solution containing 2 mM HAuCl$_4$; **b** Spherical nano-gold deposited at +0.5 V versus Ag/AgCl in 0.2 M H$_2$SO$_4$ solution containing 0.2 mM HAuCl$_4$; **c** Convex polyhedron nano-gold prepared in same conditions as A, except at an applied potential of − 0.1 V versus Ag/AgCl. (Bottom Row) In situ tapping mode AFM images of cobalt grown on a BDD surface from 10 mm Co(II) at potentials of **e** −1.05 V and **f** −1.15 V. Reprinted from Li et al. [82] with permission from Elsevier (Copyright © 2006). Reprinted from Simm et al. [64] with permission from Wiely-VCH (Copyright © 2006 WILEY-VCH Verlag GmbH & Co. KGaA, Weinheim)

The nature of the solution in which electrodeposition occurs is an important consideration. The concentration of metal ions in solutions seems to affect the nucleation pathway of Ag and Hg deposition [142]. In addition, concentration affects nanoparticle shape. This is shown in Fig. 9.3, where a 10-fold dilution of HAuCl$_4$ leads to spherical deposition, compared to "flower-like" shape at higher concentrations (see Fig. 9.3a, b). [82]. Adjusting concentration can have unusual effects. For instance, the density of electrodeposited Ag nanoparticles is inversely proportional to size, except at very dilute concentrations of the metal salt [31]. Other adjustments to the type of solution and pH can be made. Additionally, the pH of the solution [83, 105] or local pH at the interface [103, 143] can affect electrodeposition. The latter has been discussed in the previous section for the case of metal oxide deposition.

Deposition temperature can also play a role in the deposition process but is rarely investigated. MacPherson's group [144] have studied electrochemical Pt nanoparticle deposition at different temperatures by utilising a pulsed laser to

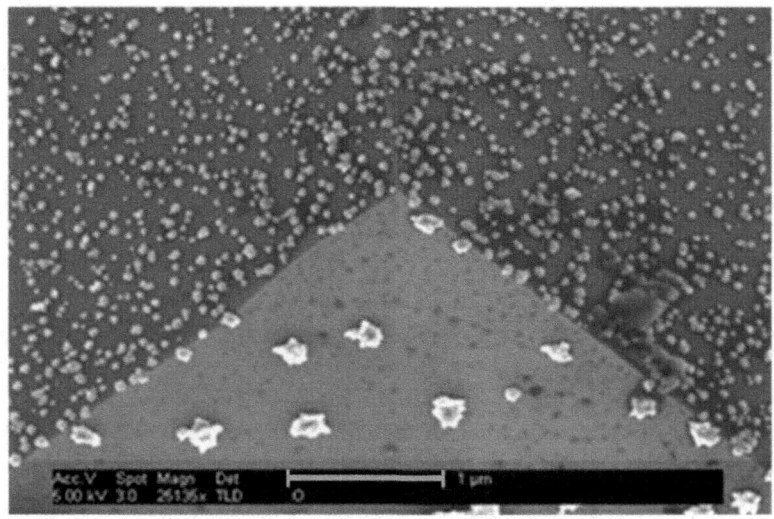

Fig. 9.4 Scanning electron microscopy image of Pt nanoparticles on boron-doped diamond deposited with potential cycling from −0.8 V to +1.6 versus Ag/AgCl in 0.5 M H₂SO₄ at a sweep rate of 500 mV/s. The (111) facet edge is clearly visible. Reprinted with permission from Gonzalez-Gonzalez et al. [120], Copyright (2009) American Chemical Society

nonisothermally heat the electrode during deposition. The pulsed laser approach results in continuous heating and cooling of the electrode substrate and leads to smaller particle size with higher index crystal facets and a more porous morphology compared to deposition at room temperature [144].

9.2.2.2 Non-electrochemical Methods

Non-electrochemical methods are used less frequently. They are often complex and require advanced equipment, unavailable in most laboratories. Two of the more common techniques are briefly discussed here.

Physical vapour deposition techniques, such as ion impact and sputtering depositions are effective in vacuo methods for depositing nanoparticles on diamond. Both techniques vaporise metal from a source, but the former accelerates the ions by 800 keV towards the sample, rather than merely coating the surface by physical deposition [20]. Often, these techniques are used to modify CVD diamond films during their creation. For instance, DC magnetron sputtering was used by the Swain group to create Pt-sandwich electrodes, in which a Pt layer was sputtered between two CVD grown layers, halfway through the CVD process [133]. Bimetallic deposition is possible if two or more metal sources are in the vacuum chamber. Uniform nanoparticle distribution is often the result of using a PVD method, as nanoparticle deposition is relatively unaffected by the inhomogeneous distribution of boron at BDE or by grain boundaries [134]. Heat treatment post-PVD deposition may be necessary, as in the case of Au nanoparticles [79, 99, 145]. Ma et al. [81] used a plasma oxidising thermal treatment after sputtering of gold film to create Au nanoparticles with average size of 8 nm and with a dense and uniform coverage. When conducting ion implantation, the modified BDD must be heated by high temperature in the presence of ambient H_2 not necessarily to modify the structure of deposited nanoparticles, but to recover metastable diamond structures produced by the ion bombardment itself [5, 47, 48, 67]. Some groups introduce sp2 carbon into diamond intentionally: Deng et al. [119] fabricated a Ni-graphite-BDE electrode by sputtering a layer of Ni onto BDE followed by thermal catalytic etching. This treatment resulted in Ni NPs embedded into the diamond, encased with graphitic layers formed at the etched surfaces [119]. By far the greatest advantage of PVD methods is that the ultra-high vacuum conditions reduce unwanted physio-chemical effects (i.e. oxidation, water, etc.) on nanoparticles due to solution or ambient conditions. Wide-spread use of PVD techniques is not possible, due to the cost and complexity of the UHV systems required.

The sol-gel method is a popular means to deposit metal oxide nanoparticles on a diamond surface [26, 20, 56, 57, 113]. Generally, a metal precursor is stabilized as a colloid in ethanol, or 2 propanol, and then coated onto the electrode by dip-coating, spin-coating, or as a paste. Afterwards, a thermal heat treatment is sufficient to ensure nanoparticle formation having the proper structure. For example, anatase TiO_2 has been deposited onto diamond via this route at thermal heat treatment at 400 °C [146]. While the sol-gel method typically leads to random and uniform distribution of nanoparticles on diamond, it suffers from the disadvantages of large particle sizes and particle aggregation. This is exemplified by the image of Fig. 9.2a, in the case of nanoparticle deposition on diamond nanoparticles. A recently developed alternative to the sol-gel method is the aforementioned electro-precipitation of the oxide, either during a single step [25] or as a coating for pre-deposited nanoparticles [103]. Results suggest that this method offers better control of nanoparticle deposition via electrochemical parameters, the effects of which are discussed in some detail later. Some of the state-of-the-art wet-chemical deposition routes are discussed in Sect. 9.5.

9.2.2.3 Substrate

Thus far, conductive diamond has been referred to simply as BDE. In actual fact, BDE are categorized by the method of their creation and by their crystallinity. More detail than will be described here concerning the manufacturing of BDE can be found in several reviews [1, 16, 21, 147, 148]. The chemical vapour deposition (CVD) method is most commonly used, and this has several consequences for the deposition of metal and metal oxide nanoparticles on diamond substrates.

CVD diamond is made by ionizing hydrogen and methane with a thermal or microwave source. The ions re-combine on a substrate—often silicon—to form a sp^3 carbon network. Boron doping can be introduced to the diamond film by having

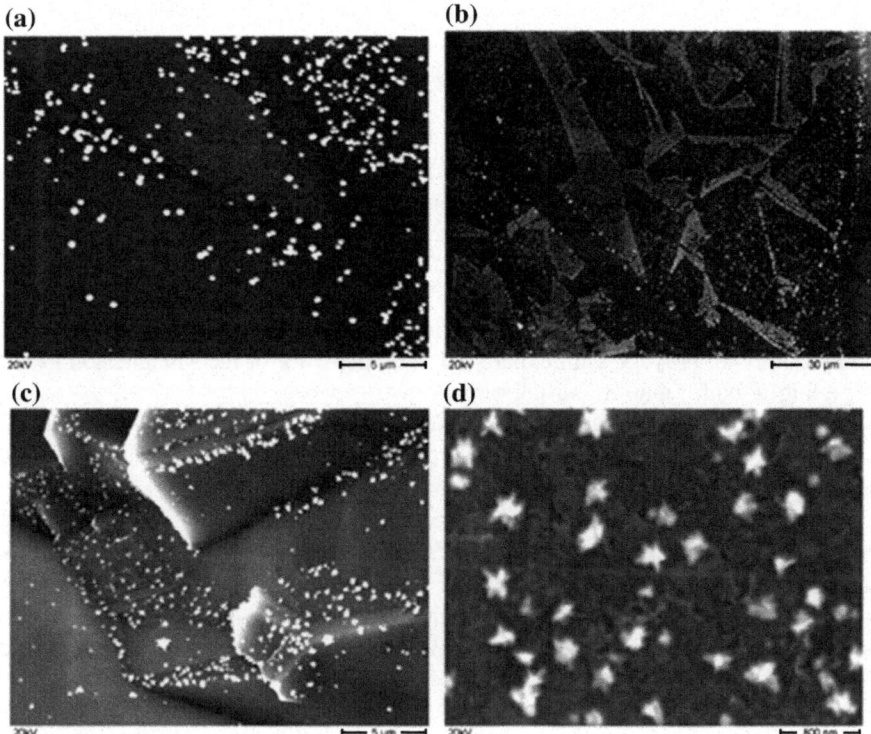

Fig. 9.5 SEM images of platinum deposited on **a, b** polished smooth boron-doped diamond electrode, **c** as-grown large grain, and **d** small grain diamond substrates. Reprinted from Hu et al. [33] with permission from Wiley-VCH, Copyright © 2009 WILEY-VCH Verlag GmbH & Co. KGaA, Weinheim

a source of boron in vapour form during the CVD process. This manufacture process can be adjusted to make a wide variety of diamond substrates, with different distributions of B doping and crystallinity. The substrate does not affect the nanoparticle growth or distribution tremendously when non-electrochemical methods are used to deposit nanoparticles. Therefore, subsequent discussion concerns itself with the affect substrate has on the electro-deposition of nanoparticles.

Metal nanoparticle nucleation occurs at the most electroactive sites on a diamond surface, such as boron centres, grain boundaries, and defects [20]. Thus, the distribution of B and the dimensions of surface crystals are important factors to be considered. The distribution of boron centres is inhomogeneous, thus partially explaining the non-uniformity of nanoparticle growth when using electrodeposition techniques [55, 135, 149]. This inhomogeneity can be seen in Fig. 9.5a–c. The grain boundaries have increased surface concentrations of electroactive sp^2 carbon and oxygen moieties; thus, denser concentrations of nanoparticles tend to be found in these regions [48, 80, 96].

Further increases in nanoparticle density can be achieved by depositing onto BDE with smaller grain sizes (i.e. macro > micro > nano), due to their increased surface area [38]. This trend can be seen by comparing Fig. 9.5c, d. It is thought that this trend explains the effectiveness of lower nanoparticle loadings on micro-crystalline diamond electrode, compared to its macro counterpart [38]. Smaller grain sizes also help to reduce aggregation, reduce particle sizes, and make distribution more uniform, as seen in Fig. 9.5d. This distribution and morphology promotes hemispherical diffusion profiles, which are electrochemically favourable [20, 38]. There is some debate whether metal nanoparticles deposit randomly or non-randomly on polycrystalline diamond, with the work of the Macpherson group suggesting is both random and uniform [42].

Pre-treatment of BDE prior to deposition has been shown to lead to improved nanoparticle stability and electrode performance. Oxygen-terminated diamond stabilizes Pt nanoparticles better than hydrogenated diamond [42, 54]. One can add oxygen functionality to the surface by a variety of means, including oxygen plasma and anodic electrochemical treatment. [87, 150]. However, passivation of the diamond film by thick overlayers of oxygen is to be avoided. Pre-roughening of the BDE surface can increase surface area, thereby improving nanoparticle adhesion and uptake; this is shown to be effective by Hu et al. [132] who used nanodiamond powder suspensions as the roughening medium.

In contrast to stronger attachment of Pt NPs on oxygen-terminated BDE, better adhesion has been reported for Au NPs on hydrogen-terminated BDE. Cathodic pretreatment and resulting lower surface oxygen content was first shown to improve Au nanoparticle deposition by cyclic voltammetry [90]. This observation was studied more closely by Svanberg-Larsson et al. [89], who compared explicitly hydrogen- and oxygen-terminated BDEs as substrates for electrochemical deposition of gold nanoparticles and found the adhesion of Au NPs to be significantly better on H-terminated BDE.

Interestingly, the diamond surface alone may not necessarily have an effect, but rather the material which supports it. Recently, Gao et al. [31] demonstrated that the spontaneous deposition of Ag, Cu, Au, Pd, and Pt on diamond could be achieved in HF, when the diamond surface was in ohmic contact with hydrogen-terminated silicon. It is thought that this type of contact increases the surface electron energy of diamond, thereby promoting the migration of excess electrons from the silicon/HF interface to the conductive diamond interface [31]. In this manner, the diamond surface is made more reductive, thereby leading to the spontaneous reduction of metal ions onto the diamond surface. Silicon dioxide was unable to achieve this same effect, thus highlighting the importance of substrate choice to deposition chemistry.

9.2.3 Surface Characteristics

9.2.3.1 Characterization Techniques

Surface characterization of NP-modified diamond films is necessary to both confirm nanoparticle deposition and understand the fundamental surface chemistry of the electrode. The size and morphology of deposited nanoparticles is best studied using atomic force microscopy (AFM), scanning electron microscopy (SEM), and transmission electron microscopy (TEM). Notably, in situ studies of nanoparticle nucleation and growth are possible using electrochemical AFM (EC-AFM) [20, 64]. The surface chemistry of the deposited nanoparticles can be effectively determined using X-ray photoelectron spectroscopy (XPS), time-of-flight secondary ion mass spectroscopy (ToF-SIMS), and Raman spectroscopy. Crystal structure of both the substrate and the nanoparticles can be determined using X-ray diffraction (XRD). Finally, the electrochemical properties and mass loading of nanoparticles is usually determined by cyclic voltammetry and stripping voltammetry.

9.2.3.2 Nanoparticle Distribution

The distribution of nanoparticles across an electrode can be modified to achieve a given electrochemical or electroanalytic purpose. Higher current densities, higher signal to noise ratios, and higher faradaic to non-faradaic current ratios, occur at surfaces comprising widely-spaced and uniformly distributed nanoparticles [151, 152]. In this case, mass transport of reactants to the nanoparticles occur via fast, three-dimensional, diffusion, as opposed to linear diffusion [38, 151]. These characteristics are known to improve electroanalysis and electrocatalysis at nanoparticle-based electrodes. In cases where macroelectrode performance must be realised at reduced cost, one might distribute nanoparticles more densely (e.g. 50–100 nm particle to particle spacing), to ensure that electron transfer is limited by mass transport (i.e. planar diffusion) of the reactant to the surface, thereby lowering electrode sensitivity [38, 151, 152]. In some cases, the kinetics of electrochemical reactions are paramount; the location and distribution of reactive nanoparticles can control or exploit the interplay between reactant mobility and adsorption [152]. Nanoparticle deposition near active sites, defects, and grain boundaries could be of benefit if such sites are important to the electrochemical system. Finally, the promotion of certain nanoparticle shapes requires preferential growth on certain grain boundaries of diamond [123]. The degree to which particles distributed uniformly can be determined by comparing experimental data to theoretical models [152, 153]

Non-uniform nanoparticle distribution on BDE is promoted by inhomogeneous distributions of boron-centres, grain boundaries, high number of surface defects, and high levels of sp^2 contamination. These surface features are highly electroactive, with nanoparticles nucleating and growing on these sites during electrodeposition [33]. This explains the difficulty of depositing homogeneous distributions of NP by electrodeposition. More uniform distributions are obtained by reducing the number of defects or grain boundaries (cf. Fig. 9.5c, d), and by making the surface less electroactive by increasing the surface density of sp^3 carbon [69]. Some reports do suggest that electrodeposition of nanoparticles can occur grain-independently [42, 69].

The uniformity of nanoparticle growth depends on the identity of the metal. Mavrokefalos et al. [95] have reported that Pd can be deposited on areas of lower B concentration, leading to a uniform coverage, whereas Sn is more likely to nucleate at grain boundaries and thus to form a non-uniform coverage with larger particle size.

Uniform nanoparticle distribution is produced, generally, by non-electrochemical methods of deposition [80]. An alternative is to functionalise the diamond and/or nanoparticle surface prior to deposition, in order to promote uniform nanoparticle distribution. For example, two dimensional arrays of Au nanoparticles have been created by covalent attachment of citrate-coated Au NPs to an amine terminated diamond surface [83]. Spătaru et al. [107] electrodeposited TiO_2 nanoparticles in the presence of a surfactant. A stable surfactant layer on hydrogen-terminated BDD surface hindered the deposition rate, thus leading to greater uniformity of the TiO_2 nanoparticle distribution [107].

Another strategy for achieving a uniform NP distribution is to control the diffusion of metal precursor towards the substrate during electrodeposition. Delivering the Pt precursor onto the electrode surface entrapped in vesicles that were allowed to adsorb onto the BDE surface prior to electrodeposition resulted in a more uniform distribution of Pt NPs, avoiding preferential deposition onto more reactive facets and grain boundaries [44]. More simply, the pre-roughening of the diamond substrates has been shown to promote uniform nanoparticle distribution [132].

9.2.3.3 Nanoparticle Adhesion and Stability

The dissolution of nanoparticles from a diamond support is not desirable, as it leads to instable electrochemical performance and increases the cost of any electrical or catalytical device. Moreover, the introduction of metal ions into biological systems may have toxic effects. Therefore, the question of particle adhesion and stability is crucial.

Dropcoating has been studied as a possible approach for metal nanoparticle deposition onto BDE. Gan et al. [114] found that dropcoated Cu(II) phthalocyanine adhered more strongly onto hydrogen-terminated diamond due to the nonpolar nature of the metal complex. Svanberg-Larsson et al. [89] showed that Au nanoparticles dropcoated from aqueous solution adhered more strongly to oxygen-terminated diamond due to favourable electrostatic interactions, but the particles were less uniformly distributed and in much poorer electrical contact with the substrate on oxygen-terminated BDE compared to hydrogen-terminated BDE.

Better adhesion and stability may be realised by using alternative materials to an initial choice. For instance, in acidic environments, metal nanoparticles can dissolve, therefore metal oxides should be considered for their stability in these environments. Examples of successful replacements include: use of RuO_2 instead of Ti, Ta, Ti-Pd by RuO_2 for chlorine evolution [101]; use of bimetallic or metal oxide nanoparticles instead of Pt nanoparticles [33, 98, 132]. A consultation of Table 9.1 may inspire the search for alternative materials for a given application.

Sandwiching nanoparticles between layers of CVD diamond seems to be the best method for retaining nanoparticles. The classic example is that of Swain et al. [133] who sputtered Pt onto CVD diamond, followed by a period of further diamond growth. The high stability of Pt nanoparticles was due to their being partially buried by diamond overlayers or anchored to the surface by diamond growth around the base of individual nanoparticles. Kondo et al. [154] fabricated micrometre-sized porous spherical diamond particles with embedded Pt nanoparticles by short CVD growth of diamond on Pt NP-DND clusters formed by spray-drying a an aqueous slurry of Pt NPs and DND particles. Ion implantation achieves a similar level of stability as high energy metal ions are partially buried into the diamond substrate, as demonstrated for Ni [5], Cu [5], Pt [47], and Ir [48] implantation into BDE. Ion implantation offers 5 month stability, in the best case scenario [5, 155]

Other methods of deposition improve particle adhesion, as noted throughout this chapter. Multi-step potentiostatic methods may be superior to single step potentiostatic methods [20, 38, 103]. Potential pulsing leads to better IrO_x adhesion and is highly recommended for the deposition of metal oxides, compared to galvanostatic or potential cycling methods [105]. Novel wet chemical routes (discussed in Sect. 9.5) also offer the potential for improved particle adhesion and stability [32].

Substrate pre-treatment can also improve particle adhesion and electrode stability. Increasing the surface density of oxygen moieties can improve Pt nanoparticle adhesion [42, 54]. This can be accomplished by pre-treatment in acid, anodic electrochemical treatment, or O_2 plasma [42, 87, 101, 150]. Sol-gel deposited metal oxide nanoparticles and sputter deposited Au nanoparticles have improved adhesion after a heat treatment [99, 145]. Thermal heat treatment can make BDE more hydrophilic, which may be the cause of improved adhesion [156]. Deliberate hydrophilic treatment of diamond lead to improved FeO_x adhesion [105, 109]. Ultrasonic pre-treatment of diamond in suspensions of diamond powders improves the adhesion of Pt nanoparticles [132]. Finally, the careful choice of the type of diamond improves particle adhesion. For instance, Pt nanoparticle modified micro-crystalline diamond outperforms its macro counterpart, providing electroanalytical performance in polluted tap and river water for 150 detection runs [38]. In the case of Au nanoparticles, hydrogen-terminated BDE surfaces have improved nanoparticle adhesion [89, 90].

Poorly adhered nanoparticles do not necessarily leave the electrode, but can agglomerate along grain boundaries during an electrochemical experiment. For instance, Pt nanoparticles are stable on the (111) facet of BDD for a few hours during methanol oxidation, but on the (110) facet, particles agglomerate [123]. Agglomeration causes particle size to increase, with the consequence of reduced electrochemical selectivity and sensitivity to analytes, as nanoparticle diffusion layers overlap [20]. It is thought that functional groups on the (111) surface, such as sp^2 carbon and oxygen functionality help to stabilize attached particles. [123].

9.2.3.4 Nucleation

The nucleation of materials using electrochemical techniques is still poorly understood and there is much controversy in the field. Hyde and Compton [157] have reviewed some of this debate and recent progress in this area. Due to existing controversy in the field, recent research is discussed in some detail.

Two types of nucleation processes are used throughout the literature on electrodeposition: instantaneous and progressive. The former assumes that nuclei grow slowly on a small number of active sites, simultaneously; the latter assumes that nuclei grow quickly on many active sites, which become activated as the electrodeposition proceeds [157, 158, 159]. There exist two methods of determining the nucleation type. The first involves analysing current-time transient curves obtained by chronoamperometry [33, 160]. The second involves the use of in situ EC-AFM [161]. In both cases, the results are most commonly compared to models by Scharifker and Mostany [162]. These models assume a constant nucleation rate and that particle growth is diffusion controlled; depending on the model chosen this can be planar or hemispherical [33, 160, 161].

Diamond is a low energy surface and nucleation is most commonly instantaneous, as exemplified by the electrodeposition of Cu [73], Bi [63], Pt [33, 36, 160] on non-polished, polycrystalline diamond. Instantaneous deposition is driven by high overpotentials and the extent of such nucleation can be controlled by varying the potential [160]. Also, instantaneous nucleation is prevalent on diamond with smaller grain sizes [31, 33] This can be seen in Fig. 9.5d. Characteristics of instantaneous deposition include: the observation that the number of nucleation sites is controlled by the applied potential [163]; an monotonically broader size distribution during deposition [73], small particle size [73], small particle distribution [33], and the formation of dendritic microstructures on grain boundaries and surface defects [98]. These characteristics helped identify the nucleation type in the case of Pt and Pt-Ru deposition by sequential potentiostatic methods, compared to deposition of Pt-Ru by simultaneous potentiostatic methods [98].

Progressive nucleation does occur on BDE as well. Platinum deposits onto smooth, polished diamond by a progressive nucleation process, as shown in Fig. 9.5a [33]. These particular nanoparticles are characterised by smooth, spherical morphology with decrease electroactive surface area. By contrast, in the same study, instantaneously nucleated Pt on small and large grain BDE had dendritic structure and higher electroactive surface area (cf. Fig. 9.5d). Others characteristics of progressive nucleation include large particle sizes, clustering, and wide size distribution [33, 64, 142]. Some deposition methods may only be capable of progressive nucleation, such as the potential sweep method, in the case of simultaneous or sequential deposition of Pt-Ru onto polycrystalline diamond [98]. Progressive nucleation should be avoided if electrochemical performance is of utmost importance, as instantaneously nucleated particles are more favourable for this application [20].

It is commonly accepted that nucleation occurs most readily at grain boundaries, surface defects, and electroactive sites on the diamond surface, as in the case of Ni and Co [64, 66], Pt [33, 36, 98, 123, 132, 160], Pt-Ru [98], as well as other metal and metal oxides [20]. This phenomenon can be seen in Figs. 9.1, 9.4, and 9.5. The position of nucleation depends on the substrate, particularly the local electroactivity of the surface [33, 132]. On smooth, polished BDE nucleation of Pt may be promoted on the most electroactive facets of diamond [33]. For example, Pt is known to deposit on the (111) facet of polycrystalline diamond (see Fig. 9.4). The size and distribution of grains within the diamond substrate can affect nucleation. Pt nanoparticles are more homogeneously distributed on small grain BDE, compared to large grain BDE (see Fig. 9.5) [33]. The authors attribute this to the higher density of electroactive sites on small grain BDE. Higher boron doping levels make the BDE more electrochemically active [149], and one surmises this would promote instantaneous nucleation; however, to our knowledge this has not been systematically studied yet. Increasing the electroactivity of the BDE substrate is an effective strategy to promote nanoparticle nucleation.

9.2.3.5 Size

There is much evidence showing that the best electrochemical performance results from the deposition of small particle sizes (5–10 nm) at low loading rates, with small size distributions [1, 20, 31]. These characteristics ensure a high electroactive surface area, hemispherical diffusion controlled electrochemistry, catalytically active quantum size effects, and the availability of low co-ordinated facets at the surfaces of small particles [20, 80, 164–166]. The above characteristics help ensure the resultant electrode is effective.

Sizes reported in the literature range from 5 nm to 5 μm, with size being controlled in a number of ways. The deposition method can affect the size. Microemulsion routes [46, 97], wet chemical deposition [31], and the various electrodeposition techniques mentioned in Table 9.1 are capable of depositing nano-sized particles on the order of 2–10 nm. Some PVD methods are capable of depositing similarly sized particles [167]. The use of sol gel methods to deposit metal oxides tends towards large particle sizes, on the order of 0.5–5μm [26, 56, 57]. However, this technique does trend towards low loading rates, on the order of 10^{13} mol^{-1} cm^2 [156], thereby minimizing diffusion zone overlap of particles [20]

Many electrodeposition methods have been evaluated for their ability to control particle size, as noted by Hu et al. [33], including: chronoamperometry [42, 111], cyclic voltammetry [37, 78], galvanostatic [54], pulsed galvanostatic [55], potentiostatic step [34], and potentiostatic cooulometry [78]. These methods were used to control the size of various metal systems, including: Cu [78, 140], Pd [93], Ag [140], Au [87, 140], Ni [111], Ir [141, 168], and Pb [112]; other systems of interest can be found from Table 1.1 and in other reports [20].

Potentiostatic methods are better able to fine-tune the growth of metal nanoparticles [33, 98, 132]. For example, for the deposition of Cu^{2+}, Ag^{2+}, and Pb^{2+}, potentiostatic coulometry is a superior method to cyclic voltammetry [78]. The latter is known to incompletely strip metal sites, thus promoting uncontrolled growth of metal nanoparticles on pre-existing metal on the surface. Metal is a more energetically favourable deposition site than BDE [78, 169]. In potentiostatic coulometry, only nanoparticles on non-electroactive sites are incompletely stripped, thus ensuring that particle growth is controlled by the electrochemistry of the diamond substrate, alone [78]. Another approach to avoid metal deposition on existing metal deposits and to control the size of NPs is to entrap the metal pre-cursor in vesicles that are then adsorbed on BDE surface prior to deposition. Pt NPs deposited this way exhibited smaller size than those deposited without the use of vesicles, and it was possible to control the size of the NPs by adjusting the vesicle size [44].

Multi-step electrodeposition can be a useful means to achieve better size control, by minimizing the disadvantages of any single method. For example, to adjust size more quickly, the Foord group utilizes a two stage process. In the first stage, nano-sized Pt nuclei are created by the slow increase of negative potential, followed by the application of a fixed potential to enlarge the nuclei to the desired size [103]. The size of metal oxides can be controlled in a third electrodeposition step, in which the oxide is precipitated at the metal nuclei, as done for the case of Pt-PrO$_x$ [103].

Many deposition parameters are reportedly able to change the size of deposited particles, including: overpotential [33, 38, 66, 78, 103], metal ion concentration in the electrolyte [31, 38, 66, 142], deposition time [33, 38, 66, 78, 103], grain size of diamond substrate [42], solution acidity [31] and number of electrodeposition steps [38]. Of these parameters, the most important is overpotential, due to its influence on the energetics of the electrodeposition process, nucleation, and mass transport to electroactive sites on the substrate [20, 98]. Adjustment of the scan rates can homogenize size distribution after initial deposition [73]. Those deposition parameters which lead to instantaneous nucleation are typically those that create the smallest particle sizes, as described earlier. Progressive nucleation leads to the largest particle sizes [66].

There are several other parameters used to control size in the case of non-electrochemical methods. Heat treatment is capable of changing the size of metals, such as Au, post-deposition [80]. Also, the facet to which the metal nanoparticle adheres may influence size. For example, it has been shown that Pt nanoparticles on (110) facets have larger particle diameter, while those on the (111) facet are smaller in size (see Fig. 9.4) [123]. In this same study it was shown that merely changing the sweep rate during deposition by cyclic voltammetry is sufficient to change the preferred facet of deposition, and thus the particle size. [123]. Figure 9.4 shows the result of using the highest scan rate in that study.

9.2.3.6 Shape

Performance can be further enhanced by changing the morphology of deposited nanoparticles. A number of different shapes and particle morphologies have been made on diamond, including: convex polyhedron (see Fig. 9.3c), flower-like or dendritic (see Fig. 9.3a), spheres (see Fig. 9.3b), clusters, and agglomerations. The size of these features varies from the nano-scale to the micro-scale, depending on the system studied. Shape can be controlled in a number of ways, including: deposition type, nucleation type, overpotential, metal ion concentration, and linear sweep rates. The shape can expose favourable facets of the nanoparticle to the solution and increase the exposed surface area, both of which are desirable features for a modified BDE [82].

There are several reports showing that deposition type can control particle shape. Deposition from sol gels and microemulsion tend to form clusters and agglomerates of metal oxides and metals at BDE [26, 31, 106]. Dendritic and smooth particles result from potentiostatic and linear potential sweep methods, respectively [33, 37]. Mesoporous Pt nanoparticles have been obtained by the deposition of a Pt–Cu alloy and subsequent anodic dissolution of Cu [117].

In multi-step methods, one might consider the effect of simultaneous versus sequential deposition. For example, they lead to spherical and dendritic structure, respectively, in the case of Pt–Ru deposition [98]. Nantaphol et al. [100] have reported on the differences in shape of bimetallic Pt–Au nanoparticles grown via simultaneous and sequential deposition, where Pt deposited onto Au NPs showed nanosheet-like structures, whereas simultaneous deposition resulted in more spherical shapes. Sequential deposition led to a higher coverage of the BDE substrate, higher catalytic activity and higher resistance to poisoning [100].

Other deposition parameters can be optimized to favour certain shapes. Flower and dendritic growth is promoted by instantaneous nucleation, while spheres, clusters and agglomerations are more prevalent during progressive nucleation [29, 63, 80, 82, 98]. The parameters which favour one or the other type of nucleation have been discussed earlier. Dendrite formation requires prismatic growth in the early stages of deposition and is related to overpotential [31, 170]. This may be due to the fact that higher overpotentials cause near maximum mass transport limits [171]. For example, at constant metal ion concentration, a simple change in overpotential is capable of changing Au deposition shape from flower-like to a convex polyhedron (cf. Fig. 9.3a, c) [82]. Also, heat treatment has been used to change the shape of Au nanoclusters and does so by exposing the nanoparticles to the very electroactive Au(111) plane, which promotes dendritic-like growth [80].

The shape of nanoparticles can be changed post-deposition. Platinum nanoparticles deposited via an electrochemical deposition method have been restructured by square wave voltammetry [40, 172]. The lower and upper potential during SWV causes repeated adsorption-desorption cycles of oxygen and hydrogen on the Pt NPs, and nucleation and growth-dissolution cycles of Pt [173], guiding the growth of the nanoparticles and affecting the particle size and shape. The potentials, frequency and electrolyte will all play a role in the resulting particle morphology.

During the spontaneous deposition of Ag, Cu, Au, Pd, Hg, and Pt onto Si supported BDE, it was found that metal ion concentration is key to particle shape [31]. The authors note that dendritic, small spheres, and large spheres were related to high, low, and dilute concentrations, respectively. In the case of dilute concentrations it is more favourable for metals to deposit onto metal rather than the underlying BDE electrode, thereby increasing particle size [31]. High and low concentrations of $HAuCl_4$ were found to lead to 'flower-like and spherical nanoparticle shapes on BDE (cf. Fig. 9.3a, b) [82].

9.3   Diamond Nanoparticles as an Electrode Material

DND is of emerging interest, with research finding it to have a diverse set of applications, including biomedicine, catalysis, quantum computing, nano-composites, and for the seeding of CVD diamond growth [10]. It has a number of attractive features, including optical transparency, large surface area, and a bio-compatible sp^3 core structure. These characteristics make DND a material of choice for use in bio-medical implants, nano-scale electrochemistry, drug delivery, cell imaging technologies, and as a substrate for cellular growth [2, 10, 174–177]. The myriad of applications of nano-scale diamond has been reviewed by Holt [10] and Schrand [2].

9.3.1   Background on Detonation Nanodiamond

Lewis et al. [178] first found diamond nanoparticles in interstellar dust and meteorites. Since then, efforts have been made to create nano-scale diamond particles, synthetically by the controlled combustion of explosives. Nanoparticles of diamond are synthesized by the controlled combustion of explosives in high oxygen conditions [2, 177, 179]. These particles undergo cleaning processes (i.e. acid treatment, oxidation by Osswald method [180] to remove metal impurities and to reduce their size to primary particles, they must undergo deaggregation [2].

The resultant material is comprised inner, sp^3 diamond core of 4–5 nm and an outer shell of non-diamond character, with complex surface chemistry [181]. This outer shell of purified DND contains a mixture of sp^2 and sp^3 carbon [182] and oxygen based functional groups (i.e. carboxylic acids, esters, lactones) [2]. If desired, this surface chemistry can be changed by fluorination [183], hydrogenation [184], and the attachment of alkyl, amino, and amino acid groups [183, 185]. The physical and chemical properties of DND depend strongly on the size of the nanoparticles [184], with size control of the DND required for the stabilization of their colloidal suspensions [186] or to ensure specific reactivity [187, 188]. The surface area of primary sized DND is approximately 270–280 m^2 g^{-1} [189], thus providing greater surface area for nanoparticle attachment than conventional diamond films.

9.3.2 Electrochemistry of Detonation Nanodiamond

DND is an undoped form of diamond. One normally associates undoped diamond as being an insulator and impractical for use in an electronic device. Unlike the bulk material, DND is electrochemically active in its undoped form. Like many materials, the nano-sized version exhibits different chemical properties than its bulk form. This is due to its high surface area, multi-faceted nature, and its outer shell, which contains reactive sp^2 carbon and oxygenated functionality [186].

Commercial DND particles of primary particle size are redox active, but not in a classical sense. The classical picture considers that H-terminated diamond is redox active, if chemical potentials are below the valance band maximum [190, 191]. Also, it considers that p-type conductivity at the surface exists due to electron transfer between H-termination and dissolved oxygen in aqueous environments [192]. This is not true of DND, as determined by Holt and her colleagues in 2008 [30] and 2009 [193].

Our understanding of DND redox behaviour has been advanced by the Holt group who have conducted extensive electrochemical studies of commercial DND (5 nm) [30, 193]. They found DND to have 'molecule-like' redox behaviour [30]. That is to say that the particles themselves undergo oxidation and reduction via surface states at specific potentials [30]. It is easier to reduce DND particles than to oxidise them [193]; this process is spontaneous and slow in the presence of certain redox species without an applied potential [30, 193]. Holt's group [194] have also studied how detonation nanodiamond interacts with redox species in solution, looking at the influence of particle size, solution pH, scan rate, ionic strength of the electrolyte and charge of the redox couple using electrochemical scanning microscopy and cyclic voltammetry and infrared spectroscopy. Two catalytic processes were identified as contributing to current enhancement: electron transfer between redox species and the surface functionalities on DND, and electron transfer mediated by adsorbed species on the DND surface [194]. The electrochemical behaviour may also be due to the presence of sp^2 carbon, delocalised electrons due to oxidation, and unsaturated bonding of surface atoms; all this gives DND surfaces a semiconductor/metallic character [193]. Although doping of DND particles is possible [195, 196], it is not required for the vast majority of its electrochemical applications, due to its intrinsic redox behaviour. The electrochemistry of detonation nanodiamond has been reviewed recently [6, 197].

9.3.3 Methods of Deposition/Incorporation into Electrode Form

Novoselova et al. [198] were the first to report the use of nanodiamond powders as electrodes. The procedure involved sintering the DND particles at high temperature and pressure to form pellets [22, 198] that were then studied by cyclic voltammetry. A DND electrode can be manufactured by grinding DND powder into the tip of a Pt wire sealed in a small pipette [199, 200].

Sintering nanodiamonds into pellets and compacting them inside pipettes result in bulk nanodiamond electrodes. Another approach is to modify electrode substrates with thin films of DND particles. A quick and facile method of preparing DND films is dropcoating [30, 193, 194, 201, 202]. DND particles have been dropcoated from ethanol suspensions onto Au [30] and BDD [193, 201] electrodes. Aqueous suspensions of DND particles have been dropcoated onto BDDE [194] and glassy carbon electrode (GCE) [200]. Electrodes prepared using the dropcoating method are commonly used for sensing applications. Biosensors for glucose and lactate determination have been prepared by dropcoating DND particles and enzymes from aqueous suspensions onto Au [203, 204]. Pyrazinamide detection was performed with DND particles dropcoated onto GCE [205]. The quantification of codeine has been studied at a DND-modified GCE, where DND particles were dropcoated from suspension with dihexadecyl phosphate surfactant [206]. Peltola et al. [202] suspended DND particles in a water-ethanol mixture and dropcoated them on tetrahedral amorphous carbon substrates for dopamine detection. The downside of the dropcoating method is non-uniform coverage and DND agglomeration [30, 202].

In order to achieve more uniform coatings on electrodes, other deposition methods have been devised. DND particles have been mixed with mineral oil and the resulting past smeared onto GCE [30]. If the paste was smeared on too thickly, electrochemical blocking effects by the DND particles were observed on the GCE, but by controlling the amount of paste, a thin uniform layer on the electrode was achieved [30]. When DND particles were deposited by low-power sonication onto gold substrate, the DND coverage was found to be continuous and uniform and no clustering at substrate grain boundaries was observed [207, 208]. It was noted that the zeta potentials of particles and substrate as well as the identity of the solvent played a role in the successful deposition of DND particles [207, 208]. DND films have been formed by spraycoating water-ethanol suspensions of DND particles onto tetrahedral amorphous carbon substrate [202] and spincoating from ethanol suspension onto gold-coated glass slides [209]. Both spraycoating and spincoating were shown to improve the uniformity of DND films compared to the dropcoating method; however, the physical stability of the films was poor as the DND particles were observed to detach from the substrate during electrochemical experiments and during sonication [202, 209]. The film stability is improved when DND is electrophoretically deposited, for instance onto silicon wafers [210, 211] and gold-coated glass slides [209] as thin uniform layers. Electrophoretically deposited films were found to be physically more stable than spincoated and dropcoated films

[209]. It has been theorised the DND can be electrodeposited onto electrodes [212], but to our knowledge this has not been evaluated in practice.

9.3.3.1 Nanodiamond Composite Materials

Nanodiamond composites are of great interest as electrode materials. Both organic compounds and metals have been incorporated into composites with DND particles.

DND-polyaniline composites have been made either by electropolymerisation of aniline onto a compacted DND electrode [213, 214], electropolymerisation of aniline with DND particles dispersed in the electrolyte [215, 216], precipitation polymerisation with DND particles dispersed in the reaction solution [217] or by in situ interfacial polymerisation with aniline dissolved in the organic phase and DND particles dispersed in the aqueous phase [218, 219]. Interestingly, the simultaneous deposition method leads to the nanostructuring of PANI in one dimension, as the PANI oligimers aggregate under the influence of pi-pi stacking [217]. This result hints that DND co-deposited with other materials can lead to novel materials with interesting chemical properties.

Other polymeric components have also been utilised to create DND nanocomposites. DND particles have been included in a sol-gel matrix with (3-mercaptopropyl)-trimethoxysilane and lactate oxidase for improved electron transfer in a lactate biosensor [220]. An in situ polymerisation process under ultrasonic irradiation was employed to achieve a polypyrrole-DND nanocomposite with improved corrosion resistance compared to polypyrrole alone [221]. DND-regioregular polyhexylthiophene hybrid films, prepared on n-type silicon and FTO glass by spincoating, have exhibited better photoelectrochemical properties than corresponding nanocomposites made with zinc oxide or titanium oxide nanoparticles [222, 223].

One example of co-deposition is the modification of diamond with metal or metal oxide nanoparticles, as discussed in this review. DND has also been modified with nanoparticles in a similar way, for similar applications (see also Table 9.1). Supporting catalysts on nanodiamond will benefit from the high surface area and increased reactivity of nanodiamond, compared to its bulk form. Thus, lower loading rates are required to make a practical electroanalytical or catalytic device, based on nanodiamond powders. This promises to reduce device costs. The inclusion of DND particles enhances the physical properties of metal coatings, thus improving device performance and prolonging device life.

DND particles have been co-deposited with Ni for improved electrothermal microactuator performance [224], disperse-strengthening of tribological coatings [225] and increased corrosion resistance [226, 227]. DND particles electrochemically co-deposited with Sn and Pb have improved the coating properties of the Sn/Pb alloy [228].

Various electroanalytical sensors have been fabricated with DND particles together with metals or other carbon materials. Ag nanoparticles were deposited onto DND and graphite nanoparticles for the voltammetric determination of

Ceftizoxime, a third-generation antibiotic [229]. A novel humidity sensor was fabricated by dropcoating a DND-graphene oxide nanocomposite onto a quartz crystal microbalance [230].

Many metals and metal oxide catalyst materials have been supported on nanodiamond powder based electrodes for improved performance. TiO_2/DND nanocomposite exhibited improved catalytic activity towards nitrite oxidation [231]. A particularly promising application for nanodiamond is in direct methanol fuel cells acting as catalyst support for Pt [27, 154, 232–237] and Pt/Ru [235, 238]. For example, Kondo et al. [154] fabricated a Pt NP embedded porous diamond microparticles by spray-drying a slurry containing Pt nanoparticles and DND. The mechanical strength of the particles was improved by a short diamond growth step by MPCVD [154]. Nanodiamond powders obtained from other sources than detonation synthesis have also been used for methanol fuel cells. BDD nanopowder obtained from mechanical crushing of CVD-grown BDD was used to construct a BDD/Pt/RuOx electrocatalyst composite that showed improved activity towards methanol oxidation compared to commercial Pt-Ru/C catalyst [26]. Two different types and sizes of nanodiamond particles were tested as Pt catalyst support for methanol fuel cells by Zang et al. [232]. When DND particles of size 5 nm were used as the catalyst support, a better electrocatalytic activity was observed compared to nanodiamond particles fabricated by mechanical crushing with particle size 100 nm [232].

Catalyst support and hybrid electrode design is aimed at reducing the amount of precious metals required in a catalytic device. A great deal of research is also devoted to developing completely metal-free catalysts, which are usually based on carbon nanomaterials [239]. Various metal-free catalysts incorporating nanodiamond have been reported. Zhu et al. [240] prepared nitrogen-doped nanodiamond for oxygen reduction reaction. DND particles were first annealed to increase sp^2 content to form so-called carbon onions, and subsequently treated with imidazole or HNO_3 and heated in ammonia to introduce nitrogen doping in the form of pyridinic, pyrrolic and quaternary N [240]. Choi and Kim [241] have also tested carbon onions as catalyst for ORR, with some modifications to the procedure described above. DND particles were annealed at a higher temperature, modified Hummer's method was used to introduce oxygen groups on the surface of the particles and nitrogen-doping was achieved by treatment with urea and pyrolysis under nitrogen [241]. The photoelectrochemical activity of DND particles has been explored in carbon nitride-nanodiamond hybrids that showed enhanced activity towards photodegradation of methylene blue [242].

Fruitful research can continue in the area of nanodiamond powder based electrodes for electrocatalysis, inspired by the work done on bulk diamond. In addition, further optimisation of particle loading, stability, and electroactivity of nanodiamond supported catalysts is necessary.

9.3.3.2 DND in Electrolytes

In addition to use as electrode material, DND particles have been proposed as additives in electrolytes. *meta*-poly(ether−ketone)-grafted DND composites generated by polymerisation of 3-phenoxybenzoic acid with DND in situ demonstrated a polyelectrolyte effect when the nanodiamond loading was higher than 10 wt% [243]. Introducing DND particles into samarium-doped ceria used as electrolyte in solid oxide fuel cell improved the mobility of oxygen ions and thereby the power density obtained from the device [244]. A recent study by Yang, Gogotsi et al. [245], motivated by the do-deposition of DND particles with metals from plating solutions to increase the uniformity of deposited metal films [246, 247], shows that DND additives in the electrolyte of a lithium ion battery have the potential to suppress dendrite growth.

9.3.4 Characterization of Diamond Nanoparticle-Based Electrodes

Nanodiamond powder films on surfaces have been characterized by SEM and AFM [233, 248]. Nanoparticle layers form porous sub-micron structures, which enhances its electroactivity compared to bulk films [233]. Holt et al. have characterized DND films by FTIR, XPS, TEM, and Raman spectroscopy [30]. Graphite G and D bands in Raman spectroscopy indicate the presence of surface sp^2 carbon [30]. XPS shows the oxygenated character of the nanodiamond surface [30]. Nanodiamond has also been characterized by scanning electrochemical microscopy (SECM) [193]. Redox behaviour at diamond nanoparticles has also been evaluated by in situ infrared spectroscopy [249].

9.4 Nanostructured CVD-Grown Diamond Electrodes

Both top-down and bottom-up techniques have been used for creating nanostructured CVD-grown diamond electrodes. These methods have been reviewed in [6] and Chapter 5 of this book, and will therefore be only briefly introduced here. Nanostructuring can be introduced into CVD-grown boron-doped diamond films by reactive ion etching. By using various nanoparticles, including gold nanodots [250], titanium [59] and nanodiamond [251] as masks, this method allows the formation of vertically aligned BDD nanopillars. In oxygen plasma based reactive ion etching, the boron dopant atoms act as the mask, thus eliminating the need for mask deposition and removal steps [252]. BDD electrode arrays fabricated this way can find uses in sensing applications [253, 254].

More recently BDD has been successfully grown on various nanostructures to achieve high surface area electrodes, including silicon wires [255], TiO2 nanotubes [256, 257] and fibres [258], polypyrrole [259], carbon nanotubes [260, 261] and glassy carbon foam [262]. Nanostructuring of diamond electrodes can drastically improve the capacitive performance of BDD, which is well known for its extremely low background current. Such nanostructured CVD BDD electrodes show promise for use in supercapacitors [255–259, 261, 263].

9.5 Interactions at the Metal-Diamond Interface

Interactions between diamond, in its various forms, and the nano-materials discussed in this review are key to the understanding of the stability and electronic performance of these composite systems. Diamond has no native oxide, and therefore the electronic behaviour of its surface is highly dependent on any surface modification. Doping of diamond films with impurities which induce p or n-type doping transforms an otherwise insulating material into a semi-conductor. Redox active species at diamond surfaces (i.e. COOH, C-O-C, lactones, ketones, etc.) are known to facilitate electron transfer between the diamond interface, an electrolyte, and metal ions [264]. The above are all well-known phenomena; however, there is a lack of systematic and fundamental research concerning other, more complex phenomena at the metal-diamond interface [124, 265]. A review of some progress in this area is presented here.

The manufacturing process, polishing, cleaning, and exposure to ambient environments can introduce impurities to the diamond surface. In the case of DND particles, the level of impurities can reach 10–12% of the total surface area [266] and comprise 55 different elements [267]. Examples of these impurities include sp^2 carbon, organic-based functionality (O, N, S, B), metal salts, and defects, among others [266]. Mitev et al. [267] identified 23 major elemental impurities that originate from either the detonation or purification processes. Various methods of removing impurities have been reported, including the use of complexing agents [268] and strong acids [269] to remove metallic impurities, and thermal oxidation treatment [269], ozone treatment [270] and plasma jet [271] to remove non-diamond carbon impurities. In some cases, sp^2 carbon is deliberately introduced to improve the performance of DND especially in oxygen reduction reaction [272–274].

Some impurities can lead to favourable interactions. For example, metals with a preference for the +2 oxidation state (e.g. Ni, Ti, Cu, Co, Fe, Al) are thought to be stabilized by interactions with oxygen groups, such as carboxyls and hydroxyls. [20, 66, 266, 275, 276]. By using ^1H and ^{13}C NMR, Panich, it was determined that metals in the +2 [276–278] and +3 oxidation state [279] participate in ion exchange with the protons of carboxylic acid groups and are chemically bound to the DND via two carboxylic groups in close proximity either on one DND particle or adjacent particles. This is evidence for non-covalent bonding between oxygen-based impurities and metal nanoparticles. Diamond surface can be terminated with amine groups via a photochemical reaction, which then allows the immobilisation of gold nanoparticles on the surface via covalent bonding [83–86]. Covalent bonding via oxygen functionalities is also possible, as in the case of Al bonding to DND nanocrystals via carboxyl terminations, thereby leading to aggregation of DND in solution [266]. Carbide bonding, such as that between diamond surfaces and Au/Ti, Al, and Ti are known to improve contact resistance and mechanical properties, as determined by the circular transmission line method (c-TLM) [280]. However, not all impurity-metal bonding has a purpose or is favourable. For instance, metal bonding to sp^2 carbon is known to interfere with the electrical properties of the diamond electrode and it reduces the strength of metal-diamond adhesion [281].

Electrostatic effects are thought to play a role in governing metal deposition and stability on diamond-based electrodes. O-terminated and H-terminated DND particles have negative and positive zeta potential over a large pH range (up to pH 12) [188, 266], and this may govern the dynamics of electropositive or negative metal ions to their surfaces. Facet dependent deposition of metal and alloy nanoparticles onto diamond thin films has been observed for a variety of systems (Au, Pt, and their alloys) [123, 265]. The reason for this behaviour seems to lie in the differing electrostatic properties of the various facets [123, 265]. This is supported by the finding that for DND particles sp^3 carbon (i.e. 100 orientation) is electropositive, while sp^2 carbon (i.e. 111 orientation) is electronegative [195].

Electronic interactions between diamond and metal nanoparticles have been observed. For example, the electrical potential of DND nanoparticles is affected by the substrate, size, height, and surface termination [282]. Stehilk et al. [282] observe that the work function diamond interfaces are altered when an electric dipoles (i.e. added metal nanoparticles) is added, thus changing the energy electrons require to reach the vacuum level.

Metallization is a useful change in the electronic properties of conductive diamond, as it improves charge transfer between diamond and any surface modifier [264]. Calculations based on Density Functional Theory (DFT) and experimentation have shown that the band gap is removed if modified by metals, O-terminated diamond, and/or sp^2 functionality [266, 283]. Also, the type of electrical contact improves charge transfer. Ohmic contacts between metals and diamond have been associated with better particle adhesion, higher bond strengths, and shorter bond lengths. [283, 284]. This is possible for Ti, V, and Ta deposition, but not for Au or Pd deposition [174, 280]. Likely, charge density on carbon is facilitating backbonding to some metals [283, 284], with metals having more unpaired d orbitals (i.e. V, Ta, Ti) and filled d orbitals (Au, Pd) having stronger and weaker electronic interactions with diamond, respectively. [283, 284].

There are other lesser known, or studied, electronic interactions between metals and diamond. These include: negative electron affinity[284–286], conductive surface protrusions [287], non-diamond sp^2 [286], conductive diamond to metal backbonding [284], intrinsic structural effects [286, 288], charge trapping/transfer [264, 282], and interpretations based on energy level confinement within quantum dots for particles < 7 nm in size [282]. Plana et al. [264] note that many of these same reasons are applicable to the metal-diamond interface, in addition to electrochemical (e.g. electrolyte, double layer, etc.) and nanostructural effects. Other subtle effects are possible, such as quantum effects and metal-vacuum properties, as noted by Tyler et al. [285] in their study on electron emission from diamond nanoparticles on metal tips. The size of the metal or diamond nanoparticle is an important parameter to consider when studying any of the above effects [282]. Quantum-based and crystallinity-based arguments may have greater explanatory weight for metal-diamond interactions involving sub-5 nm nanoparticle sizes [152, 187, 282]. More research in this sub-field is necessary to truly isolate the electronic effects at the metal-diamond interface.

Some metal-diamond interactions have physical consequences. The catalytic etching of nano-sized features in diamond surfaces occurs in H_2 environments at elevated temperatures (>700 °C) [289]. Annealing diamond thin films after it has been modified with metal nanoparticles by ion bombardment can lead to the diffusion of metal nanoparticles from the bulk to the surface [20]. Metal adatoms could have preferred deposition sites, as suggested by DFT calculations of Ti on diamond (100), whereby Ti prefers pedestal sites on top of carbon dimer rows [281]. Annealing of surfaces leads to metal-diamond aggregates involving DND particles, which may be the result of covalent bonding between DND particles with metal ions acting as linkers [266].

9.6 Modern State-of-the-Art and Outlook

Research on the modification of diamond electrodes with metal and metal oxides continues apace. Research in the past three years has been motivated by interest in fuel cell technologies, photocatalysis and water-splitting, and biological applications.

To our knowledge, two understudied areas of research concern the nanoparticle-diamond interface: the effect of doping levels and the use of epitaxial diamond. Most researchers use commercial BDE and do not investigate how doping levels, doping elements, or type of doping affect nanoparticle deposition. Secondly, studies of nanoparticle deposition on epitaxial diamond could prove fruitful. It is known that the diamond (111) facet is preferred for nanoparticle deposition and such deposition leads to enhanced electrochemistry. The fundamental chemistry of that interface is important, but little understood. One group has attempted to study the role of doping on epitaxial grown diamond [196]. However, more systematic studies, such as that done by Holt et al. [149] on the effect of doping levels for unmodified (i.e. H-terminated diamond) need to be made for the case of diamond electrodes modified with nanoparticles.

Renewable energy technologies are of high interest in the 21st century and new materials are sought to address challenges in environmental and energy technologies. The use of organo-metallic ligands to either aid in the deposition process or to support nanoparticles at the surface, has led to novel electrodes. Phthalocyanine and porphyrin rings stabilize the deposition of various metals, including Co and Ru, known to be effective in dye-sensitized solar cells, CO_2 reduction and fuel cells [114, 290–292]. The deposition of TiO_2 nanoparticles on BDE and DNDE is considered a promising approach for photocatalysis, water-splitting, and water treatment based on solar energy [108, 231, 293–295]. Nanodiamond has been incorporated into devices for photoelectrochemical applications with conducting polymers [222, 223] and graphitic carbon nitride [242]. The ability of diamond itself to act as a solid-state source of electrons when irradiated with UV light has recently been reported [296, 297], leading to interest in photoelectrochemical CO_2 reduction using metal nanoparticle-modified diamond electrodes [72]. Nanodiamond has been incorporated into devices for photoelectrochemical applications [222, 223, 242]. Electrochemical reduction of CO_2 has also been reported at Cu-modified BDE [76, 77]. Bimetallic nanoparticles deposition remains important for alcohol oxidation in fuel cells (Pt-Ru [74, 95, 98, 233, 235]) and nitrate reduction (Cu-Sn [298], Cu-Pd [298]). Nanoparticles which equal the effectiveness of Pt nanoparticles, with better stability, have yet to be discovered. Therefore, much recent effort has been expended in realizing the increased stability and reduction of fouling on Pt nanoparticles for electrocatalytic applications [32, 33, 98, 103, 123, 132, 160, 196, 232, 233, 235, 299].

Recently, there has been a resurgence of interest in wet-chemical routes towards depositing metal nanoparticles on carbon-based materials. In particular, the reduction of metal precursors to metal nanoparticles at a reducing agent attached to an underlying carbon-based surface has emerged as a promising deposition route. [233, 235, 300, 301]. This method was first attempted at diamond thin films by the Nebel group who seeded the substrate with Pt by reducing H_2PtCl_6 at diamond modified by $NaBH_4$ [32]. Surface coverage and size can be controlled by the number of repetitions of the above process. Further control of size was possible via electrochemical growth processes [32]. The advantage of this approach is the realisation of homogenous distribution of nanoparticles on diamond, which is difficult using non-physical deposition techniques. This approach is likely to have wider applications, as exemplified by the modification of nanodiamond with nano-sized (2–5 nm), and homogenously distributed, Pt and Pt-Ru particles [233]. Wet-chemical assisted deposition of nanoparticles could be easily developed further, considering the existing literature on wet-chemical routes for modifying diamond [183, 302, 303].

In the past ten years, many have recognized the potential for diamond-based devices to be integrated into biological systems. This might be imagined as a direct-connection (i.e. neurological stimulation [176, 304]). Neural stimulation with electrode arrays implanted directly on or near the retina has already been realised and restores partial vision for the patient [305]. Ultrananocrystalline diamond has been shown to reduce debris particle release from implants, improve neuron–electrode communication and lead to more stable neural implants [306]. Various microelectrode arrays fabricated from diamond are currently explored for retinal implants, such as nitrogen-doped nano- and ultrananocrystalline diamond [307–309], undoped diamond [310, 311] and boron-doped diamond grown on carbon nanotubes [312, 313]. Boron-doped diamond microelectrode arrays have also been used for in vitro measurements of neural signals in brain cortical slices [314], cortical cultures [315] and cardiomyocyte-like cells [316], and even real-time measurement of neurochemicals [317].

Another aspect of the use of diamond in biological systems includes the stimulation of bio-molecules (e.g. redox of proteins [318, 319]) and application in biosensing [25, 104, 118, 299, 320–322]. Zirconia has been used to modify diamond for DNA detection by the Foord group [323]. Cholesterol detection has been realised at Ag NP-modified BDE [71]. Several reports exist showing the effectiveness of nickel and nickel hydroxide nanoparticles for glucose oxidase sensing [25, 119, 324]. Magnetic nanoparticles formed of Fe_3O_4-Au deposited on BDE have been used to detect a cancer biomarker [104]. A Pt NP modified BDE was integrated into a lab-on-chip together with an enzymatic immunoassay for the detection and degradation of a pesticide [45]. An array of self-assembled Au nanoparticles has been able to engage in direct electrochemistry with a microbe, thus suggesting more exotic use of diamond within microbial fuel cells [325]. Many reviews exist on this topic, and a selection is provided here [2, 17, 21, 23, 148, 326].

The principle of introducing nano-scale materials to diamond is also applied when proteins [318, 319, 327, 328], enzymes [320, 329] or aptamers [330–332] are attached to the surface. This aids the detection of redox events and molecules at the diamond surface. For example, the attachment of cytochrome c on BDE enables cyanide and arsenic to be detected [327]; haemoglobin attached onto BDE improves electrochemical detection of hydrogen peroxide at BDE [70]; and zanamivir, a neuraminidase inhibitor, deposited onto Au-modified BDE allows the indirect detection of neuraminidase [85]. The applicability of bio-modified diamond-based electrodes to clinical HIV detection was recently shown by Rahim Ruslinda et al. [330]. Diamond is poised to be a material of choice for the integration of solid-state and biological phenomena.

Nanoparticles of metals, metal oxides, or diamond can be easily damaged by physical means. Moreover, their preparation time can be lengthy. Various attempts are being made to address one or both of these issues by the following approaches: nano-structuring diamond [333–336], synthesizing new diamond-based nanoparticles (e.g. nanowires [337–339], nanograss [253]), grafting diamond nanoparticles [243, 340], micro-contact printing of diamond nanoparticles to surfaces [341], and the self-assembly of nanoparticles [325, 342–344]. The stabilization of nanoparticles with electropolymers continues to be a fruitful area of research, particularly for bio-applications [118, 213, 214, 216, 217].

This chapter reviewed the methods used to construct boron-doped diamond electrodes. The methods and principles discussed here are applicable to the realisation of practical diamond-based electrodes which exploit the nano-scale, either via the deposition of nanoparticles or by having nanoparticle form. By understanding the chemical and physical consequences of any given methodology, one should be able better optimise diamond electrodes to address the biological, environmental, and industrial challenges of the 21st century.

References

1. R.G. Compton, J.S. Foord, F. Marken, Electroanalysis at diamond-like and doped-diamond electrodes. Electroanalysis **15**(17), 1349–1363 (2003). https://doi.org/10.1002/elan.200302830
2. A.M. Schrand, S.A.C. Hens, O.A. Shenderova, Nanodiamond particles: properties and perspectives for bioapplications. Crit. Rev. Solid State Mater. Sci. **34**(1–2), 18–74 (2009). https://doi.org/10.1080/10408430902831987
3. S. Fierro, Y. Einaga, Advances in electrochemical biosensing using boron doped diamond microelectrode, in *Novel Aspects of Diamond: From Growth to Applications*, vol. 121, ed. by N. Yang (Springer, Berlin, 2015), pp. 295–318
4. X.F. Chen, W.J. Zhang, Diamond nanostructures for drug delivery, bioimaging, and biosensing. Chem. Soc. Rev. **46**(3), 734–760 (2017). https://doi.org/10.1039/c6cs00109b
5. T.A. Ivandini, R. Sato, Y. Makide, A. Fujishima, Y. Einaga, Electroanalytical application of modified diamond electrodes. Diam. Relat. Mater. **13**(11), 2003–2008 (2004). https://doi.org/10.1016/j.diamond.2004.07.004
6. N. Yang, J.S. Foord, X. Jiang, Diamond electrochemistry at the nanoscale: a review. *Carbon* **99**(Supplement C), 90–110 (2016). DOI:https://doi.org/10.1016/j.carbon.2015.11.061
7. J. Radjenovic, D.L. Sedlak, Challenges and opportunities for electrochemical processes as next-generation technologies for the treatment of contaminated water. Environ. Sci. Technol. **49**(19), 11292–11302 (2015). https://doi.org/10.1021/acs.est.5b02414
8. H. Sarkka, A. Bhatnagar, M. Sillanpaa, Recent developments of electro-oxidation in water treatment—A review. J. Electroanal. Chem. **754**, 46–56 (2015). https://doi.org/10.1016/j.jelechem.2015.06.016
9. P.R.F. da Costa, E.V. dos Santos, J.M. Peralta-Hernandez, G.R. Salazar-Banda, D.R. da Silva, C.A. Martinez-Huitle, modified diamond electrodes for electrochemical systems for energy conversion and storage, in *Novel Aspects of Diamond: From Growth to Applications*, vol. 121, ed. by N. Yang (Springer, Berlin, 2015), pp. 205–235
10. K.B. Holt, Diamond at the nanoscale: applications of diamond nanoparticles from cellular biomarkers to quantum computing. Philos. Trans. R. Soc. Math. Phys. Eng. Sci. **365**(1861), 2845–2861 (2007). https://doi.org/10.1098/rsta.2007.0005
11. J. Wolters, G. Kewes, A.W. Schell, N. Nüsse, M. Schoengen, B. Löchel, T. Hanke, R. Bratschitsch, A. Leitenstorfer, T. Aichele, O. Benson, Coupling of single nitrogen-vacancy defect centers in diamond nanocrystals to optical antennas and photonic crystal cavities. Physica Status Solidi (b) **249**(5), 918–24 (2012). https://doi.org/10.1002/pssb.201100156
12. R.B. Liu, W. Yao, L.J. Sham, Quantum computing by optical control of electron spins. Adv. Phys. **59**(5), 703–802 (2010). https://doi.org/10.1080/00018732.2010.505452
13. X. Rong, D.W. Lu, X. Kong, J.P. Geng, Y. Wang, F.Z. Shi, C.K. Duan, J.F. Du, Harnessing the power of quantum systems based on spin magnetic resonance: from ensembles to single spins. Adv. Phys. **2**(1), 125–68 (2017). https://doi.org/10.1080/23746149.2016.1266914
14. V.Y. Dolmatov, Detonation synthesis ultradispersed diamonds: properties and applications. Russ. Chem. Rev. **70**(7), 607 (2001). https://doi.org/10.1070/RC2001v070n07ABEH000665
15. K.I.B. Eguiluz, J.M. Peralta-Hernández, A. Hernández-Ramírez, J.L. Guzmán-Mar, L. Hinojosa-Reyes, C.A. Martínez-Huitle, G.R. Salazar-Banda, The use of diamond for energy conversion system applications: a review. Int. J. Electrochem. **2012**, 20 (2012). https://doi.org/10.1155/2012/675124
16. A. Kraft, Conductive diamond layers. Production, properties, and possible uses of new electrode materials. Jahrb. Oberflaechentech. **61**, 109–20 (2005)
17. H. Yuen Yung, C. Chia-Liang, C. Huan-Cheng, Nanodiamonds for optical bioimaging. J. Phys. D Appl. Phys. **43**(37), 374021 (2010). https://doi.org/10.1088/0022-3727/43/37/374021

18. J. Zang, L. Dong, Y.-H. Wang, Review on electrochemical property and surface modifications of nanodiamond powders. Yanshan da xue xue bao **2**, 002 (2012)
19. N. Fujimori, T. Imai, A. Doi, Characterization of conducting diamond films. Vacuum **36**(1), 99–102 (1986). https://doi.org/10.1016/0042-207X(86)90279-4
20. K.E. Toghill, R.G. Compton, Metal nanoparticle modified boron doped diamond electrodes for use in electroanalysis. Electroanalysis **22**(17–18), 1947–1956 (2010). https://doi.org/10.1002/elan.201000072
21. A. Kraft, Doped diamond electrodes. New trends and developments. Jahrb. Oberflaechentech. **63**, 85–95 (2007)
22. I. Novoselova, E. Fedorishena, E. Panov, Electrodes from diamond and diamond-like materials for electrochemical applications. J. Superhard Mater. **29**(1), 24–39 (2007). https://doi.org/10.3103/S1063457607010042
23. Y. Zhou, J. Zhi, The application of boron-doped diamond electrodes in amperometric biosensors. Talanta **79**(5), 1189–1196 (2009). https://doi.org/10.1016/j.talanta.2009.05.026
24. D. Yamada, T.A. Ivandini, M. Komatsu, A. Fujishima, Y. Einaga, Anodic stripping voltammetry of inorganic species of as 3 + and as 5 + at gold-modified boron doped diamond electrodes. J. Electroanal. Chem. **615**(2), 145–153 (2008). https://doi.org/10.1016/j.jelechem.2007.12.004
25. L.A. Hutton, M. Vidotti, A.N. Patel, M.E. Newton, P.R. Unwin, J.V. Macpherson, Electrodeposition of nickel hydroxide nanoparticles on boron-doped diamond electrodes for oxidative electrocatalysis. J. Phys. Chem. **115**(5), 1649–1658 (2010). https://doi.org/10.1021/jp109526b
26. G.R. Salazar-Banda, K.I. Eguiluz, L.A. Avaca, Boron-doped diamond powder as catalyst support for fuel cell applications. Electrochem. Commun. **9**(1), 59–64 (2007). https://doi.org/10.1016/j.elecom.2006.08.038
27. L. Bian, Y. Wang, J. Zang, F. Meng, Y. Zhao, Detonation-synthesized nanodiamond as a stable support of Pt electrocatalyst for methanol electrooxidation. Int. J. Electrochem. Sci. **7**(8), 7295–303 (2012)
28. S. Szunerits, R. Boukherroub, Investigation of the electrocatalytic activity of boron-doped diamond electrodes modified with palladium or gold nanoparticles for oxygen reduction reaction in basic medium. C. R. Chim. **11**(9), 1004–1009 (2008). https://doi.org/10.1016/j.crci.2008.01.015
29. S.R. Belding, F.W. Campbell, E.J. Dickinson, R.G. Compton, Nanoparticle-modified electrodes. Phys. Chem. Chem. Phys. **12**(37), 11208–11221 (2010). https://doi.org/10.1039/C0CP00233J
30. K.B. Holt, C. Ziegler, D.J. Caruana, J. Zang, E.J. Millán-Barrios, J. Hu, J.S. Foord, Redox properties of undoped 5 nm diamond nanoparticles. Phys. Chem. Chem. Phys. **10**(2), 303–310 (2008). https://doi.org/10.1039/B711049A
31. J.-S. Gao, T. Arunagiri, J.-J. Chen, P. Goodwill, O. Chyan, J. Perez, D. Golden, Preparation and characterization of metal nanoparticles on a diamond surface. Chem. Mater. **12**(11), 3495–3500 (2000). https://doi.org/10.1021/cm000465o
32. F. Gao, N. Yang, W. Smirnov, H. Obloh, C.E. Nebel, Size-controllable and homogeneous platinum nanoparticles on diamond using wet chemically assisted electrodeposition. Electrochim. Acta **90**, 445–451 (2013). https://doi.org/10.1016/j.electacta.2012.12.050
33. J. Hu, X. Lu, J.S. Foord, Q. Wang, Electrochemical deposition of Pt nanoparticles on diamond substrates. Physica Status Solidi (a) **206**(9), 2057–62 (2009). https://doi.org/10.1002/pssa.200982226
34. F. Montilla, E. Morallon, I. Duo, C. Comninellis, J. Vazquez, Platinum particles deposited on synthetic boron-doped diamond surfaces. Appl. Methanol Oxid. Electrochim. Acta **48**(25), 3891–3897 (2003). https://doi.org/10.1016/S0013-4686(03)00526-7
35. G. Sine, I. Duo, B.E. Roustom, G. Foti, C. Comninellis, Deposition of clusters and nanoparticles onto boron-doped diamond electrodes for electrocatalysis. J. Appl. Electrochem. **36**(8), 847–862 (2006). https://doi.org/10.1007/s10800-006-9159-2

36. O. Enea, B. Riedo, G. Dietler, AFM study of Pt clusters electrochemically deposited onto boron-doped diamond films. Nano Lett. **2**(3), 241–244 (2002). https://doi.org/10.1021/nl015666l
37. I. Gonzalez-Gonzalez, D. Tryk, C.R. Cabrera, Polycrystalline boron-doped diamond films as supports for methanol oxidation electrocatalysts. Diam. Relat. Mater. **15**(2), 275–278 (2006). https://doi.org/10.1016/j.diamond.2005.08.037
38. S. Hrapovic, Y. Liu, J.H. Luong, Reusable platinum nanoparticle modified boron doped diamond microelectrodes for oxidative determination of arsenite. Anal. Chem. **79**(2), 500–507 (2007). https://doi.org/10.1021/ac061528a
39. B. Rismetov, T.A. Ivandini, E. Saepudin, Y. Einaga, Electrochemical detection of hydrogen peroxide at platinum-modified diamond electrodes for an application in melamine strip tests. Diam. Relat. Mater. **48**(Supplement C), 88–95 (2014). https://doi.org/10.1016/j.diamond.2014.07.003
40. Y. Hernández-Lebrón, C.R. Cabrera, Square wave voltammetry restructuring of platinum nanoparticle at boron doped diamond electrode for enhanced ammonia oxidation. J. Electroanal. Chem. **793**(Supplement C), 174–83 (2017). DOI:https://doi.org/10.1016/j.jelechem.2016.12.036
41. Y. Hernández-Lebrón, L. Cunci, C.R. Cabrera, Ammonia oxidation at electrochemically platinum-modified microcrystalline and polycrystalline boron-doped diamond electrodes. Electrocatalysis **7**(2), 184–192 (2016). https://doi.org/10.1007/s12678-015-0295-5
42. L. Hutton, M.E. Newton, P.R. Unwin, J.V. Macpherson, Amperometric oxygen sensor based on a platinum nanoparticle-modified polycrystalline boron doped diamond disk electrode. Anal. Chem. **81**(3), 1023–1032 (2008). https://doi.org/10.1021/ac8020906
43. L.C.S. Figueiredo-Filho, E.R. Sartori, O. Fatibello-Filho, Electroanalytical determination of the linuron herbicide using a cathodically pretreated boron-doped diamond electrode: comparison with a boron-doped diamond electrode modified with platinum nanoparticles. Anal. Methods **7**(2), 643–649 (2015). https://doi.org/10.1039/C4AY02182G
44. A.I. Căciuleanu, T. Spătaru, L. Preda, M. Anastasescu, P. Osiceanu, C. Munteanu, R.D. Bărăţoiu, A.A. Iovescu, N. Spătaru, Platinum–carbon electrocatalytic composites via liposome-directed electrodeposition at conductive diamond. Int. J. Hydrog. Energy **41**(47), 22529–22537 (2016). https://doi.org/10.1016/j.ijhydene.2016.05.226
45. M. Medina-Sánchez, C.C. Mayorga-Martinez, T. Watanabe, T.A. Ivandini, Y. Honda, F. Pino, K. Nakata, A. Fujishima, Y. Einaga, A. Merkoçi, Microfluidic platform for environmental contaminants sensing and degradation based on boron-doped diamond electrodes. Biosens. Bioelectron. **75**, 365–374 (2016). https://doi.org/10.1016/j.bios.2015.08.058
46. G. Siné, D. Smida, M. Limat, G. Foti, C. Comninellis, Microemulsion synthesized pt/ru/sn nanoparticles on bdd for alcohol electro-oxidation. J. Electrochem. Soc. **154**(2), B170–B174 (2007). https://doi.org/10.1149/1.2400602
47. T.A. Ivandini, R. Sato, Y. Makide, A. Fujishima, Y. Einaga, Pt-implanted boron-doped diamond electrodes and the application for electrochemical detection of hydrogen peroxide. Diam. Relat. Mater. **14**(11), 2133–2138 (2005). https://doi.org/10.1016/j.diamond.2005.08.022
48. T.A. Ivandini, R. Sato, Y. Makide, A. Fujishima, Y. Einaga, Electrochemical detection of arsenic (III) using iridium-implanted boron-doped diamond electrodes. Anal. Chem. **78**(18), 6291–6298 (2006). https://doi.org/10.1021/ac0519514
49. K. Panda, K.J. Sankaran, E. Inami, Y. Sugimoto, N.H. Tai, I.-N. Lin, Direct observation and mechanism for enhanced field emission sites in platinum ion implanted/post-annealed ultrananocrystalline diamond films. Appl. Phys. Lett. **105**(16), 163109 (2014). https://doi.org/10.1063/1.4898571
50. K.J. Sankaran, P. Kalpataru, S. Balakrishnan, N.-H. Tai, I.N. Lin, Catalytically induced nanographitic phase by a platinum-ion implantation/annealing process to improve the field electron emission properties of ultrananocrystalline diamond films. J. Mater. Chem. C **3**(11), 2632–2641 (2015). https://doi.org/10.1039/C4TC02334J

51. K. Panda, E. Inami, Y. Sugimoto, K.J. Sankaran, I.N. Lin, Straight imaging and mechanism behind grain boundary electron emission in Pt-doped ultrananocrystalline diamond films. Carbon 111(Supplement C), 8–17 (2017). https://doi.org/10.1016/j.carbon.2016.09.062

52. D.K. Belghiti, M. Zadeh-Habchi, E. Scorsone, P. Bergonzo, Boron doped diamond/metal nanoparticle catalysts hybrid electrode array for the detection of pesticides in tap water, in *Proceedings of the 30th Anniversary Eurosensors Conference—Eurosensors*, vol. 168, ed. by I. Barsony, Z. Zolnai, G. Battistig (Elsevier Science Bv, Amsterdam, 2016), pp. 428–31

53. X. Lyu, J.P. Hu, J.S. Foord, C.S. Lou, W.Q. Zhang, Synthesis and electrocatalytic performance of BDD-Supported platinum nanoparticles. J. Mater. Eng. Perform. 24(2), 1031–1037 (2015). https://doi.org/10.1007/s11665-014-1317-9

54. J. Wang, G.M. Swain, Fabrication and evaluation of platinum/diamond composite electrodes for electrocatalysis preliminary studies of the oxygen-reduction reaction. J. Electrochem. Soc. 150(1), E24–E32 (2003). https://doi.org/10.1149/1.1524612

55. Bennett, J. A.; Show, Y.; Wang, S.; Swain, G. M., Pulsed galvanostatic deposition of Pt particles on microcrystalline and nanocrystalline diamond thin-film electrodes I. Characterization of as-deposited metal/diamond surfaces. J. Electrochem. Soc. 152(5), E184–E92 (2005). https://doi.org/10.1149/1.1890745

56. G. Salazar-Banda, H. Suffredini, L. Avaca, Improved stability of PtOx sol-gel-modified diamond electrodes covered with a Nafion® film. J. Braz. Chem. Soc. 16(5), 903–906 (2005). https://doi.org/10.1590/S0103-50532005000600003

57. H.B. Suffredini, G.R. Salazar-Banda, S.T. Tanimoto, M.L. Calegaro, S.A. Machado, L.A. Avaca, AFM studies and electrochemical characterization of boron-doped diamond surfaces modified with metal oxides by the Sol-Gel method. J. Braz. Chem. Soc. 17(2), 257–264 (2006). https://doi.org/10.1590/S0103-50532006000200007

58. G.R. Salazar-Banda, H.B. Suffredini, L.A. Avaca, S.A.S. Machado, Methanol and ethanol electro-oxidation on Pt-SnO₂ and Pt-Ta2O5 sol-gel-modified boron-doped diamond surfaces. Mater. Chem. Phys. 117(2–3), 434–442 (2009). https://doi.org/10.1016/j.matchemphys.2009.06.027

59. F. Gao, R. Thomann, C.E. Nebel, Aligned Pt-diamond core-shell nanowires for electrochemical catalysis. Electrochem. Commun. 50, 32–35 (2015). https://doi.org/10.1016/j.elecom.2014.11.006

60. J. Kim, Y.S. Chun, S.K. Lee, D.S. Lim, Improved electrode durability using a boron-doped diamond catalyst support for proton exchange membrane fuel cells. RSC Adv. 5(2), 1103–1108 (2015). https://doi.org/10.1039/c4ra13389g

61. K.E. Toghill, L. Xiao, G.G. Wildgoose, R.G. Compton, Electroanalytical determination of cadmium (II) and lead (II) using an antimony nanoparticle modified boron-doped diamond electrode. Electroanalysis 21(10), 1113–1118 (2009). https://doi.org/10.1002/elan.200904547

62. C.W. Foster, A.P. de Souza, J.P. Metters, M. Bertotti, C.E. Banks, Metallic modified (bismuth, antimony, tin and combinations thereof) film carbon electrodes. Analyst 140(22), 7598–7612 (2015). https://doi.org/10.1039/C5AN01692D

63. K.E. Toghill, G.G. Wildgoose, A. Moshar, C. Mulcahy, R.G. Compton, The fabrication and characterization of a bismuth nanoparticle modified boron doped diamond electrode and its application to the simultaneous determination of cadmium (II) and lead (II). Electroanalysis 20(16), 1731–1737 (2008). https://doi.org/10.1002/elan.200804277

64. A.O. Simm, X. Ji, C.E. Banks, M.E. Hyde, R.G. Compton, AFM studies of metal deposition: instantaneous nucleation and the growth of cobalt nanoparticles on boron-doped diamond electrodes. ChemPhysChem 7(3), 704–709 (2006). https://doi.org/10.1002/cphc.200500557

65. T.-L. Wee, B.D. Sherman, D. Gust, A.L. Moore, T.A. Moore, Y. Liu, J.C. Scaiano, Photochemical synthesis of a water oxidation catalyst based on cobalt nanostructures. J. Am. Chem. Soc. 133(42), 16742–16745 (2011). https://doi.org/10.1021/ja206280g

66. N.R. Stradiotto, K.E. Toghill, L. Xiao, A. Moshar, R.G. Compton, The fabrication and characterization of a nickel nanoparticle modified boron doped diamond electrode for electrocatalysis of primary alcohol oxidation. Electroanalysis **21**(24), 2627–2633 (2009). https://doi.org/10.1002/elan.200900325

67. S. Treetepvijit, A. Preechaworapun, N. Praphairaksit, S. Chuanuwatanakul, Y. Einaga, O. Chailapakul, Use of nickel implanted boron-doped diamond thin film electrode coupled to HPLC system for the determination of tetracyclines. Talanta **68**(4), 1329–1335 (2006). https://doi.org/10.1016/j.talanta.2005.07.047

68. V. Sáez, J. González-García, F. Marken, Active catalysts of sonoelectrochemically prepared iron metal nanoparticles for the electroreduction of chloroacetates. Phys. Procedia **3**(1), 105–109 (2010). https://doi.org/10.1016/j.phpro.2010.01.015

69. L.A. Hutton, M.E. Newton, P.R. Unwin, J.V. Macpherson, Factors controlling stripping voltammetry of lead at polycrystalline boron doped diamond electrodes: new insights from high-resolution microscopy. Anal. Chem. **83**(3), 735–745 (2011). https://doi.org/10.1021/ac101626s

70. L.Y. Jiang, J.P. Hu, J.S. Foord, Electroanalysis of hydrogen peroxide at boron doped diamond electrode modified by silver nanoparticles and haemoglobin. Electrochim. Acta **176**, 488–496 (2015). https://doi.org/10.1016/j.electacta.2015.07.013

71. S. Nantaphol, O. Chailapakul, W. Siangproh, A novel paper-based device coupled with a silver nanoparticle-modified boron-doped diamond electrode for cholesterol detection. Anal. Chim. Acta **891**, 136–143 (2015). https://doi.org/10.1016/j.aca.2015.08.007

72. N. Roy, Y. Hirano, H. Kuriyama, P. Sudhagar, N. Suzuki, K.I. Katsumata, K. Nakata, T. Kondo, M. Yuasa, I. Serizawa, T. Takayama, A. Kudo, A. Fujishima, C. Terashima, Boron-doped diamond semiconductor electrodes: efficient photoelectrochemical CO_2 reduction through surface modification. Sci. Rep. **6**, 9 (2016). https://doi.org/10.1038/srep38010

73. C.M. Welch, C.E. Banks, G. Richard, The detection of nitrate using in-situ copper nanoparticle deposition at a boron doped diamond electrode. Anal. Sci. **21**(12), 1421–30 (2005). https://doi.org/10.2116/analsci.21.1421

74. C.K. Mavrokefalos, G.W. Nelson, C.G. Poll, R.G. Compton, J.S. Foord, Electrochemical aspects of Pt-Cu and Cu modified boron-doped diamond. Physica Status Solidi A-Appl. Mat. **212**(11), 2559–2567 (2015). https://doi.org/10.1002/pssa.201532163

75. K.R. Saravanan, M. Chandrasekaran, V. Suryanarayanan, Efficient electrocarboxylation of benzophenone on silver nanoparticles deposited boron doped diamond electrode. J. Electroanal. Chem. **757**, 18–22 (2015). https://doi.org/10.1016/j.jelechem.2015.08.033

76. H.S. Panglipur, T.A. Ivandini, R. Wibowo, Y. Einaga, Electroreduction of CO_2 using copper-deposited on boron-doped diamond (BDD). AIP Conf. Proc. **1729**(1), 020047 (2016). https://doi.org/10.1063/1.4946950

77. N. Roy, Y. Shibano, C. Terashima, K. Katsumata, K. Nakata, T. Kondo, M. Yuasa, A. Fujishima, Ionic-liquid-assisted selective and controlled electrochemical CO_2 reduction at Cu-modified boron-doped diamond electrode. ChemElectroChem **3**(7), 1044–1047 (2016). https://doi.org/10.1002/celc.201600105

78. C.M. Welch, A.O. Simm, R.G. Compton, Oxidation of electrodeposited copper on boron doped diamond in acidic solution: manipulating the size of copper nanoparticles using voltammetry. Electroanalysis **18**(10), 965–970 (2006). https://doi.org/10.1002/elan.200603493

79. B. El Roustom, G. Fóti, C. Comninellis, Preparation of gold nanoparticles by heat treatment of sputter deposited gold on boron-doped diamond film electrode. Electrochem. Commun. **7**(4), 398–405 (2005). https://doi.org/10.1016/j.elecom.2005.02.014

80. I. Yagi, T. Ishida, K. Uosaki, Electrocatalytic reduction of oxygen to water at Au nanoclusters vacuum-evaporated on boron-doped diamond in acidic solution. Electrochem. Commun. **6**(8), 773–779 (2004). https://doi.org/10.1016/j.elecom.2004.05.025

81. Y. Ma, J. Liu, H. Li, Diamond-based electrochemical aptasensor realizing a femtomolar detection limit of bisphenol A. Biosens. Bioelectron. **92**(Supplement C), 21–5 (2017). https://doi.org/10.1016/j.bios.2017.01.041

82. M. Li, G. Zhao, R. Geng, H. Hu, Facile electrocatalytic redox of hemoglobin by flower-like gold nanoparticles on boron-doped diamond surface. Bioelectrochemistry **74**(1), 217–221 (2008). https://doi.org/10.1016/j.bioelechem.2008.08.004

83. R.-H. Tian, T.N. Rao, Y. Einaga, J.-F. Zhi, Construction of two-dimensional arrays gold nanoparticles monolayer onto boron-doped diamond electrode surfaces. Chem. Mater. **18**(4), 939–945 (2006). https://doi.org/10.1021/cm0519481

84. T.A. Ivandini, Harmesa, E. Saepudin, Y. Einaga, Yeast-based biochemical oxygen demand sensors using Gold-modified boron-doped diamond electrodes. Anal. Sci. 31(7), 643–649 (2015). https://doi.org/10.2116/analsci.31.643

85. W.T. Wahyuni, T.A. Ivandini, E. Saepudin, Y. Einaga, Development of neuraminidase detection using gold nanoparticles Boron-Doped diamond electrodes. Anal. Biochem. **497**, 68–75 (2016). https://doi.org/10.1016/j.ab.2015.12.003

86. T.A. Ivandini, E. Saepudin, H. Wardah, Harmesa, N. Dewangga, Y. Einaga, Development of a biochemical oxygen demand sensor using Gold-Modified boron doped diamond electrodes. Anal. Chem. **84**(22), 9825–9832 (2012). https://doi.org/10.1021/ac302090y

87. Y. Zhang, V. Suryanarayanan, I. Nakazawa, S. Yoshihara, T. Shirakashi, Electrochemical behavior of Au nanoparticle deposited on as-grown and O-terminated diamond electrodes for oxygen reduction in alkaline solution. Electrochim. Acta **49**(28), 5235–5240 (2004). https://doi.org/10.1016/j.electacta.2004.07.005

88. L. Rassaei, M. Sillanpää, R.W. French, R.G. Compton, F. Marken, Arsenite determination in phosphate media at electroaggregated gold nanoparticle deposits. Electroanalysis **20**(12), 1286–1292 (2008). https://doi.org/10.1002/elan.200804226

89. J. Svanberg-Larsson, G.W. Nelson, S.E. Steinvall, B.F. Leo, E. Brooke, D.J. Payne, J.S. Foord, A comparison of explicitly-terminated diamond electrodes decorated with gold nanoparticles. Electroanalysis **28**(1), 88–95 (2016). https://doi.org/10.1002/elan.201500442

90. K.B. Holt, G. Sabin, R.G. Compton, J.S. Foord, F. Marken, Reduction of tetrachloroaurate (III) at Boron-Doped diamond electrodes: gold deposition versus gold colloid formation. Electroanalysis **14**(12), 797–803 (2002). https://doi.org/10.1002/1521-4109(200206)14:12%3c797:AID-ELAN797%3e3.0.CO;2-M

91. Á.I. López-Lorente, J. Izquierdo, C. Kranz, B. Mizaikoff, Boron-doped diamond modified with gold nanoparticles for the characterization of bovine serum albumin protein. Vib. Spectrosc. **91**(Supplement C), 147–56 (2017). https://doi.org/10.1016/j.vibspec.2016.10.010

92. S. Chai, Y. Wang, Y.-N. Zhang, M. Liu, Y. Wang, G. Zhao, Selective electrocatalytic degradation of odorous mercaptans derived from S-Au bond recognition on a dendritic gold/ boron-doped diamond composite electrode. Environ. Sci. Technol. **51**(14), 8067–8076 (2017). https://doi.org/10.1021/acs.est.7b00393

93. C. Batchelor-McAuley, C.E. Banks, A.O. Simm, T.G. Jones, R.G. Compton, The electroanalytical detection of hydrazine: a comparison of the use of palladium nanoparticles supported on boron-doped diamond and palladium plated BDD microdisc array. Analyst **131**(1), 106–110 (2006). https://doi.org/10.1039/B513751A

94. C. Batchelor-McAuley, C.E. Banks, A.O. Simm, T.G. Jones, R.G. Compton, Nano-Electrochemical detection of hydrogen or protons using palladium nanoparticles: distinguishing surface and bulk hydrogen. ChemPhysChem 7(5), 1081–1085 (2006). https://doi.org/10.1002/cphc.200500571

95. C.K. Mavrokefalos, M. Hasan, W. Khunsin, M. Schmidt, S.A. Maier, J.F. Rohan, R.G. Compton, J.S. Foord, Electrochemically modified boron-doped diamond electrode with Pd and Pd-Sn nanoparticles for ethanol electrooxidation. Electrochimica Acta **243**(Supplement C), 310–319 (2017). https://doi.org/10.1016/j.electacta.2017.05.039

96. G. Siné, G. Foti, C. Comninellis, Boron-doped diamond (BDD)-supported Pt/Sn nanoparticles synthesized in microemulsion systems as electrocatalysts of ethanol oxidation. J. Electroanal. Chem. **595**(2), 115–124 (2006). https://doi.org/10.1016/j.jelechem.2006.07.012

97. G. Siné, C. Comninellis, Nafion®-assisted deposition of microemulsion-synthesized platinum nanoparticles on BDD: activation by electrogenerated OH radicals. Electrochim. Acta **50**(11), 2249–2254 (2005). https://doi.org/10.1016/j.electacta.2004.10.008

98. X. Lu, J. Hu, J.S. Foord, Q. Wang, Electrochemical deposition of Pt–Ru on diamond electrodes for the electrooxidation of methanol. J. Electroanal. Chem. **654**(1), 38–43 (2011). https://doi.org/10.1016/j.jelechem.2011.01.034

99. B. El Roustom, G. Sine, G. Foti, C. Comninellis, A novel method for the preparation of bi-metallic (Pt–Au) nanoparticles on boron doped diamond (BDD) substrate: application to the oxygen reduction reaction. J. Appl. Electrochem. **37**(11), 1227–1236 (2007). https://doi.org/10.1007/s10800-007-9359-4

100. S. Nantaphol, T. Watanabe, N. Nomura, W. Siangproh, O. Chailapakul, Y. Einaga, Bimetallic Pt-Au nanocatalysts electrochemically deposited on boron-doped diamond electrodes for nonenzymatic glucose detection. Biosens. Bioelectron. **98**, 76–82 (2017). https://doi.org/10.1016/j.bios.2017.06.034

101. S. Ferro, A. De Battisti, Electrocatalysis and chlorine evolution reaction at ruthenium dioxide deposited on conductive diamond. J. Phys. Chem. B **106**(9), 2249–2254 (2002). https://doi.org/10.1021/jp012195i

102. T. Spătaru, L. Preda, P. Osiceanu, C. Munteanu, M. Marcu, C. Lete, N. Spătaru, A. Fujishima, Electrochemical deposition of Pt-RuO (x) a < ...nH(2)O composites on conductive diamond and its application to methanol oxidation in acidic media. Electrocatalysis **7**(2), 140–148 (2016). https://doi.org/10.1007/s12678-015-0292-8

103. L. Chen, J. Hu, J.S. Foord, Electrodeposition of a Pt–PrO2 − x electrocatalyst on diamond electrodes for the oxidation of methanol. Physica Status Solidi (a) **209**(9), 1792–1796 (2012). https://doi.org/10.1002/pssa.201200049

104. M. Braiek, Y. Yang, C. Farre, C. Chaix, F. Bessueille, A. Baraket, A. Errachid, A.D. Zhang, N. Jaffrezic-Renault, Boron-doped diamond electrodes modified with Fe3O4@Au magnetic nanocomposites as sensitive platform for detection of a cancer biomarker, Interleukin-8. Electroanalysis **28**(8), 1810–1816 (2016). https://doi.org/10.1002/elan.201600060

105. C. Terashima, T.N. Rao, B.V. Sarada, N. Spataru, A. Fujishima, Electrodeposition of hydrous iridium oxide on conductive diamond electrodes for catalytic sensor applications. J. Electroanal. Chem. **544**, 65–74 (2003). https://doi.org/10.1016/S0022-0728(03)00066-4

106. F. Marken, A.S. Bhambra, D.-H. Kim, R.J. Mortimer, S.J. Stott, Electrochemical reactivity of TiO$_2$ nanoparticles adsorbed onto boron-doped diamond surfaces. Electrochem. Commun. **6**(11), 1153–1158 (2004). https://doi.org/10.1016/j.elecom.2004.09.006

107. T. Spătaru, L. Preda, C. Munteanu, A.I. Căciuleanu, N. Spătaru, A. Fujishima, Influence of boron-doped diamond surface termination on the characteristics of titanium dioxide anodically deposited in the presence of a surfactant. J. Electrochem. Soc. **162**(8), H535–H540 (2015). https://doi.org/10.1149/2.0741508jes

108. F. Espinola-Portilla, R. Navarro-Mendoza, S. Gutiérrez-Granados, U. Morales-Muñoz, E. Brillas-Coso, J.M. Peralta-Hernández, A simple process for the deposition of TiO$_2$ onto BDD by electrophoresis and its application to the photoelectrocatalysis of Acid Blue 80 dye. J. Electroanal. Chem. **802**(Supplement C), 57–63. (2017). https://doi.org/10.1016/j.jelechem.2017.08.041

109. K.J. McKenzie, F. Marken, Electrochemical characterization of hydrous ruthenium oxide nanoparticle decorated boron-doped diamond electrodes. Electrochem. Solid-State Lett. **5**(9), E47–E50 (2002). https://doi.org/10.1149/1.1497515

110. G.C. Sedenho, J.L. da Silva, M.A. Beluomini, A.C. de Sá, N.R. Stradiotto, Determination of electroactive organic acids in sugarcane vinasse by high performance anion-exchange chromatography with pulsed amperometric detection using a nickel nanoparticle modified boron-doped diamond. Energy Fuels **31**(3), 2865–2870 (2017). https://doi.org/10.1021/acs. energyfuels.6b02783

111. G.C. Sedenho, P.T. Lee, H.S. Toh, C. Salter, C. Johnston, N.R. Stradiotto, R.G. Compton, Nanoelectrocatalytic oxidation of lactic acid using nickel nanoparticles. J. Phys. Chem. C **119**(12), 6896–6905 (2015). https://doi.org/10.1021/acs.jpcc.5b00335

112. A.J. Saterlay, S.J. Wilkins, K.B. Holt, J.S. Foord, R.G. Compton, F. Marken, Lead dioxide deposition and electrocatalysis at highly boron-doped diamond electrodes in the presence of ultrasound. J. Electrochem. Soc. **148**(2), E66–E72 (2001). https://doi.org/10.1149/1. 1339874

113. C.K. Mavrokefalos, M. Hasan, J.F. Rohan, R.G. Compton, J.S. Foord, Electrochemically deposited Cu$_2$O cubic particles on boron doped diamond substrate as efficient photocathode for solar hydrogen generation. Appl. Surf. Sci. **408**, 125–134 (2017). https://doi.org/10. 1016/j.apsusc.2017.02.148

114. P. Gan, J.S. Foord, R.G. Compton, Surface modification of boron-doped diamond with microcrystalline copper phthalocyanine: oxygen reduction catalysis. ChemistryOpen **4**(5), 606–612 (2015). https://doi.org/10.1002/open.201500075

115. F. Shang, J.D. Glennon, J.H. Luong, Glucose oxidase entrapment in an electropolymerized poly (tyramine) film with sulfobutylether-β-cyclodextrin on platinum nanoparticle modified boron-doped diamond electrode. J. Phys. Chem. C **112**(51), 20258–20263 (2008). https:// doi.org/10.1021/jp807482a

116. M.-J. Song, J.H. Kim, S.K. Lee, J.-H. Lee, D.S. Lim, S.W. Hwang, D. Whang, Pt-polyaniline nanocomposite on boron-doped diamond electrode for amperometic biosensor with low detection limit. Microchim. Acta **171**(3–4), 249–255 (2010). https://doi.org/10. 1007/s00604-010-0432-z

117. H.F. Cui, Y.F. Bai, W.W. Wu, X.Y. He, J.H.T. Luong, Modification with mesoporous platinum and poly(pyrrole-3-carboxylic acid)-based copolymer on boron-doped diamond for nonenzymatic sensing of hydrogen peroxide. J. Electroanal. Chem. **766**, 52–59 (2016). https://doi.org/10.1016/j.jelechem.2016.01.026

118. M.-J. Song, S.-K. Lee, D.-S. Lim, Dopamine sensor based on a boron-doped diamond electrode modified with a polyaniline/Au nanocomposites in the presence of ascorbic acid. Anal. Sci. **28**(6), 583–587 (2012). https://doi.org/10.2116/analsci.28.583

119. Z. Deng, H. Long, Q. Wei, Z. Yu, B. Zhou, Y. Wang, L. Zhang, S. Li, L. Ma, Y. Xie, J. Min, High-performance non-enzymatic glucose sensor based on nickel-microcrystalline graphite-boron doped diamond complex electrode. Sens. Actuators B: Chem. **242** (Supplement C), 825–34 (2017). https://doi.org/10.1016/j.snb.2016.09.176

120. I. Duo, S. Ferro, A. De Battisti, C. Comninellis, Conductive metal-oxide nanoparticles on synthetic boron-doped diamond surfaces, in *Catalysis and Electrocatalysis at Nanoparticle Surfaces*, ed. by A. Wieckowski, E.R. Savinova, C.G. Vayenas (Marcel Dekker Inc, NY, 2003), pp. 877–906

121. K.-W. Park, J.-H. Choi, B.-K. Kwon, S.-A. Lee, Y.-E. Sung, H.-Y. Ha, S.-A. Hong, H. Kim, A. Wieckowski, Chemical and electronic effects of Ni in Pt/Ni and Pt/Ru/Ni alloy nanoparticles in methanol electrooxidation. J. Phys. Chem. B **106**(8), 1869–1877 (2002). https://doi.org/10.1021/jp013168v

122. M.A. Watanabe, S. Motoo, Electrocatalysis by ad-atoms: Part II. Enhancement of the oxidation of methanol on platinum by ruthenium ad-atoms. J. Electroanal. Chem. Interfacial Electrochem. **60**(3), 267–73 (1975). https://doi.org/10.1016/S0022-0728(75)80261-0

123. I. González-González, E.R. Fachini, M.A. Scibioh, D.A. Tryk, M. Tague, H.C.D. Abruña, C.R. Cabrera, Facet-selective platinum electrodeposition at free-standing polycrystalline boron-doped diamond films. Langmuir **25**(17), 10329–10336 (2009). https://doi.org/10. 1021/la8035055

124. T. Kondo, S. Aoshima, K. Hirata, K. Honda, Y. Einaga, A. Fujishima, T. Kawai, Crystal-Face-Selective Adsorption of Au Nanoparticles onto Polycrystalline diamond surfaces. Langmuir **24**(14), 7545–7548 (2008). https://doi.org/10.1021/la800782r

125. M. Wei, L.G. Sun, Z.Y. Xie, J.F. Zhii, A. Fujishima, Y. Einaga, D.G. Fu, X.M. Wang, Z.Z. Gu, Selective determination of dopamine on a boron-doped diamond electrode modified with gold nanoparticle/polyelectrolyte-coated polystyrene colloids. Adv. Func. Mater. **18**(9), 1414–1421 (2008). https://doi.org/10.1002/adfm.200701099

126. M. Wei, Z. Xie, L. Sun, Z.Z. Gu, Electrochemical properties of a boron-doped diamond electrode modified with gold/polyelectrolyte hollow spheres. Electroanalysis **21**(2), 138–143 (2009). https://doi.org/10.1002/elan.200804411

127. M. Osawa, K.-I. Ataka, K. Yoshii, Y. Nishikawa, Surface-Enhanced infrared spectroscopy: the origin of the absorption enhancement and band selection rule in the infrared spectra of molecules adsorbed on fine metal particles. Appl. Spectrosc. **47**(9), 1497–1502 (1993). https://doi.org/10.1366/0003702934067478

128. J.K. Zak, J.E. Butler, G.M. Swain, Diamond optically transparent electrodes: demonstration of concept with ferri/ferrocyanide and methyl viologen. Anal. Chem. **73**(5), 908–914 (2001). https://doi.org/10.1021/ac001257i

129. H.B. Martin, P.W. Morrison, Application of a diamond thin film as a transparent electrode for in situ infrared spectroelectrochemistry. Electrochem. Solid-State Lett. **4**(4), E17–E20 (2001). https://doi.org/10.1149/1.1353162

130. D. Neubauer, J. Scharpf, A. Pasquarelli, B. Mizaikoff, C. Kranz, Combined in situ atomic force microscopy and infrared attenuated total reflection spectroelectrochemistry. Analyst **138**(22), 6746–6752 (2013). https://doi.org/10.1039/C3AN01169K

131. J. Izquierdo, B. Mizaikoff, C. Kranz, Surface-enhanced infrared spectroscopy on boron-doped diamond modified with gold nanoparticles for spectroelectrochemical analysis. Physica Status Solidi (a) 213(8), 2056–2062 (2016). https://doi.org/10.1002/pssa.201600222

132. J. Hu, X. Lu, J. Foord, Nanodiamond pretreatment for the modification of diamond electrodes by platinum nanoparticles. Electrochem. Commun. **12**(5), 676–679 (2010). https://doi.org/10.1016/j.elecom.2010.03.004

133. Wang, J.; Swain, G.; Tachibana, T.; Kobashi, K., The incorporation of Pt nanoparticles into boron-doped diamond thin-films: dimensionally stable catalytic electrodes. J. New Mater. Electrochem. Syst. 3(1), 75–82 (2000)

134. J. Wang, G.M. Swain, T. Tachibana, K. Kobashi, Electrocatalytic diamond thin film electrodes with incorporated PT. Electrochem. Soc. Inc: Pennington **2002**, 157–167 (2001)

135. N.R. Wilson, S.L. Clewes, M.E. Newton, P.R. Unwin, J.V. Macpherson, Impact of grain-dependent boron uptake on the electrochemical and electrical properties of polycrystalline boron doped diamond electrodes. J. Phys. Chem. B **110**(11), 5639–5646 (2006). https://doi.org/10.1021/jp0547616

136. F. Bottari, K. De Wael, Electrodeposition of gold nanoparticles on boron doped diamond electrodes for the enhanced reduction of small organic molecules. J. Electroanal. Chem. **801** (Supplement C), 521–526 (2017). https://doi.org/10.1016/j.jelechem.2017.07.053

137. K.P. Loh, S.L. Zhao, W. De Zhang, Diamond and carbon nanotube glucose sensors based on electropolymerization. Diam. Relat. Mater. **13**(4), 1075–1079 (2004). https://doi.org/10.1016/j.diamond.2003.11.009

138. C. Martínez-Huitle, N.S. Fernandes, S. Ferro, A. De Battisti, M. Quiroz, Fabrication and application of Nafion®-modified boron-doped diamond electrode as sensor for detecting caffeine. Diam. Relat. Mater. **19**(10), 1188–1193 (2010). https://doi.org/10.1016/j.diamond.2010.05.004

139. P.R. Roy, M.S. Saha, T. Okajima, S.G. Park, A. Fujishima, T. Ohsaka, Selective detection of dopamine and its metabolite, DOPAC, in the presence of ascorbic acid using diamond electrode modified by the polymer film. Electroanalysis **16**(21), 1777–1784 (2004). https://doi.org/10.1002/elan.200303026

140. A.O. Simm, C.E. Banks, S. Ward-Jones, T.J. Davies, N.S. Lawrence, T.G. Jones, L. Jiang, R.G. Compton, Boron-doped diamond microdisc arrays: electrochemical characterisation and their use as a substrate for the production of microelectrode arrays of diverse metals (Ag, Au, Cu) via electrodeposition. Analyst **130**(9), 1303–1311 (2005). https://doi.org/10.1039/b506956d

141. A. Salimi, M.E. Hyde, C.E. Banks, R.G. Compton, Boron doped diamond electrode modified with iridium oxide for amperometic detection of ultra trace amounts of arsenic (III). Analyst **129**(1), 9–14 (2004). https://doi.org/10.1039/B312285A

142. N. Vinokur, B. Miller, Y. Avyigal, R. Kalish, Cathodic and anodic deposition of mercury and silver at boron-doped diamond electrodes. J. Electrochem. Soc. **146**(1), 125–130 (1999). https://doi.org/10.1149/1.1391574

143. H. Terashima, T. Tsuji, Adsorption of bovine serum albumin onto mica surfaces studied by a direct weighing technique. Colloids Surf. B **27**(2), 115–122 (2003). https://doi.org/10.1016/S0927-7765(02)00044-9

144. H.E.M. Hussein, H. Amari, J.V. Macpherson, Electrochemical synthesis of nanoporous platinum nanoparticles using laser pulse heating: application to methanol oxidation. ACS Catal. **7**(10), 7388–7398 (2017). https://doi.org/10.1021/acscatal.7b02701

145. M. Limat, B. El Roustom, H. Jotterand, G. Fóti, C. Comninellis, Electrochemical and morphological characterization of gold nanoparticles deposited on boron-doped diamond electrode. Electrochim. Acta **54**(9), 2410–2416 (2009). https://doi.org/10.1016/j.electacta.2008.02.050

146. C. Zhang, L. Gu, Y. Lin, Y. Wang, D. Fu, Z. Gu, Degradation of X-3B dye by immobilized TiO$_2$ photocatalysis coupling anodic oxidation on BDD electrode. J. Photochem. Photobiol. A **207**(1), 66–72 (2009). https://doi.org/10.1016/j.jphotochem.2009.01.014

147. F. Celii, J. Butler, Diamond chemical vapor deposition. Annu. Rev. Phys. Chem. **42**(1), 643–684 (1991). https://doi.org/10.1146/annurev.pc.42.100191.003235

148. J.H. Luong, K.B. Male, J.D. Glennon, Boron-doped diamond electrode: synthesis, characterization, functionalization and analytical applications. Analyst **134**(10), 1965–1979 (2009). https://doi.org/10.1039/B910206J

149. K.B. Holt, A.J. Bard, Y. Show, G.M. Swain, Scanning electrochemical microscopy and conductive probe atomic force microscopy studies of hydrogen-terminated boron-doped diamond electrodes with different doping levels. J. Phys. Chem. B **108**, 15117–15127 (2004). https://doi.org/10.1021/jp048222x

150. H. Notsu, I. Yagi, T. Tatsuma, D.A. Tryk, A. Fujishima, Introduction of oxygen-containing functional groups onto diamond electrode surfaces by oxygen plasma and anodic polarization. Electrochem. Solid-State Lett. **2**(10), 522–524 (1999). https://doi.org/10.1149/1.1390890

151. D.W. Arrigan, Nanoelectrodes, nanoelectrode arrays and their applications. Analyst **129**(12), 1157–1165 (2004). https://doi.org/10.1039/b415395m

152. F. Maillard, M. Eikerling, O. Cherstiouk, S. Schreier, E. Savinova, U. Stimming, Size effects on reactivity of Pt nanoparticles in CO monolayer oxidation: the role of surface mobility. Faraday Discuss. **125**, 357–377 (2004). https://doi.org/10.1039/b303911k

153. S.R. Belding, E.J. Dickinson, R.G. Compton, Diffusional cyclic voltammetry at electrodes modified with random distributions of electrocatalytic nanoparticles: theory. J. Phys. Chem. C **113**(25), 11149–11156 (2009). https://doi.org/10.1021/jp901664p

154. T. Kondo, T. Morimura, T. Tsujimoto, T. Aikawa, M. Yuasa, Platinum nanoparticle-embedded porous diamond spherical particles as an active and stable heterogeneous catalyst. Sci Rep. **7**, 10 (2017). https://doi.org/10.1038/s41598-017-08949-0

155. U. Griesbach, D. Zollinger, H. Pütter, C. Comninellis, Evaluation of boron doped diamond electrodes for organic electrosynthesis on a preparative scale*. J. Appl. Electrochem. **35**(12), 1265–1270 (2005). https://doi.org/10.1007/s10800-005-9038-2

156. D. Bavykin, E. Milsom, F. Marken, D. Kim, D. Marsh, D. Riley, F. Walsh, K. El-Abiary, A. Lapkin, A novel cation-binding TiO_2 nanotube substrate for electro-and bioelectro-catalysis. Electrochem. Commun. **7**(10), 1050–1058 (2005). https://doi.org/10.1016/j.elecom.2005.07.010

157. M.E. Hyde, R.G. Compton, A review of the analysis of multiple nucleation with diffusion controlled growth. J. Electroanal. Chem. **549**, 1–12 (2003). https://doi.org/10.1016/S0022-0728(03)00250-X

158. D. Grujicic, B. Pesic, Iron nucleation mechanisms on vitreous carbon during electrodeposition from sulfate and chloride solutions. Electrochim. Acta **50**(22), 4405–4418 (2005). https://doi.org/10.1016/j.electacta.2005.02.013

159. D. Grujicic, B. Pesic, Reaction and nucleation mechanisms of copper electrodeposition from ammoniacal solutions on vitreous carbon. Electrochim. Acta **50**(22), 4426–4443 (2005). https://doi.org/10.1016/j.electacta.2005.02.012

160. S. Jones, K. Tedsree, M. Sawangphruk, J.S. Foord, J. Fisher, D. Thompsett, S.C.E. Tsang, Promotion of direct methanol electro-oxidation by Ru Terraces on Pt by using a reversed spillover mechanism. ChemCatChem **2**(9), 1089–1095 (2010). https://doi.org/10.1002/cctc.201000106

161. M.E. Hyde, R. Jacobs, R.G. Compton, In situ AFM studies of metal deposition. J. Phys. Chem. B **106**(43), 11075–11080 (2002). https://doi.org/10.1021/jp0213607

162. B. Scharifker, J. Mostany, Three-dimensional nucleation with diffusion controlled growth: Part I. Number density of active sites and nucleation rates per site. J. Electroanal. Chem. Interfacial Electrochem. 177(1–2), 13–23 (1984). https://doi.org/10.1016/0022-0728(84)80207-7

163. Z.D. Wei, S.H. Chan, Electrochemical deposition of PtRu on an uncatalyzed carbon electrode for methanol electrooxidation. J. Electroanal. Chem. **569**(1), 23–33 (2004). https://doi.org/10.1016/j.jelechem.2004.01.034

164. M. Mavrikakis, P. Stoltze, J.K. Nørskov, Making gold less noble. Catal. Lett. **64**(2), 101–106 (2000). https://doi.org/10.1023/A:1019028229377

165. A. Sanchez, S. Abbet, U. Heiz, W.-D. Schneider, H. Häkkinen, R. Barnett, U. Landman, When gold is not noble: nanoscale gold catalysts. J. Phys. Chem. A **103**(48), 9573–9578 (1999). https://doi.org/10.1021/jp9935992

166. M. Valden, X. Lai, D.W. Goodman, Onset of catalytic activity of gold clusters on titania with the appearance of nonmetallic properties. Science 281(5383), 1647–1650 (1998). https://doi.org/10.1126/science.281.5383.1647

167. O. Niwa, Electroanalytical chemistry with carbon film electrodes and micro and nano-structured carbon film-based electrodes. Bull. Chem. Soc. Jpn. **78**(4), 555–571 (2005). https://doi.org/10.1246/bcsj.78.555

168. I. Duo, C. Comninellis, W. Haenni, Perret A, in *Deposition of Nanoparticles of Iridium Dioxyde on a Synthetic Boron-Doped Diamond Surface, Diamond Materials Vii, Proceedings, Pennington*, ed. by G.M. Swain, J.L. Davidson, J.C. Angus, T. Ando, W.D. Brown (Electrochemical Society Inc, Pennington, 2001), pp. 147–156

169. M.E. Hyde, C.E. Banks, R.G. Compton, Anodic stripping voltammetry: an AFM study of some problems and limitations. Electroanalysis **16**(5), 345–354 (2004). https://doi.org/10.1002/elan.200302863

170. J. Barton, J.M. Bockris, The electrolytic growth of dendrites from ionic solutions, in *Proceedings of the Royal Society of London A: Mathematical, Physical and Engineering Sciences, The Royal Society*, pp. 485–505 (1962)

171. M. Paunovic, M. Schlesinger, Kinetics and mechanism of electrodeposition, in *Fundamentals of Electrochemical Deposition*, 2nd edn. (Wiley, Hoboken, NJ, 2006), pp. 77–112

172. I. González-González, Y. Hernández-Lebrón, E. Nicolau, C.R. Cabrera, Ammonia oxidation enhancement at square-wave treated platinum particle modified boron-doped diamond electrodes. ECS Trans. **33**(1), 201–209 (2010). https://doi.org/10.1149/1.3484517

173. Z.-Y. Zhou, N. Tian, J.-T. Li, I. Broadwell, S.-G. Sun, Nanomaterials of high surface energy with exceptional properties in catalysis and energy storage. Chem. Soc. Rev. **40**(7), 4167–4185 (2011). https://doi.org/10.1039/C0CS00176G

174. R. Lam, M. Chen, E. Pierstorff, H. Huang, E. Osawa, D. Ho, Nanodiamond-embedded microfilm devices for localized chemotherapeutic elution. ACS Nano **2**(10), 2095–2102 (2008). https://doi.org/10.1021/nn800465x

175. B. Guan, F. Zou, J. Zhi, Nanodiamond as the pH-Responsive vehicle for an anticancer drug. Small **6**(14), 1514–1519 (2010). https://doi.org/10.1002/smll.200902305

176. A. Thalhammer, R.J. Edgington, L.A. Cingolani, R. Schoepfer, R.B. Jackman, The use of nanodiamond monolayer coatings to promote the formation of functional neuronal networks. Biomaterials **31**(8), 2097–2104 (2010). https://doi.org/10.1016/j.biomaterials.2009.11.109

177. Y. Wang, J. Zhi, Y. Liu, J. Zhang, Electrochemical detection of surfactant cetylpyridinium bromide using boron-doped diamond as electrode. Electrochem. Commun. **13**(1), 82–85 (2011). https://doi.org/10.1016/j.elecom.2010.11.019

178. R.S. Lewis, T. Ming, J.F. Wacker, E. Steel, Interstellar diamonds in meteorites. Nature **326**, 160–162 (1987). https://doi.org/10.1038/326160a0

179. V. Danilenko, Shock-wave sintering of nanodiamonds. Phys. Solid State **46**(4), 711–715 (2004). https://doi.org/10.1134/1.1711456

180. S. Osswald, G. Yushin, V. Mochalin, S.O. Kucheyev, Y. Gogotsi, Control of sp2/sp3 carbon ratio and surface chemistry of nanodiamond powders by selective oxidation in air. J. Am. Chem. Soc. **128**(35), 11635–11642 (2006). https://doi.org/10.1021/ja063303n

181. A. Krueger, M. Ozawa, G. Jarre, Y. Liang, J. Stegk, L. Lu, Deagglomeration and functionalisation of detonation diamond. Physica Status Solidi (a) **204**(9), 2881–2887 (2007). https://doi.org/10.1002/pssa.200776330

182. B. Palosz, C. Pantea, E. Grzanka, S. Stelmakh, T. Proffen, T. Zerda, W. Palosz, Investigation of relaxation of nanodiamond surface in real and reciprocal spaces. Diam. Relat. Mater. **15** (11), 1813–1817 (2006). https://doi.org/10.1016/j.diamond.2006.09.001

183. Y. Liu, Z. Gu, J.L. Margrave, V.N. Khabashesku, Functionalization of nanoscale diamond powder: Fluoro-, Alkyl-, Amino-, and amino acid-nanodiamond derivatives. Chem. Mater. **16**(20), 3924–3930 (2004). https://doi.org/10.1021/cm048875q

184. I. Kulakova, Surface chemistry of nanodiamonds. Phys. Solid State **46**(4), 636–643 (2004). https://doi.org/10.1134/1.1711440

185. A. Härtl, E. Schmich, J.A. Garrido, J. Hernando, S.C. Catharino, S. Walter, P. Feulner, A. Kromka, D. Steinmüller, M. Stutzmann, Protein-modified nanocrystalline diamond thin films for biosensor applications. Nat. Mater. **3**(10), 736–742 (2004). https://doi.org/10.1038/nmat1204

186. F. Neugart, A. Zappe, F. Jelezko, C. Tietz, J.P. Boudou, A. Krueger, J. Wrachtrup, Dynamics of diamond nanoparticles in solution and cells. Nano Lett. **7**(12), 3588–3591 (2007). https://doi.org/10.1021/nl0716303

187. P.-H. Chung, E. Perevedentseva, C.-L. Cheng, The particle size-dependent photoluminescence of nanodiamonds. Surf. Sci. **601**(18), 3866–3870 (2007). https://doi.org/10.1016/j.susc.2007.04.150

188. O.A. Williams, J. Hees, C. Dieker, W. Jäger, L. Kirste, C.E. Nebel, Size-dependent reactivity of diamond nanoparticles. ACS Nano **4**(8), 4824–4830 (2010). https://doi.org/10.1021/nn100748k

189. V. Bondar', I. Pozdnyakova, A. Puzyr', Applications of nanodiamonds for separation and purification of proteins. Phys. Solid State **46**(4), 758–760 (2004). https://doi.org/10.1134/1.1711468

190. C. Nebel, H. Kato, B. Rezek, D. Shin, D. Takeuchi, H. Watanabe, T. Yamamoto, Electrochemical properties of undoped hydrogen terminated CVD diamond. Diam. Relat. Mater. **15**(2), 264–268 (2006). https://doi.org/10.1016/j.diamond.2005.08.012

191. D. Shin, H. Watanabe, C.E. Nebel, Insulator−Metal transition of intrinsic diamond. J. Am. Chem. Soc. **127**(32), 11236–11237 (2005). https://doi.org/10.1021/ja052834t

192. V. Chakrapani, J.C. Angus, A.B. Anderson, S.D. Wolter, B.R. Stoner, G.U. Sumanasekera, Charge transfer equilibria between diamond and an aqueous oxygen electrochemical redox couple. Science **318**(5855), 1424–1430 (2007). https://doi.org/10.1126/science.1148841

193. K.B. Holt, D.J. Caruana, E.J. Millán-Barrios, Electrochemistry of undoped diamond nanoparticles: accessing surface redox states. J. Am. Chem. Soc. **131**(32), 11272–11273 (2009). https://doi.org/10.1021/ja902216n

194. T.S. Varley, M. Hirani, G. Harrison, K.B. Holt, Nanodiamond surface redox chemistry: influence of physicochemical properties on catalytic processes. Faraday Discuss. **172**, 349–364 (2014). https://doi.org/10.1039/C4FD00041B

195. A.S. Barnard, M. Sternberg, Crystallinity and surface electrostatics of diamond nanocrystals. J. Mater. Chem. **17**(45), 4811–4819 (2007). https://doi.org/10.1039/B710189A

196. T. Brülle, A. Denisenko, H. Sternschulte, U. Stimming, Catalytic activity of platinum nanoparticles on highly boron-doped and 100-oriented epitaxial diamond towards HER and HOR. Phys. Chem. Chem. Phys. **13**(28), 12883–12891 (2011). https://doi.org/10.1039/C1CP20852G

197. K.B. Holt, *CHAPTER 6 Electrochemistry of Nanodiamond Particles* (The Royal Society of Chemistry, In Nanodiamond, 2014), pp. 128–150

198. I.A. Novoselova, E.N. Fedoryshena, É.V. Panov, A.A. Bochechka, L.A. Romanko, Electrochemical properties of compacts of nano-and microdisperse diamond powders in aqueous electrolytes. Phys. Solid State **46**(4), 748–750 (2004). https://doi.org/10.1134/1.1711465

199. L.H. Chen, J.B. Zang, Y.H. Wang, L.Y. Bian, Electrochemical oxidation of nitrite on nanodiamond powder electrode. Electrochim. Acta **53**(8), 3442–3445 (2008). https://doi.org/10.1016/j.electacta.2007.12.023

200. J. Zang, Y. Wang, L. Bian, J. Zhang, F. Meng, Y. Zhao, S. Ren, X. Qu, Surface modification and electrochemical behaviour of undoped nanodiamonds. Electrochim. Acta **72**, 68–73 (2012). https://doi.org/10.1016/j.electacta.2012.03.169

201. I. Shpilevaya, J.S. Foord, Electrochemistry of methyl viologen and anthraquinonedisulfonate at diamond and diamond powder electrodes: the influence of surface chemistry. Electroanalysis **26**(10), 2088–2099 (2014). https://doi.org/10.1002/elan.201400310

202. E. Peltola, N. Wester, K.B. Holt, L.-S. Johansson, J. Koskinen, V. Myllymäki, T. Laurila, Nanodiamonds on tetrahedral amorphous carbon significantly enhance dopamine detection and cell viability. Biosens. Bioelectron. **88**(Supplement C), 273–282 (2017). https://doi.org/10.1016/j.bios.2016.08.055

203. M. Briones, E. Casero, M.D. Petit-Dominguez, M.A. Ruiz, A.M. Parra-Alfambra, F. Pariente, E. Lorenzo, L. Vazquez, Diamond nanoparticles based biosensors for efficient glucose and lactate determination. Biosens. Bioelectron. **68**, 521–528 (2015). https://doi.org/10.1016/j.bios.2015.01.044

204. M. Briones, M.D. Petit-Dominguez, A.M. Parra-Alfambra, L. Vazquez, F. Pariente, E. Lorenzo, E. Casero, Electrocatalytic processes promoted by diamond nanoparticles in enzymatic biosensing devices. Bioelectrochemistry **111**, 93–99 (2016). https://doi.org/10.1016/j.bioelechem.2016.05.007

205. N.B. Simioni, T.A. Silva, G.G. Oliveira, O. Fatibello, A nanodiamond-based electrochemical sensor for the determination of pyrazinamide antibiotic. Sens. Actuator B-Chem. **250**, 315–323 (2017). https://doi.org/10.1016/j.snb.2017.04.175

206. N.B. Simioni, G.G. Oliveira, F.C. Vicentini, M.R.V. Lanza, B.C. Janegitz, O. Fatibello-Filho, Nanodiamonds stabilized in dihexadecyl phosphate film for electrochemical study and quantification of codeine in biological and pharmaceutical samples. Diam. Relat. Mater. **74**(Supplement C), 191–196 (2017). https://doi.org/10.1016/j.diamond.2017.03.007

207. W. Zhang, K. Patel, A. Schexnider, S. Banu, A.D. Radadia, Nanostructuring of biosensing electrodes with nanodiamonds for antibody immobilization. ACS Nano **8**(2), 1419–1428 (2014). https://doi.org/10.1021/nn405240g

208. N. Hasan, W. Zhang, A.D. Radadia, Characterization of nanodiamond seeded interdigitated electrodes using impedance spectroscopy of pure water. Electrochim. Acta **210**, 375–382 (2016). https://doi.org/10.1016/j.electacta.2016.05.053

209. Y. Goto, F. Ohishi, K. Tanaka, H. Usui, Formation of diamond nanoparticle thin films by electrophoretic deposition. Jpn. J. Appl. Phys. **55**(3), 6 (2016). https://doi.org/10.7567/jjap.55.03dd10

210. L. La-Torre-Riveros, K. Soto, M.A. Scibioh, C.R. Cabrera, Electrophoretically fabricated diamond nanoparticle-based electrodes. J. Electrochem. Soc. **157**(6), B831–B836 (2010). https://doi.org/10.1149/1.3374403

211. S. Su, J. Wang, J. Wei, J. Qiu, S. Wang, Thermal conductivity studies of electrophoretically deposited nanodiamond arrays. Mater. Sci. Eng.: B **225**(Supplement C), 54–59 (2017). https://doi.org/10.1016/j.mseb.2017.08.010

212. G.L. Bilbro, Theory of electrodeposition of diamond nanoparticles. Diam. Relat. Mater. **11** (8), 1572–1577 (2002). https://doi.org/10.1016/S0925-9635(02)00104-8

213. X. Zhao, J. Zang, Y. Wang, L. Bian, J. Yu, Electropolymerizing polyaniline on undoped 100 nm diamond powder and its electrochemical characteristics. Electrochem. Commun. **11** (6), 1297–1300 (2009). https://doi.org/10.1016/j.elecom.2009.04.029

214. J. Zang, Y. Wang, X. Zhao, G. Xin, S. Sun, X. Qu, S. Ren, Electrochemical synthesis of polyaniline on nanodiamond powder. Int. J. Electrochem. Sci **7**(2), 1677–1687 (2012)

215. H. Ashassi-Sorkhabi, M. Es'haghi, Electro-Synthesis of Nano-Colloidal PANI/ND composite for enhancement of Corrosion-Protection effect of PANI coatings. J. Mater. Eng. Perform. **22**(12), 3755–3761 (2013). https://doi.org/10.1007/s11665-013-0638-4

216. E. Tamburri, S. Orlanducci, V. Guglielmotti, G. Reina, M. Rossi, M.L. Terranova, Engineering detonation nanodiamond–Polyaniline composites by electrochemical routes: structural features and functional characterizations. Polymer **52**(22), 5001–5008 (2011). https://doi.org/10.1016/j.polymer.2011.09.003

217. E. Tamburri, V. Guglielmotti, S. Orlanducci, M.L. Terranova, D. Sordi, D. Passeri, R. Matassa, M. Rossi, Nanodiamond-mediated crystallization in fibers of PANI nanocomposites produced by template-free polymerization: conductive and thermal properties of the fibrillar networks. Polymer **53**(19), 4045–4053 (2012). https://doi.org/10.1016/j.polymer.2012.07.014

218. V. Kumar, R. Mahajan, D. Bhatnagar, I. Kaur, Nanofibers synthesis of ND:PANI composite by liquid/liquid interfacial polymerization and study on the effect of NDs on growth mechanism of nanofibers. Eur. Polym. J. **83**(Supplement C), 1–9 (2016). https://doi.org/10.1016/j.eurpolymj.2016.07.025

219. V. Kumar, R. Mahajan, I. Kaur, K.-H. Kim, Simple and Mediator-Free urea sensing based on engineered nanodiamonds with polyaniline nanofibers synthesized in situ. ACS Appl. Mater. Interfaces **9**(20), 16813–16823 (2017). https://doi.org/10.1021/acsami.7b01948

220. M. Briones, E. Casero, L. Vazquez, F. Pariente, E. Lorenzo, M.D. Petit-Dominguez, Diamond nanoparticles as a way to improve electron transfer in sol-gel L-lactate biosensing platforms. Anal. Chim. Acta **908**, 141–149 (2016). https://doi.org/10.1016/j.aca.2015.12.029

221. H. Ashassi-Sorkhabi, R. Bagheri, B. Rezaei-Moghadam, Corrosion protection properties of PPy-ND composite coating: sonoelectrochemical synthesis and design of experiment. J. Mater. Eng. Perform. **25**(2), 611–622 (2016). https://doi.org/10.1007/s11665-016-1886-x

222. M.K. Ram, H. Gomez, F. Alvi, E. Stefanakos, Y. Goswami, A. Kumar, Novel nanohybrid structured regioregular polyhexylthiophene blend films for photoelectrochemical energy applications. J. Phys. Chem. C **115**(44), 21987–21995 (2011). https://doi.org/10.1021/jp205297n

223. N. Giambrone, M. McCrory, A. Kumar, M.K. Ram, Comparative photoelectrochemical studies of regioregular polyhexylthiophene with microdiamond, nanodiamond and hexagonal boron nitride hybrid films. Thin Solid Films **615**, 226–232 (2016). https://doi.org/10.1016/j.tsf.2016.07.028

224. L.-N. Tsai, G.-R. Shen, Y.-T. Cheng, W. Hsu, Performance improvement of an electrothermal microactuator fabricated using Ni-diamond nanocomposite. J. Microelectromech. Syst. **15**(1), 149–158 (2006). https://doi.org/10.1109/JMEMS.2005. 863737

225. E. Levashov, P. Vakaev, E. Zamulaeva, A. Kudryashov, V. Kurbatkina, D. Shtansky, A. Voevodin, A. Sanz, Disperse-strengthening by nanoparticles advanced tribological coatings and electrode materials for their deposition. Surf. Coat. Technol. **201**(13), 6176–6181 (2007). https://doi.org/10.1016/j.surfcoat.2006.08.134

226. M. Sajjadnejad, H. Omidvar, M. Javanbakht, Influence of pulse operational parameters on electrodeposition, morphology and microstructure of Ni/nanodiamond composite coatings. Int. J. Electrochem. Sci. **12**(5), 3635–3651 (2017). https://doi.org/10.20964/2017.05.52

227. M. Sajjadnejad, H. Omidvar, M. Javanbakht, A. Mozafari, Textural and structural evolution of pulse electrodeposited Ni/diamond nanocomposite coatings. J. Alloy. Compd. **704**, 809–817 (2017). https://doi.org/10.1016/j.jalicom.2016.12.318

228. T. Fujimura, V.Y. Dolmatov, G. Burkat, E. Orlova, M. Veretennikova, Electrochemical codeposition of Sn–Pb–metal alloy along with detonation synthesis nanodiamonds. Diam. Relat. Mater. **13**(11), 2226–2229 (2004). https://doi.org/10.1016/j.diamond.2004.06.009

229. S. Shahrokhian, S. Ranjbar, M. Ghalkhani, Modification of the electrode surface by ag nanoparticles decorated nano Diamond-graphite for voltammetric determination of ceftizoxime. Electroanalysis **28**(3), 469–476 (2016). https://doi.org/10.1002/elan.201500377

230. Y. Yao, Y.J. Xue, Impedance analysis of quartz crystal microbalance humidity sensors based on nanodiamond/graphene oxide nanocomposite film. Sens. Actuator B-Chem. **211**, 52–58 (2015). https://doi.org/10.1016/j.snb.2014.12.134

231. L. Bian, Y. Wang, J. Lu, J. Zang, Synthesis and electrochemical properties of TiO₂/nanodiamond nanocomposite. Diam. Relat. Mater. **19**(10), 1178–1182 (2010). https://doi.org/10.1016/j.diamond.2010.05.007

232. L.Y. Bian, Y.H. Wang, J.B. Zang, J.K. Yu, H. Huang, Electrodeposition of Pt nanoparticles on undoped nanodimond powder for methanol oxidation electrocatalysts. J. Electroanal. Chem. **644**(1), 85–88 (2010). https://doi.org/10.1016/j.jelechem.2010.04.001

233. L. La-Torre-Riveros, E. Abel-Tatis, A.E. Méndez-Torres, D.A. Tryk, M. Prelas, C.R. Cabrera, Synthesis of platinum and platinum–ruthenium-modified diamond nanoparticles. J. Nanopart. Res. **13**(7), 2997–3009 (2011). https://doi.org/10.1007/s11051-010-0196-8

234. L.Y. Bian, Y.H. Wang, J.B. Zang, F.W. Meng, Y.L. Zhao, Microwave synthesis and characterization of Pt nanoparticles supported on undoped nanodimond for methanol electrooxidation. Int. J. Hydrog. Energy **37**(2), 1220–1225 (2012). https://doi.org/10.1016/j.ijhydene.2011.09.118

235. L. La-Torre-Riveros, R. Guzman-Blas, A.N.E. Méndez-Torres, M. Prelas, D.A. Tryk, C.R. Cabrera, Diamond nanoparticles as a support for Pt and PtRu catalysts for direct methanol fuel cells. ACS Appl. Mater. Interfaces **4**(2), 1134–11347 (2012) https://doi.org/10.1021/am2018628

236. J. Zang, Y. Wang, L. Bian, J. Zhang, F. Meng, Y. Zhao, R. Lu, X. Qu, S. Ren, Graphene growth on nanodiamond as a support for a Pt electrocatalyst in methanol electro-oxidation. Carbon **50**(8), 3032–3038 (2012). https://doi.org/10.1016/j.carbon.2012.02.089

237. Y.L. Zhao, Y.H. Wang, J.B. Zang, J. Lu, X.P. Xu, A novel support of nano titania modified graphitized nanodiamond for Pt electrocatalyst in direct methanol fuel cell. Int. J. Hydrog. Energy **40**(13), 4540–4547 (2015). https://doi.org/10.1016/j.ijhydene.2015.02.041

238. Y. Zhang, Y.H. Wang, L.Y. Bian, R. Lu, J.B. Zang, Functional separation of oxidation-reduction reaction and electron transport: PtRu/undoped nanodiamond and acetylene black as a hybrid electrocatalyst in a direct methanol fuel cell. Int. J. Hydrog. Energy **41**(8), 4624–4631 (2016). https://doi.org/10.1016/j.ijhydene.2016.01.082

239. L. Dai, Y. Xue, L. Qu, H.-J. Choi, J.-B. Baek, Metal-free catalysts for oxygen reduction reaction. Chem. Rev. **115**(11), 4823–4892 (2015). https://doi.org/10.1021/cr5003563

240. Y.S. Zhu, Y.M. Lin, B.S. Zhang, J.F. Rong, B.N. Zong, D.S. Su, Nitrogen-doped annealed nanodiamonds with varied sp(2)/sp(3) ratio as metal-free electrocatalyst for the oxygen reduction reaction. Chemcatchem **7**(18), 2840–2845 (2015). https://doi.org/10.1002/cctc. 201402930

241. E.Y. Choi, C.K. Kim, Fabrication of nitrogen-doped nano-onions and their electrocatalytic activity toward the oxygen reduction reaction. Sci Rep. **7**(1), 4178 (2017). https://doi.org/10. 1038/s41598-017-04597-6

242. L. Zhou, H. Zhang, X. Guo, H. Sun, S. Liu, M.O. Tade, S. Wang, Metal-free hybrids of graphitic carbon nitride and nanodiamonds for photoelectrochemical and photocatalytic applications. J. Colloid Interface Sci. **493**(Supplement C), 275–80 (2017). https://doi.org/10. 1016/j.jcis.2017.01.038

243. D.H. Wang, L.-S. Tan, H. Huang, L. Dai, E. Ōsawa, In-situ nanocomposite synthesis: Arylcarbonylation and grafting of primary diamond nanoparticles with a poly(ether − ketone) in polyphosphoric acid. Macromolecules **42**(1), 114–124 (2009). https://doi.org/10. 1021/ma8019078

244. K. Pei, H.D. Li, G.T. Zou, R.C. Yu, H.F. Zhao, X. Shen, L.Y. Wang, Y.P. Song, D.C. Qiu, Detonation nanodiamond introduced into samarium doped ceria electrolyte improving performance of solid oxide fuel cell. J. Power Sources **342**, 515–520 (2017). https://doi.org/ 10.1016/j.jpowsour.2016.12.051

245. X.-B. Cheng, M.-Q. Zhao, C. Chen, A. Pentecost, K. Maleski, T. Mathis, X.-Q. Zhang, Q. Zhang, J. Jiang, Y. Gogotsi, Nanodiamonds suppress the growth of lithium dendrites. Nat. Commun. **8**(1), 336 (2017). https://doi.org/10.1038/s41467-017-00519-2

246. V. Medeliene, V. Stankevič, G. Bikulčius, The influence of artificial diamond additions on the formation and properties of an electroplated copper metal matrix coating. Surf. Coat. Technol. **168**(2), 161–168 (2003). https://doi.org/10.1016/S0257-8972(03)00224-X

247. N.K. Shrestha, T. Takebe, T. Saji, Effect of particle size on the co-deposition of diamond with nickel in presence of a redox-active surfactant and mechanical property of the coatings. Diam. Relat. Mater. **15**(10), 1570–1575 (2006). https://doi.org/10.1016/j.diamond.2005.12. 040

248. L. Cunci, C.R. Cabrera, Preparation and electrochemistry of boron-doped diamond nanoparticles on glassy carbon electrodes. Electrochem. Solid-State Lett. **14**(3), K17–K19 (2011). https://doi.org/10.1149/1.3532943

249. J. Scholz, A.J. McQuillan, K.B. Holt, Redox transformations at nanodiamond surfaces revealed by in situ infrared spectroscopy. Chem. Commun. **47**(44), 12140–12142 (2011). https://doi.org/10.1039/C1CC14961J

250. Y.S. Zou, Y. Yang, W.J. Zhang, Y.M. Chong, B. He, I. Bello, S.T. Lee, Fabrication of diamond nanopillars and their arrays. Appl. Phys. Lett. **92**(5), 053105 (2008). https://doi.org/ 10.1063/1.2841822

251. N. Yang, H. Uetsuka, E. Osawa, C.E. Nebel, Vertically aligned nanowires from boron-doped diamond. Nano Lett. **8**(11), 3572–3576 (2008). https://doi.org/10.1021/nl801136h

252. P. Subramanian, S. Kolagatla, S. Szunerits, Y. Coffinier, W.S. Yeap, K. Haenen, R. Boukherroub, A. Schechter, Atomic force microscopic and raman investigation of boron-doped diamond nanowire electrodes and their activity toward oxygen reduction. J Phys. Chem. C **121**(6), 3397–3403 (2017). https://doi.org/10.1021/acs.jpcc.6b11546

253. M. Wei, C. Terashima, M. Lv, A. Fujishima, Z.-Z. Gu, Boron-doped diamond nanograss array for electrochemical sensors. Chem. Commun. **24**, 3624–3626 (2009). https://doi.org/ 10.1039/B903284C

254. M. Lv, M. Wei, F. Rong, C. Terashima, A. Fujishima, Z.-Z. Gu, Electrochemical detection of catechol based on as-grown and nanograss array boron-doped diamond electrodes. Electroanalysis **22**(2), 199–203 (2010). https://doi.org/10.1002/elan.200900296

255. F. Gao, G. Lewes-Malandrakis, M.T. Wolfer, W. Müller-Sebert, P. Gentile, D. Aradilla, T. Schubert, C.E. Nebel, Diamond-coated silicon wires for supercapacitor applications in ionic liquids. Diam. Relat. Mater. **51**(Supplement C), 1–6 (2015). https://doi.org/10.1016/j. diamond.2014.10.009

256. K. Siuzdak, R. Bogdanowicz, M. Sawczak, M. Sobaszek, Enhanced capacitance of composite TiO2 nanotube/boron-doped diamond electrodes studied by impedance spectroscopy. Nanoscale **7**(2), 551–558 (2015). https://doi.org/10.1039/C4NR04417G
257. M. Sobaszek, K. Siuzdak, M. Sawczak, J. Ryl, R. Bogdanowicz, Fabrication and characterization of composite TiO2 nanotubes/boron-doped diamond electrodes towards enhanced supercapacitors. Thin Solid Films **601**(Supplement C), 35–40 (2016). https://doi.org/10.1016/j.tsf.2015.09.073
258. V. Petrak, Z.V. Zivcova, H. Krysova, O. Frank, A. Zukal, L. Klimsa, J. Kopecek, A. Taylor, L. Kavan, V. Mortet, Fabrication of porous boron-doped diamond on SiO$_2$ fiber templates. Carbon **114**, 457–464 (2017). https://doi.org/10.1016/j.carbon.2016.12.012
259. C. Hébert, E. Scorsone, M. Mermoux, P. Bergonzo, Porous diamond with high electrochemical performance. Carbon **90**(Supplement C), 102–109. (2015) https://doi.org/10.1016/j.carbon.2015.04.016
260. B.C. Lourencao, R.A. Pinheiro, T.A. Silva, E.J. Corat, O. Fatibello-Filho, Porous boron-doped diamond/CNT electrode as electrochemical sensor for flow-injection analysis applications. Diam. Relat. Mater. **74**(Supplement C), 182–190 (2017). https://doi.org/10.1016/j.diamond.2017.03.006
261. A.A. Silva, R.A. Pinheiro, C.D.A. Razzino, V.J. Trava-Airoldi, E.J. Corat, Thin-film nanocomposites of BDD/CNT deposited on carbon fiber. Diam. Relat. Mater. **75** (Supplement C), 116–122 (2017). https://doi.org/10.1016/j.diamond.2017.02.017
262. M. Varga, S. Stehlik, O. Kaman, T. Izak, M. Domonkos, D.S. Lee, A. Kromka, Templated diamond growth on porous carbon foam decorated with polyvinyl alcohol-nanodiamond composite. Carbon **119**(Supplement C), 124–132. https://doi.org/10.1016/j.carbon.2017.04.022
263. T. Kondo, K. Yajima, T. Kato, M. Okano, C. Terashima, T. Aikawa, M. Hayase, M. Yuasa, Hierarchically nanostructured boron-doped diamond electrode surface. Diam. Relat. Mater. **72**(Supplement C), 13–19 (2017) https://doi.org/10.1016/j.diamond.2016.12.004
264. D. Plana, J. Humphrey, K. Bradley, V. Celorrio, D. Fermín, Charge transport across high surface area metal/diamond nanostructured composites. ACS Appl. Mater. Interfaces. **5**(8), 2985–2990 (2013). https://doi.org/10.1021/am302397p
265. T. Kondo, K. Hirata, T. Kawai, M. Yuasa, Self-assembled fabrication of a polycrystalline boron-doped diamond surface supporting Pt (or Pd)/Au-shell/core nanoparticles on the (111) facets and Au nanoparticles on the (100) facets. Diam. Relat. Mater. **20**(8), 1171–1178 (2011). https://doi.org/10.1016/j.diamond.2011.06.033
266. V. Plotnikov, B. Dem'yanov, S. Makarov, Effects of aluminum on the interaction of detonation diamond nanocrystals during high-temperature annealing. Tech. Phys. Lett. **35** (5), 473–475 (2009). https://doi.org/10.1134/s1063785009050265
267. D.P. Mitev, A.T. Townsend, B. Paull, P.N. Nesterenko, Screening of elemental impurities in commercial detonation nanodiamond using sector field inductively coupled plasma-mass spectrometry. J. Mater. Sci. **49**(10), 3573–3591 (2014). https://doi.org/10.1007/s10853-014-8036-3
268. V.Y. Dolmatov, A. Vehanen, V. Myllymäki, K.A. Rudometkin, A.N. Panova, K.M. Korolev, T.A. Shpadkovskaya, Purification of detonation nanodiamond material using high-intensity processes. Russ. J. Appl. Chem. **86**(7), 1036–1045 (2013). https://doi.org/10.1134/s1070427213070161
269. V. Pichot, M. Comet, E. Fousson, C. Baras, A. Senger, F. Le Normand, D. Spitzer, An efficient purification method for detonation nanodiamonds. Diam. Relat. Mater. **17**(1), 13–22 (2008). https://doi.org/10.1016/j.diamond.2007.09.011
270. O. Shenderova, A. Koscheev, N. Zaripov, I. Petrov, Y. Skryabin, P. Detkov, S. Turner, G. Van Tendeloo, Surface chemistry and properties of ozone-purified detonation nanodiamonds. J. Phys. Chem. C **115**(20), 9827–9837 (2011). https://doi.org/10.1021/jp1102466
271. S.P. Hong, T.H. Kim, S.W. Lee, Plasma-assisted purification of nanodiamonds and their application for direct writing of a high purity nanodiamond pattern. Carbon **116**(Supplement C), 640–647 (2017). https://doi.org/10.1016/j.carbon.2017.02.040

272. N. Kannari, T. Itakura, J.-I. Ozaki, Electrochemical oxygen reduction activity of intermediate onion-like carbon produced by the thermal transformation of nanodiamond. Carbon **87**(Supplement C), 415–417 (2015). https://doi.org/10.1016/j.carbon.2015.02.050

273. J. Koh, S.H. Park, M.W. Chung, S.Y. Lee, S.I. Woo, Diamond@carbon-onion hybrid nanostructure as a highly promising electrocatalyst for the oxygen reduction reaction. RSC Adv. **6**(33), 27528–27534 (2016). https://doi.org/10.1039/c5ra28066d

274. X.X. Liu, Y.H. Wang, L. Dong, X. Chen, G.X. Xin, Y. Zhang, J.B. Zang, One-step synthesis of shell/core structural boron and nitrogen co-doped graphitic carbon/nanodiamond as efficient electrocatalyst for the oxygen reduction reaction in alkaline media. Electrochim. Acta **194**, 161–167 (2016). https://doi.org/10.1016/j.electacta.2016.02.002

275. K.E. Toghill, L. Xiao, N.R. Stradiotto, R.G. Compton, The determination of methanol using an electrolytically fabricated nickel microparticle modified boron doped diamond electrode. Electroanalysis **22**(5), 491–500 (2010). https://doi.org/10.1002/elan.200900523

276. A. Panich, A. Altman, A. Shames, V.Y. Osipov, A. Aleksenskiy, A.Y. Vul, Proton magnetic resonance study of diamond nanoparticles decorated by transition metal ions. J. Phys. D Appl. Phys. **44**(12), 125303 (2011). https://doi.org/10.1088/0022-3727/44/12/125303

277. A.I. Shames, A.M. Panich, V.Y. Osipov, A.E. Aleksenskiy, A.Y. Vul', T. Enoki, K. Takai, Structure and magnetic properties of detonation nanodiamond chemically modified by copper. J. Appl. Phys. **107**(1), 014318 (2010). https://doi.org/10.1063/1.3273486

278. A. Panich, A. Shames, O. Medvedev, V.Y. Osipov, A. Aleksenskiy, A.Y. Vul, Magnetic resonance study of detonation nanodiamonds with surface chemically modified by transition metal ions. Appl. Magn. Reson. **36**(2–4), 317 (2009). https://doi.org/10.1007/s00723-009-0028-0

279. A.M. Panich, A.I. Shames, N.A. Sergeev, V.Y. Osipov, A.E. Alexenskiy, A.Y. Vul', Magnetic resonance study of gadolinium-grafted nanodiamonds. J. Phys. Chem. C **120**(35), 19804–19811 (2016). https://doi.org/10.1021/acs.jpcc.6b05403

280. H.J. Looi, L.Y. Pang, M.D. Whitfield, J.S. Foord, R.B. Jackman, Engineering low resistance contacts on p-type hydrogenated diamond surfaces. Diam. Relat. Mater. **9**(3), 975–981 (2000). https://doi.org/10.1016/S0925-9635(00)00240-5

281. Y. Jia, W. Zhu, E. Wang, Y. Huo, Z. Zhang, Initial stages of Ti growth on diamond (100) surfaces: from single adatom diffusion to quantum wire formation. Phys. Rev. Lett. **94**(8), 086101 (2005). https://doi.org/10.1103/PhysRevLett.94.086101

282. S. Stehlik, T. Petit, H.A. Girard, J.-C. Arnault, A. Kromka, B. Rezek, Nanoparticles assume electrical potential according to substrate, size, and surface termination. Langmuir **29**(5), 1634–1641 (2013). https://doi.org/10.1021/la304472w

283. I. Motochi, N. Makau, G. Amolo, Metal–semiconductor ohmic contacts: An ab initio Density Functional Theory study of the structural and electronic properties of metal–diamond (111) − (1 × 1) interfaces. Diam. Relat. Mater. **23**, 10–17 (2012). https://doi.org/10.1016/j.diamond.2011.12.021

284. M. Geis, J. Twichell, T. Lyszczarz, Diamond emitters fabrication and theory. J. Vac. Sci. Technol. B: Microelectron. Nanometer Struct. Process. Meas. Phenom. **14**(3), 2060–2067 (1996). https://doi.org/10.1116/1.588986

285. T. Tyler, V. Zhirnov, A. Kvit, D. Kang, J. Hren, Electron emission from diamond nanoparticles on metal tips. Appl. Phys. Lett. **82**(17), 2904–2906 (2003). https://doi.org/10.1063/1.1570498

286. N. Xu, Y. Tzeng, R. Latham, Similarities in the 'cold' electron emission characteristics of diamond coated molybdenum electrodes and polished bulk graphite surfaces. J. Phys. D Appl. Phys. **26**(10), 1776 (1993). https://doi.org/10.1088/0022-3727/26/10/035

287. V.V. Zhirnov, E.I. Givargizov, P.S. Plekhanov, Field emission from silicon spikes with diamond coatings. J. Vac. Sci. Technol. B: Microelectron. Nanometer Struct. Process. Meas. Phenom. **13**(2), 418–421 (1995). https://doi.org/10.1116/1.587960

288. A. Karabutov, V. Frolov, V. Konov, Diamond/sp 2-bonded carbon structures: quantum well field electron emission? Diam. Relat. Mater. **10**(3), 840–846 (2001). https://doi.org/10.1016/S0925-9635(00)00569-0

289. Y. Takasu, S. Konishi, W. Sugimoto, Y. Murakami, Catalytic formation of nanochannels in the surface layers of diamonds by metal nanoparticles. Electrochem. Solid-State Lett. 9(7), C114–C117 (2006). https://doi.org/10.1149/1.2201995
290. I.G. Casella, M. Contursi, Cobalt oxide electrodeposition on various electrode substrates from alkaline medium containing Co–gluconate complexes: a comparative voltammetric study. J. Solid State Electrochem. 16(12), 3739–3746 (2012). https://doi.org/10.1007/s10008-012-1794-4
291. S.A. Yao, R.E. Ruther, L. Zhang, R.A. Franking, R.J. Hamers, J.F. Berry, Covalent attachment of catalyst molecules to conductive diamond: CO_2 reduction using "smart" electrodes. J. Am. Chem. Soc. 134(38), 15632–15635 (2012). https://doi.org/10.1021/ja304783j
292. I. Zegkinoglou, P.L. Cook, P.S. Johnson, W. Yang, J. Guo, D. Pickup, R.N. González-Moreno, C. Rogero, R.E. Ruther, M.L. Rigsby, Electronic structure of diamond surfaces functionalized by Ru(tpy)2. J. Phys. Chem. C 116(26), 13877–13883 (2012). https://doi.org/10.1021/jp304016t
293. T. Ochiai, K. Nakata, T. Murakami, A. Fujishima, Y. Yao, D.A. Tryk, Y. Kubota, Development of solar-driven electrochemical and photocatalytic water treatment system using a boron-doped diamond electrode and TiO_2 photocatalyst. Water Res. 44(3), 904–910 (2010). https://doi.org/10.1016/j.watres.2009.09.060
294. P. Wang, M. Cao, Y. Ao, C. Wang, J. Hou, J. Qian, Investigation on Ce-doped TiO_2-coated BDD composite electrode with high photoelectrocatalytic activity under visible light irradiation. Electrochem. Commun. 13(12), 1423–1426 (2011). https://doi.org/10.1016/j.elecom.2011.09.009
295. T. Zhao, J. Wang, L. Jiang, T. Cheng, Preparation method of titanium dioxide and boron-doped diamond compounded photoelectric-synergetic electrode. CN101875007 A, 2010
296. D. Zhu, L. Zhang, R.E. Ruther, R.J. Hamers, Photo-illuminated diamond as a solid-state source of solvated electrons in water for nitrogen reduction. Nat. Mater. 12(9), 836–841 (2013). https://doi.org/10.1038/nmat3696
297. R.J. Hamers, J.A. Bandy, D. Zhu, L. Zhang, Photoemission from diamond films and substrates into water: dynamics of solvated electrons and implications for diamond photoelectrochemistry. Faraday Discuss. 172, 397–411 (2014). https://doi.org/10.1039/C4FD00039K
298. J.T. Matsushima, A.B. Couto, N.G. Ferreira, M.R. Baldan, Study of the electrochemical deposition of Cu/Sn alloy nanoparticles on boron doped diamond films for electrocatalytic nitrate reduction. MRS Proc. 1511 (2013). https://doi.org/10.1557/opl.2013.16
299. M.-J. Song, S.-K. Lee, D.-S. Lim, Fabrication of Pt nanoparticles-decorated CVD diamond electrode for biosensor applications. Anal. Sci. 27(10), 985–985 (2011). https://doi.org/10.2116/analsci.27.985
300. N. Yang, F. Gao, C.E. Nebel, Diamond decorated with copper nanoparticles for electrochemical reduction of carbon dioxide. Anal. Chem. 85(12), 5764–5769 (2013)
301. P. Kim, J.B. Joo, W. Kim, J. Kim, I.K. Song, J. Yi, NaBH4-assisted ethylene glycol reduction for preparation of carbon-supported Pt catalyst for methanol electro-oxidation. J. Power Sources 160(2), 987–990 (2006). https://doi.org/10.1016/j.jpowsour.2006.02.050
302. A. Barras, S. Szunerits, L. Marcon, N. Monfilliette-Dupont, R. Boukherroub, Functionalization of diamond nanoparticles using "Click" chemistry. Langmuir 26(16), 13168–13172 (2010). https://doi.org/10.1021/la101709q
303. A. Krueger, D. Lang, Functionality is key: recent progress in the surface modification of nanodiamond. Adv. Func. Mater. 22(5), 890–906 (2012). https://doi.org/10.1002/adfm.201102670
304. Sung, C.-M. Diamond neural devices and associated methods. US20110282421A1, 2011
305. A. Barriga-Rivera, L. Bareket, J. Goding, U.A. Aregueta-Robles, G.J. Suaning, Visual prosthesis: interfacing stimulating electrodes with retinal neurons to restore vision. Front. Neurosci. 11(620) (2017). https://doi.org/10.3389/fnins.2017.00620

306. Y.-C. Chen, D.-C. Lee, T.-Y. Tsai, C.-Y. Hsiao, J.-W. Liu, C.-Y. Kao, H.-K. Lin, H.-C. Chen, T.J. Palathinkal, W.-F. Pong, N.-H. Tai, I.N. Lin, I.-M. Chiu, Induction and regulation of differentiation in neural stem cells on ultra-nanocrystalline diamond films. Biomaterials 31(21), 5575–5587 (2010). https://doi.org/10.1016/j.biomaterials.2010.03.061

307. A.E. Hadjinicolaou, R.T. Leung, D.J. Garrett, K. Ganesan, K. Fox, D.A.X. Nayagam, M.N. Shivdasani, H. Meffin, M.R. Ibbotson, S. Prawer, B.J. O'Brien, Electrical stimulation of retinal ganglion cells with diamond and the development of an all diamond retinal prosthesis. Biomaterials 33(24), 5812–5820 (2012). https://doi.org/10.1016/j.biomaterials.2012.04.063

308. A. Ahnood, H. Meffin, D.J. Garrettm, K. Fox, K. Ganesan, A. Stacey, N.V. Apollo, Y.T. Wong, S.G. Lichter, W. Kentler, O. Kavehei, U. Greferath, K.A. Vessey, M.R. Ibbotson, E. L. Fletcher, A.N. Burkitt, S. Prawer, Diamond devices for high acuity prosthetic vision. Adv. Biosyst. 1(1–2), 1600003-n/a (2017). https://doi.org/10.1002/adbi.201600003

309. K. Ganesan, D.J. Garrett, A. Ahnood, M.N. Shivdasani, W. Tong, A.M. Turnley, K. Fox, H. Meffin, S. Prawer, An all-diamond, hermetic electrical feedthrough array for a retinal prosthesis. Biomaterials 35(3), 908–915 (2014). https://doi.org/10.1016/j.biomaterials.2013. 10.040

310. A. Bendali, L. Rousseau, G. Lissorgues, E. Scorsone, M. Djilas, J. Dégardin, E. Dubus, S. Fouquet, R. Benosman, P. Bergonzo, J.-A. Sahelm, S. Picaud, Synthetic 3D diamond-based electrodes for flexible retinal neuroprostheses: Model, production and in vivo biocompatibility. Biomaterials 67(Supplement C), 73–83 (2015). https://doi.org/10. 1016/j.biomaterials.2015.07.018

311. A. Bendali, C. Agnès, S. Meffert, V. Forster, A. Bongrain, J.-C. Arnault, J.-A. Sahel, A. Offenhäusser, P. Bergonzo, S. Picaud, Distinctive glial and neuronal interfacing on nanocrystalline diamond. PLoS ONE 9(3), e92562 (2014). https://doi.org/10.1371/journal. pone.0092562

312. G. Piret, C. Hébert, J.-P. Mazellier, L. Rousseau, E. Scorsone, M. Cottance, G. Lissorgues, M.O. Heuschkel, S. Picaud, P. Bergonzo, B. Yvert, 3D-nanostructured boron-doped diamond for microelectrode array neural interfacing. Biomaterials 53(Supplement C), 173– 83 (2015). https://doi.org/10.1016/j.biomaterials.2015.02.021

313. C. Hébert, J.P. Mazellier, E. Scorsone, M. Mermoux, P. Bergonzo, Boosting the electrochemical properties of diamond electrodes using carbon nanotube scaffolds. Carbon 71(Supplement C), 27–33 (2014). https://doi.org/10.1016/j.carbon.2013.12.083

314. F. Vahidpour, L. Curley, I. Biró, M. McDonald, D. Croux, P. Pobedinskas, K. Haenen, M. Giugliano, Z.V. Živcová, L. Kavan, M. Nesládek, All-diamond functional surface micro-electrode arrays for brain-slice neural analysis. Physica Status Solidi (a) 214(2), 1532347-n/a (2017). https://doi.org/10.1002/pssa.201532347

315. M. McDonald, A. Monaco, F. Vahidpour, K. Haenen, M. Giugliano, M. Nesladek, Diamond microelectrode arrays for in vitro neuronal recordings. MRS Commun. 7(3), 683–690 (2017). https://doi.org/10.1557/mrc.2017.62

316. V. Maybeck, R. Edgington, A. Bongrain, J.O. Welch, E. Scorsone, P. Bergonzo, R.B. Jackman, A. Offenhäusser, Boron-Doped nanocrystalline diamond microelectrode arrays monitor cardiac action potentials. Adv. Healthc. Mater. 3(2), 283–289 (2014). https://doi. org/10.1002/adhm.201300062

317. D. Gaurab, T. Chao, S. Shabnam, U.A. Prabhu, Enabling long term monitoring of dopamine using dimensionally stable ultrananocrystalline diamond microelectrodes. Mater. Res. Express 3(9), 094001 (2016). https://doi.org/10.1088/2053-1591/3/9/094001

318. N. Yang, R. Hoffmann, W. Smirnov, C.E. Nebel, Interface properties of cytochrome c on a nano-textured diamond surface. Diam. Relat. Mater. 20(2), 269–273 (2011). https://doi.org/ 10.1016/j.diamond.2010.12.012

319. N. Yang, W. Smirnov, A. Kriele, R. Hoffmann, C.E. Nebel, Diamond nanotextured surfaces for enhanced protein redox activity. Physica Status Solidi (a) 207(9), 2069–72 (2010). https://doi.org/10.1002/pssa.201000085

320. B.C. Janegitz, R.A. Medeiros, R.C. Rocha-Filho, O. Fatibello-Filho, Direct electrochemistry of tyrosinase and biosensing for phenol based on gold nanoparticles electrodeposited on a boron-doped diamond electrode. Diam. Relat. Mater. **25**, 128–133 (2012). https://doi.org/10. 1016/j.diamond.2012.02.023

321. A. Liu, Q. Ren, T. Xu, M. Yuan, W. Tang, Morphology-controllable gold nanostructures on phosphorus doped diamond-like carbon surfaces and their electrocatalysis for glucose oxidation. Sens. Actuators B: Chem. **162**(1), 135–142 (2012). https://doi.org/10.1016/j.snb. 2011.12.050

322. Y. Yu, Y. Zhou, L. Wu, J. Zhi, Electrochemical biosensor based on boron-doped diamond electrodes with modified surfaces. Int. J. Electrochem. **2012**, 10 (2012). https://doi.org/10. 1155/2012/567171

323. B. Liu, J. Hu, J.S. Foord, Electrochemical detection of DNA hybridization by a zirconia modified diamond electrode. Electrochem. Commun. **19**, 46–49 (2012). https://doi.org/10. 1016/j.elecom.2012.03.007

324. A. Zeng, C. Jin, S.-J. Cho, H.O. Seo, Y.D. Kim, D.C. Lim, D.H. Kim, B. Hong, J.-H. Boo, Nickel nano-particle modified nitrogen-doped amorphous hydrogenated diamond-like carbon film for glucose sensing. Mater. Res. Bull. **47**(10), 2713–2716 (2012). https://doi. org/10.1016/j.materresbull.2012.04.041

325. W. Wu, R. Xie, L. Bai, Z. Tang, Z. Gu, Direct electrochemistry of shewanella loihica PV-4 on gold nanoparticles-modified boron-doped diamond electrodes fabricated by layer-by-layer technique. J. Nanosci. Nanotechnol. **12**(5), 3903–3908 (2012). https://doi. org/10.1166/jnn.2012.6175

326. C.-C. Wu, C.-C. Han, H.-C. Chang, Applications of surface-functionalized diamond nanoparticles for mass-spectrometry-based proteomics. J. Chin. Chem. Soc. **57**(3B), 583– 594 (2010). https://doi.org/10.1002/jccs.201000082

327. X. Fuku, F. Iftikar, E. Hess, E. Iwuoha, P. Baker, Cytochrome c biosensor for determination of trace levels of cyanide and arsenic compounds. Anal. Chim. Acta **730**, 49–59 (2012). https://doi.org/10.1016/j.aca.2012.02.025

328. R. Hoffmann, A. Kriele, S. Kopta, W. Smirnov, N. Yang, C.E. Nebel, Adsorption of cytochrome c on diamond. physica Status Solidi (a) **207**(9), 2073–2077 (2010). https://doi. org/10.1002/pssa.201000043

329. Y. Zou, D. Lou, K. Dou, L. He, Y. Dong, S. Wang, Amperometric tyrosinase biosensor based on boron-doped nanocrystalline diamond film electrode for the detection of phenolic compounds. J. Solid State Electrochem. **20**(1), 47–54 (2016). https://doi.org/10.1007/ s10008-015-3003-8

330. A. Rahim Ruslinda, K. Tanabe, S. Ibori, X. Wang, H. Kawarada, Effects of diamond-FET-based RNA aptamer sensing for detection of real sample of HIV-1 Tat protein. Biosens. Bioelectron. **40**(1), 277–282 (2013). https://doi.org/10.1016/j.bios.2012. 07.048

331. M.-J. Song, S.-K. Lee, J.-Y. Lee, J.-H. Kim, D.-S. Lim, Electrochemical sensor based on Au nanoparticles decorated boron-doped diamond electrode using ferrocene-tagged aptamer for proton detection. J. Electroanal. Chem. **677–680**, 139–144 (2012). https://doi.org/10.1016/j. jelechem.2012.05.019

332. D.T. Tran, V. Vermeeren, L. Grieten, S. Wenmackers, P. Wagner, J. Pollet, K.P.F. Janssen, L. Michiels, J. Lammertyn, Nanocrystalline diamond impedimetric aptasensor for the label-free detection of human IgE. Biosens. Bioelectron. **26**(6), 2987–2993 (2011). https:// doi.org/10.1016/j.bios.2010.11.053

333. O. Babchenko, E. Verveniotis, K. Hruska, M. Ledinsky, A. Kromka, B. Rezek, Direct growth of sub-micron diamond structures. Vacuum **86**(6), 693–695 (2012). https://doi.org/ 10.1016/j.vacuum.2011.08.011

334. K. Honda, M. Yoshimura, T.N. Rao, D.A. Tryk, A. Fujishima, K. Yasui, Y. Sakamoto, K. Nishio, H. Masuda, Electrochemical properties of Pt-modified nano-honeycomb diamond electrodes. J. Electroanal. Chem. **514**(1–2), 35–50 (2001). https://doi.org/10.1016/S0022- 0728(01)00614-3

335. A. Kriele, O.A. Williams, M. Wolfer, J.J. Hees, W. Smirnov, C.E. Nebel, Formation of nano-pores in nano-crystalline diamond films. Chem. Phys. Lett. **507**(4–6), 253–259 (2011). https://doi.org/10.1016/j.cplett.2011.03.089

336. F. Weigl, S. Fricker, H.-G. Boyen, C. Dietrich, B. Koslowski, A. Plettl, O. Pursche, P. Ziemann, P. Walther, C. Hartmann, M. Ott, M. Möller, From self-organized masks to nanotips: a new concept for the preparation of densely packed arrays of diamond field emitters. Diam. Relat. Mater. **15**(10), 1689–1694 (2006). https://doi.org/10.1016/j.diamond.2006.02.007

337. C.E. Nebel, N. Yang, H. Uetsuka, E. Osawa, N. Tokuda, O. Williams, Diamond nano-wires, a new approach towards next generation electrochemical gene sensor platforms. Diam. Relat. Mater. **18**(5–8), 910–917 (2009). https://doi.org/10.1016/j.diamond.2008.11.024

338. P. Subramanian, Y. Coffinier, D. Steinmüller-Nethl, J. Foord, R. Boukherroub, S. Szunerits, Diamond nanowires decorated with metallic nanoparticles: A novel electrical interface for the immobilization of histidinylated biomolecuels. Electrochim. Acta **110**, 4–8 (2013). https://doi.org/10.1016/j.electacta.2012.11.010

339. N. Yang, W. Smirnov, C.E. Nebel, Three-dimensional electrochemical reactions on tip-coated diamond nanowires with nickel nanoparticles. Electrochem. Commun. **27**, 89–91 (2013). https://doi.org/10.1016/j.elecom.2012.10.044

340. H.A. Girard, S. Perruchas, C. Gesset, M. Chaigneau, L. Vieille, J.-C. Arnault, P. Bergonzo, J.-P. Boilot, T. Gacoin, Electrostatic grafting of diamond nanoparticles: a versatile route to nanocrystalline diamond thin films. ACS Appl. Mater. Interfaces. **1**(12), 2738–2746 (2009). https://doi.org/10.1021/am900458g

341. H. Zhuang, B. Song, T. Staedler, X. Jiang, Microcontact printing of monodiamond nanoparticles: an effective route to patterned diamond structure fabrication. Langmuir **27** (19), 11981–11989 (2011). https://doi.org/10.1021/la2024428

342. O. Babchenko, A. Kromka, K. Hruska, M. Michalka, J. Potmesil, M. Vanecek, Nanostructuring of diamond films using self-assembled nanoparticles. Cent. Eur. J. Phys. **7**(2), 310–314 (2009). https://doi.org/10.2478/s11534-009-0026-8

343. G. Powch, A. Jain, In *Directed Self Assembly: a Novel, High Speed Method of Nanocoating Ultra-thin Films and Monolayers of Particles, 2012 NSTI Nanotechnology Conference and Expo*, Santa Clara, CA (CRC Press: Santa Clara, CA, 2012), pp. 474–7

344. H. Sim, S.-I. Hong, S.-K. Lee, D.-S. Lim, J.-E. Jin, S.-W. Hwang, Fabrication of boron-doped nanocrystalline diamond nanoflowers based on 3D Cu(OH)2 dendritic architectures. J. Korean Phys. Soc. **60**(5), 836–841 (2012). https://doi.org/10.3938/jkps.60.836

Chapter 10
Diamond Nanowires: Theoretical Simulation and Experiments

Yuan Yu and Jinfang Zhi

Abstract The unique physicochemical properties of diamond nanowires, such as negative electron affinity, chemical inertness, high Young's modulus, the highest hardness and room-temperature thermal conductivity, etc., have generated wide interest for their use as fillers in nanocomposites, as light detectors and emitters, as substrates for nanoelectronic devices, as tips for scanning probe microscopy as well as for sensing applications. Although theoretical comparisons with carbon nanotubes have shown that diamond nanowires are energetically and mechanically viable structures, reproducibly synthesizing the crystalline diamond nanowires has remained challenging, due to its high hardness and chemical inertness. However, many efforts have been made to realize various structures of DNWs both theoretically and experimentally. Ab initio modeling of DNW structures including dodecahedral, cubic and cylindrical shapes were theoretically studied. The density functional theory (DFT) study of the surface structure of diamond nanowires with different morphology and cross-sections were also carried out. A number of experimental results have also been reported in fabrication of DNWs with different shapes. In this chapter, we present a comprehensive, up-to-date review for the diamond nanowires, wherein we will give a discussing for their synthesis along with their structures, properties and applications both from theoretical simulation and experiments.

Y. Yu
Institute of Atomic and Molecular Science, Shaanxi University of Science & Technology, Xi An, China

Y. Yu · J. Zhi (✉)
Key Laboratory of Photochemical Conversion and Optoelectronic Materials,
Technical Institute of Physics and Chemistry, Chinese Academy of Sciences,
Beijing 100190, People's Republic of China
e-mail: zhi-mail@mail.ipc.ac.cn

© Springer Nature Switzerland AG 2019
N. Yang (ed.), *Novel Aspects of Diamond*, Topics in Applied Physics 121,
https://doi.org/10.1007/978-3-030-12469-4_10

10.1   Introduction

Diamond is a material of extremes. It is the hardest, the most conductive thermally, the highest hardness, the greatest in atomic density of any known material and stable in single-photon emission even at room temperature [1, 2]. Diamond can be doped n- and p-type, from insulating, semiconducting, to metallic, respectively. Boron-doped diamond (BDD) has been recognized as the best electrode material for electrochemistry, since diamond electrodes have unique features like (a) high chemical stability, (b) no easy fouling, (c) good biocompatibility, (d) low background current, and (e) wide potential window [3], etc. The electrochemical background current of BDD in phosphate buffer is 10 times lower than that of gold electrode and 400 times lower than that of glassy carbon electrode. In addition, diamond shows strongest bonding stability to deoxyribonucleic acid (DNA) [4]. Applications of diamond electrodes in electrochemistry [5–7] and in biosensors [8–10] have been well demonstrated.

Diamond nanowires (DNWs) have motivated considerable interest in the applications of Micro-Electro-Mechanical Systems [11], quantum information processing applications [12], electrochemical gene sensing platforms [13], etc. DNWs can also be used as a single photon source with greater flux and less power compared to that of bulk diamond [14, 15]. Moreover, single crystal DNW tips can be used for ultrasensitive force microscopy [16]. Due to its high hardness and chemical inertness of diamond materials, it is difficult to fabricate DNW of an arbitrary shape. However, many efforts have been made to realize various structures of DNWs both theoretically and experimentally. Ab initio modeling of DNW structures including dodecahedral, cubic and cylindrical shapes were theoretically studied [17, 18]. The density functional theory (DFT) study of the surface structure of diamond nanowires with cubic morphology and two distinct rectangular cross-sections were studied by Barnard [18]. Garuma [19] investigated the effect of geometrical shape of diamond nanowire on its mechanical properties using the finite element modeling (FEM) software. A number of experimental results have also been reported in fabrication of DNWs with different shapes. The DNWs precursor of porous diamond films was successfully fabricated by reactive ion etching in O_2 and CF_4 plasma [20], thereafter, nanostructured diamond honeycomb films were prepared by oxygen plasma etching with porous anodic alumina mask [21]. DNWs of 80–100 nm in length were synthesized in ultra-nanocrystalline diamond (UNCD) films which were deposited by microwave plasma enhanced chemical-vapor deposition (MPECVD) system with an introduction of nitrogen in the mixture gas [22, 23]. Further to note, the hybrid graphite-DNWs (G-DNWs) were already known as promising candidate in nanoelectronics [24–26]. Wherein, the insertion of sp^2 graphene carbon on the surface of insulating diamond or metal nanowires may certainly enhance their conductivity [22, 27]. However, developing such G-DNWs majorly involved the chemical vapor deposition (CVD) techniques, which requires expensive instruments to maintain the high temperature. Very recently, Shellaiah reported an affordable wet chemical route for the reproducible hybrid graphite-diamond nanowires (G-DNWs) growth from cysteamine functionalized diamond nanoparticles (ND-Cys) via pH induced self-assembly [28].

As a result of the numerous efforts, many synthetic methods of diamond nanowires have being developed. However, the synthesis of these diamond nanowires has proven to be a low probability event and very difficult to reproduce, despite the many attempts We and others have put in. This challenge is an attractive one although reproducibly synthesizing the crystalline diamond nanowires has remained challenging. This review is presented in the hope that it will help generate more interest and more effort to address this challenge. We will present a comprehensive, up-to-date review of the diamond nanowires, wherein we discuss their synthesis along with their structures, properties and applications. We highlight some of the most important synthetic routes that underpin the synthesis of diamond nanowires, then discuss the structures and properties of the diamond nanowires from the reported theoretical and experimental results, finally, summarize the applications of the diamond nanowires [29, 30].

10.2 Synthetic Strategies of Diamond Nanowires

Numerous efforts have been directed to develop the synthetic methods of diamond nanowires, such as reactive-ion-etching (RIE) [16, 31–45], plasma post-treatment carbon nanotubes [46, 47], transfer fullerene to diamond nanowires at high temperature and high pressure [48], template or catalyst assisted CVD method [49–55], etc.

10.2.1 Plasma-Assisted Reactive Ion Etching(RIE)Route

Reactive-ion etching (RIE) is an etching technology used in top-down micro- and nanofabrication. As shown in Fig. 10.1, it uses chemically reactive plasma to remove material deposited on wafers. High-energy ions from plasma attack the surface of the sample and react with it forming desired nanostructures. The "output" of plasma etching depends on a nonlinear way on a multitude of adjustable input parameters: power density, frequency, pressure, dc bias, gas composition, flow rate and so forth [56]. The first realization of diamond nanowires by RIE was performed by Shiomi [20], who demonstrated the formation of porous diamond film through oxygen plasma RIE. Later, the plasma-assisted RIE technologies have been widely developing for the top-down fabrication of diamond nanowires. In the process, a planar nanocrystalline or microcrystalline diamond film was first deposited, and then nanowires or nanorods arrays film was fabricated by the etching of planar film. Depending on whether the mask is used, the plasma-assisted RIE technologies for diamond nanowires consist of two types, mask-need and maskless ones.

Fig. 10.1 A diagram of a common RIE setup. An RIE consists of two electrodes (**1** and **4**) that create an electric field (**3**) which can accelerate the ions (**2**) toward the surface of the samples (**5**)

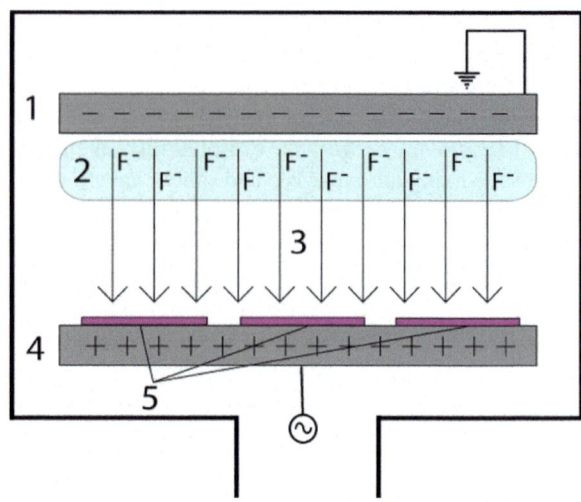

10.2.1.1 Mask-Needed Plasma-Assisted RIE Technology

Diamond nanowires are generally obtained by etching of various planar diamond films through various masks including metal nanoparticles, oxides nanoparticles, and diamond nanoparticles. These nanoparticles used as masks were a few nanometers in diameter, and the density of the as-prepared diamond nanowires was depended on the size of the masks.

10.2.1.2 Metal-Masked Plasma-Assisted RIE Technology

The first realization of diamond nanowires was performed in 1997 by Shiomi [20], who demonstrated the formation of porous diamond film through oxygen plasma RIE by using Al as mask. Subsequently, Hatta [57] and Ando [33] achieved aligned diamond nanowires through radio frequency (RF) RIE of planar diamond film (obtained from CVD) in Ar/O_2 or CF_4/O_2 plasma by the assistance of the patterned Al mask. The patterned Al masks were formed on diamonds by conventional photolithography. The densities of diamond nanowires, roughness of the etched diamond surface, and the etching rate were dependent on the type, ratio and pressure of the mixture gas plasma. As mentioned above, diamond nanowires obtained from Al-masked RIE are polycrystalline, which have the disadvantages of the existence of grain boundaries, impurities, and large stress in the films. Through Al-masked RIE method, Liao [58] and his coworkers had successfully fabricated single-crystal diamond nanowires from the high-pressure high-temperature (HPHT) type Ib-diamond substrate, which was implanted with carbon ions [59]. Very recently, Kentaro reported single-crystal diamond nanowires on a bulk, single crystalline diamond near an edge of aluminum coating using inductively coupled plasma reactive ion etching [37]. The nanowire sizes realized in the dense (sparse) area are 858 ± 22 nm (876 ± 25 nm) in height and 126 ± 6 nm (124 ± 7 nm) in diameter. Besides Al, other metals, such as Mo [34, 60], Ni [35, 36], Au [61, 62], Cr [39, 41] have been successfully used as masks to oxygen plasma etch undoped or boron-doped diamond (BBD) nanowires. Gu fabricated arrays of diamond nanostructures, differing in both diameter and top end shape, with HSQ and Cr as the etching mask materials, aiming toward large scale fabrication of single-photon sources with enhanced collection efficiency made of nitrogen vacancy (NV) embedded diamond [39]. Gheeraert systematically investigated the effect of metal film types, i.e., Al, Ti, Co, Ni, Cu, Pd, Pt and Au film, and thicknesses on surface density, shape and size of the resulting diamond nanowires [63].

10.2.1.3 Oxides-Masked Plasma-Assisted RIE Technology

A. Fujishima et al. reported a technique for the preparation of periodic diamond nanowires arrays by means of reactive ion etching with oxygen plasma using two-dimensional (2D) arrays and monodisperse solid SiO_2 particles as masks [64, 65]. Afterward, some oxides nanoparticles, such as SiO_2 and Al_2O_3, have been successfully used as mask for realizing synthesis of diamond nanowires in both single crystal and polycrystalline diamond by Hausmann [66]. Very recently, Degen fabricated single-crystal diamond nanowires using Al_2O_3 as mask, and these nanowires possessed typical lengths of a few micrometers and diameters around 100 nm [16]. As shown in Fig. 10.2, tip radii of the nanowires were of order 10 nm, making them suitable probes for scanning probe applications.

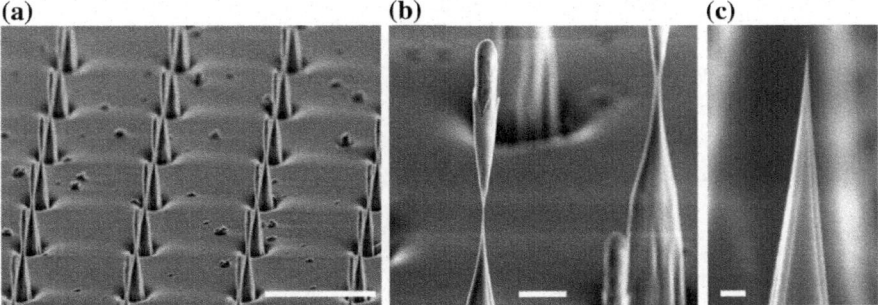

Fig. 10.2 Lithographically defined single crystal diamond tips. **a** Regular array of tips formed after 30 min of ICP etching in an oxygen plasma. The scale bar is 10 μm. **b** Zoom-in view of a nanowire showing the thin waist before cleavage. The scale bar is 1 μm. **c** Cleavage results in sharp tips with radii of about 10 nm. The scale bar is 100 nm

10.2.1.4 *Diamond Nanoparticles-Masked Plasma-Assisted RIE Technology*

Yang et al. realized vertically aligned diamond nanowires from metallically boron doped single crystalline CVD diamond film by using of diamond nanoparticles as mask [32, 67–69]. The fabrication process of vertically aligned diamond nanowires was illustrated in Fig. 10.3. First, metallically boron-doped (p-type) diamonds with atomically smooth surfaces are grown by homoepitaxy on Ib diamond substrates, using a microwave-assisted chemical vapor deposition technique [70]. Then, an etching mask from diamond nanoparticles is deposited. Diamond nanoparticles used as mask can be produced with well-defined size and quality accoring to the ref. [71]. The diamond nanoparticles of about 8–10 nm in diameter are dissolved in water by ultrasonification (200 W, 20 kHz, 12 h) to form a pseudo-stable suspension [72, 73]. The concentration of this suspension is crucial and pretreatment of diamond powder affects the stability of the suspension. Then the planar diamond film is immersed into the suspension and sonificated (100 W, 10 min) to seed diamond nanoparticles on the diamond surface. The diamond nanoparticle layer is dense and depends on suspension quality and time of sonification. After deposition of diamond nanoparticles, reactive ion etching (RIE) in an O_2 (97%)/CF_4 (3%) gas mixture is applied for typical time between 2 and 60 s. Vertically aligned diamond nanowires arise where diamond nanoparticles have been deposited. The as-fabricated diamond nanowires were bio-functionalizated and used for DNA sensing [68, 69].

10.2.1.5 Maskless Plasma-Assisted RIE Technology for Highly Doped Diamond Nanowires

As mentioned above, 1D nanostructured diamonds are generally fabricated by etching through various masks such as anodic alumina, SiO_2 ordered arrays, Au nanodots, molybdenum, nanodiamond particles, and other materials. However, these methods suffer from certain limitations and are unfavorable for large-scale fabrication. For example, some masks need be removed by additional chemical or

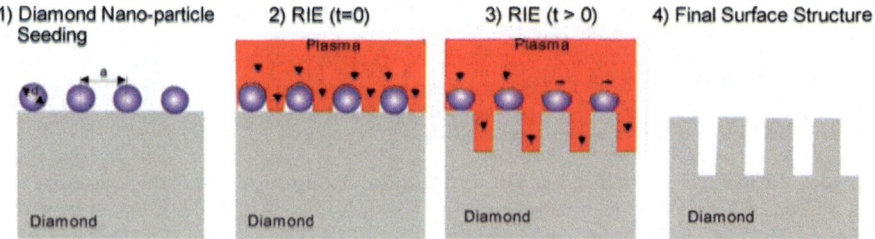

Fig. 10.3 Schematic plots of fabrication of diamond nanowires. Here d is the diameter of nanodiamond particles, and a is the distance between particles

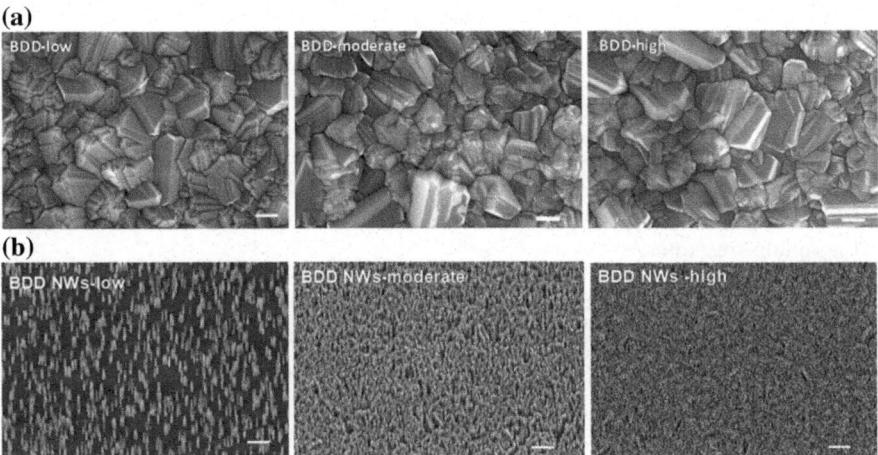

Fig. 10.4 **a** SEM images of diamond interfaces grown on p-type doped (100) silicon wafers from methane/hydrogen mixtures in the presence of trimethyl borane gas: BDD-low (1000 B ppm; $N_A = 8 \times 10^{17}$ B cm^{-3}), BDD-moderate (5000 B ppm; $N_A = 1.5 \times 10^{19}$ B cm^{-3}), and BDD-high (10,000; $N_A = 2 \times 10^{20}$ B cm^{-3}) (scale bar = 200 nm). **b** SEM images of the different BDD interfaces after treatment with oxygen plasma (pressure = 150 mT; flow speed = 20 sccm; plasma power = 350 W) forming BDD NWs of different doping levels (scale bar = 1 μm)

physical processes, or an etching mask needs be intentionally deposited by pre-preparation processes. Fujishima et al. reported a RIE technology for heavily B-doped diamond (the boron doping level is 2.1×10^{21} B cm^{-3}) nanowires using oxygen plasma without any additional mask assistance [74]. The boron dopant atoms in the diamond act as the mask during plasma etching, so avoiding the complicated pre-preparation processes involved in using an intentional mask or removal of the template by using additional processes. Very recently, Szunerits et al. also reported Reactive ion etching of diamond interfaces using oxygen plasma approach for the formation of diamond nanowires. Especially, they highlighted the influence of the doping level of the etched diamond substrate on the density of the resulting nanowires. The BDD NWs were prepared through RIE of BDD film with different boron concentrations, including 8×10^{17} B cm^{-3} (1000 ppm, BDD-low), 1.5×10^{19} B cm^{-3} (5000 ppm, BDD-moderate), and 2×10^{20} B cm^{-3} (10,000 ppm, BDD-high), and as shown in Fig. 10.4, the heavily boron-doped diamond interfaces result in very dense diamond nanowires, while etching of low boron-doped diamond substrates results in sparsely formed nanostructures, as boron dopant atoms in the diamond act as masks during the etching process [38]. The formation sites of the nanowire array depend greatly on the boron distribution. This straightforward method displays several advantages as it does not require any complicated processing steps such as mask deposition and removal. Until now, this maskless RIE method has been widely used to realize diamond nanowires [31, 40, 42–44, 45, 75–77].

10.2.2 Chemical Vapor Deposition Method (CVD)

Vapor phase growth is extensively useful bottom-up method to produce nanowires. Starting with the simple evaporation technique in an appropriate atmosphere to produce elemental or oxide nanowires, vapour–liquid–solid, vapour–solid and other processes are also used [78–81]. Depending on whether the template is used, the CVD technologies for diamond nanowires consist of two types, template-assisted and template-free ones.

10.2.2.1 Template-Assisted CVD Method

Template-assisted CVD synthesis is a convenient and versatile method for generating 1D nanostructures [82–85]. The template serves as a scaffold against which other materials with similar morphologies are synthesized. The in-situ generated material is shaped into a nanostructure with morphology complementary to that of the template. Templates could be nanoscale channels within mesoporous materials, such as porous alumina or polycarbonate membranes, etc.

10.2.2.2 Nanowires-Templated CVD for Diamond Nanowires

One of the most important methods for inorganic nanowire synthesis is template-assisted CVD. The semiconductor nanowires derived from non-colloidal synthesis makes them convenient templates for gas phase synthesis including both physical coating as well as chemical transformation. Nanowires-templated CVD method for diamond nanowires usually consists of two steps. One is the synthesis of various nanowires templates, and the other is conformal coating of the nanowires templates with nanodiamond forming polycrystalline diamond nanowires by CVD method. The size of the as-prepared diamond nanowires is dependent on the size of the nanowires templates. The possibility of deposition of layered microdiamond coatings onto tungsten wires by the CVD method was demonstrated for the first time by May et al. [86]. Afterward, many others reported successful the coating of microdiamond onto a variety of substrate wires including silicon carbide, copper, tungsten, and titanium [87–89]. Singh et al. reported two-step method for synthesis of high-density nanocrystalline diamond fibers (nanowires). This method includes the synthesis of templates (silica (α-SiO$_2$) nanofibers) by a conventional vapor–liquid–solid method and the conformal coating of the nanofibers with 15–20 nm sized NCD grains by a microwave plasma enhanced chemical vapor deposition technique in hydrogen-deficient conditions [90–92]. The as–prepared diamond nanowires showed good electron field emission properties [91, 92]. Furthermore, our group also reported a Si nanowires–templated hot filament chemical vapor deposition method for fabrication boron-doped diamond nanowires arrays and 2D diamond nanowires network [93, 94]. Recently, Nebel reported a Si nanowire–

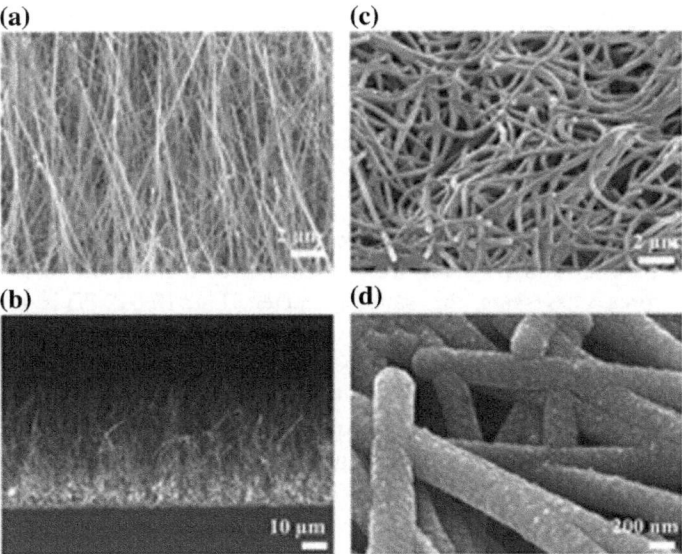

Fig. 10.5 a SEM image of the morphology of Si NWs recorded at 45° tilted angle.
b Cross-sectional view of Si NWs. **c** and **d** Low and high resolution SEM images of
diamond-coated Si NWs

templated CVD method for diamond nanowires. Theas-grown functionalized diamond nanowires were employed as electrodes in a symmetric micro-supercapacitor
(MSC) using a protic ionic liquid electrolyte [51, 95–97]. The Si NWs were first
grown on ⟨111⟩ highly doped silicon substratesby chemical vapor deposition
(CVD) via gold catalyst (4 nm evaporated thin film). Then, Si NWs were immersed
into an aqueous nanodiamond colloid and treated in an ultrasonic bath for 1 min at
a power of 26 W to obtain a uniform distribution of nanodiamond particles on
Si NW surface. Lately, the samples were introduced into a microwave chemical
vapor deposition (MWCVD) system and exposed to H_2/CH_4 gas mixtures. The
boron doping on diamond was carried out by adding trimethylborane to the H_2/CH_4
mixture. Briefly, the growth parameters can be summarized as follow: 4% methane
in H_2, a microwave power of 2750 W, a gas pressure of 40 mbar and a substrate
temperature of approximately 675 °C. Figure 10.5 shows the as-grown Si nanowires and diamond coated Si nanowires.

10.2.2.3   AAO-Templated CVD

Porous anodic aluminum oxide templates played a dominant role in the preparation
of highly ordered nanowires [98, 99] with controllable size. Masuda et al. reported
the preparation of well-aligned polycrystalline diamond nanocylinders (nanowires)
and diamond-like carbon (DLC) nanotubes on anodic aluminum oxide templates by

microwave plasma-assisted chemical vapor deposition (CVD) [100]. The alumina templates [101, 102] for diamond deposition were prepared by electrochemical anodization of aluminum sheet (thickness 0.15 mm) in 0.3 M phosphoric acid at 1 °C under a constant voltage of 190 V for 70 min. The resulted well-ordered nanoporous through-hole membranes were used as templates for diamond deposition. As shown in Fig. 10.6, prior to deposition, the alumina templates were nucleated with 50 and 5 nm diamond particles for the deposition of diamond nanocylinders and diamond-like nanotubes, respectively. For the growth of nanowires arrays, the 50 nm particles were nucleated at the bottom of the alumina membrane pores possessing the same diameter (Fig. 10.6a). This enables the growth of diamond cylinders along the length of the pores. For the growth of nanotubes, smaller diamond nanoparticles (5 nm) were ultrasonically dispersed on the pore walls by keeping the templates in particle suspensions (Fig. 10.6b). The growth of diamond nanowires or nanotubes was carried out by plasma-assisted CVD. The deposition of diamond initially proceeds along the alumina pores and continues to grow on the membrane to yield a continuous film, which acts as a

Fig. 10.6 Schematic diagram of fabrication process for cylindrical diamond and tubular structures. **a** The cylindrical structures were prepared by nucleating with 50 nm diamond particles at the bottom of the membrane holes, followed by microwave plasma CVD for 3 ± 4 h using acetone as the carbon source. **b** Tubular structures were fabricated by nucleating with 5 nm particles on the pore walls of the membrane followed by microwave plasma CVD

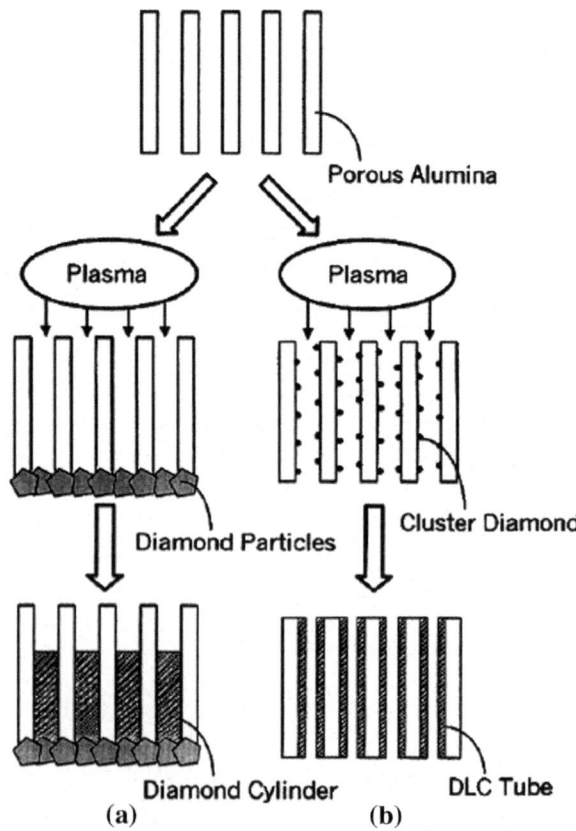

support for the nanostructures. The resulting nano-fibers were released from the alumina membrane by immersion in concentrated phosphoric acid at 250 °C.

10.2.2.4 Template-Free CVD for Diamond Nanowires

The growth of nanowires via a gas phase reaction involving the vapor–liquid–solid (VLS) and vapor–solid (VS) processes has been widely studied. CVD synthesis of nanowires is essentially a two-step process, consisting of a catalyst preparation step followed by synthesis of the nanowires. The experimental observations suggest that there are three growth stages: metal alloying, crystal nucleation and axial growth in the CVD growth of a nanowire. More recently, the enhanced CVD methods, such as microwave or plasma enhanced CVD methods for diamond nanowires were widely investigated.

10.2.2.5 *Microwave Plasma Enhanced CVD (MPCVD) Method*

MPCVD is an effective method for fabrication of hybrid diamond-graphite nanowires [103, 104] Valsov et al. reported on the MPCVD synthesis and characterization of new hybrid diamond-graphite nanowires, formed via ultra-nanocrystalline diamond (UNCD) film deposited with adding nitrogen gas [103]. The UNCD films were deposited on Si substrates in a microwave plasma CVD reactor in $Ar/CH_4/H_2/N_2$ mixtures [104]. Kamatchi [53] reported DNWs with length of 50−200 nm with a few nanometers in diameter by N_2-based microwave plasma enhanced chemical vapor deposition (94% N_2/6%CH_4). Figure 10.7 displays an HRTEM image of a DNW, exhibiting a clear core−shell microstructure. Each DNW is found to be encased by the graphite phase, which probably forms during the growth of DNWs. The thickness of the graphitic layer varies from a few atomic layers to more than 5 nm. The electron diffraction pattern, display the coexistence of sp^3 diamond and sp^2 graphitic phases in DNWs films. Sankaran [54] found that the hydrogen plasma treatment could result in microstructural and chemical modification of the diamond nanowires obtained from N_2-based MPECVD. Moreover, they also investigated [55] the effect of hydrogen plasma treatment on the electrical conductivity and electron field emission (EFE) properties for diamond nanowire (DNW) films. Transmission electron microscopy depicted that DNW films mainly consist of wire-like diamond nanocrystals encased in a nanographitic sheath, which formed conduction channels for efficient electron transport and hence lead to excellent electrical conductivity and EFE properties for these films. Hydrogen plasma treatment initially enhanced the electrical conductivity and EFE properties of DNW films and then degraded with an increase in treatment time. Shang et al. [105] reported the growth of ultrathin diamond nanorods (UDNRs) by a MPCVD method (in N_2/CH_4 plasma). As-deposited UDNRs have a length of 50–300 nm and a thin diameter of 2.1 nm, less than the theoretical minimum value (2.7 nm) for energetically stable UDNRs [106, 107]. The growth of UDNRs is suggested to follow a

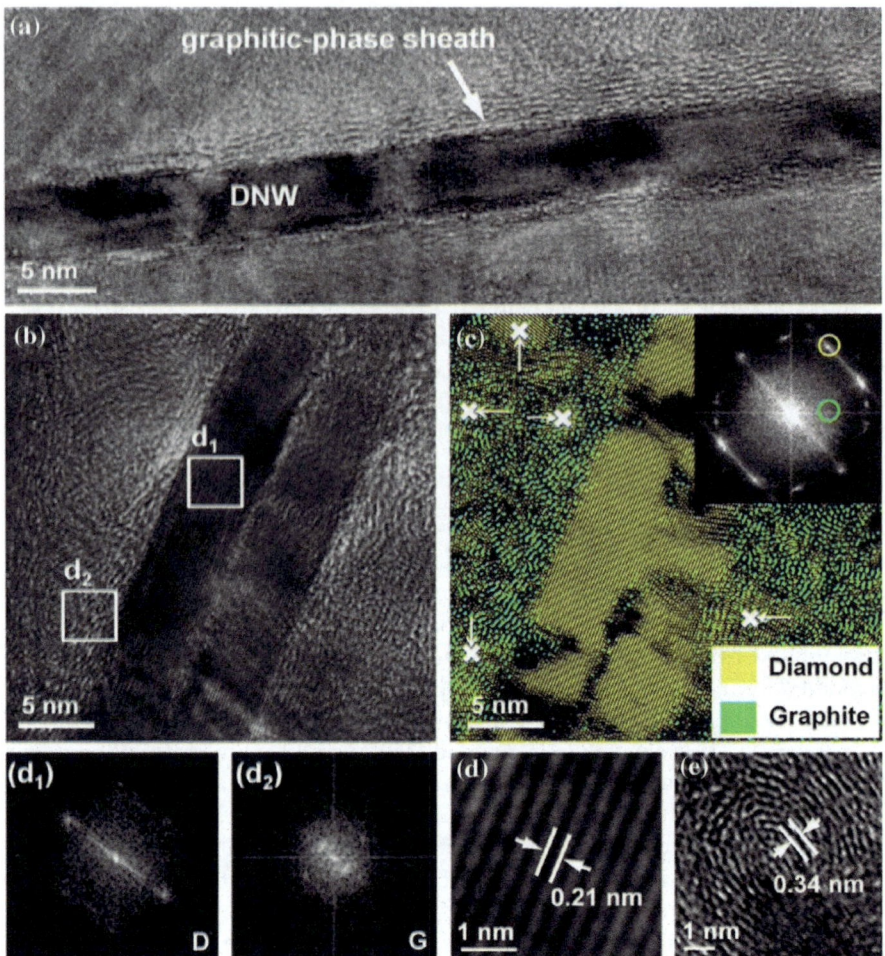

Fig. 10.7 **a** HRTEM image of a single DNW disclosing the core–shell microstructure. **b** HRTEM image with its corresponding FT pattern displayed in the inset. **c** Inverse FT image from the inset of (**b**) to mark the diamond and graphite regimes. FT patterns of (**d₁**) and (**d₂**) areas in TEM image **b** are shown in figures (**d₁**) and (**d₂**), respectively. **e** Crystalline structure of the (**d₁**) area in (**b**) presenting a clear (111) planes of the diamond structure with a lattice spacing of 0.21 nm

heterogeneous self-catalytic vapor liquid mechanism. Together with diamond nanoclusters and multilayer graphene nanowires/nano-onions, UDNRs are self-assembled into isolated electron-emitting spherules and exhibit a low-threshold, high current-density (flat panel display threshold: 10 mA/cm² at 2.9 V/μm) field emission performance. Recently, Sobia et al. reported the effect of nitrogen on hydrogen incorporation in diamond nanorods thin films obtained through MPCVD method(in Ar/H₂/CH₄/N₂ plasma). The results showed that the addition of nitrogen in gas phase has been utilized to increase the nondiamond content in the films. The incorporation of hydrogen in the diamond nanowires was found to increase with increasing the addition of nitrogen in the feed gases in deposition chamber. The increase in incorporated amount of hydrogen is related to the low crystalline quality of the film due to increase in non-diamond content presented in the samples as is supported by Raman spectroscopy measurements [108]. More recently, Shalini et al. reported the synthesis of diamond nanowires films using the MPCVD (in Ar-N_2/CH_4 plasma) [52, 109]. The experimental results showed that the electrical conductivity and the fraction of sp^2 bonding in the diamond nanowires film grain boundary (πstates) increased proportionally with the amount of nitrogen incorporation [109]. They used the diamond nanowires films realized by N_2-rich MPCVD method (6% CH_4/94%N_2) as electrode to detect the dopamine.

During the DNW growth process of MPCVD, the N_2 play an important role on the growth of wire-like diamond grains. MPCVD induced high N_2-incorporated DNW usually display good electrochemical properties, which is accounted for by increase in sp^2 content, new C–N bonds at the diamond grains, and increase in the electrical conductivity at the diamond grain boundary [52]. Recently, Feng et.al. found that the methane could induce electrical property change of nitrogen doped ultrananocrystalline diamond (UNCD) nanowires. The experimental data indicated a significant decrease in the resistance caused by the presence of the methane impurity at room temperature. Therefore, it makes N-doped DNWs an excellent material platform for methane gas sensor applications [110].

10.2.2.6   Hot Cathode Direct Current Plasma Chemical Vapor Deposition Method (HCDC-PCVD)

Nanocrystalline diamond films with different grain sizes, surface smoothness and quality have been prepared on Si substrate by changing the composition of CH_4/Ar/H_2 gas mixture using hot cathode direct current chemical vapor deposition (HCDC-PCVD) method [111]. HCDC-PCVD method is an effective method for the deposition of diamond films (including nanocrystalline, microcrystalline diamond films, etc.), and diamond films can be deposited uniformly with large area and high growth rate. Zeng et al. reported (111) diamond microcrystals and (100) diamond microcrystals and nanorods were synthesized on Si substrate by HCDC-PCVD method in CH_4/H_2 gas mixture [112]. The results showed that the high-temperature (1223 K) and low CH_4 concentration led to the growth of (111) microcrystalline diamond films, but the low temperature (1098 K) and high CH_4 concentration can lead to the growth of (100) diamond microcrystals and nanorods. Furthermore, the low reactor temperature and high CH_4 concentration decreased the grain sizes, purity, and quality of diamond films but increased the transmittance of diamond films. All the mentioned diamond were grown on Si substrates. Very recently, Motahari reported on the synthesis of DNRs by means of the bottom-up and template-free method from NCD films by the hot filament chemical vapor deposition system. The substrate materials used for diamond deposition were stainless steel (AISI 316) and chromium nitride-coated stainless steel. Their orientations are almost random in the diamond thin-film surface. In addition, the density of DNRs on the NCD film for the CrN interlayer is more than for the stainless-steel substrate. The NCD/DNR films are dense, adhesive, continuous, and almost uniform on the CrN-coated stainless-steel substrate [113].

10.2.2.7 Catalyst-Assisted Atmospheric-Pressure Chemical Vapor Deposition

As mentioned above, techniques for growing crystalline diamond have evolved from the high-temperature high-pressure (HTHP) method [114] to plasma enhanced chemical vapor deposition (PECVD) techniques [52, 109]. Diamond microwires with 25 μm diameter and 400 μm length were synthesized in 1968 using a radiation heating unit developed from a superhigh- pressure xenon tube [115]. However, the fabrication of long, single crystalline diamond nanowires using conventional thermal CVD methods has so far proven elusive, despite the potential benefits. Until to 2010, by chance, Hsu et al. synthesized diamond nanowires in an atmospheric pressure chemical vapor deposition method without plasma or energy radiation used during the CVD growth [49]. The growth process began with methane and hydrogen flow over the Fe catalyst solution dispersed on a silicon substrate under conventional chemical vapor deposition (CVD) conditions at 900 °C. After this process was completed, and without pumping the residual methane from the chamber, pure hydrogen was flowed through the quartz tube chamber at the rate of 200 sccm while the temperature was slowly lowered to ambient at the rate of ~ 1.2 °C/min over a period of 12 h. As shown in Fig. 10.8a, b, the as-prepared diamond were straight, thin and long, and longitudinally uniform in exterior diameter (60–90 nm) along the entire lengths of tens of micrometers and the structure of the nanowire indeed consists of a diamond core is encased within a graphitic shell. HRTEM image (Fig. 10.8c) of one of these nanowires reveals a crystalline diamond wire structure with a lattice constant of 0.21 nm, corresponding to the (111) orientation of diamond. Micro-Raman spectra (Fig. 10.8d) taken with a 532 nm laser beam focused on randomly selected microspots of individual isolated nanowires further verified the core-shell structure of the diamond nanowires. Hydrogen plays an important role in the formation of the diamond nanowires since it can facilitate the transformation of sp and sp^2 bonds into sp^3 bonds [116]. When the final step of cooling under hydrogen flow was omitted, carbon nanotubes and amorphous carbon were synthesized but no diamond nanowires. Other combinations of temperature, pressure, gas contents, flow rates, and process times, etc., were attempted but did not yield any diamond content. Meanwhile, transition metals (Fe in this experiment) are known to facilitate the dissociation of hydrogen molecules into atomic hydrogen at significant low hydrogen dissociation barrier [117, 118]. Such atomic hydrogen was demonstrated to enhance diamond nucleation as well as

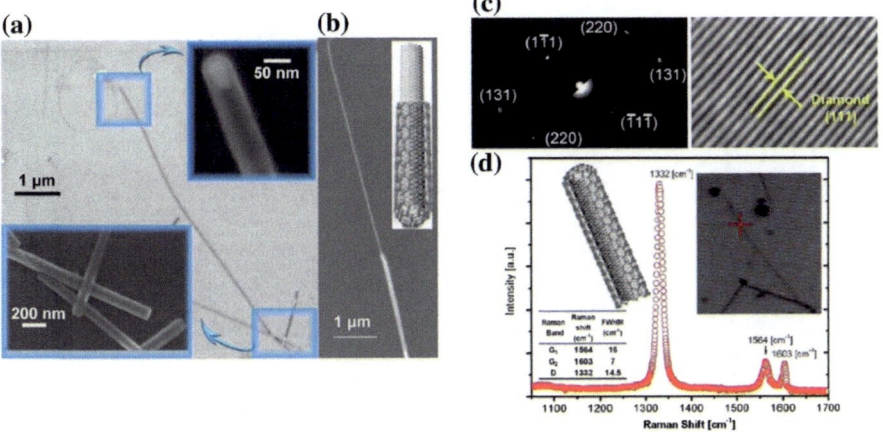

Fig. 10.8 **a** Electron microscopy of diamond nanowires encased within a carbon nanotube shell. Diamond core enclosed in a CNT sheath, typically tens of microns in length and 60–90 nm in exterior diameter. **b** An SEM micrograph showing laser-stripping of a portion of the graphitic shell of a CVD grown core–shell diamond nanowire. **c** High resolution transmission electron microscope (HRTEM) images and selective area electron diffraction (SAED) pattern of a single diamond nanowire indicating a crystalline cubic diamond (c-diamond) structure. **d** Raman spectroscopy of the diamond nanowire

graphite etching and thus has been widely applied in the synthetic diamond growth processes. The authors summarized their achievement in a mini-review, and gave a possible formation process of the diamond nanowires [50]. They rationalized that the diamond nanowire growth process begins with the conventional synthesis of either diamond stud or graphitic tubes with a vapor–liquid– solid (VLS) process and is then followed by in-situ nucleation and growth of diamond wires, fuelled by the subsequent hydrogen flow in the presence of solid and vapor carbon feedstock. This process is schematically illustrated in Fig. 10.9a–d. The tubular enclosure provides an environment where the capillary pressure (which scales inversely with the radius of the tube) can reach the order of GPa in a nanotube with an interior radius on the order of 10 nm.

10.2.3   Diamond Nanowires Realized from Sp² Carbon and Sp³ Diamondoid

The transformation of graphite to diamond nanocrystals has been one of the most challenging problems of material science for many decades. Generally, high pressures and high temperatures are needed to induce this transformation, and catalysts are used to increase the yield of diamond. Research has revealed that carbon

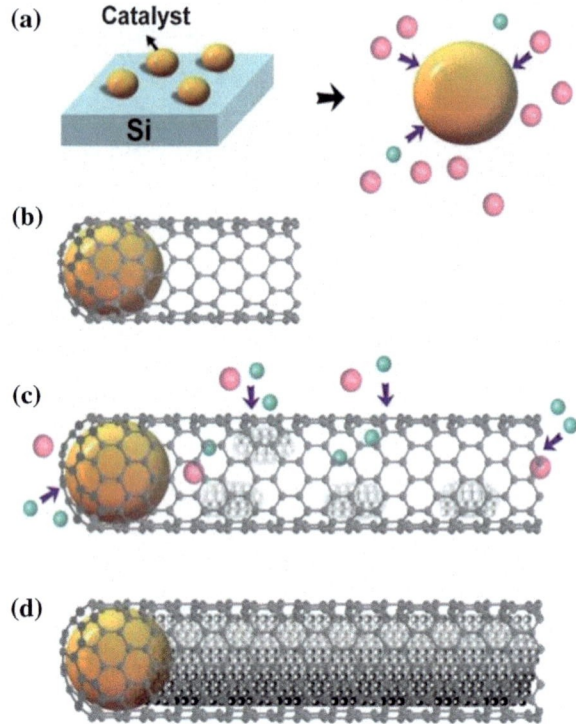

Fig. 10.9 Schematic diagram showing a possible formation process of diamond nanowires: **a** catalytic particles are formed from the evaporated or deposited thin film on the substrate as the temperature rises; **b** carbon-containing radicals reach the surface of the catalysts, leading to the growth of either a diamond stud or a graphitic tube via the VLS mechanism. The size is determined by the catalyst. **c** Hydrogen assists in the growth process by either preferentially etching sp and sp2 bonds or transforming them into sp3 bonds. With the higher capillary pressure at smaller diameters, the diamond phase could be more stable but the apillary pressure rapidly decreases with diameter, leaving the shell more stable in the graphitic phase. **d** The grain growth, boundary healing and structural reorganization take place in the slow cooling period in the presence of a pure hydrogen flow

nanotubes could also transform to diamond nanocrystals under different processing conditions—laser induced transformation [119], chemical vapor deposition by nanotube coating [120], shock wave [121], and direct transformation under high pressures and high temperatures (HPHT) [122–124]. The mechanism of transformation from carbon nanotubes to diamond was proposed to be nanotubes–carbon onions–diamond [125]. The last step had been identified to be critical in the nucleation and growth of diamond nanocrystals in the centers of spherical carbon onions under intense electron irradiation at high temperatures. However, transferring sp^2 carbon to 1D diamond nanowires is still a challenging.

10.2.3.1 Hydrogen Plasma Post-treatment of Multiwalled Carbon Nanotubes (MWCNTs) for Diamond Nanowires

Sun et al. provided a simple way for the transformation from CNTs to nanocrystalline diamond, that is, hydrogen plasma induced structural transformation from CNTs to nanodiamond [126]. Only by prolonging the hydrogen plasma treatment time, diamond nanowires with diameters of 4–8 nm and with lengths up to several hundreds of nanometers were obtained [25, 26]. The high-magnification TEM image shows that the nanowires form a core-sheath structure. Moreover, the vapor-liquid-solid (VLS) growth mechanism of 1D nanomaterials seems to be unlikely for the present diamond nanowires growth, since there is no metal catalysts used in the synthesis process. The author proposed a three-step process for diamond formation and growth under the hydrogen plasma treatment of MWCNTs, including clustering, crystallization, growth, and faceting, which is similar to that proposed by Singh [127] for diamond nucleation, crystallization, and growth from amorphous carbon precursors. Figure 10.10 shows the proposed model for the formation of nanodiamonds, and the growth of diamond nanowires. Amorphous carbon clusters are formed in step I, the crystallization of diamond begins in the core of the carbon clusters(stepII), followed by the diamond growth and faceting stage (stepIII), after the diamond nanocrystallites are faceted, diamond nanowires begin to grow at the nanowire tips (stepIV). The amorphous carbon layers that sheathe both the diamond nanoparticles and nanowires are important for the 1D growth of diamond nanowires, by preventing the lateral growth of diamond and providing the carbon source for diamond nanowires growth.

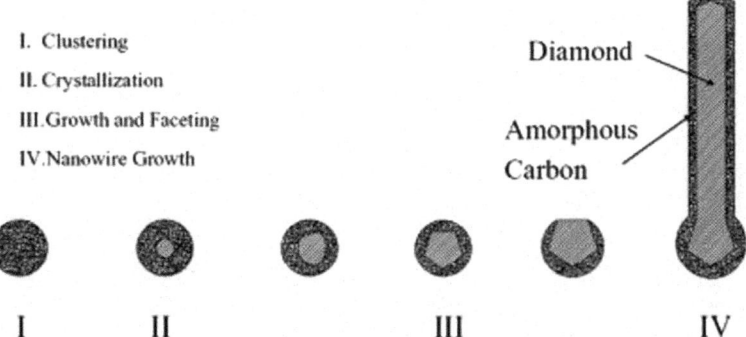

I. Clustering
II. Crystallization
III. Growth and Faceting
IV. Nanowire Growth

Diamond

Amorphous Carbon

I   II   III   IV

Fig. 10.10 The proposed model for the formation of nanodiamonds, and the growth of diamond nanowires under hydrogen plasma irradiation of multiwalled carbon nanotubes at high temperatures

10.2.3.2 Diamond Nanowires Grown from Fullerence (C_{60})

Dubrovinskaia et al. reported a bulk sample of nanocrystalline cubic diamond with crystallite sizes of 5–12 nm which was synthesized from fullerene C_{60} at 20 Gpa and 2000 °C using a multi-anvil apparatus [128]. The new material is at least as hard as single crystal diamond. It was found that nanocrystalline diamond at high temperature and ambient pressure kinetically is more stable with respect to graphitization than usual diamond. Almost at the same time, they synthesized aggregated diamond nanorods (ADNRs) from fullerene C_{60} at 20(1) GPa and 2200 °C using a multi-anvil apparatus [27]. Individual diamond nanoroads are of 5–20 nm in diameter and longer than 1 μm. The measured density of ADNRs is about of 0.2–0.4% higher than that of usual diamond.

10.2.3.3 Diamond Nanowires from Diamonoids

Diamondoids have been found to have strong affinity towards compatible host structures, such as cyclodextrins [129] and CNTs [130]. Analogous to the fabrication of 1D sp^2 CNTs from 0D sp^2 fullerenes [131, 132], the diamondoids may also fuse and transform into 1D sp^3 diamond nanowires. Templated growth of these nanowires from 1D diamondoid assemblies confined in CNTs provides a probable pathway through a "face-fused" reaction. However, the fusion reaction of adamantane into diamond nanowires has been shown to be energetically unfavorable [130]. Zhang et al. presented theoretical and experimental evidence for the feasibility of a fusion reaction of diamondoid derivatives containing relatively reactive functional groups, diamantane-4,9-dicarboxylic acid to 1D diamond nanowires inside CNTs [133]. The bisapical diamondoid diacid is more reactive than the pristine diamondoid, requiring milder reaction conditions. Unlike in 3D space, the diamantine dicarboxylic acid molecules are pulled inside a CNT by an effective "capillary force" that originates in the stabilization of the molecule inside the surrounding nanotube. The fusion of diamantane-4,9-dicarboxylic acid under the confinement of CNTs may be a promising choice to yield diamond nanowires. The diamantane-4,9-dicarboxylic acid was sublimed and self-assembled into the quasi 1D space of double-wall CNTs (DWCNTs) by a vapor phase reaction. Since the sublimation temperature of diamantane-4,9-dicarboxylic acid was at 300 °C under air atmosphere, An encapsulation temperature of 280 °C was chosen for it at 10^{-6} Torr owing to its relatively low sublimation temperature in vacuo. Similar to adamantane, the encapsulation of diamantane-4,9-dicarboxylic acid is highly selective with respect to the CNT inner diameter (D_{inner}). There is no encapsulated diamantane-4,9-dicarboxylic acid in DWCNTs which have $D_{inner} < 0.8$ nm (Fig. 10.11a). However, the diamantane-4,9-dicarboxylic acid well aligned along the axis of DWCNTs with compatible $D_{inner} \approx 1.0$ nm (Fig. 10.11b), consistent with computational results for diamondoid encapsulation [134]. Multiple arrays of encapsulated molecules can often be found in CNTs with much wider D_{inner} (Fig. 10.11c). As demonstrated in the diamond formation by chemical vapor

Fig. 10.11 HRTEM and simulated images, as well as model structures of **a** empty DWCNTs ($_{Dinner}$ < 0.8 nm), **b** linear diamondoid arrays inside DWCNTs ($_{Dinner}$ ≈ 1 nm), **c** multiple diamondoid arrays inside DWCNTs ($_{Dinner}$ ≈ 1.3 nm), and **d** optimized structure of the diamantane-4, 9-dicarboxylic acid molecule; C yellow, O red, H white

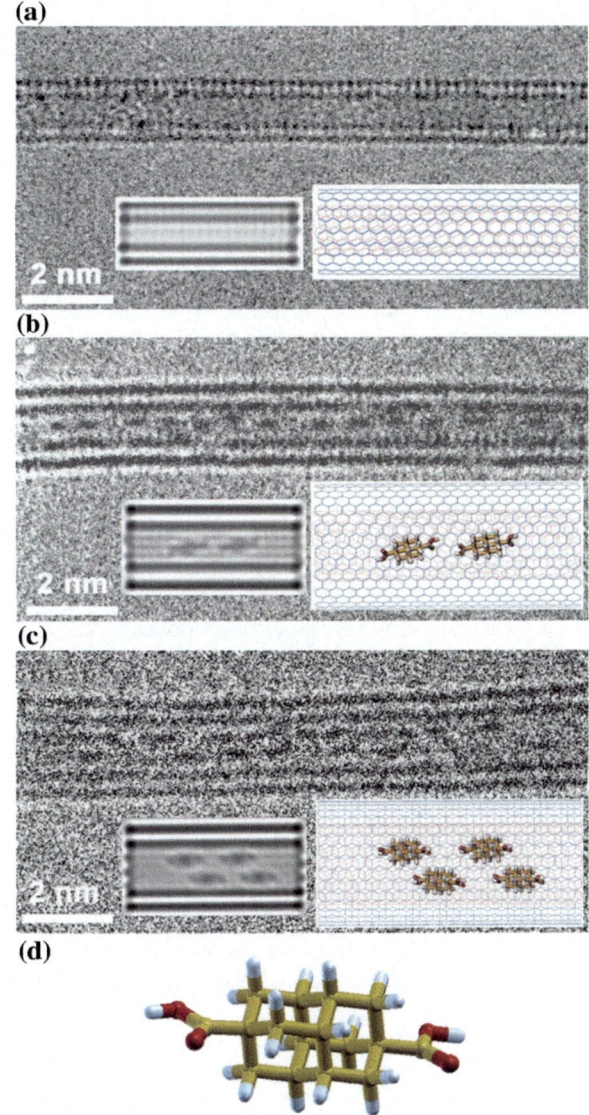

deposition (CVD) [135] and thermal annealing of adamantine inside CNTs [130], hydrogen plays a crucial role in retaining the sp^3 diamond cages. The as-produced diamondoid 1D assemblies in DWCNTs were annealed at 600 °C for 12 h under a flow of hydrogen. Diamond nanowires (CNWs) are then found inside DWCNTs with D_{inner} ≈ 1.3 nm. The rod-shaped product has a diameter about 0.78 nm, and it is very stiff and straight. No nanowire structures have been observed in DWCNTs with a smaller D_{inner} of around 1.0 nm.

10.3 Structures and Properties

10.3.1 Structural Stability of Diamond Nanowires

Diamond-based materials have been suggested to be the optimal choice for nanomechanical designs because of their high elastic modulus and strength-to-weight ratio [136]. This has prompted a number of theoretical studies investigating various aspect of diamond on the nanoscale. Results of these investigations have shown that dehydrogenated C (111) octahedral nanodiamond surfaces are structurally unstable, with their presence inducing phase transitions from the sp^3 structure of nanodiamonds to the sp^2 structure of carbon onions. However, the presence of cubic surface facets has been found to promote stability. For example, whereas cuboctahedral nanodiamond structures have exhibited preferential exfoliation of C (111) surfaces over lower-index surfaces, increasing the C (100) surface area produces a more stable nanodiamond structure and reduced surface graphitization.

Attention is now turning to 1-D diamond nanostructures diamond nanostructures. Barnard et al. studied the structural properties of dehydrogenated diamond nanowires using the Vienna ab initio simulation package (VASP) [107]. Three kinds of diamond wires, including dodecahedral, cubic and cylindrical nanowires have been considered. The dodecahedral structures are bounded by (110) surfaces in all lateral directions, with a square cross section, and have a principal axis in the [99] direction. The cubic diamond nanowires are bounded by two C (100) surfaces and two C (110) surfaces in the lateral directions, with a rectangular cross section, and have a principal axis in the [109] direction. Finally, the three cylindrical nanowires considered here are bounded by two C (100) surfaces and two C (110) surfaces in the lateral directions, with a circular cross section, and have a principal axis in the [99] direction. Changes in energy per atom (ΔE), cross-sectional area (ΔA), and nanowire segment lengths (ΔL) resulted from the relaxation of each nanowire's morphology were shown in Table 10.1. The structural changes ΔA and ΔL are also shown in terms of the percentage change in

Table 10.1 Changes in Energy Per Atom (ΔE), Cross-Sectional Area (ΔA), and Nanowire Segment Lengths (ΔL) Resulting from the Relaxation of Each Nanowire's Morphology[a]

Morphology	Atoms	ΔE (eV)	ΔL (nm)	ΔA (nm^2)
Dodecahedral	75	−0.2271	+0.0883	−0.0428
Dodecahedral	144	−0.2150	+0.1051	−0.0578
Dodecahedral	196	−0.2057	+0.0722	−0.0586
Cubic[a]	84	−0.9812	−0.0222	
Cubic	132	−0.4847	−0.0094	−0.0265
Cubic	240	−0.4339	−0.0038	−0.0615
Cylindrical	63	−0.7063	+0.0199	−0.0336
Cylindrical	128	−0.5687	+0.0182	−0.0421
Cylindrical	228	−0.2676	+0.0017	−0.0448

[a]Nonclassical nanotube

Fig. 10.12 Percentage change in cross-sectional area ΔA (top) and in periodic segment length ΔL (bottom) for the respective morphologies

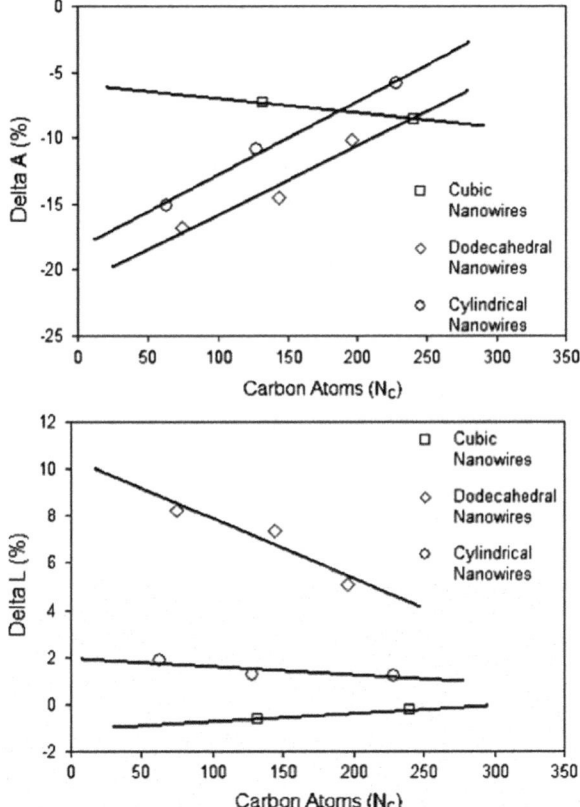

Fig. 10.12. In theory, each plot in Fig. 10.12 should converge to zero as the number of atoms increases (the macroscopic limit), but there are too few data points to deduce anything but the overall trend. The smallest cubic nanowire has been excluded from this comparison because of the conversion to a nonclassical nanotube. It is apparent from Table 10.1 that the remaining cubic nanowires still exhibit unusual structural changes. The slopes for the cubic nanowires are positive, whereas the cylindrical and dodecahedral nanowire slopes are negative. This is thought to be a product of the cubic nanowire having a principal axis in the [109] direction, suggesting that this is not an optimal choice for diamond nanowire structures. It has been shown from the ab initio relaxation of diamond nanowires that nanocrystalline diamond may be structurally stable in one dimension. Diamond nanowires with dodecahedral and cubododecahedral morphology retained the diamond structure upon relaxation but did exhibit significant relaxation involving changes in the length and cross-sectional area. The stability, characterized by the variation in these structural properties from that of bulk diamond, has been found to be dependent on both the surface morphology and the crystallographic direction of the principal axis of the nanowire. For example, nanowires having a principal axis

in the [109] direction do not represent an optimal choice for diamond nanowire structures.

Barnard et al. presented a theoretical investigation of the relative phase stability of 1D carbon nanostructures, including carbon nanotubes and diamond nanowires, using an atomic heat of formation model previously used successfully to compare the phase stability of diamond nanocrystals and fullerenes [106]. The results of this study indicate that carbon nanotubes represent the most energetically preferred form for fine 1D carbon nanostructures, and the diamond nanowires occupy a "window" of stability. This window ranges from approximately 2.7 to 3.7–9 nm in (lateral) diameter, beyond which graphite is energetically preferred. The limits of this range are sensitive both to the nanowire morphology, and the method used to scale the graphite structures (required to ensure dimensional consistency). These results are considered to be useful in estimating the size range for which diamond nanowires may be expected during synthesis, and as a guide to the relative stability of some sp^2 and sp^3 carbon in 1D nanostructures.

Tanskanen et al. studied the structural stability of the polyicosahedral diamond nanowires obtained from icosahedral diamondoids and conventional diamond nanowires superimposable with bulk diamond [137]. Molecular structures of three icosahedral diamondoids, $C_{20}H_{20}$, $C_{20}@C_{80}H_{60}$, and $C_{20}@C_{80}@C_{180}H_{120}$, together with the corresponding polyicosahedral diamond nanowires are illustrated in Fig. 10.13. The strain energies suggest the 1D diamond nanowires (DNWs) to be favored over the dodecahedral $C_{20}H_{20}$. The low structural strain of the DNWs derived from the dodecahedrane is in agreement with previous studies on short oligomers of dodecahedrane [138, 139]. To take the effect of thermodynamics into account, the authors calculated the Gibbs corrected strain energies at $T = 298.15$ K for dodecahedrane and its 1D counterparts. The Gibbs corrected strain energies were determined by

$$\Delta E = \left[E\left(C_x(CH)_y(CH_2)_z \right) - xE(C) - yE(CH) - zE(CH_2) \right] (x + y + z)^{-1}$$

(10.1)

all energy quantities replaced by the corresponding Gibbs free energies. The Gibbs corrections make the dodecahedrane slightly favored over its 1D counterparts. The differences are small, however, suggesting the polyicosahedral DNWs to be thermodynamically viable. With the exclusion of dodecahedrane, the strain energies increase, while only slightly, as a function of length of the polyicosahedral diamondoid, suggesting the preference for icosahedral structures. Combining the icosahedral diamondoids to polyicosahedral DNRs reduces the number of strain-inducing pentagons. However, connecting the cages introduce additional strain to the interface region due to fused pentagons. Nevertheless, the strain energies are systematically lower for the polyicosahedral DNRs than for the experimentally known dodecahedrane. The relative proportion of the fused pentagons becomes reduced as a function of the diameter, decreasing the strain energies of the polyicosahedral DNRs derived from the larger icosahedral diamondoids

Fig. 10.13 Icosahedral
diamondoids (top left) and
their polyicosahedral
Diamond nanowires (top right
and bottom): **a** $C_{20}H_{20}$,
b $C_{20}@C_{80}H_{60}$, and
c $C_{20}@C_{80}@C_{180}H_{120}$

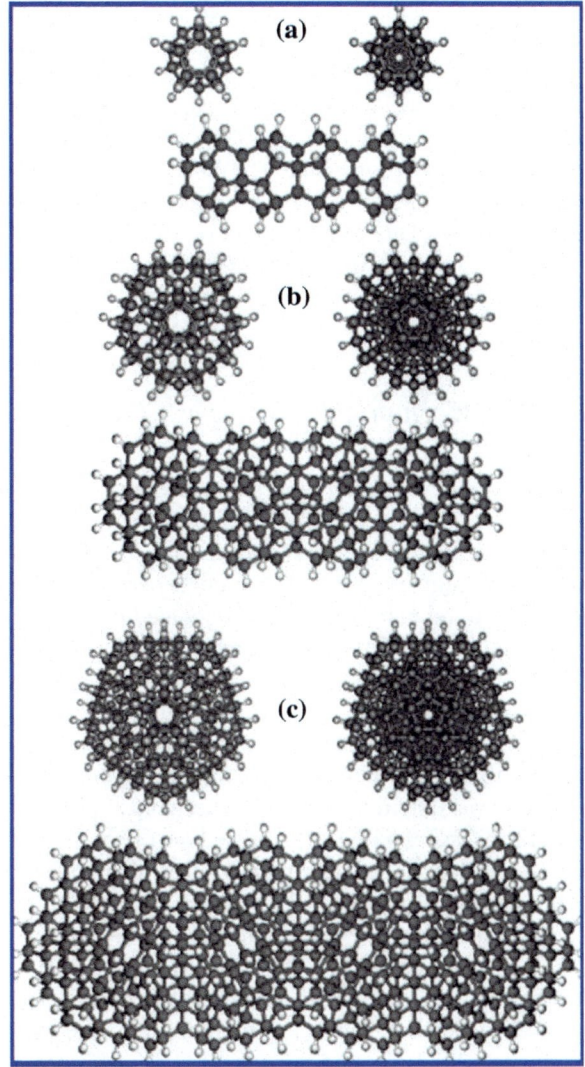

[137]. Comparisons between the 1D diamond nanowires of $C_{20}@C_{80}@C_{180}H_{120}$ and the corresponding hollow $C_{80}H_{20}@C_{180}H_{120}$ show the filled structures to be energetically favored.

Representative examples of the hydrogenated conventional DNWs, i.e., DNWs that are superimposable with bulk diamond, are investigated. The calculated strain energies are clearly the lowest for those conventional DNWs that are parallel to the [109] direction of bulk diamond. The highest stability of the [109] DNWs can be understood to originate from surface hydrogenation. The H-H distance between the surface hydrogen is around 2.5 Å for the [109] DNWs, while they are down to 2.0

Å for the [99] and [52] DNWs. The repulsion between the surface hydrogen thus somewhat destabilizes the [99] and [52] DNWs, whereas the repulsion is negligible for the [109] DNWs. The described preference for hydrogenated [109] DNWs also has been experimentally observed, DNRs synthesized by hydrogen plasma post-treatment of multi-walled CNTs preferring the [109] growth direction [60]. Previously, dehydrogenated DNWs have been shown to prefer structures with principal axis parallel to the [99] direction, while [109] DNWs have been reported unstable [140]. Due to the impact of H–H interactions at the surface, the presence (or absence) of hydrogen in the synthesis process of DNWs and DNRs may have an effect on the orientation of the products.

10.3.2   Mechanical Properties of Diamond Nanowires

A few results have been reported on the mechanical property of DNWs. The effect of cross-sectional area, length and temperature on thermal conductivity of DNWs has been studied using non-equilibrium molecular dynamics methods [141]. The strain rate, temperature and size dependent mechanical properties of [001] orientation diamond nanowires are investigated by Guo et al. using molecular dynamics (MD) simulations [142]. The mechanical properties of DNWs, including Poisson's ratios, Young's moduli, and shear moduli, were investigated by Tanskanen et al. [19, 137].

Shenderova et al. compared the stiffness and fracture force of hydrogenated diamond nanorods with those of single-walled and multi-walled carbon nanotubes [143]. It was determined that the mechanical properties of the nanorods depend on both the diameter of the nanorod and the orientation of the principal axis. The results of their molecular models indicate that diamond nanorods are energetically competitive with nanotubes of a similar diameter and possess desirable mechanical properties, making them a viable target for synthesis. The mechanical properties of the polyicosahedral and the conventional DNWs, including Poisson's ratios, Young's moduli, and shear moduli, were investigated by Tanskanen et al. [137]. In general, the Young's and shear moduli follow the same trends. The calculations give the highest Young's modulus for the CNT, included as a reference, a value of about 900 GPa being in agreement with previous theoretical and experimental work [144]. The studied conventional DNWs have Young's moduli around 360–680 GPa, the moduli increasing as a function of the wire diameter. For approximately the same cross-sectional area of the wire, the [99] DNWs have clearly lower moduli than the [110] and [111] DNWs. At a cross-sectional area of 1.4 nm^2 and beyond, the [111] DNWs have the highest Young's moduli of the studied conventional DNWs. This in agreement with previous calculations, suggesting the [111] direction having the highest Young's modulus for the low-index orientations of bulk diamond [48]. The Young's moduli of both hydrogenated conventional DNWs and the polyicosahedral

Fig. 10.14 Young's moduli (Y) of the hydrogenated conventional and the polyicosahedral DNWs as a function of the cross-sectional area of the stucture

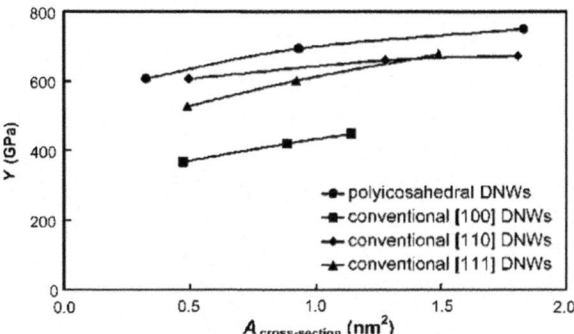

DNWs are presented in Fig. 10.14. Similar to the conventional DNWs, the elastic moduli of the polyicosahedral DNWs increases as a function of the cross-sectional area. Furthermore, their Young's moduli are even higher than those of the conventional DNWs. Accordingly, the polyicosahedral DNWs, while being somewhat more strained than the conventional DNWs, could turn out to be valuable in nanomechanical designs.

The strain rate, temperature and size dependent mechanical properties of [001] orientation diamond nanowires are investigated by Guo et al. using molecular dynamics simulations [142]. It is found that, for the same cross-sectional areas, strain rates have almost no effect on yield strength and Young's modulus, provided strain rates are within the range from 0.001 to 0.025 ps^{-1}. The calculated results have also indicated that, at the temperature ranging from 100 to 500 K, diamond nanowires' yield strength, Young's modulus, fracture strength and fracture strain are all decreasing with increasing temperature.

Very recently, Garuma [19] discussed the effect of cross-sectional area of a CVDDNW on its mechanical property. A bending test under concentrated load applied at one of the free ends is simulated using the finite element modeling (FEM) software. Displacement from the neutral central axis, the second moment, von Mises stress distribution, and strain were discussed. Figure 10.9 shows the force response of a rectangular and triangular cantilever of length 10 μm, width 2 μm and thickness 1 μm. The cantilever is fixed at one end and uniformly distributed load, shown in log scale in Fig. 10.15, is applied. The simulation results confirmed that a triangular cross-section is more flexible than the rectangular one as shown in Fig. 10.15. Moreover, in the comparisons with other cross-sections, such as circular, hexagonal, square or rectangular cross-sections, the triangular shaped DNW has highest bending capability, and the bending capability of a triangular cross-section is about three times than that of the conventional rectangular shape for most cantilever applications. Both FEM result and COMSOL Multiphysics simulations result showed that largest deformation is obtained for nanowire with triangular cross-section and lowest values for circular and hexagonal cross-sections.

Fig. 10.15 Deflection, Z_{max} versus log(10) of applied uniformly distributed force comparing a rectangular cross-section with that of triangular case. The length (L), width (w) and thickness (t) are $L \times w \times t = 10 \times 2 \times 1$ μm for both rectangular and triangular cross-sections and compared with that of Kupfer et al. [145]

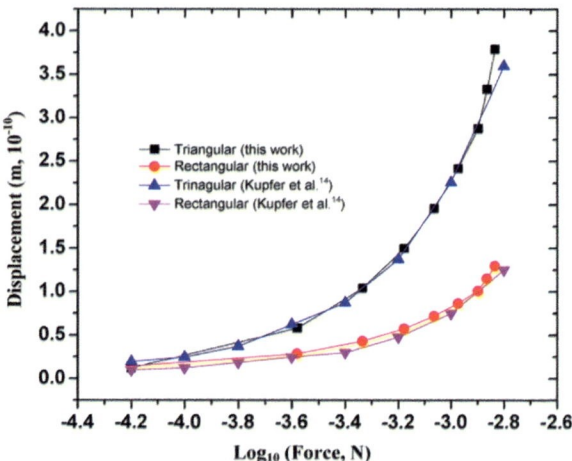

10.3.3 Density and Compressibility Properties of Diamond Nanowires

Density and compressibility properties of diamond nanowires were also investigated via both theoretical calculations and experiments. Dubrovinskaia et al. reported the synthesis of aggregated diamond nanorods (ADNRs) from fullerene C60 at 20(1) GPa and 2200 °C using a multianvil apparatus [48]. Individual aggregated diamond nanorod is of 5–20 nm in diameter and longer than 1 μm. The X-ray density of diamonds with natural abundances of isotopes is 3.515–3.519 g/cm³. The lattice parameter of ADNRs obtained from X-ray powder diffraction is 3.5617(3) Å, and consequently, the X-ray density of the material is 3.528(1) g/cm³, ~0.2–0.4% higher than that of usual diamond [146, 147]. The samples synthesized in a multianvil apparatus have a cylindrical shape that allows us to determine their volumes. The measured density of the ADNRs bulk sample was found to be 3.532 (5) g/cm³ in good correspondence with X-ray data [146, 147]. This result is in agreement with theoretical calculations of the structural relaxations of the diamond nanowires [107]. Although exact crystallographic configuration realized for individual rods in ADNRs was not theoretically considered in [107], the main conclusion, that the outerlayer contraction, characteristic for diamond nanowires, causes shortening of the C–C bonds, may explain higher density of ADNRs. Both diamond and ANDRs were compressed simultaneously in Ar pressure medium to over 27 GPa, until the samples were bridged between diamond anvils and further compression was not possible. The experimental pressure-volume data were fitted (Fig. 10.16) using the third order Birch-Murnaghan equation of state. ADNRs are more than by 11% less compressible than normal diamond that makes it, an incompressible form of carbon. It has also the lowest experimentally determined compressibility [148]. (compare with the highest K_T measured so far: 462 GPa for Os, 420 GPa for WC, 383 GPa for Ir, 380 GPa for cubic BN, 306 GPa for HfN).

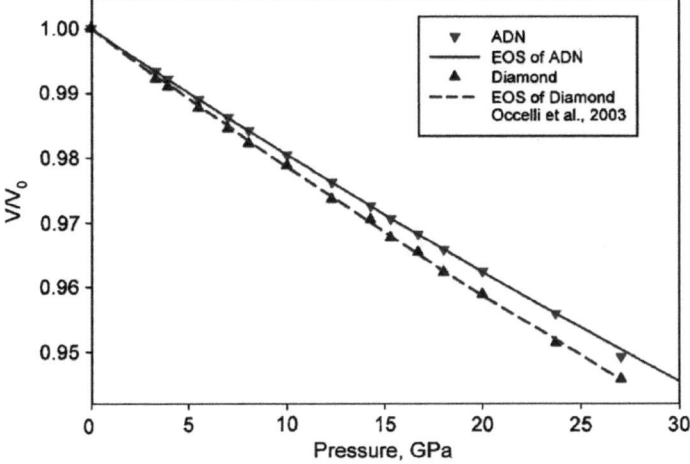

Fig. 10.16 Pressure dependence of reduced volume (V/V_0) of diamond (triangles) and aggregate diamond nanoroads (inversed triangles). Dashed line shows third-order Birch-Murnaghan equation of state (EOS) with parameters $K_T = 446(3)$ GPa, $K' = 3.0$, $V_0 = 3.4170(5)$ cm^3/mol. Continues line is the fit to our data with $K_T = 491(3)$ GPa, $K' = 3.1(2)$, $V_0 = 3.4014(5)$ cm^3/mol

10.3.4 Phonon Optical Mode and Electronic Structure of Diamond Nanowires

The electronic structure of the polyicosahedral diamond nanowires and conventional diamond nanowires were determined by Tanskanen through quantum chemical calculations [137]. Comparing with the diamondoids, the HOMO-LUMO gaps of polyicosahedral diamond nanowires is smaller. The gaps are generally smaller for the polyicosahedral diamond nanowires with larger diameters and longer length. For conventional diamond nanowires, the band gaps have the same trend as the band gaps calculated for the polyicosahedral DNWs.

It is well-known that phonon properties play an important role in the considered systems because of their significance for the analysis of various physical processes, such as, charge and thermal transport, and optical transitions, hence phonon band structures of Si and Ge nanowires have been investigated through DFPT (density functional perturbation theory) [149, 150]. Recently, A local bond-polarization model based on the displacement–displacement Green's function and the Born potential are applied to study the confined optical phonons and Raman scattering of diamond nanowires (DNWs) by Trejo et al. [151]. Also, the electronic band structure of DNWs are investigated by means of a semi-empirical tight binding (TB) approach and compared with density functional theory (DFT) within local density approximation (LDA). The results show that the highest-frequency Raman peak shifts to lower frequencies as the nanowire width increases, due to phonon confinement, in concordance with the experimental data [152].

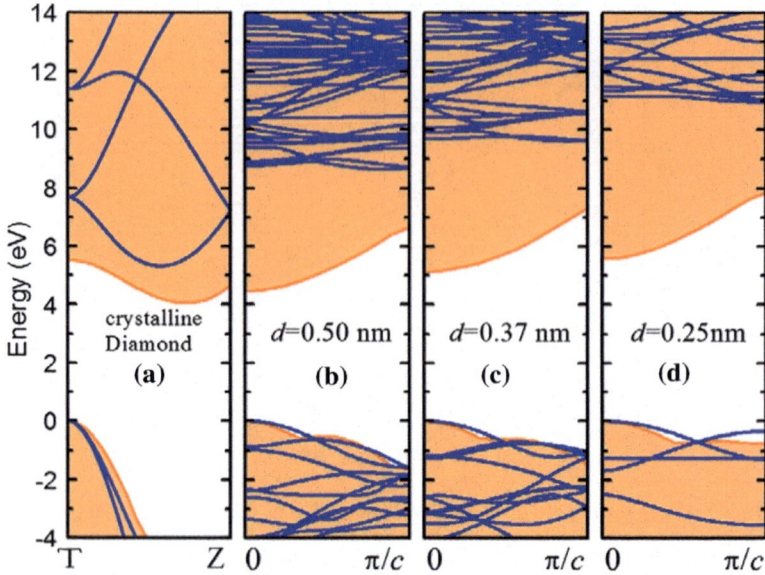

Fig. 10.17 Electronic band structure for: **a** c–D and **b**–**d** DNWs calculated from TB (blue lines) and DFT–LDA (orange area)

The authors described the electronic properties of H passivated, free standing diamond nanowires oriented along [001] direction with widths ranging from 0.25 to 2.52 nm, contained in corresponding supercells from 9 to 441 carbon atoms. Figure 10.17a–d shows the electronic band structure of c–D and DNWs, obtained from TB (blue lines) and DFT–LDA (orange area) calculations. The calculations of electronic band structure from TB for DNWs with $d = 0.50$, 0.37, and 0.25 nm have been solidly shifted by $d = 4.58$, 3.43, and 2.69 eV, respectively in order to adjust its valence band maximum (VBM) energy to the DFT–LDA one. It can be seen that the good agreement continues for the VBM but the LDA band gap underestimation becomes greater when reducing size of DNWs. It is worth to observe that DFT–LDA band gap is always direct. Meanwhile, the TB one is indirect for all cases. The difference between the conduction bands could be due to that the s* orbital has non d-wave symmetry and additionally the geometry optimization is absent in the TB calculations. However, in both approaches the band gap show a clear broadening due to quantum confinement. The calculation results also showed that the almost linear decrease in the band gap when the width of the diamond nanowires is increasing, and the important contribution of hydrogen atoms to the electronic states around the conduction band minimum (CBM).

Advances in single photon emission have been achieved experimentally by introducing nitrogen-vacancy (N-V) centers on diamond nano-structures [39, 153, 154]. However, theoretical modeling of the anisotropic effects on the electronic properties of these materials is almost nonexistent. Solano [155] analyzed the electronic band structure and density of states of diamond nanowires with N-V defects through first principles approach using the density functional theory and the supercell scheme. The nanowires were modeled on two growth directions [001] and [52]. All surface dangling bonds were passivated with hydrogen (H) atoms. The results show that the N–V introduces multiple trap states within the energy band gap of the diamond nanowire. The energy difference between these states is influenced by the growth direction of the nanowires, which could contribute to the emission of photons with different wavelengths. The presences of these trap states could reduce the recombination rate between the conduction and the valence band, thus favoring the single photon emission. Zhang and his coworkers [156] have investigated the electronic, magnetic, and optical properties of hydrogenated diamond nanowires (DNWs) with nitrogen-vacancy (NV) centers using density functional theory. The strong localization of defect states results in the formation of local magnetic moments with a spin-polarization energy that is close to those for transition-metal atoms. Such spin-polarized defect states are found to be stable well above room temperature, in agreement with previous experimental reports. In addition, they found that a semiconductor-metal transition can be triggered upon applying a transverse electric field. Furthermore, an enhanced optical absorption in the visible-light region is predicted in DNWs with NV centers. The strength and the position of the absorption can be tuned or optimized by an external electric field and/or the nanowire diameter.

10.3.5   Thermal Conductivity of Diamond Nanowires

On the basis of the high thermal conductivity of bulk diamond, DNWs may have thermal conductivities that rival CNTs. Moreover, the thermal conductivity of DNWs may be less sensitive to surface functionalization than modeling suggests for CNTs [157], thereby providing a potentially important mode for enhancing heat transfer within a nanocomposite via cross-linking. On the other hand, the nanometer-scale dimensions of DNWs may severely reduce their thermal conductivity compared to bulk diamond. Experiments and theoretical analysis by Novikov et al. [158], for example, show that thermal conductivity in polycrystalline diamond thin films is severely reduced as grain sizes approach the nanometer scale due to phonon scattering. Moreland et al. used simulations to characterize the thermal conductivity of a (10, 10) CNT and a diamond nanowire [159]. They showed that the conductivity of the diamond nanowire is significantly less than that of the CNT but that the calculated values for the thermal conductivity for both structures depends on the choice of thermostat. Padgeet et al. reported thermal conductivity calculations using classical trajectories of hydrogen-terminated and functionalized DNRs with a [109] long axis, and cross sectional radii and lengths ranging from

0.578 to 1.606 nm and from 0.016 to 0.128 μm, respectively [160]. The simulations predict that thermal conductivities for DNRs with hydrogen surface termination are about a factor of 4 less than previously calculated values for pristine (10, 10) CNTs. To study the effect of surface functionalization on thermal conductivity, structures on which attached phenyl groups replace surface hydrogen have been modeled. The simulations indicate that the thermal conductivities of DNRs are much less influenced by surface functionalization than are thermal conductivities of CNTs, suggesting that DNRs are a viable alternative to CNTs for thermal management in nanocomposites. The simulation results show a strong dependence of thermal conductivity on length and radius of the DNRs for both the hydrogen-terminated and surface functionalized structures. Guo et al. presented non-equilibrium molecular dynamics simulations on geometry and temperature dependent thermal conductivities of diamond nanowires [141]. The calculated results indicated that at temperature 300 K, diamond nanowire thermal conductivities increase with increase in length and cross-sectional areas. It is also found that at the same length, temperature and cross-sectional area, thermal conductivities of the [011] crystal orientation DNWs are larger than those in the [001] and [111] crystal orientation DNWs. All diamond nanowires thermal conductivities considered in this work are smaller than those of the corresponding orientations bulk diamond. Additionally, in the temperature range 0–1000 K, diamond nanowire's thermal conductivities initially increase with increasing temperature, and then decrease. Geometry and temperature effects on diamond nanowires thermal conductivities can be explained well by calculated densities of phonon states.

10.4 Application of Diamond Nanowires

10.4.1 Field Emission from Diamond Nanowire

Diamond is a good candidate for solid-state electronics emitters because of its negative or very low electron affinity [161]. Electron emitters can be used in vacuum microelectronics such as Spindt-type emitters. Diamond has a negative electron affinity (NEA) when the diamond [52] surface is terminated by hydrogen [161, 162]. Nishimura et al. reported that the diamond [99] surface also exhibits the NEA property [163]. NEA permits diamond to be a superior field emitter [164]. Defects at the diamond tip may further assist the emission by providing a focal point for both field and current [165]. There are many reports concerning field emission from chemical vapor deposited (CVD) diamond nanocrystals or diamond-like carbon thin films. Up to now, a few examples on synthesis and field emission studies from 1D nanowire diamond-based systems have appeared.

10.4.1.1 Electron Field Emission (EFE) from Planar Diamond Nanowire Film and a Single DNW

Electron field emission of nanodiamond emitter arrays was reported by Madaleno and his coworkers [92]. Emission starts abruptly when the applied field reaches 5.5 V/μm. This type of phenomenon was also reported by Jiang et al. [166] and Gupta et al. [167] in the case of NCD films. A small plateau can be observed in the current density, between 5.5 and 7 V/μm. The emission does not increase even with the increase of the applied field. A similar behavior was reported by Gan et al. [168] in the electron emission from hydrogenated natural diamond. A high emission current density of 10 mA cm^{-2} at 12 V/μm has been obtained from NCD microtip arrays. This value is comparable to those of high quality NCD diamond films deposited on silicon substrates. Shang et al. reported the growth of ultrathin diamond nanorods (DNRs) by a microwave plasma assisted chemical vapor deposition method, using a mixture gas of nitrogen and methane [105]. DNRs have a diameter as thin as 2.1 nm, which is not only smaller than reported one-dimensional diamond nanostructures (4–300 nm) but also smaller than the theoretical value for energetically stable DNRs. Together with diamond nanoclusters and multilayer graphene nanowires/nano-onions, DNRs are self-assembled into isolated electron emitting spherules and exhibit a low-threshold, high current-density (flat panel display threshold: 10 mA/cm^2 at 2.9 V/μm) field emission performance, better than that of all other conventional (Mo and Si tips, etc.) and popular nanostructural (ZnO nanostructure and nanodiamond, etc.) field emitters except for oriented CNTs. Sankaran et al. reported the synthesis of conducting diamond nanowires (DNWs) films encased by graphite by N$_2$-based microwave plasma enhanced chemical vapor deposition [53]. The EFE properties of the DNWs encased by the highly conducting graphite were tested. The results show emission current density, J, as a function of the applied electric field, E. It reveals a low turn-on field of 4.35 V/μm at a current density of 3 μA/cm^2. In addition, the current density increases rapidly and reaches the large magnitude of ~ 3.42 mA/cm^2 at the field of 9.1 V/μm. The DNWs film exhibits far more efficient EFE properties of lowest E_0 and highest J_e values than that of other diamond related materials reported previously [169–171]. Very recently, the effects of hydrogen plasma treatment on the electrical conductivity and electron field emission (EFE) properties for diamond nanowire (DNW) films were systematically investigated [55]. Figure 10.18a shows the J versus E characteristics of pristine and hydrogen treated DNW films, whereas the inset of the figure is the F–N plots of the corresponding J–E curves. For the DNW0 films (curve I, Fig. 10.18a), E_0 shows a higher value of 6.8 V/μm with a lower J value of 0.4 mA/cm^2 at an electric field of 8.5 V/μm. The E_0 value decreases consistently with the hydrogen plasma treatment time. The best EFE properties are observed for DNW10 films with a lower E_0 value of 4.2 V/μm and a higher J value of 5.1 mA/cm^2 at an electric field of 8.5 V/μm. However, increasing the hydrogen treatment time further to 15 min surprisingly degrades the EFE properties, i.e., E_0 value increased to 10.1 V/μm with a sudden decrease in J value to 0.01 mA/cm^2 at an applied field of 8.5 V/μm. Figure 10.1b shows the electrical conductivity values

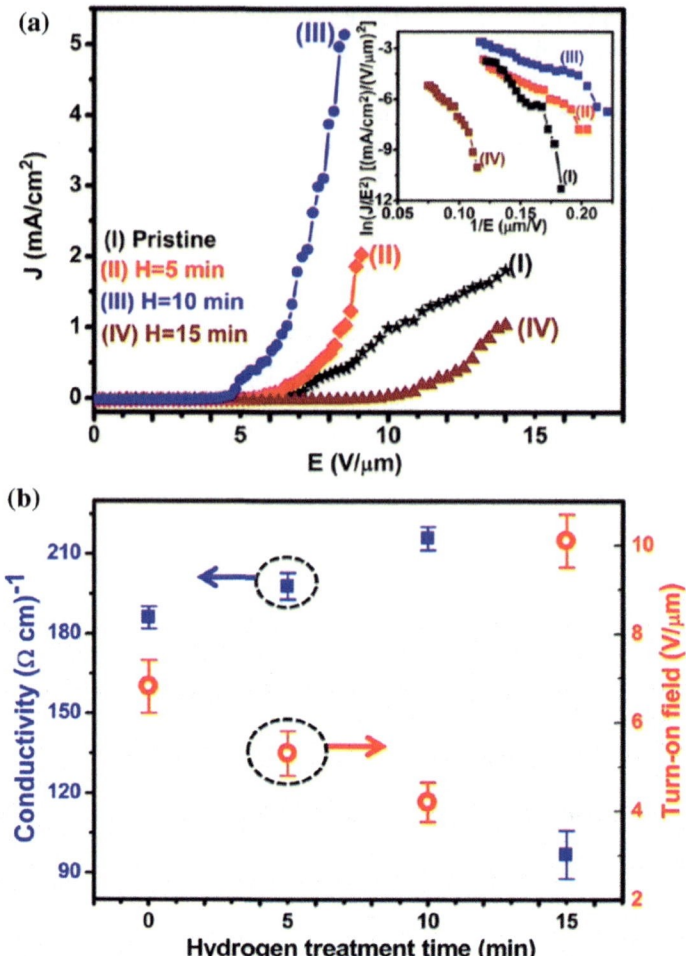

Fig. 10.18 a Electric field emission current density (J) as a function of applied electric field (E) for (I) pristine, (II) 5 min, (III) 10 min, and (IV) 15 min hydrogen plasma-treated DNW films. The F−N plot $\ln(J/E^2)$ −$1/E$ are presented in the inset. **b** Variation in the electrical conductivity (solid squares) and turn-on field (open circles) with hydrogen treatment time

of these films as a function of hydrogen treatment time along with the corresponding E_0 values for each sample. Interestingly, the conductivity for DNW films increases monotonously upon hydrogen plasma treatment time up to 10 min and suddenly dropped upon longer hydrogen plasma treatment for DNW15 films. Scanning tunneling spectroscopy in current imaging tunneling spectroscopy mode clearly shows significant increase in local emission sites in 10 min hydrogen plasma treated diamond nanowire (DNW10) films as compared to the pristine films that is ascribed to the formation of graphitic phase around the DNWs due to the hydrogen plasma treatment process. The degradation in EFE properties and electrical conductivity of extended (15 min) hydrogen plasma-treated DNW films was explained by the removal of nanographitic phase surrounding the DNWs.

Hsu and Xu reported for the first time the measurement of the field emission from a single diamond nanowire from the samples synthesized by catalyst-assisted atmospheric-pressure chemical vapor deposition method [50]. The liquid phase Gallium–Indium eutectic was used to achieve better contact between the manipulator tip and the nanowire throughout the test. They measured the field emission data for a single isolated diamond nanowire. The result is an I–V plot with a near-linear Fowler–Nordheim (F–N) relationship and reveals an ultra-low threshold electron field emission. When taking $\varphi = 5$ eV for diamond as Shiraishi suggested [172], it was found that the field enhancement factor b reached $\sim 60{,}000$. This is almost two orders of magnitude larger than the factor reported for carbon nanotube field emitters (~ 1000) [173]. It is also worth noting the low threshold field (ETH). The threshold field is defined as the applied field required for an emission of a certain current, which was set to be 100 nA. The authors then compared their diamond nanowire to the field emission characteristic obtained by a single 80 nm diameter carbon nanotube reported by Smith and Silva [174]. The diamond nanowire's threshold field of 1.25 V μm^{-1} is four times lower than the carbon nanotube's 5 V μm^{-1}. The possible reasons for such a remarkable high field enhancement factor (sixty times higher) and low threshold field (four times lower) of the diamond nanowire over carbon nanotubes are the negative electron affinity, the greatly intensified field at the tip of a large curvature and the possible defects at the nanowire tip, which itself is a 'geometric singularity' [175]. Furthermore, the chemical inertness, high mechanical strength and high thermal conductivity of the diamond nanowires allow them to endure a longer and more intense emission. This excellent EFE feature of DNWs films and a single DNW evidence a high potential for flat panel display applications [176].

10.4.2   Photonic Quantum Applications from DNWs Embedded with Nitrogen-Vacancy (NV) Centers

Solid-state quantum emitters, including semiconductor quantum dots and color centers in crystals, are indispensable elements in various quantum applications [177]. Among the solid-state quantum emitters, negatively-charged nitrogen-vacancy (NV) centers in diamond play a leading role due to their distinctive spin and optical properties, e.g. long coherence times of electron and nuclear spin states [178, 179], single-photon emission even at room temperature [180], spin state initialization and read-out by incoherent excitation [181]. These properties of the NV centers are desirable for quantum information and sensing applications, such as quantum network and highly-sensitive quantum magnetometer, and are also potentially useful in quantum key distribution and linear optical quantum computation.

Since NV centers in diamond are promising solid-state quantum emitters that can be utilized for photonic quantum applications, various diamond nanophotonic devices have been fabricated for efficient extraction of single photons emitted from NV centers to a single guided mode. However, for constructing scalable quantum networks, further efficient coupling of single photons to a guided mode of a single-mode fiber (SMF) is indispensable and a difficult challenge. For example, The tapered diamond nanowires, fabricated by the top-down approach, i.e. e-beam lithography, can also readily be detached from the parent substrate using a micro-manipulator. Although the maximum coupling efficiency of 75% was predicted for such a system, experimentally observed coupling efficiency still remains to be 16–37%, possibly due to the difficulties in fabrication of the tapered diamond structures themselves [182]. Very recently, Yonezu [153] have proposed a novel efficient hybrid system between the nanofiber and the diamond nanowire. A main advantage of our proposed coupling system is that the fabrication of the diamond nanowire could be easier than the other techniques, e.g. the adiabatically tapered diamond waveguide [183]. Assuming a cylindrical diamond nanowire with flat facets (i.e. no tapered-structures), the maximum coupling efficiency as high asb75% for the sum of both fiber ends is numerically obtained by optimizing the system geometry (the nanofiberbradius $r_f = 240$ nm, the nanowire radius $r_d = 85$ nm, and the nanowire length $L_d = 3.6$ μm). The numerical results indicate that the optimization of the two physical effects is important to maximize the coupling efficiency: (1) the interference between the two supermodes and (2) the Fabry-Perot resonance due to the reflection from the nanowire facets.

10.4.3 Diamond Nanowires for Highly Sensitive Matrix-Free Mass Spectrometry Analysis of Small Molecules

Mass spectrometry (MS) is widely accepted as a 'gold-standard' method for the identification of chemicals or biological products. It is, nowadays, applied in highly diversified domains like those directly or indirectly tied to healthcare or regulation-driven demand such as drug development, diagnostics, food and environmental safety testing. Diamond-like carbon (DLC), an amorphous carbon with mixed levels of sp^3 and sp^2 hybridized carbons, has been successfully used for matrix-free laser desorption/ionization mass spectrometry (LDI-MS) particularly for the analysis of small metabolites such as carbohydrates, lipids, and low-molecular-weight peptides [184]. Coffinier et al. reported, for the first time, on the use of boron-doped diamond nanowires (BDD NWs) as an inorganic substrate for matrix-free LDI-MS analysis of small molecules [76]. The diamond nanowires are prepared by reactive ion etching (RIE) with oxygen plasma of highly boron-doped (the boron level is

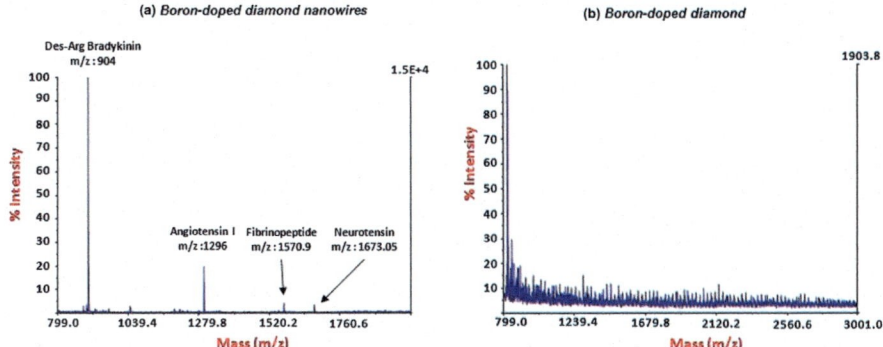

Fig. 10.19 MS spectra obtained for a peptide mixture (50 fmol uL⁻¹) on BDD NWs (**a**) and NcBDD starting material (**b**) substrates

10^{19} B cm⁻³) or undoped nanocrystalline diamond substrates. The resulting diamond nanowires are coated with a thin silicon oxide layer that confers a super-hydrophilic character to the surface. To minimize droplet spreading, the nanowires were chemically functionalized with octadecyltrichlorosilane (OTS) and then UV/ ozone treated to reach a final water contact angle of 120°. For LDI-MS experiments: a 1 μL droplet containing 50 fmol μL⁻¹ of a peptide mixture, Des-Arg1-bradykinin (904 Da), angiotensin I (1296 Da), fibropeptide (1571 Da) and neurotensin (1672 Da, 10 fmol μL⁻¹), was deposited on the diamond nanowire surface, dried and then exposed to a pulsed nitrogen laser (λ = 337 nm, 3.68 eV). The diamond nanowires act as an energy absorbing material. It should be noticed that the SiO$_x$ overlayer was kept during the LDI-MS experiments to favor heat confinement inside the diamond nanowires. Figure 10.19a shows the mass spectrum of the peptide mixture obtained on BDD NWs interface. As it can be seen, all peptides have been detected with relatively high signal intensity. As a control, the same experiment was performed on a planar nanocrystalline BDD, i.e. the same interface without any RIE step process etching. The absence of peaks in the MS spectrum (Fig. 10.19b) clearly indicates that the presence of nanostructures on the BDD substrate is mandatory for achieving D/I of biomolecules. Table 10.2 displays the signal to noise (S/N) ratio values for the detected peptides by LDI-MS on BDD and UDD NWs. The S/N ratios obtained on BDD nanowires are 1078, 431, 55 and 59.3 for [Des-Arg1]-bradykinin, angiotensin I, [Glu1]-fibrinopeptide B, and neurotensin, respectively, whereas those obtained on UDD nanowires are 115, 44, 8.3 and 4.3. We can notice that the S/N ratio values obtained using the UDD NWs are very low in comparison with those of BDD NWs, suggesting that to reach an efficient D/I on diamond, the surface should be nanostructured with a dense layer of wires and display antireflective properties. Boron-doped diamond nanowire substrate was successfully used as an energy absorbing material under UV laser irradiation, allowing D/I of small molecules and their mass spectrometry analysis. The sub-bandgap absorption and heat confinement inside the wires lead to thermal

Table 10.2 Different small molecules analyzed by LDI-MS on B-doped (BDD) and undoped (UDD) diamond nanowires

Analytes	Concentrations/ μL	Main peak (m/z)	Signal/ noise BDD NW	Signal/ noise UDD NW	Fragments/ adducts (m/z)
Des-Arg1 bradykinin	50 fmol	904	1078	115	–
Angiotension I	50 fmol	1296	431	44	–
[Glu1]-fibrinopeptide B	50 fmol	1570.9	55	8.3	–
Neurotensin	10 fmol	1673.05	59.3	4.3	–
Verapamil	2 pmol, 200 amol and 200 zmol	455	1187, 457, and 41[a]	–	303
Betaine	1 pmol	118.1	318	–	–
Histidine	1 pmol	156	453	–	−178 [M + Na]$^+$
Cortisone	1 pmol	361	20	–	–

[a]2 pmol, 200 amol, and 200 zmol, respectively

ionization of analytes. The sensitivity is at about the fmol level for the tested peptides. A set of small molecules have also been assessed with m/z ranging from 100 to 460 including an amino acid and a corticosteroid. A detection limit of 200 zmol μL^{-1} for verapamil was attained using these substrates. The requirements for the realization of the LDI process using diamond interface are: the diamond should be nanostructured (dense black layer of nanowires), small sized, and highly doped and should provide a thermal confinement inside the wires through SiOx overlayer coating. More recently, Szunerits reviewed the application of diamond nanowires in mass spectrometry [185].

10.4.4 Suspended Single-Crystal Diamond Nanowires (SCD) for High-Performance Nano-Electromechanical Switches

The nano- or micro-electro-mechanical (N/MEM) switch is one of the most attractive fields due to its zero-current leakage, low power consumption, and no necessity for the formation of a p–n junction, and can be applied in various fields such as digital logic circuits, memory, and radio-frequency wireless communication. The current N/MEM switches based on silicon technologies are suffering from problems of stiction, abrasion, and poor mechanical and tribological properties, leading to the poor reliability of the electric contacts [186, 187]. In particular, silicon-based switches are not able to work under extreme conditions such as high temperature, environments with corrosive chemicals, high-speed/high-power radiofrequency switches, etc. Diamond is the ideal material for high-performance N/MEMS devices due to its outstanding properties such as its extremely high Young's modulus, the highest hardness, hydrophobic surface, low mass density, the highest thermal conductivity, high corrosion resistance upon caustic chemicals, and biocompatibility. However, all the reported diamond N/MEMS devices are based on polycrystalline or nanocrystalline films, which have the disadvantages of the existence of grain boundaries, impurities, and large stress in the films; difficulty in electrical conductivity control; and poor reproducibility [188, 189]. Therefore, N/MEMS devices made from these diamonds suffer from degradation in performance and poor reproducibility.

Liao et al. demonstrate high-performance all-SCD NEM switches by developing a batch fabrication process for suspended SCD nanowires with well-controlled dimensions and by proposing a lateral device concept [58]. The SCD NEMS switches utilize a unique concept of diamond (conductor)-on-diamond (insulator), different from those of silicon-based technology. It is revealed that the SCD NEM switches exhibit no stiction, high controllability, high reproducibility, high-temperature stability above 500 K, and repeated switching. This work not only opens a novel class of NEM switches with all diamond, based on the concept of diamond-on-diamond, but also has an extensive impact on the field of NEMS.

10.4.5 Diamond Nanowires for Sensors

Planar boron-doped diamond (BDD) electrodes have been recognized as one of the most promising electrode materials and sensing platforms due to the unique physical properties of the diamond film. The diamond interface is chemically stable, exhibits favourable biocompatibility and shows an enlarged potential window together with a low background current, as compared to other electrode materials such as gold or glassy carbon [3]. The electronic and chemical properties can be tuned by changing the surface termination of diamond [186]. BDD electrodes have been consequently used for a wide range of electrochemical applications [3, 5]. In the past years, diamond nanowires was motivated by increasing the surface area of the electrode to achieve higher sensitivity and selectivity, as compared to planar BDD interfaces [67, 68].

Yang et al. introduced for the first time the electrochemical application of vertically aligned diamond nanowires for DNA sensing. Nanowires separated by approximately 11 nm were selected because anchoring DNA molecules onto these wires will result in a density of DNA of about 10^{12} cm^{-2}, which is promising for DNA sensing with high efficiency [68, 69, 32]. As DNA self-aligns with the phenyl linker groups, functionalization of the nanowire tips produces a pattern of dispersed DNA bonding governed by the nanowires' structure. Sensitivity curves for DNA hybridization were measured by varying the concentration of complementary target

DNA from 1 μM to 10 pM. The overall performance of the diamond nanowire
DNA sensor is compared with published data for Au [190–192], and polycrystalline
diamond using comparable DNA structures in Fig. 10.20a. Sensing with diamond
nanowires is about 100–1000 times better than with smooth surfaces of Au or
polycrystalline diamond. To identify the sensitivity limit exactly, experiments with
between 0 and 10 pm of complementary DNA were performed, and the results
(Fig. 10.20b) indicate a sensitivity limit of around 2 pM.

Diamond nanowires electrodes allow also the direct electrochemical detection of
glucose under strong basic conditions [36, 43, 93]. Our group [93] reported synthesis of boron-doped diamond nanowires arrays film by hot filament chemical
vapor deposition (HFCVD) method using Si nanowires as template. This BDDNF
electrode exhibits very attractive electrochemical performance compared to conventional planar boron-doped diamond (BDD) electrodes, notably improved sensitivity and selectivity for glucose detection. It is clear that the electrochemical
response to glucose at the BDDNF electrode is very fast in reaching a dynamic
equilibrium upon each addition of the sample solution, generating a steady-state
current signal within a short time (less than 20 s). The calibration curve for the
electrochemical responses of the BDDNF electrode to glucose is at 0.7 V (vs. SCE)
in the concentration range between 0 and 15 mM. The response to glucose displays
a linear range from 0 to 7 mM with a correlation coefficient (R) of 0.993 and a
sensitivity of 8.1 $\mu AmM^{-1} cm^{-2}$ (slope). The limit of detection was estimated at a
signal-to-noise ratio of 3 to be 0.2 ± 0.01 μM. Moreover, BDDNF electrode is very
favorable for the selective determination of glucose in the presence of AA and UA.

Diamond nanowires are also adapted for the sensitive electrochemical detection
of aromatic amino acids such as tryptophan and tyrosine, two important precursors
of adrenaline, dopamine or melatonine [43]. Szunerits et al. reported on the fabrication and electrochemical investigation of boron-doped diamond nanowires
(BDD NWs) electrodes [75]. The interface with the most favourable electrochemical response is investigated for the detection of tryptophan using differential
pulse voltammetry. As shown in Fig. 10.21, a detection limit of 5×10^{-7} M was

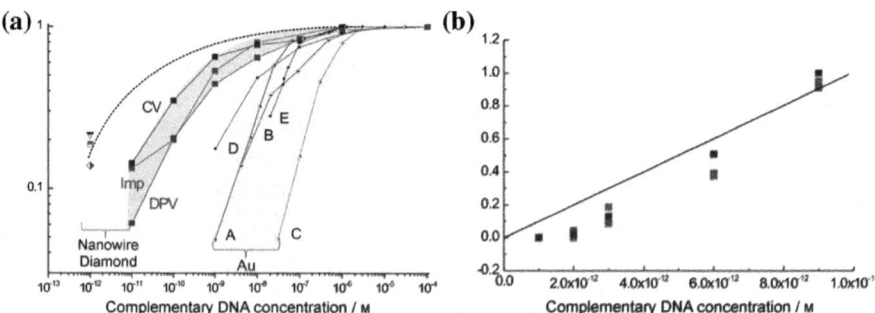

Fig. 10.20 **a** Sensitivity curves of diamond nanowires compared with those of gold electrodes
(from [14] (A), [15] (B), [16] (C), and [17] (D)) and diamond [13] (E). **b** Detection limit measured
on diamond nanowires

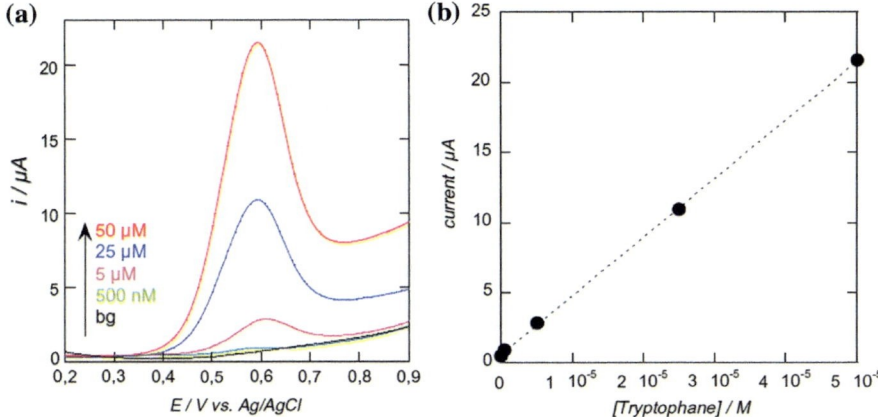

Fig. 10.21 **a** Differential pulse voltammograms of different concentrations of tryptophan (50, 25, 5, 0.5 μM) in Na_2PO_4/NaOH (pH 11) on oxidized BDD NWs; **b** calibration curve

obtained on BDD NWs. This is significantly lower than on planar polycrystalline BDD [193], but an order of magnitude higher than reported on glassy carbon electrodes modified with single-walled carbon nanotube films [194]. The oxidation of other small molecules such as dopamine, uric acid and ascorbic acid was reported by Shalini et al. using nitrogen-doped diamond nanowire electrodes [52, 195].

Up to date, the preliminary success on the diamond nanowires based gas sensor has been achieved. Feng [196] synthesized well-arranged nitrogen-doped ultra-nanocrystalline diamond (UNCD) nanowires (NWs), which has excellent sensing performances to CH_4. UNCD differs from NCD and MCD in its small grain sizes ranging from 3 to 5 nm and a nanoscaled root mean square (RMS) roughness of 4–7 nm, which provides more adsorption sites for gas molecules resulting in optimized sensing performances. Recently, the CO sensor arrays consist of three groups of well-defined B-UNCD NWs have been successfully prepared by Peng on a single Si wafer via MPCVD, followed by EBL/RIE processes [197]. The three groups of B-UNCD NWs all have high selectivity and responses to CO. The nanosizing and the grain boundaries of UNCD crystal are ascribed as the main reason for high response to CO.

10.4.6 Other Applications of Diamond Nanowires

In addition to the application mentioned above, diamond nanowires are also used in many other fields, such as electrochemical catalysis [95, 198, 199], supercapacitor [96, 97, 200], drug delivery [45, 201, 202], plasma illumination [203], degradation of organic pollutants [204], and so on. It is precisely because of its good performance and wide application of diamond nanowires that attract more and more interest in research.

10.5 Conclusions and Outlook

Numerous efforts have been directed to develop the various synthetic methods to prepare diamond nanowires, including reactive-ion-etching, plasma post-treatment carbon nanotubes, transfer fullerene to diamond nanowires at high temperature and high pressure, template or catalyst assisted CVD method, and so on. Template-assisted synthesis may be an effective approach for realization of reproducible diamond nanowires due to its unparalleled ability to produce highly controlled nanostructures.

On other hand, simulation and experiments have verified that 1D diamond nanowires possess excellent properties, including mechanical, electron field emission, structural stability, good electrochemical properties and so on. Many practical applications, such as EFE device, high-performance nano-electromechanical switches, and electrochemical biosensor etc., have been exploited. However, the promise of the application of the diamond nanowires has not been fulfilled. For example, despite the impressive progress in the development of BDD-based amperometric biosensors, the promise of the application of these biosensors in real biological systems has not been fulfilled, and there are still many challenges and obstacles related to the achievement of a highly stable and reliable continuous biomolecules monitoring.

Future developments will rely upon large-scale synthesis of diamond nanowires, and the close collaboration of analytical technology, electrochemistry, biological engineering, nanoelectronics and other relative technologies to ensure its effective application. Continued investigation into this area will likely yield many new synthetic approaches and open up larger application space of diamond nanowires.

Acknowledgements The authors appreciate the supports of the International Science & Technology Cooperation Program of China (no. 2013DFG50150), the Natural Foundation of Sciences of the People's Republic of China (Grant no. 21,175,144, and 20,903,111) and the Key Project of Beijing Natural Science Foundation (Grant No. 2,120,002).

References

1. V.V. Brazhkin, Interparticle interaction in condensed media: some elements are 'more equal than others'. Phys.-Usp. **52**, 369–376 (2009)
2. R.F. Davis, *Diamond Films and Coatings* (Noyes Publications, New Jersey, 1992)
3. R.L. McCreery, Advanced carbon electrode materials for molecular electrochemistry. Chem. Rev. **108**, 2646–2687 (2008)

4. W. Yang, O. Auciello, J.E. Butler, W. Cai, J.A. Carlisle, J.E. Gerbi, D.M. Gruen, T. Knickerbocker, T.L. Lasseter, J.N. Russell, J. Smith, R.J. Hamers, DNA-modified nanocrystalline diamond thin-films as stable, biologically active substrates. Nature Mater. **1**, 253–257 (2002)
5. Y. Zhou, J. Zhi, Y. Zou, W. Zhang, S.T. Lee, Direct electrochemistry and electrocatalytic activity of cytochrome c covalently immobilized on a boron-doped nanocrystalline diamond electrode. Anal. Chem. **80**, 4141–4146 (2008)
6. T. Watanabe, T.A. Ivandini, Y. Makide, A. Fujishima, Y. Einaga, Selective detection method derived from a controlled diffusion process at metal-modified diamond electrodes. Anal. Chem. **78**, 7857–7860 (2006)
7. A. Suzuki, T.A. Ivandini, K. Yoshimi, A. Fujishima, G. Oyama, T. Nakazato, N. Hattori, S. Kitazawa, Y. Einaga, Fabrication, characterization, and application of boron-doped diamond microelectrodes for in vivo dopamine detection. Anal. Chem. **79**, 8608–8615 (2007)
8. H. Gu, X.D. Su, K.P. Loh, Electrochemical impedance sensing of DNA hybridization on conducting polymer film-modiried diamond. J. Phys. Chem. B **109**, 13611–13618 (2005)
9. W. Yang, J.E. Butler, J.N. Russell, R.J. Hamers, Interfacial electrical properties of DNA-modified diamond thin films: Intrinsic response and hybridization-induced field effects. Langmuir **20**, 6778–6787 (2004)
10. B. Rezek, D. Shin, C.E. Nebel, Properties of hybridized DNA arrays on single-crystalline undoped and boron-doped (100) diamonds studied by atomic force microscopy in electrolytes. Langmuir **23**, 7626–7633 (2007)
11. J.L. Davidson, R. Ramesham, C. Ellis, J. Electrochem. Soc. **137**, 3206 (1990)
12. B.J.M. Hausmann, M. Khan, Y. Zhang, T.M. Babinec, K. Martinick, M. McCutcheon, P.R. Hemmer, M. Loncar, Diam. Relat. Mater. **19**, 621 (2010)
13. C.E. Nebel, N. Yang, H. Uetsuka, E. Osawa, N. Tokuda, O. Williams, Diam. Relat. Mater. **18**, 910 (2009)
14. T.M. Babinec, B.J.M. Hausmann, M. Khan, Y. Zhang, J.R. Maze, P.R. Hemmer, M. Loncar, Nat. Nanotechnol. **5**, 195 (2010)
15. I. Friedler, C. Sauvan, J.P. Hugonin, P. Lalanne, J. Claudon, J.M. Gérard, Opt. Express **17**, 2095 (2009)
16. Y. Tao, C.L. Degen, Single-crystal diamond nanowire tips for ultrasensitive force microscopy. Nano Lett. **15**, 7893–7897 (2015)
17. A.S. Barnard, S.P. Russo, I.K. Snook, Nano Lett. **3**, 1323 (2003)
18. A.S. Barnard, S.P. Russo, I.K. Snook, Surf. Sci. **538**, 204 (2003)
19. G.A. Denu, Z.C. Liu, J. Fu, H. Wang, X: A finite element analysis of the effects of geometrical shape on the elastic properties of chemical vapor deposited diamond nanowire. AIP Adv. **7**, 015025 (2017)
20. H. Shiomi, Reactive ion etching of diamond in O-2 and CF4 plasma, and fabrication of porous diamond for field emitter cathodes. Jpn. J. Appl. Phys. **36**, 7745–7748 (1997)
21. H. Masuda, M. Watanabe, K. Yasui, D. Tryk, T. Rao, A. Fujishima, Fabrication of a Nanostructured Diamond Honeycomb Film. Adv. Mater. **12**, 444–447 (2000)
22. R. Arenal, P. Bruno, D.J. Miller, M. Bleuel, J. Lal, D.M. Gruen, Diamond nanowires and the insulator-metal transition in ultrananocrystalline diamond films. Phys. Rev. B **75**, 195431 (2007)
23. R. Arenal, G. Montagnac, P. Bruno, D.M. Gruen, Multiwavelength Raman spectroscopy of diamond nanowires present in n-type ultrananocrystalline films. Phys. Rev. B **76**, 245316 (2007)
24. I. Vlasov et al., Hybrid diamond-graphite nanowires produced by microwave plasma chemical vapor deposition. Adv. Mater. **19**, 4058–4062 (2007)
25. K. Panda, K.J. Sankaran, B.K. Panigrahi, N.-H. Tai, I.N. Lin, Direct observation and mechanism for enhanced electron emission in hydrogen plasma-treated diamond nanowire films. ACS Appl. Mater. Interfaces **6**, 8531–8541 (2014)
26. K.J. Sankaran et al., Structural and electrical properties of conducting diamond nanowires. ACS Appl. Mater. Interfaces **5**, 1294–1301 (2013)

27. I.N. Kholmanov et al., Improved electrical conductivity of graphene films integrated with metal nanowires. Nano Lett. **12**, 5679–5683 (2012)
28. M. Shellaiah, T.H. Chen, T. Simon, L.C. Li, K.W. Sun, F.H. Ko, An Affordable Wet Chemical Route to Grow Conducting Hybrid Graphite-Diamond Nanowires: Demonstration by A Single Nanowire Device. Sci. Reports **7**, 11243 (2017)
29. Q.X. Liu, C.X. Wang, S.W. Li, J.X. Zhang, G.W. Yang, Nucleation stability of diamond nanowires inside carbon nanotubes: A thermodynamic approach. Carbon **42**, 629–633 (2004)
30. A.S. Barnard, S.P. Russo, I.K. Snook, Surface structure of cubic diamond nanowires. Surf. Sci. **538**, 204–210 (2003)
31. C. Terashima, K. Arihara, S. Okazaki, T. Shichi, D.A. Tryk, T. Shirafuji, N. Saito, O. Takai, A. Fujisima, Fabrication of vertically aligned diamond whiskers from highly boron-doped diamond by oxygen etching. ACS Appl. Mater. Interfaces **3**, 177–182 (2011)
32. N. Yang, H. Uetsuka, E. Osawa, C.E. Nebel, Vertically aligned diamond nanowires for DNA sensing. Angew. Chem. Int. Ed. **47**, 5183–5185 (2008)
33. Y. Ando, Y. Nishibayashi, K. Kobashi, T. Hirao, K. Oura, Smooth and high-rate reactive ion etching of diamond. Diam. Relat. Mater. **11**, 824–827 (2002)
34. E.S. Baik, Y.J. Baik, S.W. Lee, D. Jeon, Fabrication of diamond nano-whiskers. Thin Solid Films **377–378**, 295–298 (2000)
35. W. Smirnov, A. Kriele, N. Yang, C.E. Nebel, Aligned diamond nano-wires: Fabrication and characterisation for advanced applications in bio- and electrochemistry. Diam. Relat. Mater. **19**, 186–189 (2010)
36. N. Yang, W. Smirnov, C.E. Nebel, Three-dimensional electrochemical reactions on tip-coated diamond nanowires with nickel nanoparticles. Electrochem. Commun. **27**, 89–91 (2013)
37. K. Wakui, Y. Yonezu, T. Aoki, M. Takeoka, K. Semba, Simple method for fabrication of diamond nanowires by inductively coupled plasma reactive ion etching. Jpn. J. Appl. Phys. **56**(5), 058005 (2017)
38. P. Subramanian, S. Kolagatla, S. Szunerits, Y. Coffinier, W.S. Yeap, K. Haenen, R. Boultherroub, A. Schechter, Atomc Force Microscopic and Raman Investigation of Boron-Doped Diamond Nanowire Electrodes and Their Activity toward Oxygen Reduction. J. Phys. Chem. C **121**(6), 3397–3403 (2017)
39. Q.Q. Jiang, W.X. Li, C.C. Tang, Y.C. Chang, T.T. Hao, X.Y. Pan, H.T. Ye, J.J. Li, C.Z. Gu, Large scale fabrication of nitrogen vacancy-embedded diamond nanostructures for single-photon source applications. Chin. Phys. B **25**(11), 118105 (2016)
40. Q. Wang, Y. Coffinier, M.S. Li, R. Boukherroub, S. Szunerits, Light-triggered release of biomolecules from diamond nanowire electrodes. Langmuir **32**, 6515–6523 (2016)
41. C.J. Widmann, C. Giese, M. Wolfer, D. Brink, N. Heidrich, C.E. Nebel, Fabrication and characterization of single crystalline diamond nanopillars with NV-centers. Diam. Relat. Mater. **54**, 2–8 (2015)
42. L. Marcon, A. Addad, Y. Coffinier, R. Boukherroub, Cell micropatterning on superhydrophobic diamond nanowires. Acta Biomater. **9**, 4585–4591 (2013)
43. Q. Wang, P. Subramanian, M.S. Li, W.S. Yeap, K. Haenen, Y. Coffinier, R. Boukherroub, S. Szunerits, Non-enzymatic glucose sensing on long and short diamond nanowire electrodes. Electrochem. Commun. **34**, 286–290 (2013)
44. Q. Wang, A. Vasilescu, P. Subramanian, A. Vezeanu, V. Andrei, Y. Coffinier, M.S. Li, R. Boukherroub, Szunerits S: simultaneous electrochemical detection of tryptophan and tyrosine using boron-doped diamond and diamond nanowire electrodes. Electrochem. Commun. **35**, 84–87 (2013)
45. P. Subramanian, Y. Coffinier, D. Steinmüller-Nethl, J. Foord, R. Boukherroub, S. Szunerits, Diamond nanowires decorated with metallic nanoparticles: A novel electrical interface for the immobilization of histidinylated biomolecuels. Electrochim. Acta **110**, 4–8 (2013)

46. L.T. Sun, J.L. Gong, Z.Y. Zhu, D.Z. Zhu, Z.X. Wang, W. Zhang, J.G. Hu, Q.T. Li, Synthesis and characterization of diamond nanowires from carbon nanotubes. Diam. Relat. Mater. **14**, 749–752 (2005)
47. L.T. Sun, J.L. Gong, D.Z. Zhu, Z.Y. Zhu, S.X. He, Diamond nanorods from carbon naotubes. Adv. Mater. **16**, 1849–1853 (2004)
48. N. Dubrovinskaia, L. Dubrovinsky, Aggregated diamond nanorods, the densest and least compressible form of carbon. Appl. Phys. Lett. **87**, 083106 (2005)
49. C.H. Hsu, S.G. Cloutier, S. Palefsky, J. Xu, Synthesis of diamond nanowires using atmospheric-pressure chemical vapor deposition. Nano Lett. **10**, 3272–3276 (2010)
50. C.H. Hsu, J. Xu, Diamond nanowire—a challenge from extremes. Nanoscale **4**, 5293–5299 (2012)
51. D. Aradilla, F. Gao, G. Lewes-Malandrakis, W. Muller-Sebert, P. Gentile S. Pouget, C.E. Nebel, G. Bidan, Powering electrodes for high performance aqueous micro-supercapacitors: diamond-coated silicon nanowires operating at a wide cell voltage of 3 V. Electrochim. Acta **242**, 173–179 (2017)
52. J. Shalini, K.J. Sankaran, C.L. Dong, C.Y. Lee, N.H. Tai, I.N. Lin, Insitu detection of dopamine using nitrogen incorporated diamond nanowire electrode. Nanoscale **5**, 1159–1167 (2013)
53. K.J. Sankaran, Y.F. Lin, W.B. Jian, H.C. Chen, K. Panda, B. Sundaravel, C.L. Dong, N.H. Tai, I.N. Lin, Structural and electrical properties of conducting diamond nanowires. ACS Appl. Mater. Interfaces **5**, 1294–1301 (2013)
54. K.J. Sankaran, N. Kumar, H.C. Chen, C.L. Dong, A. Bahuguna, S. Dash, A.K. Tyagi, C.Y. Lee, N.H. Tai, I.N. Lin, Sci. Adv. Mater. **5**, 687–698 (2013)
55. K. Panda, K.J. Sankaran, B.K. Panigrahi, N.H. Tai, I.N. Lin, Direct observation and mechanism for enhanced electron emission in hydrogen plasma-treated diamond nanowire films. ACS Appl. Mater. Interfaces **6**, 8531–8541 (2014)
56. G.S. Oehrlein, Reactive-ion etching. Phys. Today **39**, 26–33 (1986)
57. C.Y. Li, Y. Hatta, Preparation of diamond whiskers using Ar/O2 plasma etching. Diam. Relat. Mater. **14**, 1780–1783 (2005)
58. M.Y. Liao, S. Hishita, E. Watanabe, S. Koizumi, Y. Koide, Suspended single-crystal diamond nanowires for high-performance nanoelectromechanical switches. Adv. Mater. **22**, 5393–5397 (2010)
59. Y. Tzeng, J. Wei, J.T. Woo, W. Lanford, Freestanding single-crystalline chemical vapor deposited diamond films. Appl. Phys. Lett. **63**, 2216–2218 (1993)
60. E.S. Baik, Y.J. Baik, D. Jeon, Aligned diamond nanowhiskers. J. Mater. Res. **15**, 923–926 (2000)
61. Y.S. Zou, T. Yang, W.J. Zhang, Y.M. Chong, B. He, I. Bello, S.T. Lee, Fabrication of diamond nanopillars and their arrays. Appl. Phys. Lett. **92**, 053105 (2008)
62. W. Smirnov, A. Kriele, R. Hoffmann, E. Sillero, J. Hees, O.A. Williams, N. Yang, C. Kranz, C.E. Nebel, Diamond-modified afm probes: from diamond nanowires to atomic force microscopy-integrated boron-doped diamond electrodes. Anal. Chem. **83**, 4936–4941 (2011)
63. W. Janssen, E. Gheeraert, Dry etching of diamond nanowires using self-organized metal droplet masks. Diam. Relat. Mater. **20**, 389–394 (2011)
64. S. Okuyama, *Matsushita SI* (Preparation of periodic microstructured diamond surfaces. chemistry letters, Fujishima A, 2000), pp. 534–535
65. S. Okuyama, S.I. Matsushita, A. Fujishima, Periodic submicrocylinder diamond surfaces using two-dimensional fine particle arrays. Langmuir **18**, 8282–8287 (2002)
66. B.J.M. Hausmann, M. Khan, Y. Zhang, T.M. Babinec, K. Martinick, M. McCutcheon, P.R. Hemmer, M. Loncar, Fabrication of diamond nanowires for quantum information processing applications. Diam. Relat. Mater. **19**, 621–629 (2010)
67. N. Yang, H. Uetsuka, E. Osawa, C.E. Nebel, Vertically aligned nanowires from Boron-doped diamond. Nano Lett. **8**, 3572–3576 (2008)
68. N. Yang, H. Uetsuka, C.E. Nebel, Biofunctionalization of Vertically aligned diamond nanowires. Adv. Funct. Mater. **19**, 887–893 (2009)

69. N. Yang, H. Uetsuka, O.A. Williams, E. Osawa, N. Tokuda, C.E. Nebel, Vertically aligned diamond nanowires: fabrication, characterization, and application for DNA sensing. Phys. Status Solidi A **206**, 2048–2056 (2009)

70. N. Tokuda, H. Umezawa, T. Saito, K. Yamabe, H. Okushi, S. Yamasaki, Surface roughening of diamond (001) films during homoepitaxial growth in heavy boron doping. Diamond Relat. Mater. **16**, 767–770 (2007)

71. A. Kruger, F. Kataoka, M. Ozawa, T. Fujino, Y. Suzuki, A.E. Aleksenskii, A.Y. Vul, E. Osawa, Unusually tight aggregation in detonation nanodiamond: Identification and disintegration. Carbon **43**, 1722–1730 (2005)

72. O.A. Williams, O. Douheret, M. Daenen, K. Haenen, E. Osawa, M. Takahashi, Enhanced diamond nucleation on monodispersed nanocrystalline diamond. Chem. Phys. Lett. **445**, 255–258 (2007)

73. O.A. Williams, M. Daenen, J. Dhaen, K. Haenen, J. Maes, V.V. Moshchalkov, M. Nesladek, D.M. Gruen, Comparison of the growth and properties of ultrananocrystalline diamond and nanocrystalline diamond. Diam. Relat. Mater.**15**, 654–658 (2006)

74. M. Wei, C. Terashima, M. Lv, A. Fujishima, Z.Z. Gu, Boron-doped diamond nanograss array for electrochemical sensors. Chem. Commun. **24**, 3624–3626 (2009)

75. S. Szunerits, Y. Coffinier, E. Galopin, J. Brenner, Boukherroub R: Preparation of boron-doped diamond nanowires and their application for sensitive electrochemical detection of tryptophan. Electrochem. Commun. **12**, 438–441 (2010)

76. Y. Coffinier, S. Szunerits, H. Drobecq, M. Oleg, R. Boukherroub, Diamond nanowires for highly sensitive matrix-free mass spectrometry analysis of small molecules. Nanosacle **4**, 231–238 (2012)

77. Y. Coffinier, E. Galopin, S. Szunerits, R. Boukherroub, Preparation of superhydrophobic and oleophobic diamond nanograss arrays. J. Mater. Chem. **20**, 10671–10675 (2010)

78. P.S. Shah, T. Hanrath, K.P. Johnston, S.A. Korgel, Nanocrystal and nanowire synthesis and dispersibility in supercritical fluids. J. Phys. Chem. B **108**, 9574–9587 (2004)

79. Y. Wu, P. Yang, Direct observation of vapor-liquid-solid nanowire growth. J. Am. Chem. Soc. **123**, 3165–3166 (2001)

80. S. Kodambaka, J. Tersoff, M.C. Reuter, F.M. Ross, Germanium nanowire growth below the eutectic temperature. Science **316**, 729–732 (2007)

81. J.L. Lensch-Falk, E.R. Hemesath, D.E. Perea, L.J. Lauhon, Alternative catalysts for VSS growth of silicon and germanium nanowires. J. Mater. Chem. **19**, 849–857 (2009)

82. Y. Xia, P. Yang, Y. Sun, Y. Wu, B. Mayers, B. Gates, Y. Yin, F. Kim, H. Yan, One-dimensional nanostructures: Synthesis, characterization, and applications. Adv. Mater. **15**, 353–389 (2003)

83. H. Ringsdorf, B. Schlarb, J. Verzmer, Molecular architecture and function of polymeric oriented systems-models for the study of organization, surface recognition, and dynamics of biomembranes. Angew. Chem. Int. Ed. **27**, 113–158 (1988)

84. C.N.R. Rao, A. Govindaraj, F.L. Deepak, N.A. Gunari, M. Nath, Surfactant-assisted synthesis of semiconductor nanotubes and nanowires. Appl. Phys. Lett. **78**, 1853–1855 (2001)

85. Y. Yin, Y. Lu, Y. Sun, Y. Xia, Silver nanowires can be directly coated with amorphous silica to generate well-controlled coaxial nanocables of silver/silica. Nano Lett. **2**, 427–430 (2002)

86. P.W. May, CVD Diamond: A new technology for future. Endeavour **19**, 101–106 (1995)

87. S.S. Lee, O. Takai, H. Itoh, Uniform coating of CVD diamond on metallic wire substrates. J. Mater. Sci. **32**, 2417–2422 (1997)

88. G. Chollon, R. Naslain, C. Prentice, R. Shatwell, P. May, High temperature properties of SiC and diamond CVD-monofilaments. J. Eur. Ceram. Soc. **25**, 1929–1942 (2005)

89. V. Baranauskas, H.J. Ceragioli, A.C. Peterlevitz, A.F. Durrant, Development of tubes of micro-crystalline diamond and diamond-like carbon. Thin Solid Film **398**, 250–254 (2001)

90. M.K. Singh, E. Titus, J.C. Madaleno, G. Cabral, J. Gracio, Novel two-step method for synthesis of high-density nanocrystalline diamond fibers. Chem. Marer. **20**, 1725–1732 (2008)

91. M.K. Singh, E. Titus, M.G. Willinger, J.C. Madaleno, J. Gracio, Microstructure and electron field emission study of diamond nanorod decorated a-SiO$_2$ nanowires by microwave Ar–CH4/H2 plasma chemical vapor deposition with addition of N2. Diamond Relat. Mater. **18**, 865–869 (2009)

92. J.C. Madaleno, M.K. Singh, E. Titus, G. Cabral, J. Gracio, Electron field emission from patterned nanocrystalline diamond coated a-SiO$_2$ micrometer-tip arrays. Appl. Phys. Lett. **92**, 023113 (2008)

93. D. Luo, L. Wu, J. Zhi, Fabrication of boron-doped diamond nanorod forest electrodes and their application in nonenzymatic amperometric glucose biosensing. ACS Nano **8**, 2121–2128 (2009)

94. D. Luo, L. Wu, J. Zhi, 2-D dimensional micro-network for boron-doped diamond film: fabrication and electrochemical sensing application. Chem. Commun. **46**, 6488–6490 (2010)

95. D. Aradilla, F. Gao, G. Lewes-Malandrakis, W. Muller-Sebert, P. Gentile, M. Boniface, D. Aldakov, B. Iliev, T.J.S. Schubert, C.E. Nebel, Designing 3D multihierarchical hetero-nanostructures for high-performance on-chip hybrid supercapacitors: poly(3,4-(ethylene-dioxy)thiophene)-coated diamond/silicon nanowire electrodes in an aprotic ionic liquid. ACS Appl. Mater. Interfaces. **8**, 18069–18077 (2016)

96. D. Aradilla, F. Gao, G. Lewes-Malandrakis, W. Muller-Sebert, D. Gaboriau, P. Gentile, B. Iliev, T. Schubert, S. Sadki, G. Bidan, C.E. Nebel, A step forward into hierarchically nanostructured materials for high performance micro-supercapacitors: Diamond-coated SiNW electrodes in protic ionic liquid electrolyte. Electrochem. Commun. **63**, 34–38 (2016)

97. F. Gao,G. Lewes-Malandrakis, M.T. Wolfer, W. Müller-Sebert, P. Gentile, D. Aradilla,T. Schubert, C.E. Nebel, Diamond-coated silicon wires for supercapacitor applications in ionic liquids. Diam. Relat. Mater. **51**, 1–6 (2015)

98. C.G. Granqvist, A. Andersson, O. Hundri, Spectrally selective surfaces of Ni-pigmented anodic Al2O3. Appl. Phys. Lett. **35**, 268–270 (1979)

99. C.A. Huber, T.E. Huber, M. Sadoqi, J.A. Lubin, S. Mannlis, C.B. Prater, Nanowire array composites. Science **263**, 800–802 (1994)

100. H. Masuda, T. Yanagishita, K. Yasui, K. Nishio, I. Yagi, T.N. Rao, A. Fujishima, synthesis of well-aligned diamond nanocylinders. Adv. Mater. **13**, 247–249 (2001)

101. F. Keller, M.S. Hunter, D.L. Robinson, Structural features of oxide coatings on aluminium. J. Electrochem. Soc. **100**, 411–419 (1953)

102. H. Masuda, K. Fukuda, Ordered metal nanohole arrays made by A two-step replication of honeycomb structures of anodic alumina. Science **268**, 1466–1468 (1995)

103. I.I. Vlasov, O.I. Lebedev, V.G. Ralchenko, E. Goovaerts, G. Bertoni, G.V. Tendeloo, V.I. Konov, Hybrid diamond –graphite nanowires produced by microwave plasma chemical vapor deposition. Adv. Mater. **19**, 4058–4062 (2007)

104. I.I. Vlasov, V.G. Ralchenko, E. Goovaerts, A.V. Saveliev, M.V. Kanzyuba, Bulk and surface-enhanced Raman spectroscopy of nitrogen-doped ultrananocrystalline diamond films. Phys. Status Solidi A Appl. Mater. Sci. **203**, 3028–3035 (2006)

105. N. Shang, P. Papakonstantinou, P. Wang, A. Zakharov, U. Palnitkar, I.N. Lin, M. Chu, A. Stamboulis, Self-assembled growth, microstructure, and field-emission high-performance of ultrathin diamond nanorods. ACS Nano **3**, 1032–1038 (2009)

106. A.S. Barnard, I.K. Snook, Phase stability of nanocarbon in one dimension: nanotubes versus diamond nanowires. J. Chem. Phys. **120**, 3817–3821 (2004)

107. A.S. Barnard, S.P. Russo, I.K. Snook, Ab initio modeling of diamond nanowire structures. Nano Lett. **3**, 1323–1328 (2003)

108. A.R. Sobia, S. Adnan, A. Mukhtiar, A.A. Khurram, A.A. Turab, A. Awais, A. Naveed, Q. J. Faisal, H. Javaid, G.J. Yu, Effect of nitrogen addition on hydrogen incorporation in diamond nanorod thin films. Curr. Appl. Phys. **12**, 712–717 (2012)

109. J. Shalini, Y.C. Lin, T.H. Chang, K.J. Sankaran, H.C. Chen, C.Y. Lee, N.H. Tai, Ultra-nanocrystalline diamond nanowires with enhanced electrochemical properties. Electrochim. Acta **92**, 9–19 (2013)

110. P. Feng, X.P. Wang, A. Aldalbahi, A.F. Zhou, Methane induced electrical property change of nitrogen doped ultrananocrystalline diamond nanowires. Appl. Phys. Lett. **107**, 233103 (2015)

111. L.Y. Zeng, H.Y. Peng, W.B. Wang, Y.Q. Chen, D. Lei, W. Qi, J.Q. Liang, J.L. Zhao, X.G. Kong, H. Zhang, Nanocrystalline diamond films deposited by the hot cathode direct current plasma chemical vapor deposition method with different compositions of CH4/Ar/H-2 gas mixture. J. Phys. Chem. C **112**, 1401–1406 (2008)

112. L.Y. Zeng, H.Y. Peng, W.B. Wang, Y.Q. Chen, D. Lei, W. Qi, J.Q. Liang, J.L. Zhao, X.G. Kong, H. Zhang, Synthesis and characterization of diamond microcrystals and nanorods deposited by hot cathode direct current plasma chemical vapor deposition method. J. Phys. Chem. C **112**, 6160–6164 (2008)

113. H. Motahari, R. Malekfar, Bottom-up diamond nanorod growth in HFCVD from nanocrystalline diamond film as a template-free method. Mater. Res. Express **4** (2017)

114. F.P. Bundy, H.T. Hall, H.M. Strong, R.H. Wentorf, Man-made diamonds. Nature **176**, 51–55 (1955)

115. B.V. Derjaguin, D.V. Fedoseev, V.M. Lukyanovich, B.V. Spitzin, V.A. Ryabov, A.V. Lavrentyev, Filamentary diamond crystals. J. Cryst. Growth **2**, 380–384 (1968)

116. W.R.L. Lambrecht, C.H. Lee, B. Segall, J.C. Angus, Z.D. Li, M. Sunkara, Diamond nucleation by hydrogenation of the edges of graphitic precursors. Nature **364**, 607–610 (1993)

117. A. Gross, Hydrogen dissociation on metal surfaces - a model system for reactions on surfaces. Appl. Phys. A Mater. Sci. Process. **67**, 627–635 (1998)

118. H.F. Berger, E. Grosslinger, K.D. Rendulic, Coupling of vibrational and translational energy in the adsorption of H_2 on FE(100)—state-resolved sticking coefficients. Surface Sci. **261**, 313–320 (1992)

119. J.H. Zhang, B.Q. Wei, J. Liang, Z.D. Gao, D.H. Wu, Synthesis of diamond from buckytubes by laser and quenching treatment. Mater. Lett. **31**, 79–82 (1997)

120. Y.Q. Hou, D.M. Zhuang, G. Zhang, M.S. Wu, J.J. Liu, Preparation of diamond films by hot filament chemical vapor deposition and nucleation by carbon nanotubes. Appl. Surf. Sci. **185**, 303–308 (2002)

121. Y.Q. Zhu, T. Sekine, T. Kobayashi, T. Takazawa, M. Terrones, H. Terrones, Collapsing carbon nanotubes and diamond formation under shock waves. Chem. Phys. Lett. **287**, 689–693 (1998)

122. B.Q. Wei, J. Liang, Z.D. Gao, J.H. Zhang, Y.Q. Zhu, Y.B. Li, D.H. Wu, The transformation of fullerenes into diamond under different processing conditions. J. Mater. Process. Technol. **63**, 573–578 (1997)

123. L.M. Cao, C.X. Gao, H.P. Sun, G.T. Zou, Z. Zhang, X.Y. Zhang, M. He, M. Zhang, Y.C. Li, J. Zhang, D.Y. Dai, L.L. Sun, W.K. Wang, Synthesis of diamond from carbon nanotubes under high pressure and high temperature. Carbon **39**, 311–314 (2001)

124. H. Yusa, Nanocrystalline diamond directly transformed from carbon nanotubes under high pressure. Diamond Relat. Mater. **11**, 87–91 (2002)

125. B. Wei, J. Zhang, J. Liang, D. Wu, The mechanism of phase transformation from carbon nanotube to diamond. Carbon **36**, 997–1001 (1998)

126. L.T. Sun, J.L. Gong, Z.Y. Zhu, D.Z. Zhu, S.X. He, Nanocrystalline diamond from carbon nanotubes. Appl. Phys. Lett. **84**, 2901–2903 (2004)

127. J. Singh, Nucleation and Growth-Mechanism of Diamond During Hot-Filament Chemical-Vapor-Deposition. J. Mater. Sci. **29**, 2761–2766 (1994)

128. N. Dubrovinskaia, L. Dubrovinsky, F. Langenhorst, S. Jacobsen, C. Liebske, Nanocrystalline diamond synthesized from C60. Diamond Relat. Mater. **14**, 16–22 (2005)

129. J. Voskuhl, M. Waller, S. Bandaru, B.A. Tkachenko, C. Fregonese, B. Wibbeling, P.R. Schreiner, B.J. Ravoo, Nanodiamonds in sugar rings: an experimental and theoretical investigation of cyclodextrin-nanodiamond inclusion complexes. Org. Biomol. Chem. **10**, 4524–4530 (2012)

130. J. Zhang, Y. Feng, H. Ishiwata, Y. Miyata, R. Kitaura, J.E.P. Dahl, R.M.K. Carlson, H. Shinohara, D. Tomanek, Synthesis and Transformation of Linear Adamantane Assemblies inside Carbon Nanotubes. ACS Nano **6**, 8674–8683 (2012)
131. D.A. Britz, A.N. Khlobystov, K. Porfyrakis, A. Ardavan, G.A.D. Briggs, Chemical reactions inside single-walled carbon nano test-tubes. Chem. Commun. **1**, 37–39 (2005)
132. J. Zhang, Y. Miyata, R. Kitaura, H. Shinohara, Preferential synthesis and isolation of (6, 5) single-wall nanotubes from one-dimensional C-60 coalescence. Nanoscale **3**, 4190–4194 (2011)
133. J. Zhang, Z. Zhu, Y.Q. Feng, H. Ishiwata, Y. Miyata, R. Kitaura, J.E.P. Dahl, R.M.K. Carlson, N.A. Fokina, P.R. Schreiner, Evidence of diamond nanowires formed inside carbon nanotubes from diamantine dicarboxylic acid. Angew. Chem. Int. Ed. **52**, 3717–3721 (2013)
134. G.C. McIntosh, M. Yoon, S. Berber, D. Tomanek, Diamond fragments as building blocks of functional nanostructures. Phys. Rev. B **70**, 045401 (2004)
135. W. Piekarczyk, How and why CVD diamond is formed: A solution of the thermodynamic paradox. J. Mater. Sci. **33**, 3443–3453 (1998)
136. O.A. Shenderova, V.V. Zhirnov, D.W. Brenner, Carbon nanostructures. Crit. Rev. Solid State Mater. Sci. **27**, 227–356 (2002)
137. J.T. Tanskanen, M. Linnolahti, A.J. Karttunen, T.A. Pakkanen, From fulleranes and icosahedral diamondoids to polyicosahedral nanowires: structural, electronic, and mechanical characteristics. J. Phys. Chem. C **112**, 11122–11129 (2008)
138. F.L. Liu, Theoretical study on the coplanar double-cage dodecahedrane C35H30. Phys. Chem. Chem. Phys. **6**, 906–909 (2004)
139. F.L. Liu, L. Peng, J.X. Zhao, S.Q. Wang, Theoretical study of two C50H40 isomers with three dodecahedrane cages sharing two pentagons. Int. J. Quantum Chem. **103**, 167–175 (2005)
140. A.S. Barnard, S.P. Russo, I.K. Snook, From nanodiamond to diamond nanowires: structural properties affected by dimension. Philos. Mag. **84**, 899–907 (2004)
141. J. Guo, B. Wen, R. Melnik, S. Yao, T. Li, Geometry and temperature dependent thermal conductivity of diamond nanowires: Anon-equilibrium molecular dynamics study. Phys. E **43**, 155–160 (2010)
142. J. Guo, B. Wen, R. Melnik, S. Yao, T. Li, Molecular dynamics study on diamond nanowires mechanical properties: strain rate, temperature and size dependent effects. Diam. Relat. Mater. **20**, 551–555 (2011)
143. O.A. Shenderova, D.W. Brenner, R.S. Ruoff, Would diamond nanorods be stronger than fullerene nanotubes? Nano Lett. **3**, 805–809 (2003)
144. J.N. Coleman, U. Khan, W.J. Blau, Gun'ko YK: small but strong: a review of the mechanical properties of carbon nanotube-polymer composites. Carbon **44**, 1624–1652 (2006)
145. B.Z. Kupfer, R.K. Ahmad, A. Zainal, R.B. Jackman, Diam. Relat. Mater. **19**, 742 (2010)
146. F. Occelli, P. Loubeyre, R. Letoullec, Properties of diamond under hydrostatic pressures up to 140 GPa. Nat. Mater. **2**, 151–154 (2003)
147. T. Yamanaka, S. Morimoto, H. Kanda, Influence of the isotope ratio on the lattice-constant of diamond. Phys. Rev. B **49**, 9341–9343 (1994)
148. H. Cynn, J.E. Klepeis, C.S. Yoo, D.A. Young, Osmium has the lowest experimentally determined compressibility. Phys. Rev. Lett. **88**, 135701 (2002)
149. H. Peelaers, B. Partoens, F.M. Peeters, Phonon band structure of Si nanowires: a stability analysis. Nano Lett. **9**, 107–111 (2009)
150. H. Peelaers, B. Partoens, F.M. Peeters, Phonons in Ge nanowires. Appl. Phys. Lett. **95**, 122110 (2009)
151. A. Trejo, A. Miranda, L. Rivera, A. Diaz-mendez, M. Cruz-Irisson, Phonon optical modes and electronic properties in diamond nanowires. Micoelectron. Eng. **90**, 92–95 (2012)
152. K.W. Sun, J.Y. Wang, T.Y. Ko, Raman spectroscopy of single nanodiamond: Phonon-confinement effects. Appl. Phys. Lett. **92**, 153115 (2008)

153. Y. Yonezu, K. Wakui, K. Furusawa, M. Takeoka, K. Semba, T. Aoki, Efficient Single-Photon Coupling from a Nitrogen-Vacancy Center Embedded in a Diamond Nanowire Utilizing an Optical Nanofiber. Sci. Rep. **7**, 12985 (2017)
154. S. Li, C.H. Li, B.W. Zhao, Y. Dong, C.C. Li, X.D. Chen, Y.S. Ge, F.W. Sun, A bright single-photon source from nitrogen-vacancy centers in diamond nanowires. Chin. Phys. Lett. **34**, 096101 (2017)
155. J.R. Solano, A.T. Banos, A.M. Duran, E.C. Quiroz, M.C. Irisson, DFT study of anisotropy effects on the electronic properties of diamond nanowires with nitrogen-vacancy center. J. Mol. Model. **23**, 292 (2017)
156. J.H. Zhang, J.X. Cao, X.Y. Chen, J.W. Ding, P.H. Zhang, W. Ren, Diamond nanowires with nitrogen vacancy under a transverse electric field. Phys. Rev. B **91**, 045417 (2015)
157. C.W. Padgett, D.W. Brenner, Influence of chemisorption on the thermal conductivity of single-wall carbon nanotubes. Nano Lett. **4**, 1051–1053 (2004)
158. N.V. Novikov, A.P. Podoba, S.V. Shmegera, A. Witek, A.M. Zaitsev, A.B. Denisenko, W.R. Fahrner, M. Werner, Influence of isotopic content on diamond thermal conductivity. Diamond Relat. Mater. **8**, 1602–1606 (1999)
159. J.F. Moreland, J.B. Freund, G. Chen, The disparate thermal conductivity of carbon nanotubes and diamond nanowires studied by atomistic simulation. Microscale Thermophys. Eng. **8**, 61–69 (2004)
160. C.W. Padgett, O. Shenderova, D.W. Brenner, Thermal conductivity of diamond nanorods: Molecular simulation and scaling relations. Nano Lett. **6**, 1827–1831 (2006)
161. F.J. Himpsel, J.A. Knapp, J.A. Vanvechten, D.E. Eastman, Quantum photoyield of diamond (111)—stable negative-affinity emitter. Phys. Rev. B **20**, 624–627 (1979)
162. B.J. Cui, J. Ristein, L. Ley, Electron affinity of the bare and hydrogen covered single crystal diamond (111) surface. Phys. Rev. Lett. **81**, 429–432 (1998)
163. T. Ito, M. Nishimura, M. Yokoyama, M. Irie, C.L. Wang, Highly efficient electron emitting diode fabricated with single-crystalline diamond. Diamond relat. Mater. **9**, 1561–1568 (2000)
164. I.L. Krainsky, V.M. Asnin, G.T. Mearini, J.A. Dayton, Negative-electron-affinity effect on the surface of chemical-vapor-deposited diamond polycrystalline films. Phys. Rev. B **53**, R7650–R7653 (1996)
165. Y.K. Chang, H.H. Hsieh, W.F. Pong, M.H. Tsai, F.Z. Chien, P.K. Tseng, L.C. Chen, T.Y. Wang, K.H. Chen, D.M. Bhusari, J.R. Yang, S.T. Lin, Quantum confinement effect in diamond nanocrystals studied by X-ray-absorption spectroscopy. Phys. Rev. Lett. **82**, 5377–5380 (1999)
166. N. Jiang, K. Eguchi, S. Noguchi, T. Inaoka, Y. Shintani, Structural characteristics and field electron emission properties of nano-diamond/carbon films. J. Cryst. Growth **236**, 577–582 (2002)
167. S. Gupta, B.L. Weiss, B.R. Weiner, G. Morell, Electron field emission from sulfur-incorporated nanocrystalline carbon thin films. Appl. Phys. Lett. **79**, 3446–3448 (2001)
168. L. Gan, E. Baskin, C. Saguy, R. Kalish, Quantization of 2D hole gas in conductive hydrogenated diamond surfaces observed by electron field emission. Phys. Rev. Lett. **96**, 196808 (2006)
169. W. Zhu, G.P. Kochanski, S. Jin, Low-field electron emission from undoped nanostructured diamond. Science **282**, 1471–1473 (1998)
170. D. Pradhan, I.N. Lin, Grain-Size-Dependent Diamond-Nondiamond Composite Films: Characterization and Field-Emission Properties. ACS Appl. Mater. Interfaces **1**, 1444–1450 (2009)
171. J.P. Thomas, H.C. Chen, N.H. Tai, I.N. Lin, Freestanding Ultrananocrystalline Diamond Films with Homojunction Insulating Layer on Conducting Layer and Their High Electron Field Emission Properties. ACS Appl. Mater. Interfaces **3**, 4007–4013 (2011)

172. M. Shiraishi, M. Ata, Work function of carbon nanotubes. Carbon **39**, 1913–1917 (2001)
173. Z. Xu, X.D. Bai, E.G. Wang, Wang Zl: Field emission of individual carbon nanotube with in situ tip image and real work function. Appl. Phys. Lett. **87**, 163106 (2005)
174. R.C. Smith, S.R.P. Silva, Interpretation of the field enhancement factor for electron emission from carbon nanotubes. J. Appl. Phys. **106**, 014314 (2009)
175. A. Mayer, N.M. Miskovsky, P.H. Cutler, Photon-stimulated field emission from semiconducting (10,0) and metallic (5,5) carbon nanotubes. Phys. Rev. B **65**, 195416 (2002)
176. Q.H. Wang, A.A. Setlur, J.M. Lauerhaas, J.Y. Dai, E.W. Seeling, R.P.H. Chang, A nanotube-based field-emission flat panel display. Appl. Phys. Lett. **72**, 2912–2913 (1998)
177. I. Aharonovich, D. Englund, M. Toth, Solid-state single-photon emitters. Nat. Photonics **10**, 631–641 (2016)
178. P.C. Maurer et al., Room-temperature quantum bit memory exceeding one second. Science **336**, 1283–1286 (2012)
179. C. Kurtsiefer, S. Mayer, P. Zarda, H. Weinfurter, Stable solid-state source of single photons. Phys. Rev. Lett. **85** 290 (2000)
180. T. Gaebel et al., Room-temperature coherent coupling of single spins in diamond. Nat. Phys. **2**, 408–413 (2006)
181. J. Maze et al., Nanoscale magnetic sensing with an individual electronic spin in diamond. Nature **455**, 644–647 (2008)
182. M. Almokhtar, M. Fujiwara, H. Takashima, Takeuchi S: Numerical simulations of nanodiamond nitrogen-vacancy centers coupled with tapered optical fibers as hybrid quantum nanophotonic devices. Opt. Express **22**, 20045–20059 (2014)
183. A. Sipahigil et al., An integrated diamond nanophotonics platform for quantum-optical networks. Science **354**, 847–850 (2016)
184. M. Najam-ul-Haq, M. Rainer, C.W. Huck, P. Hausberger, H. Kraushaar, G.K. Bonn, Nanostructured diamond-like carbon on digital versatile disc as a matrix-free target for laser desorption/ionization mass spectrometry. Anal. Chem. **80**, 7467–7472 (2008)
185. S. Szunerits, Y. Coffinier, R. Boukherroub, Diamond nanowires: a novel platform for electrochemistry and matrix-free mass spectrometry. Sensors **15**, 12573–12593 (2015)
186. A. Socoliuc, Atomic-scale control of friction by actuation of nanometer-sized contacts. Science **313**, 207–210 (2006)
187. J.M. Kim, J.H. Park, C.W. Baek, Y.K. Kim, The SiOG-based single-crystalline silicon (SCS) RF MEMS switch with uniform characteristics. J. Microelectromech. Syst. **13**, 1036–1042 (2004)
188. M. Adamschik, J. Kuserer, P. Schmid, K.B. Schad, D. Grobe, E. Kohn, Diamond microwave micro relay. Diam. Relat. Mater. **11**, 672–676 (2002)
189. V.P. Adiga, A.V. Sumant, S. Suresh, C. Gudeman, O. Auiello, J.A. Carlisle, R.W. Carpick, Mechanical stiffness and dissipation in ultrananocrystalline diamond microresonators. Phys. Rev. B **79**, 245403 (2009)
190. G. Carpini, F. Lucarelli, G. Marrazza, M. Mascini, Oligonucleotide-modified screen-printed gold electrodes for enzyme-amplified sensing of nucleic acids. Biosens. Bioelectron. **20**, 167–175 (2004)
191. O. Paenke, A. Kirbs, F. Lisdat, Voltammetric detection of single base-pair mismatches and quantification of label-free target ssDNA using. Biosens. Bioelectron. **22**, 2656–2662 (2007)
192. H. Aoki, H. Tao, Gene sensors based on peptide nucleic acid (PNA) probes: Relationship between sensor sensitivity and probe/target duplex stability. Analyst **130**, 1478–1482 (2005)
193. G. Zhao, Y. Qi, Y. Tian, Simultaneous and direct determination of tryptophan and tyrosine at boron-doped diamond electrode. Electroanalysis **18**, 830–834 (2006)
194. W. Huang, G. Mai, Y. Liu, C. Yang, W. Qua, Voltammetric Determination of Tryptophan at a Single-Wall Carbon Nanotubes Modified Electrode. J. Nanosci. Nanotechnol. **4**, 423–427 (2004)
195. J. Shalini, K.J. Sankaran, C.Y. Lee, N.H. Tai, I.N. Lin, An amperometric urea bisosensor based on covalent immobilization of urease on N-2 incorporated diamond nanowire electrode. Biosens. Bioelectron. **56**, 64–70 (2014)

196. P. Feng, X. Wang, A. Aldalbahi, A.F. Zhou, Methane induced electrical property change of nitrogen doped ultrananocrystalline diamond nanowires. Appl. Phys. Lett. **107**, 233103 (2015)
197. X.Y. Peng, J. Chu, L.D. Wang, S.K. Duan, P. Feng, Boron-doped diamond nanowires for CO gas sensing application. Sens. Actuators B Chem. **241**, 383–389 (2017)
198. F. Gao, R. Thomann, C.E. Nebel, Aligned Pt-diamond core-shell nanowires for electrochemical catalysis. Electrochem. Commun. **50**, 32–35 (2015)
199. N.J. Yang, J.S. Foord, X. Jiang, Diamond electrochemistry at the nanoscale: a Rev. Carbon **99**, 90–110 (2016)
200. D. Aradilla, F. Gao, G. Lewes-Malandrakis, W. Muller-Sebert, P. Gentile, S. Pouget, C.E. Nebel, G. Bidan, Powering electrodes for high performance aqueous micro-supercapacitors: diamond-coated silicon nanowires operating at a wide cell voltage of 3 V. Electrochim. Acta **242**, 173–179 (2017)
201. X.F. Chen, W.J. Zhang, Diamond nanostructures for drug delivery, bioimaging, and biosensing. Chem. Soc. Rev. **46**, 734–760 (2017)
202. X.Y. Zhu, S.Y. Kwok, M.F. Yuen, L. Yan, W. Chen, Y. Yang, Z.G. Wang, K.N. Yu, G.Y. Zhu, W.J. Zhang, Dense diamond nanoneedle arrays for enhanced intracellular delivery of drug molecules to cell lines. J. Mater. Sci. **50**, 7800–7807 (2015)
203. K.J. Sankaran, C.J. Yeh, S. Drijkoningen, P. Pobedinskas, M.K. Van Bael, K.C. Leou, I.N. Lin, K. Haenen, Enhancement of plasma illumination characteristics of few-layer graphene-diamond nanorods hybrid. Nanotechnology **28**, 065701 (2017)
204. C.H. Lee, E.S. Lee, Y.K. Lim, K.H. Park, H.D. Park, D.S. Lim, Enhanced electrochemical oxidation of phenol by boron-doped diamond nanowire electrode. RSC Adv. **7**, 6229–6235 (2017)

Chapter 11
Spectroscopy of Nanodiamond Surface: Investigation and Applications

Ashek-I-Ahmed, Elena V. Perevedentseva, Artashes Karmenyan and Chia-Liang Cheng

Abstract Nanodiamond is one of the carbon families that has attracted much attention recently for its versatile spectroscopic properties; render it promising potential in optoelectronic, quantum computation, and bio/medical applications. With sizes ranging from several hundred down to single digit nanometers, nanodiamond represents a group of nanomaterial with rich surfaces both in its physical and chemical properties. Its surface and bulk consist of carbon of bonding of different nature. The sp^2/sp^3 hybridization forming diamond, graphite, disorder/amorphous carbons and defects are easily detectable using infrared and Raman spectroscopy and allow both easy characterization and the surface modification or conjugation with molecules of interest. Recently nanodiamond is considered as one of the most biocompatible nanomaterials. With the newly discussed defects-originated fluorescence, renders nanodiamond suitable for bio-labeling, bio-sensing and drug delivery. In this chapter, the authors review and discuss the spectroscopic studies of nanodiamond surfaces focusing on the infrared spectroscopy, Raman spectroscopy, including SERS, photoluminescence spectroscopy and lifetime analysis. Applications of these methods to detect and analyze surface structural features, surface molecular groups and macromolecules conjugations, interactions between nanodiamond surface moieties and its environments, etc. are discussed. These open new possibilities for bio-medical applications, for multimodal imaging, sensing, and controllable drug delivery of nanodiamond.

11.1 Introduction

Nanodiamond (ND) has been demonstrated as a promising material with various applications in multidisciplinary fields due to its unique properties such as super hardness, high thermal conductivity, stable photoluminescence with visible light

Ashek-I-Ahmed · E. V. Perevedentseva · A. Karmenyan · C.-L. Cheng (✉)
Department of Physics, National Dong Hwa University, Hualien 97401, Taiwan
e-mail: clcheng@gms.ndhu.edu.tw

E. V. Perevedentseva
P.N. Lebedev Physics Institute of Russian Academy of Science, Moscow 119991, Russia

© Springer Nature Switzerland AG 2019
N. Yang (ed.), *Novel Aspects of Diamond*, Topics in Applied Physics 121,
https://doi.org/10.1007/978-3-030-12469-4_11

active nitrogen-vacancy center (with zero-phonon lines, ZPL; NV^- ~ 636 nm, NV^o ~ 575 nm), phonon vibrational signature (1332 cm^{-1}), tailorable surface and innate biocompatibility, etc. [1]. Amongst those, nanodiamond attracts intensive attentions lately in bio and medical applications such as bio-sensor, bio-labeling, imaging and drug delivery [2–9].

Recently, a number of researches on ND in fundamental science and on its applications demonstrate rapid growth of the importance of the diamond and its applications. Review papers in journals and books try to systemize and to structure the knowledge considering the information from different points of view [3–8]. While vibrational (IR and Raman) spectroscopy of ND are well-presented not only in original research works but also in reviews papers, much information of the surface and ND photoluminescence, from our point of view, is not yet sufficiently understood. Along with other approaches that can not be neglected when considering the properties of the ND surface, in this chapter we focus on the spectroscopic studies of nanodiamond surfaces and the influence of surface on the ND's applications. Other issues such as aggregation [10]; surface treatment with laser light; laser-color centers interaction (including some observed nonlinear effects) [11, 12] are of important to the applications of nanodiamond, but will not be the focus of this chapter.

11.2 Diamond Crystal Surface and Surface Spectroscopy

Diamond is a highly crystalline structured material in the group of carbon allotropes, having unique physical/chemical properties such as super hardness, high thermal conductivity, optical transparency, high index of refraction, and surface inertness at room temperature, etc. The crystal structure of the diamond is usually formed as a tetrahedral sp^3-covalent bonding of carbon atoms with four nearest neighbors in a cubic lattice which allows to form the {111}, {110} and {100} facets in a single crystal (Fig. 11.1). Due to the cubic arrangement of the atoms in diamond structure, there can have many facets belonging to a cube, octahedron, dodecahedron and hexahedron etc., therefore the crystals could be shaped with rounded off or roughed edges [13].

Diamond is a wide bandgap semiconductor with bandgap corresponding to the deep ultraviolet (UV) wavelength of 225 nm (5.5 eV). Naturally it is a visibly transparent material in the range of ~400–750 nm; however they exhibit a variety of different colorations such as yellow, green, blue, pink, etc., due to the impurity/defects appearing in the crystal lattice [14]. Diamond is mostly found in a combination with other carbon allotropes, the identification of their structure was initially a complex task. Diffraction techniques are used frequently to characterize the crystalline structure of carbon-based materials. However, the complex combination of diamond, graphite, and amorphous phases in natural or synthesized diamond makes them difficult to be resolved even by electron microscopy and selected area diffraction (SAD). Additionally, electron diffraction patterns of the polycrystalline diamond are similar to those of basal-plane oriented polycrystalline graphite

Fig. 11.1 Crystal structure of diamond in a cubic lattice system

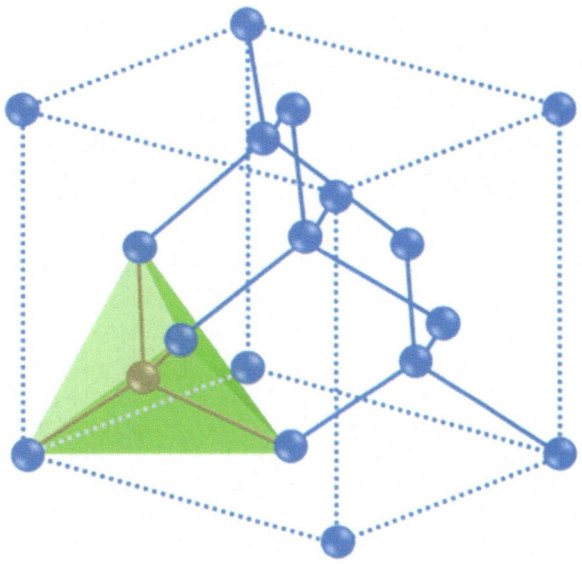

resulting difficulty in separating one pattern to the other. An alternative way of identifying the diamond structure is to use Raman spectroscopy. This technique determines with great accuracy of the bonding states of carbon atoms such as sp^2 for graphite or sp^3 for diamond by exciting their vibrational energy states. The core sp^3 hybridized structure for diamond is distinguished with a single sharp Raman peak at 1332 cm^{-1}, compare to the disordered carbon (D-band) at 1350 cm^{-1}, and graphitic structure (sp^2 bonded carbon) with a band (G-band) centered at 1580 cm^{-1} [15].

For the understanding of diamond surface structure and diamond synthesis an investigation of how hydrogen atoms interact with diamond surfaces is important. Although diamond surface is chemically inert at room temperature, the surfaces are reactive natured at high temperature with gas medium [16]. Figure 11.2A illustrates in a smooth C(111) surface, single CH stretching can be observed via in situ Fourier Transform Infrared Spectroscopy (FTIR). The surface structure of the C(111) surface allows single CH dipole vertically aligned (Fig. 11.3) and therefore reflected in a sharp CH bonding frequency as showed. This spectrum is obtained with single crystal diamond C(111) surface subjected to atomic hydrogen reaction generated by hot filament cracking of the hydrogen molecules (H$_2$). The atomic hydrogen generated is of high energy and also reactive to the surface. This function created CH bonds on the C(111) surface and as an effect; it also etched away the disordered carbon atoms residing on the surface. Therefore, there is an equilibrium state one can achieve with maximum CH on the surface to from domains of hydrogenated surfaces. Further analysis of anisotropic diamond etching dynamics via detection of CH stretches on C(111), C(110), and C(100) surfaces using FTIR

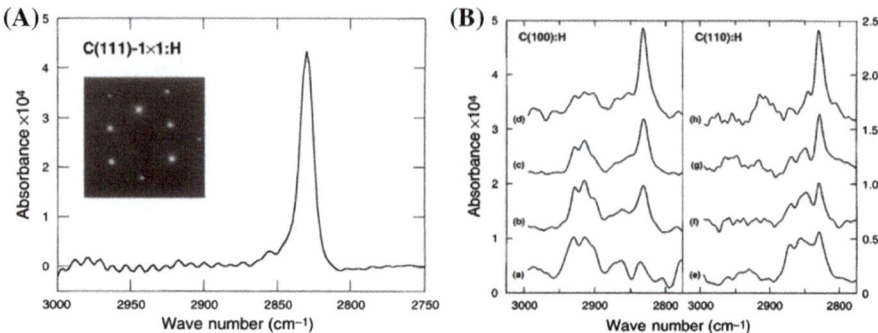

Fig. 11.2 **A** Infrared polarization spectrum of CH stretching on C(111)−(1 × 1) at 800 K. Inset: low-energy electron diffraction pattern of the C(111)−(1 × 1):H surface. **B** Evolution of the polarization spectra of C–H stretches as a function of the hydrogen etching time of 3, 4, 5, and 8 h (**a–d**) for C(100), and 3, 7, 8, and 9 h (**e–h**) for C(110). Reprinted with the permission from [16], C.-L. Cheng et al., Phys. Rev. Lett. 78(9) 1997, © (1997) by the American Physical Society

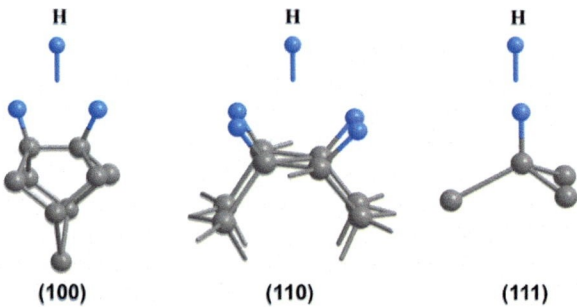

Fig. 11.3 Schematic of three low index planes of diamond surfaces and the relation of surfaces subject to incoming atomic hydrogen. Both C(100) and C(110) planes are with their surface C–C bonds facing the atomic hydrogen

Spectroscopy (Fig. 11.2B) indicates that, in the presence of atomic hydrogen, {111}-oriented facets formation irreversibly occurs on both C(110) and C(100), whereas C(111) remains intact.

The anisotropic etching of atomic hydrogen demonstrates different surfaces are with different physical strength when subject to atomic collision. This effect of hydrogen etching is interpreted as physical and non-reversible. For the C(110) and C (100) planes, extensive atomic etching result in {111} facets forming hydrogenated surfaces domains as evidenced from the observed CH at ∼2835 cm^{-1}, typical of CH on C(111) surface. A schematic of difference low index surfaces is shown in Fig. 11.3. When atomic hydrogen approaches, C(110) and C(100) surfaces are more likely to subject to physical collision, due to its geometric arrangement, while C(111) surface can be more resistant to atomic hydrogen etching.

With the decreasing of the diamond crystal size its properties change. Decreasing size up to less than 1 μm and in the nanoscale range, a different class of diamond named nanodiamond (ND) appears. In this order, diamond properties start to vary due to its structure and size. Its physical/chemical properties and a wide variability of these properties are currently considered as model system for physics studies and also as a promising nano-system for various applications. In this chapter we will discuss the surface properties of ND, their investigation using spectroscopy, methods of the surface modification, effects of the physical properties and some biomedical applications.

11.3 Characterization of Nanodiamond Surface and Structure

Nanodiamond usually refers to diamond crystals in the nanoscale, ranging from single digit nanometers to several hundred nanometers. Its physical and chemical properties are, in a large degree, determined by surface properties due to large fraction of the surface-localized carbon atoms. It has been estimated that a Nanodiamond of 4.2 nm in diameter consists of totally about 6728 carbon atoms, and among them 16% reside on the surface [17]. With increase diamond nanoparticle (NP) size, the part of surface atoms decreases correspondingly and the properties of bulk diamond start to compete with surface properties. Understanding the ND surface properties and pre-determined varying surface properties allows therefore to control the physical/chemical properties of nanodiamond and hence their use for different applications [18]. There are numerous spectroscopic methods developed for the analysis of surface and structure of complex nanomaterial in past decades which includes Infrared (IR), Raman, X-ray Photoelectron (XPS)/X-ray diffraction spectroscopy (XRD etc.) However, in this section the methods have been limited to IR and Raman spectroscopy.

11.3.1 Infrared Spectroscopy and ND Surface

Nanodiamond usually contains a composite structure of hybridized carbon bonding with sp^3/sp^2 ratio corresponding to the diamond/graphitic carbons on the surface. It also contains large number of surface molecular and ionic groups such as C–O–C, –COOH, –OH, –CH$_x$ (x = 1 − 3), etc. (Fig. 11.4 [1, 19]) due to their purification process; for example acid treatment (with 3:1 = H_2SO_4:HNO_3). Among the optical methods of spectroscopy, Fourier transform infrared spectroscopy (FTIR) is a well-developed technique for direct analysis of surface molecular dipole vibration in situ. In order for further applications of the as produced ND, surface functionalization with different methods such as wet-chemical and gas phase treatment are

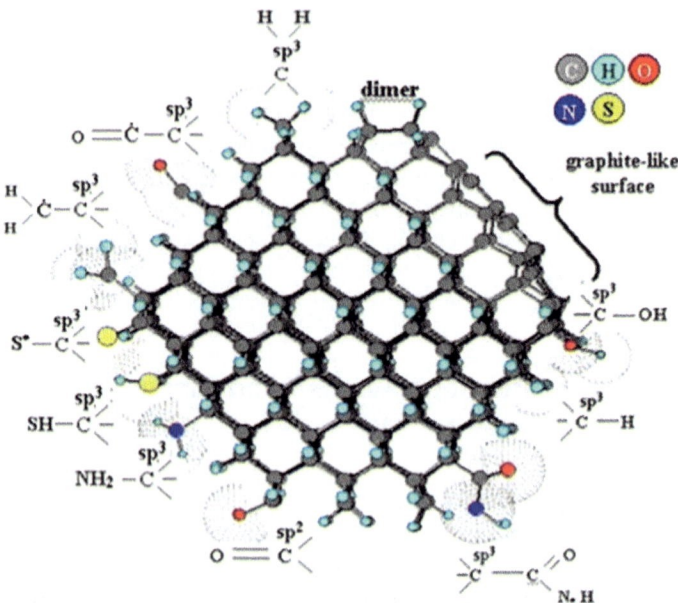

Fig. 11.4 A schematic nanodiamond particle with different surface groups and reconstructions. The structure is optimized by the semi-empirical quantum chemical PM3 method. Reprinted with the permission from [19], Springer Nature, "Synthesis, Properties and Applications of Ultrananocrystalline Diamond; Physical and Chemical Problems of Modification of Detonation Nanodiamond Surface Properties". vol 192, by V. Yu. Dolmatov, T. Fujimura, © 2005

applied, and these can be characterized using FTIR to observe the surface functional groups vis the surface dipole vibrations.

Figure 11.5 displays typical FTIR spectra of a commercially produced ND of grain size 100 nm, explaining the surface molecular compositions. It reveals the presence of surface groups and shows influence of the measurements conditions on the spectra (on the surface). Spectrum-I is a general feature of surface groups observed in ambient condition while (II) recorded in vacuum. The band near 1630 cm^{-1} is attributed to the O–H bending vibration of physically adsorbed water which is absent in the vacuum (base pressure $\sim 10^{-6}$ Torr). The C=O stretching mode appeared in wide range from 1750 to 1800 cm^{-1} and even above corresponds to the carboxylic acid group, sometimes a shoulder band observed near 1800–1850 cm^{-1} can be originated from carboxylic anhydride. O–H stretching from surface –COOH group appears at about 3710 cm^{-1}; the broad band in range 3000–3600 cm^{-1} can be attributed to the hydrogen-bonded-OH of physically adsorbed water on the surface; in vacuum the band narrowing corresponds to their desorption. The band at 2800–3000 cm^{-1} is the CH stretching vibrational feature from various origins [20]. The overlapped bands in the range 700–1450 cm^{-1} usually are attributed to the ether/ester-like (C–O–C/C–O) groups on the diamond powders [21–24].

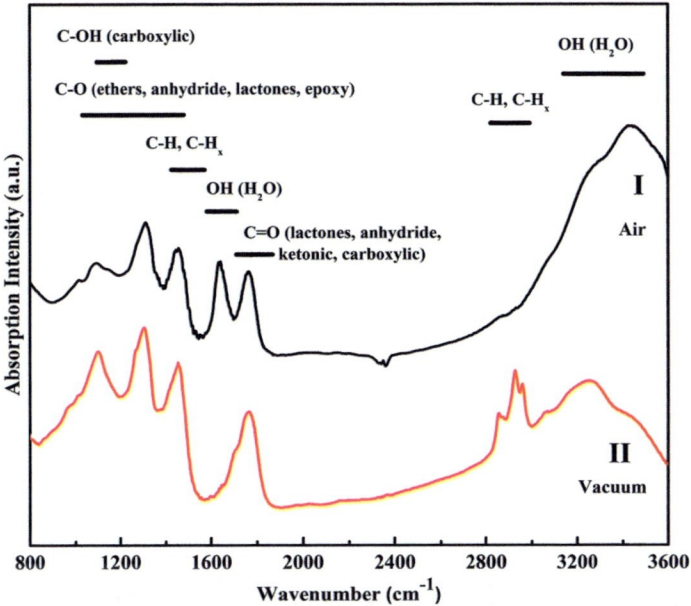

Fig. 11.5 The IR spectra of 100 nm nanodiamond measured at ambient conditions (**I**) and in vacuum (10^{-6} torr) (**II**)

In addition to the analysis of chemical compositions, infrared (IR) spectroscopy can be used as a unique tool to visualize the molecular interactions on the studied surface. For example in situ observation of H_2 gas interaction with nanodiamond, resulting in C–H bonding on the different facets [16]. Although IR spectroscopy is a useful and nondestructive method to characterize the surface, the attribution and characterization of IR active molecular groups on the nanostructure surfaces is somehow complicated due to their spectral complexity such as overlapping, shifting, etc. which are dependent on the surface area to volume ratio of the materials. In the other sense, this complexity can provide a general understanding of the size and shape of the nanostructured diamond. Chen et al. has showed the spectral shape of the C–H stretching frequency on diamond is dependent on the size of nanocrystals (Fig. 11.6I) and it is interpreted in terms of a domain size effect which affects the surface dipoles. The size of domains of the {111}-type facets were decreasing with decreasing ND particles size, and dipole-dipole interaction was used to explain the broadening of the spectral features [25].

The surface C=O stretching frequency also has found size-dependent [18]. It was analyzed using ND particles of sizes from 5 to 500 nm. The C=O stretching frequency observed at 1820 cm^{-1} for large particles of size 500 nm, is down shifted to 1725 cm^{-1} with decreasing ND crystallites size to 5 nm (Fig. 11.6II). This shift is attributed to a result of hydrogen bond formation between the COOH groups of the carboxylated nanodiamond surfaces with neighboring water molecules,

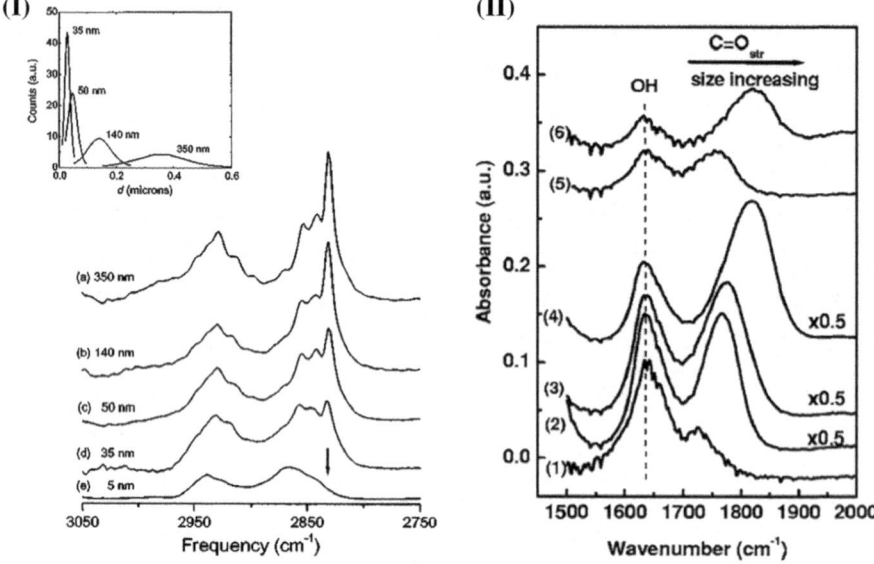

Fig. 11.6 **I** The FTIR spectra of C–H feature of hydrogen treated nanodiamond with average sizes of 350, 140, 50, 35, and 5 nm. The arrow mark indicates the diminishing of the characteristic 2834 cm^{-1} peak with decreasing the size of crystallite from 350 to 5 nm. The inset is the size distribution of corresponding nanocrystals. **II** The characteristic FTIR spectra of different size nanodiamond with C=O surface functional group for (1) 5 nm; (2)100 nm; (3) 200 nm; (4) 300 nm; (5) 400 nm; (6) 500 nm. Reprinted Figure (I) and (II) with the permission from [25] Y.-R. Chen et al., J. Chem. Phys, 199(20), 2003, © 2003 and [18], J.-S. Tu et al., J. Chem. Phys, 125(17), 2006, © 2006 AIP publishing respectively

or neighboring nanodiamond. The C=O stretching vibration is extremely sensitive to electrochemical environment and can be considered to probe of surface molecular transformation of nanodiamond. The H-bonds formation in C=O stretching frequency observation is also found to be temperature-dependent [26]. Additionally, electrochemical states of the functional groups can be affected by the surface structural defects and surface curvature, and rarely by symmetry violation and lattice constant changes [27]. The temperature and also gas treatments usually modify the surface states and molecular interactions affecting the hydrogen bonds network, facilitating desorption of the carbons/physisorbed molecules and decomposition of some surface molecular groups, and stimulating partial graphitization of the upper layer of diamond [26, 28].

Therefore, FTIR spectroscopic analysis of ND allows realizing the fundamental profile of the surface characterization and their further applications. However, the explanation of the obtained spectra can sometimes be complicate due to the complexity of the surfaces and surface dipole interaction with the environment and with among surfaces functional groups.

11.3.2 Raman Spectroscopy

Raman spectroscopy is vibrational spectroscopic techniques based on detection of non-elastic light scattering. Obtained spectra are characteristic fingerprints of molecular and crystal vibrations. It is a routinely used tool to characterize various carbon phases via analysis of phonon modes of sp^2 and sp^3 bounded carbon [29–32]. In ND it allows observation of diamond sp^3 hybridized carbon, graphite (sp^2) phase, presence of disordered sp^2 hybridized carbon and amorphous sp^3/sp^2 carbon; to analyze their relation between the phases content quantitatively and their spatial distribution in the system, and, correspondingly, surface predominant phases; to estimate size distribution of crystallites using phonon confinement model [33, 34]. Additionally, some works discuss using Raman not only for analysis of the surface physical structure/lattice, but surface chemistry via analysis of functional molecular and ionic groups [35–37].

11.3.2.1 Raman in ND Surface Vibrational Study

Raman spectroscopic analysis of ND currently is not limited to the identification of the carbon phases; it is a complementary technique of IR spectroscopy to provide information about the phase composition as well as the surface molecular terminations of the material [38].

In Fig. 11.7 the Raman spectra are compared for detonation nanodiamond (DND) with high and low content of non-diamond carbon and correspondingly with low and high content of surface groups, marked UD50 and UD90, respectively. The UD90OxHCl is air oxidized, HCl treated ND90, with high content –OH and C=O

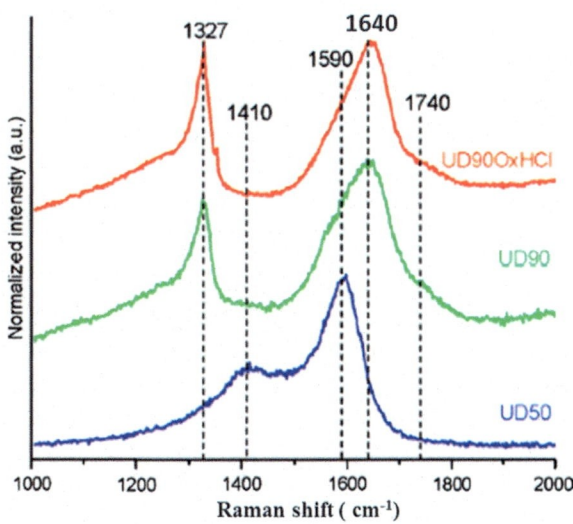

Fig. 11.7 Raman spectra of nanodiamond powders, with 325 nm wavelength laser excitation at room temperature. The nanodiamond, UD50 contains low content of diamond phase whereas UD90 contains higher content of diamond structures. Reprinted with the permission from [35], V. Mochalin et al., Chem. Mater. 21 (2), 2009, 273–279, © 2009 American Chemical Society

groups [35]. The spectra were excited with $\lambda_{exc} = 325$ nm wavelength laser. DND with low content of diamond phase (UD50) exhibits characteristic Raman features of graphitic carbon: the G-band centered at 1590 cm^{-1} and the broader disorder D-band at ~ 1400 cm^{-1}. The diamond Raman signal is not detected because the diamond core is shielded by covering graphitic layers and amorphous carbon [39]. The Raman spectra of ND with high diamond content (UD90 and UD90OxHCl) reveal an asymmetrically broadened sharp diamond peak centered near 1327 cm^{-1} with a shoulder toward lower wavenumbers, and a broad, asymmetric peak with a maximum at 1640 cm^{-1}. The Raman peak observed for DND close to 1640 cm^{-1}, the authors attributed as the superposition of sp^2 carbon band at 1590 cm^{-1} with the peak of O–H bending vibration at 1640 cm^{-1}. Additionally, C=O stretching vibrations, positioned at 1740 cm^{-1} is observed in the Raman spectrum of oxidized ND, as a shoulder on the 1640 cm^{-1} peak. The input of the surface functional groups (–OH and C=O) to the Raman scattering is demonstrated together with confinement effect in smallest ND particles. Thus ignoring their contributions may result in a significant overestimation of the graphite sp^2 phase and underestimation of the crystalline diamond sp^3 phase [40]. Due to the contribution of surface group vibrations in the Raman spectrum, Raman spectroscopy can be used for the analysis of surface modification of ND [36] and for the study of the surface features' interactions with environment, for example, interaction with water.

Water dispersions of ND with various functional/poly-functional surface terminations were compared using FTIR, Raman, X-ray absorption, and Photoluminescence spectroscopy [41, 42]. Mutual influences of ND surface and water environment are discussed. The hydrogen bond network in dispersion of hydrogenated nanodiamond (ND-H), analyzed via Raman spectra in aqueous, and is differed from the dispersions of nanodiamond with other surface terminations [42]. In case of hydrogenated nanodiamond, no H-bonds are formed between water molecules and hydrogenated surface groups, however a long-range disruption of the water H-bond network has been observed in dispersion medium. It has been proposed this unusual hydration structure observed is resulted from electron accumulation at the diamond-water interface. The modification of water H-bonds network around ND-H should be considered for further studies on the catalytic and chemical reactivity of ND-H in aqueous media.

11.3.2.2 Surface Enhanced Raman Scattering on ND

Surface-enhanced Raman scattering (SERS) arises due to the excitation of surface plasmon and photon states density modification of a noble metal's nanostructured surface [43]. Particularly, SERS has been observed at target molecules adsorption on nano-sized roughened surfaces with an enhancement of spectral intensity by orders of magnitude, theoretically, up to 10^{14} times. A second considered mechanism is originated from the formation of new electronic states as a result of adsorbate-substrate bonding interactions; this effect also can significantly increase the Raman scattering cross-section [44]. Used initially to study organic molecules [45, 46],

SERS also can help in understanding the surface, structure and the physical/chemical properties of nanocrystalline materials, such as carbon-containing nanocrystals TiO_2 [47] as well as other carbon nanostructures. Studies for ND [48–50], fullerenes [50, 51], and diamond-like carbon (DLC) [52–54], have demonstrated the sensitivity of SERS to local structural features of carbons in nanoscale.

The characteristic features of carbon contents using SERS in various nanocarbons such as amorphous, graphitic, diamond-like carbon, etc. were well studied [50, 51, 53]. The enhancement of graphite signal in the range of 1580–1600 cm^{-1}, corresponding to the characteristic feature of sp^2-hybridized carbon (G-band), and disordered carbon (D-band) nearly at 1350 cm^{-1}, "selective enhancement" of the signal from local nano- or sub-nanocarbon structures and "blinking effect" for the set of selectively enhanced peaks were demonstrated [53, 54]. The SERS application on ND in close contact with Ag (Fig. 11.8) and with Au has been shown by several groups [55, 56]. In SERS of diamond nanocrystal, the Raman signals of surface graphite and disordered carbon usually predominate over the signal of diamond sp^3 bounded carbon at 1332 cm^{-1} due to their large Raman scattering cross-section [53].

Additionally, Raman study of diamond film suggests that the presence of nano-phase structure on the film surface that affects the change of diamond spectral shape toward asymmetric order could be observed under SERS investigation [52]. Thus, the difference in the SERS efficiency from sp^3 and non-diamond surface features determines the shape of the nanostructured diamond spectra, facilitating detail analysis of their structure and surface properties [48–50]. Being sensitive to local structural properties of carbon SERS can serve for probing carbon atom

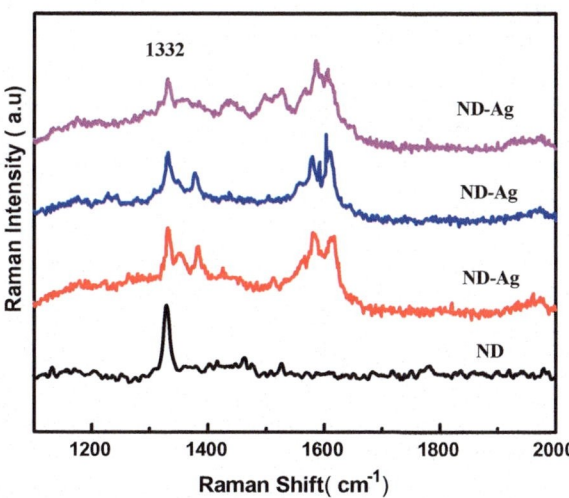

Fig. 11.8 Raman spectra of ND and ND on Ag film as a SERS substrate. The ND has been laser-positioned on the Ag like described in [48]. The spectra are collected with 532 excitation wavelength, laser power at Raman detection is 0.5 mW in focal spot. Duration of spectrum measurements is 10 s. The Raman spectrum from the ND on Si substrate measured at the same conditions and with comparable ND particles number in focal spot is presented for comparison

hybridization and for analysis of the spatial distribution of the different kinds of hybridization along the sample surface. The presence of different carbon phases on the surface and their distribution, including the studies of mechanical effects on the carbon surface and plastic deformation in the carbon can be observed and studied [57–61].

11.4   Methods of Surface Functionalization of Nanodiamond

For the use of ND in a number of applications, it is necessary to modify the physical and chemical properties of the ND surface and produce materials with the pre-determined properties. Surface functionalization becomes a required process to develop such properties of ND, as surface chemistry determines the particles nature such as hydrophilicity/hydrophobicity and their electrochemistry, surface homogeneity, physical/chemical adsorption/desorption as well as colloidal ability, etc. [62, 63]. The efficient and homogeneous surface modification of ND can be initially achieved with several methods such as wet-chemical (strong acids treatment), gas treatments (for example, hydrogen, oxygen and ozone) [64, 65], and even cold plasma treatment using CF_4 or SF_6 gas, etc. [66].

Using ND for bio and chemical applications, the particles exhibit some major drawbacks such as aggregation, nonhomogeneous surface reactivity etc. [62, 67]. These are due to the presence of graphitic sp^2, amorphous carbon and oxide groups on the surface [68]; or simply the size effect. Surface functionalization via wet-chemical and gas phase methods is usually applied to eliminate the obtained drawbacks in optimal condition to modify ND' surface for further applications [69].

Surface hydrogenation of nanodiamond using gas treatment is considered to be a direct method to yield the least surface contamination with maximum homogeneity, narrowest colloidal particle size distribution and highest zeta potential with negative electron affinity. The hydrogenation corresponding the C–H bonding on the surface of nanodiamond provides a primary bio-chemical linker for further modification [70, 71]. Cost-effective and facile hydrogenation process of NDs is developed by many groups, but high temperature (>800 °C) and atomic hydrogen are usually involved. Amongst those recently it has been shown that the successful hydrogenation of NDs is possible at low temperature annealing (<500 °C) using molecular hydrogen treatment [72, 73]. Using spectroscopic methods, it is demonstrated that the hydrogenation at such low temperature is dependent on the presence of amorphous/sp^2 carbon fragment leading to a catalytic reaction to dissociate molecular hydrogen. Although gas phase hydrogenation has been achieved much earlier [16], and practice routinely, the mechanism leading to the hydrogenation is still not conclusive; mostly due to the complexity of the ND surfaces. For the low temperature hydrogenation using H_2 molecules, the proposed mechanism suggests during annealing of ultra-disperse diamond (UDD) nanoparticle, the surface carbons (sp^2 fragments/amorphous) desorb as a radical form from the surface, mostly C_3 and they are reactive to molecular hydrogen dissociation which leads C–H adsorption on the diamond surface; the reaction steps (a–e) in schematic diagram are shown in Fig. 11.9I. The mass spectroscopic result of C_3 radical analysis during annealing and corresponding C–H feature, characterized using in situ FTIR, on the diamond are displayed in the Fig. 11.9II–IV.

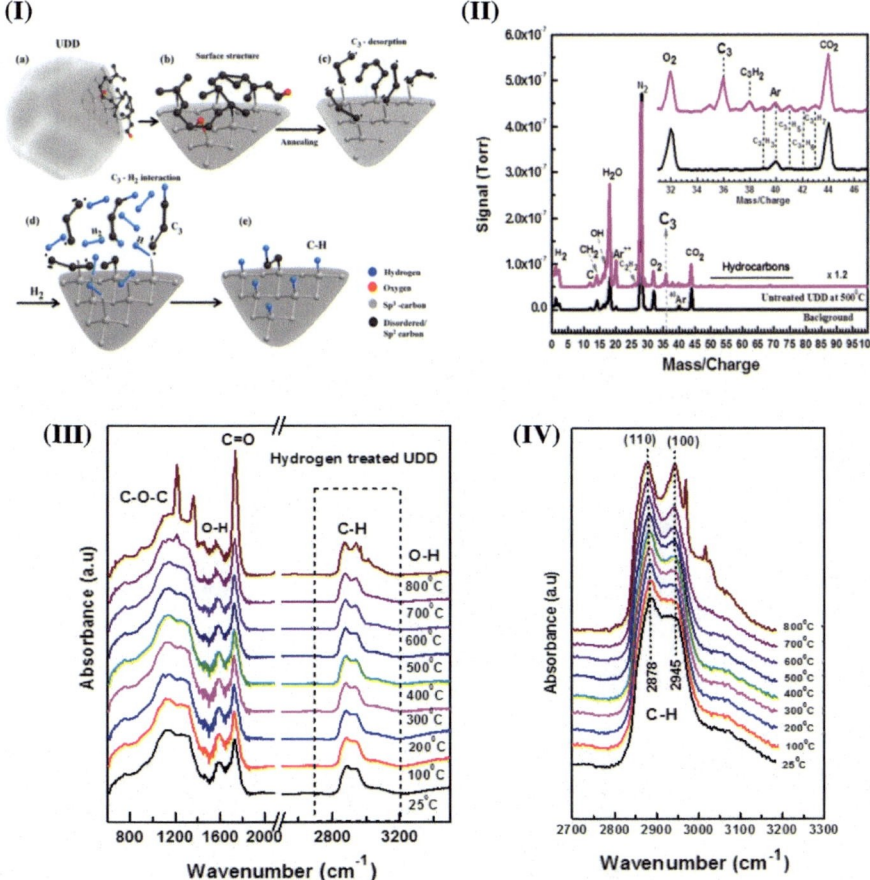

Fig. 11.9 **I** Schematic diagrams of the process of hydrogenation of UDD. **a** Nanodiamond particle (<10 nm), **b** a portion of the surface containing disordered/amorphous carbon (black color), **c** formation and desorption of carbon radical as C_3, **d** interaction between C_3 and molecular hydrogen (H_2), resulting in the production of atomic hydrogen, **e** C–H-bond formation on surface sp^3 dangling carbon. **II** The mass spectrum of untreated-UDD annealed at 500 °C for 30 min in UHV (base pressure $\sim 4.0 \times 10^{-8}$ Torr). Inset is the detailed mass analysis of desorbed gas from the sample in a specific range. **III** FTIR spectra of hydrogenated UDD annealed at various temperatures (100–800 °C) for 30 min to each in vacuum. **IV** Corresponding C–H stretching feature in detail. Reprinted with the permission from [72], A.-I. Ahmed et al., Carbon 110 (438–442), 2016, © (2016) Elsevier

The FTIR result of remarkably stable C–H bonding on hydrogenated nanodiamond reannealed up to 800 °C, Fig. 11.9III, IV, meaning the C–H bonds are formed with dangling carbon of core sp^3 diamond planes. The low temperature hydrogenation could be considered the cleanest surface modification as this temperature range doesn't make any substantial change of the surface structural state such as graphitization, oxidization etc. Along with hydrogenation, oxidation of the NDs in air was demonstrated as a tool to optimize the crystalline quality of the NDs via removing non-diamond carbons including surface molecular groups [74, 75].

In wet chemical process, the method of carboxylation of ND using acid treatment is prominent for further modification [18, 21, 28]. Besides, several kinds of functionalization on ND surface with suitable linker or moieties such as hydroxyl (–OH), amine (N–H, NH_2), thiol (SH) groups including other covalent and non-covalent immobilization of different functional moieties are studied. The reaction pathways are displayed in Fig. 11.10 [62].

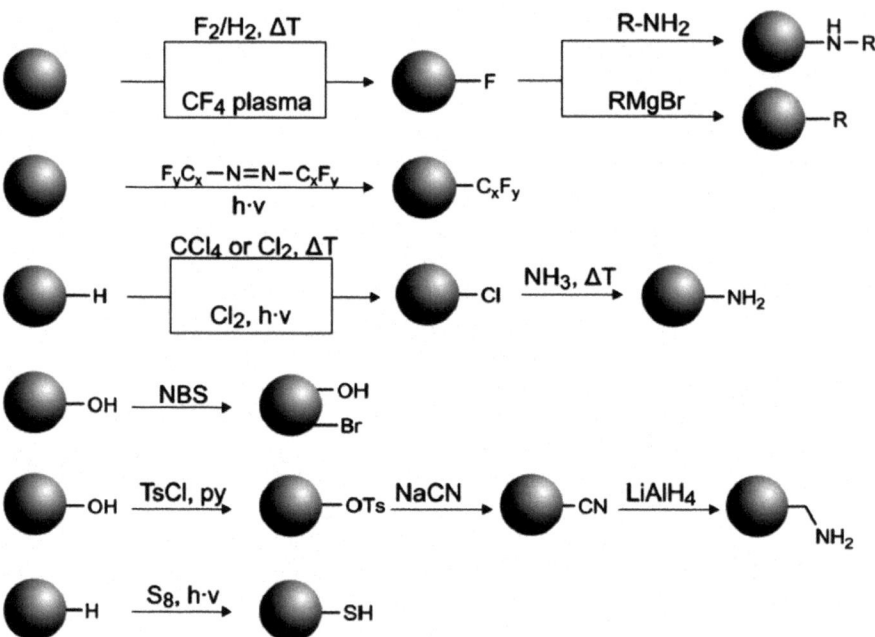

Fig. 11.10 Schemes of several functionalization methods of as received or pre-modified nanodiamond (with H, OH, F, Cl etc.) using wet-chemical synthesis. The used methods of subsequent reactions open the possibilities of homogeneous surface termination. Reprinted with the permission from [62], A. Krueger et al., Adv. Funct. Mater. 22 (5), 890–906, 2012, © 2012 WILEY-VCH Verlag GmbH & Co. KGaA, Weinheim

11.5 Role of the Surface, Surface Interactions and Their Effects on the Photoluminescence of Nanodiamond

Among the spectroscopic methods, IR and Raman spectroscopy are quite established tools to study ND surface chemistry and interactions; vibrational spectroscopic data are analyzed in detail in many original and review works. NDs containing multicolor (blue, green and red) centers can be applied as markers for optical bio-imaging [6, 12, 76, 77]. Among a hundred color centers catalogued in diamond, the negatively charged nitrogen-vacancy center, $(NV)^-$, is of particular interest since its emission is at the window for convenient bio-imaging. The fluorescence of ND is stable and exhibits no photobleaching as compared to molecular dyes commonly used for bio-imaging; therefore it has become a popular subject of study in the past decade. However, the photoluminescence of nanodiamond can depend on its surface structures, moieties, and functional groups; this section is devoted to the effect of the surface properties to the photoluminescence (PL) of nanodiamond.

11.5.1 Surface Functional Groups and Moieties on Photoluminescence

Although the PL origin of nanostructured diamond is mainly the defects and admixtures in the diamond lattice, so called color centers, and the ND's fluorescence is emitted predominantly from the color centers in diamond core [12, 76], surface also can play important roles. Among the color defect centers, the most well-studied and practically utilized is the nitrogen vacancy (NV) color center which appears as neutral (NV^0) and negative (NV^-) charge states [77]. NV color centers emit stable emission with narrow band width at room temperature and are recently used for bio-imaging, bio-sensing [1–6] and single photon emission [78–80]. Color centers based luminescence of ND can be significantly increased to achieve the so called fluorescent ND (FND), ND with enhanced fluorescence [81–83].

The surface contribution in the PL is originated from surface localized defects (including defects in sp^2 shell of nanodiamond) and impurities [37]. Factors such as ND size, graphitization degree, surface moieties of the nanodiamond host and bandgap structure of the dielectric environment in different degree and by different ways can affect the surface state and, correspondingly, have been found affecting the PL of ND [84–88]. The mechanisms are discussed. The recombination of donor-acceptor pairs can determine the surface role in ND luminescence [89, 90]. Delocalized π electrons in aromatic hydrocarbons can create conditions for photons' strong absorption and re-emission [91]. Conjugated aromatic molecules-like structures in nanoparticulate structures are considered to explain new-observed optical properties of nanocarbons [92, 93], involve either stable charge traps on the surface or surface band bending [94–96].

Among the surface interactions influence, the study of diamond-water coupling effects in colloidal suspensions of nanodiamond of different surface functionalization has revealed dependence of ND luminescence intensity on the type of functional groups on the ND surface. Observed dependence has demonstrated a relationship between the broadband luminescence typical for NDs and the chemical state of ND surface. In the water solution the increase in the PL intensity of ND was found to correlate with the hydrogen bonds strengthening in ND water suspension [41]. The most intense luminescence was detected for the ND functionalized with OH groups, which forms the strongest hydrogen bonds among the studied samples of different surface terminations. However, in the study of the methanol-water and ethanol-water DND solutions it has been shown that polarity of a solvent strongly affects the DND–COOH fluorescence by other way. The greater the polarity of a solvent is, the weaker the fluorescence of DND–COOH in the suspension is. Thus the weaker and less dense the hydrogen bonds in the solvent are, the less impact they have on the fluorescence, that is, the weaker the solvent quenches the fluorescence of the DND–COOH [97]. Direct effect of water adsorption on the PL of ND of different origin has been observed [98]. It was found that the ND luminescence related to surface defects can increase almost by order of magnitude upon heating to 200–250 °C. To explain these phenomena, it was suggested that water adsorbed on ND surfaces efficiently quenches PL so desorption of water during heating leads to dramatic increase of the radiative de-excitation. These studies reveal high selectivity of the termination of ND surface interactions and hence potential for development of sensing applications.

More studies of the contribution of the surface on the optical properties of ND controlled by surface termination were conducted with calculations on the surface structure and theoretical considerations [86–88, 98]. Detonation nanodiamond has been found exhibiting significant luminescence dependent on excitation wavelength from the visible to the near-infrared spectral region above 800 nm and on the surface functionality of the DND particles [37]. The effect of surface functionalization with different kinds of molecular groups (of various sizes) on the PL characteristics of DND is shown in the Fig. 11.11. In addition, to create surface molecular groups, during the functionalization processes the chemical structure of the outermost carbon layers is changed and can result in a change of non-diamond carbon fraction. The brightest fluorescence among the samples is observed from octadecylamine-functionalized particles and is more than 100 times brighter than the least fluorescent particles, carboxylated DNDs [37]. In this case mostly the photons emitted by functionalized particles originated from non-diamond carbon. However, emission from NV centers in hydrogenated, hydroxylated, and carboxylated DND also has been observed increased. Therefore the surface chemistry of DND is shown to be critical for their fluorescence properties. Two possible mechanisms of the surface influence on the fluorescence were suggested. First, the creation of new fluorescent defects on the particle surface or in its close proximity via the local reconfiguration of carbon atoms as a result of the functionalization

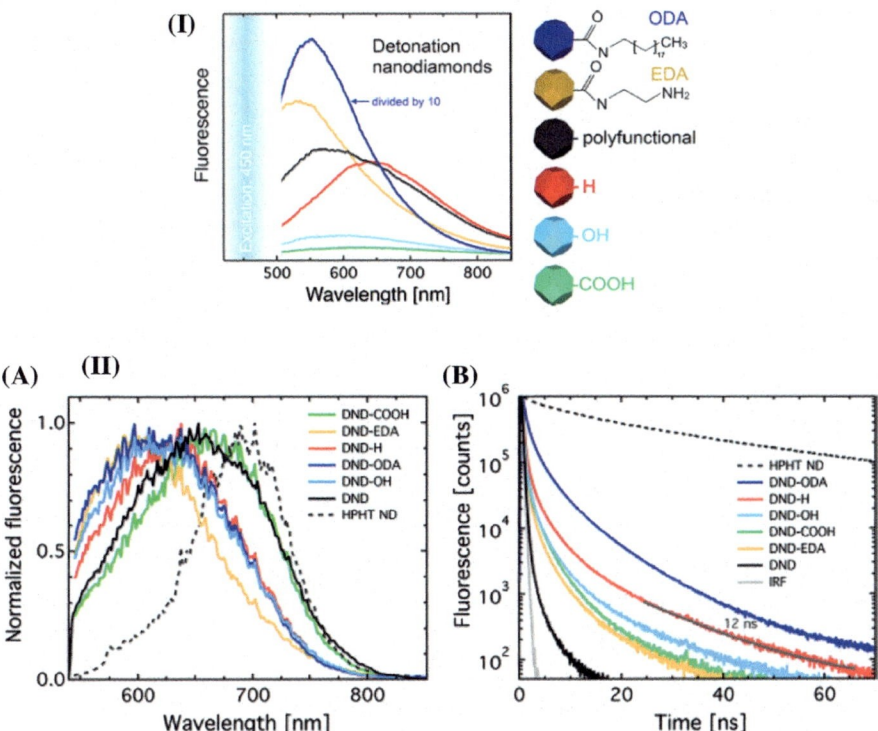

Fig. 11.11 I Effect of surface functionalization with different kinds of molecular groups on the intensity of DND PL. **II-A** The normalized fluorescence spectra illustrating the peak dependence on the surface functionalization. **II-B** Lifetime fluorescence spectra of different DNDs with typical fluorescence spectrum of high pressure-high temperature (HPHT) NDs revealing NV⁻ center for comparison. Reprinted with permission from [37], P. Reineck et al., ACS Nano, 2017, 11 (11), 10924–10934, © 2017 American Chemical Society

process; and second, through the modified interaction of the particle surface with the optical defects and color centers inside the particles' diamond fraction.

Most of studies are performed using detonation ND. However surface effects on the high temperature-high pressure synthesized (HTHP) ND, owing to its better diamond structure and usually high content of color centers also are subjects of discussion. A number of works reported that the luminescence of ND (and FND) can be controlled by changing the ND surface chemistry [99–101]. The surface termination with H^-, OH^- and O^- causes a changes in the charge state of the nitrogen-vacancy centers and, correspondingly, in the occupancy of the NV^- and NV^0 energy states. Consequently the change of NV^-/NV^0 relation affects the shape of the ND (FND) fluorescence spectra (which includes inputs from different defects). The luminescence was found lower for NV^- localized near a hydrogen-terminated surface compared to an oxygen-terminated surface. Probably, it can be explained by a de-charging of the defect, as the accumulation of holes in

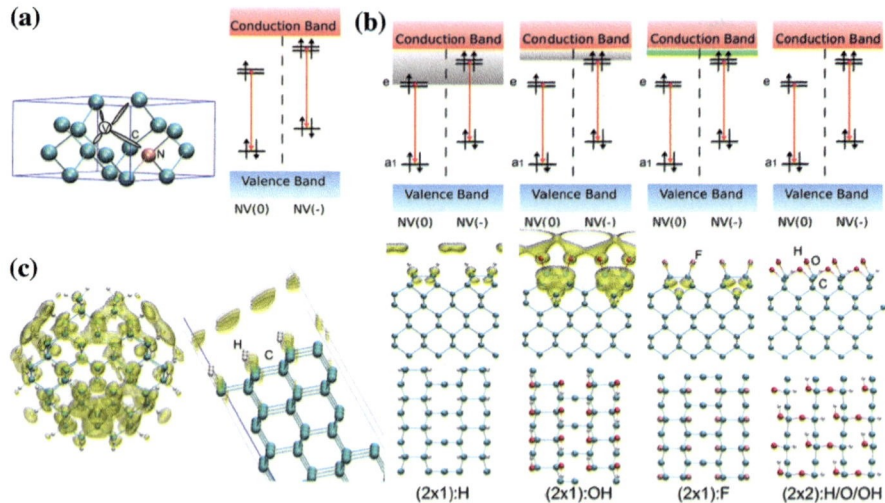

Fig. 11.12 The NV defect centers in bulk diamond and at close proximity to diamond surfaces of different structures and terminations. **a** Represents the schematic view of the NV geometry and its energy bandgap levels in the neutral and in the negatively charge state. The absorption/emission for the neutral and negatively charged defects is dependent on the excitation/de-excitation from/to the level a1 to/from e level. The defect levels are rearranged upon charging the defect with an electron donated by substitutional nitrogen. **b** Diagram of NV^0 and NV^- energy levels (spin-averaged) close to different surfaces (the gray and green regions depict the surface bands of delocalized image states and the localized acceptor surface states, respectively) and the corresponding charge density isosurfaces (on structure) of each surface related state. The probable direct photoionization of NV^- on the (2×1): H and (2×1): OH surfaces are shown. The (2×2): H/O/OH model surface (no surface related states in the bandgap) is proposed as ideal to host the NV^- defect. **c** The lowest unoccupied state on an H-terminated tetrahedral ND with primarily (111) facets and on the surface of an H-terminated (111) slab. Reprinted with permission from [103], M. Kaviani et al., Nano Lett., 2014, 14 (8), pp 4772–4777, © 2014 American Chemical Society

the H-terminated diamond surface is not compatible with a negatively charged NV center [102].

Manipulation of emission state of the NV center is shown by controlling surface moieties of the nanodiamond host and bandgap structure of the dielectric environment [84]. The modified surface with oxygen or hydrogen termination was found to promote or suppress the nearest centers emission, respectively. Additionally, the observed blinking phenomenon in diamond NV centers was analyzed and interpreted in terms of tunneling of the photoexcited electron from the NV center to a proximal electron acceptor site in the substrate.

Quantum mechanical simulations reveal that typical model diamond surfaces possess image states with sub-bandgap energies, and compromise the photostability of NV centers placed within a few nm of the surface, due to the mixture of the NV-related gap states and the surface image states (Fig. 11.12). It allows identifying a combination of surface terminators that is optimal for NV center based nanoscale sensing [103]. Studies of the influence of local electromagnetic environment at different refractive indexes on ND-environment interface on the parameters of NV color center luminescence in nanodiamond are presented [104].

Studying the dependence of ND PL on the excitation wavelength, the vital role outside functional groups and surface states on the ND fluorescence are of important [105]. Each functional group has its optimum excitation and excitation wavelengths, when the excitation wavelength changes, the corresponding emission wavelength shifts accordingly and shows an excitation-dependent fluorescence. If the surface groups were reduced to a single type (by surface functionalization), the fluorescence should be excitation-independent. The fluorescence origins of nanodiamond influenced by OH, ketone C=O, and ester C=O groups are illustrated in Fig. 11.13. The effect of the excitation-dependent fluorescence determined by the relative intensity of these three groups, and the low-lying effects are different: the lowest unoccupied molecular orbital (LUMO) of the blue band changes more than LUMO of the green or yellow bands.

Effect of pH of medium on the PL of ND (for the cases of detonation ND [11] and FND [106]) in suspension and transition between molecular and ionic surface groups is discussed. A strong pH dependence of the ND PL is observed and compared to the pH dependent fluorescence of aromatic hydrocarbons (AHs). The strongest PL of DND is found at a pH of 11.8. It decreases by almost one order of magnitude when the pH is decreased to 3.7. This change coincides with a protonation of COO– groups as the pH changes from 11.8 (COO–) to 3.7 (mostly COOH). For comparison, the fluorescence of simple AHs reveals a qualitatively similar pH dependence as the one observed for DND particles. Additionally, from pH 6 to 12 the DND particles show a relatively long fluorescence lifetime of close to 5 ns, making fluorescence lifetime imaging (FLIM) for bio-imaging applications feasible.

11.5.2 Effects of Surface-Attached Macromolecules on the PL of Nanodiamond

In addition to surface groups, chemically or physically attached macromolecules can affect the PL of nanodiamond. These phenomena are studied, using macromolecules with different optical properties to attach. Significant (in comparison with other ways of surface functionalization) increase in the luminescence intensity from octadecylamine-functionalized diamond nanoparticles has been reported [37]. Using time-resolved and time-integrated photoluminescence spectroscopy, it has been shown that optical and opto-electronic properties of H-ND film can be significantly and reversibly modified by a thin polypyrrole (PPy) layer grafted on the film surface [107]. A model was proposed, where the PPy layer on the film surface promotes spatial separation of photo-generated charge carriers both in non-diamond carbon phase and in bulk diamond.

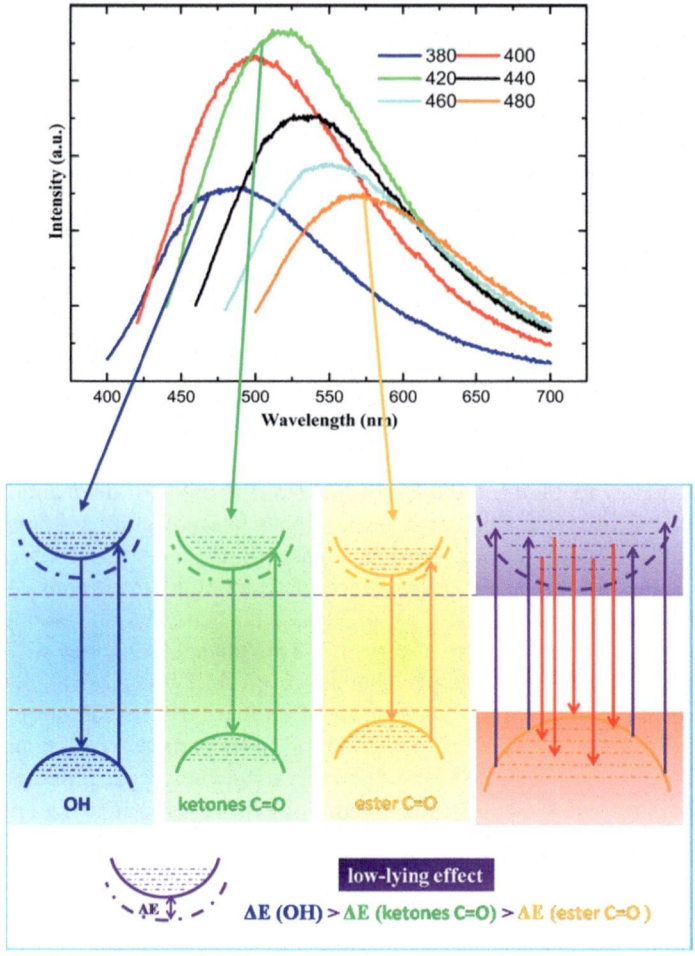

Fig. 11.13 Origin of excitation-dependent fluorescence in ND. The blue, green, and yellow emissions are from OH, ketone C=O, and ester C=O respectively. Reprinted with the permission from [105], J. Xiao et al., J. Phys. Chem. C, 2015, 119 (4), 2239–2248, © 2015 American Chemical Society

Supramolecular donor-acceptor assemblies composed of ND and porphyrin macromolrcules are constructed via interensemble hydrogen bonding and π–π interactions [108]. Formation of the supramolecular clusters composed of ND and porphyrin enhanced the photoelectrochemical performance and broader the photocurrent response in the visible region due to photoinduced electron transfer from the photoexcited porphyrin to the ND and photoenergy conversion.

The relation between the physical/chemical properties of point perturbations in ND, including surface, and the optical properties of the whole system, ND lattice and color defects of different structure and admixtures, is theoretically discussed using the density functional theory (DFT) technique to obtain the main influencing parameters [86]. Among them structure, lowest-lying excitation energy, charge distribution, density of states, components of the main orbitals involved in the excitation processes and interatomic interactions properties are considered.

A surface-charge-induced modulation of FND PL is observed at noncovalent interaction of the FND surface with charged molecules [106]. FND surface charge density was found reversal upon binding cationic polymers as a result of deprotonation of the polymer containing amino groups. It led to the depletion of NV$^-$ occupancy and a subsequent decrease in its emission intensity and allowed monitoring of molecular binding events in proximity to the ND surface.

It has also been shown that strong fluorescence arises and has been observed at passivation of non-fluorescent nanocarbons with predominantly sp^2 structure (e.g. so called carbon dots, CD; or DND graphitized surface) with different organic non-fluorescent macromolecules. The emission range was depending on the macromolecules properties [109]. The effect of influence of surface-attached macromolecules on fluorescence of CD or on ND color centers involves the electron transfer between the color center, surface carbon (non-diamond) nanostructures (nanoclusters, carbon dots) and attached macromolecule [110, 111]. Similar effect with participation of surface-localized non-diamond carbon was observed also for HTHP ND of size 100–300 nm at their surface interaction with protein macromolecules [112]. PL spectra of ND changed their shape and position of the spectral peaks at adsorption of non-fluorescent proteins lysozyme and albumin on their surface. The effect was found depended on excitation wavelength. The red shift of PL was observed for ND-protein complex. To explain this phenomenon a model was suggested based on the energy transfer between graphite-like nanoclusters on ND surface through adsorbed macromolecules to luminescent centers of ND (defect or color centers) (Fig. 11.14).

The influence on photoluminescence of CD and ND with adsorbed macromolecules has been repeatedly observed for surface-attached non-fluorescent molecules [90, 109, 112], chromophore-functionalized ND are also considered for it allows easy tuning of luminescence through an appropriate choice of the fluorophore [38]. The energy or electron transfer processes allow the use of optical and electronic properties of ND and the chromophore in synchronization, rendering these hybrid materials as potent candidates for light harvesting and other optoelectronic applications. Recently blue fluorescence has been reported in nanodiamond by simple chemical functionalization with octadecylamine. The origin of this kind of the fluorescence is not clearly understood, but probably it also originates from the graphitic part of the nanodiamond [112, 113]. The study of the effect of surface state, termination and surface interactions on PL of ND (and nanocarbons at whole) opens new perspectives in the field of nanophotonics, including new applications. Note however, some studies didn't observe the effect of adsorption on PL properties in ND [114, 115]. In the interaction of transferrin adsorbed on ND with transferrin receptor, the fluorescence spectra and lifetimes of carboxylated FND, FND-Tf bioconjugates and FND-Tf–TfR complexes were measured.

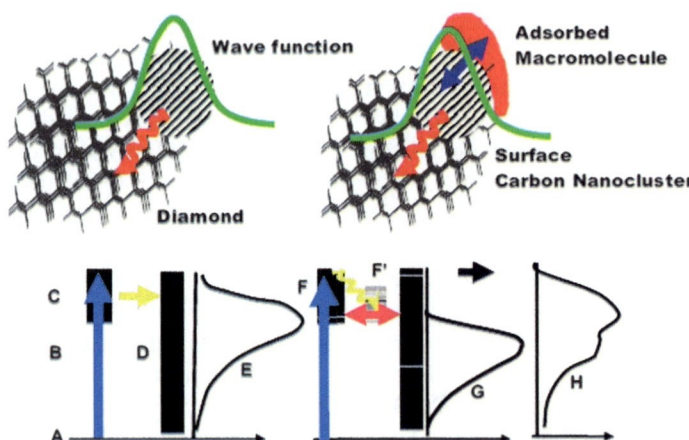

Fig. 11.14 A schematic of the energy transfer mechanism model among carbon nanocluster, adsorbed protein molecule to diamond structure in protein modified diamond system: A—energy ground state, B—excitation, C—excited states of surface carbon nanocluster. The excitation from the surface carbon nanocluster transfers to D—energy levels of the excited states of defects and admixtures in the diamond; emission is realized as wide luminescence band E. F—excited states of surface carbon nanocluster with adsorbed macromolecule; F'—energy levels of protein molecule; the energy transfers between the nanocluster and macromolecule and between nanocluster and the diamond defects and admixtures; the emission from these states reveals the shifted luminescence band G. As a result, the luminescence is observed as the superposition of G and E, shown as band H. Reprinted with the permission from [112], E. Perevedentseva et al., J. App. Phys. 109, 034704 (2011), © 2011 American Institute of Physics

No changes were observed, the measured lifetimes and fluorescence spectra are comparable [114]. Some weak effect of adsorption can be seen for adsorption of polypeptides, or "protein-based polymer" with FND, but the authors don't discuss it [116].

11.5.3 Fluorescence Lifetime for the Surface State and Surface Interactions Analysis

The fluorescence lifetime of ND is even more sensitive than the spectral intensity and shape to the interactions between fluorophore and its environment. ND color centers' luminescence lifetime can be measured to analyze the ND color center environment, role of surface structure chemistry, and the interaction with the environment. On the other hand, the methods to modify lifetime have been considered. ND with modified fluorescence lifetime can be used, for example, in applications such as fluorescence lifetime imaging microscopy (FLIM) and time-gate imaging, and some sensing applications.

Measuring ND luminescence lifetime of different color centers is currently applied to analyze and characterize comprehensively the centers and their interactions with surrounding crystal lattice, with some surface features or with particle's close environment [117–120]. For this, different color centers exhibit different lifetimes, for example, NV⁻ (with ZPL at 637 nm, lifetime of 25 ns), SiV (with ZPL at 736 nm, lifetime of 1.2 ns), NE8 (with ZPL at 794 nm, lifetime of 2 ns), GRI (with ZPL at 742 nm, lifetime of 2–4 ns), H3 (with ZPL at 503 nm, lifetime of 17.5), H4 (with ZPL at 496 nm, lifetime of 19 ns), etc. [119, 120]. As functionalized surface with high content of non-diamond carbon contributes significantly in PL of ND, the photoluminescence lifetime is also found depending on the functionalization. It is clearly seen in the Fig. 11.11c where time-resolved fluorescence decay traces are presented [37]. All types of functionalized DND show multi-exponential (in these cases bi-exponential) fluorescence decay. This implies the presence of more than one fluorescence decay pathway for optically excited states and allows discussing corresponding mechanisms.

Figure 11.15 shows the effects of surface termination combined with temperature treatment on the lifetime decays. In this case temperature treatment modifies both the ND structure and surface. For HTHP ND with high content of well-ordered diamond structure the PL is mostly determined by color centers. Color center's PL in such kind of ND can be increased using high energy treatment. The interactions between color centers and terminated surface are observed both for ND (Fig. 11.15a) and FND (Fig. 11.15b).

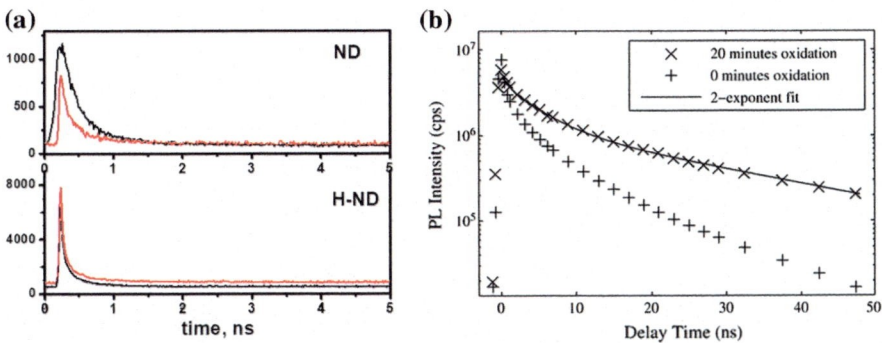

Fig. 11.15 a The fluorescence lifetimes estimated using 2-exponential approximation (fitting model $F(t) = a_1 e - t/t_1 + a_2 e - t/t_2$): for ND $t_1 = 0.55$, $t_2 = 0.1$ ns; for H-ND $t_1 = 0.05$, $t_2 = 0.1$ ns. 2-photon excitation is with femtosecond laser at 800 nm; **b** effect of oxidation in air at 400 °C on PL from NV centers of 55 nm fluorescent nanodiamond (FND) and corresponding two-exponent approximation of the decays after oxidation shown as solid line. For the FND, the slowest decay component increases from 13.3 to 24.7 ns after 20 min, and then stabilizes. Short-lived luminescence intensity decreases by 33%, while NV PL intensity increases by 2-fold. Reprinted (**a**) with the permission from [121], J. Mona et al., J. Appl. Phys. 113, 114907 (2013), © 2013 American Institute of Physics. Reprinted (**b**) with the permission from [117], B.R. Smith et al., Diam. Relat. Mater 19 (4), 314–318, 2010, © 2010 Elsevier B.V

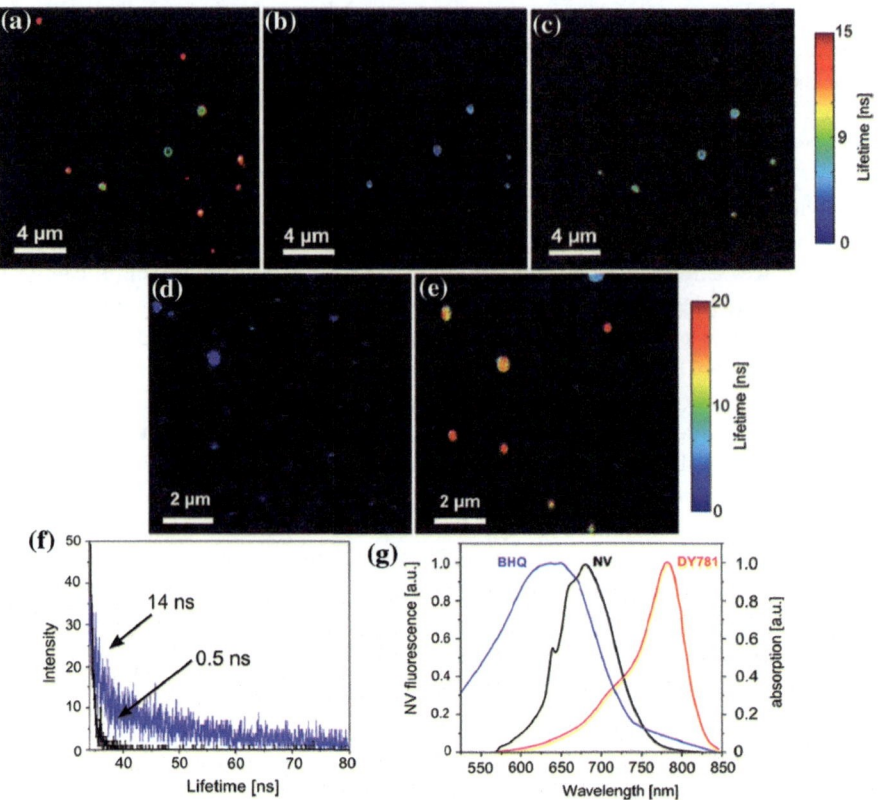

Fig. 11.16 The FLIM images of ND with Black Hole Quencher (BHO) and DY781 dye. **a** Co-localization of (**b**) and (**c**), where **b** is the image of DY781 attached to nanodiamond, **c** is the image of NV center fluorescence from the same ND shown in (**b**). ND with BHQ before (**d**) and after (**e**) bleaching. Both lifetime and intensity increased after the quencher bleaching. **f** Lifetimes of ND before (black curve) and after bleaching (blue curve). About 30% of all nanocrystals contained a NV centers and were suitable for the PL measurements. **g** Emission spectra of ND, BHQ and DY781. Reprinted with permission from [123], J. Tisler et al., ACS Nano, 2011, 5 (10), 7893–7898, © 2011 American Chemical Society

Interaction between color centers, surface and environment involving electron or energy transfer determines the lifetime decay sensitivity to FND surface interaction with fluorescent molecule or quencher molecule [122, 123]. Highly efficient Förster resonance energy transfer (FRET) between negatively charged nitrogen-vacancy centers deeply embedded in diamond as donor and surface-attached dyes or quenchers molecules as acceptor by comparing fluorescence lifetime images (FLIM) of NDs before and after bleaching of acceptor has been observed (Fig. 11.16). It has been shown that FRET may significantly enhance the fluorescence signal observed from a single FND; and the ability of organic fluorophores attached to the FND surface to transfer efficiently the excitation energy to NV⁻

centers is confirmed. The dyes positioned on the ND facile surface can serve as harvesting antennas. Functional groups on the FND surface are used both to attach linker and then to localize the fluorophore on calibrated distance [124].

11.5.4   Nanodiamond Hybrid Structures with Controlled Photoluminescence

The interaction of the ND surface with plasmonic materials and forming hybrid systems including ND particles and films and nanostructured noble metals can be considered as one of the promising ways to affect and control ND's photoluminescence. In this case, surface plasmon is involved in the interaction between ND with embedded color centers and noble metal at close contact. The PL of nanodiamond (or nanocarbon structure, for example, DLC) affecting by surface contact with plasmonic materials is originated from nanostructured surfaces of noble metals which can increase or decrease the incident local electric field, and correspondingly increase or decrease the radiative decay rate of the fluorophore in near-field area [125–127]. These modifying effects are based on surface plasmons, collective oscillations of the electrons of conductivity. The effects of local field enhancement and nonradiative energy transfer from fluorophore to the particle are competing in such system. In this case, the local field enhancement leads to an increased excitation rate while nonradiative energy transfer leads to decreasing the quantum yield and the luminescence quenching. The resulting enhancement or quenching depends on the distance between fluorophore molecule and the metal nanostructured surface. Changing this distance the continuous transition from luminescence enhancement to luminescence quenching was observed [128].

The noble metal-modified fluorescence was observed for nanodiamond. Using various modifications of the systems based on this effect development of plasmonic devices is discussed, e.g. light emitter diodes, optical nanoantennas, etc. [129]. The increase in photoluminescence of ND together with SERS of surface non-diamond fractions (see also Sect. 11.3.2.2) are observed for structures of nanodiamond combined with Ag (obtained using the laser deposition of the nanodiamond particles) at 488 and 532 nm wavelength laser excitations, while for the ND interacting with Au film characteristic SERS effect was observed at 785 nm wavelength laser excitation [49, 56]. The laser deposition creates conditions for strong interaction between the surface plasmon, arising in Ag or in Au at corresponding excitations, and ND surface and provides simultaneous enhancement of the Raman signal of ND surface (SERS of D- and G-bands, see also Sect. 11.3.2.2) and defect luminescence of the ND. The enhancement of luminescence prevails in ND due to the color centers in ND are not on the surface, but are embedded into diamond lattice and some of them can be found on the optimal distance for the enhancement (Fig. 11.17). The enhanced fluorescence emissions of ND color centers coupled to surface plasmons was shown for the ND of an average size of about 20 nm

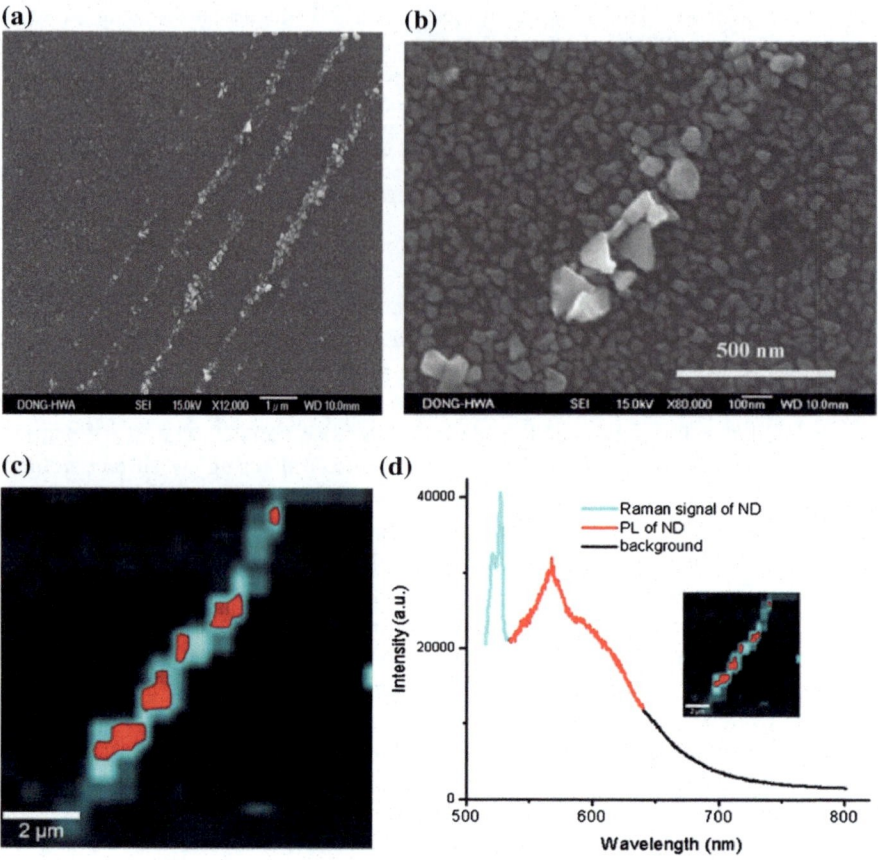

Fig. 11.17 The images of ND of average size 100-nm positioned by laser on Ag island film on Si: SEM images, **a** ×12,000; **b** ×80,000; **c** the emission intensity mapping, at 488 nm. Spatial distribution of Raman signal intensity is depicted at the integration in range 515–530 nm (cyan); PL signal is depicted at the integration in range 540–640 nm (red) according to spectral distribution on (**d**); **d** emission from ND laser positioned on Ag film (corresponding to spatial distribution on **c**, inset). Reprinted with the permission from [56], A. V. Karmenyan, et al., Plasmonics 8(2), 2012, © 2012 Springer Nature

positioned on an Au film with roughness of 1–2 nm, excited with 532 nm wavelength laser. Individual NV^0 and NV^- centers in the ND were operating like a single photon-emitting, stable solid-state quantum emitter systems, even at room temperature [130].

The above mentioned works not only demonstrate the effect of noble metal on ND fluorescence, but suggest the possibilities to control the PL in such hybrid system via controlling the distance between the metals layer and color centers in the ND (or in FND). Observed fluorescence enhancements of FNDs fluorescence positioned on silver island film surfaces depended on the nanoparticles size.

The enhancement observed for the 140 nm FNDs was significantly less than for 5 nm quantum dots due to only $\sim 10\%$ of the fluorophores in FND (like NV centers) are estimated to be located inside the diamond lattices within the enhancement region (on the distance of ~ 10 nm from the metal surface). Also, the enhancement can be controlled by adjusting the distance between silver islands and FND surface (FND enhancement region) by nanoparticles overcoating macromolecules [131].

Using the atomic force microscope, a single nanodiamond (of size 20–35 nm) containing a single NV center were combined with gold nanospheres with plasmon resonance at 540 nm wavelength laser excitation for direct contact with detection of maximum PL emission. The emission from the neutral NV^0 defects can be significantly enhanced in contrast to the NV centers in negative state as PL of NV^0 is closer to the plasmon resonance of gold. The luminescence enhancement effect was observed diminished at increasing the distance between the Ag nanoparticles and the color center emitters inside the FND of size 35 or 140 nm which were wrapped by DNA molecules [132]. This observation as well as the PL enhancement at laser deposition of ND of metal surface confirms [56] that no additional buffer layers between ND particles and metal surface are required for the PL enhancement of PL emitted by the NV centers inside the diamond lattice which are separated from the metal by non-diamond surface layer.

Covalently functionalized nanodiamond with rare earth (RE) complexes as other kind of luminescent nanodiamond-based hybrid material was prepared which may find potential application in optical tools. The emission colors of these composites can be tuned by adjusting the molar ratio of the RE. Additionally, the PL of the hybrid composites exhibits high photostability under ultraviolet irradiation [133]. Together with various plasmonic coupling with NV centers using noble metal, excitonic coupling has been realized using semiconductor quantum dots with well-defined excitonic features. Both variants of the hybrid systems allow to control a wide range of ND optical properties, such as fluorescence lifetime [134]. The effect Ag and Au and ND hybrids on fluorescence lifetime is shown in the Fig. 11.18.

Formation of ND-based hybrid materials using the self-assembling capabilities of biological systems which serve as a structural scaffold for controlled ordering of surface functionalized ND in one-, two- and three-dimensional system is discussed [135]. Such system was formed by interconnecting ND using self-assembling protein scaffolds allowing the precise positioning of surface functionalized NDs. The expected dipolar coupling interaction with neighboring NV centers and the expected decoherence time were evaluated for the high-fidelity creation of entanglement, cluster states and quantum simulation applications.

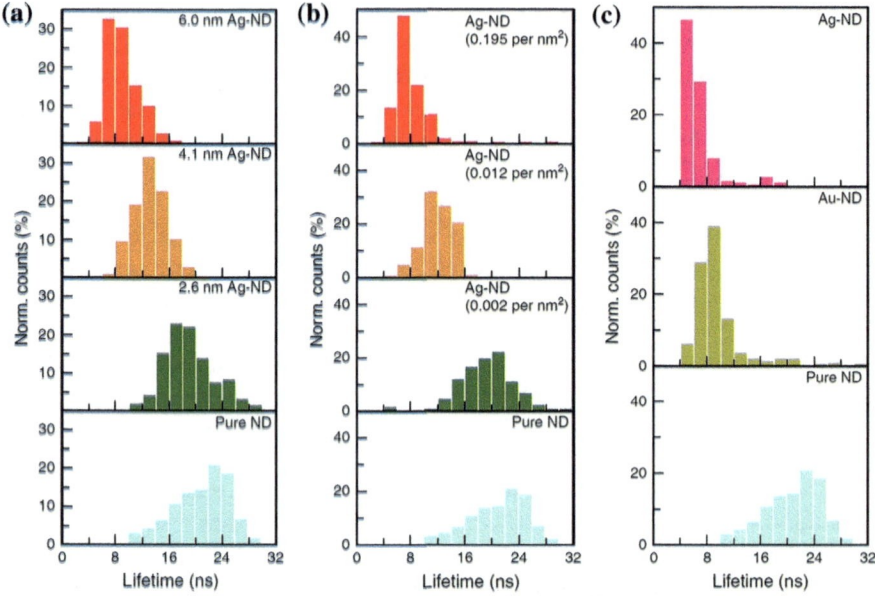

Fig. 11.18 Effect of plasmon-NV coupling in various ND-metal nanostructures. **a** Fluorescence lifetime of NV centers versus the size of Ag nanoparticles in a ND-Ag hybrid nanostructure. All ND-Ag nanostructures possess same surface coverage of Ag subunits (0.004 particles per nm^2). **b** Fluorescence lifetime of NV centers versus the coverage of Ag nanoparticles in a hybrid ND-Ag nanostructure. Mean size of Ag subunits in all hybrid ND-Ag nanostructures is 4.5 nm. **c** Fluorescence lifetime of NV centers versus the composition of metal nanoparticles in a hybrid ND based nanostructure. Surface coverage density of both ND–metal nanostructures is 0.008 particles per nm^2. Reprinted with the permission from [134], J. Gong, et al., Nature Comm. 7, 11820 (2016), © 2016 Springer Nature

11.6 Bioapplications Using Surface Spectroscopic Properties of Nanodiamond

Nanocarbon materials, such as nanodiamond, graphene, fullerene, carbon nanotube, carbon dots, etc., presently attract much attention as prospective technological materials. They are highly suitable for numerous possible applications, for example, quantum information processing, optoelectronics applications, including topical investigations in field of terahertz applications, optical markers for cellular imaging, chemical and bio-chemical sensing [1, 78, 79]. It is important to understand the physical/chemical processes of these materials for further application development. In this chapter our focus is on bio-applications. Therefore, examples are chosen to illustrate the applicability of the surface spectroscopic properties.

11.6.1 Spectroscopic Analysis of ND Surface Interaction with Bio-active Molecules in Bio-systems

The surface modification of nanodiamond with biologically active molecules is developed mainly for applications in drug delivery, gene delivery, targeted delivery of the nano-probes and nano-sensors [3–5, 9] or bio-engineering materials [136, 137]. The aims of ND for all these purposes, particularly for drug delivery are, (a) to enhance drug and delivery efficiency with minimal drug toxicity and side-effects [8, 9, 138], including the effects on chemoresistant tumors [139] and direct delivery across blood-brain barrier [140]; (b) to utilize ND's multifunctionality [8, 9, 141], including the ability of bi-functionalization with different bio-active molecules [142], simultaneous delivery tracing and imaging using various imaging modalities and numerical estimations [3–8, 143]; and (c) to achieve stimuli-responsive delivery and release [144–146]. The tunable functionalization properties of the surface with various chemical/physical treatments make nanodiamond possible to graft with macro-molecules such as polymers, drugs, biomolecules etc. Pre-modified nanodiamond with functional groups such as COOH, C–O/C=O, OH, H, NH_2/CO–NH_2, and some halides (Cl, F) etc., shown in Fig. 11.19, has been used as the precursor linker of further termination of larger molecules [1, 3]. Among the functional groups that are used for further functionalizaiton, amine/amide (NH_2/CO–NH_2) is found to be an active linker of mostly used drugs/bio-molecules/gene [147]. Amino-functionalized

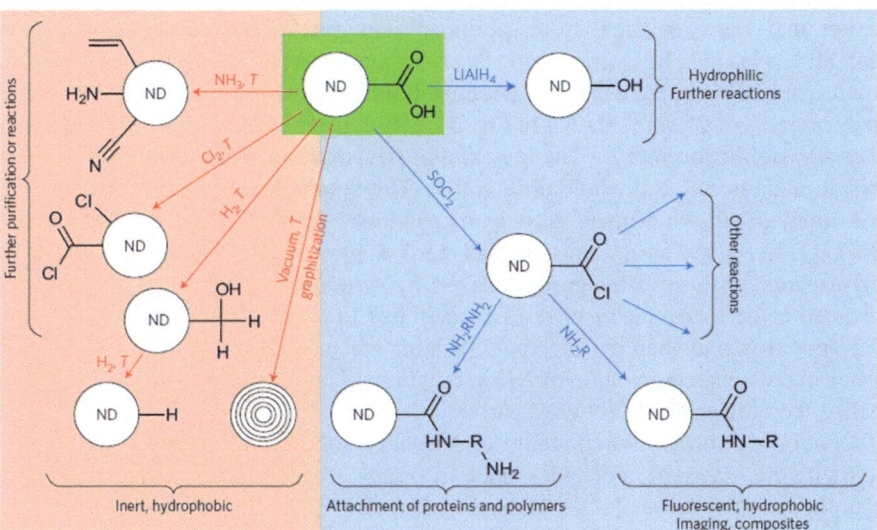

Fig. 11.19 Scheme of the frequently used chemical transformations on the nanodiamond surface, relevant for various applications. Reprinted with the permission from [1], V. N. Mochalin et al., Nat. Nanotechnol 7, 11–23 (2012), © 2011 Springer Nature

ND provides a positive surface zeta-potential allowing macromolecular adsorption as well as chemical substation.

The chemical approach of clinical used anticancer drugs paclitaxel/doxorubicin conjugation with ND utilizing its surface properties relating to a cancer treatment model has been studied [1, 148]. Using amine modified nanodiamond to paclitaxel substation; ND-conjugated paclitaxel by covalent bonding is illustrated in Fig. 11.20A. The linkage of ND (3–5 nm) and paclitaxel and the intermediate derivatives were examined using FTIR spectroscopy. The data analysis confirmed the successful functionalization of ND at various steps. The IR frequencies of the specific functionalities of the intermediate compounds, like SO_2 (1203 and 1315 cm^{-1}), NH_2 (3390 cm^{-1}) (Fig. 11.20B-b and inset (II)) and CONH (1700 cm^{-1}) and their disappearances in the subsequent stages were clearly observed. For further evidence of the conjugation of paclitaxel on the ND surface, deuterated paclitaxel-2-succinate was analyzed. The FTIR spectra for the deuterated compound displayed intense bands around $\sim$2131 and 2219 cm^{-1}; attributed to the characteristic of the C–D (carbon-deuterium) stretching frequency (Fig. 11.20B (c–e)) and inset (III)). The C–D and N–H observed in FTIR ensure the products of each step are formed according to the designed scheme. This method synthesized a novel covalent linkage of ND and paclitaxel, which can be delivered into lung carcinoma cells. More importantly, the covalent bonding of ND-paclitaxel still preserves its anticancer activities on the mitotic blockage, apoptosis induction and anti-tumorigenesis in human lung carcinoma cells. A functional ND-paclitaxel conjugate developed for a novel strategy on cancer drug delivery and therapy.

The effect of ND-paclitaxel on anti-tumorigenesis of lung carcinoma tumor xenograft in SCID mice is observed for 70 days incubation and the visible lung tumors that were separated from sacrificed xenograft SCID mice is displayed in Fig. 11.21a [148]. The average growth of the tumors were found 300–400 mm^3 in control and ND groups, while ND-paclitaxel dramatically reduced the tumor size to an average of $\sim$25 mm^3 shown in Fig. 11.21b. It is observed the ND alone doesn't play any significant role on altering the tumorigenesis of A549 cells (lung cancer cell) in mice during the whole observation. The uptake ability of ND-paclitaxel to cell correlates the anticancer activity of paclitaxel leading cell death and mitotic blockage in lung cancer cells. It has been suggested the chemical linkage of ND-paclitaxel via an ester group cleaved by esterase of lung carcinoma cells is assumed to be a key factor to release paclitaxel to execute microtubule inhibition and apoptosis and thereby eventually inhibit the tumor growth. As a result, the spectroscopic characterization of ND in functionalizing/modifying surface is vital to realize the efficiency of ND-drug carrier.

Controlling surface functionality of ND allows the strong electrostatic interaction with the macro/biomolecules, such as protein, nucleic acids, DNA, RNA, etc. in colloidal suspensions [149]. Delivery of nucleic acids into the cells is a highly precise and efficient treatment for many devastating diseases. Nanodiamond has been used as highly effective carrier for the cell transfection with plasmid DNA, siRNA and miRNA and has also opened a window of tracking the transfection directly using fluorescent nitrogen vacancy centers [150]. Using the effect of

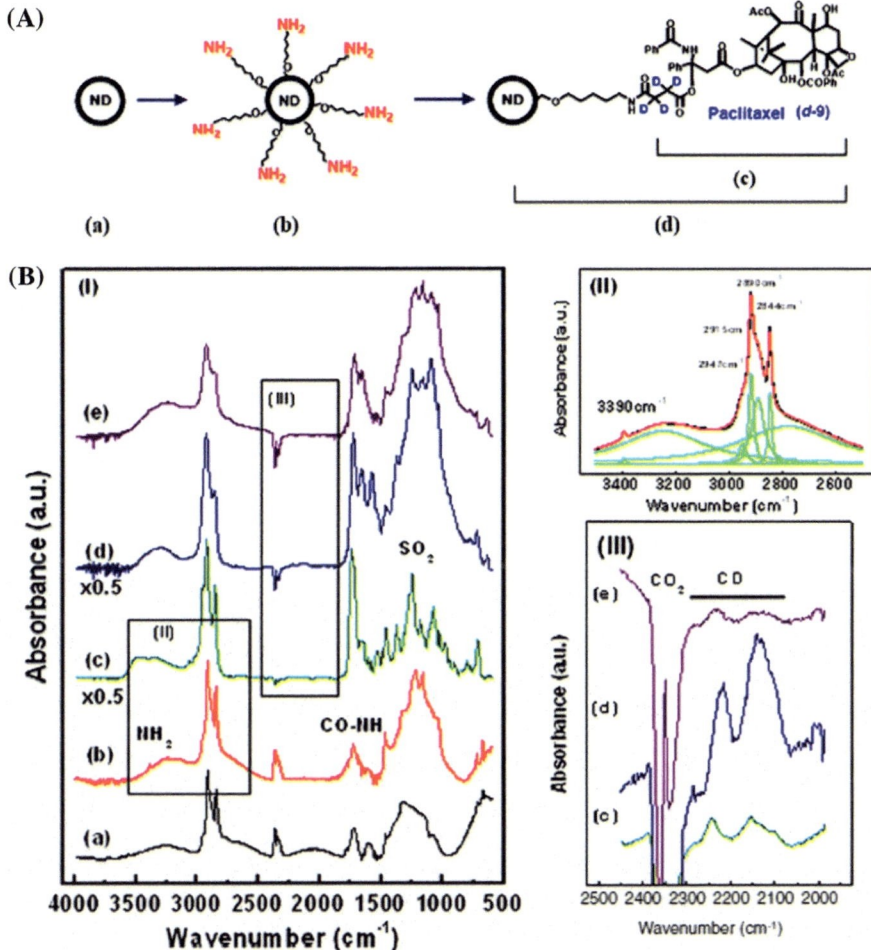

Fig. 11.20 Illustrates **A** Scheme of the ND-paclitaxel conjugation process: **a** ND of 3–5 nm; **b** ND derivative, ND–(CH$_2$O–(CH2)$_6$–NH$_2$)x; **c** deuterated paclitaxel; **d** ND-deuterated paclitaxel. **B** FTIR spectra measured on each step of the synthesis process in vacuum ($\sim 10^{-6}$ Torr), **I** FTIR spectra of the ND and its derivatives: **a** 3–5 nm ND; **b** ND–(CH2O–(CH$_2$)$_6$–NH$_2$)x and the C–H peaks are in the range 2800–3000 cm^{-1}; **c** spectrum of deuterated paclitaxel; **d** spectrum of ND-deuterated paclitaxel; **e** spectrum of ND-deuterated paclitaxel treated with 1 M NaOH. **II** The extended view of spectrum (**b**) in the 2700–3500 cm^{-1} range; identifying N–H stretching (~ 3390 cm^{-1}) frequency from ND–(CH$_2$O–(CH$_2$)$_6$–NH$_2$)x, **III** the extended view of C-D stretching frequency lying in the 1920–2520 cm^{-1} to observe from the deuterated paclitaxel in spectra (**c**, **d**). The broad peaks around 2350 cm^{-1} are for atmospheric CO$_2$ outside the vacuum chamber. Reprinted with the permission from [148], K.-K. Liu et al., Nanotechnology 21(2010) 315106, © 2010 IOP Publishing Ltd

Fig. 11.21 The Comparative effect of the paclitaxel and ND-paclitaxel on tumorigenesis of lung carcinoma tumor xenograft in CB17/Icr-Prkdc^{scid}/Crl mice. **a** the tumors treated with ND and ND-paclitaxel which are collected from five different (for each group) sacrificed xenograft mice after the incubation for 70 days, **b** the estimated volume of tumor growth measured using digital caliper every four days until 70 days. The observed results are from five mice per each group. Here the bar represents the mean ± S.E. **p < 0.01 indicates significant difference between the effect of controls and ND-paclitaxel. Reprinted with the permission from [148], K.-K. Liu et al., Nanotechnology 21(2010) 315106, © 2010 IOP Publishing Ltd

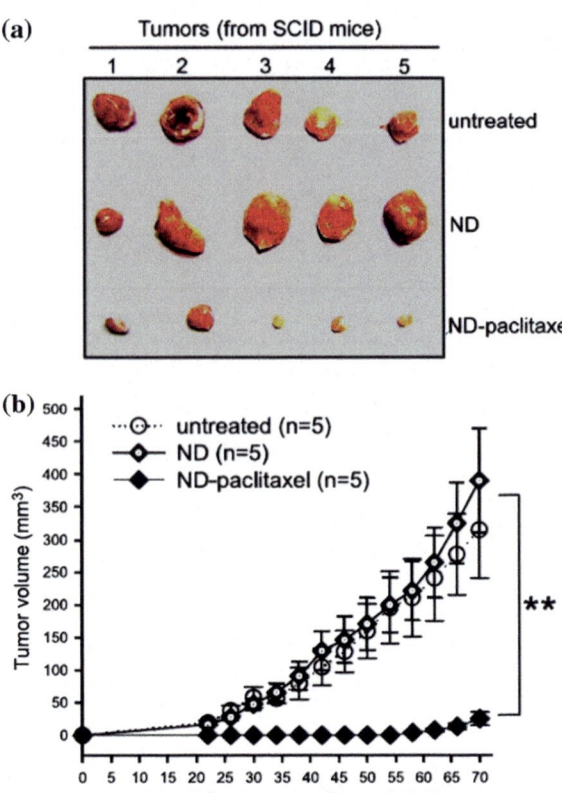

surface electrostatic charge Poly-ethyleneimine (PEI) modified FND (Fluorescent nanodiamond) has been shown as a labeling sensor for gene (DNA) delivery. After the conjugation of Poly-ethyleneimine on a negatively charged oxidized FND, the surface turns to be a positively charged character that effects on the surface contained nitrogen vacancy luminescence, leading dramatically reduced photoluminescence of NV^- ZPL and enhanced NV^0 ZPL luminescence, while the NV^- ZPL luminescence is found completely restored upon the formation of the FND–PEI–DNA complex. The surface interaction of FND with non-labeled DNA causing the dynamic changes in the NV states provides a unique optical visualization of the molecular events occurring on the FND surface; the NV^-/NV^0 ratio changes the luminescence shows a color shift shown in Fig. 11.22a. It is shown that the electrostatic immobilization of the charged molecules on the surface of FND manipulates the Fermi level position (E_F) by the surface band bending of energetic levels shown in Fig. 11.22b which corresponds to the occupation of the NV^0 and NV^- states and consequently their strength of the luminescence. Besides, hydrogenated ND, for the labeling as well as optimum rate of transfection can be used, for example as a potential carrier of electrostatically negative charged siRNA/DNA [3].

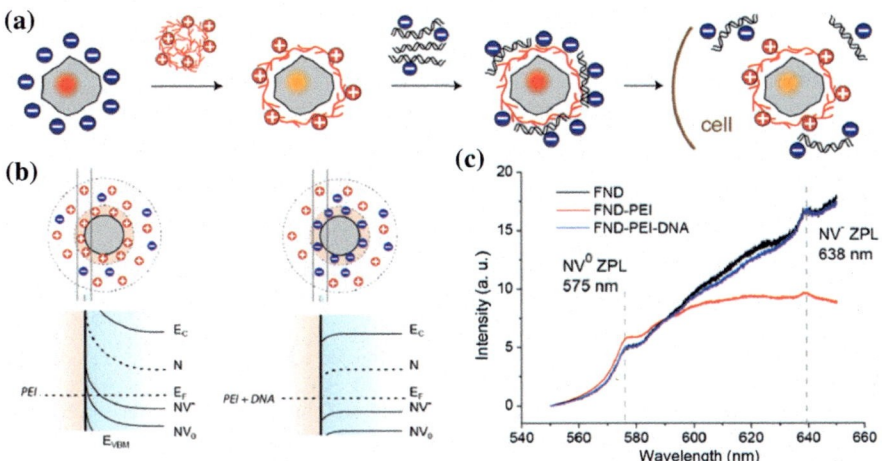

Fig. 11.22 a The schematic diagram of the FND-PEI-DNA complex formation by electrostatic interactions and DNA release after entering into the cell. FND particles' colors in different stages depict the expected changes in FND emission colors upon interaction with PEI and non-labeled DNA. **b** The surface charge density in the proximity of an FND for FND-PEI (left) and FND-PEI-DNA (right) complexes and the corresponding band bending of energetic levels in the diamond. **c** Photoluminescence spectra of oxidized FNDs and FND-PEI and FND-PEI-DNA complexes in aqueous solution (FND concentration 0.2 mg ml^{-1}) at 514 nm excitation wavelength. NV$^-$ luminescence is increased significantly in the formation of FND-PEI complex compared to oxidized FND while increases again upon binding DNA over FND-PEI. Reprinted with the permission from [150], V. Petrakova et al., Nanoscale, 2016, 8, 12002–12012, © 2016 The Royal Society of Chemistry

To understand the effect of surface modified nanodiamond to the larger bio-system such as red blood cell (RBC), bacteria, etc., several spectroscopic methods have been developed. The surface interaction of carboxylated nanodiamond (cND) with RBC is studied using Raman and UV-Visible spectroscopy [151]. Raman spectroscopic analysis of the oxygenation and deoxygenation state of RBCs during the interaction of 100 nm carboxylated nanodiamond is shown in Fig. 11.23a. The distinguished spectral peaks for the oxygenated and deoxygenated states of hemoglobin (Hb) are marked by the dotted lines. Raman feature showed that the principal bands used at 1604 to 1608 cm^{-1} (for deoxygenated RBCs); at 1638–1640 cm^{-1} and 1586–1588 cm^{-1} (for oxygenated cells). The bands at 1358–1372 cm^{-1} and at 1212–1224 cm^{-1} remain unchanged for the cND interacted RBCs. The intensity of the band at 755 cm^{-1} (a pyrrole breathing mode) remaining independent for different states is shown as the reference point. The observation suggested the surface interaction of ND doesn't change any structural conformations of Hb. As an alternative approach, similar results are also confirmed using UV-visible spectroscopy shown in Fig. 11.23b. However, it is demonstrated higher concentration of ND can affect some RBC properties such as deformability and aggregation. [151].

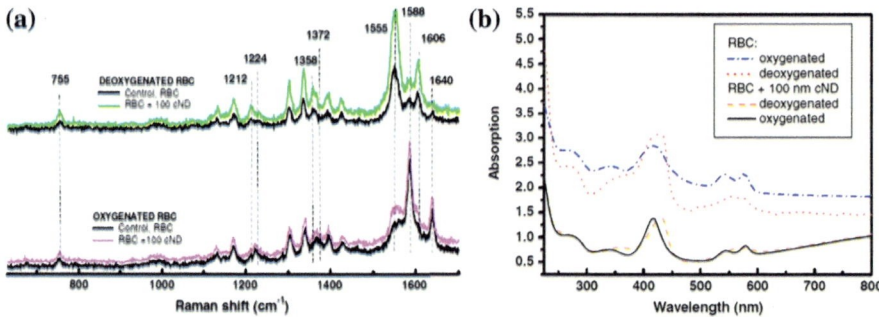

Fig. 11.23 **a** Raman spectra of oxygenated and deoxygenated RBC during their interaction with 100 nm carboxilated ND. The major Raman peaks of RBC are marked by dotted lines. **b** UV-visible absorption spectra of RBC interacting with 100 nm carboxilated ND and RBC in PBS suspension (as control) at their oxygenation and deoxygenation in situ. The oxygenation and deoxygenation was performed using N_2 purging in medium. The concentration of RBC used was 1 μl/ml and 100 cND was 100 μl/ml. Reprinted with the permission from [151], Y.-C. Lin et al., The influence of nanodiamond on the oxygenation states and micro rheological properties of human red blood cells in vitro, J. of Biomedical Optics, 17(10), 101512 (2012), © 2012 Society of Photo-Optical Instrumentation Engineers (SPIE)

ND surface is usually considered chemically inert and biocompatible [7, 8], however ND shows toxicity to the bacteria due to their surface functional groups. It is observed that carboxylated detonation nanodiamond (cDND) has an obvious antibacterial activity toward the gram-negative bacteria. Chatterjee et al. showed that in the presence of DND (~ 5 nm cDND), the surface of *E. coli* can be damaged causing the bacterial death. It was found after the interaction of cDND the maximal absorbance of *E. coli* in UV-visible region is shifted from 260 to 285 nm and additional peak is generated at around 300 nm, (Fig. 11.24), revealing the change of the cell wall structure of the *E. coli* [152]. Using Raman spectroscopy the spectral changes were observed in the range of 1400–1700 cm^{-1} corresponding to amino groups signal appeared from the proteins of destroyed cell wall, as shown in Fig. 11.24II-c [152]. A small shift of the amide-I peak (of protein) from 1654 to 1667 cm^{-1} indicates the changes in the structure of the cell wall components in the presence of cDND nanoparticles. Raman mapping shows a strong evidence of cDND attachment at the bacterial cell wall surface; the bright signal originated from the cDND (point 1) revealed cDND adsorption at the bacteria surface (Fig. 11.24II-b). The less intense spectra have been observed for the non- cDND adsorbed part of the bacterial cell wall (near point 2). This report also demonstrated the consistent results from optical and SEM images. Additionally, the interaction of lysozyme with *E. coli* is also shown for the comparative analysis. As of the findings, it has been concluded the surface chemistry of NDs plays as the driving force of the ND antibacterial effect.

In a recent related study [153], several types of surface modified nanodiamond (ND) with different purity grades (GO1, GO2 as in Figs. 11.25 and 11.26) which has been annealed in air (negatively charged ND-, ND$_{\text{pure}-}$) and in an H$_2$

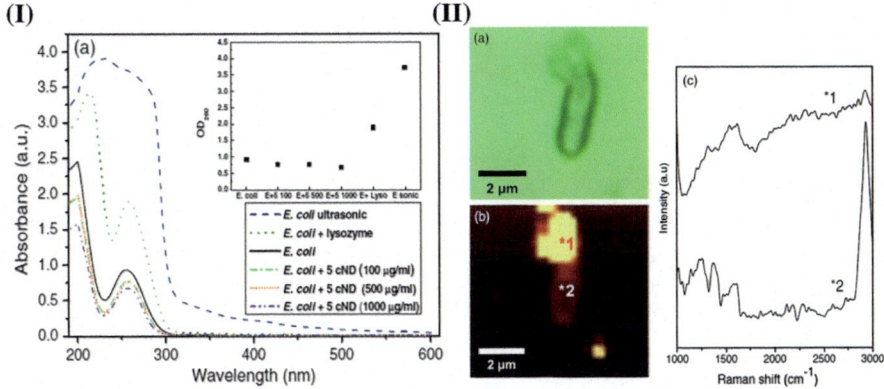

Fig. 11.24 I UV-visible spectra of the growing bacteria *E. coli* in medium treated with 5 nm cDND with the concentrations of 100, 500, 1000 µg/ml, *E. coli* treated with 50 µg/ml of lysozyme for 4 h of interaction, and *E. coli* (negative control) ultrasonic destruction of the bacterial membrane. The medium was separated from the *E. coli* and ND using centrifugation. Inset shows the values of Optical Density$_{260\ nm}$ for all the samples. **II** Depicts **a** optical image **b** the Raman mapping and **c** Raman spectra of a single *E. coli* interacted with 5 nm cDND (25 µg/ml) for 30 s in aqueous medium (488 nm wavelength laser excitation and 6 mW laser power at the 100× (N. A = 0.90) objective output). In Raman spectra, point 1 as marked in (**b**) corresponding to the place where 5 nm cDNDs are attached on bacterial cell wall and for *E. coli* Raman signal (C–H, 2820–3030 cm^{-1}) corresponding to point 2 as marked in (**b**). Reprinted with the permission from [152], A. Chatterjee et al., Antibacterial effect of ultrafine nanodiamond against gram-negative bacteria *Escherichia coli* J. of Biomedical Optics 20(5), 051014 (2015), © 2015 Society of Photo-Optical Instrumentation Engineers (SPIE)

atmosphere (positively charged ND$_+$, ND$_{pure+}$) were applied to observe the bacterial cell viability for Gram-positive (Bacillus subtilis) and Gram-negative (*E. Coli*). It is found the negatively charged ND (annealed) and raw ND/ultrasonicated raw ND exhibit strong antibacterial properties under aqueous conditions. However, hydrogen treated surface (ND$_+$) causes bacterial death only at high ND concentrations whereas highly purified ND does not show any bactericidal effects. The modified surface and their activity on bacteria are schematically displayed in Fig. 11.25.

The FTIR spectroscopic measurement of the modified ND are shown in Fig. 11.26. This work reported annealing of ultra-dispersed nanodiamond (~5 nm size) in air (oxidized) results in the formation of carboxylic anhydride groups on the surface; the anhydride structural band of carboxyl group is observed at around 1750–1850 cm^{-1}, which plays key role in the bacterial death at ambient conditions [153]. In summary, it can be concluded the surface spectroscopic analysis of nanodiamond is vital to the understanding of their fundamental characteristics and applications in various fields.

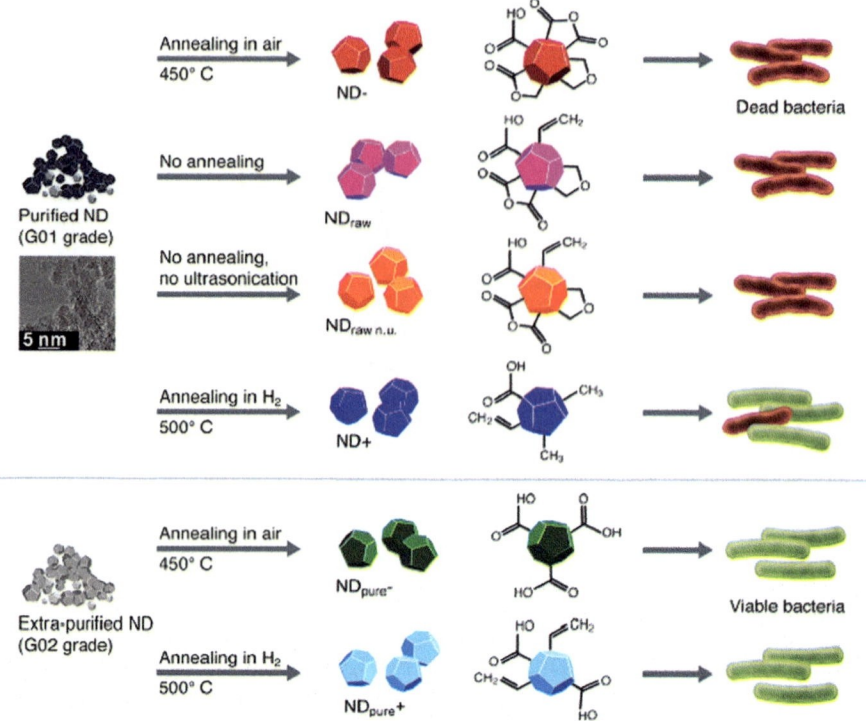

Fig. 11.25 Several surface treated nanodiamonds corresponding to neutral, negative and positively surface charged and their effects on bacterial cell viability. Grade GO1 air annealed ND and raw ND exhibit strong antibacterial activity in aqueous conditions, whereas ND$_+$ causes bacterial death only at high ND concentrations. No bactericidal effect is observed for purified ND, independent of its charge. Reprinted with the permission from [153], J. Wehling et al., ACS Nano, 2014, 8 (6), 6475–6483, © 2014 American Chemical Society

11.6.2 Perspectives of the Use of ND's Spectroscopic Properties for Bio-sensing

It is possible to combine ND's properties of multiple functionalities to develop multimodality bio applications. Among the applications, the most mentioned are drug delivery, bio-imaging and bio-sensing. For this matter, wide variability of methods of detection is employed for detection based on ND's optical/ spectroscopic properties. In this section, we consider development of bio-sensing.

Of the semiconductor nanostructures which are considered for the sensors development, carbon-based nanostructures are the most bio-compatible. For bio-sensors, carbon-based nanostructured materials are studied and used in the forms of nanoparticles. Nanotubes and carbon dots (CD) [11, 154, 155], particles and structures based on graphene and graphene oxide, including their composites

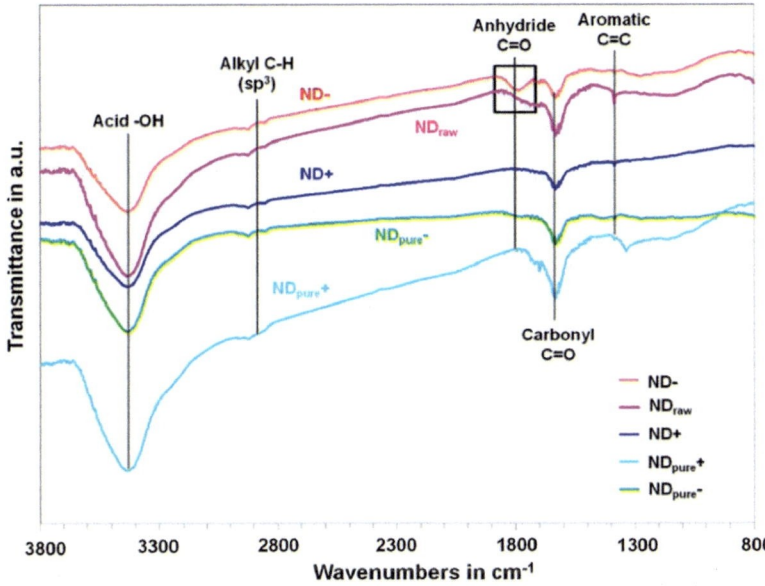

Fig. 11.26 The FTIR spectra of several types of nanodiamond (GO1, GO2) with positive and negatively charged surface character (ND−, ND$_{pure−}$, ND+, ND$_{pure+}$). In negatively charged (air annealed and raw) ND, the absorption band of acid anhydride group is at around 1850–1750 cm^{-1}. Reprinted with the permission from [153], J. Wehling et al., ACS Nano, 2014, 8 (6), 6475–6483, © 2014 American Chemical Society

with other materials, considered as quasi-two-dimensional carbon structures [154, 156], nanodiamond self-assembly arrays and thin layers [156] are of the focus. Particularly, bio-sensors on the base of carbon nanotubes are discussed owing to their unique structure-dependent electronic and mechanical properties. Due to their high surface-to-volume ratio and electron transport properties, the electronic conductance of such nanostructures can be influenced by surface perturbations associated with binding of macromolecules. Thus, these materials are considered for development of methods of rapid (in real-time) and sensitive label-free bio-electronic detection [157]. Surface modification (termination, adsorption of macromolecules, etc.) has been found modifying the fluorescence properties not only in ND but also in other kinds of nanocarbons, like CD [154, 158]. Since the first reported nanoscale bright photoluminescence of carbon nanoparticles upon simple surface passivation [159], fluorescent CDs have shown attractive properties for their use in bio-imaging and bio-sensing due to their high photostability, excellent cell membrane permeability, water solubility and biocompatibility [160]. The newest and also promising type of CD are graphene and graphene oxide dots, now extremely wide studied and developed for bio-sensing [161] and bio-imaging [162] demonstrate tunable fluorescence properties by varying the structure and surface.

As for ND, based on the above discussion of the surface state and interactions with closest environment here we want to focus mostly on the abilities and perspectives of bio-sensing using optical-spectroscopic methods. However we need to note that large number of other methods and approaches have been used, e.g. utilizing paramagnetic properties of NV centers or NV-based magnetometry, a measure of electronic paramagnetic resonance (EPR) frequency shifts caused by external perturbation or detection and three-dimensional mapping of nuclear spins in nanoscale samples (Nuclear Magnetic Resonance). The most common approaches are the direct detection of the EPR spectrum using continuous-wave spectroscopy, the measurement of the time evolution in (pulsed) pump-probe experiments, and measurements of spin relaxation times, chemical analysis of surfaces and associated dynamics [163] magnetic noise detection [164, 165]. Nuclear Spin Imaging with high spatial resolution is suggested for sensing individual spins in a complex biomolecule, and for unraveling couplings among them, revealing local characteristics of the molecule structure [166]; developing hybrid sensor combining a layer of nitrogen color centers with piezomagnetic elements for study small forces and pressures at protein folding or DNA stretching on the nanoscale [167]; impedance spectroscopy using bio-sensing electrodes nanostructured with ND [168], etc.

Petrakova et al. [106] have shown fluorescent nanodiamond (FND) as nanosensors that enable direct optical detection of noncovalent binding molecular events. The sensing mechanism is based on switching between the negatively charged and neutral states of NV centers which occurs at the noncovalent interaction of the FND surface with charged molecules, attached on the surface or being close environment in aqueous solution. It is hypothesized that these interactions cause modulation of fluorescence accordingly to changes in the FND surface charge. Correspondingly, in the model system gradual changes in the charge density on the FND surface were optically detected. These changes were caused by deprotonation of a polymer containing amino groups due to reversal ND surface charge upon binding cationic polymers. This, in turn, should lead to the depletion of NV^- occupancy and a decrease in its emission intensity. This finding opens new possibilities for construction of biocompatible photostable two-color fluorescent bio-nanosensors with potential applications in biomedical imaging and chemical sensing.

As it's shown in Sect. 11.5.3 surface state and interactions with environment affect parameters of ND PL such as lifetime decay due to energy transfer between the surface features and environment (color centers can be involved). For non-organic ND-QD system it has been demonstrated in for ND-CdSe hybrid nanostructures [134]. Förster resonance energy transfer (FRET) in such structures occurred showing clear dependence of the excitonic energy of semiconductor quantum dots. The energy transferring properties of such system can be uniquely tuned by tailoring the size of the quantum dots. This effect can be considered and used for bio-sensing. Thus, efficient FRET between negatively charged NV^- centers deeply embedded in ND as donor and surface-attached dyes or quenchers molecules as acceptor by comparing fluorescence lifetime images (FLIM) of NDs before and after bleaching of acceptor has been observed [124, 169]. Additionally, bio-compatible and photostable ND potentially is considered as convenient FRET label, for example ND of size of 3–5 nm with single color center has been suggested for FRET-based monitoring of rotation of molecular nanomotor (some kind of ATP synthase); ND was attached to the rotating subunit, and fluorophore EGFP was contained by the stator subunit. As the FRET efficiency depends on distance between donor and acceptor, this distance dependence can be used to map the spacing between the subunits of the molecule during rotation (at least in the case of monoexponential decay) [170, 171]. While previously the subunits rotation has been monitored by single-molecule FRET, limited by the photostability of the fluorophores, fluorescent ND particles with single NV-center advance these FRET measurements to long time scales which also can be varied with combination of FND with FRET-donor or FRET-acceptor fluorophore.

Moreover, as the optical-spectroscopic properties of NV centers are affected by surface state, in the following applications, surface may contribute significantly. The most developed sensing applications of ND utilize along with luminescent and magnetic properties of NV centers. The NV$^-$ center in diamond can be considered as an atom-like quantum system with a total electron spin of 1. The ground states of the spins show a crystal field splitting of 2.87 GHz, separating the 0 and ± 1 sublevels. The transitions between the spin sublevels can be optically detected and manipulated by microwave radiation. This technique is known as optically detected magnetic resonance (ODMR) [172]. Applied for imaging, ODMR provides a way to selectively modulate the intensity of emission of NV-containing nanodiamond using relatively weak microwave fields, while the other fluorescent moieties can remain unaffected by the microwaves. Thus, ODMR allows to selectively identi-fying the fluorescence from ND in the presence of environment's auto-fluorescence, scattering, and other spectrally overlapped background signals. In addition to imaging in biological systems (Fig. 11.27) [173 and references herein] ODMR can also be used for measurements of local magnetic fields [174], local electric fields [175], temperature [4, 176–178]. Thus, ODMR in nanodiamond has great potential as a quantum-based analytical tool for bio applications.

ODMR spectroscopy has been performed in the intracellular environment of primary cortical neurons. Simpson et al. demonstrated intracellular temperature mapping with high spatial resolution using FND probes (Fig. 11.28) and discuss a correlations between temperature and neuronal activity (or fertilization processes, analyzed for bovine embryo [176]) and to show that the sensing can be achieved on a time scale relevant to cellular processes [178].

Development and using hybrid materials utilize both fluorescence properties of ND and its surface properties for the hybrid creation [134, 135, 179, 180]. As an example, hybrid nanomaterial (peroxynanosensor, PNS) consisting of an organic fluorescent probe bound to a ND can be considered. This system allows concurrent and extended cell imaging of the ND and ratiometric detection of hydrogen per-oxide in the ND environment. Its use has been demonstrated for analysis of the cell cytoplasm. For that, ND's red fluorescence is detected for the imaging, while the green fluorescence of the organic fluorescent probe attached to the nanodiamond

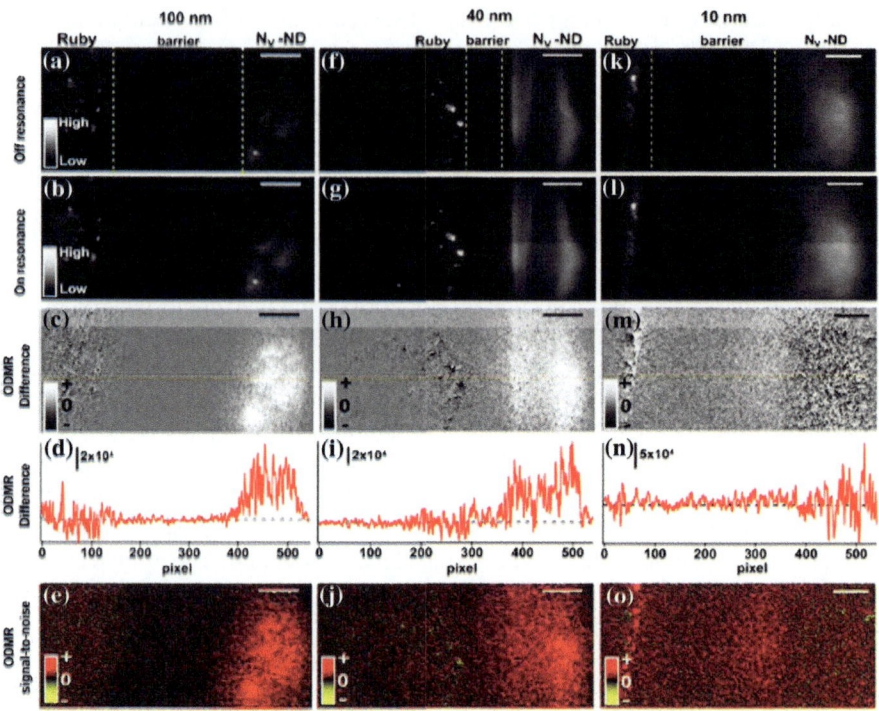

Fig. 11.27 The selective imaging of ND of sizes 100 (**a–e**), 40 (**f–j**), and 10 nm (**k–o**) and ruby particles. The Images **a**, **f**, and **k** were collected with an off-resonance microwave frequency (2.93 GHz) and **b**, **g**, and **l** were collected with the on resonance, microwave frequency (2.87 GHz). Grayscale ranges are ranging from zero to $\sim 6 \times 10^6$ counts per pixel. Images **c**, **h**, and **m** are the difference image of I(2.93 GHz)–I(2.87 GHz), representing the microwave-induced modulation in intensity where ND gives rise to positive signals (white in the image). Panels **d**, **i**, and **n** present cross-sections through the ODMR difference images and **e**, **j**, and **o** represent the ODMR signal-to-noise ratio. The vertical scale ranges are S/N = (−100, 100) for panels **e** and **j** and S/N = (−30, +30) for panel **o**, horizontal scale bars are 25 μm for all images. Reprinted with the permission from [173], M. E. Robinson et al., Anal. Chem., 2018, 90 (1), 769–776, © 2017 American Chemical Society

surface indicates hydrogen peroxide [179]. In another example, a stimuli-responsive metastasis detection complex of ND functionalized with fluorescent-labeled peptide which is specific to metastasis biomarker matrix metalloproteinase (MMP9). Using this hybrid biosensor, protease activity of MMP9 correlated to MMP9 expression can be numerically measured. It allows identifying potential metastatic cancer sites [180].

Finally, chemically modified DND can exhibit different non-linear optical response to laser excitation compared to non-modified one. It has been shown in studies of aminated DND, which seems to be very interesting for their possible interaction with hydroxyl groups of some polymers. For the aminated DND more pronounce non-linear optical effects were observed (non-linear scattering and

Fig. 11.28 The intracellular temperature mapping of NDs inside primary cortical neurons overlaid with the ND fluorescence image at 37.3 °C. Reprinted with the permission from [178], D. A. Simpson et al., ACS Nano, 2017, 11 (12), 12077–12086, © 2017 American Chemical Society

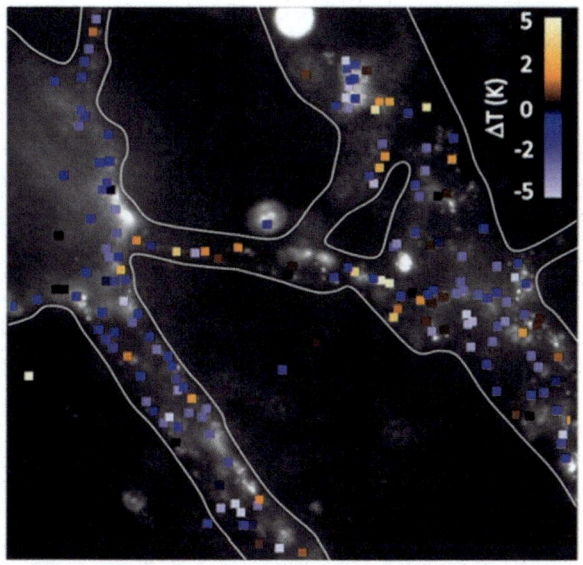

Kerr effects) and the founding depended on wavelength [12]. Affecting the ND spectroscopic properties both by surface manipulation and laser can be achieved and considered for applications.

Thus, the role of surface in ND interaction with laser light is that NV centers' charge and emission states manipulating by control of ND surface state [101] (see also Sect. 11.5) can interfere with effects occurring at ND interaction with laser light. Further example, when the ND levitated in a free-space optical trap excited by a 532 nm pulsed laser aligned confocally with the 1064 nm CW trap laser, the NV's PL demonstrates a high degree of suppression of PL from the NV^0 defect state while the trap laser is on, without affecting NV^- [181].

11.7 Conclusions

In this chapter, the spectroscopic studies of nanodiamond surface structure, composition and the surface interactions are reviewed. The methods of characterization of the nanodiamond surface and modification methods are introduced. The reviewing in optical spectroscopy of the nanodiamond surface is focused on infrared spectroscopy, Raman and photoluminescence and their applications both for nanodiamond characterization and for development of the ways of nanodiamond bio-applications. Based on wide variety of the approaches, important role of the surface and surface interactions on the vibrational and luminescence spectroscopic properties of nanodiamond is revealed. Surface related bio/medical researches are reviewed indicating promising potential of nanodiamond for further development of new bio/medical applications, particularly, bio-sensing in nanoscale.

Acknowledgements The authors appreciate the financial support of this research by the Ministry of Science and Technology (MOST) of Taiwan, Grant No. MOST 106-2112-M-259-009-MY3.

References

1. V.N. Mochalin, O. Shenderova, D. Ho, Y. Gogotsi, The properties and applications of nanodiamonds. Nat. Nanotechnol. **7**(1), 11–23 (2012)
2. E. Perevedentseva, Y.-C. Lin, M. Jani, C.-L. Cheng, Biomedical applications of nanodiamonds in imaging and therapy. Futur. Med. **8**(12), 2041–2060 (2013)
3. K. Turcheniuk, V.N. Mochalin, Biomedical applications of nanodiamond (Review). Nanotechnology **28**, 252001 (2017)
4. M. Chipaux, K.J. van der Laan, S.R. Hemelaar, M. Hasani, T. Zheng, R. Schirhagl, Nanodiamonds and their applications in cells. Small **14**(24), e1704263 (2018). https://doi.org/10.1002/smll.201704263
5. G. Hong, S. Diao, A.L. Antaris, H. Dai, Carbon nanomaterials for biological imaging and nanomedicinal therapy. Chem. Rev. **115**(19), 10816–10906 (2015)
6. T. Plakhotnik, H. Aman, NV-centers in nanodiamonds: how good they are. Diam. Relat. Mater. **82**, 87–95 (2018)
7. Y. Zhang, K.Y. Rhee, D. Hui, S.-J. Park, A critical review of nanodiamond based nanocomposites: synthesis, properties and applications. Compos. B **143**, 19–27 (2018)
8. K.J. van der Laan, M. Hasani, T. Zheng, R. Schirhagl, Nanodiamonds for in vivo applications. Small **14**(19), e1703838 (2018)
9. D.G. Lim, R.E. Prim, K.H. Kim, E. Kang, K. Park, S.H. Jeong, Combinatorial nanodiamond in pharmaceutical and biomedical applications. Int. J. Pharm. **514**, 41–51 (2016)
10. C. Bradac, I. Das Rastogi, N.M. Cordina, A. Garcia-Bennett, L.J. Brown, Influence of surface composition on the colloidal stability of ultra-small detonation nanodiamonds in biological media. Diam. Relat. Mater. **83**, 38–45 (2018)
11. P. Reineck, D.W.M. Lau, E.R. Wilson, N. Nunn, O.A. Shenderova, B.C. Gibson, Visible to near-IR fluorescence from single-digit detonation nanodiamonds: excitation wavelength and pH dependence. Sci. Rep. **8**, 2478 (2018)
12. V. Pichot, O. Muller, A. Seve, A. Yvon, L. Merlat, D. Spitzer, Optical properties of functionalized nanodiamonds. Sci. Rep. **7**, 14086 (2017)
13. R. Tappert, M.C. Tappert, *Diamond in nature; A guide to rough diamonds* (Chapter-1 & 2, Springer, Berlin, Heidelberg, 2011), pp. 1–40. https://doi.org/10.1007/978-3-642-12572-0
14. H.O. Pierson, (ed.), Structure and properties of diamond and diamond polytypes, in *Handbook of Carbon, Graphite, Diamond and Fullerenes; Properties, Processing and Applications* (Chapter-11, William Andrew Publishing, Oxford, 1993), pp. 244–277. doi: https://doi.org/10.1016/B978-0-8155-1339-1.50016-5
15. R.P. Mildren, J.R. Rabeau, (eds.), *Optical engineering of diamond* (Chapter-5, Wiley-VCH Verlag GmbH & Co. KGaA, Boschstr.12, 69469 Weinheim, Germany, 2013), pp. 143–145 [https://doi.org/10.1002/9783527648603]
16. C.-L. Cheng, H.-C. Chang, J.-C. Lin, K.-J. Song, J.-K. Wang, Direct observation of hydrogen etching anisotropy on diamond single crystal surfaces. Phys. Rev. Lett. **78**(19), 3713–3716 (1997)
17. A.P. Jones, L.I. D'Hendecourt, Interstellar nanodiamonds: the carriers of mid-infrared emission bands? Astron. Astrophys. **355**, 1191–1200 (2000)
18. J.-S. Tu, E. Perevedentseva, P.-H. Chung, C.-L. Cheng, Size-dependent surface CO stretching frequency investigations on nanodiamond particles. J. Chem. Phys. **125**, 174713 (2006)

19. V. Dolmatov, T. Fujimura, Physical and chemical problems of modification of detonation nanodiamond surface properties, in *Synthesis, Properties and Applications of Ultrananocrystalline Diamond*, ed. by D.M. Gruen, O.A. Shenderova, A.Y. Vul'. NATO Science Series (Series II: Mathematics, Physics and Chemistry), vol. 192 (Springer, Dordrecht, 2005), pp. 217–230

20. C.-L. Cheng, C.-F. Chen, W.-C. Shaio, D.-S. Tsai, K.-H. Chen, The CH stretching features on diamonds of different origins. Diam. Relat. Mater. **14**(9), 1455–1462 (2005)

21. L.C.L. Huang, H.-C. Chang, Adsorption and immobilization of cytochrome c on nanodiamonds. Langmuir **20**(14), 5879–5884 (2004)

22. T. Ando, K. Yamamoto, M. Ishii, M. Kamo, Y. Sato, J. Chem, Vapour-phase oxidation of diamond surfaces in O_2 studied by diffuse reflectance Fourier-transform infrared and temperture-programmed desorption spectroscopy. Soc. Faraday Trans. **89**, 3635–3640 (1993)

23. T. Tsubota, O. Hirabayashi, S. Ida, S. Nagaoka, M. Nagata, Y. Matsumoto, Chemical modification of hydrogenated diamond surface using benzoyl peroxides. Phys. Chem. Chem. Phys. **4**, 806–811 (2002)

24. P.H. Chung, E. Perevedentseva, J.S. Tu, C.C. Chang, C.-L. Cheng, Spectroscopic study of bio-functionalized nanodiamonds. Diam. Relat. Mater. **15**, 622–625 (2006)

25. Y.-R. Chen, H.-C. Chang, C.-L. Cheng, C.-C. Wang, J. C. Jiang, Size dependence of C-H stretching features on diamond nanocrystal surfaces: infrared spectroscopy and density functional theory calculations. J. Chem. Phys. **119**, 10626 (2003)

26. C.-D. Chu, E. Perevedentseva, V. Yeh, S.-J. Cai, J.-S. Tu, C.-L. Cheng, Temperature-dependent surface C=O stretching frequency investigation of functionalized ND particles. Diam. Relat. Mater. **19**, 76–81 (2009)

27. I.I. Kulakova, Surface chemistry of nanodiamonds. Phys. Solid State **46**, 636–643 (2004)

28. M. Jani, J.-S. Tu, T.-Y. Kang, C.-Y. Tsai, E. Perevedentseva, C.-L. Cheng, Surface modification of nanodiamond: photoluminescence and Raman studies. Diam. Relat. Mater. **24**, 134–138 (2012)

29. A.M. Rao, E. Richter, S. Bandow, B. Chase, P.C. Eklund, K.A. Williams, S. Fang, K.R. Subbaswamy, M. Menon, A. Thess, R.E. Smalley, G. Dresselhaus, M.S. Dressehaus, Diameter-selective Raman scattering from vibrational modes in carbon nanotubes. Science **275**, 187–191 (1997)

30. A.C. Ferrari, J. Robertson, Interpretation of Raman spectra of disordered and amorphous carbon. J. Phys. Rev. B **61**, 14095–14107 (2000)

31. A.C. Ferrari, J. Robertson, Origin of the 1150 cm^{-1} Raman mode in nanocrystalline diamond. Phys. Rev. B **63**, 121405(R) (2001)

32. R. Pfeiffer, H. Kuzmany, N. Salk, B. Gunther, Evidence for trans-polyacetylene in nanocrystalline diamond films from H-D isotropic substitution experiments. Appl. Phys. Lett. **82**, 4149–4150 (2003)

33. S. Prawer, K.W. Nugent, D.N. Jamieson, J.O. Orwa, L.A. Bursill, J.L. Peng, The Raman spectrum of nanocrystalline diamond. Chem. Phys. Lett. **332**(1–2), 93–97 (2000)

34. V.I. Korepanov, H.O. Hamaguchi, E. Osawa, V. Ermolenkov, I.K. Lednev, B.J.M. Etzold, O. Levinson, B. Zousman, C. Prakash Epperla, H.-C. Chang, Carbon structure in nanodiamonds elucidated from Raman spectroscopy. Carbon **121**, 322e329 (2017)

35. V. Mochalin, S. Osswald, Y. Gogotsi, Contribution of functional groups to the Raman spectrum of nanodiamond powders. Chem. Mater. **21**(2), 273–279 (2009)

36. M. Mermoux, A. Crisci, T. Petit, H.A. Girard, J.-C. Arnault, Surface modifications of detonation nanodiamonds probed by multiwavelength Raman spectroscopy. J. Phys. Chem. C **118**, 23415–23425 (2014)

37. P. Reineck, D.W.M. Lau, E.R. Wilson, K. Fox, M.R. Field, C. Deeleepojananan, V.N. Mochalin, B.C. Gibson, Effect of surface chemistry on the fluorescence of detonation nanodiamonds. ACS Nano **11**, 10924–10934 (2017)

38. V.N. Mochalin, Y. Gogotsi, Wet chemistry route to hydrophobic blue fluorescent nanodiamond. J. Am. Chem. Soc. **131**(13), 4594–4595 (2009)

39. S. Osswald, G. Yushin, V. Mochalin, S.O. Kucheyev, Y. Gogotsi, Control of sp2/sp3 carbon ratio and surface chemistry of nanodiamond powders by selective oxidation in Air. J. Am. Chem. Soc. **128**(35), 11635–11642 (2006)

40. V.Y. Osipov, A.M. Panich, A.V. Baranov, Comment on "Carbon structure in nanodiamonds elucidated from Raman spectroscopy" by V.I. Korepanov et al. Carbon **127**, 193–194 (2018)

41. T.A. Dolenko, S.A. Burikov, J.M. Rosenholm, O.A. Shenderova, I.I. Vlasov, Diamond −water coupling effects in Raman and photoluminescence spectra of nanodiamond colloidal suspensions. J. Phys. Chem. C **116**, 24314–24319 (2012)

42. T. Petit, L. Puskar, T. Dolenko, S. Choudhury, E. Ritter, S. Burikov, K. Laptinskiy, Q. Brzustowski, U. Schade, H. Yuzawa, M. Nagasaka, N. Kosugi, M. Kurzyp, A. Venerosy, H. Girard, J.-C. Arnault, E. Osawa, N. Nunn, O. Shenderova, E.F. Aziz, Unusual water hydrogen bond network around hydrogenated nanodiamonds. J. Phys. Chem. C **121**, 5185–5194 (2017)

43. J. Zheng, Y. Ding, B. Tian, Z.L. Wang, X. Zhuang, Luminescent and Raman active silver nanoparticles with polycrystalline structure. J. Am. Chem. Soc. **130**, 10472–10473 (2008)

44. M. Moskovits, Surface-enhanced spectroscopy. Rev. Mod. Phys. **57**, 783–826 (1985)

45. S. Nie, S.R. Emory, Probing single molecules and single nanoparticles by surface-enhanced Raman scattering. Science **275**, 1102–1106 (1997)

46. H. Xu, E.J. Bjerneld, M. Kall, L. Borjesson, Spectroscopy of single hemoglobin molecules by surface enhanced Raman scattering. Phys. Rev. Lett. **83**, 4357–4360 (1999)

47. Z.-C. Hong, E. Perevedentseva, S. Treschev, J.-B. Wang, C.-L. Cheng, Surface enhanced Raman scattering of nanodiamond using visible-light-activated TiO$_2$ as a catalyst to photo-reduce nanostructured silver from AgNO$_3$ as SERS-active substrate. J. Raman Spectrosc. **40**(8), 1016–1022 (2009)

48. A. Karmenyan, E. Perevedentseva, A. Chiou, C.-L. Cheng, Positioning of carbon nanostructures on metal surfaces using laser acceleration and the Raman analyses of the patterns. Eur. J. Phys. **61**, 513–517 (2007)

49. M. Veres, E. Perevedentseva, A.V. Karmenyan, S. Tóth, S. Koós, Catalytic activity of gold on nanocrystalline diamond support. Phys. Status Solidi C **7**(3–4), 1211–1214 (2010)

50. E. Perevedentseva, A. Karmenyan, P.H. Chung, Y.T. He, C.-L. Cheng, Surface enhanced Raman spectroscopy of carbon nanostructures. Surf. Sci. **600**, 3723–3728 (2006)

51. S.J. Chase, W.S. Bacsa, M.G. Mitch, L.J. Pilione, J.S. Lannin, Surface-enhanced Raman scattering and photoemission of C60 on noble-metal surfaces. Phys. Rev. B **46**, 7873–7877 (1992)

52. M. Roy, V.C. George, A.K. Dua, P. Raj, S. Schulze, D.A. Tenne, G. Salvan, D.R.T. Zahn, Detection of nanophase at the surface of HFCVD grown diamond films using surface enhanced Raman spectroscopic (SERS) technique. Diam. Relat. Mater. **11**, 1858–1862 (2002)

53. M. Veres, M. Fule, S. Toth, M. Koos, I. Pocsik, Surface enhanced Raman scattering (SERS) investigation of amorphous carbon. Diam. Relat. Mater. **13**, 1412–1415 (2004)

54. E.C. Le Ru, P.G. Etchegoin, Sub-wavelength localization of hot-spots in SERS. Chem. Phys. Lett. **396**, 393–397 (2004)

55. M. Veres, S. Tóth, E. Perevedentseva, A. Karmenyan, M. Koós, Detection of structural units of nanocrystalline diamond surfaces using surface-enhanced Raman scattering, in *Nanotechnological Basis for Advanced Sensors* (NATO Science for Peace and Security Series B: Physics and Biophysics), ed. by J.P. Reithmaier, P. Paunovic (Springer, Netherlands, 2011), pp. 111–120

56. A.V. Karmenyan, E. Perevedentseva, M. Veres, C.-L. Cheng, Simultaneous PL and SERS observation of ND at laser deposition on noble metals. Plasmonics **8**, 325–333 (2012)

57. H. Chacham, L. Kleinman, Instabilities in diamond under high shear stress. Phys. Rev. Lett. **85**, 4904 (2000)

58. J. Qian, C. Pantea, G. Voronin, T.W. Zerda, Partial graphitization of diamond crystals under high-pressure and high-temperature conditions. J. Appl. Phys. **90**, 1632 (2001)

59. R.M. Erasmus, R.D. Daniel, J.D. Comins, Three-dimensional mapping of stresses in plastically deformed diamond using micro-Raman and photoluminescence spectroscopy. J. Appl. Phys. **109**, 013527 (2011)
60. Y. Gogotsi, A. Kailer, K.G. Nickel, Transformation of diamond to graphite. Nature **40**, 663 (1999)
61. D.S. Knight, W.B. White, Characterization of diamond films by Raman spectroscopy. J. Mater. Res. **4**, 385–393 (1989)
62. A. Krueger, D. Lang, Functionality is key: recent progress in the surface modification of nanodiamond. Adv. Func. Mater. **22**(5), 890–906 (2012)
63. I. Kulakova, V.V. Korol'kov, R.Y. Yakovlev, G.V. Lisichkin, The structure of chemically modified detonation synthesized nanodiamond particles. Nanotechnol. Russ. **5**(7–8), 474–485 (2010)
64. B.V. Spitsyn, J.L. Davidson, M.N. Gradoboev, T.B. Galushko, N.V. Serebryakova, T.A. Karpukhina, I.I. Kulakova, N.N. Melnik, Inroad to modification of detonation nanodiamond. Diam. Relat. Mater. **15**, 296–299 (2006)
65. I. Petrov, O. Shenderova, V. Grishko, V. Grichko, T. Tyler, G. Cunningham, G. McGuire, Detonation nanodiamonds simultaneously purified and modified by gas treatment. Diam. Relat. Mater. **16**, 2098–2103 (2007)
66. M.A. Ray, O. Shenderova, W. Hook, A. Martin, V. Grishko, T. Tyler, G.B. Cunningham, G. McGuire, Cold plasma functionalization of nanodiamond particles. Diam. Relat. Mater. **15**, 1809–1812 (2006)
67. G.A. Chiganova, Aggregation of particles in ultradispersed diamond hydrosols. Colloid J. **62**(2), 238–243 (2000)
68. A. Krüger, F. Kataoka, M. Ozawa, T. Fujino, Y. Suzuki, A.E. Aleksenskii, A.Y. Vul', E. Ōsawa, Unusually tight aggregation in detonation nanodiamond: identification and disintegration. Carbon **43**(8), 1722–1730 (2005)
69. A. Krueger, Y. Liang, G. Jarre, J. Stegk, Surface functionalisation of detonation diamond suitable for biological applications. J. Mater. Chem. **16**(24), 2322–2328 (2006)
70. H.A. Girard, T. Petit, S. Perruchas, T. Gacoin, C. Gesset, J.C. Arnault, P. Bergonzo, Surface properties of hydrogenated nanodiamonds: a chemical investigation. Phys. Chem. Chem. Phys. **13**(24), 11517–11523 (2011)
71. V.V. Korolkov, I.I. Kulakova, B.N. Tarasevich, G.V. Lisichkin, Dual reaction capacity of hydrogenated nanodiamond. Diam. Relat. Mater. **16**(12), 2129–2132 (2007)
72. A.-I. Ahmed, S. Mandal, L. Gines, O.A. Williams, C.-L. Cheng, Low temperature catalytic reactivity of nanodiamond in molecular hydrogen. Carbon **110**, 438–442 (2016)
73. O.A. Williams, J. Hees, C. Dieker, W. Jager, L. Kirste, C.E. Nebel, Size-dependent reactivity of diamond nanoparticles. ACS Nano **4**(8), 4824–4830 (2010)
74. S. Osswald, G. Yushin, V. Mochalin, S.O. Kucheyev, Y. Gogotsi, Control of sp2/sp3 carbon ratio and surface chemistry of nanodiamond powders by selective oxidation in air. J. Am. Chem. Soc. **128**(35), 11635–11642 (2006)
75. E. Neu, F. Guldner, C. Arend, Y. Liang, S. Ghodbane, H. Sternschulte, D. Steinmüller-Nethl, A. Krueger, C. Becher, Low temperature investigations and surface treatments of colloidal narrowband fluorescent nanodiamonds. J. Appl. Phys. **113**, 203507 (2013)
76. I. Aharonovich, A.D. Greentree, S. Prawer, Diamond photonics. Nat. Photonics **5**, 397–405 (2011)
77. M.W. Doherty, N.B. Manson, P. Delaney, F. Jelezko, J. Wrachtrup, L.C.L. Hollenberg, The nitrogen-vacancy colour centre in diamond. Phys. Rep. **528**(1), 1–45 (2013)
78. I. Aharonovich, S. Castelletto, D.A. Simpson, C.-H. Su, A.D. Greentree, S. Prawer, Diamond-based single-photon emitters. Rep. Prog. Phys. **74**, 076501 (2011)
79. J. Wrachtrup, F. Jelezko, Processing quantum information in diamond J. Phys. Condens. Matter **18**, S807 (2006)
80. H.S. Knowles, D.M. Kara, M. Atatüre, Observing bulk diamond spin coherence in high-purity nanodiamonds. Nat. Mater. **13**, 21–25 (2014)

81. A.W. Schell, G. Kewes, T. Hanke, A. Leitenstorfer, R. Bratschitsch, O. Benson, T. Aichele, Single defect centers in diamond nanocrystals as quantum probes for plasmonic nanostructures. Opt. Express **19**, 7914–7920 (2011)

82. C.C. Fu, H.-Y. Lee, K. Chen, T.-S. Lim, H.-Y. Wu, P.-K. Lin, P.-K. Wei, P.-H. Tsao, H.-C. Chang, W. Fann, Characterization and application of single fluorescent nanodiamonds as cellular biomarkers. Proc. Natl. Acad. Sci. U.S.A. **104**(3), 727–732 (2007)

83. J. Narayan, A. Bhaumik, Novel synthesis and properties of pure and NV-doped nanodiamonds and other nanostructures. Mater. Res. Lett. **5**(4), 242–250 (2016)

84. C. Bradac, T. Gaebel, C.I. Pakes, J.M. Say, A.V. Zvyagin, J.R. Rabeau, Effect of the nanodiamond host on a nitrogen-vacancy color-centre emission state. Small **9**, 132 (2013)

85. G.S. Gildenblat, S.A. Grot, A. Badzian, The electrical properties and device applications of homoepitaxial and polycrystalline diamond films. Proc. IEEE **79**, 647–668 (1991)

86. I. Kratochvılova, A. Kovalenko, F. Fendrych, V. Petráková, S. Záliš, M. Nesládek, Tuning of nanodiamond particles' optical properties by structural defects and surface modifications: DFT modelling. J. Mater. Chem. **21**, 18248 (2011)

87. A. Kovalenko, V. Petráková, P. Ashcheulov, S. Záliš, M. Nesládek, I. Kraus, I. Kratochvílová, Parameters affecting the luminescence of nanodiamond particles: quantum chemical calculations. Phys. Status Solidi A **209**, 1769–1773 (2012)

88. V. Petrakova, A. Taylor, I. Kratochvílová, F. Fendrych, J. Vacík, J. Kučka, J. Štursa, P. Cígle, M. Ledvina, A. Fišerová, P. Kneppo, M. Nesládek, Luminescence of nanodiamond driven by atomic functionalization: towards novel detection principles. Adv. Funct. Mater. **22**, 812–819 (2012)

89. K. Yakoubovskii, Luminescence excitation spectra in diamond. Phys. Rev. B **61**(15), 010174 (2000)

90. S.Y. Lim, W. Shen, Z. Gao, Carbon quantum dots and their applications. Chem. Soc. Rev. **44**, 362–381 (2015)

91. P. Reineck, A. Francis, A. Orth, D.W.M. Lau, R.D.V. Nixon-Luke, I.D. Rastogi, W.A.W. Razali, L.M. Parker, V.K.A. Sreenivasan, L.J. Brown, B.C. Gibson, Brightness and photostability of emerging red and near-IR fluorescent nanomaterials for bioimaging. Adv. Opt. Mater. **4**, 1549–1557 (2016)

92. M. Fu, F. Ehrat, Y. Wang, K.Z. Milowska, C. Reckmeier, A.L. Rogach, J.K. Stolarczyk, A. S. Urban, J. Feldmann, Carbon dots: a unique fluorescent cocktail of polycyclic aromatic hydrocarbons. Nano Lett. **15**, 6030–6035 (2015)

93. G. Eda, Y.Y. Lin, C. Mattevi, H. Yamaguchi, H.A. Chen, I.S. Chen, C.W. Chen, M. Chhowalla, Blue photoluminescence from chemically derived graphene oxide. Adv. Mater. **22**, 505–509 (2010)

94. R.G. Ryan, A. Stacey, K.M. O'Donnell, T. Ohshima, B.C. Johnson, L.C.L. Hollenberg, P. Mulvaney, D.A. Simpson, Impact of surface functionalisation on the quantum coherence of nitrogen vacancy centres in nanodiamond. ACS Appl. Mater. Interfaces (2018). https://doi.org/10.1021/acsami.7b19238

95. F. Maier, M. Riedel, B. Mantel, J. Ristein, L. Ley, Origin of surface conductivity in diamond. Phys. Rev. Lett. **85**, 3472–3475 (2000)

96. C. Bradac, T. Gaebel, N. Naidoo, M.J. Sellars, J. Twamley, L.J. Brown, A.S. Barnard, T. Plakhotnik, A.V. Zvyagin, J.R. Rabeau, Observation and control of blinking nitrogen-vacancy centres in discrete nanodiamonds. Nat. Nanotechnol. **5**, 345–349 (2010)

97. A.M. Vervald, S.A. Burikov, O.A. Shenderova, N. Nunn, D.O. Podkopaev, I.I. Vlasov, T.A. Dolenko, Relationship between fluorescent and vibronic properties of detonation nanodiamonds and strength of hydrogen bonds in suspensions. J. Phys. Chem. C **120**, 19375–19383 (2016)

98. A.A. Khomich, O.S. Kudryavtsev, T.A. Dolenko, A.A. Shiryaev, A.V. Fisenko, V.I. Konov, I.I. Vlasov, Anomalous enhancement of nanodiamond luminescence upon heating. Laser Phys. Lett. **14**, 025702 (2017)

99. V. Petráková, M. Nesladek, A. Taylor, Luminescence properties of engineered nitrogen vacancy centers in a close surface proximity. Phys. Status Solidi A **208**(9), 2051–2056 (2011)
100. L. Rondin, G. Dantelle, A. Slablab, F. Grosshans, F. Treussart, P. Bergonzo, S. Perruchas, T. Gacoin, M. Chaigneau, H.-C. Chang, V. Jacques, J.-F. Roch, Surface-induced charge state conversion of nitrogen-vacancy defects in nanodiamonds. Phys. Rev. B Condens. Matter **82**, 115449 (2010)
101. A.N. Newell, D.A. Dowdell, D.H. Santamore, Surface effects on nitrogen vacancy centers neutralization in diamond. J. Appl. Phys. **120**, 185104 (2016)
102. M.V. Hauf, B. Grotz, B. Naydenov, M. Dankerl, S. Pezzagna, J. Meijer, F. Jelezko, J. Wrachtrup, M. Stutzmann, F. Reinhard, J.A. Garrido, Chemical control of the charge state of nitrogen-vacancy centers in diamond. Phys. Rev. B Condens. Matter **83**, 081304 (2011)
103. M. Kaviani, P. Deak, B. Aradi, T. Frauenheim, J.-P. Chou, A. Gali, Proper surface termination for luminescent near-surface NV centers in diamond. Nano Lett. **14**(8), 4772–4777 (2014)
104. A. Khalid, K. Chung, R. Rajasekharan, D.W.M. Lau, T.J. Karle, B.C. Gibson, S. Tomljenovic-Hanic, Lifetime reduction and enhanced emission of single photon color centers in nanodiamond via surrounding refractive index modification. Sci. Rep. **5**, 11179 (2015)
105. J. Xiao, P. Liu, L. Li, G. Yang, Fluorescence origin of nanodiamonds. J. Phys. Chem. C **119**, 2239–2248 (2015)
106. V. Petrakova, I. Rehor, J. Stursa, M. Ledvina, M. Nesladeka, P. Cigler, Charge-sensitive fluorescent nanosensors created from nanodiamonds. Nanoscale **7**, 12307 (2015)
107. P. Galar, J. Čermák, P. Malý, A. Kromka, B. Rezek, Electrochemically grafted polypyrrole changes photoluminescence of electronic states inside nanocrystalline diamond. J Appl. Phys. **116**, 223103 (2014)
108. M. Ohtani, P.V. Kamat, S. Fukuzumi, Supramolecular donor-acceptor assemblies composed of carbon nanodiamond and porphyrin for photoinduced electron transfer and photocurrent generation. J. Mater. Chem. **20**, 582–587 (2010)
109. S. Zhu, J. Shao, Y. Song, X. Zhao, J. Du, L. Wang, H. Wang, K. Zhang, J. Zhang, B. Yang, Investigating the surface state of graphene quantum dots. Nanoscale **7**, 7927–7933 (2015)
110. O. Shenderova, S. Hens, I. Vlasov, S. Turner, Y.-G. Lu, G. Van Tendeloo, A. Schrand, S.A. Burikov, T.A. Dolenko, Carbon-dot-decorated nanodiamonds. Part. Part. Syst. Charact. **31**, 580–590 (2014)
111. U. Maitra, A. Jain, S.J. George, C.N.R. Rao, Tunable fluorescence in chromophore-functionalized nanodiamond induced by energy transfer. Nanoscale **3**, 3192–3197 (2013)
112. E. Perevedentseva, N. Melnik, C.-Y. Tsai, Y.-C. Lin, M. Kazaryan, C.-L. Cheng, Effect of surface adsorbed proteins on the photoluminescence of nanodiamond. J. Appl. Phys. **109**, 034704 (2011)
113. K.S. Subrahmanyam, P. Kumar, A. Nag, C.N.R. Rao, Blue light emitting graphene-based materials and their use in generating white light. Solid State Commun. **150**(37–38), 1774–1777 (2010)
114. M.-F. Weng, S.-Y. Chiang, N.-S. Wang, H. Niu, Fluorescent nanodiamonds for specifically targeted bioimaging. Diam. Relat. Mater. **18**, 587–591 (2009)
115. Z. Wang, C. Xu, C. Liu, Surface modification and intrinsic green fluorescence emission of a detonation nanodiamond. Mater. Chem. C **1**, 6630 (2013)
116. T. Zheng, F.P. Martínez, I.M. Storm, W. Rombouts, J. Sprakel, R. Schirhagl, R. de Vries, Recombinant protein polymers for colloidal stabilization and improvement of cellular uptake of diamond nanosensors. Anal. Chem. **89**(23), 12812–12820 (2017)
117. B.R. Smith, D. Gruber, T. Plakhotnik, The effects of surface oxidation on luminescence of nano diamonds. Diam. Relat. Mater. **19**, 314 (2010)
118. L.A. Stewart, C. Bradac, J. M. Dawes, M. J. Steel, J. R. Rabeau, M.J. Withford, Characterization of emission lifetime of nitrogen-vacancy centres in *Nanodiamonds Conference on Lasers and Electro-Optics (CLEO) and Quantum Electronics and Laser Science Conference (QELS)*, IEEE JWF24 (2010)

119. S. Pezzagna, D. Rogalla, D. Wildanger, J. Meijer, A. Zaitsev, Creation and nature of optical centres in diamond for single-photon emission-overview and critical remarks. New J. Phys. **13**, 035024 (2011)

120. J.-H. Hsu, W.-D. Su, K.-L. Yang, Y.-K. Tzeng, H.-C. Chang, Nonblinking green emission from single H3 color centers in nanodiamonds. Appl. Phys. Lett. **98**, 193116 (2011)

121. J. Mona, E. Perevedentseva, A. Karmenyan, H.-M. Liou, T.-Y. Kang, C.-L. Cheng, Tailoring of structure, surface, and luminescence properties of nanodiamonds using rapid oxidative treatment. J. Appl. Phys. **113**, 114907 (2013)

122. J. Tisler, G. Balasubramanian, B. Naydenov, R. Kolesov, B. Grotz, R. Reuter, J.-P. Boudou, P.A. Curmi, M. Sennour, A. Thorel, M. Börsch, K. Aulenbacher, R. Erdmann, P.R. Hemmer, F. Jelezko, J. Wrachtrup, Fluorescence and spin properties of defects in single digit nanodiamonds. ACS Nano **3**, 1959–1965 (2009)

123. J. Tisler, R. Reuter, A. Lämmle, F. Jelezko, G. Balasubramanian, P.R. Hemmer, F. Reinhard, J. Wrachtrup, Highly efficient FRET from a single Nitrogen-vacancy center in nanodiamonds to a single organic molecule. ACS Nano **5**, 7893 (2011)

124. R. Fudala, S. Raut, B.P. Maliwal, T.W. Zerda, I. Gryczynski, E. Simanek, J. Borejdo, R. Rich, I. Akopova, Z. Gryczynski, FRET enhanced fluorescent nanodiamonds. Curr. Pharm. Biotechnol. **14**, 1127 (2013)

125. C.D. Geddes, J.R. Lakowicz, Metal-enhanced fluorescence. J. Fluoresc. **12**(2), 121–129 (2002)

126. K.L. Kelly, E. Coronado, L.L. Zhao, G.C. Schatz, The optical properties of metal nanoparticles: the influence of size, shape, and dielectric environment. J Phys. Chem. B **107**, 668–677 (2003)

127. J.R. Lakowicz, Plasmonics in biology and plasmon controlled fluorescence. Plasmonics **1**, 5–33 (2006)

128. P. Anger, P. Bharadwaj, L. Novotny, Enhancement and quenching of single-molecule fluorescence. Phys. Rev. Lett. **96**, 113002 (2006)

129. E. Ozbay, Plasmonics: merging photonics and electronics at nanoscale dimensions. Science **311**, 189–193 (2006)

130. Y. Chi, G. Chen, F. Jelezko, E. Wu, H. Zeng, Enhanced photoluminescence of single-photon emitters in nanodiamonds on a gold film. IEEE Photonics Technol. Lett. **23**(6), 374–376 (2011)

131. T.S. Lim, C.C. Fu, K.C. Lee, H.Y. Lee, K. Chen, W.F. Cheng, W.W. Pai, H.C. Chang, W. Fann, Fluorescence enhancement and lifetime modification of single nanodiamonds near a nanocrystalline silver surface. Phys. Chem. Chem. Phys. **11**, 1508–1514S (2009)

132. S. Schietinger, M. Barth, T. Aichele, O. Benson, Plasmonenhanced single photon emission from a nanoassembled metaldiamond hybrid structure at room temperature. Nano Lett **9**(4), 1694–1698 (2009)

133. D. Zhang, Q Zhao, J. Zang, Y.-J. Lu, L. Dong, C.-X. Shan, Luminescent hybrid materials based on nanodiamonds. Carbon **127**, 170–176 (2018)

134. J. Gong, N. Steinsultz, M. Ouyang, Nanodiamond-based nanostructures for coupling nitrogen-vacancy centers to metal nanoparticles and semiconductor quantum dots. Nat. Comm. **7**, 11820 (2016)

135. A. Albrecht, G. Koplovitz, A. Retzker, F. Jelezko, S. Yochelis, D. Porath, Y. Nevo, O. Shoseyov, Y. Paltiel, M.B. Plenio, Self-assembling hybrid diamond–biological quantum devices. New J. Phys. **16**, 093002 (2014)

136. D.-K. Lee, T. Kee, Z. Liangd, D. Hsiou, D. Miya, B. Wu, E. Osawa, E.K.-H. Chow, E.C. Sungi, M.K. Kang, D. Ho, Clinical validation of a nanodiamond-embedded thermoplastic biomaterial. Proc. Natl. Acad. Sci. U.S.A. **114**(45), E9445–E9454 (2017)

137. T.-K. Ryu, R.-H. Kang, K.-Y. Jeong, D.-R. Jun, J.-M. Koh, D. Kim, S.K. Bae, S.-W. Choi, Bone-targeted delivery of nanodiamond-based drug carriers conjugated with alendronate for potential osteoporosis treatment. J. Control. Release **232**, 152–160 (2016)

138. G. Xi, E. Robinson, B. Mania-Farnell, E.F. Vanin, K.-W. Shim, T. Takao, E.V. Allender, C. S. Mayanil, M.B. Soares, D. Ho, T. Tomita, Convection-enhanced delivery of nanodiamond drug delivery platforms for intracranial tumor treatment. Nanomed. Nanotechnol. Biol. Med. **10**, 381–391 (2014)

139. T.-B. Toh, D.-K. Lee, W. Hou, L.N. Abdullah, J. Nguyen, D. Ho, E.K.-H. Chow, Nanodiamond−Mitoxantrone complexes enhance drug retention in chemoresistant breast cancer cells. Mol. Pharm. **11**, 2683–2691 (2014)

140. O. Shimoni, B. Shi, P.A. Adlard, A.I. Bush, Delivery of fluorescent nanoparticles to the brain. J. Mol. Neurosci. **60**(3), 405–409 (2016)

141. J. Whitlow, S. Pacelli, P. Arghya, Multifunctional nanodiamonds in regenerative medicine: recent advances and future directions. J. Control. Release **261**, 62–86 (2017)

142. T. Meinhardt, D. Lang, H. Dill, A. Krueger, Pushing the functionality of diamond nanoparticles to new horizons: orthogonally functionalized nanodiamond using click chemistry. Adv. Funct. Mater. **21**, 494–500 (2011)

143. S. Haziz, N. Mohan, Y. Loe-Mie, A.-M. Lepagnol-Bestel, S. Massou, M.-P. Adam, X. Loc Le, J. Viard, C. Plancon, R. Daudin, P. Koebel, E. Dorard, C. Rose, F.-J. Hsieh, C.-C. Wu, B. Potier, Y. Herault, C. Sala, A. Corvin, B. Allinquant, H.-C. Chang, F. Treussart, M. Simonneau, Fluorescent nanodiamond tracking reveals intraneuronal transport abnormalities induced by brain-disease-related genetic risk factors. Nat. Nanotechnol. **12**(4), 322–328 (2017)

144. R.A. Shimkunas, E. Robinson, R. Lam, S. Lu, X. Xu, X.-Q. Zhang, H. Huang, E. Osawa, D. Ho, Nanodiamond–insulin complexes as pH-dependent protein delivery vehicles. Biomaterials **30**, 5720–5728 (2009)

145. A.H. Smith, E.M. Robinson, X.Q. Zhang, E.K. Chow, Y. Lin, E. Osawa, J. Xi, D. Ho, Triggered release of therapeutic antibodies from nanodiamond complexes. Nanoscale **3**(7), 2844–2848 (2011)

146. V.N. Mochalin, A. Pentecost, X.-M. Li, I. Neitzel, M. Nelson, C. Wei, T. He, F. Guo, Y. Gogotsi, Adsorption of drugs on nanodiamond: toward development of a drug delivery platform. Mol. Pharm. **10**, 3728–3735 (2013)

147. G. Jarre, S. Heyer, E. Memmel, T. Meinhardt, A. Krueger, Synthesis of nanodiamond derivatives carrying amino functions and quantification by a modified Kaiser test. Beilstein J. Org. Chem. **10**, 2729–2737 (2014)

148. K.-K. Liu, W.-W. Zheng, C.-C. Wang, Y.-C. Chiu, C.-L. Cheng, Y.-S. Lo, C. Chen, J.-I. Chao, Covalent linkage of nanodiamond paclitaxel for drug delivery and cancer therapy. Nanotechnology **21**, 315106 (2010)

149. O.A. Shenderova, G.E. McGuire, Science and engineering of nanodiamond particle surfaces for biological applications. Biointerphases **10**, 030802 (2015)

150. V. Petrakova, V. Benson, M. Buncek, A. Fiserova, M. Ledvina, J. Stursa, P. Cigler, M. Nesladek, Imaging of transfection and intracellular release of intact, non-labeled DNA using fluorescent nanodiamonds. Nanoscale **8**, 12002 (2016)

151. Y.-C. Lin, L.-W. Tsai, E. Perevedentseva, H.-H. Chang, C.-H. Lin, D.-S. Sun, A.E. Lugovtsov, A. Priezzhev, J. Mona, C.-L. Cheng, The influence of nanodiamond on the oxygenation states and micro rheological properties of human red blood cells in vitro. J. Biomed. Optics **17**(10), 101512 (2012)

152. A. Chatterjee, E. Perevedentseva, M. Jani, C.-Y. Cheng, Y.-S. Ye, P.-H. Chung, C.-L. Cheng, Antibacterial effect of ultrafine nanodiamond against gram-negative bacteria Escherichia coli. J. Biomed. Optics **20**(5), 051014 (2015)

153. J. Wehling, R. Dringen, R.N. Zare, M. Maas, K. Rezwan, Bactericidal activity of partially oxidized nanodiamonds. ACS Nano **8**(6), 6475–6483 (2014)

154. Z. Zhu, An overview of carbon nanotubes and graphene for biosensing applications. Nano-Micro Lett. **9**, 25 (2017). https://doi.org/10.1007/s40820-017-0128-6

155. M. Tuerhong, Y. Xu, X.-B. Yin, Review on carbon dots and their applications. Chinese J Anal. Chem. **45**(1), 139–150 (2017)

156. K.S. Novoselov, V.I. Falko, L. Colombo, P.R. Gellert, M.G. Schwab, K. Kim, A roadmap for graphene. Nature **490**, 192–200 (2012)
157. D. Fu, L.-J. Li, Label-free electrical detection of DNA hybridization using carbon nanotubes and graphene. Nano Rev. **1**, 5354 (2010)
158. Y.-P. Sun, B. Zhou, Y. Lin, W. Wang, K.A.S. Fernando, P. Pathak, M.J. Meziani, B.A. Harruff, X. Wang, H. Wang, P.G. Luo, H. Yang, M.E. Kose, B. Chen, L.M. Veca, S.-Y. Xie, Quantum-sized carbon dots for bright and colorful photoluminescence. J. Am. Chem. Soc. **128**, 7756–7757 (2006)
159. L. Wang, Y. Yin, A. Jain, H.S. Zhou, Aqueous phase synthesis of highly luminescent, nitrogen-doped carbon dots and their application as bioimaging agents. Langmuir **30**, 14270 (2014)
160. S. Liu, N. Zhao, Z. Cheng, H. Liu, Amino-functionalized green fluorescent carbon dots as surface energy transfer biosensors for hyaluronidase. Nanoscale **7**, 6836–6842 (2015)
161. Z.S. Qian, X.Y. Shan, L.J. Chai, J.J. Ma, J.R. Chen, H. Feng, DNA nanosensor based on biocompatible graphene quantum dots and carbon nanotubes. Biosens. Bioelectron. **60**, 64–70 (2014)
162. N. Prabhakar, T. Näreoja, E. von Haartman, D.Ş. Karaman, S.A. Burikov, T.A. Dolenko, T. Deguchi, V. Mamaeva, P.E. Hänninen, I.I. Vlasov, O.A. Shenderova, J.M. Rosenholm, Functionalization of graphene oxide nanostructures improves photoluminescence and facilitates their use as optical probes in preclinical imaging. Nanoscale **7**, 10410 (2015)
163. R. Schirhagl, K. Chang, M. Loretz, C.L. Degen, Nitrogen-vacancy centers in diamond: nanoscale sensors for physics and biology. Annu. Rev. Phys. Chem. **65**, 83–105 (2014)
164. A. Ermakova, G. Pramanik, J.-M. Cai, G. Algara-Siller, U. Kaiser, T. Weil, Y.-K. Tzeng, H. C. Chang, L.P. McGuinness, M.B. Plenio, B. Naydenov, F. Jelezko, Detection of a few metallo- protein molecules using color centers in nanodiamonds. Nano Lett. **13**, 3305–3309 (2013)
165. A. Albrecht, G. Koplovitz, A. Retzker, F. Jelezko, S. Yochelis, D. Porath, Y. Nevo, O. Shoseyov, Y. Paltiel, M.B. Plenio, Self-assembling hybrid diamond–biological quantum devices. New J. Phys. **16**, 093002 (2014)
166. A. Ajoy, U. Bissbort, M.D. Lukin, R.L. Walsworth, P. Cappellaro, Atomic-scale nuclear spin imaging using quantum-assisted sensors in diamond. Phys. Rev. X **5**, 011001 (2015)
167. J.M. Cai, F. Jelezko, M.B. Plenio, Hybrid sensor based on colour centres in diamond and piezoactive layers. Nat. Commun. **5**, 4065 (2014)
168. W. Zhang, K. Patel, A. Schexnider, S. Banu, A.D. Radadia, Nanostructuring of biosensing electrodes with nanodiamonds for antibody immobilization. ASC Nano **8**(2), 1419–1428 (2014)
169. M. Börsch, R. Reuter, G. Balasubramanian, R. Erdmann, F. Jelezko, J. Wrachtrup, Fluorescent nanodiamonds for FRET-based monitoring of a single biological nanomotor FoF1-ATP synthase. Proc. SPIE **7183**, 71832N (2009)
170. M. Borsch, J. Wrachtrup, Fluorescent nanodiamonds for FRET-based monitoring of a single biological nanomotor FoF1-ATP synthase. Chem. Phys. Chem. **12**(3), 542 (2011)
171. H. Pinto, R. Jones, D.W. Palmer, J.P. Goss, P.R. Briddon, S. Öberg, Theory of the surface effects on the luminescence of the NV⁻ defect in nanodiamond. Phys. Status Solidi A **208**, 2045 (2011)
172. W.W.-W. Hsiao, Y.Y. Hui, P.-C. Tsai, H.-C. Chang, Fluorescent nanodiamond: a versatile tool for long-term cell tracking, super-resolution imaging, and nanoscale temperature sensing. Acc. Chem. Res. **49**, 400–407 (2016)
173. M.E. Robinson, J.D. Ng, H. Zhan, J.T. Buchman, O.A. Shenderova, C.L. Haynes, Z. Ma, R. H. Goldsmith, R.J. Hamers, Optically detected magnetic resonance for selective imaging of diamond nanoparticles. Anal. Chem. **90**, 769–776 (2018)
174. G. Balasubramanian, I.Y. Chan, R. Kolesov, M. Al-Hmoud, J. Tisler, C. Shin, C. Kim, A. Wojcik, P.R. Hemmer, A. Krueger, T. Hanke, A. Leitenstorfer, R. Bratschitsch, F. Jelezko, J. Wrachtrup, Nanoscale imaging magnetometry with diamond spins under ambient conditions. Nature **455**, 648–651 (2008)

175. F. Dolde, H. Fedder, M.W. Doherty, T. Nobauer, F. Rempp, G. Balasubramanian, T. Wolf, F. Reinhard, C.L. Hollenberg, F. Jelezko, J. Wrachtrup, Electric-field sensing using single diamond spins. Nat. Phys. **7**, 459–463 (2011)

176. M. Alkanti, L. Jiang, R. Brick, P. Hemmer, M. Scully, Nanometer-scale luminescent thermometry in bovine embryos. Opt. Lett. **42**(23), 4812–4815 (2017)

177. G. Kucsko, P.C. Maurer, N.Y. Yao, M. Kubo, H.J. Noh, P.K. Lo, H. Park, M.D. Lukin, Nanometer scale thermometry in a living cell. Nature **500**(7460), 54–58 (2013)

178. D.A. Simpson, E. Morrisroe, J.M. McCoey, A.H. Lombard, D.C. Mendis, F. Treussart, L.T. Hall, S. Petrou, L.C.L. Hollenberg, Non-neurotoxic nanodiamond probes for intraneuronal temperature mapping. ACS Nano **11**, 12077–12086 (2017)

179. M.S. Purdey, P.K. Capon, B.J. Pullen, P. Reineck, N. Schwarz, P.J. Psaltis, S.J. Nicholls, B. C. Gibson, A.D. Abell, An organic fluorophore-nanodiamond hybrid sensor for photostable imaging and orthogonal, on-demand biosensing. Sci. Rep. **7**, 15967 (2017)

180. X. Wang, M. Gu, T.B. Toh, N.L. Binti Abdullah, E.K.-H. Chow, Stimuli-responsive nanodiamond-based biosensor for enhanced metastatic tumor site detection. SLAS Technol., 1–13 (2017)

181. L.P. Neukirch, J. Gieseler, R. Quidant, L. Novotny, A.N. Vamivakas, Observation of nitrogen vacancy photoluminescence from an optically levitated nanodiamond. Opt. Lett. **38** (16), 2976–2979 (2013)

Chapter 12
Surface Modifications of Nanodiamonds and Current Issues for Their Biomedical Applications

J. C. Arnault

Abstract Combining numerous unique assets, nanodiamonds are promising nanoparticles for biomedical applications. The present chapter focuses on the current knowledge of their properties. It shows how the control of their surface chemistry governs their colloidal behavior. This allows a fine tuning of their surface charge. Developments of bioapplications using nanodiamonds are summarized and further promising challenges for biomedicine are discussed.

12.1 Introduction

Nanoparticles (NPs) are currently designed and developed for nanomedecine applications which constitute an emerging research field [1]. Among the main stakes, cancer therapy expands with three main paradigms: drug delivery, biomarkers for diagnosis and therapeutics.

The main interest to use NPs for drug delivery is the ability to target cancer cells. Indeed, some tumors present a higher vascularisation which could favor a passive targeting of NPs via the EPR effect (Enhanced Permeability and Retention) [2]. The targeting can also be active when a specific grafting of NPs with specific markers is performed [3]. The use of NPs can enhance the treatment efficiency allowing the delivery of higher doses while it reduces significantly side effects. Polymeric NPs [4] and liposomes are currently the most used NPs to encapsulate drugs. The polymer degradation permits their progressive delivery.

Second, NPs can also act as biomarkers for tumor imaging or tracking via their optical properties as quantum dots [5] or gold NPs via surface plasmon resonance [6]. However, some of these luminescent probes are limited either by their photostability or their cytotoxicity. Inorganic NPs can be used as contrast agents for magnetic resonance imaging [7, 8]. Lastly, C^{14} grafting on carbon nanotubes was

J. C. Arnault (✉)
CEA, LIST, Diamond Sensors Laboratory, 91191 Gif sur Yvette, France
e-mail: jean-charles.arnault@cea.fr

© Springer Nature Switzerland AG 2019
N. Yang (ed.), *Novel Aspects of Diamond*, Topics in Applied Physics 121,
https://doi.org/10.1007/978-3-030-12469-4_12

reported conferring radiolabeling properties for exploration of the in vivo biodistribution and pharmacokinetics [9].

Third, NPs can have therapeutic roles if they could be activated by an external stimulus. For example, under an external magnetic field, metallic NPs can generate temperature elevation called hyperthermia leading to cell death [10]. This activation can also be performed thanks to optical properties i.e. phototherapy for graphene oxide nanosheets [11, 12] or metallic NPs [13]. Efficiency of radiotherapy treatments can be enhanced by coupling with NPs. This constitutes a promising way to treat resistant tumors. Such approach may permit the delivery of lower doses for the same efficacy. A radiosensitization effect can be obtained under irradiation producing Reactive Oxygen Species (ROS) in cells. For this application, gold [14], platinum [15] or rare earths [16] NPs are used due to their high ratio of Auger electrons generated under irradiation.

The current trend which associates therapeutic and diagnostic properties is called theranostics [17, 18]. For this purpose, hybrid nanoparticles are designed combining properties of their core and shell: for example, superparamagnetic NPs embedded in mesoporous silica [19] or gold NPs in liposomes [20].

Nanodiamonds (NDs) constitute promising NPs for biomedicine applications [21–23]. They possess many unique assets. They are scalable with sizes ranging from 100 nm down to 5 nm. One can expect the elimination by kidney for the smaller NDs (diameter lower than 6 nm) [24]. Nanodiamonds are also available in large quantities at reasonable prices (<$1/g). Several studies reported their very low toxicity related to the high chemical inertia of diamond [25–29]. Recent research demonstrates they are even biocompatible with advantages for drug delivery or fluorescence labels [30, 31]. NDs enable covalent grafting of various chemical moieties on their surfaces [32] resulting in stable colloidal aqueous solutions. Covalent chemistry works similar way on NDs as for organics.

Their surface charge can be tuned allowing the electrostatic adsorption of biomolecules such as siRNA [31, 33, 34]. They have been used for drug delivery applications [35–39].

Moreover, NDs can act as biomarkers for diagnosis. Indeed, color centers can be generated from nitrogen present in the diamond core by electron or helium irradiation [40] followed by a thermal annealing. Under green illumination, produced nitrogen-vacancy (NV) centers emits in the red wavelength region, they are highly photostable presenting neither photoblinking nor photobleaching [41]. Fluorescent NDs were used as probes for tracking in cells [42].

Potential therapeutic behaviors of NDs related to their specific surface properties are currently under investigation. Recently, an in vitro radiosensitizing effect of hydrogenated detonation NDs was reported under gamma rays turning radioresistant human cancer cells into senescence [43]. Physical and chemical mechanisms involved in the production of free radicals by hydrogenated nanodiamonds suspended in water under irradiation were investigated (Sect. 12.6) [44].

Investigations of nanodiamonds for biomedical applications started quite recently. Numerous parameters remain partially controlled such as their different surface chemistry, their surface charge, their colloidal stability and the role of facets in their surface properties. The present chapter summarizes the current knowledge and the assets of NDs for nanomedecine. First, the different ways to produce nanodiamonds will be briefly presented (Sect. 12.2). Then, the characterization tools of their core and surface will be detailed focusing on the most relevant characterization methods (Sect. 12.3). The main surface modifications performed using different approaches (chemical, physical or plasma treatments) will be reviewed (Sect. 12.4). Electronic properties of modified nanodiamonds will be presented as well as their colloidal behaviors in water (Sect. 12.5). Finally, main assets of NDs for biomedical applications will be discussed before giving some future promising challenges (Sect. 12.6). A recent book reported in 2017 a detailed state of art in terms of production, characterization techniques and applications fields of nanodiamonds [45].

12.2 Production of Nanodiamonds

According to thermodynamics, graphite is the stable carbon form at ambient conditions while diamond is metastable [46, 47]. Since 1950s, synthetic diamond can be obtained by high pressure high temperature (HPHT) method using hydraulic press which reproduces the conditions of natural diamond formation [48]. In 1980s, chemical vapor deposition (CVD) diamond was discovered and developed [49]. Different techniques are presently used to fabricate nanodiamonds: either NDs could be obtained by milling of HPHT/CVD diamond or NDs could be synthetized using explosives or laser ablation of graphite precursors [50, 51].

Monocrystalline nanodiamonds with size down to 20 nm can be obtained by ball milling of micron-sized HPHT diamonds. High Resolution Transmission Electron Microscopy (HRTEM) revealed faceted shapes with sharp edges corresponding to fractured crystallographic planes. In few cases, quasi spherical nanodiamonds smaller than 10 nm were obtained by HPHT milling [52, 53]. More recently, 70–80 nm nanodiamonds were also produced from polycrystalline CVD diamond by bead assisted sonic disintegration (BASD) [54]. Boron doped NDs milled from CVD diamond were used for colloidal suspensions [55].

Alternatively, nanodiamonds could be synthetized by several dynamic processes from molecules of explosives and different carbon precursors [56] (Fig. 12.1). Three methods could be distinguished: the direct transformation of graphite by an external shock wave, the detonation of graphite mixed with explosives (hexogen/RDX) or the detonation of high energy explosives (TNT/hexogen/RDX). For detonation synthesis, the key parameters governing the yield of nanodiamonds are the heat capacity, the content and the reactivity of the medium, the cooling of detonation products [57]. Nanodiamonds produced by detonation usually exhibit smaller diameters ranging typically from 2 to 20 nm. Several models were proposed to explain their formation by detonation synthesis [57]. More details about ND synthesis by detonation methods could be found in previous reports [57, 58].

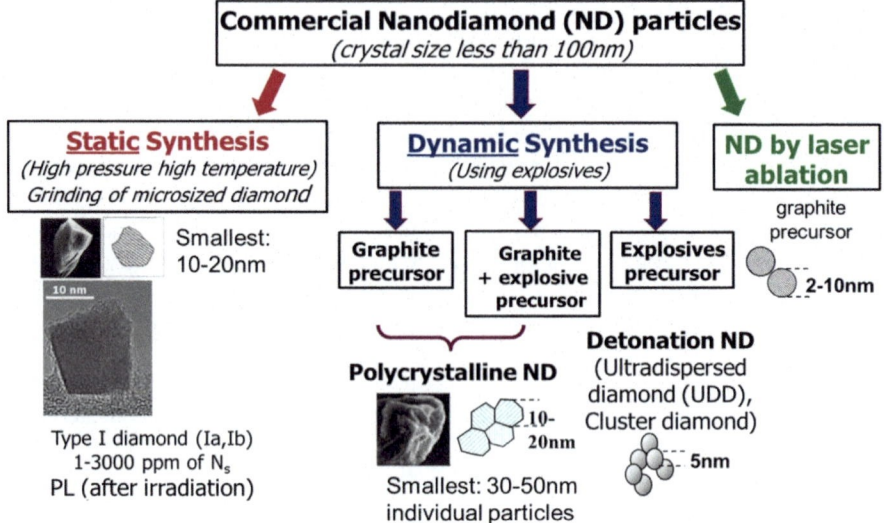

Fig. 12.1 The different natures of NDs [56] courtesy of O. Shenderova, International Technology Center, USA

After detonation, the collected soot contains metals, metal oxides and carbides. In addition to carbon, its elemental composition reveals hydrogen, nitrogen and oxygen species. Indeed, nitrogen is initially present in HPHT diamond and in explosives used in the detonation synthesis. In most cases, a-C and/or sp^2 carbon are surrounding the diamond core. Detonation NDs are extracted after purification procedures using liquid oxidants to remove non-diamond carbon. Indeed, diamond and non-diamond carbon have different stability to oxidants. The efficiency of an ozone treatment was also reported to eliminate the non-diamond carbon [58]. Then, liquid-phase oxidation using HCl or other high boiling acids is applied to eliminate the non-carbon impurities. The resulting surface chemistry of detonation nanodiamonds will be described in the next section (Sect. 12.3).

Diamond nanoparticles have been classified according to their size [56]. Nanocrystalline particles (10–100 nm) could be either monocrystalline (HPHT milling) or polycrystalline after shock-wave compression of graphite or detonation of carbon/explosives mixture. Ultrananocrystalline NDs (2–10 nm) are mainly produced by detonation or laser ablation [59]. Smallest particles behave to the family of diamondoids. Higher diamondoids (1–2 nm) are extracted from petroleum with hydrogen termination [60]. The smallest specie of cubic diamond, the adamantane, contains 10 carbon atoms ($C_{10}H_{16}$).

12.3 Characterization Tools

Depending on their synthesis methods (Sect. 12.2), nanodiamonds can behave different core-shell structures, morphologies and surface chemistry. Experimental and theoretical works reported multi-shell structures based on a diamond core which could be surrounded by defective diamond, amorphous carbon and/or several onion like outer-shells [61, 62]. According to calculations, this structure constitutes the most stable energetic form for sp^3 nanometric clusters [63]. Characterization methods were developed to specifically probe the crystalline structure as well as the chemical composition of diamond core, outer shells and surface of nanodiamonds. The most relevant techniques to assess these characteristics are presented in Table 12.1.

12.3.1 Diamond Core

Its crystalline structure can be accurately investigated by high resolution transmission electron microscopy (HRTEM) in bright field mode. The signature of diamond lattice is also confirmed in the diffraction mode. Moreover, structural defects as twins or dislocations within diamond lattice can be resolved especially for nanodiamonds produced by detonation method [64]. The spherical or faceted shape of nanodiamonds which could greatly influence their reactivity could be characterized at the nanoscale using HRTEM. As an illustration, (111) facets with (100) truncations were identified even for detonation NDs of 2 nm [64] (Fig. 12.2). More recently, modern HRTEM instruments were developed to operate at low electron energy down to 40 keV. Lens aberrations are reduced allowing image resolution down to 50 pm [65]. In these experimental conditions, the graphitization of NDs which may occur under electron beam is completely avoided. Diamond core structure and size of nanodiamonds can be also investigated by X-ray Diffraction (XRD) [66] or Small Angle X-ray Scattering (SAXS) [67]. Contrary to XRD which

Table 12.1 Characterization tools of nanodiamonds

Diamond core	
Particle size, crystalline structure	HRTEM, XRD, SAXS
Morphology and shape	HRTEM
Structural defects	HRTEM
Chemical impurities	XPS, EELS/STEM, ICP-MS
Local environment	NEXAFS, ESR, NMR, ELNES
NV centers	Time resolved PL, confocal PL
Outer shells and surface	
a-C, sp^2 carbon	Raman, HRTEM, XRD, XPS
Surface chemistry and grafting	FTIR, XPS, TDMS

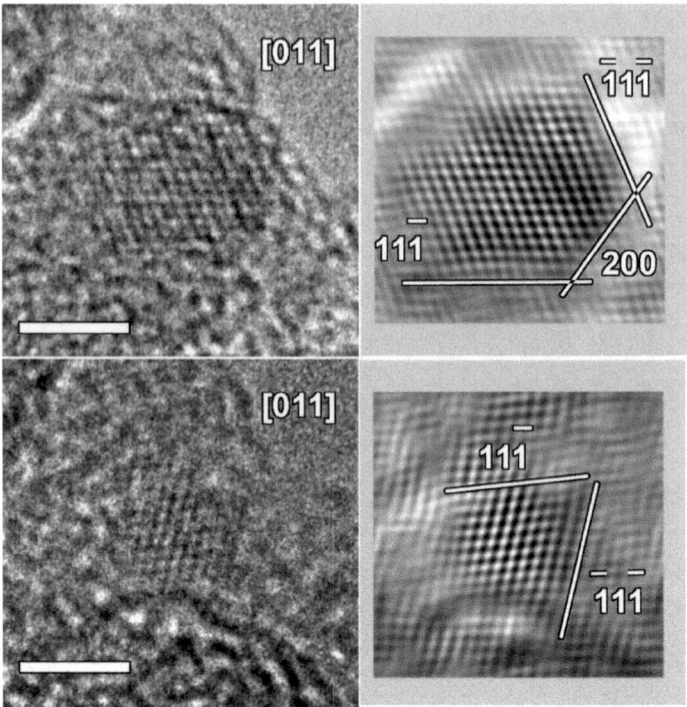

Fig. 12.2 HRTEM pictures of detonation nanodiamonds, scale bar 2 nm [64] courtesy of S. Turner, Aurubis Belgium Research

is limited by diffraction effects leading to physical broadening, SAXS give access to sizes very close to HRTEM. Size distributions can even be extracted from SAXS data assuming a shape for NDs.

The concentration of chemical impurities (N, O, H, metals remaining from synthesis) within diamond core and their location in the diamond lattice is another important issue. For example, the nitrogen presence could confer optical properties to nanodiamonds via the formation of luminescent nitrogen vacancy (NV) centers. Atomic concentrations of impurities (excepted hydrogen) could be assessed from X-ray Photoemission Spectroscopy (XPS). Typically, the detection limit for oxygen and nitrogen is close to 0.5 at.% [68]. A recent review focuses on the assets of XPS for analysis of nanodiamonds [69]. Very few techniques such as SIMS or NMR are sensitive to hydrogen. During synthesis of nanodiamonds, hydrogen is likely to be inserted into diamond core. Indeed, hydrogen diffusion was previously reported in bulk diamond [70]. There is a crucial need to probe hydrogen into diamond core because it could confer new properties to nanodiamonds.

Electron spin resonance (ESR) is the most sensitive method to detect nitrogen (10^{12} spins) [71, 72]. Indeed, single nitrogen in substitution in diamond lattice is paramagnetic [73]. Core loss electron energy loss spectroscopy (EELS) profiles can also probe light elements as nitrogen, oxygen, boron or carbon. Energy loss near edge structures (ELNES) technique provides a signature of the local binding of impurities. For example, the tetrahedral configuration can be deduced from the nitrogen K edge structure [64, 74]. Recent work shows this technique permits to confirm the tetrahedral insertion of boron atoms in diamond leading to p-doping [75, 76]. EELS combined with scanning TEM (STEM) can even provide mapping or profiles of impurities with an atomic resolution [77–79]. It was recently applied to determine the sites occupied by boron atoms in milled doped nanodiamonds [55]. Near-edge X-ray absorption fine structure (NEXAFS) technique could probe the local bonding environment [80]. Nuclear Magnetic Resonance permits to probe hydrogen (^{1}H), nitrogen (^{15}N) and carbon (^{13}C) spins [81].

The concentration of chemical impurities present in NDs especially metallic impurities can be assessed by inductively coupled plasma hyphenated methods like Inductively Coupled Plasma-Mass Spectrometry (ICP-MS) which behave a very low detection limit (5×10^{-6} wt%) compared to other techniques [82].

Photoluminescence (PL) spectroscopy can reveal signature of structural imperfection of diamond lattice (defects, impurities). Luminescent NV centers composed by a nitrogen atom in substitution combined with a neighbouring vacancy are characterized by PL. Under excitation at 531 nm, emission wavelengths at 575 and 638 nm are detected corresponding to different charge states NV^0 and NV^-, respectively [71]. NV centers exhibit a high stability without any photo-blinking or photo-bleaching. Recently, the influence of NDs surface chemistry on NV luminescence was shown, especially a quenching of the NV^- luminescence when nanodiamond surface is hydrogenated [83].

12.3.2 Outer Shells and Surface Chemistry

Surface chemistry of nanodiamonds is strongly dependent on their synthesis method and the purification treatments (Sect. 12.2). For pristine NDs, the diamond core is usually coated with non sp^3 carbon. These carbon outer-shells could be structured as onion-like shells [84] or could exhibit an amorphous structure (a-C) [85]. The nature of outer shells must be well characterized because it strongly affects the surface reactivity of nanodiamonds. Raman spectroscopy gives access to composition and structure for bulk carbon phases [86]. A signature at 1332 cm^{-1} is obtained for diamond while D and G bands at 1410 and 1590 cm^{-1} originate from amorphous and graphitic carbon, respectively. Nevertheless, a specific Raman interpretation is needed for detonation nanodiamonds (size lower than 6 nm) taking into account a phonon confinement leading to a downshift and an asymmetry of the diamond Raman peak [87]. Some features like the one located at 1640 cm^{-1} are still a matter of debate [47]. A recent work of Mermoux et al. focuses on a detailed investigation of detonation NDs using a multi-wavelength approach [88]. XPS is also sensitive to sp^2 and sp^3 hybridization of carbon, it was recently used to monitor the surface graphitization of nanodiamonds [68]. However, artefacts due to nano-materials especially the high fraction of surface versus bulk must be taken into account in the data interpretation [89, 90]. The inputs of XPS to characterize the surface chemistry and the in situ reactivity of nanodiamonds were recently reviewed [69]. Auger Electron Spectroscopy (AES) is even more sensitive to carbon hybridization because two valence electrons are involved in the C KVV transition [85]. Using low loss EELS, a-C or graphite could also be discriminated [91]. HRTEM imaging provides the structure of the outer-shells at atomic scale [64, 90]. For nanodiamonds smaller than 3 nm, surface reconstructions dependent on size and shape have been evidenced by HRTEM and calculations of electronic structure [92].

After their synthesis, nanodiamonds underwent oxidizing treatments to clean metallic impurities (Sect. 12.2). This leads to a heterogeneous surface chemistry involving different carbon-oxygen groups. To investigate the surface chemistry of nanodiamonds after purification or functionalization, Fourier Transformed Infra-Red (FTIR) Spectroscopy is a powerful technique as diamond core is trans-parent to IR. The FTIR technique is described in a recent review [93]. It is sensitive to the vibrations of carbon-hydrogen (C–H$_x$) [94] and the different carbon-oxygen bonds (O–H, C–O, C–O–C, C=O). Moreover, it could detect C–N or N–H vibra-tions [95]. Although FTIR is usually done in transmission mode, a diamond-coated ATR prism for infrared absorption spectroscopy was recently used to characterize surface-modified diamond nanoparticles [96]. Raman spectroscopy performed under UV light could provide signature of C–H surface bonds for hydrogenated diamond surface [97, 98]. High resolution electron energy loss spectroscopy (HREELS) can probe the surface bonds of the first atomic plane with a resolution of 5 meV. It was recently used to characterize the surface chemistry of hydrogenated nanodiamonds [91]. NMR experiments are sensitive to ozone-treated or hydroxy-lated nanodiamonds as well as functionalization with fluorine (^{19}F) [81].

Finally, as an indirect surface chemistry probe, let cite the thermal desorption mass spectrometry (TDMS) analysis which is a continuous measurement of the mass spectra of gases released from a sample under programmed heating [99, 100]. This technique permits a fine identification of functional groups according to their specific desorption temperature and gives information concerning their thermal stability. As an illustration, TDMS spectra can help to understand the anhydrides desorption from ozone purified NDs [58] (Fig. 12.3).

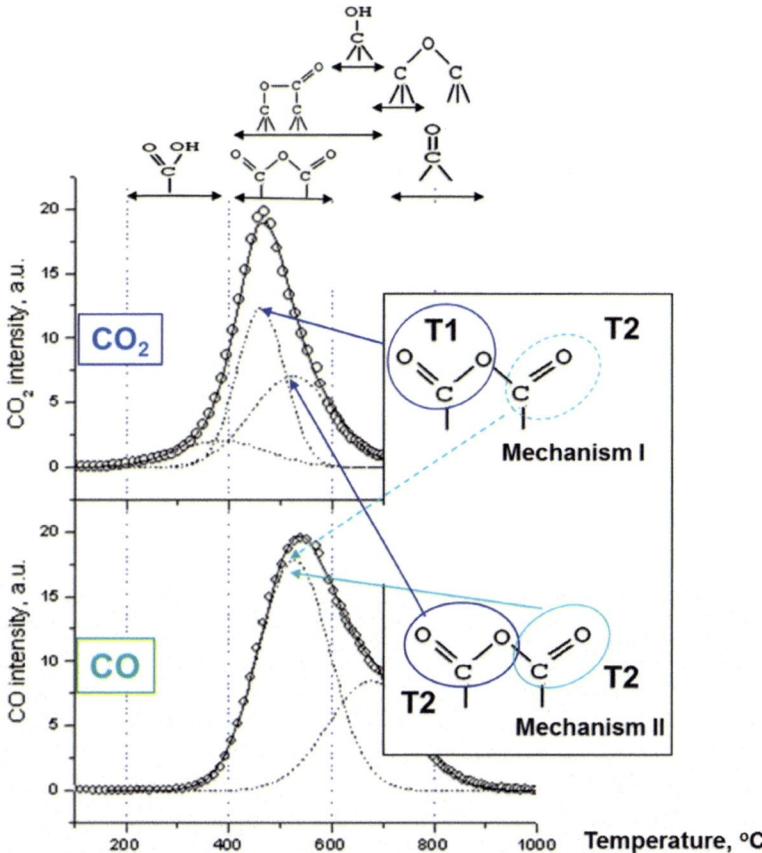

Fig. 12.3 Two possible desorption mechanisms of anhydrides on ozone treated NDs for two temperatures from TDMS [58] courtesy of O. Shenderova, International Technology Center, USA

12.4 Surface Modifications of Nanodiamonds

Nanodiamond surface appears highly heterogeneous exhibiting different carbon-oxygen functional groups (ether, hydroxyl, carbonyl, carboxyl,...), a-C and/or sp^2 carbon. This scattered surface chemistry is mainly related to purification steps (Sect. 12.3). Surface treatments were developed to get homogeneous chemistry allowing more efficient further graftings [32] either by chemical, thermal or plasma approaches. In addition, these controlled surface chemistries help to stabilize NDs in aqueous suspension limiting aggregation (Sect. 12.5). Indeed, detonation NDs tend to aggregate while many applications require isolated primary NDs [47]. Milling or BASD procedures [101] were successfully applied for deaggregation. Nevertheless, these treatments can induce either chemical contaminations or surface graphitization. The efficiency of surface treatments such as oxidation in air [102] or

Table 12.2 Surface terminations of NDs versus treatments

Surface terminations		Treatments	Objectives
Hydrogenated		• H$_2$ plasma [85, 104] • Annealing at 500 °C under H$_2$ [105, 106]	• Positive ZP with SCL • Grafting
Oxidized	Carboxylated	• Different acid treatments [32] • Air annealing at 400–430 °C [107] • Ozone treatment at 150–200 °C [58]	• Negative ZP • Disaggregation • Grafting of peptides
	Hydroxylated	• Borane reduction [108, 109] • Fenton reagent [110] • Milling with beads [84, 101] • Photochemistry [111]	• Grafting • Silanisation
Fluorinated		• F$_2$/H$_2$ exposure at 150–470 °C [112] • CF$_4$ plasma [113] • Ag$^+$ Catalysed substitution [114]	• Grafting
Aminated		• Cl-term NDs in gaseous ammonia [115] • Covalent grafting of amine derivative [101, 116, 117]	• Positive ZP • Grafting
Surface graphitised		• Annealing at 750 °C under vacuum [68, 118, 119] • Long beads milling [120]	• Positive ZP by oxygen hole doping • Hybrid properties

plasma hydrogenation [103] was reported. Indeed, tuning the surface chemistry can confer intrinsic electronic properties to NDs surface having consequences on their aggregation and colloidal behavior (Sect. 12.5). Thermal treatments permit to create hybrid nanoparticles constituted by a diamond core surrounded by sp^2 reconstructions [90]. Table 12.2 presents the major reported terminations and their uses.

12.4.1 Surface Hydrogenation of Nanodiamonds

The aim is to generate homogeneous C–H bonds at nanodiamond surface. Hydrogenated intrinsic diamond surface behave specific electronic properties characterized by a negative electron affinity (NEA) [121]. After air exposure, it presents a surface conductivity by a transfer doping involving adsorbed species [122]. Hydrogen atoms may also diffuse into the diamond lattice where they are preferentially trapped on structural defects or chemical impurities [70].

The hydrogenation of nanodiamonds was successfully achieved by two surface treatments: one performed by annealing under hydrogen flow, the second using a hydrogen microwave plasma. These different studies have been recently reviewed [123]. Annealing treatments under H$_2$ flow at 850–900 °C were reported for diamond particles bigger than 100 nm [124, 125]. Nevertheless, this temperature range is very close to the one leading to graphitization for smaller nanodiamonds. Recent reports demonstrated that surface graphitization started at 750 °C for 5 nm detonation NDs [68, 118]. Similar treatments performed at lower temperature (typically 500 °C) seem to produce an efficient deaggregation of detonation NDs after centrifugation. Nevertheless, HRTEM reveals that onion like shells are still present after the annealing [105]. FTIR shows a partial coverage by C–H bonds conferring a hydrophobic behavior to treated nanodiamonds [105, 126]. These results suggest hydrogen is more likely bonded to sp^2 carbon. The formation of C–H bonds at the surface of detonation nanodiamonds under molecular hydrogen flow at 500 °C was explained by the desorption of C_3 radicals leading to atomic hydrogen production [106]. Another method for hydrogenation would use organic chemistry with metal catalyst such as Pd, Pt or Ni [127, 128]. However, in that case, catalysts must be further eliminated.

An interesting alternative consists to expose NDs to H_2 CVD plasma. Under microwave field or thermal cracking [116, 129], dihydrogen molecules are efficiently dissociated into atomic hydrogen. This specie has a high reactivity towards carbon and etches sp^2 carbon easier than sp^3 carbon [130]. Thus, its interaction with nanodiamonds may allow the loss of non-diamond carbon, the reduction of oxygen species at the surface and the formation of C sp^3-H bonds. This surface hydrogenation by CVD plasma is widely used for diamond layers to get hydrogen-terminated boron doped diamond for electrochemistry [131, 132]. Optimized CVD conditions for hydrogenation were determined for detonation nanodiamonds using a Microwave Plasma Chemical Vapor Deposition (MPCVD) reactor connected to a UHV set-up [85]. Nanodiamonds were deposited on multilayer Silicon On Insulator (SOI) substrates by drop-casting and samples were analyzed without air exposure. A XPS sequential study allows the monitoring of the oxygen removal after MPCVD exposure. A temperature threshold of 700 °C is needed to fully remove oxygen from detonation NDs. The hydrogenation efficiency is closely related to the concentration of atomic hydrogen produced in the MPCVD plasma [85].

Proper conditions were then used in a home-made set-up allowing plasma hydrogenation of NDs in powder (Fig. 12.4). Nanodiamonds located in a quartz tube are directly treated in the gas phase [104]. A rotation of this tube allows the exposure of their whole surface to atomic hydrogen. This method is able to treat simultaneously a large amount of NDs (hundreds mg) which is a prerequisite to a meaningful surface functionalization. Both HPHT and detonation nanodiamonds which differ by their size and the graphitic shells surrounding the sp^3 core, have been hydrogenated.

The surface chemistry of hydrogenated nanodiamonds (H-NDs) was then characterized using XPS and FTIR (Fig. 12.4b). For dried H-NDs, FTIR spectra well exhibit signals located around 2900 cm^{-1} [104]. These bands have been previously assigned to C–H stretching on hydrogenated NDs surfaces [94].

The surface reactivity of HPHT H-NDs (50 nm) was investigated applying several surface modifications [111]: a selective oxidation under UV exposure and functionalization with alkenes and diazonium moieties. These reactions are

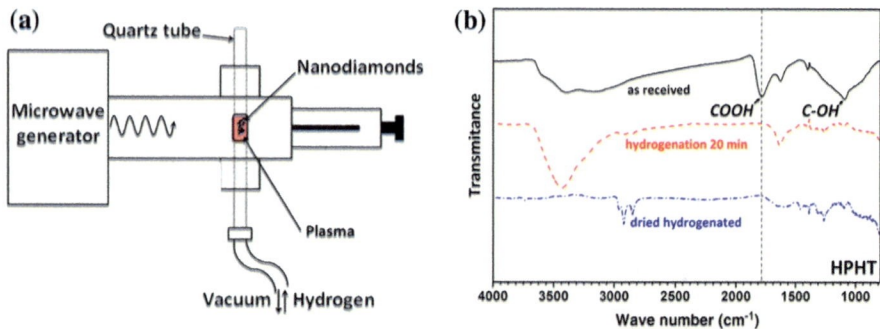

Fig. 12.4 **a** Set-up for plasma hydrogenation; **b** FTIR spectra of hydrogenated HPHT NDs [104]

common on hydrogenated diamond films towards efficient biological functionalizations (DNA, proteins grafting) [133–135] so their use on NDs may also lead to promising developments. However, the chemical mechanisms involved in these grafting routes require specific surface properties, involving charge transfer, which is only provided to the diamond surface by C–H hydrogen terminations [136].

On H-NDs, the oxidation procedure under UV gives rise to a surface hydroxylation as shown by the FTIR bands at 1050 and 3300 cm^{-1}. In addition, a new component appears in the XPS C1s core level [111]. A similar modification of the C1s peak was reported for a diamond surface after UV oxidation [137]. This oxidation treatment appears selective on H-NDs as neither carboxyl nor carbonyl groups were evidenced by XPS or FTIR. An inversion of the Zeta potential was measured after UV oxidation +30 mV compared to −45 mV for as received NDs. The origin of this positive Zeta potential will be discussed in the Sect. 12.5.

HPHT NDs were also reacted with pure undecylenic acid under 254 nm UV irradiation. This alkene was chosen for its long alkyl chain and its acid group, both functionalities easily identifiable by FTIR analysis with characteristic structures at 2800 cm^{-1} and 1700 cm^{-1} respectively (Fig. 12.5). Furthermore, the vinyl function

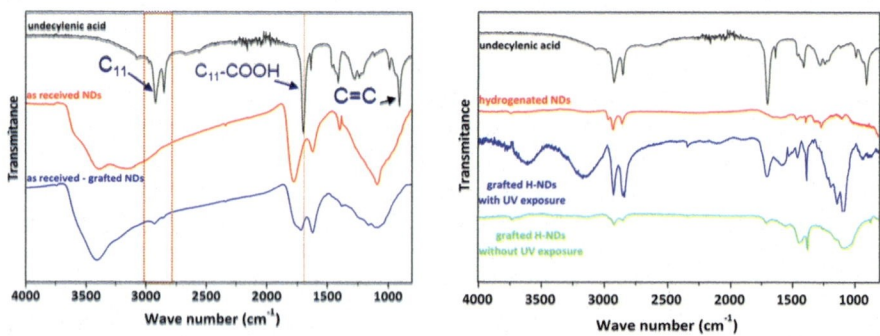

Fig. 12.5 FTIR spectra on (left) as received HPHT NDs (right) HPHT H-NDs [111]

is evidenced by a peak located around 900 cm^{-1}. For comparison, the same reaction was also conducted on as-received NDs. The different signatures of undecylenic acid are observed only on the FTIR spectra for hydrogenated HPHT NDs (Fig. 12.5). Moreover, the essential role of UV light in the mechanisms of grafting was demonstrated. On diamond layers, numerous studies report that C–H strongly enhances chemical reaction [136] leading to the formation of covalent C–C bonds between the diamond surface and the grafted moiety. Indeed, hydrogenated terminations confer to the diamond layers specific electronic surface properties, such as a superficial conductive layer (SCL) [138] and a negative electron affinity (NEA) [121, 139]. The latter is usually highlighted to explain the reactivity of hydrogenated diamond layers towards alkenes, by the mean of photo-excited electrons allowed to propagate out of the diamond into the nearby liquid phase, even with sub-bandgap excitation [140, 141]. Authors agree that hydrogen terminations are required for this reaction, and that electron transfer arises from sp^3 surfaces and not from grain boundaries and/or sp^2 species at the interface. This constitutes a strong indirect proof of the efficiency of hydrogen plasma on NDs.

The spontaneous grafting of a diazonium salt on hydrogenated HPHT NDs was also demonstrated [111]. On a diamond surface, this arises from specific electronic properties conferred by C–H terminations [142]. The covalent bonding goes through the creation of a phenyl radical by the mean of an electron transfer from the diamond surface to the aryldiazonium salt. This electron transfer is related to the presence of a superficial conductive p-type layer. The commonly accepted mechanism, so-called transfer doping [138] describes an accumulation of holes due to an electron transfer from the valence band and a redox couple adsorbed on the surface when the diamond surface is exposed to wet atmosphere or immersed in an electrolyte. This phenomenon is only possible on hydrogenated layers which exhibit a low ionization potential, matching with the chemical potentials of the adsorbates.

These different surface modifications of hydrogenated HPHT NDs suggest their chemical reactivity is very similar to the one of hydrogenated diamond surfaces. The previous results were supported by the report of Bolker et al. who measured a negative electronic affinity for plasma hydrogenated nanodiamonds of sizes included between 4 and 15 nm by Scanning Tunneling Spectroscopy (STS) [143]. In the Sect. 12.5, surface properties of hydrogenated nanodiamonds will be discussed.

12.4.2 Oxidation of Nanodiamonds

The surface chemistry of pristine nanodiamonds is highly heterogeneous (Sect. 12.3). The selective formation of C–O bonds at the surface by applying specific treatments is a major challenge to obtain more efficient functionalisations.

Among C–O groups, carboxylic acid termination has a specific interest. Indeed, its basic form (COO$^-$) confers a negative charge to the NDs over a wide pH range in water [90]. This ensures stable colloidal suspensions (Sect. 12.4). Surface of NDs could be saturated with carboxyl functions using nitric (HNO$_3$), sulfuric (H$_2$SO$_4$) or perchloric (HCl) acids. A mixture of these acids in equal amounts reveals powerful [57, 107, 144]. "Piranha" water (sulfuric acid and hydrogen peroxide) constitutes an efficient alternative [145]. Air annealing of NDs at 400–450 °C leads to the formation of acid functions mainly carboxylic and to the etching of non-diamond carbon [107, 146, 147]. A quantitative analysis of COOH groups was performed using Boehm titration after acid treatment [148]. The mean amount corresponds to 0.85 COOH groups par nm^2 of ND surface. Recently, ozone purification treatment was reported leading to the formation of anhydrides [58].

Starting from pristine nanodiamonds, several procedures inducing hydroxyl terminations were reported. First, chemical reduction can be performed with borane (BH$_3$) or hydrides (LiAlH$_4$) starting from COOH [108, 109, 149]. Second, C–OH bonds can also be formed using the "Fenton reaction" which takes place in a mixture of hydrogen peroxide and iron sulfate in strong acid solution [110, 150]. Third, a significant amount of hydroxyl groups can be generated during milling [84] or BASD [101] of NDs in water. More recently, a photochemical oxidation under UV was performed on NDs previously hydrogenated [111].

A recent review provides more details on the different oxidation procedures [32].

12.4.3 Amination, Fluorination or Chlorination of Nanodiamonds

Several attempts to generate fluorine, amine or chlorine terminations at NDs surface by annealing under controlled atmosphere were reported [47, 151, 152, 153]. Nevertheless, annealing temperatures are usually limited to avoid surface graphitization (Sect. 12.3). As a consequence, some oxidized groups and non-diamond carbon may remain at the surface. The selectivity of these gaseous treatments is often limited: after heating in ammonia flow, aminogroups, C≡N and C=N-containing groups were identified at ND surface [153, 112]. In order to create more reactive species, a CF$_4$ atmospheric pressure plasma was used [113]. A partial fluorination of oxidised HPHT nanodiamonds was recently reported occurring by a Ag$^+$-catalyzed substitution of carbonyl groups [114]. The ratio of C–F bonds represent 28% of the available carbon atoms according to XPS analysis. Chlorination of NDs was also successfully achieved by photochemical reaction of gaseous chlorine with NDs previously hydrogenated [115]. The interaction of these chlorinated NDs with ammonia led to the formation of NH$_2$ groups.

12.4.4 Surface Graphitization of Nanodiamonds

Diamond graphitization can be induced by different mechanisms: via thermal treatment under vacuum [154, 155] or exposure to reactive gas under ambient atmosphere [156–158]. It can be also obtained by beam irradiation (electron, ion, laser, gamma-ray) [159–161]. Finally, the graphitization could be catalyzed by metals as Fe or Co [162–164].

A recent review was devoted to diamond phase transitions at nanoscale [165]. For nanodiamond, the graphitization process can lead to a full transformation into onion like carbon (OLC) structures formed by concentric closed graphitic shells. Corresponding kinetics were extensively studied experimentally by annealing at temperatures included between 1100 and 1900 °C in different atmospheres (vacuum, oxygen, argon, hydrogen) [118, 119, 166, 167, 168, 169, 170]. Formations of curved graphite like structures and concentric-shells fullerenes on nanodiamonds were theoretically investigated [171–173]. For nanodiamonds 4–5 nm in size, the evolution of carbon bulk density measured by a gamma-ray attenuation method drops from 3.1 to 2.2 g cm^{-3} in the 900–1300 °C temperature range [174]. By comparison, the diamond/graphite transition occurs above 1600 °C for bulk diamond [174, 175]. Starting from detonation nanodiamonds, HRTEM revealed the formation of fullerene like shells after an annealing at 1100–1300 °C [166]. At 1500 °C, nanodiamonds are completely converted into OLC structures while polygonal hollow onions are formed at temperatures higher than 1870 °C (Fig. 12.6). OLC have promising properties for energy storage, catalysis or composites [165, 119, 166, 176].

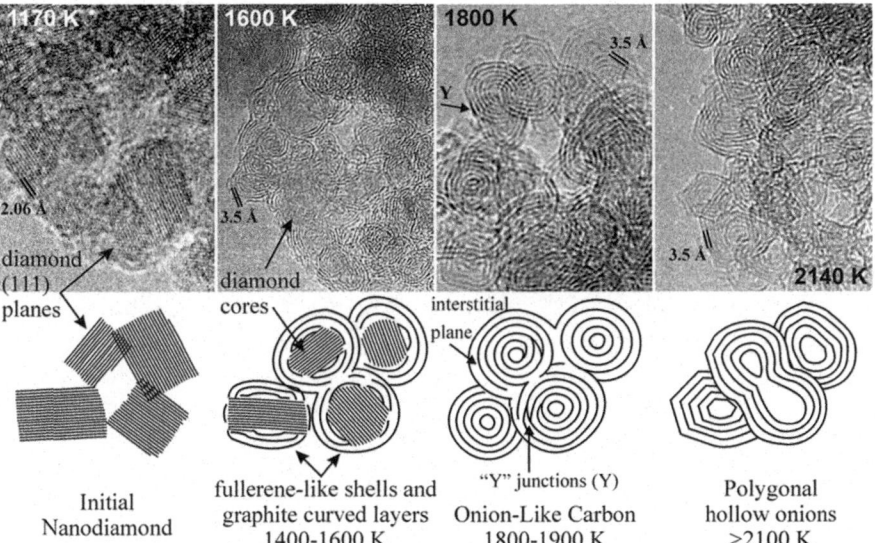

Fig. 12.6 Transformation of NDs into onion like carbon [119] courtesy of Y.V. Butenko, ESTEC, The Netherlands

The graphitization mechanism is strongly sensitive to the presence of non-diamond carbon at the NDs surface [165]. Indeed, detonation NDs underwent purification procedures which can differ from a supplier to another (Sect. 12.2). This can explain the wide dispersion for the onset temperature of nanodiamonds graphitization reported in the literature, it varies from 670 to 1000 °C [177, 178]. The selective desorption of oxygen groups has also a significant effect on the graphitization onset. For HPHT NDs, a size effect is expected with a higher temperature suitable to initiate graphitization. Graphitization kinetics for submicron diamond and NDs were compared, different graphitization rates were determined [174]. Moreover, the previous report demonstrates that activation energies differ for {110} and {111} crystallographic facets [165, 179].

Recently, several groups focused on the early stages of surface graphitization of detonation NDs using XPS, HRTEM, Raman or NMR characterizations [68, 90, 113, 180, 181]. The aim is to control the formation of hybrid nanoparticles with a sp^2 organized surface covering a sp^3 core, thus avoiding the graphitization of the diamond core. Such hybrid NPs may have promising properties linked to fullerene or graphene assets as radiosensitization or photothermal therapy [182, 183]. In addition, surface graphitized nanodiamonds exhibit electrical conductivity sufficient for electrochemical applications [184, 185].

Surface graphitization of detonation NDs by annealing under UHV was monitored using XPS [68]. The evolution of the carbon core level (C1s) shows the existence of two regimes according to annealing temperature T (Fig. 12.7). For

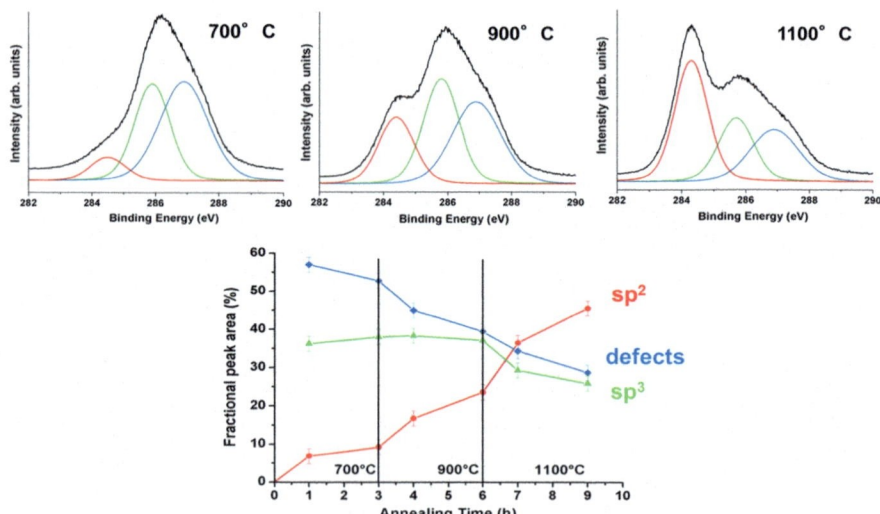

Fig. 12.7 C1s XPS spectra of NDs after sequential annealing treatments of 3 h at 700, 900, 1100 °C. Fitting components CI (red), CII (green), CIII (blue), related to sp^2 carbon-carbon bonds, sp^3 carbon-carbon bonds, and defects, respectively, are plotted under the experimental curves. Evolution of the fractional peak areas versus annealing time [68]

T included between 700 and 900 °C, surface graphitization is initiated from non-diamond carbon present at ND surface. Fullerene like reconstructions (FLRs) are formed as shown by HRTEM [90]. On the other hand, for $T > 900$ °C, the diamond core starts to graphitize leading to the formation of bucky diamond. It is thus possible to generate FLRs at the surface without altering the diamond core. The stability of FLRs at the NDs surface was previously demonstrated by ab initio calculations confirmed by X-ray absorption and emission [171]. The selective synthesis of a thin graphitic layer on the ND surface by annealing under vacuum at temperature lower than 900 °C gives rise to hybrid nanocarbons combining the intrinsic core properties of diamond with the surface reactivity of sp^2-based nanomaterials. These results were recently confirmed by a Raman study of the sp^3 to sp^2 conversion for detonation NDs [180]. NMR and EPR investigations detect the early stages of graphitization for annealing temperatures included between 600 and 800 °C [181].

Proper conditions determined for surface graphitization were then reported in a set-up where nanodiamonds are annealed under vacuum in a crucible. Annealed NDs are dispersable in water, their colloidal properties will be discussed in the Sect. 12.5. HRTEM observations confirm the structure modifications previously described showing FLRs formation [90]. A specific chemical reactivity was reported for detonation NDs after annealing at 750 °C exhibiting FLRs structures allowing arylation reactions [119].

In conclusion, surface chemistry of nanodiamonds can be tuned by thermal annealing, plasma exposure or chemical reactions. Hydrogenated and carboxylated NDs are currently well controlled whereas hydroxylation remains more difficult to achieve. These terminations are particularly suitable for the grafting of biomolecules or drugs on NDs [32, 47]. Sp^2 reconstructions can be generated at detonation NDs surface without altering the diamond core [68]. Arylation reactions allow their functionalization with complex organic moieties for bioapplications [119]. On the other hand, further progresses are needed to control the selectivity of amination, fluorination and chlorination of NDs.

The control of NDs surface charge is essential for electrostatic loading of drugs [31, 36]. Moreover, it greatly influences the interaction with cell membrane and the internalization pathways of NDs into cells [186]. The control of surface terminations allows a tuning of NDs surface charge in suspension from negative to positive versus pH. Colloidal properties of modified nanodiamonds in water will be now presented (Sect. 12.5). The link between their Zeta potential and their surface electronic properties will be investigated in details.

12.5 Colloidal Properties of Modified Nanodiamonds

The colloidal stability of nanodiamonds in biologic media is of major interest for their biomedical applications. It strongly depends on their surface charge which controls hydrophilic/hydrophobic interactions with other NDs and moieties.

Moreover, in the biological media, proteins adsorption leads to the formation of a corona [187] surrounding nanoparticles. The nature of proteins forming the corona is a key parameter for cell internalization pathways which occur by their specific recognition. In liquid media, nanodiamonds tend to form aggregates, this may be favored by opposite charged facets [188] or by graphitic carbon present at NDs surface [84, 108]. Several experimental methods were developed to reduce aggregation and obtain monodisperse colloidal suspensions. Bead assisted sonic disintegration (BASD) was performed using ceramic microbeads, nevertheless, it can produce chemical contaminations or surface graphitization [101, 189, 190]. Dry milling using salts and sugars avoiding contamination was also reported [191]. Modifications in NDs surface chemistry like plasma hydrogenation [103], surface graphitization [119], oxidation in air [102] or borane reduction [108] have also proved their efficiency for de-aggregation. For example, a thermal deagglomeration protocol of detonation nanodiamonds was reported involving a vacuum annealing at a temperature higher than 450 °C [192]. The sp^3-sp^2 transformation between faceted NDs was monitored by XPS, XAS, FTIR and Raman. This method allows to get colloidal suspensions of NDs in various liquid media.

The behavior of nanoparticles in liquid media could be investigated by Dynamic Light Scattering (DLS). From diffused intensity signal, this technique provides the measurement of the hydrodynamic diameter of stable NPs. This diameter is higher than the real one because it includes the solvated layer of adsorbed ions which is solvent and ionic force dependent [193]. The Zeta Potential (ZP) is defined from the double layer model (Stern and Gouy-Chapman layers) which describes the counter-ions adsorption. ZP is the potential at the surface of the electrical double layer called slipping plane. It is obviously related to the surface charge of NDs, very different values are obtained versus surface chemistry and size [27]. Zeta Potential is calculated from the electrophoretic mobility measured in DLS. Its evolution versus pH must be investigated. Indeed, functional groups present at NDs surface can switch from protonated to deprotonated states according to their pKa value. The colloidal stability will be then dependent on pH value. Indeed, the ZP governs the electrostatic repulsive interactions in liquid medium. Stable NPs in colloids usually exhibit ZP potential absolute value higher than 30 mV [194, 195].

The control of ND surface charge is essential for drug or biological moieties adsorption [31, 33, 34, 36] it can also play a major role in internalization pathways and interactions with negatively charged cell membrane [186]. In this part, the surface reactivity of modified NDs will be discussed. Especially, stable NDs in aqueous suspensions exhibiting positive or negative ZP will be considered, mechanisms such as charge transfer leading to these specific charges will be also discussed.

12.5.1 Surface Reactivity of Modified NDs

The hydrophilicity of detonation nanodiamonds from the same origin was investigated versus their surface chemistry after plasma hydrogenation, air oxidation and surface graphitization [103]. Water adsorption isotherms measured by BET revealed a significantly higher hydrophilicity for H-NDs and sp^2-NDs compared to COOH-NDs. Taking into account the specific surface area and assuming a monomolecular adsorption, hydrophilic sites were estimated to 2.2, 1.7 and 1.4 sites / nm^2 for sp^2-NDs, H-NDs and COOH-NDs, respectively. These hydrophilic sites on H-NDs and sp^2-NDs are likely to facilitate the dispersion in water of otherwise hydrophobic NDs (Sect. 12.3).

This high affinity of hydrogenated and surface graphitized NDs with oxygen is supported by XPS analysis performed after air exposure. Oxygen atomic concentrations of 6.5 at.% and 4.0 at.% were measured for sp^2-NDs and H-NDs after air exposure [103] while it was completely removed after plasma hydrogenation or annealing under vacuum (Sect. 12.4) [68, 85].

12.5.2 Solubility, Stability in Colloids

Stable suspensions of individual detonation NDs in water were achieved by successive sonication and centrifugation [105, 196]. Sonication allows the breaking of Van der Waals interactions between aggregated NDs. This is an important issue for biomedical applications because detonation nanodiamonds (<10 nm) are expected to be easily eliminated by kidney (Sect. 12.6) [24]. The colloidal stability is closely related to surface charge of NDs because of electrostatic repulsive interactions. Indeed, these interactions can also promote aggregation as discussed previously. The impact of the colloidal stability on the electronic properties of NDs has started to be investigated using soft X-ray absorption and emission spectroscopies with Synchrotron facilities. For oxidized NDs, these studies revealed a vanishing of $\pi*$ transitions originating from sp^2 carbon as soon as NDs are in suspension in water [197]. Holes are detected in the valence band for colloids. X-ray absorption measurements of these suspensions through a micro-jet allowed to distinguish the interfacial water from the bulk one [198]. Indeed, the oxygen atoms behave different unoccupied electronic states.

Among the literature, positive [149, 199] or negative [57, 105, 117, 200, 201] Zeta Potentials were reported for nanodiamonds in suspension in water depending on their surface chemistry. These colloidal properties constitute an important asset for nanodiamonds compared to other carbon nanomaterials. Fullerenes or carbon nanotubes exhibit hydrophobic behavior and their stability in water suspension can only be obtained after their functionalization [202–204]. These aqueous colloids of NDs were also used to obtain homogeneous seeding of substrates for CVD growth of thin diamond films [205, 206]. Some groups used surfactants like sodium oleate

[199] or oleylamine [207] to improve the colloidal stability of NDs. Nevertheless, aqueous suspensions are more suitable for bioapplications such as drug delivery.

12.5.3   Negatively Charged NDs

Carboxylated nanodiamonds obtained after annealing in air (Sect. 12.4) exhibit a strong negative Zeta potential included between −30 and −50 mV when dispersed in water for pH > 5 [57, 90]. The origin of this negative potential is related to the presence at the surface of carboxylate groups (COO⁻) the basic form of COOH. Below pH = 5, the formation of mainly carboxylic groups (COOH) leads to a drop of Zeta potential. This threshold pH value is close to the pKa of COOH/COO– couple [208].

Several studies [84, 105, 209] reported a similar behavior i.e. negative ZP for pristine NDs. This has been explained by the presence of carboxylate groups at the NDs surface formed during purification treatments (Sect. 12.4).

12.5.4   Positively Charged NDs

Other "as received" nanodiamonds from different suppliers were stabilized in water suspensions exhibiting a positive Zeta Potential (> + 30 mV) [57, 200]. The origin of this positive ZP is not yet clearly understood. In the literature, colloidal aqueous suspensions of positively charged NDs (ZP > 30 mV) were also reported after different surface modifications: reduction with borane or $LiAlH_4$ [18, 149], UV hydroxylation [111], annealing under hydrogen flow [105], plasma hydrogenation [103] or surface graphitization [90]. These charge modifications may be related to graphitic carbon surrounding NDs core or to protonation of some chemical functional groups or lastly to intrinsic properties of hydrogenated diamond surface. Different likely mechanisms have been previously proposed in the literature and will be now discussed.

12.5.4.1   Presence of Graphitic Carbon at the NDs Surface

Gibson et al. reported positive Zeta potential for detonation nanodiamonds after two oxidization treatments CrO_2/H_2SO_4 followed by $NaOH/H_2O_2$ [201]. Authors attributed this positive ZP to electrostatic interactions between pyrone groups and π-electrons from sp^2-carbon present at NDs surface corresponding to protonation of pyrone structures. Theoretical investigations supported this effect leading to carbon basicity [210, 211].

Williams et al. obtained a positive Zeta potential for detonation NDs exposed to hydrogen flow at 500 °C then dispersed into water solution [105]. Authors claimed

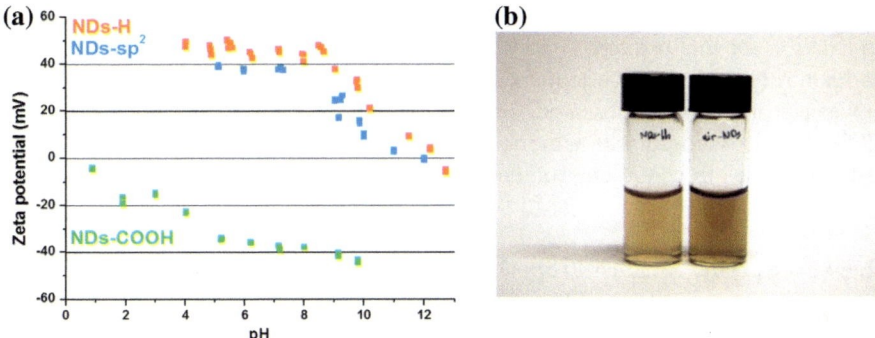

Fig. 12.8 a Zeta potential of H-NDs, COOH-NDs and sp²-NDs dispersed in water versus pH. **b** Colloidal suspensions of sp²-NDs and COOH-NDs in water

it originates from the interaction of onion like carbon (OLC) structures (π bonding) remaining at NDs surface with oxonium ions (H_3O^+) present in water suspension at acidic pH. Their HRTEM pictures support the presence of OLC on hydrogen treated NDs. By analogy with carbon black, authors proposed an electron—donor —acceptor complex may be formed [212]. In both previous studies, a charge exchange between species from aqueous solution and sp² carbon remaining at NDs surface is involved.

More recently, positive Zeta potential was measured at pH < 8 on detonation NDs on which Fullerene like reconstructions (FLRs) were intentionally generated by annealing under vacuum before their dispersion into water (Fig. 12.8) [90]. The origin of the positive charge was linked to the formation of endoperoxide groups by cycloaddition of O_2 molecules on FLRs. BET and XPS measurements well emphasize the high reactivity of sp²-NDs toward oxygen species after air exposure [103]. In situ XPS analysis after UHV annealing suggests a reversible oxidation similar to the one observed on graphene [213]. Moreover, after in situ desorption, the FTIR spectrum exhibits an intense band at 1100 cm^{-1} well corresponding to C–O stretching of ether or epoxy groups [214]. This last result supports the formation of endoperoxide groups on FLRs structures. The protonation of these endoperoxides may be the explanation for oxygen hole doping [90].

12.5.4.2 Chemically Induced Positive ZP

Ozawa et al. reported on a positive Zeta potential switching after hydroxylation of carboxylated detonation NDs by a borane reduction [215] Osawa et al. [195] Hydroxyl groups probed on their NDs by FTIR measurements, allowed further silanization coupling. Shenderova et al. obtained positive ZP for detonation NDs initially carboxylated after LiAlH₄ reduction [149]. Girard et al. also observed a ZP inversion (+30 mV at pH = 7 compared to—45 mV initially) after UV photochemical treatment of HPHT nanodiamonds [111]. Hydroxyl groups were identified from FTIR and XPS investigations. Nevertheless, any previous work provides an explanation for this positive charge and no link between positive ZP and hydroxylation has been currently established. Finally, positive ZP potential (25 mV for pH between 2 and 3) was measured for nanodiamonds grafted with ethylenediamine (EDA) [117]. According to authors, protonation of amino groups may explain the basic properties in acidic environment.

12.5.4.3 Specific Surface Properties of Hydrogenated Nanodiamonds

Stable aqueous suspensions of detonation nanodiamonds previously plasma hydrogenated were recently reported [103]. Measured Zeta Potential is included between + 5−50 mV for pH values in the range of 4–9 (Fig. 12.8). The experiment was carried out on initially negatively or positively charged NDs. According to authors, a transfer doping occurs onto 5 nm diamond nanoparticles suspended in water based on the semi-conductive behavior of the NDs core. These experiments show that an electrochemical exchange between residual CO_2 takes place between aqueous solution and hydrogenated diamond surface leading to holes accumulation similarly to the one observed for micro-sized diamond particles [216]. The key role played by oxygen was underlined by nitrogen flushing which provokes flocculation of H-NDs. A transfer doping similar to the one occurring on diamond surfaces also occurs onto 5 nm NDs suspended in water [103]. This effect was not obvious at nanoscale. Indeed, the band bending inducing the hole accumulation layer is of the same order of magnitude than the particle diameter. This transfer doping at the hydrogenated ND surface was recently confirmed by resistivity measurements [217]. Resistivity drops from 10^7 to 10^5 Ω cm for detonation nanodiamonds heated in hydrogen gas (600–900 °C). Oxidation of the hydrogenated sample at 300 °C recovers resistivity to its original value.

The understanding of the specific colloidal properties of hydrogenated NDs in water must take into account their interactions with water molecules. By a combined investigation (thermal analysis, FTIR and Kelvin probe microscopy), the hydrophobic interaction of hydrogenated NDs with water was evidenced with a weaker binding energy compared to oxidized NDs [218]. These results support the formation of water droplets included in hydrogenated NDs suspension (Fig. 12.9). A recent report also underlines that the structure of water molecules surrounding NDs is strongly dependent on their surface chemistry [219]. X-ray absorption, infra-red and Raman spectroscopies well showed a specific water structure around hydrogenated NDs characterized by modification of the long range order. This effect was not observed for hydroxylated or carboxylated NDs.

To conclude, Zeta potential of nanodiamonds can be tuned over a wide pH range applying different surface modifications. This behaviour is highly suitable for bio-applications. According to the literature, ZP switching can be explained by charge exchanges between NDs surface and species from aqueous suspensions. Some works emphasized the role played by sp^2 carbon at NDs surface. Other studies were performed using surface cleaned nanodiamonds i.e. Fenton reaction

Fig. 12.9 Scheme of interaction of water with **a** hydrogenated and **b** oxidized detonation NDs [218]

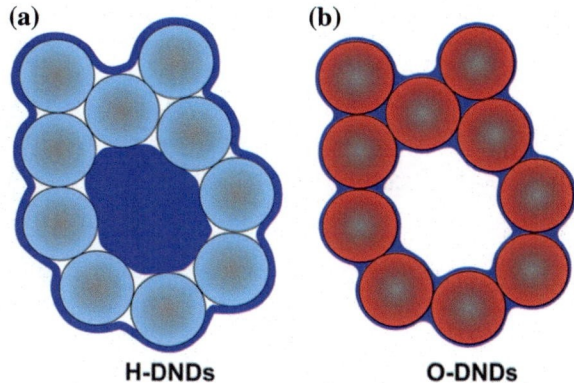

which clearly consumes a part of NDs or plasma hydrogenation. In the latter case, positive Zeta potential of H-NDs is due to specific surface properties of hydrogenated diamond surface. Neither chemical functional groups nor graphitic structures seem to be involved. This suggests mechanisms of different natures could explain this positive surface charge.

12.6 Nanodiamonds and Biomedical Applications

12.6.1 NDs Assets

Nanodiamonds (NDs) combined many required assets for biomedical applications: extremely low cytoxicity and genotoxicity, carbon chemistry allowing covalent grafting, photoluminescent color centers, tunable size (down to 5 nm). Let us discuss these advantages.

12.6.1.1 NDs Toxicity and Biodistribution

Several long-term in vitro [25, 220] and in vivo [221–223] previous studies demonstrated that NDs are non-cytotoxic, better tolerated by cells than other nanocarbon materials [224]. NDs introduced into *Caenorhabditis elegans* worm did not cause any detectable stress to worms [224]. In mice, intravenous injection of 50 nm NDs led to long-term entrapment in the liver and the lung [221], but no mice showed any abnormal symptoms. Similarly, subcutaneous and intraperitoneal injection of 100 nm NDs in rats led to accumulation of NDs in retention organs over one month with no impact on the rats' viability [223]. However, a slight surface-dependent genotoxicity of NDs was recently reported on embryonic stem cells [225].

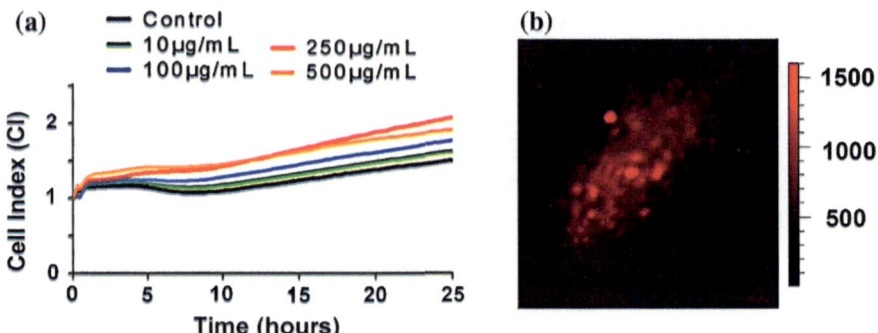

Fig. 12.10 **a** Cell index real-time monitoring of HepG2 cells exposed 24 h to 20 nm NDs; **b** Raman/photoluminescence image on a HepG2 cell [28]

Until now, no extensive study combining cytotoxicity and genotoxicity of NDs was reported. A recent study [28] focuses on in vitro cytotoxicity and genotoxicity of HPHT NDs perfectly characterized in terms of size (20 and 100 nm) and surface chemistry. The cellular induced effects of two sets of NDs were investigated in six human cell lines: HepG2 and Hep3B (liver), Caki-1 and Hek-293 (kidney), HT29 (intestine) and A549 (lung). The screening of ND cytotoxicity was carried out by measuring cell impedance (xCELLigence). This technique permits real-time monitoring of NPs effects on cell morphology, proliferation, adhesion and membrane potential [226]. Flow cytometry allowed the discrimination of viable cells, containing or not NDs, and dead cells. Finally, using immunofluorescence detection of nuclear γ-H2Ax foci, considered the most sensitive method for detecting DNA double-strand breaks [227], genotoxicity was also analyzed. According to the results, NDs do not induce any significant toxic effect on the six cell lines up to an exposure dose of 250 mg/mL (Fig. 12.10). As a comparison, 25 nm SiO_2 NPs are 9 times more genotoxic at a 10 times lower concentration [28].

The previous results were confirmed by conventional in vitro genotoxicity tests carried out on two human cell lines. No genotoxicity activity was detected for carboxylated HPHT NDs for sizes (20, 50 and 100 nm) and concentrations investigated [29]. This demonstrates that carboxylated NDs could be used as a negative reference in nanotoxicology.

This low toxicity can be exploited in drug delivery applications. The delivery of daunorubicin by NDs was found less toxic than the direct treatment by daunorubicin [228]. This demonstrates the high potential of NDs to improve drug tolerance.

12.6.1.2 Radiosensitization of Hydrogenated NDs

An in vitro study showed that plasma hydrogenated detonation NDs behave a radiosensitizing effect on radioresistant cancer cell lines under gamma irradiation [43]. A detailed investigation of the biological effects concluded that radioresistant

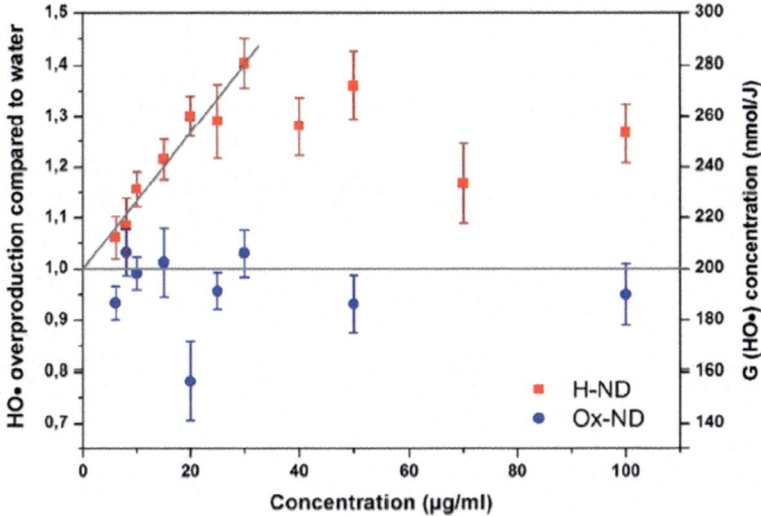

Fig. 12.11 Hydroxyl radical overproduction versus ND concentration for hydrogenated NDs (red) and oxidized NDs (blue). The horizontal line represents production in water [44]

cancer cell lines are turned into senescence which is an irreversible ageing state. This asset provides potential therapeutic abilities to hydrogenated NDs as radiosensitizing agents. Nevertheless, the mechanisms involved are not currently well understood.

In the Sect. 12.4, the electronic properties of hydrogenated NDs were discussed. Especially, their negative electron affinity which has been measured by STS [143]. As a consequence, hydrogenated NDs may emit electrons under illumination more easily than oxidized NDs. The behaviour of NDs suspended in water under ionizing radiations (X-ray and Gamma) was studied to investigate the production of reactive oxygen species (ROS), in particular hydroxyl radicals (HO•). The detection of HO• radicals was realized in the presence of a fluorescence probe, the 7 OH-coumarin [229]. An overproduction of HO• was measured for hydrogenated detonation NDs under X-ray [44]. In the same irradiation conditions, no effect was detected for oxidized NDs (Fig. 12.11).

12.6.1.3   Grafting of Biological Moieties

We have previously underlined surface charge of NDs suspended in aqueous solution can be tuned over a wide pH range (Sect. 12.5) thanks to specific surface modifications (Sect. 12.4). Efficient surface treatments also permit to conjugate ND surfaces with fluorescent molecules [230–233], with DNA [234], siRNA [31], proteins [235, 236], lysozymes [236], growth hormones [237], antibodies [238, 239], anti-cancer drugs [35–37], as well as dopamine derivatives [240].

Either covalent or non-covalent graftings were performed depending on expected bio-applications. These attempts have been recently reviewed in details [32]. The non-covalent grafting of large biomolecules based on electrostatic interactions was performed from well controlled NDs surface charge (Sect. 12.5). Cytochrome c [199] and bovine insulin [241] were immobilized on nanodiamonds. Furthermore, lysozymes [41, 242, 243] and apoobelin [244] were attached. Anticancer drugs like doxorubicine [245, 246], paclitaxel [35] or HCPT [37] were non-covalently adsorbed and then delivered. Some polymer coatings like polyethylenimine (PEI) have been also carried out on NDs to confer cationic surface charge for siRNA adsorption and delivery for gene therapy (Ewing's sarcoma) [31, 33, 233, 247]. More recently, the siRNA delivery was achieved using detonation NDs which have been rendered cationic by plasma hydrogenation [34]. The expression of the targeted gene was efficiently inhibited by these siRNA- NDs.

In most cases, the functionality of biomolecules is preserved [241, 248] although significantly less efficiency has sometimes been observed. Cumulative effects were even obtained as lysozyme adsorption is combined with cytochrome c [242]. Depending on the functionality of the biologic unit, covalent grafting via organic chemistry could be suitable to get a specific binding site on the ND or to avoid modifications of the biomolecule conformation. Grafting routes which may involve a linker molecule could be performed from different surface terminations of NDs [32].

12.6.1.4 Photoluminescent Centers

Color centers could be efficiently generated in HPHT nanodiamonds by high-energy particle beam irradiation (He^+ or electrons) leading to vacancy creation [145, 230]. During annealing, vacancies combined with nitrogen atoms to form NV centers which could be characterized by Photoluminescence (Sect. 12.3). NV centers hosted by HPHT NDs exhibited stable photoluminescence without bleaching or blinking in the red and near-infrared region corresponding to transmission window of tissues for size down to 7 nm [249]. These properties allowed the long-term tracking of fluorescent NDs into cells localized at subcellular scale by fluorescence and TEM microscopies. The intraneuronal transport in transgenic mouse lines was monitored by tracking of fluorescent NDs [250]. In that way, abnormalities which may be induced by brain-disease were detected. Expected applications of fluorescent NDs concern biomedical imaging [230] and fluorescence energy transfer (FRET) with other fluorophores [251]. For detonation NDs, the limited nitrogen amount, the nanometric size (<5 nm) and structural defects present at the surface and in the diamond core prevent the observation of stable photoluminescence [79, 83]. Recently, the effect of the surface chemistry on photoluminescence of detonation nanodiamonds was studied. This photoluminescence can be enhanced by 100 times for NDs grafted with octadecylamine [252].

The optical properties of color centers based on other impurities are currently intensively investigated. Silicon-vacancy (SiV) centers attract attention of research groups. Its molecular structure is depicted on Fig. 12.12 [253]. Indeed, SiV centers

Fig. 12.12 Molecular
structure of the SiV center
[253]

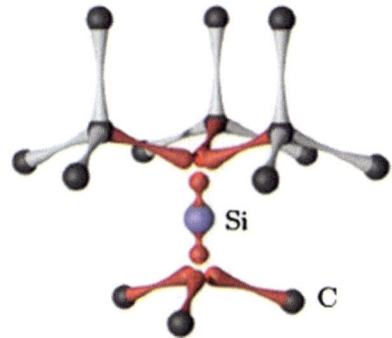

exhibit a narrower zero-phonon emission line (ZPL) compared to NV, at a wavelength of 738 nm. Such SiV center can be hosted by very small NDs (1.6 nm)
[254]. The Si incorporation can be achieved either by CVD [255] or by HPHT
[256]. According to their emission wavelength, SiV can act as biomarkers in the
near infrared which corresponds to the transmission window of biological tissues
[255]. SiV defects were used as effective biomarkers of primary neural precursor
cells) isolated from the adult mouse brain [257].

12.6.1.5 Tunable Size

Nanodiamonds are scalable nanoparticles down to few nanometers [196]. The main
size of NDs produced by detonation synthesis was decreased down to 2.8 nm by
nanostructuration of the explosive particles [258, 259]. For smallest NDs, in vivo
clearance or kidney elimination could be expected according to the size of kidney
capillarities. This constitutes an essential advantage for drug delivery or biomarkers
applications. In addition, size effects on photoluminescence were reported which
tune the wavelength of emitted photons [260].

12.6.2 Some Current Challenges

According to the previous assets, several challenges appear highly relevant and
promising for the development of future bio-applications involving nanodiamonds:
their labeling, their use for safer by design or their developments as a multifunctional platform for drug delivery.

12.6.2.1 Labeling

Labeling is inescapable for biodistribution and pharmacokinetics studies, for
development of biomarkers or targeted drug delivery. Although NV centers are

promising for in vitro imaging applications [231, 261, 262], their limited photo-luminescence intensity is an obstacle to their detection in the body. In the literature, several groups reported the efficient tracking of NDs into cells by grafting of fluorophores [27, 199, 230, 263]. Bright blue luminescence was even obtained by octadecylamine covalently attached to NDs-COOH [264]. The main stake concerns the stability of fluorophores in cellular environment. An interesting alternative for tracking consists to combine NDs with contrast agents currently used for magnetic resonance imaging (MRI). The rich surface chemistry of NDs (Sect. 12.4) allows the covalent grafting of amine-functionalized Gd(III). Gd(III)-NDs conjugates exhibited ten times higher relaxivity and can be well detected by MRI [265]. The radioactive labeling was reported for carbon nanotubes using ^{14}C [9]. The tritium labeling of detonation NDs has demonstrated by Girard et al. using the MPCVD plasma with a set-up previously used for hydrogenation (Fig. 12.4) [266]. Tritium is a β^- emitter with a half-life of 12.32 years. A quantification of ^3H was achieved by liquid scintillation. The total amount of tritium corresponds to 9120 µCi mg^{-1} with 93% of ^3H atoms bonded to the surface. A significant part of tritium is trapped in the diamond core (7%). Such labelled NDs which could be further functionalized will keep their radioactivity during their use in cells for bio-distribution and pharmacokinetics studies of NPs.

12.6.2.2 Safer by Design

An important issue for drug delivery concerns the conception of safer nanoparticles [267, 268]. For this purpose, NPs physicochemical determinants such as size, shape, composition, surface chemistry or coating nature which may be critical for cytotoxic and genotoxic effects must be investigated. In vitro screening for assessing the possible reactivity, biomarkers of inflammation and cellular uptake can provide precious indications to chemists and physicists to design "safe" nanomaterials. According to their extremely weak in vitro toxicity [28, 29], nan-odiamonds constitute excellent candidates to study the specific effects of physical and chemical parameters on toxicity.

12.6.2.3 Multifunctional Platform for Drug Delivery

Nanodiamonds combine biocompatibility, scalability and stability in aqueous solution (Sect. 12.5). Moreover, several studies reported their ability to carry and deliver different classes of drugs [35–37] or nucleic acids like siRNA [30, 31, 33, 245]. To build such platform, a stable labeling is required to allow tracking of NDs. Another important issue is related to the specific targeting of proteins or DNA. A recent study [209] reports on 20 nm HPHT NDs covalently grafted with peptide nucleic acids (PNA). The original functionalization route is based on an optimized amidation of ND carboxylic acids groups. ND-PNA conjugates were validated through a successful recognition of complementary DNA in a mixture, showing

Fig. 12.13 Fluorescence of
ND-PNA conjugates
compared to ND-COOH after
DNA hybridization observed
at λ_{exc} = 550 nm and
λ_{em} = 570 nm [209]

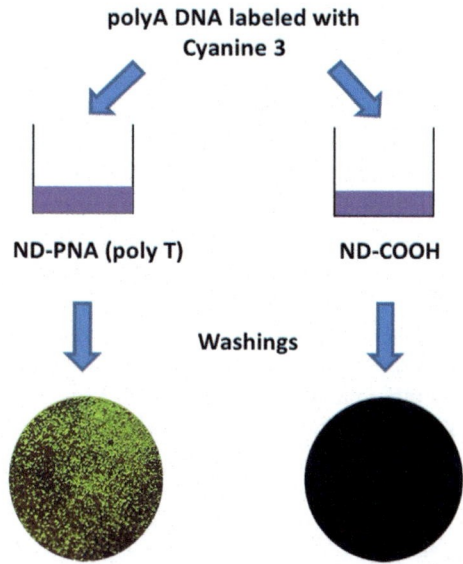

their efficiency toward nucleic acid detection (Fig. 12.13). Such nucleic acid
functionalized NDs open the way to a wide range of biomedical applications
towards genetic diseases, genomic research or early cancer diagnosis.

12.7 Conclusion

Nanodiamonds can behave very different surface properties according to their
tunable surface chemistry. The current state of art of controlled surface terminations
was summarized in this chapter (Sect. 12.4). The link between surface chemistry,
surface charge and colloidal properties of NDs was particularly emphasized.
Indeed, it appears highly relevant for biomedical applications because it is a key
parameter for drug adsorption or interactions with cell membrane. Among the
potential surface chemistries, hydroxyl and halogen terminations (F, Cl, Br) offer
specific reactivity. However, experimental protocols must be further improved to
reach a higher selectivity.

Some NDs characteristics related to the diamond core remain also partially
understood at the present state. In particular, it is essential to further investigate the
impact of structural defects or chemical impurities on NDs properties. The presence
of different crystallographic facets on HPHT and some detonation NDs should also
significantly influence their surface reactivity. Moreover, consequences of hydrogen
incorporated in nanodiamonds especially during detonation synthesis (Sect. 12.2)
must be better understood. According to previous works performed on bulk diamond, hydrogen should be trapped by structural defects or chemical impurities [70].

The tritium labeling of NDs has shown than a significant amount of tritium is trapped in the diamond core [266].

Among further main challenges for biomedecine, the radioactive labeling appears as a priority to investigate in vivo the biodistribution and pharmacokinetics on NDs. Such achievement has not been yet reported although tritium labeling was reported [266, 268]. Mechanisms of drug adsorption/desorption on NDs must be deeper understood to achieve a controlled release. These phenomena are highly sensitive to size and surface chemistry. The combination of NDs with NV or SiV centers can permit novel developments in biology like the probing of ion channels in cells or the neuron imaging. Finally, the synthesis of hybrid NDs is also an essential challenge. For example, surface graphitized NDs may possess surface properties linked to fullerene or graphene assets required for radiosensitization or photothermal therapy. In conclusion, according to their multiple assets, nanodiamonds constitute excellent candidates to be used as active NPs with therapeutic behaviors for biomedical applications.

Acknowledgements J.C. Arnault would like to thank H. A. Girard, T. Petit, M. Kurzyp, C. Gesset, C. Sicard-Roselli, E. Brun and M. Mermoux for fruitful discussions. He also acknowledges the different coworkers which contribute to studies dealing with surface modified nanodiamonds. The author also thanks Professor E. Osawa for providing detonation nanodiamonds.

References

1. M.L. Etheridge, S.A. Campbell, A.G. Erdman, C.L. Haynes, S.M. Wolf, J. McCullough, The big picture on nanomedicine: the state of investigational and approved nanomedicine products. Nanomed. Nanotechnol. Biol. Med. **9**, 1–14 (2013). https://doi.org/10.1016/j.nano.2012.05.013
2. Y. Matsumura, H. Maeda, A New concept for macromolecular therapeutics in cancer chemotherapy: mechanism of tumoritropic accumulation of proteins and the antitumor agent smancs. Cancer Res. **46**, 6387–6392 (1986) (WOS:A1986E976600069)
3. J.D. Byrne, T. Betancourt, L. Brannon-Peppas, Active targeting schemes for nanoparticle systems in cancer therapeutics. Adv. Drug Deliv. Rev. **60**, 1615–1626 (2008). https://doi.org/10.1016/j.addr.2008.08.005
4. E.V. Batrakova, A.V. Kabanov, Pluronic block copolymers: evolution of drug delivery concept from inert nanocarriers to biological response modifiers. J. Control. Release **130**, 98–106 (2008). https://doi.org/10.1016/j.jconrel.2008.04.013
5. X. Michalet, F.F. Pinaud, L.A. Bentolila, J.M. Tsay, S. Doose, J.J. Li, G. Sundaresan, A.M. Wu, S.S. Gambhir, S. Weiss, Quantum dots for live cells, in vivo imaging, and diagnostics. Science **307**, 538–544 (2005). https://doi.org/10.1126/science.1104274
6. T.A. Taton, C.A. Mirkin, R.L. Letsinger, Scanometric DNA array detection with nanoparticle probes. Science **289**, 1757–1760 (2000). https://doi.org/10.1126/science.289.5485.1757
7. H.B. Na, I.C. Song, T. Hyeon, Inorganic nanoparticles for MRI contrast agents. Adv. Mater. **21**, 2133–2148 (2009). https://doi.org/10.1002/adma.200802366
8. D. Yoo, J.H. Lee, T.H. Shin, J. Cheon, Theranostic magnetic nanoparticles. Acc. Chem. Res. **44**, 863–874 (2011). https://doi.org/10.1021/ar200085c

9. D. Georgin, B. Czarny, M. Botquin, M. Mayne-L'hermite, M. Pinault, B. Bouchet- Fabre, M. Carriere, J.L. Poncy, Q. Chau, R. Maximilien, V. Dive, F. Taran, Preparation of (14) C-labeled multiwalled carbon nanotubes for biodistribution investigations. J. Am. Chem. Soc. **131**, 14658–14659 (2009). https://doi.org/10.1021/ja906319z

10. C.S.S.R. Kumar, F. Mohammad, Magnetic nanomaterials for hyperthermia based therapy and controlled drug delivery. Adv. Drug Deliv. Rev. **63**, 789–808 (2011). https://doi.org/10.1016/j.addr.2011.03.008

11. K. Yang, S. Zhang, G. Zhang, X. Sun, S.T. Lee, Z. Liu, Graphene in mice: ultrahigh in vivo tumor uptake and efficient photothermal therapy. Nano Lett. **10**, 3318–3323 (2010). https://doi.org/10.1021/nl100996u

12. K. Yang, J. Wan, S. Zhang, B. Tian, Y. Zhang, Z. Liu, The influence of surface chemistry and size of nanoscale graphene oxide on photothermal therapy of cancer using ultra-low laser power. Biomaterials **33**, 2206–2214 (2012). https://doi.org/10.1016/j.biomaterials.2011.11.064

13. P. Cherukuri, E.S. Glazer, S.A. Curley, Targeted hyperthermia using metal nanoparticles. Adv. Drug Deliv. Rev. **62**, 339–345 (2010). https://doi.org/10.1016/j.addr.2009.11.006

14. J.F. Hainfeld, D.N. Slatkin, H.M. Smilowitz, The use of gold nanoparticles to enhance radiotherapy in mice. Phys. Med. Biol. **49**, N309–N315 (2004). https://doi.org/10.1088/0031-9155/49/18/N03

15. E. Porcel, S. Liehn, H. Remita, N. Usami, K. Kobayashi, Y. Furusawa, C. Le Sech, S. Lacombe, Platinum nanoparticles: a promising material for future cancer therapy? Nanotechnology **21**, 85103 (2010). https://doi.org/10.1088/0957-4484/21/8/085103

16. L. Maggiorella, G. Barouch, C. Devaux, A. Pottier, E. Deutsch, J. Bourhis, E. Borghi, L. Levy, Nanoscale radiotherapy with hafnium oxide nanoparticles. Futur. Oncol. **8**, 1167–1181 (2012). https://doi.org/10.2217/FON.12.96

17. M.J. Sailor, J.H. Park, Hybrid nanoparticles for detection and treatment of cancer. Adv. Mater. **24**, 3779–3802 (2012). https://doi.org/10.1002/adma.201200653

18. W.T. Al-Jamal, K. Kostarelos, Liposomes: from a clinically established drug delivery system to a nanoparticle platform for theranostic nanomedicine. Acc. Chem. Res. **44**, 1094–10104 (2011). https://doi.org/10.1021/ar200105p

19. M. Liong, J. Lu, M. Kovochich, T. Xia, S.G. Ruehm, A.E. Nel, F. Tamanoi, J.I. Zink, Multifunctional inorganic nanoparticles for imaging, targeting, and drug delivery. ACS Nano 2, 889–896 (2008). https://doi.org/10.1021/nn800072t

20. G. Wu, A. Mikhailovsky, H.A. Khant, C. Fu, W. Chiu, J.A. Zasadzinski, Remotely triggered liposome release by near-infrared light absorption via hollow gold nanoshells. J. Am. Chem. Soc. **130**, 8175–8177 (2008). https://doi.org/10.1021/ja802656d

21. E. Perevedentseva, Y.C. Lin, M. Jani, C.L. Cheng, Biomedical applications of nanodiamonds in imaging and therapy. Nanomedecine **8**, 2041–2060 (2013). https://doi.org/10.2217/NNM.13.183

22. O.A. Shenderova, G.E. McGuire, Science and engineering of nanodiamond particle surfaces for biological applications. Biointerphases **10**(030802), 1–24 (2015). https://doi.org/10.1116/1.4927679

23. K. Turcheniuk, V.N. Mochalin, Biomedical applications of nanodiamond. Nanotechnology **28**, 252001 (2017). https://doi.org/10.1088/1361-6528/aa6ae4

24. H.S. Choi, W. Liu, P. Misra, E. Tanaka, J.P. Zimmer, B. Itty Ipe, M.G. Bawendi, J.V. Frangioni, Renal clearance of quantum dots. Nat. Biotechnol. **25**, 1165–1170 (2007). https://doi.org/10.1038/nbt1340

25. V. Vyjayanthimala, Y.K. Tzeng, H.C. Chang, C.L. Li, The biocompatibility of fluorescent nanodiamonds and their mechanism of cellular uptake. Nanotechnology **20**, 425103 (2009). https://doi.org/10.1088/0957-4484/20/42/425103

26. Y. Yuan, X. Wang, G. Jia, J.H. Liu, T. Wang, Y. Gu, S.T. Yang, S. Zhen, H. Wang, Y. Liu, Pulmonary toxicity and translocation of nanodiamonds in mice. Diam. Relat. Mater. **19**, 291–299 (2010). https://doi.org/10.1016/j.diamond.2009.11.022

27. A.M. Schrand, S.A.C. Hens, O.A. Shenderova, Nanodiamond particles: properties and perspectives for bioapplications. Crit. Rev. Solid State Mater. Sci. **34**, 18–74 (2009). https://doi.org/10.1080/10408430902831987

28. V. Paget, J.A. Sergent, R. Grall, S. Altmeyer-Morel, H.A. Girard, T. Petit, G. Gesset, M. Mermoux, P. Bergonzo, J.C. Arnault, S. Chevillard, Nanodiamonds are neither cytotoxic nor genotoxic on kidney, intestine, lung and liver human cell lines. Nanotoxicology. https://doi.org/10.3109/17435390.2013.855828

29. H. Moche, V. Paget, D. Chevalier, E. Lorge, N. Claude, H.A. Girard, J.C. Arnault, S. Chevillard, F. Nesslany, Carboxylated nanodiamonds can be used as negative reference in in vitro nanogenotoxicity studies. J. Appl. Toxicol. **37**, 954–961 (2017). https://doi.org/10.1002/jat.3443

30. B. Zhang, Y. Li, C.Y. Fang, C.C. Chang, C.S. Chen, Y.Y. Chen, H.C. Chang, Receptor-mediated cellular uptake of folate-conjugated fluorescent nanodiamonds: a combined ensemble and single-particle study. Small **5**, 2716–2721 (2009). https://doi.org/10.1002/smll.200900725

31. A. Alhaddad, M.P. Adam, J. Botsoa, G. Dantelle, S. Perruchas, T. Gacoin, C. Mansuy, S. Lavielle, C. Malvy, F. Treussart, J.R. Bertrand, Nanodiamond as a vector for siRNA delivery to Ewing sarcoma cells. Small **7**, 3087–3095 (2011). https://doi.org/10.1002/smll.201101193

32. A. Krueger, D. Lang, Functionality is key: recent progress in the surface modification of nanodiamond. Adv. Func. Mater. **22**, 890–906 (2012). https://doi.org/10.1002/adfm.201102670

33. A. Alhaddad, C. Durieu, G. Dantelle, E. Le Cam, C. Malvy et al., Influence of the internalization pathway on the efficacy of siRNA delivery by cationic fluorescent nanodiamonds in the Ewing sarcoma cell model. PLoS ONE **7**(12), e52207 (2012). https://doi.org/10.1371/journal.pone.0052207

34. J.R. Bertrand, C. Pioche-Durieu, J. Ayala, T. Petit, H.A. Girard, C. Malvy, E. Le Cam, F. Treussart, J.C. Arnault, Plasma hydrogenated cationic detonation nanodiamonds efficiently deliver to human cells in culture functional siRNA targeting the Ewing sarcoma junction oncogene. Biomaterials **45**, 93–98 (2015). https://doi.org/10.1016/j.biomaterials.2014.12.007

35. K.K. Liu, W.W. Zheng, C.C. Wang, Y.C. Chiu, C.L. Cheng, Y.S. Lo, C. Chen, J.I. Chao, Covalent linkage of nanodiamond-paclitaxel for drug delivery and cancer therapy. Nanotechnology **21**, 315106 (2010). https://doi.org/10.1088/0957-4484/21/31/315106

36. E.K. Chow, X.Q. Zhang, M. Chen, R. Lam, E. Robinson, H. Huang, D. Schaffer, E. Osawa, A. Goga, D. Ho, Nanodiamond therapeutic delivery agents mediate enhanced chemoresistant tumor treatment. Sci. Transl. Med. **3**, 73ra21(2011). https://doi.org/10.1126/scitranslmed.3001713

37. J. Li, Y. Zhu, W. Li, X. Zhang, P. Peng, Q. Huang, Nanodiamonds as intracellular transporters of chemotherapeutic drug. Biomaterials **31**, 8410–8418 (2010). https://doi.org/10.1016/j.biomaterials.2010.07.058

38. L. Moore, V. Grobarova, H. Shen, H.B. Man, J. Mičova, M. Ledvina, J. Štursa, M. Nesladek, A. Fišerova, D. Ho, Comprehensive Interrogation of the Cellular Response to Fluorescent. Detonation and Functionalized Nanodiamonds. Nanoscale **6**, 11712–11721 (2014). https://doi.org/10.1039/c4nr02570a

39. D. Ho, A. Zappinpar, E.K. Chow, Diamonds, digital health, and drug development: optimizing combinatorial nanomedicine. ACS Nano **10**, 9087–9092 (2016). https://doi.org/10.1021/acsnano.6b06174

40. Y.R. Chang, H.Y. Lee, K. Chen, C.C. Chang, D.S. Tsai, C.C. Fu, T.S. Lim, Y.K. Tzeng, C.Y. Fang, C.C. Han, H.C. Chang, W. Fann, Mass production and dynamic imaging of fluorescent nanodiamonds. Nat. Nanotech. **3**, 284–288 (2008). https://doi.org/10.1038/nnano.2008.99

41. J.I. Chao, E. Perevedentseva, P.H. Chung, K.K. Liu, C.Y. Cheng, C.C. Chang, C.L. Cheng, Nanometer-sized diamond particle as a probe for biolabeling. Biophys. J. **93**, 2199–2208 (2007). https://doi.org/10.1529/biophysj.107.108134

42. Y.Y. Hui, W. Wei-Wen Hsiao, S. Haziza, M. Simonneau, F. Treussart, H.C. Chang, Single particle tracking of fluorescent nanodiamonds in cells and organisms. Curr. Opin. Solid State Mater. Sci. **21**, 35–42 (2016). https://doi.org/10.1016/j.cossms.2016.04.002

43. R. Grall, H.A. Girard, L. Saad, T. Petit, C. Gesset, M. Combis-Schlumberger, V. Paget, J. Delic, J.C. Arnault, S. Chevillard, Impairing the radioresistance of cancer cells by hydrogenated nanodiamonds. Biomaterials **61**, 290–298 (2015). https://doi.org/10.1016/j.biomaterials.2015.05.034

44. M. Kurzyp, H.A. Girard, Y. Cheref, E. Brun, C. Sicard-Roselli, S. Saada, J.C. Arnault, Hydroxyl radical production induced by plasma hydrogenated nanodiamonds under X-ray irradiation. Chem. Commun. **53**, 1237–1240 (2017). https://doi.org/10.1039/c6cc08895c

45. J.C. Arnault (ed.), *Nanodiamonds: Advanced Material Analysis, Properties and Applications* (Elsevier, 2017). ISBN: 978-0-323-43029-6

46. F.P. Bundy, W.A. Bassett, M.S. Weathers, R.J. Hemley, H.K. Mao, A.F. Goncharov, The pressure-temperature phase and transformation diagram for carbon; updated through 1994. Carbon **34**, 141–153 (1996). https://doi.org/10.1016/0008-6223(96)00170-4

47. V.N. Mochalin, O. Shenderova, D. Ho, Y. Gogotsi, The properties and applications of nanodiamonds. Nature Nanotech. **7**, 11–23 (2012). https://doi.org/10.1038/NNANO.2011.209

48. F.P. Bundy, H.T. Hall, H.M. Strong, R.H. Wentorf, Man-Made Diamonds. Nature **176**, 51–55 (1955). https://doi.org/10.1038/176051a0

49. J.C. Angus, C.C. Hayman, Low-pressure, metastable growth of diamond and "diamondlike" phases. Science **241**, 913–921 (1988). https://doi.org/10.1126/science.241.4868.913

50. G.W. Yang, J.B. Wang, Q.X. Liu, Preparation of nano-crystalline diamonds using pulsed laser induced reactive quenching. J. Phys.: Condens. Matter **10**, 7923–7928 (1998). https://doi.org/10.1088/0953-8984/10/35/024

51. J. Sun, S.L. Hu, X.W. Du, Y.W. Lei, Ultrafine diamond synthesized by long-pulse-width laser. Appl. Phys. Lett. **89**, 183115 (2006). https://doi.org/10.1063/1.2385210

52. J.P. Boudou, P.A. Curmi, F. Jelezko, J. Wrachtrup, P. Aubert, M. Sennour, G. Balasubramanian, R. Reuter, A. Thorel, E. Gaffet, High yield fabrication of fluorescent nanodiamonds. Nanotechnology **20**, 235602 (2009). https://doi.org/10.1088/0957-4484/20/35/359801

53. J.P. Boudou, J. Tisler, R. Reuter, A. Thorel, P.A. Curmi, F. Jelezko, J. Wrachtrup, Fluorescent nanodiamonds derived from HPHT with a size of less than 10 nm. Diam. Relat. Mater. **37**, 80–86 (2013). https://doi.org/10.1016/j.diamond.2013.05.006

54. E. Neu, C. Arend, E. Gross, F. Guldner, C. Hepp, D. Steinmetz, E. Zscherpel, S. Ghodbane, H. Sternschulte, D. Steinmüller-Nethl, Y. Liang, A. Krueger, C. Becher, Narrowband fluorescent nanodiamonds produced from chemical vapor deposition films. Appl. Phys. Lett. **98**, 243107 (2011). https://doi.org/10.1063/1.3599608

55. S. Heyer, W. Janssen, S. Turner, Y.G. Lu, W.S. Yeap, J. Verbeeck, K. Haenen, A. Krueger, Toward deep blue nano hope diamonds: heavily boron-doped diamond nanoparticles. ACS Nano **8**, 5757 (2014). https://doi.org/10.1021/nn500573x

56. V. Danilenko, O. Shenderova, Advances in synthesis of nanodiamond particles, in *Ultrananocrystalline Diamond: synthesis, properties and applications*, 2nd edn., by O. Shenderova, D.M. Gruen (Elsevier, 2012)

57. V.Y. Dolmatov, Detonation-synthesis nanodiamonds: synthesis, structure, properties and applications. Russ. Chem. Rev. **76,** 339–360 (2007) (WOS:000247118100004)

58. O. Shenderova, A. Koscheev, N. Zaripov, I. Petrov, Y. Skryabin, P. Detkov, T. Turner, G. Van Tendeloo, Surface chemistry and properties of ozone-purified detonation nanodiamonds. J. Phys. Chem. C **115**, 9827–9837 (2011). https://doi.org/10.1021/jp1102466

59. B. Zousman, O. Levinson, Pure nanodiamonds produced by laser-assisted technique, in *RSC Nanoscience and Nanotechnology*, vol. 31, ed. by O.A. Williams (2014), pp. 112–127
60. J.E. Dahl, S.G. Liu, R.M.K. Carlson, Isolation and structure of higher diamondoids, nanometer-sized diamond molecules. Science **299**, 96–102 (2003). https://doi.org/10.1126/science.1078239
61. O.O. Mykhaylyk, Y.M. Solonin, D.N. Batchelder, R. Brydson, Transformation of nanodiamond into carbon onions: A comparative study by high-resolution transmission electron microscopy, electron energy-loss spectroscopy, x-ray diffraction, small-angle x-ray scattering, and ultraviolet Raman spectroscopy. J. Appl. Phys. **97**, 074302 (2005). https://doi.org/10.1063/1.1868054
62. E. Osawa, D. Ho, Nanodiamond and its application to drug delivery. J. Med. Allied. Sci. **2**, 31–40 (2012) (P r i n t I S S N: 2 2 3 1 1 6 9 6 O n l i n e I S S N: 2 2 3 1 1 7 0 X)
63. A.S. Barnard, M. Sternberg, Crystallinity and surface electrostatics of diamond nanocrystals. J. Mater. Chem. **17**, 4811–4819 (2007). https://doi.org/10.1039/b710189a
64. S. Turner, O.I. Lebedev, O. Shenderova, I.I. Vlasov, J. Verbeeck, G. Van Tendeloo, Determination of size, morphology, and nitrogen impurity location in treated detonation nanodiamond by transmission electron microscopy. Adv. Funct. Mater. **19**, 2116–2124 (2009)
65. D.C. Bell, C.J. Russo, D.V. Kolmykov, 40 keV atomic resolution TEM. Ultramicroscopy **114**, 38–45 (2012). https://doi.org/10.1016/j.ultramic.2011.12.001
66. B. Palosz, S. Stelmakh, E. Grzanka, S. Gierlotka, W. Palosz, Application of apparent lattice parameter to determination of core-shell structure of nanocrystals. Z. Kristallogr. **222**, 580–594 (2007). https://doi.org/10.1524/zkri.2007.222.11.580
67. V.L. Kuznetsov, M.N. Aleksandrov, I.V. Zagoruiko, A.L. Chuvilin, E.M. Moroz, V.N. Kolomiichuk, V.A. Lizholobov, P.M. Brylyakov, G.V. Sakovitch, Study of ultradispersed diamond powders obtained using explosion energy. Carbon **29**, 665–668 (1991). https://doi.org/10.1016/0008-6223(91)90135-6
68. T. Petit, J.C. Arnault, H.A. Girard, M. Sennour, P. Bergonzo, Early stages of surface graphitization on nanodiamond probed by x-ray photoelectron spectroscopy. Phys. Rev. B **84**, 233407 (2011). https://doi.org/10.1103/PhysRevB.84.233407
69. J.C. Arnault, X-ray photoemission spectroscopy applied to nanodiamonds: From surface chemistry to in situ reactivity Diam. Relat. Mater. **84**, 157–168 (2018) https://doi.org/10.1016/j.diamond.2018.03.015
70. D. Ballutaud, F. Jomard, T. Kociniewski, E. Rzepka, H.A. Girard, S. Saada, Sp(3)/sp(2) character of the carbon and hydrogen configuration in micro- and nanocrystalline diamond Diam. Relat. Mater. **17**, 451–456 (2008). https://doi.org/10.1016/j.diamond.2007.10.004
71. B.R. Smith, D. Inglis, B. Sandnes, J. Rabeau, A.V. Zvyagin, D. Gruber, C.J. Noble, R. Vogel, E. Osawa, T. Plakhotnik, Five-nanometer diamond with luminescent nitrogen-vacancy defect centers. Small **5**, 1649–1653 (2009). https://doi.org/10.1002/smll.200801802
72. A.V. Fionov, A. Lund, W.M. Chen, N.N. Rozhkova, I.A. Buyanova, G.I. Emel'yanova, L.E. Gorlenko, E.V. Golubina, E.S. Lokteva, E. Osawa, V.V. Lunin, Paramagnetic centers in detonation nanodiamonds studied by CW and pulse EPR. Chem. Phys. Lett. **493**, 319–322 (2010). https://doi.org/10.1016/j.cplett.2010.05.050
73. J.H.N. Loubser, J.A. Van Wyk, Electron spin resonance in the study of diamond. Rep. Progr. Phys. **41**, 1201–1248 (1978). https://doi.org/10.1088/0034-4885/41/8/002
74. V. Pichot, O. Stephan, M. Comet, E. Fousson, J. Mory, K. March, D. Spitzer, High nitrogen doping of detonation nanodiamonds. J. Phys. Chem. C **114**, 10082–10087 (2010). https://doi.org/10.1021/jp9121485
75. Y.G. Lu, S. Turner, J. Verbeeck, S.D. Janssens, P. Wagner, K. Haenen, G. Van Tendeloo, Direct visualization of boron dopant distribution and coordination in individual chemical vapor deposition nanocrystalline B-doped diamond grains. Appl. Phys. Letters **101**, 041907 (2012). https://doi.org/10.1063/1.4738885

76. S. Turner, Y.G. Lu, S.D. Janssens, F. Da Pieve, D. Lamoen, J. Verbeeck, K. Haenen, P. Wagner, G. Van Tendeloo, Local boron environment in B-doped nanocrystalline diamond films. Nanoscale **4**, 5960–5964 (2012). https://doi.org/10.1039/c2nr31530k

77. A.V. Kvit, V.V. Zhirnov, T. Tyler, J.J. Hren, Aging effect and nitrogen distribution in diamond nanoparticles. Comput. Part B Eng. **35**, 163–166 (2004). https://doi.org/10.1016/j.compositesb.2003.08.003

78. I.I. Vlasov, Hydrid diamond-graphite nanowires produced by microwave plasma chemical vapor deposition. Adv. Mater. **19**, 4058–4062 (2007). https://doi.org/10.1002/adma.200700442

79. O.A. Shenderova, I.I. Vlasov, S. Turner, G. Van Tendeloo, S.B. Orlinskii, A.A. Shiryaev, A. A. Khomich, S.N. Sulyanov, F. Jelezko, J. Wrachtrup, Nitrogen control in nanodiamond produced by detonation shock-wave-assisted synthesis. J. Phys. Chem. C **115**, 14014–14024 (2011). https://doi.org/10.1021/jp202057q

80. T. Berg, E. Marosits, J. Maul, P. Nagel, U. Ott, F. Schertz, S. Schuppler, C. Sudek, G. Schonhense, Quantum confinement observed in the x-ray absorption spectrum of size distributed meteoritic nanodiamonds. J. Appl. Phys. **104**, 064303 (2008). https://doi.org/10.1063/1.2978217

81. A.M. Panich, Nuclear magnetic resonance studies of nanodiamonds. Crit. Rev. Solid State Mater. Sci. **37**, 276–303 (2012). https://doi.org/10.1080/10408436.2011.606930

82. P.N. Nesterenko, D. Mitev, B. Paull, Elemental analysis of nanodiamonds by inductively coupled plasma hyphenated methods, in *Nanodiamonds: Advanced Material Analysis, Properties and Applications,* ed. by J.C. Arnault (Elsevier, 2017). ISBN: 978-0-323-43029-6

83. C. Bradac, T. Gaebel, N. Naidoo, M.J. Sellars, J. Twamley, L.J. Brown, A.S. Barnard, T. Plakhotnik, A.V. Zvyagin, J.R. Rabeau, Observation and control of blinking nitrogen-vacancy centres in discrete nanodiamonds. Nat. Nanotech. **5**, 345–349 (2010). https://doi.org/10.1038/NNANO.2010.56

84. A. Krüger, F. Kataoka, M. Ozawa, T. Fujino, Y. Suzuki, A.E. Aleksenskii, A. Ya, A. Vul, E. Osawa, Unusually tight aggregation in detonation nanodiamond: Identification and disintegration. Carbon **43**, 1722–1730 (2005). https://doi.org/10.1016/j.carbon.2005.02.020

85. J.C. Arnault, T. Petit, H.A. Girard, A. Chavanne, C. Gesset, M. Sennour, M. Chaigneau, Surface chemical modifications and surface reactivity of nanodiamonds hydrogenated by CVD plasma. Phys. Chem. Chem. Phys. **13**, 11481 (2011). https://doi.org/10.1039/c1cp20109c

86. M. Mermoux, B. Marcus, G.M. Swain, J.E. Butler, A confocal raman imaging study of an optically transparent boron-doped diamond electrode. J. Phys. Chem. B **106**, 10816–10827 (2002). https://doi.org/10.1021/jp0202946

87. S. Osswald, V.N. Mochalin, M. Havel, G. Yushin, Y. Gogotsi, Phonon confinement effects in the Raman spectrum of nanodiamond. Phys. Rev. B **80**, 075419 (2009). https://doi.org/10.1103/PhysRevB.80.075419

88. M. Mermoux, A. Crisci, T. Petit, H.A. Girard, J.C. Arnault, Surface modifications of detonation nanodiamonds probed by multiwavelength Raman spectroscopy. J. Phys. Chem. C **118**, 23415–23425 (2014). https://doi.org/10.1021/jp507377z

89. D.R. Baer, M.H. Engelhard, XPS analysis of nanostructured materials and biological surfaces. J. Electron Spectrosc. Relat. Phenom. 178–179: 415–432 (2010). https://doi.org/10.1016/j.elspec.2009.09.003

90. T. Petit, J.C. Arnault, H.A. Girard, M. Sennour, T.Y. Kang, C.L. Cheng, P. Bergonzo, Oxygen hole doping of nanodiamond. Nanoscale **4**, 6792 (2012). https://doi.org/10.1039/c2nr31655b

91. S. Michaelson, A. Stacey, R. Akhvlediani, S. Prawer, A. Hoffman, High resolution electron energy loss spectroscopy surface studies of hydrogenated detonation nano-diamond spray-deposited films. Surf. Sci. **604**, 1326–1330 (2010). https://doi.org/10.1016/j.susc.2010.04.022

92. S.L.Y. Chang, C. Dwyer, E. Osawa, A.S. Barnard, Size dependent reconstruction in detonation nanodiamonds. Nanoscale Horizons (2017). https://doi.org/10.1039/c7nh00125h

93. T Petit, L Puskar, FTIR spectroscopy of nanodiamonds: Methods and interpretation. Diam. Relat. Mater. **89**, 52–66. https://doi.org/10.1016/j.diamond.2018.08.005

94. C.L. Cheng, C.F. Chen, W.C. Shaio, D.S. Tsai, K.H. Chen, The CH stretching features on diamonds of different origins. Diam. Relat. Mater. **14**, 1455–1462 (2005). https://doi.org/10.1016/j.diamond.2005.03.003

95. P.H. Chung, E. Perevedentseva, J.S. Tu, C.C. Chang, C.L. Cheng, Spectroscopic study of bio-functionalized nanodiamonds. Diam. Relat. Mater. **15**, 622–625 (2006). https://doi.org/10.1016/j.diamond.2005.11.019

96. Z. Remes, H. Kozak, B. Rezek, E. Ukraintsev, O. Babchenko, A. Kromka, H.A. Girard, J.C. Arnault, P. Bergonzo, Diamond-coated ATR prism for infrared absorption spectroscopy of surface-modified diamond nanoparticles. Appl. Surf. Sci. **270**, 411–417 (2013). https://doi.org/10.1016/j.apsusc.2013.01.039

97. S. Ghodbane, A. Deneuville, D. Tromson, P. Bergonzo, E. Bustarret, D. Ballutaud, Sensitivity of Raman spectra excited at 325 nm to surface treatments of undoped polycrystalline diamond films. Phys. Status Solidi (a) **203**, 2397–2402 (2006). https://doi.org/10.1002/pssa.200521462

98. A. Crisci, M. Mermoux, B. Saubat-Marcus, Deep ultra-violet Raman imaging of CVD boron-doped and non-doped diamond films. Diam. Relat. Mater. **17**, 1207–1211 (2008). https://doi.org/10.1016/j.diamond.2008.01.025

99. F. Cataldo, A. Koscheev, A study of the action of ozone and on the thermal stability of nanodiamond. Fuller. Nanotub. Carbon Nanostructures **11**, 201 (2003). https://doi.org/10.1081/FST-120024039

100. A. Koshcheev, Thermodesorption mass spectrometry in the light of solution of the problem of certification and unification of the surface properties of detonation nano-diamonds. Russ. J. Gener. Chem. **79**, 2033–2044 (2009). https://doi.org/10.1134/S1070363209090357

101. A. Krueger, M. Ozawa, G. Jarre, Y. Liang, J. Stegk, L. Lu, Deagglomeration and functionalisation of detonation diamond. Phys. Status Solidi (a) **204**, 2881–2887 (2007). https://doi.org/10.1002/pssa.200776330

102. A.E. Aleksenskiy, E.D. Eydelman, A.Y. Vul, Deagglomeration of detonation nanodiamonds. Nanosci. Nanotechnol. Lett. **3**, 68–74 (2011). https://doi.org/10.1166/nnl.2011.1122

103. T. Petit, H.A. Girard, A. Trouve, I. Batonneau-Genner, P. Bergonzo, J.C. Arnault, Surface transfer doping can mediate both colloidal stability and self-assembly of nanodiamonds. Nanoscale **5**, 8958–8962 (2013). https://doi.org/10.1039/c3nr02492j

104. H.A. Girard, J.C. Arnault, S. Perruchas, S. Saada, T. Gacoin, J.P. Boilot, P. Bergonzo, Hydrogenation of nanodiamonds using MPCVD: A new route toward organic functionalization. Diam. Relat. Mater. **19**, 1117–1123 (2010). https://doi.org/10.1016/j.diamond.2010.03.019

105. O.A. Williams, J. Hees, C. Dieker, W. Jäger, L. Kirste, C.E. Nebel, Size-dependent reactivity of diamond nanoparticles. ACS Nano **4**, 4824–4830 (2010). https://doi.org/10.1021/nn100748k

106. A.I. Ahmed, S. Mandal, L. Gines, O.A. Williams, C.L. Cheng, Low temperature catalytic reactivity of nanodiamond in molecular hydrogen. Carbon **110**, 438–442 (2016). https://doi.org/10.1016/j.carbon.2016.09.019

107. S. Osswald, G. Yushin, V. Mochalin, S.O. Kucheyev, Y. Gogotsi, Control of sp2/sp3 carbon ratio and surface chemistry of nanodiamond powders by selective oxidation in air. J. Am. Chem. Soc. **128**, 11635–11642 (2006). https://doi.org/10.1021/ja063303n

108. A. Krueger, The structure and reactivity of nanoscale diamond. J. Mater. Chem. **18**, 1485–1492 (2008). https://doi.org/10.1039/b716673g

109. A. Krüger, Y. Liang, G. Jarre, J. Stegk, Surface functionalisation of detonation diamond suitable for biological applications. J. Mater. Chem. **16**, 2322–2328 (2006). https://doi.org/10.1039/b601325b

110. R. Martin, P.C. Heydorn, M. Alvaro, H. Garcia, General strategy for high-density covalent functionalization of diamond nanoparticles using fenton chemistry. Chem. Mater. **21**, 4505–4514 (2009). https://doi.org/10.1021/cm9012602

111. H.A. Girard, T. Petit, S. Perruchas, J.C. Arnault, P. Bergonzo, Surface properties of hydrogenated nanodiamonds: a chemical investigation. Phys. Chem. Chem. Phys. **13**, 11511–11516 (2011). https://doi.org/10.1039/c1cp20424f

112. V.N. Mochalin, S. Osswald, C. Portet, G. Yushin, C. Hobson, M. Havel, Gogotsi Y high temperature functionalization and surface modification of nanodiamond powders. MRS Proc. **1039**, 201–211 (2007). https://doi.org/10.1557/PROC-1039-P11-03

113. M.A. Ray, T. Tyler, B. Hook, A. Martin, G. Cunningham, O. Shenderova, J.L. Davidson, M. Howell, W.P. Kang, G. McGuire, Cool plasma functionalization of nano-crystalline diamond films. Diam. Relat. Mater. **16**, 2087–2089 (2007). https://doi.org/10.1016/j.diamond.2007.07.016

114. J. Havlik, H. Raabova, M. Gulka, V. Petrakova, M. Krecmarova, V. Masek, P. Lousa, J. Stursa, G. BoyenH, M. Nesladek, P. Cigler, Benchtop fluorination of fluorescent nanodiamonds on a preparative scale: toward unusually hydrophilic bright particles. Adv. Funct. Mater. **26**, 4134–4142 (2017). https://doi.org/10.1002/adfm.201504857

115. K.I. Sotowa, T. Amamoto, A. Sobana, K. Kusakabe, T. Imato, Effect of treatment temperature on the amination of chlorinated diamond. Diam. Relat. Mater. **13**, 145–150 (2004). https://doi.org/10.1016/j.diamond.2003.10.029

116. W.S. Yeap, S. Chen, K.P. Loh, Detonation nanodiamond: an organic platform for the suzuki coupling of organic molecules. Langmuir **25**, 185–191 (2009). https://doi.org/10.1021/la8029787

117. L.C.L. Huang, H.C. Chang, Adsorption and immobilization of cytochrome c on nanodiamonds. Langmuir **20**, 5879–5884 (2004). https://doi.org/10.1021/la0495736

118. Y. Liang, T. Meinhardt, G. Jarre, M. Ozawa, P. Vrdoljak, A. Schöll, F. Reinert, A. Krueger, Deagglomeration and surface modification of thermally annealed nanoscale diamond. J. Colloid Interface Sci. **354**, 23–30 (2011). https://doi.org/10.1016/j.jcis.2010.10.044

119. J. Chen, S.Z. Deng, J. Chen, Z.X. Yu, N.S. Xu, Graphitization of nanodiamond powder annealed in argon ambient. Appl. Phys. Lett. **74**, 3651 (1999). https://doi.org/10.1063/1.123211

120. Y. Liang, M. Ozawa, A. Krueger, A general procedure to functionalize agglomerating nanoparticles demonstrated on nanodiamond. ACS Nano 3, 2288–2296 (2009). https://doi.org/10.1021/nn900339s

121. J.B. Cui, J. Ristein, L. Ley, Electron affinity of the bare and hydrogen covered single crystal diamond (111) surface. Phys. Rev. Lett. **81**, 429–432 (1998). https://doi.org/10.1103/PhysRevLett.81.429

122. L. Ley, J. Ristein, F. Meier, M. Riedel, P. Strobel, Surface conductivity of the diamond: a novel transfer doping mechanism. Phys. B **376**, 262–267 (2006). https://doi.org/10.1016/j.physb.2005.12.068

123. J.C. Arnault, H.A. Girard, Hydrogenated nanodiamonds: synthesis and surface properties Curr. Opin. Solid State Mater. Sci. **21**, 10–16 (2017). https://doi.org/10.1016/j.cossms.2016.06.007

124. B.V. Spitsyn, S.A. Denisov, N.A. Skorik, A.G. Chopurova, S.A. Parkaeva, L.D. Belyakova, O.G. Larionov, The physical-chemical study of detonation nanodiamond application in adsorption and chromatography. Diam. Relat. Mater. **19**, 123–127 (2010). https://doi.org/10.1016/j.diamond.2009.10.020

125. S. Ida, T. Tsubota, O. Hirabayashi, M. Nagata, Y. Matsumoto, A. Fujishima, Chemical reaction of hydrogenated diamond surface with peroxide radical initiators. Diam. Relat. Mater. **12**, 601–605 (2003). https://doi.org/10.1016/S0925-9635(02)00334-5

126. I.I. Obraztsova, N.K. Eremenko, Physicochemical modification of nanodiamonds. Russ. J. Appl. Chem. **81**, 603–608 (2008). https://doi.org/10.1134/S107042720804006X

127. M.B. Smith, J. March, *March's Advanced Organic Chemistry*, 6th edn. (Wiley, Hoboken, 2007)

128. D Ager, Hydrogenation of carbon-carbon double bonds in *Science of Synthesis, Stereoselective Synthesis*, vol. 1 ed. by J.G. De Vries, G.A. Molander, P.A. Evans (2011), pp. 185–256

129. M. Yeganeh, P. Coxon, A. Brieva, V. Dhanak, L. Šiller, Y. Butenko, Atomic hydrogen treatment of nanodiamond powder studied with photoemission spectroscopy. Phys. Rev. B **75**, 1–8 (2007). https://doi.org/10.1103/PhysRevB.75.155404

130. J. Angus, H.A. Will, W.S. Stanko, Growth of diamond seed crystals by vapor deposition. J. Appl. Phys. **39**, 2915–2922 (1968). https://doi.org/10.1063/1.1656693

131. E. Van Hove, J. De Sanoit, J.C. Arnault, S. Saada, C. Mer, P. Mailley, P. Bergonzo, M. Nesladek, Stability of H-terminated BDD electrodes: an insight into the influence of the surface preparation. Phys. Status Solidi (a) **204**, 2931–2939 (2007). https://doi.org/10.1002/pssa.200776340

132. R. Kiran, E. Scorsone, J. De Sanoit, J.C. Arnault, P. Mailley, P. Bergonzo, Boron doped diamond electrodes for direct measurement in biological fluids: an in situ regeneration approach. J. Electrochem. Soc. **160**, H67–H73 (2013). https://doi.org/10.1149/2.014302jes

133. W.S. Yang, O. Auciello, J.E. Butler, W. Cai, J.A. Carlisle, J. Gerbi, D.M. Gruen, T. Knickerbocker, T.L. Lasseter, J.N. Russell, L.M. Smith, R.J. Hamers, DNA-modified nanocrystalline diamond thin-films as stable, biologically active substrates. Nat. Mater. **1**, 253–257 (2002). https://doi.org/10.1038/nmat779

134. T. Strother, T. Knickerbocker, J. Russell, J. Butler, L. Smith, R. Hamers, Photochemical functionalization of diamond films. Langmuir **18**, 968–971 (2002). https://doi.org/10.1021/la0112561

135. A. Hartl, E. Schmich, J.A. Garrido, J. Hernando, S.C.R. Catharino, S. Walter, P. Feulner, A. Kromka, D. Steinmuller, M. Stutzmann, Protein-modified nanocrystalline diamond thin films for biosensor applications. Nat. Mater. **3**, 736–742 (2004). https://doi.org/10.1038/nmat1204

136. Y. Zhong, K. Loh, The chemistry of C-H bond activation on diamond. Chem.–Asian J. **5**, 1532–1540 (2010). https://doi.org/10.1002/asia.201000027

137. S. Szunerits, R. Boukherroub, Different strategies for functionalization of diamond surfaces. J. Solid State Electrochem. **12**, 1205–1218 (2008). https://doi.org/10.1007/s10008-007-0473-3

138. F. Maier, M. Riedel, B. Mantel, J. Ristein, L. Ley, Origin of surface conductivity in diamond. Phys. Rev. Lett. **85**, 3472–3475 (2000). https://doi.org/10.1103/PhysRevLett.85.3472

139. C. Bandis, B.B. Pate, electron-emission due to exciton breakup from negative electron-affinity diamond. Phys. Rev. Lett. **74**, 777–780 (1995). https://doi.org/10.1103/PhysRevLett.74.777

140. B.M. Nichols, J.E. Butler, J.N. Russell, R.J. Hamers, Photochemical functionalization of hydrogen-terminated diamond surfaces: a structural and mechanistic study. J. Phys. Chem. B **109**, 20938–20947 (2005). https://doi.org/10.1021/jp0545389

141. D. Shin, B. Rezek, N. Tokuda, D. Takeuchi, H. Watanabe, T. Nakamura, T. Yamamoto, C. E. Nebel, Photo- and electrochemical bonding of DNA to single crystalline CVD diamond. Phys. Status Solidi A **203**, 3245–3272 (2006). https://doi.org/10.1002/pssa.200671402

142. S. Lud, M. Steenackers, R. Jordan, P. Bruno, D. Gruen, P. Feulner, J. Garrido, M. Stutzmann, Chemical grafting of biphenyl self-assembled monolayers on ultrananocrystalline diamond. J. Am. Chem. Soc. **128**, 16884–16891 (2006). https://doi.org/10.1021/ja0657049

143. A. Bolker, C. Saguy, R. Kalish, Transfer doping of single isolated nanodiamonds, studied by scanning probe micros copy techniques. Nanotechnology **25**(385702), 1–7 (2014). https://doi.org/10.1088/0957-4484/25/38/385702

144. Y.V. Butenko, V.L. Kuznetsov, E.A. Paukshtis, A.I. Stadnichenko, I.N. Mazov, S.I. Moseenkov, A.I. Boronin, S.V. Kosheev, The thermal stability of nanodiamond surface groups and onset of nanodiamond graphitization. Fuller. Nanotub. Carbon Nanostructures **14**, 557–564 (2006). https://doi.org/10.1080/15363830600666779

145. L. Rondin, G. Dantelle, A. Slablab, F. Grosshans, F. Treussart, P. Bergonzo, S. Perruchas, T. Gacoin, M. Chaigneau, H.C. Chang, V. Jacques, J.F. Roch, Surface-induced charge state conversion of nitrogen-vacancy defects in nanodiamonds. Phys. Rev. B **82**, 115449 (2010). https://doi.org/10.1103/PhysRevB.82.115449

146. O. Shenderova, I. Petrov, J. Walsh, V. Grichko, T. Tyler, G. Cunningham, Modification of detonation nanodiamonds by heat treatment in air. Diam. Relat. Mater. **15**, 1799–1803 (2006). https://doi.org/10.1016/j.diamond.2006.08.032

147. D. Mitev, R. Dimitrova, M. Spassova, C. Minchev, S. Stavrev, Surface peculiarities of detonation nanodiamonds in dependence of fabrication and purification methods. Diam. Relat. Mater. **16**, 776–780 (2007). https://doi.org/10.1016/j.diamond.2007.01.005

148. M. Comet, V. Pichot, B. Siegert, F. Britz, D. Spitzer, Detonation Nanodiamonds for Doping Kevlar. J. Nanosci. Nanotechnol. **10**, 4286–4292 (2010). https://doi.org/10.1166/jnn.2010.2186

149. O. Shenderova, A.M. Panich, S. Moseenkov, S.C. Hens, V. Kuznetsov, H.M. Vieth, Hydroxylated detonation nanodiamond : FTIR, XPS, and NMR studies. J. Phys. Chem. C **115**, 19005–19011 (2011). https://doi.org/10.1021/jp205389m

150. R. Martín, M. Álvaro, J.R. Herance, H. García, Fenton-treated functionalized diamond nanoparticles as gene delivery system. ACS Nano **4**, 65–74 (2010). https://doi.org/10.1021/nn901616c

151. Y. Liu, Z. Gu, J.L. Margrave, V.N. Khabashesku, Functionalization of nanoscale diamond powder: Fluoro-, alkyl-, amino-, and amino acid-nanodiamond derivatives. Chem. Mater. **16**, 3924–3930 (2004). https://doi.org/10.1021/cm048875q

152. G. Lisichkin, V. Korol'kov, B. Tarasevic, I. Kulakova, A. Karpukhin, Photochemical chlorination of nanodiamond and interaction of its modified surface with C-nucleophiles. Russ. Chem. Bull. **55**, 2212–2219 (2006). https://doi.org/10.1007/s11172-006-0574-7

153. B.V. Spitsyn, J.L. Davidson, M.N. Graboboev, T.B. Galushko, N.V. Serebryakova, T.A. Karpukhina, I.I. Kulakova, N.N. Melnik, In road to modifications of detonation nanodiamond. Diam. Relat. Mater. **15**, 296–299 (2006). https://doi.org/10.1016/j.diamond.2005.07.033

154. V. Ralchenko, L. Nistor, E. Pleuler, A. Khomich, I. Vlasov, R. Khmelnitskii, Structure and properties of high-temperature annealed CVD diamond. Diam. Relat. Mater. **12**, 1964–1970 (2003). https://doi.org/10.1016/S0925-9635(03)00214-0

155. S. Ogawa, T. Yamada, S. Ishizduka, A. Yoshigoe, M. Hasegawa, Y. Teraoka, Y. Takakuwa, Vacuum annealing formation of graphene on diamond C(111) surfaces studied by real-time photoelectron spectroscopy. Jap. J. Appl. Phys. **51**, 11PF02 (2012). https://doi.org/10.1143/jjap.51.11pf02

156. T. Evans, Changes produced by high temperature treatment of diamond, in *The Properties of Natural and Synthetic Diamonds*, ed. by J.E. Field (Academic Press, London, 1979), pp. 403–425

157. K.S. Uspenskaya, Y.N. Tolmachev, D.V. Fedoseev, Oxidation and graphitization of diamond at low pressures. Zh. Fiz. Khim. **56**, 495 (1982) (in Russian). WOS: A1982ND19900073

158. D.V. Fedoseev, S.P. Vnusov, V.L. Bukhovets, B.A. Anikin, Surface graphitization of diamond at high temperatures. Surf. Coat. Technol. **28**, 207–214 (1986). https://doi.org/10.1016/0257-8972(86)90059-9

159. G. Davies, *Properties and Growth of Diamond* (INSPEC, London, 1994)

160. J.F. Prins, Ion implantation of diamond for electronics applications. Semicond. Sci. Technol. **18**, S27 (2003). https://doi.org/10.1088/0268-1242/18/3/304

161. F. Banhart, Irradiation effects in carbon nanostructures. Rep. Prog. Phys. **62**, 1181 (1999). https://doi.org/10.1088/0034-4885/62/8/201

162. J.E. Field (ed.), *The Properties of Natural and Synthetic Diamonds* (Academic Press, London, 1977)

163. O.P. Krivoruchko, V.I. Zaikovski, K.I. Zamaraev, Formation of unsual liquid-like FeC particles and dynamics of their nehavior on amorphous carbon surface at 920–1170 K. Dkl. Akad. Nauk. **329**, 744 (1993) (WOS:A1993LR07400017)
164. M.S. Dresselhaus, G. Dresselhaus, P.C. Eklund, *Science of Fullerenes and Carbon Nanotubes* (Academic Press, San Diego, 1996)
165. V.L. Kuznetsov, Y.V. Butenko, Diamond phase transitions at nanoscale, in *Ultrananocrystalline Diamond: Synthesis, Properties and Applications*, 2nd edn., ed. by O. Shenderova, D.M. Gruen (Elsevier, 2012)
166. Y.V. Butenko, S. Krishnamurthy, A.K. Chakraborty, V.L. Kuznetsov, V.R. Dhanak, M.R.C. Hunt, L. Scaroniller, L. Šiller, Photoemission study of onionlike carbons produced by annealing nanodiamonds. Phys. Rev. B **71**, 75420 (2005). https://doi.org/10.1103/PhysRevB.71.075420
167. D. Pech, M. Brunet, H. Durou, P.H. Huang, V. Mochalin, Y. Gogotsi, Ultra-high-power micrometre-sized supercapacitors based on onion-like carbon. Nat. Nanotechnol. **5**, 651–654 (2010). https://doi.org/10.1038/nnano.2010.162
168. O. Shenderova, C. Jones, V. Borjanovic, S. Hens, G. Cunningham, S. Moseenkov, Detonation nanodiamond and onion-like carbon: applications in composites. Phys. Status Solidi a **205**, 2245–2251 (2008). https://doi.org/10.1002/pssa.200879706
169. O. Shenderova, T. Tyler, V. Borjanovic, G. Cunningham, M. Ray, J. Walsh, M. Casulli, Nanodiamond and onion-like carbon polymer nanocomposites. Diam. Relat. Mater. **16**, 1213–1217 (2007). https://doi.org/10.1016/S0925-9635(07)00337-8
170. V.L. Kuznetsov, A.L. Chuvilin, Y.V. Butenko, I.L. Malkov, V.M. Titov, Onion-like carbon from ultradisperse diamond. Chem. Phys. Lett. **222**, 343 (1994). https://doi.org/10.1016/0009-2614(94)87072-1
171. F. Fugaciu, H. Hermann, G. Seifert, Concentric-shell fullerenes and diamond particles: a molecular-dynamics study. Phys. Rev. B **60**, 10711 (1999). https://doi.org/10.1103/PhysRevB.60.10711
172. J.Y. Raty, G. Galli, C. Bostedt, T.W. van Buuren, L.J. Terminello, Quantum confinement and fullerenelike surface reconstructions in nanodiamonds. Phys. Rev. Lett. **90**, 37401 (2003). https://doi.org/10.1103/PhysRevLett.90.037401
173. V.L. Kuznetsov, I.L. Zilberberg, Y.V. Butenko, A.L. Chuvilin, B. Segall, Theoretical study of the formation of closed curved graphite-like structures during annealing of diamond surface. J. Appl. Phys. **86**, 863 (1999). https://doi.org/10.1063/1.370816
174. Y.V. Butenko, V.L. Kuznetsov, A.L. Chuvilin, V.N. Kolomiichuk, S.V. Stankus, R.A. Khairulin, The kinetics of the graphitization of dispersed diamonds at low temperatures. J. Appl. Phys. **88**, 4380 (2000). https://doi.org/10.1063/1.1289791
175. G. Davies, T. Evans, Graphitization of diamond at zero temperature and a high pressure. Proc. R. Soc. **328**, 413 (1972). https://doi.org/10.1098/rspa.1972.0086
176. D.S. Su, N.I. Maksimova, G. Mestl, V.L. Kuznetsov, V. Keller, R. Schlogl, N. Keller, Oxidative dehydrogenation of ethylbenzene to styrene over ultra-dispersed diamond and onion-like carbon. Carbon **45**, 2145–2151 (2007). https://doi.org/10.1016/j.carbon.2007.07.005
177. K. Xu, Q. Xue, A new method for deaggregation of nanodiamond from explosive detonation: graphitization-oxidation method. Phys. Solid State **46**, 649–650 (2004). https://doi.org/10.1134/1.1711442
178. O.E. Anderson, B.L.V. Prasad, H. Sato, T. Enoki, Y. Hishiyama, Y. Kaburagi, Structure and electronic properties of graphite nanoparticles. Phys. Rev. B **58**, 16387–16395 (1998)
179. J. Qian, C. Pantea, J. Huang, T.W. Zerda, Y. Zhao, Graphitization of diamond powders of different sizes at high pressure-high temperature. Carbon **42**, 2691 (2004). https://doi.org/10.1016/j.carbon.2004.06.017
180. J. Cebik, J.K. McDonough, F. Peerally, R. Medrano, I. Neitzel, Y. Gogotsi, S. Osswald, Raman spectroscopy study of the nanodiamond-to-carbon onion transformation. Nanotechnology **24**, 205703 (2013). https://doi.org/10.1088/0957-4484/24/20/205703

181. A. Panich, A.I. Shames, N.A. Sergeev, M. Olszewski, J.K. McDonough, V.N. Mochalin, Y. Gogotsi, Nanodiamond graphitization: a magnetic resonance study. J. Phys.: Condens. Matter **25**, 245303 (2013). https://doi.org/10.1088/0953-8984/25/24/245303

182. Z. Markovic, V. Trajkovic, Biomedical potential of the reactive oxygen species generation and quenching by fullerenes (C60). Biomaterials **29**, 3561–3573 (2008). https://doi.org/10.1016/j.biomaterials.2008.05.005 (ref 156)

183. K. Yang, J. Wan, S. Zhang, B. Tian, Y. Zhang, Z. Liu, The influence of surface chemistry and size of nanoscale graphene oxide on photothermal therapy of cancer using ultra-low laser power. Biomaterials **33**, 2206–2214 (2012). https://doi.org/10.1016/j.biomaterials.2011.11.064 (ref 157)

184. C. Portet, G. Yushin, Y. Gogotsi, Electrochemical performance of carbon onions, nanodiamonds, carbon black and multiwalled nanotubes in electrical double layer capacitors. Carbon 45: 2511–2518 (2007). https://doi.org/10.1016/j.carbon.2007.08.024 (ref 158)

185. J. Zang, Y. Wang, L. Bian, J. Zhang, F. Meng, Y. Zhao, S. Ren, X. Qu, Surface modification and electrochemical behaviour of undoped Nanodiamonds. Electrochem. Acta **72**, 68–73 (2012). https://doi.org/10.1016/j.electacta.2012.03.169 (ref 159)

186. G. Su, H. Zhou, Q. Mu, Y. Zhang, L. Li, P. Jiao, G. Jiang, B. Yan, Effective surface charge density determines the electrostatic attraction between nanoparticles and cells. J. Phys. Chem. C **116**, 4993–4998 (2012). https://doi.org/10.1021/jp211041m

187. Y.Y. Liu, H. Miyoshi, M. Nakamura, Nanomedicine for drug delivery and imaging: a promising avenue for cancer therapy and diagnosis using targeted functional nanoparticles. Int. J. Cancer **120**, 2527–2537 (2007). https://doi.org/10.1002/ijc.22709

188. A.S. Barnard, Self-assembly in nanodiamond agglutinates. J. Mater Chem. **18**, 4038–4041 (2008). https://doi.org/10.1039/b809188a

189. E.D. Eidelman, V.I. Siklitsky, L.V. Sharonova, A stable suspension of single ultrananocrystalline diamond particles. Diam. Relat. Mater. **14**, 1765–1769 (2005). https://doi.org/10.1016/j.diamond.2005.08.057

190. E. Osawa, Recent progress and perspectives in single-digit nanodiamond. Diam. Relat. Mater. **16**, 2018–2022 (2007). https://doi.org/10.1016/j.diamond.2007.08.008

191. A. Pentecost, S. Gour, V. Mochalin, I. Knoke, Y. Gogotsi, Deaggregation of nanodiamond powders using salt- and sugar-assisted milling. ACS Appl. Mater. Interfaces **2**, 3289–3294 (2010). https://doi.org/10.1021/am100720n (ref 165)

192. A.T. Dideikin, A.E. Aleksenskii, M.V. Baidakova, P.N. Brunkov, M. Brzhezinskaya, V.Y. Davydov, V.S. Levitskii, S.V. Kidalov, Y.A. Kukushkina, D.A. Kirilenko, V.V. Shnitov, A. V. Shvidchenko, B. Senkovskiy, M.S. Shestakov, A.Y. Vul, Rehybridization of carbon on facets of detonation diamond nanocrystals and forming hydrosols of individual particles. Carbon **122**, 737–745 (2017). https://doi.org/10.1016/j.carbon.2017.07.013

193. R.J. Hunter, *Zeta Potential in Colloids Science* (Academic Press, NY, 1981)

194. T.M. Riddick, *Zeta-Meter Operating Manual zm-75* (Zeta-Meter Inc., New York, 1968)

195. A.V. Delgado, F. González-Caballero, R.J. Hunter, L.K. Koopal, J. Lyklema, Measurement and interpretation of electrokinetic phenomena (IUPAC Technical Report). Pure Appl. Chem. **77**, 1753–1805 (2005). https://doi.org/10.1351/pac200577101753

196. M. Ozawa, M. Inakuma, M. Takahashi, F. Kataoka, A. Krueger, E. Osawa, Preparation and behavior of brownish, clear nanodiamond colloids. Adv. Mater. **19**, 1201–1206 (2007). https://doi.org/10.1002/adma.200601452

197. T. Petit, M. Pflüger, D. Tolksdorf, J. Xiao, E.F. Aziz, Valence holes observed in nanodiamonds dispersed in water. Nanoscale **7**, 2987–2991 (2015). https://doi.org/10.1039/C4NR06639A

198. T. Petit, H. Yuzawa, M. Nagasaka, R. Yamanoi, E. Osawa, N. Kosugi, E.F. Aziz, Probing interfacial water on nanodiamonds in colloidal dispersion. J. Phys. Chem. Lett. **6**, 2909–2912 (2015). https://doi.org/10.1021/acs.jpclett.5b00820

199. V.N. Mochalin, I. Neitzel, B. Etzold, A.M. Peterson, G. Palmese, Y. Gogotsi, Covalent incorporation of aminated nanodiamond into an epoxy polymer network. ACS Nano **9**, 7494–7502 (2011). https://doi.org/10.1021/nn2024539

200. Y. Morita, T. Takimoto, H. Yamanaka, K. Kumekawa, S. Morino, S. Aonuma, T. Kimura, N. Komatsu, A facile and scalable process for size-controllable separation of nanodiamond particles as small as 4 nm. Small **4**, 2154–2157 (2008). https://doi.org/10.1002/smll. 200800944

201. N. Gibson, O. Shenderova, T.J.M. Luo, S. Moseenkov, V. Bondar, A. Puzyr, K. Purtov, Z. Fitzgerald, D.W. Brenner, Colloidal stability of modified nanodiamond particles. Diam. Relat. Mater. **18**, 620–626 (2009). https://doi.org/10.1016/j.diamond.2008.10.049

202. K. Kokubo, K. Matsubayashi, H. Tategaki, H. Takada, T. Oshima, Facile synthesis of highly water-soluble fullerenes more than half-covered by hydroxyl groups. ACS Nano **2**, 327–333 (2008). https://doi.org/10.1021/nn700151z

203. Y.F. Li, C.I. Hung, C.C. Li, W. Chin, B.Y. Wei, W.K. Hsu, A gas-phase hydrophilization of carbon nanotubes by xenon excimer ultraviolet irradiation. J. Mater. Chem. **19**, 6761 (2009). https://doi.org/10.1039/b905995d

204. L. Pospíšil, M. Gál, M. Hromadová, J. Bulíčková, V. Kolivoška, J. Cvacka, K. Nováková, L. Kavan, M. Zukalová, L. Dunsch, Search for the form of fullerene C(60) in aqueous medium. Phys. Chem. Chem. Phys. **12**, 14095–14101 (2010). https://doi.org/10.1039/c0cp00986e

205. H.A. Girard, S. Perruchas, C. Gesset, M. Chaigneau, L. Vieille, J.C. Arnault, P. Bergonzo, J. P. Boilot, T. Gacoin, Electrostatic grafting of diamond nanoparticles: a versatile route to nanocrystalline diamond thin films. ACS Appl. Mater. Interfaces **1**, 2738–2746 (2009). https://doi.org/10.1021/am900458g

206. J. Hees, A. Kriele, O.A. Williams, Electrostatic self-assembly of diamond nanoparticles. Chem. Phys. Lett. **509**, 12–15 (2011). https://doi.org/10.1016/j.cplett.2011.04.083

207. C.C. Li, C.L. Huang, Preparation of clear colloidal solutions of detonation nanodiamond n organic solvents. Coll. Surf. Physicochem. Eng. Asp. **353**, 52–56 (2010). https://doi.org/10. 1016/j.colsurfa.2009.10.019

208. A.I. Shames, A.M. Panich, V.Y. Osipov, A.E. Aleksenskiy, A.Y. Vul', T. Enoki, K. Takai, Structure and magnetic properties of detonation nanodiamond chemically modified by copper. J. Appl. Phys. **107**, 014318 (2010). https://doi.org/10.1063/1.3273486

209. C. Gaillard, H.A. Girard, C. Falck, V. Paget, V. Simic, N. Ugolin, P. Bergonzo, S. Chevillard, J.C. Arnault, RSC Adv. (2013) https://doi.org/10.1039/c3ra45158e

210. M.A. Montes-Moran, D. Suarez, J.A. Menendez, E. Fuente, On the nature of basic sites on carbon surfaces: an overview. Carbon **42**, 1219–1225 (2004). https://doi.org/10.1016/j. carbon.2004.01.023

211. C. Leon, J.M. Solar, V. Calemma, L.R. Radovic, Evidence for the protonation of basal-plane sites on carbon. Carbon **30**, 797–811 (1992). https://doi.org/10.1016/0008-6223(92)90164-R

212. V.L. Kuznetsov, Y.V. Butenko, A.L. Chuvilin, A.I. Romanenko, A.V. Okotrub, Electrical resistivity of graphitized ultra-disperse diamond and onion-like carbon. Chem. Phys. Lett. **336**, 397–404 (2001). https://doi.org/10.1016/S0009-2614(01)00135-X

213. S. Biniak, G. Szymanski, J. Siedlewskia, A. Swiatkowskib, The characterization of activated carbons with oxygen and nitrogen surface groups. Carbon **35**, 1799–1810 (1997). https://doi. org/10.1016/S0008-6223(97)00096-1

214. A. Krueger, J. Stegk, Y.J. Liang, L. Lu, G. Jarre, Biotinylated nanodiamond: Simple and efficient functionalization of detonation diamond. Langmuir **24**, 4200–4204 (2008). https:// doi.org/10.1021/la703482v

215. E. Fuente, J.A. Menendez, D. Suarez, M.A. Montes-Moran, Basic surface oxides on carbon materials: a global view. Langmuir **19**, 3505–3511 (2003). https://doi.org/10.1021/ la026778a

216. V. Chakrapani, J.C. Angus, A.B. Anderson, S.D. Wolter, B.R. Stoner, G.U. Sumanasekera, Charge transfer equilibria between diamond and an aqueous oxygen electrochemical redox couple. Science **318**, 1424–1430 (2007). https://doi.org/10.1126/science.1148841

217. T. Kondo, I. Neitzel, V.N. Mochalin, J. Urai, M. Yuasa, Y. Gogotsi, Electrical conductivity of thermally hydrogenated nanodiamond powders. J. Appl. Phys. **113**, 214307 (2013). https://doi.org/10.1063/1.4809549

218. S. Stehlik, T. Glatzel, V. Pichot, R. Pawlak, E. Meyer, D. Spitzer, B. Rezek, Water interaction with hydrogenated and oxidized detonation nanodiamonds - microscopic and spectroscopic analyses. Diam. Relat. Mater. **63**, 97–102 (2015). https://doi.org/10.1016/j. diamond.2015.08.016

219. T. Petit, L. Puskar, T. Dolenko, S. Choudhury, E. Ritter, S. Burikov, K. Laptinskiy, Q. Brzustowski, U. Schade, H. Yuzawa, N. Nagasaka, N. Kosugi, M. Kurzyp, A. Venerosy, H. A. Girard, J.C. Arnault, E. Osawa, N. Nunn, O. Shenderova, E.F. Aziz, Unusual Water Hydrogen Bond Network around Hydrogenated Nanodiamonds. J. Phys. Chem. C **121**, 5185–5194 (2017). https://doi.org/10.1021/acs.jpcc.7b00721

220. K.K. Liu, C.C. Wang, C.L. Cheng, J.I. Chao, Endocytic carboxylated nanodiamond for the labeling and tracking of cell division and differentiation in cancer and stem cells. Biomaterials **30**, 4249–4259 (2009). https://doi.org/10.1016/j.biomaterials.2009.04.056

221. Y. Yuan, X. Wang, G. Jia, J.H. Liu, T. Wang, Y. Gu, S.T. Yang, S. Zhen, H. Wang, Y. Liu, Pulmonary toxicity and translocation of nanodiamonds in mice. Diam. Relat. Mater. **19**, 291–299 (2009). https://doi.org/10.1016/j.diamond.2009.11.022

222. N. Mohan, C.S. Chen, H.H. Hsieh, Y.C. Wu, H.C. Chang, In vivo imaging and toxicity assessments of fluorescent nanodiamonds in Caenorhabditis elegans. Nano Lett. **10**, 3692–3699 (2010). https://doi.org/10.1021/nl1021909

223. V. Vaijayanthimala, P.Y. Cheng, S.H. Yeh, K.K. Liu, C.H. Hsiao, J.I. Chao, H.C. Chang, The long-term stability and biocompatibility of fluorescent nanodiamond as an in vivo contrast agent. Biomaterials **33**, 7794–7802 (2012). https://doi.org/10.1016/j.biomaterials. 2012.06.084

224. S.J. Yu, M.W. Kang, H.C. Chang, K.M. Chen, Y.C. Yu, Bright fluorescent nanodiamonds: no photobleaching and low cytotoxicity. J. Am. Chem. Soc. **127**, 17604–17605 (2005). https://doi.org/10.1021/ja0567081

225. Y. Xing, W. Xiong, L. Zhu, E. Osawa, S. Hussin, L. Dai, DNA damage in embryonic stem cells caused by nanodiamonds. ACS Nano **5**, 2376–2384 (2011). https://doi.org/10.1021/ nn200279k

226. J.A. Sergent, V. Paget, S. Chevillard, Toxicity and genotoxicity of nano-SiO2 on human epithelial intestinal HT-29 cell line. Ann. Occup. Hyg. **56**, 622–630 (2012). https://doi.org/ 10.1093/annhyg/mes005

227. L.J. Mah, A. El-Osta, T.C. Karagiannis, gammaH2AX: a sensitive molecular marker of DNA damage and repair. Leukemia **24**, 679–686 (2010). https://doi.org/10.1038/leu.2010.6

228. L. Moore, B. Grobarova, E. Shen, H.B. Man, J. Micova, M. Ledvina, J. Stursa, M. Nesladek, A. Fiserova, D. Ho, Comprehensive interrogation of the cellular response to fluorescent, detonation and functionalized Nanodiamonds **6**, 11712–11721 (2014). https://doi.org/10. 1039/c4nr02570a

229. C. Sicard-Roselli, E. Brun, M. Gilles, G. Baldacchino, C. Kelsey, H. McQuaid, C. Polin, N. Wardlow, F. Currell, A new mechanism for hydroxyl radical production in irradiated nanoparticle solutions. Small **10**, 3338–3346 (2014). https://doi.org/10.1002/smll. 201400110

230. A.M. Schrand, J.B. Lin, S.C. Hens, S.M. Hussain, Temporal and mechanistic tracking of cellular uptake dynamics with novel surface fluorophore-bound nanodiamonds. Nanoscale **3**, 435–445 (2011). https://doi.org/10.1039/c0nr00408a

231. I.P. Chang, K.C. Hwang, C.S. Chiang, Preparation of fluorescent magnetic nanodiamonds and cellular imaging. J. Am. Chem. Soc. **130**, 15476–15481 (2008). https://doi.org/10.1021/ ja804253y

232. U. Maitra, A. Jain, S.J. George, C.N. Rao, Tunable fluorescence in chromophore-functionalized nanodiamond induced by energy transfer. Nanoscale **3**, 3192–3197 (2011). https://doi.org/10.1039/c1nr10295h

233. Q. Zhang, V.N. Mochalin, I. Neitzel, I.Y. Knoke, J. Han, C.A. Klug, J.G. Zhou, P.I. Lelkes, Y. Gogotsi, Fluorescent PLLA-nanodiamond composites for bone tissue engineering. Biomaterials **32**, 87–94 (2011). https://doi.org/10.1016/j.biomaterials.2010.08.090

234. X.Q. Zhang, M. Chen, R. Lam, X.Y. Xu, E. Osawa, D. Ho, Polymer-functionalized nanodiamond platforms as vehicles for gene delivery. ACS Nano **3**, 2609–2616 (2009). https://doi.org/10.1021/nn900865g

235. H.D. Wang, Q. Yang, C.H. Niu, I. Badea, Protein-modified nanodiamond particles for Layer-by-Layer assembly. Diam. Relat. Mater. **20**, 1193–1198 (2011). https://doi.org/10.1016/j.diamond.2011.06.015

236. Y.K. Tzeng, O. Faklaris, B.M. Chang, Y. Kuo, J.H. Hsu, H.C. Chang, Superresolution imaging of albumin-conjugated fluorescent nanodiamonds in cells by stimulated emission depletion. Angew. Chem. Int. Ed. Eng. **50**, 2262–2265 (2011). https://doi.org/10.1002/anie.201007215

237. C.Y. Cheng, E. Perevedentseva, J.S. Tu, P.H. Chung, C.L. Cheng, K.K. Liu, J.I. Chao, P.H. Chen, C.C. Chang, Direct and in vitro observation of growth hormone receptor molecules in A549 human lung epithelial cells by nanodiamond labeling. Appl. Phys. Lett. **90**, 163903 (2007). https://doi.org/10.1063/1.2727557

238. D.T. Tran, V. Vermeeren, L. Grieten, S. Wenmackers, P. Wagner, J. Pollet, K.P. Janssen, L. Michiels, J. Lammertyn, Nanocrystalline diamond impedimetric aptasensor for the label-free detection of human IgE. Biosens. Bioelectron. **26**, 2987–2993 (2011). https://doi.org/10.1016/j.bios.2010.11.053

239. A.H. Smith, E.M. Robinson, X.Q. Zhang, E.K. Chow, Y. Lin, E. Osawa, J. Xi, Ho D (2011) triggered release of therapeutic antibodies from nanodiamond complexes. Nanoscale **3**, 2844 (2011). https://doi.org/10.1039/c1nr10278h

240. A. Barras, J. Lyskawa, S. Szunerits, P. Woisel, R. Boukherroub, Direct functionalization of nanodiamond particles using dopamine derivatives. Langmuir **27**, 12451–12457 (2011). https://doi.org/10.1021/la202571d

241. R.A. Shimkunas, E. Robinson, R. Lam, S. Lu, X. Xu, X.Q. Zhang, H. Huang, E. Osawa, D. Ho, Nanodiamond-insulin complexes as pH-dependent protein delivery vehicles. Biomaterials **30**, 5720–5728 (2009). https://doi.org/10.1016/j.biomaterials.2009.07.004

242. E. Perevedentseva, P.J. Cai, Y.C. Chiu, C.L. Cheng, Characterizing protein activities on the lysozyme and nanodiamond complex prepared for bio applications. Langmuir **27**, 1085–1091 (2011). https://doi.org/10.1021/la103155c

243. T.T.B. Nguyen, H.C. Chang, V.W.K. Wu, Adsorption and hydrolytic activity of lysozyme on diamond nanocrystallites. Diam. Relat. Mater. **16**, 872–876 (2007). https://doi.org/10.1016/j.diamond.2007.01.030

244. V.S. Bondar, I.O. Pozdnyakova, A.P. Puzyr, Applications of nanodiamonds for separation and purification of proteins. Phys. Solid State **46**, 758–760 (2004). https://doi.org/10.1134/1.1711468

245. R. Lam, M. Chen, E. Pierstorff, H. Huang, E. Osawa, D. Ho, Nanodiamond-embedded microfilm devices for localized chemotherapeutic elution. ACS Nano **2**, 2095–2102 (2008). https://doi.org/10.1021/nn800465x

246. H. Huang, E. Pierstorff, E. Osawa, D. Ho, Active nanodiamond hydrogels for chemotherapeutic delivery. Nano Lett. **7**, 3305–3314 (2007). https://doi.org/10.1021/nl0715210

247. M. Chen, X.Q. Zhang, H.B. Man, R. Lam, E.K. Chow, D. Ho, Nanodiamond vectors functionalized with polyethylenimine for siRNA delivery. J. Phys. Chem. Lett. **1**, 3087–3095 (2010). https://doi.org/10.1021/jz1013278

248. H. Huang, E. Pierstorff, E. Osawa, D. Ho, Protein-mediated assembly of nanodiamond hydrogels into a biocompatible and biofunctional multilayer nanofilm. ACS Nano **2**, 203–212 (2008). https://doi.org/10.1021/nn7000867

249. J. Tisler, R. Reuter, A. Lammle, F. JElezko, G. Balasubramanian, P.R. Hemmer, F. Reinhard, J. Wrachtrup, Highly efficient FRET from single NV center in nanodiamonds to single organic molecule. ACS Nano **5**, 7893–7898 (2011). https://doi.org/10.1021/nn2021259

250. S. Haziza, N. Mohan, Y. Loe-Mie, A.M. Lepagnol-Bestel, S. Massou, P. AdamM, X.L. Le, J. Viard, C. Plancon, R. Daudin, P. Koebel, E. Dorard, C. Rose, F.J. Hsieh, C.C. Wu, B. Potier, Y. Herault, C. Sala, A. Corvin, B. Allinquant, H.C. Chang, F. Treussart, M. Simonneau, Fluorescent nanodiamond tracking reveals intraneuronal transport abnormalities induced by brain-disease-related genetic risk factors. Nat. Nanotechnol. **12**, 322–328 (2016). https://doi.org/10.1038/NNANO.2016.260

251. N. Mohan, Y.K. Tzeng, L. Yang, Y.Y. Chen, Y.Y. Hui, C.Y. Fang, H.C. Chang, Sub-20-nm fluorescent nanodiamonds as photostable biolabels and fluorescence resonance energy transfer donors. Adv. Mater. **22**, 843–847 (2010). https://doi.org/10.1002/adma.200901596

252. P. Reineck, D.W.M. Lau, E.R. Wilson, K. Fox, M.R. Field, C. Deeleepojananan, V.N. Mochalin, B.C. Gibson, Effect of surface chemistry on the fluorescence of detonation nanodiamonds. ACS Nano **11**, 10924–10934 (2017). https://doi.org/10.1021/acsnano.7b04647

253. L.J. Rogers, K.D. Jahnke, M.H. Metsch, A. Sipahigil, J.M. Binder, T. Teraji, H. Sumiya, J. Isoya, M.D. Lukin, P. Hemmer, F. Jelezko, All-optical initialization, readout, and coherent preparation of single silicon-vacancy spins in diamond. Phys. Rev. Lett. **113**, 263602 (2014). https://doi.org/10.1103/physrevlett.113.263602

254. I.I. Vlasov, A.A. Shiryaev, T. Rendler, S. Steinert, S.Y. Lee, D. Antonov, Vörös M, J. Jelezko, A.V. Fisenko, L.F. Semjonova, J. Biskupek, U. Kaiser, O.I. Lebedev, I. Sildos, P. R. Hemmer, V.I. Konov, A. Gali, J. Wrachtrup, Molecular-sized fluorescent nanodiamonds. Nat. Nanotechnol. **9**, 54–58 (2014). https://doi.org/10.1038/NNANO.2013.255

255. H. Zhang, I. Aharonovich, D.R. Glenn, R. Schalek, A.P. Magyar, J.W. Lichtman, E.L. Hu, R.L. Walsworth silicon-vacancy color centers in nanodiamonds: cathodoluminescence imaging markers in the near infrared. Small **10** 1908–1913 (2014). https://doi.org/10.1002/smll.201303582

256. V.A. Davydov, A.V. Rakhmanina, S.G. Lyapin, I.D. Ilichev, K.N. Boldyrev, A.A. Shiryaev, V.N. Agafonov, Production of nano and microdiamonds with Si–V and N-V luminescent centers at high pressures in systems based on mixtures of hydrocarbon and fluorocarbon compounds. JETP Lett. **99**, 585–589 (2014). https://doi.org/10.1134/S002136401410004X

257. T.D. Merson, S. Castelletto, I. Aharonovitch, A. Turbic, T.J. Kilpatrick, A.M. Turnley, Nanodiamonds with silicon vacancy defects for nontoxic photostable fluorescent labeling of neural precursor cells. Opt. Lett. **38**, 4170–4172 (2013). https://doi.org/10.1364/OL.38.004170

258. V. Pichot, B. Risse, F. Schnell, J. Mory, D. Spitzer, Understanding ultrafine nanodiamond formation using nanostructured explosives. Sci. Rep. **3**, 2159 (2013). https://doi.org/10.1038/srep02159

259. V. Pichot, M. Comet, B. Risse, D. Spitzer, Detonation of nanosized explosive: new mechanistic model for nanodiamond formation. Diam. Relat. Mater. **54**, 59–63 (2015). https://doi.org/10.1016/j.diamond.2014.09.013

260. V. Grichko, T. Tyler, V.I. Grishko, O. Shenderova, Nanodiamond particles forming photonic structures. Nanotechnology **19**, 225201 (2008). https://doi.org/10.1088/0957-4484/19/22/225201

261. O. Faklaris, V. Joshi, T. Irinopoulou, P. Tauc, M. Sennour, H. Girard, C. Gesset, J.C. Arnault, A. Thorel, J.P. Boudou, P.A. Curmi, F. Treussart, Photoluminescent diamond nanoparticles for cell labeling: Study of the uptake mechanism in mammalian cells. ACS Nano **3**, 3955–3962 (2009). https://doi.org/10.1021/nn901014j

262. L.P. McGuinness, Y. Yan, A. Stacey, D.A. Simpson, L.T. Hall, D. Maclaurin, S. Prawer, P. Milvaney, J. Wrachtrup, F. Caruso, R.E. Scholten, L.C.L. Hollenberg, Quantum measurement and orientation tracking of fluorescent nanodiamonds inside living cells. Nat. Nanotech. **6**, 358–363 (2011). https://doi.org/10.1038/nnano.2011.64

263. S.C. Hens, G. Cunningham, T. Tyler, S. Moseenkov, V. Kuznetsov, O. Shenderova, Nanodiamond bioconjugate probes and their collection by electrophoresis. Diam. Relat. Mater. **17**, 1858–1866 (2008). https://doi.org/10.1016/j.diamond.2008.03.020

264. V.N. Mochalin, Y. Gogotsi, Wet chemistry route to hydrophobic blue fluorescent nanodiamond. J. Am. Chem. Soc. **131**, 4594–4595 (2009). https://doi.org/10.1021/ja9004514

265. L.M. Manus, D.J. Mastarone, E.A. Waters, X.Q. Zhang, E.A. Schultz-Sikma, K.W. MacRenaris, D. Ho, T.J. Meade, Gd(III)-nanodiamond conjugates for MRI contrast enhancement. NanoLett. **10**, 484–489 (2010). https://doi.org/10.1021/nl903264h

266. H.A. Girard, A. El Kharbachi, S. Garcia-Argote, T. Petit, P. Bergonzo, B. Rousseau, J.C. Arnault, Tritium labeling of detonation nanodiamonds. Chem. Commun. **50**, 2916–2918 (2014). https://doi.org/10.1039/c3cc49653h

267. S.S. Tinkle, Maximizing safe design of engineered nanomaterials: the NIH and NIEHS research perspective. Wiley Interdiscip. Rev.: Nanomedicine Nanobiotechnology **2,** 88–98 (2010). https://doi.org/10.1002/wnan.63

268. D.B. Warheit, P.J.A. Borm, C. Hennes, J. Lademann, Testing strategies to establish the safety of nanomaterials: conclusions of an ECETOC workshop. Inhal. Toxicol. **19**, 631–643 (2007). https://doi.org/10.1080/08958370701353080

Chapter 13
Surface-Modification of Nanodiamond by Amphiphilic Materials: Formation of Single Particle Layer and Polymer-Based Nanocomposite

Takeru Yunoki and Atsuhiro Fujimori

Abstract In this chapter, it focusses on nanodiamond. Nanodiamond is a type of diamond, which is a type of nano-carbon materials. In other words, nanodiamond is an allotrope of nano-carbon materials as same of fullerenes, carbon nanotubes, and graphene. The nanodiamond have universal properties of other nano-carbons typified by like an excellent thermal conductivity and also have antimicrobial properties and high refractive index properties at the same time. On the other hand, like other nanocarbons, it has extremely remarkable aggregation properties. Although it is a substance of a sphere with a minimum primary particle diameter at 5 nm prepared by detonation method, its outermost surface is covered with an adsorbed water nanolayer in order to stabilize the structure. As a result, it is dispersed in an aqueous solution, but it is difficult to disperse in an organic solvent or an organic polymer. Therefore, in this chapter, it will introduce a variety of research examples by surface modification to change nanodiamond into a easier-to-use material.

13.1 Introduction

In this chapter, it presents nine topics on surface-modification of nanodiamond by amphiphilic materials. Surface-modification of nanodiamond by organic chain improves affinity and miscibility with organic solvent and resin. At this time, the partial exposure of the hydrophilic surface of the nanodiamond imparts amphiphilicity to the particle at the same time. That is to say, organo-modification of nanoparticles will change the organization behavior of particles and the positive potential for creation of polymer-based composite materials (Fig. 13.1). By the

T. Yunoki · A. Fujimori (✉)
Graduate School of Science and Engineering, Saitama University, 255 Shimo-okubo, Sakura-ku, Saitama 338-8570, Japan
e-mail: fujimori@fms.saitama-u.ac.jp

© Springer Nature Switzerland AG 2019
N. Yang (ed.), *Novel Aspects of Diamond*, Topics in Applied Physics 121,
https://doi.org/10.1007/978-3-030-12469-4_13

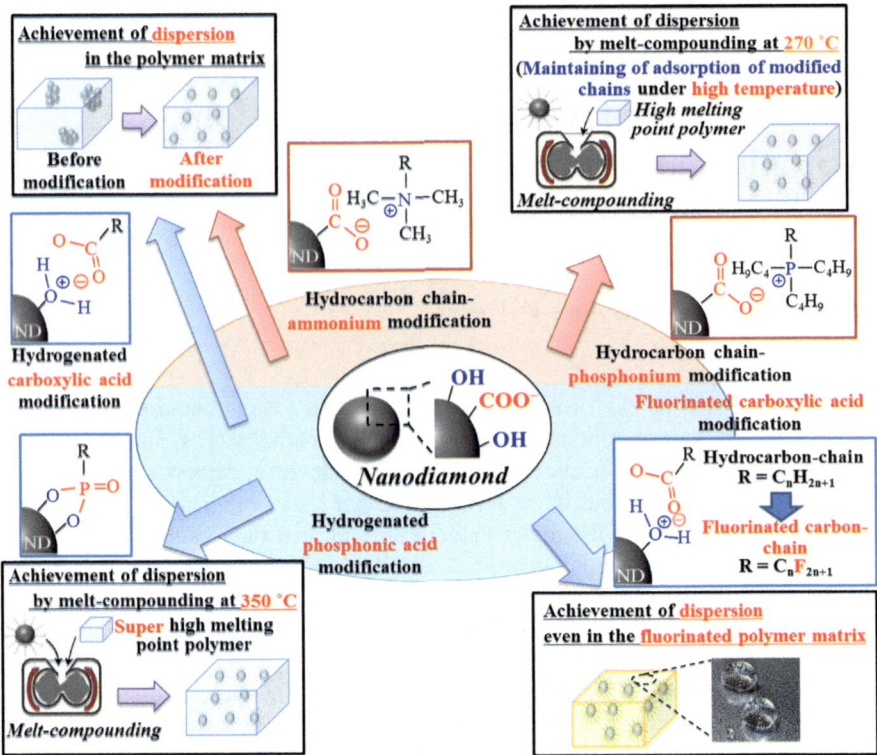

Fig. 13.1 Diagrams of application development by nanohybridization of several organo-modified nanodiamonds and their polymer-based composites

way, although there are examples in which surface-modification reaction and mesoscopic organization of organo-modified particle itself are discussed at the same time, caution should be exercised in such discussions [1, 2]. There is a hierarchical barrier [3, 4] here, which cannot be thought directly linked. To help visualize the situation, one can imagine a human with an approximate height of 1 m and an adjacent person standing 0.5 m from the first. For a molecule with a length of 1 nm and an intermolecular distance of 0.5 nm, a morphology of 1 μm as observed by electron microscopic observation is analogous to a row of people 1 or 2 km long. For the 1 mm or 1 cm long fibers that directly contribute to the development of physical properties, the corresponding row of people would be 10 or 100 km in length. It is extremely difficult to arrange humans uniformly at regular intervals and in an ordered structure over 1 km, and defects and distortions will easily occur. If a person at the end of the row twists their body, it is unlikely that this change will propagate throughout the entire 1 km row of people. In the field of crystalline polymers [5], the shape varies depending on the size, because they form a hierarchical organizational structure [6] of polymer chain → folded chain [7–9] → lamellae [10] → spherulite [11] → solid material. In order to directly link molecular chirality and mesoscopic helical fiber structure growth, this hierarchy must be followed, and cooperative molecular behavior [12, 13] without defects or distortions is necessary. In other words, organization and nanocomposite formation of nanodiamond by surface-modification derived from organic chain will be considered to have a cooperative effect beyond the hierarchy.

13.2 Fabrication of Organo-Modified Nanodiamond

It will outline the surface-modification reaction of nanodiamond by amphiphiles which correspond the core technology in this chapter. Modification reaction was carried out with long-chain carboxylic acid, fluorinated carboxylic acid, or long-chain phosphonic acid derivatives aiming at the positively charged adsorbed water nanolayer in the outermost layer of nanodiamond. Modifier molecules in the oil phase were introduced into the nano diamond dispersion medium in the aqueous phase, and reaction was carried out at the oil/water interface to obtain organo-modified nanodiamond (Fig. 13.2). Organo-nanodiamond migrated to the oil phase, and impurities were removed to isolate and purify. Confirmation of formation was made by infrared absorption spectrum, dynamic light scattering, thermogravimetric analysis, and surface pressure area-isotherms.

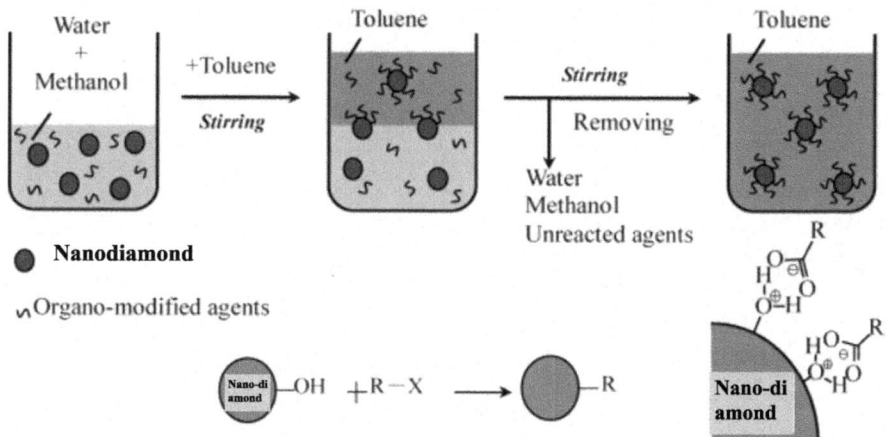

Fig. 13.2 Schematic illustration of preparation of organo-modified nanodiamond

13.3 Topics of Research on Organo-Modification of Nanodiamond by Amphiphilic Materials

13.3.1 Mono-"Particle" Dispersion of Organo-Modified Nanodiamond in Fluoropolymer Matrix of Crystalline Transparent Films of Semifluorinated Polymer/Filler Nanocomposite [14]

First of all, it introduces the research that was reported in 2014. In this study, the partially fluorinated poly(vinylidenefluoride)-based polymer/organo-modified nanodiamond composite were prepared by the melt-compounding process using fluorinated crystalline polymer and organo-nanodiamond. The structure of the polymer and the structures of dispersed organo—nanodiamond particles in the polymer matrix were investigated by using an atomic force microscope (AFM). The crystallization temperature was determined using a Differential Scanning Calorimetry (DSC). From the AFM image, the organo-nanodiamond disperses in the polymer matrix. These results suggest that organo-nanodiamond with hydrocarbon chain can be dispersed in a fluorinated copolymer. From DSC thermograms (first cooling), the crystallization peak of DSC thermograms of the composite appeared at slightly higher temperature than that of the neat fluorinated polymer. These results suggested that the polymer crystallizes epitaxially from organo—nanodiamond surface (Fig. 13.3).

13.3.2 The Role of Modifying Molecular Chains in the Formation of Organized Molecular Films of Organo-Modified Nanodiamond [15]

This topic reported in 2015 is advanced. The role of organo-modified molecular chain was went a discussion through the interfacial chemical particle-integration of

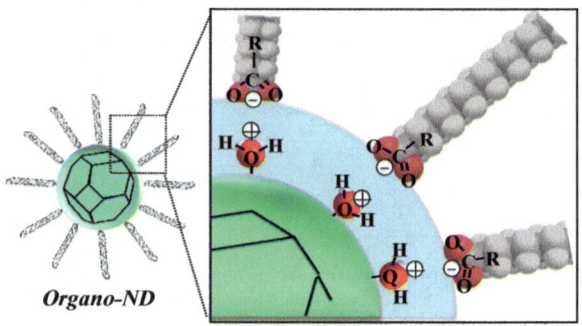

Fig. 13.3 Schematic illustration of the research concept of corresponding study

organo-modified nanodiamond with 5 nm particle size. Nanodiamond surface is covered with adsorbed water nano-layer. By the performing of organo-modification with long-chain fatty acids through the adsorption water nano-layer, nanodiamond nano-dispersed in general organic solvent as if particles were soluble in that solvent. The organo-modified nanodiamond dispersion solution was used as a spreading solution of mono-"particle" layer on the water surface, and Langmuir particle layer was integrated at the air/water interface. Next, a multi-"particle" layers by Langmuir-Blodgett technique are formed, and its fine structural analysis were performed. The effect of organo-modification become possible to that like integration and multilayers formation of inorganic nano-particles, and these origins corresponds the addition of strong van der Waals interaction between the chains of organo-modification. That is to say, by an "encounter" of a chain of organo-modification with the inorganic particles, the inorganic particles were solubilized, and were enhanced interactions between the particles. It can be said to have induced new function of organic molecules. In the case of the performing of the baking process to these particle organization, maintenance of morphology of single-particle layer was possible, and regularity of layered period was disordered. Organic chains of modification is essential for the maintenance of the layered structure (Fig. 13.4).

13.3.3 Fabrication of Transparent Nanohybrids with Heat Resistance Using High-Density Amorphous Formation and Uniform Dispersion of Nanodiamond [16]

A transparent and flexible crystalline polymer nanohybrid film containing well-dispersed nanodiamond filler with antibacterial properties, high refractive index, and high thermal conductivity was fabricated. This partially-fluorinated copolymer with a matrix composed of switchboard-type lamellae is transparent due to "high-density amorphous regions" created by drawing the film just below the melting temperature of the polymer. Although the formation of nanohybrid materials composed of fluorinated-polymer/organo-modified nanocarbon is generally difficult, we confirmed the formation, via melt-compounding, using AFM and wide-angle XRD. The nanodiamond surface had an adsorbed water layer, which we organo-modified with stearic acid. Hence, although the nanodiamond/polymer matrix nanohybrid has remarkable aggregation properties, a well-dispersed state was achieved because of improvement in wettability obtained through organo-modification. The resulting nanohybrid demonstrates transparency, increased thermal degradation temperature, and enhanced mechanical properties, which seem to be derived from the nucleation effect caused by the adsorption of the terminal polymer chain onto the organic modifier (Fig. 13.5).

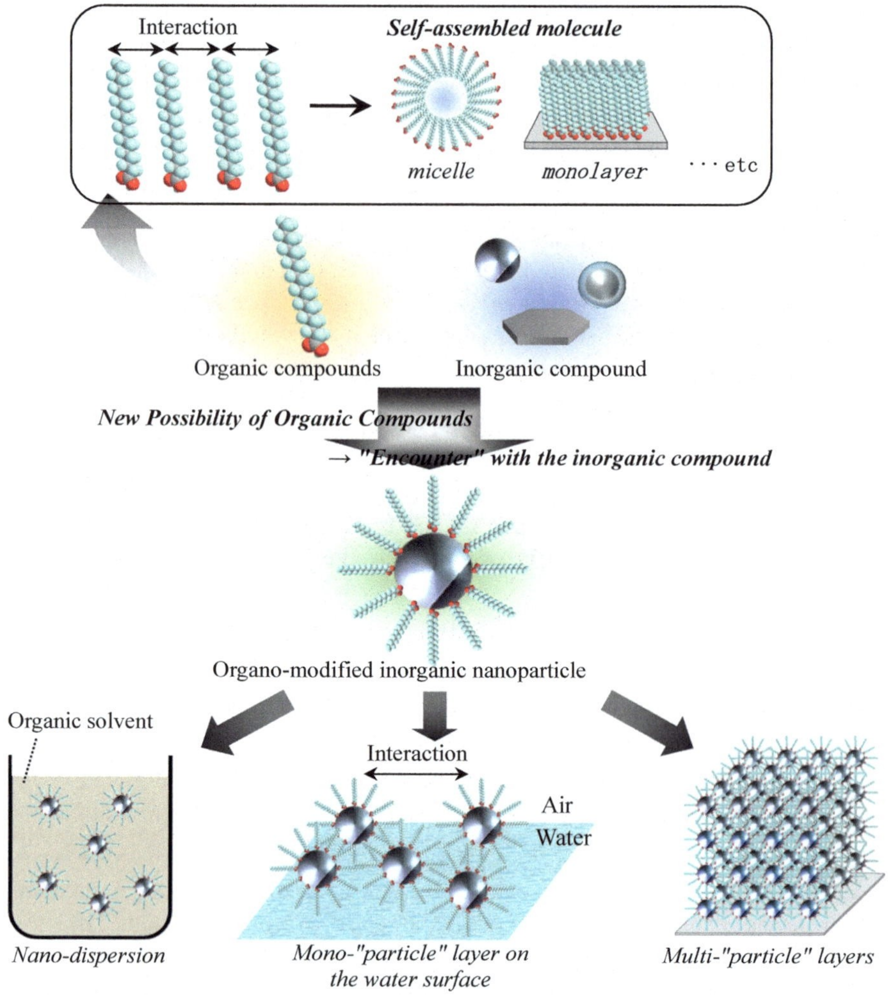

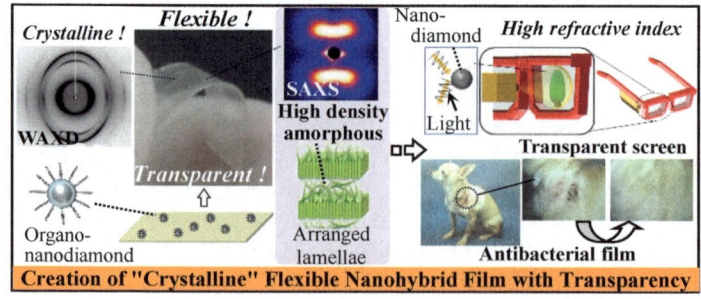

Fig. 13.4 Schematic illustration of the research concept of corresponding study

Fig. 13.5 Schematic illustration of the research strategy of corresponding study

13.3.4   Spherulitic Formation and Characterization of Partially Fluorinated Copolymers and Their Nanohybrids with Functional Fillers [17]

The next is a research example based on spherulite observation of polymer composite material using nanodiamond and clay as composite-nanofiller. Poly[vinylidenefluoride-*co*-(tetrafluoroethylene)] (P(VDFTeFE)) exhibited clear spherulitic texture with negative birefringence. The number and growth rates of the spherulites decreased at high crystallization temperature than at low crystallization temperature. Nonetheless, overall larger spherulites were found at high crystallization temperature and the brightness of the spherulites increased very fast at low crystallization temperature, thereafter it seemed as diminution of birefringence. AFM was used to investigate the impact of organo modified nanodiamond on spherulitic textures, lamellar thickness, and thickness distribution of P(VDF-TeFE) copolymer. Poly [ethylene-*co*-(tetrafluoroethylene)] (ETFE) also confirmed spherulites structure and the boundaries could be clearly observed. By incorporation of the organo modified nanodiamond and organo-modified montmorillonite in fluropolymer matrix, it was found that spherulitic texture was seriously disordered and their nanohybrids was found only to have poorly developed spherulite structure. Both of the nanohybrids samples show better crystallization temperature as compared to their neat copolymer samples. Furthermore, the incorporation of nanoparticles decreased the size of the spherulites (Fig. 13.6).

13.3.5   Dependency of Nanodiamond Particle Size and Outermost-Surface Composition on Organo-Modification [18]

The formation behavior of organized molecular film and nanocomposite with polymers have been investigated by using organo-modified nanodiamond with several particle size and outermost-surface composition. Nanodiamond used in this study indicates 3 and 5 nm particle size and contains –OH and –COOH groups in the outermost surface. Organo-modification of nanodiamond is performed to –OH$_2^+$ cation and –COO$^-$ anion on outermost surface by carboxylic anion of fatty acid and long-chain phosphonium cation, respectively. The surface of nanodiamond is known to be covered with a nano-layer of adsorbed water. This water nano-layer was exploited for organo-modification of nanodiamond with long-chain fatty acids via adsorption, leading to nano-dispersion of nanodiamond in general organic solvents as a mimic of solvency. Multi-"particle" layers were then formed via the Langmuir-Blodgett technique, and were subjected to fine structural analysis. The effect of organo-modification enabled integration and multilayer formation of inorganic nano-particles due to enhancement of the van der Waals interactions

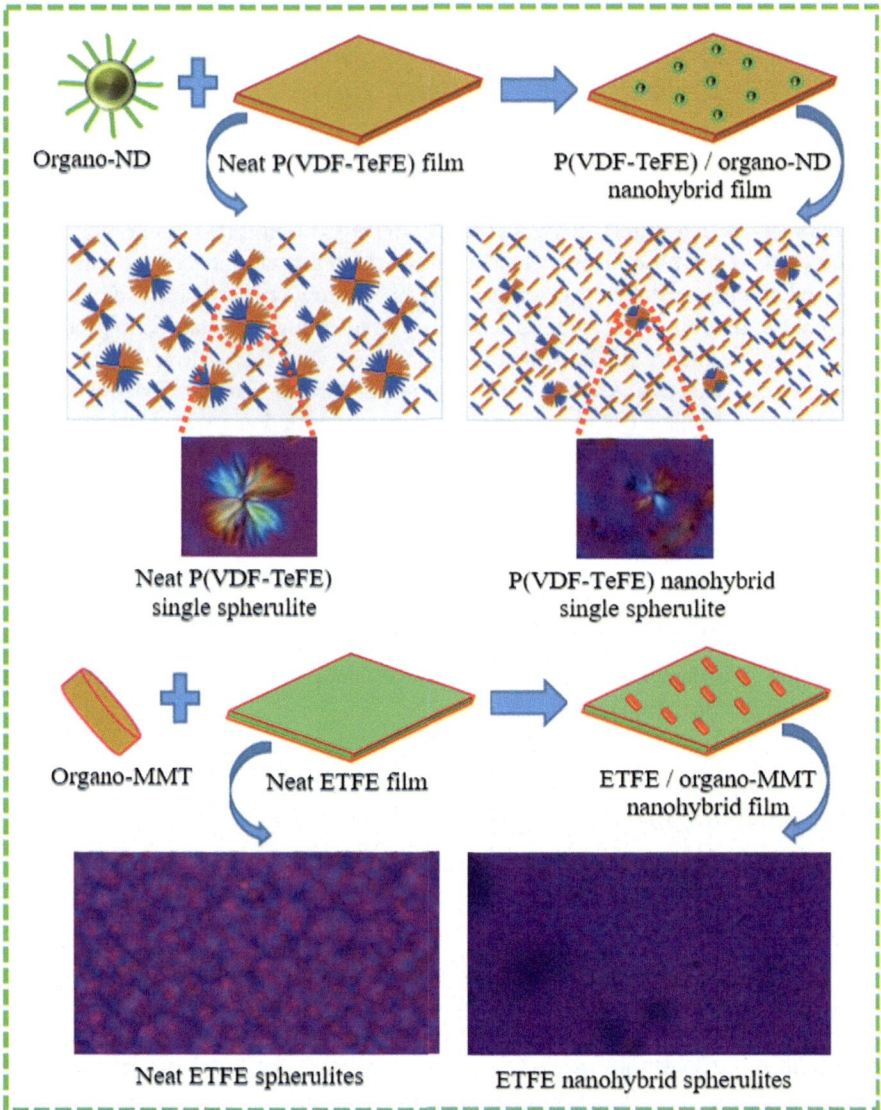

Fig. 13.6 Schematic illustration of the research strategy of corresponding study

between the chains. Therefore, the "encounter" between the organo-modifying chain and the inorganic particles led to solubilization of the inorganic particles and enhanced interactions between the particles, which can be regarded as imparting new function to the organic molecules. Nanocomposite fabrication with transparent crystalline polymer based on the nano-dispersion of nanodiamond into the matrix has also attained by this organo-modification. The resulting transparent

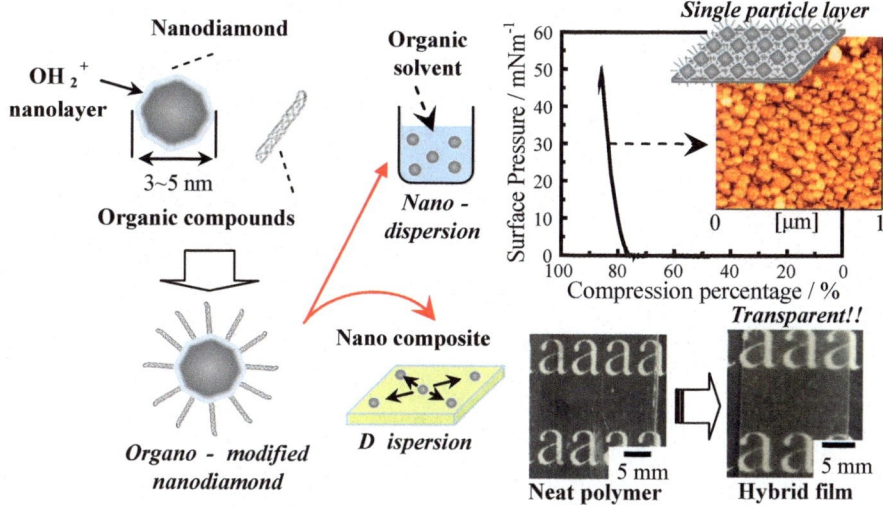

Fig. 13.7 Schematic illustration of the research strategy of corresponding study

nanocomposite indicates the enhancement of crystallization degree and improved crystallization temperature based on the nucleation effect (Fig. 13.7).

13.3.6 Nanodispersion in Transparent Polymer Matrix with High Melting Temperature Contributing to the Hybridization of Heat-Resistant Organo-Modified Nanodiamond [19]

Surface modification of outermost surface of nanodiamond was achieved by long-chain phosphonic acid, where thermal desorption of the modified chain was suppressed until 350 °C. The phosphonic acid-modified nanodiamond formed a nanohybrid by melt-compounding with transparent polymer having a high melting point over 230 °C. This nanohybrid containing nanodiamond was maintained in transparency, and the nanoparticle aggregation size on the surface was indicated at the range of 40–70 nm. The melting temperature of the nanohybrid raised with respect to the matrix polymer, and the D_{200} crystallite size was also improved. In addition, mechanical properties and thermal degradation temperature are also increased, it is thought to be based on the effect of well-dispersed state. Furthermore, the nanohybrid showed image projection ability derived from nanodiamond with high refractive index. Darkening due to carbonization was observed in the nanohybrid with the crystalline fluorinated polymers, but it was solved by complexing with modified nanodiamond with fluorinated long-chain phosphonic

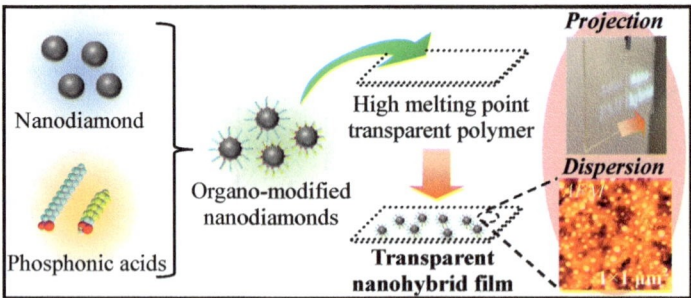

Fig. 13.8 Schematic illustration of the research strategy of corresponding study

acid. It is assumed that the desorbed modified chain was incorporated into the fluoropolymer with high molten viscosity and caused carbonization, and it seems that the miscible fluorinated-modified chain solved this problem (Fig. 13.8).

13.3.7 Nanodispersion of Fluorinated Phosphonate-Modified Nanodiamond in Crystalline Fluoropolymer Matrix to Achieve a Transparent Polymer/Nanofiller Composites [20]

Creation of polymer/organo-modified nanodiamond "well-dispersed" composite by using crystalline fluorinated polymers as matrix was investigated. All of the crystalline fluorinated polymers used in this study are heat-resistant ones that become transparent by high temperature drawing. The polyvinylidene fluoride-based copolymer used in this study had a relatively low melting point as a fluororesin and was easy to handle. Both fluorocarbon-modified and hydrocarbon-modified nanodiamond formed well-dispersed nanocomposites from the similarity of the crystal structure and the affinity between the terminal end of polymer chain and the modified chain. However, the creation of nanocomposite by combination between hydrocarbon-modified fine particles and fluorinated polymer was not universal, and this system was special. Ethylene-tetrafluoroethylene and perfluoroalkoxyalkane polymers showing high melting point cannot form a nanohybrid with hydrocarbon-modified inorganic nanoparticles. In addition, in order to form a nanocomposite by melt-compounding method, it was necessary to remarkably increase the desorption temperature of the modified chain of the outermost surface of organo-nanodiamond. Phosphonic acid containing Fluorocarbon chain-modified nanodiamond satisfied all these requirements. It was confirmed that the obtained nanocomposite had enhanced physical properties and imparted unique characteristics of nanodiamond (Fig. 13.9).

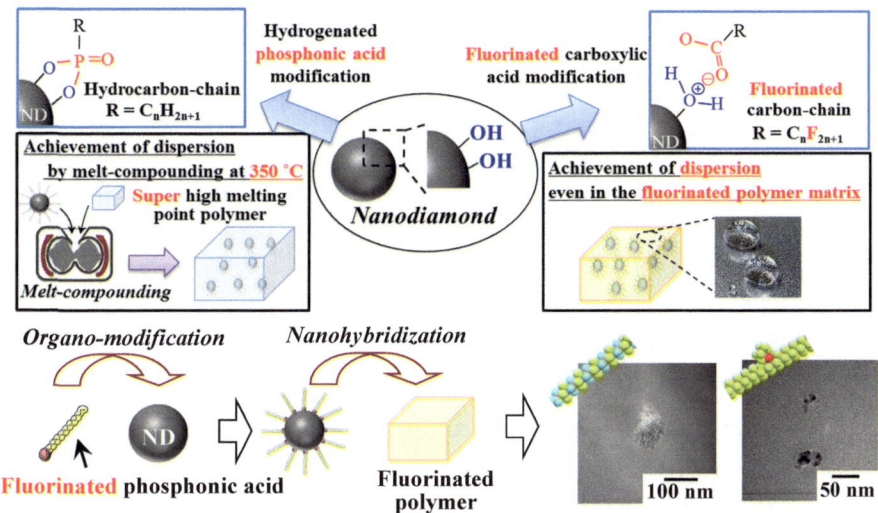

Fig. 13.9 Schematic illustration of the research strategy of corresponding study

13.3.8 Thermal Stability of Ordered Multi-particle Layers of Long-Chain Phosphonate-Modified Nanodiamond with Superior Heat-Resistance [21]

Maintenance properties of layered regularity under heating in the organized film of heat-resistant organo-modified nanodiamond were investigated. In long-chain phosphonic acid-modified nanodiamond, desorption of the modified-molecular chain by the temperature elevation was suppressed until 350 °C. It was innovative improvement compared the desorption behavior at 155 °C of the chain of stearic acid-modified nanodiamond. When layered organization is formed by technique of two-dimensional molecular film, clear layered period is shown, and it has been simultaneously confirmed an existence of unmodified-chain of impurities which cannot be removed. Temperature controlled X-ray diffraction and IR measurements showed that impurities can be fired and removed. Due to annealing effect based on heating above 100 °C, layered order has been enhanced by rearrangement of organo-nanodiamonds. Layered regularity of the organo-particle layers under heating was maintained up to 175 °C, far exceeding the value of 50 °C of the organized film of stearic acid-modified nanodiamond. This technique is expected to be developed for technology of single layer coating at the outermost layer with nanodiamond which is excellent in antibacterial property, high refractive index, and thermal conductivity (Fig. 13.10).

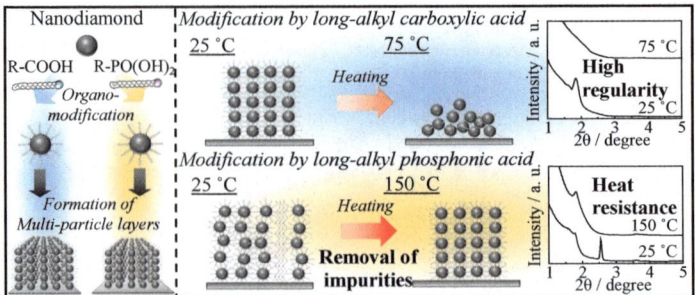

Fig. 13.10 Schematic illustration of the research strategy of corresponding study

13.3.9 Correlation Between Nanodispersion of Organo-Modified Nanodiamond in Solvent and Condensed Behavior of Their Organized Particle Films [22]

Organo-nanodiamond by surface modification using several surfactants have achieved nanodispersion in an organic solvent as if these nanoparticles were dissolved. A high surface coverage resulted in a reduction in the particle aggregation size of the organo-nanodiamond particles in the solvent and their modification by fluorocarbon chain resulted in achieving nanodispersion in a nonpolar solvent. The resulting dispersion medium could be utilized as a spreading solvent for single particle layers on the water surface of various organo-nanodiamonds. From this estimation, it was found that a close relation was found between the dispersion characteristics of the organo-nanoparticles in the solvent and the condensation behavior in their two-dimensional single particle layer. Organo-nanodiamond with high dispersion ability in solvent is inferior in two-dimensional condensability as single particle layers. Retention time is necessary for forming a homogeneous single particle layers on the water surface. The actual meaning of this property was

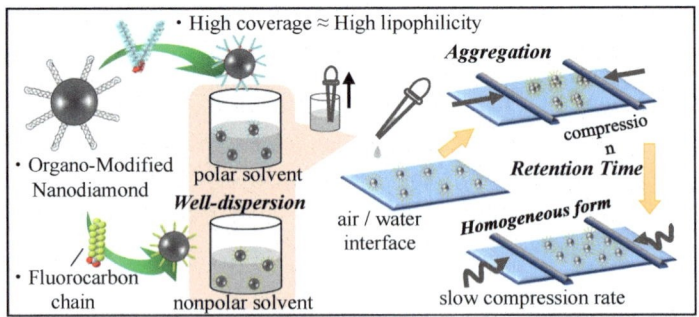

Fig. 13.11 Schematic illustration of the research strategy of corresponding study

found to be secondary aggregation of the organo-modified nanoparticles at the air/water interface. Since lipophilicized nanodiamond is not necessarily stable on the water surface, increasing of aggregated particle size and three dimensionalization were confirmed in the case that the particle compression rate during two-dimensional integration was relative fast (Fig. 13.11).

13.4   Summary

In this chapter, it presented the latest nine topics on formation of interfacial single particle layer and polymer-based nanocomposites by surface-modification of nanodiamond. Nanodiamond exhibits properties of antibacterial property and high refractive index in addition to the universal characteristics of nano-carbons, so it is expected to create practical materials (Fig. 13.12). The surface modification technology introduced in this chapter will be a core technology contributing to this realization.

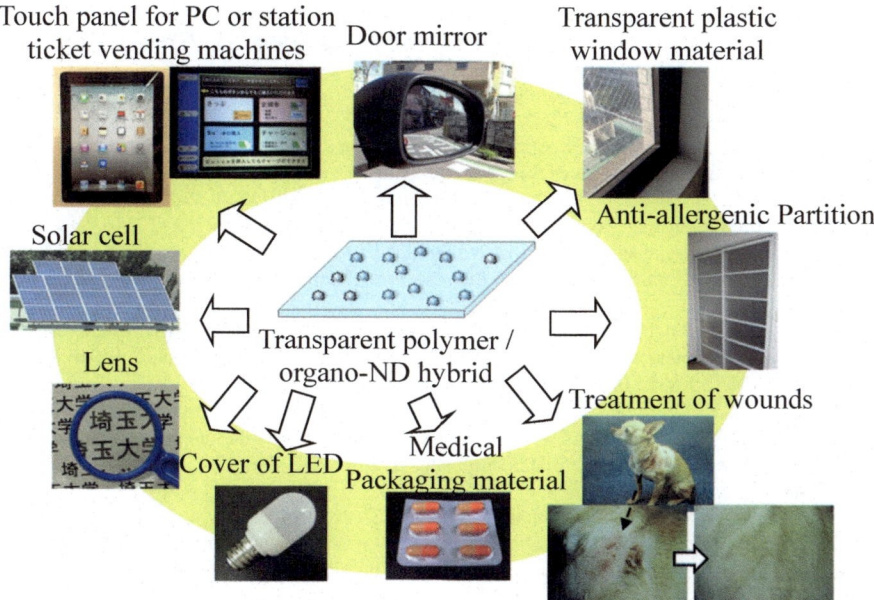

Fig. 13.12 Schematic illustration of varied applications of an antibacterial, transparent, and flexible nanohybrid with heat-resistance

Acknowledgements The authors appreciate Prof. Shuichi Akasaka, Tokyo Institute of Technology for research collaboration in many related works. Further, authors also thank Dr. Yuji Shitara, Mr. Akira Tada, Mr. Takumi Yamamoto, Mr. Tatsuki Nakajima, JXTG Energy Co. Ltd., for useful discussion. In addition, authors appreciate Mr. Koichi Umemoto, and Dr. Daisuke Shiro, DAICEL Co. Ltd., for the providing of nanodiamond samples. Finally, authors thank Prof. Tomofumi Ugai, Bio-nano electronics research center, Toyo University, for the lecture of DLS measurement.

References

1. S. Abele, D. Seebach, Preparation of achiral and of enantiopure geminally disubstituted beta-amino acids for beta-peptide synthesis. Eur. J. Org. Chem. **1**, 1–15 (2000)
2. Y. Zheng, L. Lin, X. Ye, F. Guo, X. Wang, Helical graphitic carbon nitrides with photocatalytic and optical activities. Angew. Chem. Int. Ed. **53**, 11926–119230 (2014)
3. H. Fenniri, B.L. Deng, A.E. Ribbe, Helical rosette nanotubes with tunable chiroptical properties. J. Am. Chem. Soc. **124**, 11064–11072 (2002)
4. W.J. Chung, J.W. Oh, K. Kwak, B.Y. Lee, J. Meyer, E. Wang, A. Hexemer, S.W. Lee, Biomimetic self-templating supramolecular structures. Nature **478**, 364–368 (2011)
5. E. Baer, A. Hiltner, H.D. Keith, Hierarchical structure in polymeric materials. Science **235**, 1015–1022 (1987)
6. J. Ruokolainen, G. Ten Brinke, O. Ikkala, Supramolecular polymeric materials with hierarchical structure-within-structure morphologies. Adv. Mater. **11**, 777–780 (1999)
7. A. Keller, Philos. Mag. **2**, 1171–1175 (1957)
8. P.H. Till, J. Polym. Sci. **24**, 301–306 (1957)
9. E.W.Z. Fischer, Naturforsch. **12**(a), 753–754 (1957)
10. J.D. Hoffman, R.L. Miller, Kinetics of crystallization from the melt and chain folding in polyethylene fractions revisited: theory and experiment. Polymer **38**, 3151–3212 (1997)
11. P.J. Barham, A. Keller, E.L. Otun, P.A. Holmes, Crystallization and morphology of a bacterial thermoplastic—poly-3-hydroxybutyrate. J. Mater. Sci. **19**, 2781–2794 (1984)
12. S.J. George, A. Ajayaghosh, Self-assembled nanotapes of oligo(p-phenylene vinylene)s: Sol-gel controlled optical properties in fluorescent π-electronic gels. Chem. Eur. J. **11**, 3217–3227 (2005)
13. S.H. Tolbert, P. Sieger, G.D. Stucky, S.M.J. Aubin, C.C. Wu, D.N. Hendrickson, Control of inorganic layer thickness in self-assembled iron oxide surfactant composites. J. Am. Chem. Soc. **119**, 8652–8661 (1997)
14. T. Kanehira, Y. Soutome, N. Honda, S. Akasaka, A. Fujimori, Mono-particle dispersion of organo-modified nanodiamond in fluoropolymer matrix of crystalline transparent films of semifluorinated polymer/filler nanocomposite. Trans. Mat. Res. Soc. Japan **39**, 231–234 (2014)
15. A. Fujimori, Y. Kasahara, N. Honda, S. Akasaka, The role of modifying molecular chains in the formation of organized molecular films of organo-modified nanodiamond. Construction of a highly-ordered low defect particle layer, and evaluation of desorption behavior of organic chains–. Langmuir **31**, 2895–2904 (2015)
16. M.A.A. Mamun, Y. Soutome, Y. Kasahara, Q. Meng, S. Akasaka, A. Fujimori, Fabrication of transparent nanohybrids with heat resistance using high-density amorphous formation and uniform dispersion of nanodiamond. ACS Appl. Mater. Interfaces. **7**, 17792–17801 (2015)
17. M.A.A. Mamun, Y. Kasahara, T. Tasaki, A. Fujimori, Spherulitic formation and characterization of partially fluorinated copolymers and their nanohybrids with functional fillers. Polym. Eng. Sci. **57**, 161–171 (2017)

18. T. Tasaki, Y. Guo, Q. Meng, M.A.A. Mamun, Y. Kasahara, S. Akasaka, A. Fujimori, Dependency of nanodiamond particle size and outermost-surface composition on organo-modification—evaluation by formation of organized molecular films and nano-hybridization with organic polymers. ACS Appl. Mater. Interfaces. **9**, 14379–14390 (2017)
19. Y. Kasahara, Y. Guo, T. Tasaki, Q. Meng, M.A.A. Mamun, M. Iizuka, S. Akasaka, A. Fujimori, Nano-dispersion in transparent polymer matrix with high melting temperature contributing to hybridization of heat-resistant organo-modified nanodiamond. Polym. Bull. **75**, 4145–4163 (2018)
20. T. Tasaki, Y. Guo, H. Machida, S. Akasaka, A. Fujimori, Nano-dispersion of fluorinated phosphonate-modified nanodiamond in crystalline fluoropolymer matrix of polymer/nanofiller transparent hybrid. Polym. Compos. **39**, (2018). in press. https://doi.org/10.1002/pc.25047
21. Y. Guo, K. Fukushi, S. Hirayama, H. Machida, Q. Meng, S. Akasaka, A. Fujimori, Thermal stability of ordered multi-particle layers of long-chain phosphonate-modified nanodiamond with superior heat-resistance. Colloids Surf. A **556**, 227–238 (2018)
22. H. Machida, Y. Abiko, S. Hirayama, Q. Meng, S. Akasaka, A. Fujimori, Correlation between nanodispersion of organo-modified nanodiamond in solvent and condensed behavior of their organized particle films. Colloids Surf. A (2018). in press. https://doi.org/10.1016/j.colsurfa.2018.11.003

Chapter 14
Electrochemical Applications of Conductive Diamond Powders

Takeshi Kondo

Abstract Heavily boron-doped diamond (BDD) has drawn attention as a functional electrode material that exhibits unique electrochemical properties such as wide potential window and low background current. A BDD electrode is usually fabricated by depositing a polycrystalline BDD thin film on a flat substrate, such as a silicon wafer, using the chemical vapor deposition (CVD) method. However, processing a BDD electrode to form an arbitrary shape is difficult, and it has limited practical applications. In contrast, BDD powder (BDDP) emerges as a promising electrode material for the creation of flexible and inexpensive BDD electrodes; the electrode ink is either painted or printed onto various types of substrates. In addition, the BDDP can be used as an electrode material with a large specific surface area. In this chapter, the preparation of BDDP and its application to printed electrodes and as fuel cell catalyst support is described.

14.1 Introduction

Although diamond is an insulating material, electric conductivity can be imparted to it by doping with boron at a high concentration. Heavily ($>2 \times 10^{20}$ cm^{-3}) boron-doped diamond (BDD) exhibits sufficient conductivity [1] and thus can be used as an electrochemical electrode material. BDD electrodes are known to exhibit unique electrochemical properties, such as wide potential window in aqueous and non-aqueous electrolytes and low background current, as well as extreme physical and chemical stability [2]. On the basis of these features, the application of BDD electrodes as highly sensitive electrodes for electrochemical analysis [3–6] and

T. Kondo (✉)
Department of Pure and Applied Chemistry, Faculty of Science and Technology,
Tokyo University of Science, Noda, Chiba 278-8510, Japan
e-mail: t-kondo@rs.noda.tus.ac.jp

© Springer Nature Switzerland AG 2019
N. Yang (ed.), *Novel Aspects of Diamond*, Topics in Applied Physics 121,
https://doi.org/10.1007/978-3-030-12469-4_14

highly efficient electrode materials for electrolytic systems [7–11] has been intensively studied. BDD electrodes are normally obtained by chemical vapor deposition
(CVD) of a polycrystalline BDD thin film on a flat conductive substrate. High
temperature (800 °C or higher) is required for BDD growth process to occur when
CVD method is used; therefore, the substrate material is virtually limited to conductive silicon wafer and niobium plate. In addition, the processability of the BDD
electrodes is poor because of their hardness; hence, it is difficult to obtain different
electrode configurations and to lower the production cost. To overcome this limitation, conductive diamond powder (DP) can be used as an electrode material [12].
BDD electrodes of various shapes and sizes can be fabricated by applying an ink
containing BDD powder (BDDP) onto a conductive substrate. However, the electrodes produced using CVD are expensive, and their size is limited to the dimension
of the CVD chamber. In contrast, the size of a painted BDD electrode is ideally
unlimited, and cheaper substrate materials can be used to suppress the production
cost. Furthermore, BDDP is useful for its application to a three-dimensional electrode, such as a catalyst support for fuel cells [13–15].

Several reports on the preparation of BDDP can be found in the literature. Heyer
et al. reported that BDD nanoparticles with the size as low as to 10–60 nm can be
obtained via the bead milling of BDD thin films [16]. Ekimov et al. showed that
BDD nanoparticles with size down to 10 nm can be synthesized via the
high-pressure high-temperature treatment of a sp^3-configurated organic compound
containing boron [17]. Although the BDD particles prepared using these methods
most likely contain boron, their electrochemical properties have not been investigated yet. In contrast, conductive DP was efficiently prepared by growing a BDD
layer on the surface of insulating DP as a substrate material using CVD (Fig. 14.1).
The insulating DP is commercially available as diamond abrasive powder in a wide
range of sizes. In addition, since only the surface of the substrate needs to be coated
with the BDD layer, a relatively large amount of the conductive DP can be produced. Fischer and Ay reported that the BDDP exhibits electrochemical characteristics similar to those of BDD thin films [18, 19]. In this chapter, the preparation,
characterization, and basic electrochemical characteristics of the BDDP produced
by BDD growth on insulating DP substrate are described. In addition, its application to screen-printed electrodes and catalyst supports for fuel cells are explored.

Fig. 14.1 Schematic of the
preparation of BDDP

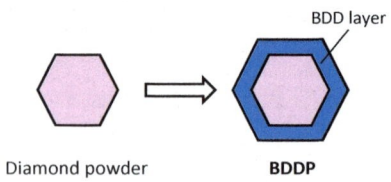

Diamond powder   **BDDP**

14.2   Preparation and Electrochemical Properties of BDDP

14.2.1   Preparation of BDDP

BDDP can be prepared by depositing a BDD layer on the surface of an insulating DP substrate using microwave plasma-assisted CVD (MPCVD). 0.8 g of DP with a size of <0.5 μm was used as the substrate. A mixture of acetone and methanol (9:1, v/v) containing trimethoxyborane was used as the carbon/boron source; a similar method was used in the preparation of conventional BDD thin films. A more detailed preparation of BDDP is described in a previous study [12]. To lower the amount of sp^2 carbon impurities generated during BDD growth in the as-deposited BDDP, the specimen was heat treated in air at 450 °C, which afforded high-quality BDDP.

14.2.2   Characterization of BDDP

The size of the newly prepared BDDP was evaluated using a laser diffraction particle-size distribution analyzer. The average size of the DP substrate was 0.23 μm; the substrate size increased from 0.23 to 0.35 μm after 8 h of BDD deposition (Fig. 14.2). This result was consistent with the scanning electron microscopy (SEM) observation (Fig. 14.3). A detailed observation of BDDP revealed that the roughness of BDDP surface increased in several tens of nanometers after CVD was performed. These results indicate that a nanocrystalline BDD layer grew on the surface of the DP substrate. The Raman spectrum of BDDP shows bands for the sp^2 carbon content, which may exists in the grain boundary of

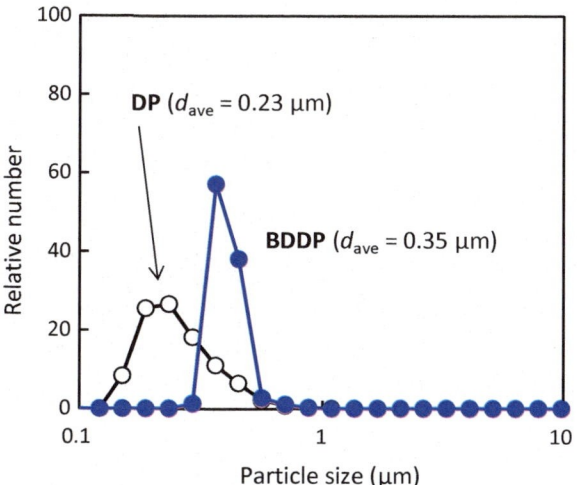

Fig. 14.2 Particle size distribution of diamond powder (DP) substrate and BDDP

Fig. 14.3 SEM images of
a DP and **b** BDDP. Panels
c and **d** are magnified images
of (**a**) and (**b**), respectively

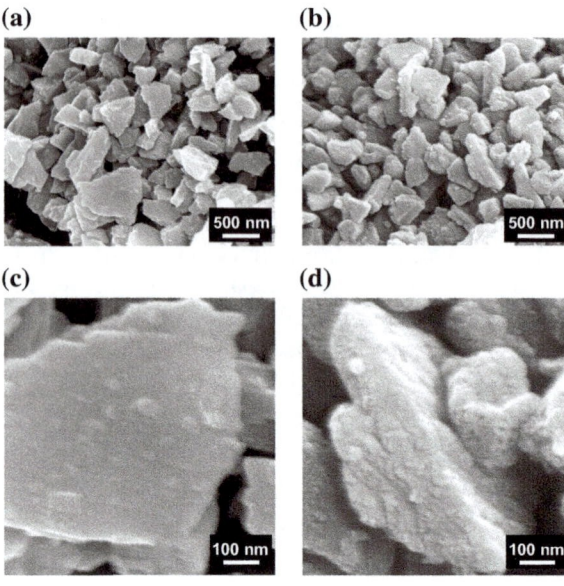

Fig. 14.4 Raman spectrum
of BDDP

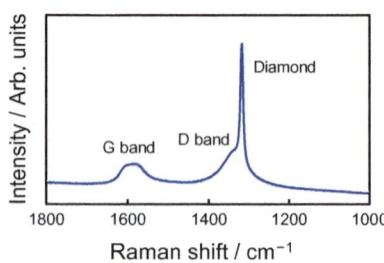

the (nanocrystalline) BDD layer (D and G bands at 1350 and 1600 cm^{-1}, respectively) (Fig. 14.4). The conductivity of BDDP was evaluated by measuring the DC electrical resistance between the ends of a packed BDDP in a glass tube (inner diameter of 1.0 mm). The calculated conductivity was typically 0.2–0.4 S cm^{-1}; such conductivity is sufficient for application to electrode material for electrochemical analysis.

The basic electrochemical properties of BDDP were assessed via cyclic voltammetry (CV) of a BDDP-paste electrode; this electrode was prepared by filling a paste containing BDDP with liquid paraffin into a cavity electrode. In addition to the quasi-reversible CV behavior of $Fe(CN)_6^{3-/4-}$, the CV showed a wide potential window and flat background current. These electrochemical properties are similar to those of conventional BDD (Fig. 14.5).

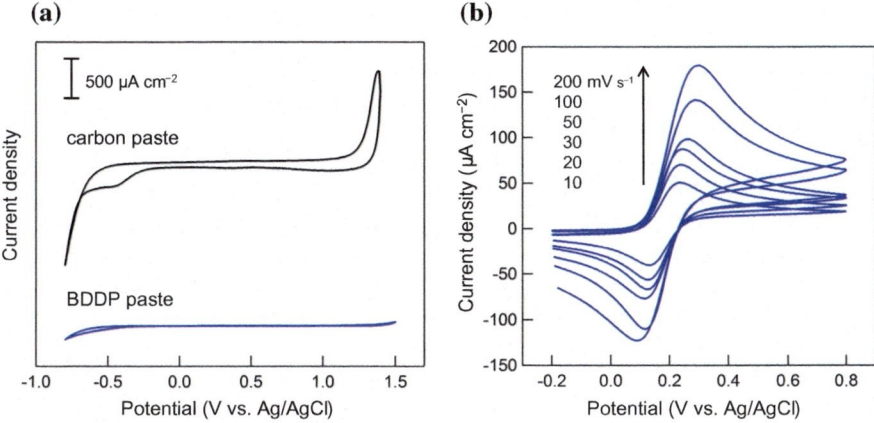

Fig. 14.5 a CV in 0.5 M Na₂SO₄ at carbon- and BDDP-paste electrodes. Potential sweep rate was 100 mV s⁻¹. **b** CV for 1 mM K₄Fe(CN)₆ in 0.5 M Na₂SO₄ at BDDP-paste electrode

14.3 Application to Screen-Printed Electrodes

14.3.1 Fabrication and Electrochemical Properties of BDDP-Printed Electrodes

As described above, BDDP is expected to be used for a variety of applications with various electrode configurations. To explore one of these possible applications, we developed a screen-printed diamond electrode as a disposable and sensitive electrochemical platform using a BDDP ink [12]. In general, screen-printed electrodes can be fabricated at low cost. When used as disposable sensors, these electrodes are characterized by a range of advantages such as versatility, miniaturization, high reproducibility, and flexibility to complex designs. In particular, screen-printed carbon electrodes, which are suitable for portable electrochemical sensors, are widely applied in clinical and biological areas [20–22], as well as in environmental [23–25] and food analysis fields [24, 26].

For the preparation of BDDP ink, an insulating polyester resin (PES) binder was dissolved in 2-butanone and isophorone. BDDP was then added and dispersed in the PES solution, affording the BDDP ink. The weight ratio of PES/BDDP, typically of 0.3, can be varied within a certain range. BDDP-printed electrodes were prepared using a screen printer. Figure 14.6 shows a representative configuration of a BDDP-printed electrode. Carbon paste (current collector), silver paste (electrical lead), resist (insulating mask), and BDDP ink were successively printed on a polyimide or polyethylene terephthalate film substrate. After each printing step, the ink was allowed to dry by heating the substrate at 120 °C for 30 min. Detailed information on the electrode preparation is described in our previous report [12].

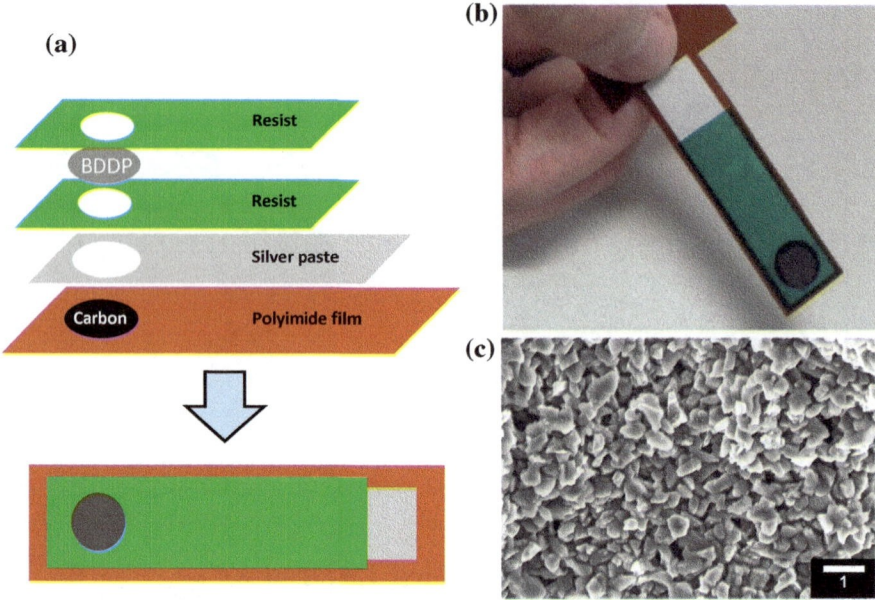

Fig. 14.6 a Schematic of the fabrication of a BDDP-printed electrode. **b** Image of a BDDP-printed electrode. **c** SEM image of a BDDP-printed electrode surface

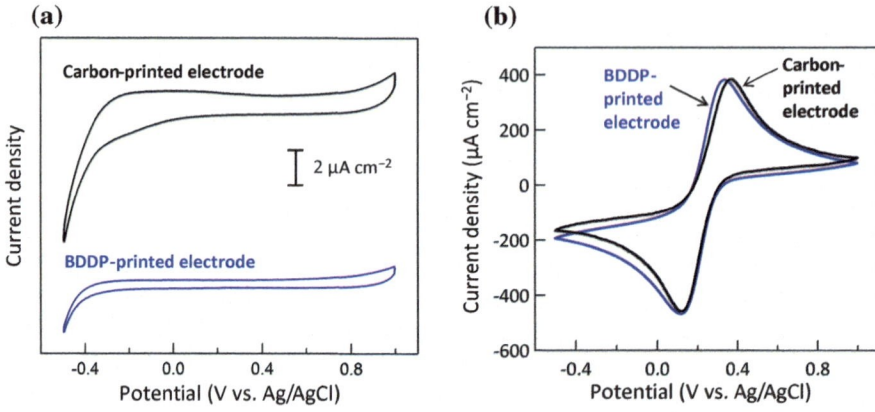

Fig. 14.7 CV in 0.5 M Na$_2$SO$_4$ at carbon- and BDDP-printed electrodes **a** in the absence and **b** in the presence of 0.5 mM K$_3$Fe(CN)$_6$. Potential sweep rate was 50 mV s^{-1}

Figure 14.7a shows the CV of the BDDP-printed electrode recorded in 0.5 M Na$_2$SO$_4$, together with that of a carbon-printed electrode for comparison. The background current was smaller for the BDDP-printed electrode than that for the conventional carbon-printed electrode. Furthermore, the potential window was slightly wider at the BDDP-printed electrode. These properties are consistent with those of a conventional BDD electrode. The CV for $Fe(CN)_6^{3-/4-}$ at the BDDP-printed electrode exhibited quasi-reversible behavior (Fig. 14.7b). The signal-to-background (S/B) ratio (defined herein by the anode peak current and background current at the peak potential, respectively) was 1909 and 436 for the BDDP- and the carbon-printed electrodes, respectively. These ratios demonstrate the potential of the BDDP-printed electrode as a disposable and sensitive electrode platform for electrochemical sensors.

14.3.2 Application to Glucose Detection

Glucose is an extremely important analyte, particularly relevant to human health. Measuring glucose levels in blood is an indispensable part of diabetes treatment; a portable glucose sensor with a disposable screen-printed electrode is used for such measurements [21, 22, 27–29]. This prompted us to fabricate a glucose-sensing electrode with a modified BDDP-printed electrode [30]. Owing to its large over-potential, glucose cannot be directly detected electrochemically using a BDD electrode. Therefore, glucose oxidase, which catalyzes the oxidation of glucose with the concomitant generation of hydrogen peroxide (H_2O_2), was used herein (14.1). Although the electrochemical oxidation of hydrogen peroxide does not occur at the bare BDD electrode because of the large overpotential, it can be promoted at lower potentials by modifying the BDD surface with a suitable electrocatalyst such as cobalt phthalocyanine (CoPc) (14.2). Therefore, we prepared a GOx-modified CoPc/BDDP-printed electrode to be used as a glucose sensor [30].

$$Glucose + O_2 \rightarrow gluconolactone + H_2O_2 \qquad (14.1)$$

$$H_2O_2 \rightarrow O_2 + 2H^+ + 2e^- \qquad (14.2)$$

For the preparation of the CoPc/BDDP-printed electrode, an ink containing 100 mg of BDDP and 6 mg of CoPc was prepared. Next, the CoPc/BDDP-printed electrode was coated with a GOx-immobilized poly(p-phenylenediamine) (PPD) layer via electropolymerization as follows: ten potential sweep cycles were performed on the CoPc/BDDP-printed electrode in a 0.1 M phosphate buffer solution (PBS) containing 0.15 mM p-phenylenediamine, 0.15 mM resorcinol, and 500 U mL^{-1} GOx from 0 to +0.6 V versus Ag/AgCl at 20 mV s^{-1}. Figure 14.8 schematizes the GOx-CoPc/BDDP-printed electrode.

The CV of 5 mM hydrogen peroxide in 0.1 M PBS at the CoPc/BDDP-printed electrode exhibited an anodic peak at +0.7 V versus Ag/AgCl. The S/B ratio for hydrogen peroxide detection at CoPc/BDDP-printed electrode was found to be higher than that at CoPc/carbon-printed electrode. A high S/B ratio demonstrates the high sensitivity of the CoPc/BDDP-printed electrode for hydrogen peroxide detection. Figure 14.9 shows the CV for 5 mM glucose in the GOx-CoPc/

Fig. 14.8 Schematic of the
GOx-CoPc/BDDP-printed
electrode

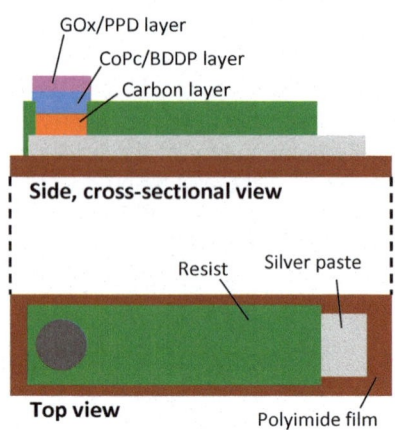

Fig. 14.9 CV for 5 mM
glucose in 0.1 M phosphate
buffer solution (PBS) at
GOx-CoPc/BDDP-printed
electrode. Potential sweep rate
was 5 mV s^{-1}

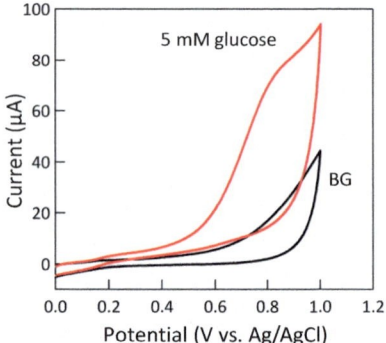

BDDP-printed electrode. An anodic peak was observed at +0.8 V versus Ag/AgCl, indicating the successful electrochemical detection of glucose at the GOx-CoPc/BDDP-printed electrode.

The amperometric detection of glucose at the GOx-CoPc/BDDP-printed electrode at a constant potential of +0.8 V versus Ag/AgCl shows a linear current response with a slope of 3.9 μA mM^{-1} (geometric electrode area was 0.28 cm^2) in the glucose concentration range of 0.2–1.0 mM (Fig. 14.10). Figure 14.11a displays the chronoamperogram of the GOx-CoPc/BDDP- and GOx-CoPc/carbon-printed electrodes at +0.8 V versus Ag/AgCl in 0.1 M PBS, which suggests that current stabilizes faster at the GOx-CoPc/BDDP-electrode than at the GOx-CoPc/carbon-printed electrode. A fast stabilization of the baseline current is essential because it contributes to obtaining reliable signal recognition and shortening the measurement time. Figure 14.11b and c show the current responses of the GOx-CoPc/carbon- and GOx-CoPc/BDDP-printed electrodes, respectively, for 50 μM glucose at +0.8 V versus Ag/AgCl. Magnetic stirring of the electrolyte produced periodic noise. An increase in current was confirmed in both electrodes after the injection of glucose. However, at the GOx-CoPc/carbon-printed electrode,

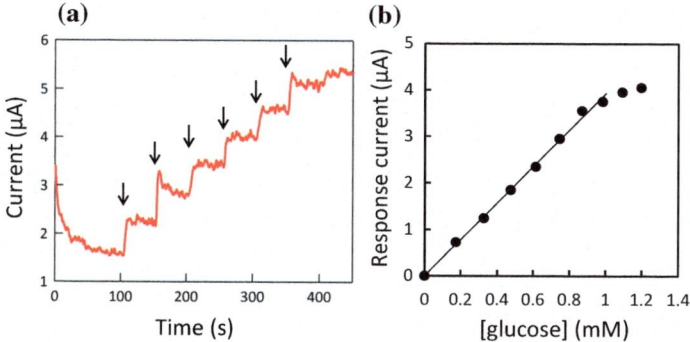

Fig. 14.10 **a** Amperometric glucose detection in 0.1 M PBS at the GOx-CoPc/BDDP-printed electrode with stirring of the electrolyte solution. Electrode potential was +0.8 V versus Ag/AgCl. Arrows indicate the time when the glucose sample was added to the solution. **b** Calibration curve for the glucose detection created from the data provided in panel (**a**)

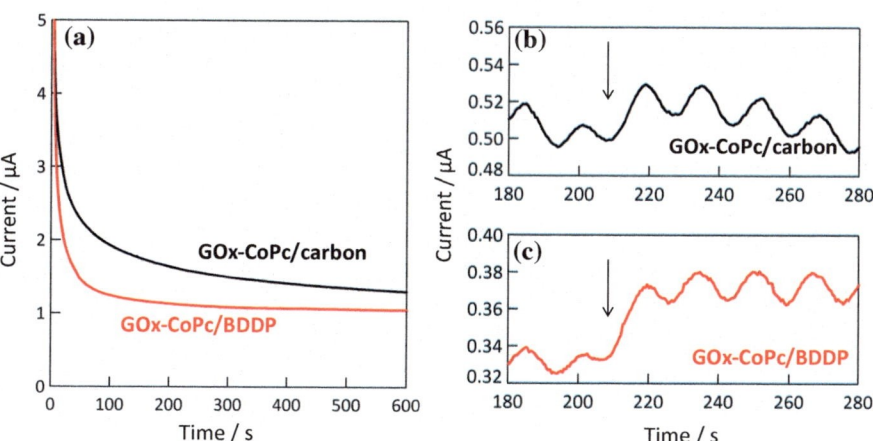

Fig. 14.11 **a** Chronoamperograms for the GOx-CoPc/carbon- and GOx-CoPc/BDDP-printed electrodes in 0.1 M PBS with a step potential of +0.8 V versus Ag/AgCl. Amperometric detection of 50 µM glucose at **b** GOx-CoPc/carbon- and **c** GOx-CoPc/BDDP-printed electrode with stirring and at an electrode potential of +0.8 V versus Ag/AgCl

the current gradually decreased and the current response was ambiguous. In contrast, at the GOx-CoPc/BDDP-printed electrode, the baseline current was rather stable and the current response was obvious. Thus, the GOx-CoPc/BDDP-printed electrode proved to be useful for glucose detection, particularly at low concentrations.

14.3.3 Random Microelectrode Array Effect for Sensitive Electrochemical Detection

Compared with conventional BDD electrodes, BDDP-printed electrodes exhibit low background current; a low background current enables sensitive electrochemical detection with a large S/B ratio. The microelectrode method is effectively used to further increase the S/B ratio. For microelectrodes, it is expected that a large faradaic current can be obtained as a result of efficient mass transport via hemispherical diffusion at the electrode/electrolyte interface with respect to a small capacitive background current. Therefore, a BDDP-printed electrode with a large PES/BDDP ratio was fabricated, and the feasibility of a highly sensitive electrochemical detection based on a random microelectrode array (or microelectrode ensemble) effect [31–35] was assessed (Fig. 14.12) [36]. For planar electrodes, the PES/BDDP ratio should be as small as possible; the optimum ratio was 0.3 herein, as described above. In contrast, to form a random microelectrode array-like structure, the exposed conductive BDDP must be dispersed in an insulating matrix (PES) on the electrode surface when the PES/BDDP ratio is significantly large.

Figure 14.13 shows the CV of $K_3Fe(CN)_6$ in a BDDP-printed electrode with various PES/BDDP ratios ranging from 0.3 to 2.0. It was found that the background current of these electrodes was flat and featureless in the potential region of 0–1.0 V versus Ag/AgCl and decreased as the PES/BDDP ratio increased (Table 14.1). This trend is most likely caused by the reduction of the total area of the BDDP surface that was exposed to the electrolyte. The CV curve in the presence of $K_3Fe(CN)_6$ showed a symmetric redox peak pair with a quasi-reversible behavior of the BDDP-printed electrode with a PES/BDDP ratio of 0.3 (Fig. 14.13a). As the PES/BDDP ratio increased, the anodic–cathodic peak separation increased, showing a decrease in the apparent electron transfer rate owing to a decrease in the total conduction area on the electrode surface. Eventually, an asymmetric CV was

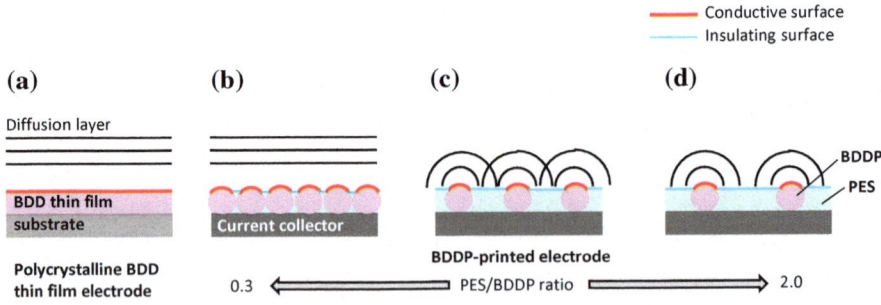

Fig. 14.12 Schematic of the diffusion layers on polycrystalline BDD thin film electrode and BDDP-printed electrodes with various PES/BDDP ratios

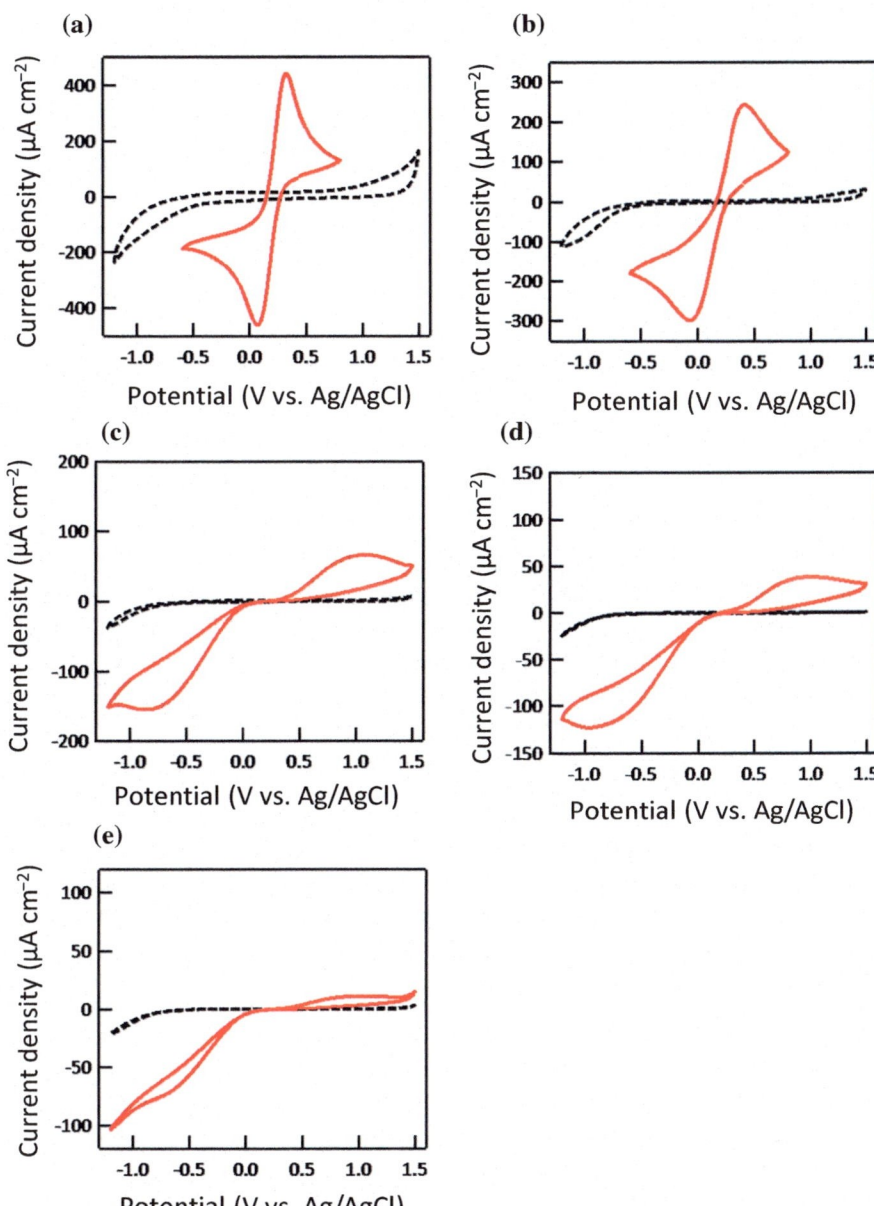

Fig. 14.13 CVs for 5 mM $K_3Fe(CN)_6$ in 0.1 M Na_2SO_4 using the BDDP-printed electrodes with various PES/BDDP ratios of **a** 0.3, **b** 0.5, **c** 1.0, **d** 1.5, and **e** 2.0. Dashed lines indicate the background current. Potential sweep rate was 100 mV s^{-1}

Table 14.1 S/B ratio of the CV current for $Fe(CN)_6^{3-}$ reduction at BDDP-printed electrode with various PES/BDDP ratios

PES/BDDP ratio	Potential (V vs. Ag/AgCl)	Signal current ($\mu A\ cm^{-2}$)	Background current ($\mu A\ cm^{-2}$)	S/B ratio
0.3	0.07	−459	−12.3	37
0.5	−0.16	−280	−3.6	78
1.0	−0.60	−134	−1.54	87
1.5	−0.66	−107	−0.83	128
2.0	−0.67	−71	−0.48	148

observed; the anode current was hardly observed at the electrode with a PES/BDDP ratio of 2.0 (Fig. 14.13e). Such a sigmoidal CV curve is typically found in microelectrodes. It was, therefore, demonstrated that the planar electrode behavior of the BDDP-printed electrode can be converted to that of a microelectrode array by increasing the PES/BDDP ratio from 0.3 to 2.0. The formation of a random microelectrode array structure in which micrometer-sized conductive domains (BDDP) are surrounded by an insulating matrix (PES) was confirmed via the elemental Cu mapping. The mapping was acquired using scanning electron microscopy with energy dispersive X-ray spectroscopy (SEM-EDS) after the electrodeposition of copper on the BDDP-printed electrode surface (Fig. 14.14).

The application of the BDDP-printed electrodes with large PES/BDDP ratio to sensitive electrochemical detection was investigated. Figure 14.15a–c show the CV curves of ascorbic acid (AA) using a conventional polycrystalline BDD thin film electrode (Fig. 14.15a) and BDDP-printed electrodes with PES/BDDP ratios of 0.3 and 1.0 (Fig. 14.15b and c, respectively). The anodic peak current of AA oxidation was nearly equivalent between the polycrystalline BDD and the BDDP-printed electrode with a PES/BDDP ratio of 0.3. However, the background current was smaller at the BDDP-printed electrode (PES/BDDP = 0.3, 1.7–1.8 $\mu A\ cm^{-2}$) compared with that at the polycrystalline BDD electrode (3.0–3.2 $\mu A\ cm^{-2}$), leading to a larger S/B ratio at the BDDP-printed electrode (Fig. 14.15d). In contrast, the CV curve at the BDDP-printed electrode with a PES/BDDP ratio of 1.0 showed an ambiguous peak, indicating an intermediate behavior between the planar electrode and the microelectrode. Although the anodic peak current was small at the BDDP-printed electrode with a PES/BDDP ratio of 1.0, the S/B ratio was much larger than that of the other two electrodes because of small background current (0.07 $\mu A\ cm^{-2}$). This result demonstrates that BDDP-printed electrodes with large PES/BDDP ratios are potentially useful as disposable, highly sensitive electrochemical sensors and exhibit larger S/B ratios than those available when using conventional BDD electrodes.

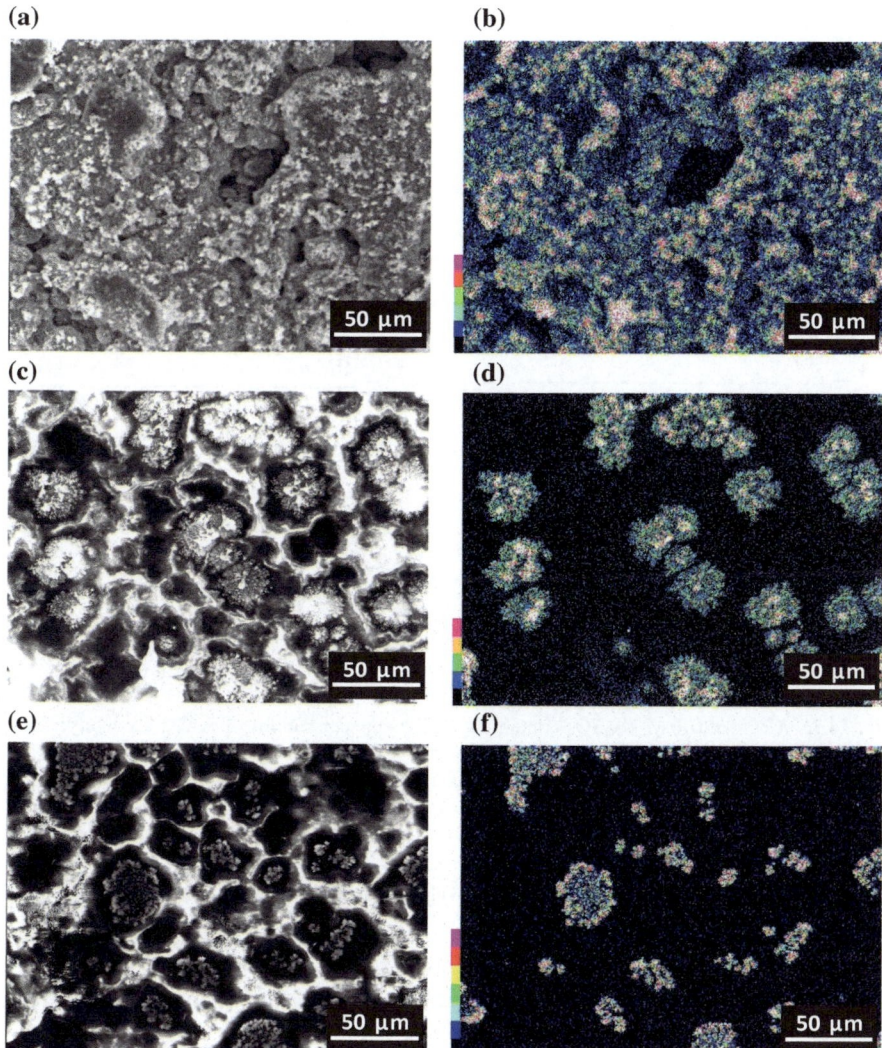

Fig. 14.14 **a, c, e** SEM images and **b, d, f** EDS elemental Cu maps of BDDP-printed electrode surfaces with PES/BDDP ratios of **a, b** 0.5, **c, d** 1.5 and **e, f** 2.0

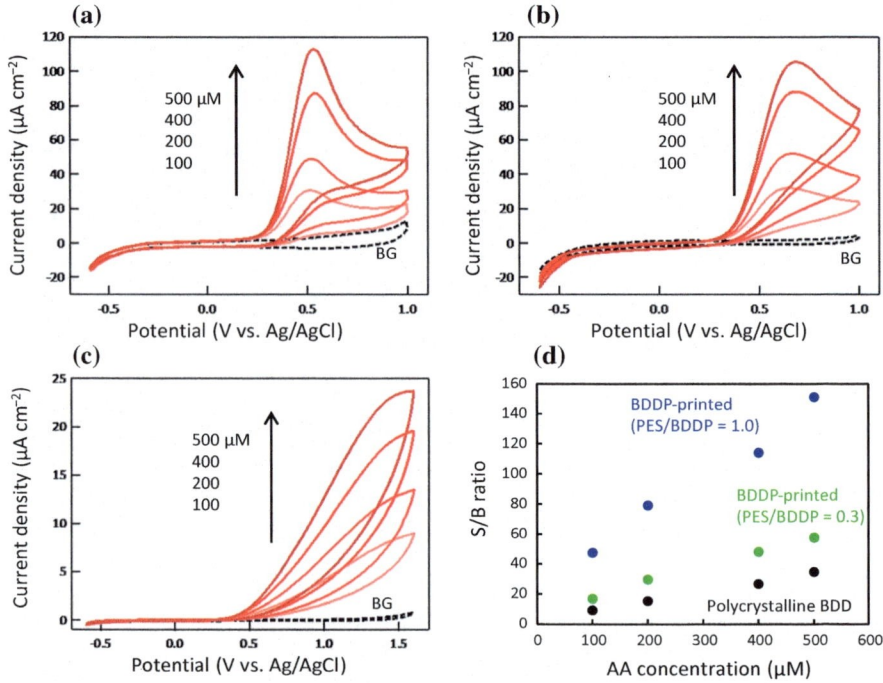

Fig. 14.15 CVs for 0.1 M Na$_2$SO$_4$ containing various concentrations of AA at **a** polycrystalline BDD thin film electrode, and BDDP-printed electrodes with PES/BDDP ratios of **b** 0.3 and **c** 1.0. Potential sweep rate was 100 mV s^{-1}. **d** S/B ratio for AA detection as a function of the concentration

14.4 Application to Catalyst Support for Fuel Cells

Polymer electrolyte fuel cell (PEFC) is known to be a promising renewable energy source with some advantages such as high energy density, low application temperature and ease of miniaturization. These properties render PEFC potentially useful as a power source for automobiles, residential applications and mobile devices [37–41]. However, the corrosion of carbon that is used as a cathode catalyst support deteriorates PEFC quality, which can be a major drawback of using PEFC in automobiles. The PEFC cathode is exposed to a highly positive potential at a start/stop operation and/or exhaustion of hydrogen gas. This exposure causes carbon corrosion and deteriorates the PEFC performance because of agglomeration and detachment of platinum [42–45]. To circumvent this problem, we considered BDDP as a possible candidate for a cathode catalyst support material that would exhibit excellent corrosion resistance and thus, render high durability under highly positive potentials. This section describes the preparation of platinum-supported BDDP (Pt/BDDP) and the investigation of its application as a cathode catalyst resistant to highly positive potentials [46].

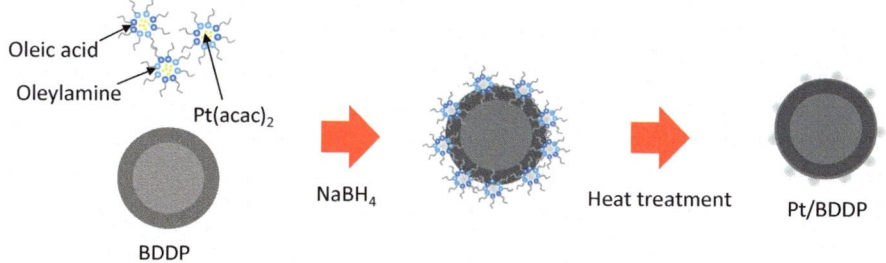

Fig. 14.16 Schematic of the preparation of Pt/BDDP using the nanocapsule method

Pt/BDDP was prepared using the nanocapsule method, which has been reported to be useful in the preparation of small Pt nanoparticle with controlled size [47]. According to this method, a surfactant is added to an organic solvent to form reverse micelles (nanocapsules), and a hydrophilic reaction substrate is incorporated therein to obtain Pt nanoparticles (Fig. 14.16). Briefly, Pt(acac)$_2$ was dissolved in a mixed solvent of 1,2-hexadecanediol and diphenyl ether, and the solution was heated and stirred. Then, oleylamine and oleic acid as surfactants and BDDP (300 nm in diameter) dispersed in ethanol were added to the solution. Further, a reducing agent (NaBH$_4$) was added to promote the deposition of Pt nanoparticles. After washing, drying, and heat treatment, Pt/BDDP was afforded. From transmission electron microscope (TEM) observations and X-ray diffraction, Pt/BDDP was confirmed to support 3–5 nm of Pt nanoparticles densely on the surface of BDDP (Fig. 14.17).

The newly prepared Pt/BDDP was then dispersed in ethanol together with Nafion, and the dispersion was cast on a graphite electrode for conducting electrochemical investigations. The oxygen reduction reaction (ORR) activity of the Pt/BDDP was estimated via linear sweep voltammetry (LSV) in O$_2$-saturated 1.0 M HClO$_4$ using the rotating disk electrode (RDE) method. The result indicated that the onset potential for ORR was substantially the same as in the case of the

Fig. 14.17 TEM image of Pt/BDDP

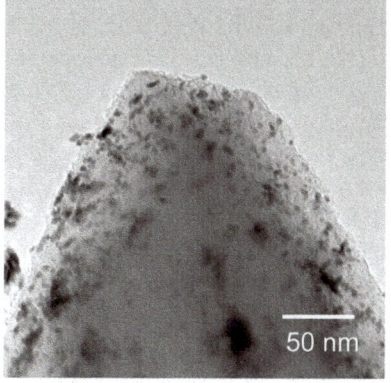

50 nm

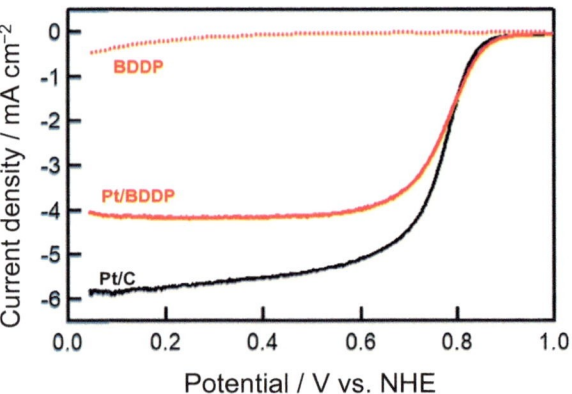

Fig. 14.18 LSV in O₂-saturated 1.0 M HClO₄ at BDDP, Pt/BDDP, and Pt/C electrodes recorded using the RDE method (1,000 rpm). Potential sweep rate was 10 mV s⁻¹

commercially available Pt/C (Vulcan XC-72), showing that BDDP could be used as a cathode catalyst support (Fig. 14.18).

A potential cycle test simulating start/stop operations of a fuel cell was performed to evaluate the durability of Pt/BDDP against application of highly positive potentials [48]. The potential cycle (2 s/cycle) in the potential range from +1.0 to +1.5 V versus NHE in 1.0 M HClO₄ was repeated, and the deterioration of the catalyst was evaluated by the change in the electrochemical active surface area (ECA) of Pt in the catalyst. The ECA can be estimated from the charge of the cathodic peak for hydrogen adsorption in the CV and by dividing this charge with that obtained for a smooth polycrystalline Pt (260 μC cm⁻²). In the case wherein the Pt/C catalyst was evaluated, a sharp decrease in the ECA was observed in the initial 1,000 cycles, followed by a gradual decrease in the subsequent cycles. In contrast, in the case wherein Pt/BDDP was evaluated, such deterioration in two stages was not observed, and the ECA decreased gently from the beginning. After 10,000 cycles, the ECA decreased to 53% of the initial value for Pt/C, whereas it remained at 61% in Pt/BDDP. This suggests that the durability of Pt/BDDP against a highly positive potential; thus, the frequent start/stop operations of a fuel cell was higher than that of Pt/C. In addition, when BDDP with a particle size of 200 nm was used, the ECA was maintained at 78% even after 10,000 cycles (Fig. 14.19). The rapid initial deterioration in Pt/C is presumably caused by the agglomeration and/or detachment of Pt nanoparticles associated with carbon corrosion, and in the following stage, the decrease in the total surface area of Pt could occur because of the agglomeration of Pt nanoparticles. In contrast, in the case wherein Pt/BDDP was evaluated, corrosion of the support (BDDP) hardly occurs, and the ECA decrease can be explained in terms of the agglomeration of Pt particles on the BDDP surface. Although further improvement in conductivity and an increase in specific surface area are desirable, from the results presented herein, Pt/BDDP can be considered as a promising candidate for its application as a highly durable cathode catalyst with an ORR activity similar to that of commercially available Pt/C catalysts.

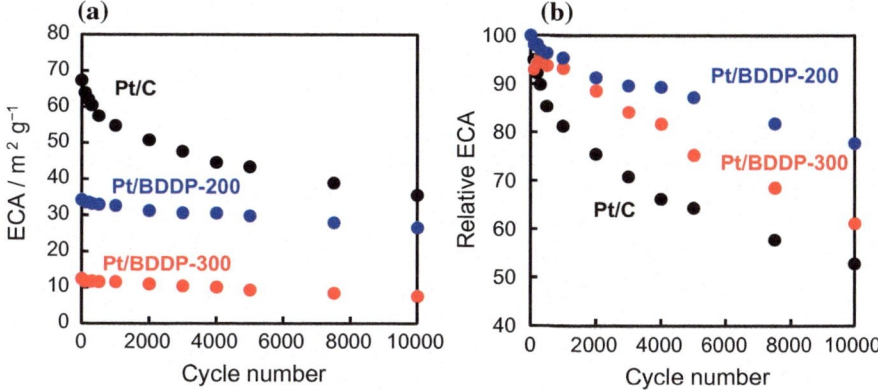

Fig. 14.19 **a** ECA and **b** normalized ECA evaluated from CV in 1.0 M HClO$_4$ at Pt/C, Pt/BDDP-300, and Pt/BDDP-200 electrodes as a function of the cycle number of the potential cycle test

14.5 Conclusions

In this chapter, the preparation of BDDP and its application as screen-printed electrode and PEFC cathode catalyst was described. Other unique applications of BDDP, such as a flexible pinpoint electrolysis unit for dental treatment [49], microfluidic paper-based analytical devices [50] and flexible sensors for plant monitoring [51] have been reported recently. BDDP is an attractive material that can provide access to inexpensive, large-sized, miniaturized, or flexible BDD electrodes, as well as BDD electrodes with large specific surface area. Moreover, BDDP could potentially form composites with various materials. Therefore, the BDDP is believed to be a promising material that can expand the application field of conductive diamond to disposable biosensors and electrochemical energy-related devices.

References

1. J.P. Lagrange, A. Deneuville, E. Gheeraert, Activation energy in low compensated homoepitaxial boron-doped diamond films1. Diam. Relat. Mater. **7**(9), 1390–1393 (1998). https://doi.org/10.1016/S0925-9635(98)00225-8
2. *Diamond Electrochemistry*, ed. by A. Fujishima, Y. Einaga, T.N. Rao, D.A. Tryk (Elsevier-BKC, Tokyo, 2005)
3. R.G. Compton, J.S. Foord, F. Marken, Electroanalysis at diamond-like and doped-diamond electrodes. Electroanalysis **15**(17), 1349–1363 (2003). https://doi.org/10.1002/elan.200302830
4. E.N. Christoph, R. Bohuslav, S. Dongchan, U. Hiroshi, Y. Nianjun, Diamond for bio-sensor applications. J. Phys. D **40**(20), 6443 (2007). https://doi.org/10.1088/0022-3727/40/20/S21

5. Y. Zhou, J. Zhi, The application of boron-doped diamond electrodes in amperometric biosensors. Talanta **79**(5), 1189–1196 (2009). https://doi.org/10.1016/j.talanta.2009.05.026
6. Y. Einaga, Diamond electrodes for electrochemical analysis. J. Appl. Electrochem. **40**(10), 1807–1816 (2010). https://doi.org/10.1007/s10800-010-0112-z
7. M. Panizza, G. Cerisola, Application of diamond electrodes to electrochemical processes. Electrochim. Acta **51**(2), 191–199 (2005). https://doi.org/10.1016/j.electacta.2005.04.023
8. M. Panizza, G. Cerisola, Direct and mediated anodic oxidation of organic pollutants. Chem. Rev. **109**(12), 6541–6569 (2009). https://doi.org/10.1021/cr9001319
9. B. Marselli, J. Garcia-Gomez, P.-A. Michaud, M.A. Rodrigo, C. Comninellis, Electrogeneration of hydroxyl radicals on boron-doped diamond electrodes. J. Electrochem. Soc. **150**(3), D79–D83 (2003). https://doi.org/10.1149/1.1553790
10. T. Kondo, Y. Tamura, M. Hoshino, T. Watanabe, T. Aikawa, M. Yuasa, Y. Einaga, Direct determination of chemical oxygen demand by anodic decomposition of organic compounds at a diamond electrode. Anal. Chem. **86**(16), 8066–8072 (2014). https://doi.org/10.1021/ac500919k
11. A. Kraft, M. Stadelmann, M. Wunsche, M. Blaschke, Electrochemical ozone production anodes and a solid polymer using diamond electrolyte. Electrochem. Commun. **8**(5), 883–886 (2006). https://doi.org/10.1016/j.elecom.2006.02.013
12. T. Kondo, H. Sakamoto, T. Kato, M. Horitani, I. Shitanda, M. Itagaki, M. Yuasa, Screen-printed diamond electrode: a disposable sensitive electrochemical electrode. Electrochem. Commun. **13**(12), 1546–1549 (2011). https://doi.org/10.1016/j.elecom.2011.10.013
13. T. Spătaru, T. Kondo, C. Anastasescu, I. Balint, P. Osiceanu, C. Munteanu, N. Spătaru, A. Fujishima, Silica veils-conductive diamond powder composite as a new propitious substrate for platinum electrocatalysts. J. Solid State Electrochem. **21**(4), 1007–1014 (2017). https://doi.org/10.1007/s10008-016-3454-6
14. L. Preda, T. Kondo, T. Spataru, M. Marin, M. Radu, P. Osiceanu, A. Fujishima, N. Spataru, Enhanced activity for methanol oxidation of platinum particles supported on iridium oxide modified boron-doped diamond powder. ChemElectroChem **4**(8), 1908–1915 (2017). https://doi.org/10.1002/celc.201700155
15. G.R. Salazar-Banda, K.I.B. Eguiluz, L.A. Avaca, Boron-doped diamond powder as catalyst support for fuel cell applications. Electrochem. Commun. **9**(1), 59–64 (2007). https://doi.org/10.1016/j.elecom.2006.08.038
16. S. Heyer, W. Janssen, S. Turner, Y.-G. Lu, W.S. Yeap, J. Verbeeck, K. Haenen, A. Krueger, Toward deep blue nano hope diamonds: heavily boron-doped diamond nanoparticles. ACS Nano **8**(6), 5757–5764 (2014). https://doi.org/10.1021/nn500573x
17. E.A. Ekimov, O.S. Kudryavtsev, A.A. Khomich, O.I. Lebedev, T.A. Dolenko, I.I. Vlasov, High-pressure synthesis of boron-doped ultrasmall diamonds from an organic compound. Adv. Mater. **27**(37), 5518–5522 (2015). https://doi.org/10.1002/adma.201502672
18. A.E. Fischer, G.M. Swain, Preparation and characterization of boron-doped diamond powder —A possible dimensionally stable electrocatalyst support material. J. Electrochem. Soc. **152**(9), B369–B375 (2005). https://doi.org/10.1149/1.1984367
19. A. Ay, V.M. Swope, G.M. Swain, The physicochemical and electrochemical properties of 100 and 500 nm diameter diamond powders coated with boron-doped nanocrystalline diamond. J. Electrochem. Soc. **155**(10), B1013–B1022 (2008). https://doi.org/10.1149/1.2958308
20. S.A. Wring, J.P. Hart, B.J. Birch, Voltammetric behaviour of screen-printed carbon electrodes, chemically modified with selected mediators, and their application as sensors for the determination of reduced glutathione. Analyst **116**(2), 123–129 (1991). https://doi.org/10.1039/AN9911600123
21. F. Darain, S.-U. Park, Y.-B. Shim, Disposable amperometric immunosensor system for rabbit IgG using a conducting polymer modified screen-printed electrode. Biosens. Bioelectron. **18**(5), 773–780 (2003). https://doi.org/10.1016/S0956-5663(03)00004-6
22. J.P. Hart, A. Crew, E. Crouch, K.C. Honeychurch, R.M. Pemberton, Some recent designs and developments of screen-printed carbon electrochemical sensors/biosensors for biomedical,

environmental, and industrial analyses. Anal. Lett. **37**(5), 789–830 (2004). https://doi.org/10. 1081/AL-120030682

23. Y.-S. Song, G. Muthuraman, Y.-Z. Chen, C.-C. Lin, J.-M. Zen, Screen printed carbon electrode modified with poly(l-lactide) stabilized gold nanoparticles for sensitive as(iii) detection. Electroanalysis **18**(18), 1763–1770 (2006). https://doi.org/10.1002/elan.200603634

24. M. Tudorache, C. Bala, Biosensors based on screen-printing technology, and their applications in environmental and food analysis. Anal. Bioanal. Chem. **388**(3), 565–578 (2007). https://doi.org/10.1007/s00216-007-1293-0

25. I. Shitanda, S. Takamatsu, K. Watanabe, M. Itagaki, Amperometric screen-printed algal biosensor with flow injection analysis system for detection of environmental toxic compounds. Electrochim. Acta **54**(21), 4933–4936 (2009). https://doi.org/10.1016/j. electacta.2009.04.005

26. M.I. Prodromidis, M.I. Karayannis, Enzyme based amperometric biosensors for food analysis. Electroanalysis **14**(4), 241–261 (2002). https://doi.org/10.1002/1521-4109(200202)14: 4<241::AID-ELAN241>3.0.CO;2-P

27. M. Albareda-Sirvent, A. Merkoçi, S. Alegret, Configurations used in the design of screen-printed enzymatic biosensors. A review. Sens. Actuat. B **69**(1), 153–163 (2000). https://doi.org/10.1016/S0925-4005(00)00536-0

28. W.-J. Guan, Y. Li, Y.-Q. Chen, X.-B. Zhang, G.-Q. Hu, Glucose biosensor based on multi-wall carbon nanotubes and screen printed carbon electrodes. Biosens. Bioelectron. **21** (3), 508–512 (2005). https://doi.org/10.1016/j.bios.2004.10.030

29. A. Heller, B. Feldman, Electrochemical glucose sensors and their applications in diabetes management. Chem. Rev. **108**(7), 2482–2505 (2008). https://doi.org/10.1021/cr068069y

30. T. Kondo, M. Horitani, H. Sakamoto, I. Shitanda, Y. Hoshi, M. Itagaki, M. Yuasa, Screen-printed modified diamond electrode for glucose detection. Chem. Lett. **42**(4), 352–354 (2013). https://doi.org/10.1246/cl.121242

31. L. Xiao, I. Streeter, G.G. Wildgoose, R.G. Compton, Fabricating random arrays of boron doped diamond nano-disc electrodes: Towards achieving maximum Faradaic current with minimum capacitive charging. Sens. Actuat. B **133**(1), 118–127 (2008). https://doi.org/10. 1016/j.snb.2008.02.003

32. S. Fletcher, M.D. Horne, Random assemblies of microelectrodes (RAM™ electrodes) for electrochemical studies. Electrochem. Commun. **1**(10), 502–512 (1999). https://doi.org/10. 1016/S1388-2481(99)00100-9

33. M. Lacroix, P. Bianco, E. Lojou, Modified random assembly of microelectrodes for the selective electrochemical detection of dopamine. Electroanalysis **11**(14), 1068–1076 (1999). https://doi.org/10.1002/(SICI)1521-4109(199910)11:14<1068::AID-ELAN1068>3.0.CO;2-V

34. E. Ilinoiu, F. Manea, P. Serra, R. Pode, Simultaneous/selective detection of dopamine and ascorbic acid at synthetic zeolite-modified/graphite-epoxy composite macro/ quasi-microelectrodes. Sensors **13**(6), 7296 (2013). https://doi.org/10.3390/s130607296

35. J. Hees, R. Hoffmann, A. Kriele, W. Smirnov, K. Obloh, K. Glorer, B. Raynor, R. Driad, N. Yang, O.A. Williams, C.E. Nebel, Nanocrystalline diamond nanoelectrode arrays and ensembles. ACS Nano **5**(4), 3339–3346 (2011). https://doi.org/10.1021/nn2005409

36. T. Kondo, I. Udagawa, T. Aikawa, H. Sakamoto, I. Shitanda, Y. Hoshi, M. Itagaki, M. Yuasa, Enhanced sensitivity for electrochemical detection using screen-printed diamond electrodes via the random microelectrode array effect. Anal. Chem. **88**(3), 1753–1759 (2016). https://doi. org/10.1021/acs.analchem.5b03986

37. B.C.H. Steele, A. Heinzel, Materials for fuel-cell technologies. Nature **414**, 345 (2001). https://doi.org/10.1038/35104620

38. O. Okada, K. Yokoyama, Development of polymer electrolyte fuel cell cogeneration systems for residential applications. Fuel Cells **1**(1), 72–77 (2001). https://doi.org/10.1002/1615-6854 (200105)1:1%3c72:AID-FUCE72%3e3.0.CO;2-P

39. H.A. Gasteiger, S.S. Kocha, B. Sompalli, F.T. Wagner, Activity benchmarks and requirements for Pt, Pt-alloy, and non-Pt oxygen reduction catalysts for PEMFCs. Appl. Cat. B **56**(1), 9–35 (2005). https://doi.org/10.1016/j.apcatb.2004.06.021

40. R. Bashyam, P. Zelenay, A class of non-precious metal composite catalysts for fuel cells. Nature **443**, 63 (2006). https://doi.org/10.1038/nature05118

41. M. Watanabe, D.A. Tryk, M. Wakisaka, H. Yano, H. Uchida, Overview of recent developments in oxygen reduction electrocatalysis. Electrochim. Acta 84 (Supplement C) 187–201 (2012). https://doi.org/10.1016/j.electacta.2012.04.035

42. L.M. Roen, C.H. Paik, T.D. Jarvi, Electrocatalytic corrosion of carbon support in PEMFC cathodes. Electrochem. Solid-State Lett. **7**(1), A19–A22 (2004). https://doi.org/10.1149/1.1630412

43. J.P. Meyers, R.M. Darling, Model of carbon corrosion in PEM fuel cells. J. Electrochem. Soc. **153**(8), A1432–A1442 (2006). https://doi.org/10.1149/1.2203811

44. S. Maass, F. Finsterwalder, G. Frank, R. Hartmann, C. Merten, Carbon support oxidation in PEM fuel cell cathodes. J. Power Source **176**(2), 444–451 (2008). https://doi.org/10.1016/j.jpowsour.2007.08.053

45. F.A. de Bruijn, V.A.T. Dam, G.J.M. Janssen, Review: durability and degradation issues of pem fuel cell components. Fuel Cells **8**(1), 3–22 (2008). https://doi.org/10.1002/fuce.200700053

46. T. Kondo, M. Kikuchi, H. Masuda, F. Katsumata, T. Aikawa, M. Yuasa, Boron-doped diamond powder as a durable support for platinum-based cathode catalysts in polymer electrolyte fuel cells. J. Electrochem. Soc. **165**(6), F3072–F3077 (2018). https://doi.org/10.1149/2.0111806jes

47. H. Yano, M. Kataoka, H. Yamashita, H. Uchida, M. Watanabe, Oxygen reduction activity of carbon-supported Pt−M (M = V, Ni, Cr, Co, and Fe) alloys prepared by nanocapsule method. Langmuir **23**(11), 6438–6445 (2007). https://doi.org/10.1021/la070078u

48. A. Ohma, K. Shinohara, A. Iiyama, T. Yoshida, A. Daimaru, Membrane and catalyst performance targets for automotive fuel cells by fccj membrane, catalyst. MEA WG. ECS Trans. **41**(1), 775–784 (2011). https://doi.org/10.1149/1.3635611

49. T. Ochiai, S. Tago, M. Hayashi, K. Hirota, T. Kondo, K. Satomura, A. Fujishima, Boron-doped diamond powder (BDDP)-based polymer composites for dental treatment using flexible pinpoint electrolysis unit. Electrochem. Commun. **68**, 49–53 (2016). https://doi.org/10.1016/j.elecom.2016.04.011

50. S. Nantaphol, R.B. Channon, T. Kondo, W. Siangproh, O. Chailapakul, C.S. Henry, Boron doped diamond paste electrodes for microfluidic paper-based analytical devices. Anal. Chem. **89**(7), 4100–4107 (2017). https://doi.org/10.1021/acs.analchem.6b05042

51. S. Tago, T. Ochiai, S. Suzuki, M. Hayashi, T. Kondo, A. Fujishima, Flexible boron-doped diamond (bdd) electrodes for plant monitoring. Sensors **17**(7), 1638 (2017). https://doi.org/10.3390/s17071638

Index

A
Abnormal growth, 7
Abrasive particles, 73
Abstraction process, 48
Acceptor, 34
Activation energy, 8
Active sites, 4
Adatoms, 8
Addressable diamond UMEAs, 235
Adsorption, 4
AFM-SECM probes, 230
AFM-SECM tips, 244
Aggregated diamond nanorods, 330, 338
Ag nanoparticle, 183
Alkaline media, 214
Alkene, 426
All-diamond, 237
Amination of naphthalene, 190
Ammonia solution, 178
Amorphous phases, 137
Amounts of nitrogen, 132
Amperometric biosensors, 352
Amperometric detection, 484
Anisotropically, 138
Anisotropic etching, 4
Anisotropic etching of atomic hydrogen, 366
Anisotropic growth, 140, 144
Annealing temperature, 114
Anodic electrochemical reactions, 176
Anodic polarization, 208
Anodic transformations, 187
Applied potential, 267
Aqueous ammonium, 179
Aqueous electrolytes, 211
Aromatic hydrocarbons, 381

Array, 224
Arrhenius equation, 47
Atomically flat diamond, 10, 14
Atomically step-free surface, 15, 17
Atomic charges, 46
Atomic force microscopy (AFM), 57, 102, 207, 389, 464
Atomic hydrogen, 142, 326
Atomic level, 6
Atom-like quantum system, 401

B
Background current, 175
B- and C-ion implantations, 129
Bandgap, 8
BDD nanoparticles, 478
BDD powder, 478
B dopant, 48
Bead assisted sonic disintegration, 432
Bead milling, 478
Bias enhanced nucleation and growth, 143
Bilayer model, 114
Bimetallic alloys, 263
Binding energy, 59, 70
Binding energy shift, 78
Bio applications, 390, 398
Biogenic substances, 244
Biologically active molecules, 391
Biological systems, 401
Biomarkers, 415
Biomedecine, 444
Biomedical applications, 416, 431, 437
Bio-sensing, 398
Bio systems, 391
Blinking effect, 373

© Springer Nature Switzerland AG 2019
N. Yang (ed.), *Novel Aspects of Diamond*, Topics in Applied Physics 121,
https://doi.org/10.1007/978-3-030-12469-4

Printed by Printforce, the Netherlands